ANTIOXIDANTS IN DISEASE MECHANISMS AND THERAPY

ADVANCES IN
PHARMACOLOGY

VOLUME 38

ADVANCES IN
PHARMACOLOGY

ANTIOXIDANTS IN DISEASE MECHANISMS AND THERAPY

Edited by

Helmut Sies

Institute of Physiological Chemistry
Heinrich-Heine-Universität Düsseldorf
Düsseldorf, Germany

ADVANCES IN
PHARMACOLOGY

VOLUME 38

ACADEMIC PRESS

San Diego London Boston New York Sydney Tokyo Toronto

RM 101
.A2
vol. 38
1997

This book is printed on acid-free paper. ∞

Academic Press, Inc.
525 B Street, Suite 1900, San Diego, California 92101-4495, USA
http://www.apnet.com

Academic Press Limited
24-28 Oval Road, London NW1 7DX, UK
http://www.hbuk.co.uk/ap/

International Standard Serial Number: 1054-3589

International Standard Book Number: 0-12-032939-5

PRINTED IN THE UNITED STATES OF AMERICA
96 97 98 99 00 01 BB 9 8 7 6 5 4 3 2 1

To Katharina and Audrey

Contents

Antioxidative and Metal-Chelating Effects of Polyamines

Erik Løvaas

Antioxidant and Chelating Properties of Flavonoids

Ludmila G. Korkina and Igor B. Afanas'ev

PART II   Synthetic Antioxidants and Enzyme Mimics

Potential Use of Iron Chelators against Oxidative Damage

Jean-Baptiste Galey

PART III Antioxidant Enzyme Induction and Pathophysiology

Antioxidant Drug Targeting

Anthony C. Allison

Antioxidant-Inducible Genes

Thomas Primiano, Thomas R. Sutter, and Thomas W. Kensler

Redox Signaling and the Control of Cell Growth and Death

Garth Powis, John R. Gasdaska, and Amanda Baker

PART IV Disease Processes

Relationships among Oxidation of Low-Density Lipoprotein, Antioxidant Protection, and Atherosclerosis
Hermann Esterbauer, Reinhold Schmidt, and Marianne Hayn

Adult Respiratory Distress Syndrome: A Radical Perspective
Samuel Louie, Barry Halliwell, and Carroll Edward Cross

Oxidative Stress in Cystic Fibrosis: Does It Occur and Does It Matter?
Albert van der Vliet, Jason P. Eiserich, Gregory P. Marelich, Barry Halliwell, and Carroll E. Cross

Oxidative Stress and Antioxidant Function in Relation to Risk for Cataract

Allen Taylor and Thomas Nowell

The Macular Pigment: A Possible Role in Protection from Age-Related Macular Degeneration

John T. Landrum, Richard A. Bone, and Mark D. Kilburn

Neurological Disease

David P. R. Muller

Role of Cysteine and Glutathione in HIV Infection and Cancer Cachexia: Therapeutic Intervention with *N*-Acetylcysteine

Wulf Dröge, Andrea Gross, Volker Hack, Ralf Kinscherf, Michael Schykowski, Michael Bockstette, Sabine Mihm, and Dagmar Galter

Role of Oxidative Stress and Antioxidant Therapy in Alcoholic and Nonalcoholic Liver Diseases

Charles S. Lieber

Antioxidant Therapy for the Prevention of Type I Diabetes

Birgit Heller, Volker Burkart, Eberhard Lampeter, and Hubert Kolb

Contributors

Numbers in parentheses indicate the pages on which the authors' contributions begin.

Igor B. Afanas'ev (151) Vitamin Research Institute, Moscow 177820, Russia

Anthony C. Allison (273) Dawa Corporation, Belmont, California 94002

Mary E. Anderson (65) Department of Biochemistry, Cornell University Medical College, New York, New York 10021

Amanda Baker (329) Arizona Cancer Center, University of Arizona, Tucson, Arizona 85724

Michel Baudry (247) Neuroscience Program, University of Southern California, Los Angeles, California 90089

Cecile Bladier (379) Commissariat a l'Energie Atomique, Centre de Cadarache, Laboratoire de Phytotechnologie, 13108 St-Paul-lez-Durance, France

David R. Blake (403) Inflammation Research Group, St. Bartholomew's and the Royal London School of Medicine and Dentistry, London E1 1AD, United Kingdom

Michael Bockstette (581) Division of Immunochemistry, Deutsches Krebsforschungszentrum, D-69120 Heidelberg, Germany

Ann M. Bode (21) Physiology Department, School of Medicine and Health Sciences, University of North Dakota, Grand Forks, North Dakota 58201

Richard A. Bone (537) Department of Physics, Florida International University, Miami, Florida 33199

Annadora Bruce (247) Neuroscience Program, University of Southern California, Los Angeles, California 90089

Volker Burkart (629) Clinical Department, Diabetes Research Institute, Heinrich-Heine University Düsseldorf, D-40225 Düsseldorf, Germany

Ian A. Cotgreave (205) Biochemical Toxicology Unit, Institute of Environmental Medicine, Karolinska Institute, S-17177 Stockholm, Sweden

Francesca Cristiano (379) Molecular Genetics and Development Group, Institute of Reproduction and Development, Monash University, Clayton Vic 3168, Australia

Carroll Edward Cross (457, 491) Division of Pulmonary and Critical Care Medicine, Department of Internal Medicine, University of California, Davis, Sacramento, California 95817

Judy B. de Haan (379) Molecular Genetics and Development Group, Institute of Reproduction and Development, Monash University, Clayton Vic 3168, Australia

Susan R. Doctrow (247) Eukarion, Inc., Bedford, Massachusetts 01730

Wulf Dröge (581) Division of Immunochemistry, Deutsches Krebsforschungszentrum, D-69120 Heidelberg, Germany

Jason P. Eiserich (491) Division of Pulmonary and Critical Care Medicine, University of California Davis Medical Center, Sacramento, California 95817

Hermann Esterbauer (425) Institute of Biochemistry, University of Graz, A-8010 Graz, Austria

Robert A. Floyd (361) Free Radical Biology & Aging Research Program, Oklahoma Medical Research Foundation, Oklahoma City, Oklahoma 73104

Jean-Baptiste Galey (167) Department of Chemistry, L'Oréal Research Center, 93600 Aulnay sous bois, France

Dagmar Galter (581) Division of Immunochemistry, Deutsches Krebsforschungszentrum, D-69120 Heidelberg, Germany

John R. Gasdaska (329) Arizona Cancer Center, University of Arizona, Tucson, Arizona 85724

Andrea Gross (581) Division of Immunochemistry, Deutsches Krebsforschungszentrum, D-69120 Heidelberg, Germany

Volker Hack (581) Division of Immunochemistry, Deutsches Krebsforschungszentrum, D-69120 Heidelberg, Germany

Barry Halliwell (3, 457, 491) Neurodegenerative Disease Research Centre, Pharmacology Group, University of London King's College, London SW3 6LX, United Kingdom

Marianne Hayn (425) Institute of Biochemistry, University of Graz, A-8010 Graz, Austria

Birgit Heller (629) Clinical Department, Diabetes Research Institute, Heinrich-Heine University Düsseldorf, D-40225 Düsseldorf, Germany

Karl Huffman (247) Eukarion, Inc., Bedford, Massachusetts 01730

Rocco Iannello (379) Molecular Genetics and Development Group, Institute of Reproduction and Development, Monash University, Clayton Vic 3168, Australia

Michael J. Kelner (379) Department of Pathology, University of California, San Diego, California 92103

Thomas W. Kensler (293) Department of Environmental Health Sciences, The Johns Hopkins School of Hygiene and Public Health, Baltimore, Maryland 21205

Mark D. Kilburn (537) Department of Chemistry, Florida International University, Miami, Florida 33199

Ralf Kinscherf (581) Division of Immunochemistry, Deutsches Krebsforschungszentrum, D-69120 Heidelberg, Germany

Ismail Kola (379) Molecular Genetics and Development Group, Institute of Reproduction and Development, Monash University, Clayton Vic 3168, Australia

Hubert Kolb (629) Clinical Department, Diabetes Research Institute, Heinrich-Heine University Düsseldorf, D-40225 Düsseldorf, Germany

Ludmila G. Korkina (151) Institute of Pediatric Hematology, Moscow 117513, Russia

Eberhard Lampeter (629) Clinical Department, Diabetes Research Institute, Heinrich-Heine University Düsseldorf, D-40225 Düsseldorf, Germany

John T. Landrum (537) Department of Chemistry, Florida International University, Miami, Florida 33199

Charles S. Lieber (601) Mount Sinai School of Medicine (CUNY), Alcohol Research and Treatment Center and G.I.-Liver-Nutrition Program, Bronx Veterans Affairs Medical Center, Bronx, New York 10468

Samuel Louie (457) Division of Pulmonary and Critical Care Medicine, Department of Internal Medicine, University of California, Davis, Sacramento, California 95817

Erik Løvaas (119) The Norwegian College of Fishery Science, University of Tromsø, 9037 Tromsø, Norway

Bernard Malfroy (247) Eukarion, Inc., Bedford, Massachusetts 01730

Catherine B. Marcus (247) Eukarion, Inc., Bedford, Massachusetts 01730

Gregory P. Marelich (491) Division of Pulmonary and Critical Care Medicine and Adult Cystic Fibrosis Program, University of California Davis Medical Center, Sacramento, California 95817

Hiroshi Masumoto (229) Institut für Physiologische Chemie I, Heinrich-Heine-Universität Düsseldorf, D-40001 Düsseldorf, Germany

Susan T. Mayne (657) Department of Epidemiology and Public Health, Yale University School of Medicine and Yale Cancer Center, New Haven, Connecticut 06520

Sabine Mihm (581) Division of Immunochemistry, Deutsches Krebsforschungszentrum, D-69120 Heidelberg, Germany

David P. R. Muller (557) Division of Biochemistry and Genetics, Institute of Child Health, London WC1N 1EH, United Kingdom

Wael Musleh (247) Neuroscience Program, University of Southern California, Los Angeles, California 90089

Thomas Nowell (515) Jean Mayer USDA Human Nutrition Research Center on Aging at Tufts University, Boston, Massachusetts 02111

Lester Packer (79) Department of Molecular and Cell Biology, University of California at Berkeley, Berkeley, California 94720

Garth Powis (329) Arizona Cancer Center, University of Arizona, Tucson, Arizona 85724

Thomas Primiano (293) Department of Environmental Health Sciences, The Johns Hopkins School of Hygiene and Public Health, Baltimore, Maryland 21205

Russel J. Reiter (103) Department of Cellular and Structural Biology, The University of Texas Health Science Center at San Antonio, San Antonio, Texas 78284

Sashwati Roy (79) Department of Molecular and Cell Biology, University of California at Berkeley, Berkeley, California 94720

Karin Scharffetter-Kochanek (639) Experimental and Clinical Photodermatology, Department of Dermatology, University of Cologne, 50931 Köln, Germany

Reinhold Schmidt (425) University Clinic of Neurology, University of Graz, A-8036 Graz, Austria

Michael Schykowski (581) Division of Immunochemistry, Deutsches Krebsforschungszentrum, D-69120 Heidelberg, Germany

Chandan K. Sen (79) Department of Molecular and Cell Biology, University of California at Berkeley, Berkeley, California 94720

Helmut Sies (229) Institut für Physiologische Chemie I, Heinrich-Heine-Universität Düsseldorf, D-40001 Düsseldorf, Germany

Thomas R. Sutter (293) Department of Environmental Health Sciences, The Johns Hopkins School of Hygiene and Public Health, Baltimore, Maryland 21205

Allen Taylor (515) Jean Mayer USDA Human Nutrition Research Center on Aging at Tufts University, Boston, Massachusetts 02111

Maret G. Traber (49) Department of Molecular and Cell Biology, University of California, Berkeley, California 94720

Albert van der Vliet (491) Division of Pulmonary and Critical Care Medicine, University of California Davis Medical Center, Sacramento, California 95817

Paul G. Winyard (403) Inflammation Research Group, St. Bartholomew's and the Royal London School of Medicine and Dentistry, London E1 1AD, United Kingdom

Ernst J. Wolvetang (379) Molecular Genetics and Development Group, Institute of Reproduction and Development, Monash University, Clayton Vic 3168, Australia

Preface

"Oxidative stress," denoting a disturbance in the prooxidant/antioxidant balance in favor of the prooxidants, leading to potential damage (see 1, 2), has been in focus for some time. The present volume brings together many of the interesting topics that have developed with regard to potential clinical implications. High hopes, voiced in the past, of spectacular effects of single antioxidants as pharmacologically active agents have not fully materialized; maybe some of the investigators hoped for too much. However, the novel role of oxidants and antioxidants as part of signaling cascades, as mediators of adaptive responses, has opened new areas of active research. Although it is still too early to speak of "free radical diseases," this volume brings together evidence for crucial roles of oxidants and antioxidants in disease states. As most diseases ultimately lead to cell injury or cell death, the accompanying oxidative damage to DNA, protein, lipids, and carbohydrates obviously is a feature of widespread interest. It is possible that the time course, and even the final outcome, of a disease can be critically modulated by strengthening the antioxidant side of the balance.

This volume focuses on some new strategies of antioxidant defense in terms of new pharmacologically active agents, presents current knowledge on known agents, and provides in-depth treatment of some disease states. It is hoped that the book will serve as a stimulus for further research.

Helmut Sies

References

1. "Oxidative Stress," Academic Press, London (1985).
2. "Oxidative Stress: Oxidants and Antioxidants," Academic Press, London (1991).

PART I

Natural Antioxidants

Barry Halliwell

Neurodegenerative Disease Research Centre
Pharmacology Group
University of London King's College
London SW3 6LX, United Kingdom

Antioxidants: The Basics—What They Are and How to Evaluate Them

I. Introduction

Words such as "antioxidant" and "oxidative stress" are widely used but surprisingly difficult to define precisely (for a discussion of the latter term see Sies, 1991). For example, the term "antioxidant" as used in the literature is often implicitly restricted to chain-breaking antioxidant inhibitors of lipid peroxidation. In particular, food scientists frequently equate antioxidants to inhibitors of lipid peroxidation because they use antioxidants largely to prevent rancidity. However, free radicals generated *in vivo* damage proteins, DNA, and other molecules in addition to lipids. Hence the author and his colleagues introduced a broader definition: an antioxidant is any substance that, when present at low concentrations compared to those of an oxidizable substrate, significantly delays or prevents oxidation of that substrate (Halliwell and Gutteridge, 1989; Halliwell, 1990). The term "oxidizable substrate" includes almost everything found in living cells, including proteins, lipids, carbohydrates, and DNA. Mechanisms of antioxidant action can include:

- Removal of O_2 (e.g., the packaging of foodstuffs under N_2)
- Scavenging reactive oxygen/nitrogen species (Table I) or their precursors
- Inhibiting ROS/RNS formation
- Binding metal ions needed for catalysis of ROS generation
- Upregulation of endogenous antioxidant defenses.

When ROS/RNS are generated *in vivo,* many antioxidants come into play. Their relative importance as protective agents depends on:

- Which ROS/RNS is generated
- How it is generated
- Where it is generated
- What target of damage is measured

For example, if human blood plasma is tested for its ability to inhibit iron ion-dependent lipid peroxidation, the proteins transferrin and caeru-

TABLE I   Reactive Oxygen and Nitrogen Species

Radicals	*Nonradicals*
Reactive oxygen species[a]	
Superoxide, $O_2^{\cdot-}$	Hydrogen peroxide, H_2O_2
Hydroxyl, $OH^\cdot$	Hypochlorous acid, HOCl
Peroxyl, $RO_2^\cdot$	Ozone, O_3
Alkoxyl, $RO^\cdot$	Singlet oxygen, $^1\Delta g$
Hydroperoxyl, $HO_2^\cdot$	
Reactive nitrogen species[b]	
Nitric oxide, $NO^\cdot$	Nitrous acid, HNO_2
Nitrogen dioxide, $NO_2^\cdot$	Dinitrogen tetroxide, N_2O_4
	Dinitrogen trioxide, N_2O_3
	Peroxynitrite, $ONOO^-$
	Peroxynitrous acid, ONOOH
	Nitronium cation, NO_2^+
	Alkyl peroxynitrates, ROONO

[a] A collective term that includes both oxygen radicals and certain nonradicals that are oxidizing agents and/or are easily converted into radicals (HOCl, O_3, $ONOO^-$, $^1O_2$, H_2O_2).

[b] A collective term including nitric oxide and nitrogen dioxide radicals, as well as such nonradicals as HNO_2 and N_2O_4. $ONOO^-$ is often included in both categories. "Reactive" is not always an appropriate term: H_2O_2, $NO^\cdot$, and $O_2^{\cdot-}$ react quickly with few molecules whereas $OH^\cdot$ reacts quickly with almost everything. $RO_2^\cdot$, $RO^\cdot$, HOCl, $NO_2^\cdot$, $ONOO^-$, and O_3 have intermediate reactivities. HOCl could also be regarded as a "reactive chlorine species."

loplasmin are the most important protective agents (Gutteridge and Quinlan, 1992). When human blood plasma is exposed to NO_2·, uric acid seems to be a major protective antioxidant (Halliwell *et al.*, 1992), whereas it appears to play little role as a scavenger of HOCl in plasma (Hu *et al.*, 1992). Similarly, if the oxidative stress is kept the same but a different target of oxidative damage is measured, different answers can result. When plasma is exposed to gas-phase cigarette smoke (CS), lipid peroxidation occurs, which is inhibited by ascorbate (Frei *et al.*, 1991), whereas ascorbate has no effect on the formation of plasma protein carbonyls by CS (Reznick *et al.*, 1992). As an extreme example, some known carcinogens (diethylstilbestrol) that aggravate oxidative DNA damage *in vivo* (Roy and Liehr, 1991) are powerful inhibitors of *in vitro* lipid peroxidation (Wiseman and Halliwell, 1993).

The previously mentioned definition emphasizes the importance of the source of stress and the target ("oxidizable substrate") measured. However, there may be cases that the definition does not include. Thus, plasma albumin may have antioxidant properties, e.g., by binding copper ions and scavenging HOCl (Halliwell, 1988). Would it therefore be beneficial to broaden the definition, perhaps to an antioxidant is any substance that inhibits oxidative damage to a target under the assay conditions being used? The risk is that every chemical in the laboratory could then be classified as either an "antioxidant" and/or a "pro-oxidant" on the basis of assays that have little biological meaning. It is easy to develop some *in vitro* test to support a postulate that compound X is an antioxidant, but it is very hard to prove that compound X actually works by an antioxidant mechanism *in vivo*.

Antioxidants are of interest to radiation chemists, food scientists, polymer chemists, and even to curators of museums (Daniels, 1989) but this chapter only discusses the antioxidants known, or proposed, to be important in aerobic organisms. Because many previous reviews as well as other chapters in the present volume cover the well-established physiological antioxidant roles of vitamin E, ascorbic acid, and such proteins as superoxide dismutase, glutathione peroxidase, catalase, and caeruloplasmin, these are not discussed in this chapter.

Many other substances have been proposed to act as antioxidants *in vivo*. They include β-carotene, other carotenoids, xanthophylls, metallothionein, carnosine and related compounds, mucus, phytic acid, taurine and its precursors, bilirubin, estrogens, creatinine, ergothioneine, dihydrolipoic acid, ovothiols, ubiquinol, polyamines, retinol, flavonoids, other phenolic compounds of plant origin, and melatonin. Some drugs already approved for administration to humans, such as nonsteroidal anti-inflammatory drugs, angiotensin converting enzyme (ACE) inhibitors, Ca^{2+} antagonists, iron-chelating agents, and thiol compounds, are frequently suggested to exert antioxidant properties *in vivo* (reviewed in Halliwell, 1991).

How can such claims be evaluated? First, one must ask how the putative antioxidant is supposed to act. Does it act directly (e.g., by scavenging ROS or inhibiting their generation)? This is the most common proposal. Does it act indirectly (e.g., by upregulating endogenous antioxidant defenses)? For example, the antioxidant properties of melatonin have received much publicity recently (see the chapter by Reiter, this volume). Our data suggest that melatonin is a mediocre direct antioxidant (Marshall *et al.,* 1996), but this does not, of course, preclude it from acting by raising the levels of endogenous antioxidant defense enzymes.

In evaluating the likelihood of direct antioxidant action *in vivo,* it is important to ask certain questions (Table II). Simple *in vitro* experiments can answer some of these questions, and the results often allow one to dismiss the proposed antioxidant role: a compound that is a poor direct antioxidant *in vitro* will not be any better *in vivo.* This chapter outlines some approaches to the characterization of direct antioxidant activity. Two obvious (but often forgotten) points:

1. A compound should be tested at concentrations achievable *in vivo.*
2. In assaying putative antioxidants, biologically relevant ROS/RNS should be used.

Interest is also growing in reactive nitrogen species in biological systems (Table I). Low molecular mass antioxidants such as GSH, other thiol compounds, albumin, urate, and ascorbate may scavenge $ONOO^-$ (Beckman *et al.,* 1994) and urate may help protect against NO_2˙ in the respiratory tract (Halliwell *et al.,* 1992).

II. Biologically Relevant ROS/RNS _____

A. Superoxide and Hydrogen Peroxide

Superoxide formed *in vivo* is largely converted by superoxide dismutase (SOD)-catalyzed or nonenzymic dismutation into H_2O_2 (Fridovich, 1989). Some enzymes (e.g., glycollate, monoamine, xanthine oxidases) produce H_2O_2 directly (Sies, 1991; Halliwell and Gutteridge, 1989). Unlike O_2˙⁻,

TABLE II Questions to Ask When Evaluating "Antioxidants" *in Vivo*

---

1. What biomolecule is the antioxidant supposed to protect? Does enough antioxidant reach that target *in vivo?*
2. How does it protect: by scavenging ROS/RNS, preventing their formation, or repairing damage?
3. If the antioxidant acts by scavenging, can the resulting antioxidant-derived radicals themselves cause damage?
4. Can the antioxidant cause damage in other biological systems (toxicological studies)?

H_2O_2 is believed to cross membranes easily (Halliwell and Gutteridge, 1989). Both $O_2 \cdot^-$ and H_2O_2 can damage a few cellular targets (Fridovich, 1989; Flint *et al.*, 1993; Cochrane, 1991; Bielski, 1985), but, in general, their reactivity is limited, i.e., few compounds react fast with $O_2 \cdot^-$ or H_2O_2 [exceptions for O_2 are $NO^\cdot$ (Huie and Padmaja, 1993) and some iron–sulfur proteins (Flint *et al.*, 1993)]. Hence it is rather difficult to design an antioxidant that scavenges $O_2 \cdot^-$ or H_2O_2 at high rates other than by developing models of such enzymes as SOD or glutathione peroxidase.

I. Measuring Superoxide Scavenging

Superoxide is easily produced by the radiolysis of water in the presence of O_2 and formate, and pulse radiolysis allows the accurate determination of reaction rate constants as well as examination of the spectrum of products formed when $O_2 \cdot^-$ reacts with scavengers/SOD mimics (Butler *et al.*, 1988). However, pulse radiolysis is unsuitable for measuring slow (rate constants $<10^5\ M^{-1}\ \text{sec}^{-1}$) reactions for $O_2 \cdot^-$ in aqueous solution. Stopped-flow methods can be used to study slower reactions (Bull *et al.*, 1983). However, approximate rate constants may be obtained using simple "test-tube" methods. For example, xanthine oxidase plus hypoxanthine (or xanthine) at pH 7.4 can be used to generate $O_2 \cdot^-$, which is detected by its ability to reduce cytochrome c or nitro blue tetrazolium (NBT) (Fridovich, 1989). If an $O_2 \cdot^-$-reactive molecule is added, it decreases the rates of cytochrome c or NBT reduction, and competition plots allow calculation of approximate rate constants from the known rate constants for reaction of $O_2 \cdot^-$ with the just-described molecules (Bielski, 1985; Halliwell, 1985). This approach can be used with other sources of $O_2 \cdot^-$ [e.g., the potassium salt of superoxide, KO_2, can be used directly (Henry *et al.*, 1976)] but anyone using it should always perform some essential controls:

1. Check that the "antioxidant" does not directly inhibit $O_2 \cdot^-$ generation. Many papers use xanthine oxidase to generate $O_2 \cdot^-$, but effects of putative $O_2 \cdot^-$ scavengers on the enzyme itself are often not reported.
2. Check that the "antioxidant" does not directly reduce cytochrome c or NBT, which will deplete their levels and hence decrease their reaction rates with $O_2 \cdot^-$.
3. Consider the possibility that a radical formed by the attack or $O_2 \cdot^-$ on an "antioxidant" interacts with cytochrome c or NBT.

2. Measuring H_2O_2 Scavenging

H_2O_2 can be sensitively measured by peroxidase-based assay systems, e.g., horseradish peroxidase uses H_2O_2 to oxidize scopoletin into a nonfluorescent product (Corbett, 1989). If an "antioxidant" is incubated with H_2O_2 and the reaction mixture is sampled at various times, the rates of H_2O_2 disappearance can be calculated. Some points to consider:

1. Check that the "antioxidant" is not a substrate for peroxidase, which could decrease the fluoresence changes by competing with scopoletin for the enzyme rather than by scavenging H_2O_2.

2. Consider adding SOD if $O_2\cdot^-$ is also being generated. $O_2\cdot^-$ inhibits peroxidase (forming compound III) and may compromise the measurement of H_2O_2 (Kettle et al., 1994).

If an "antioxidant" interferes with peroxidase-based systems, other assays for H_2O_2 can be used, including titration with acidified $KMnO_4$, measuring the O_2 release (1 mol per 2 mol of H_2O_2) when a sample of the reaction mixture is injected into an O_2 electrode containing catalase and buffer solution, or measuring the release of $^{14}CO_2$ from ^{14}C-labeled 2-oxoglutarate (Varma, 1989).

B. Hydroxyl Radical

Much of the damage done by $O_2\cdot^-$ and H_2O_2 in vivo is thought to be due to their conversion into more reactive species, including hydroxyl radical (OH˙) (Halliwell and Gutteridge, 1989, 1990). Formation of OH˙ in vivo occurs by at least four mechanisms:

1. Transition metal ion catalysis, especially by iron and copper (Halliwell and Gutteridge, 1990). Reactive species additional to OH˙ [e.g., perferryl, ferryl or Cu(III)] may also be formed, although direct chemical evidence for their existence in Fenton-type systems is lacking.

2. Exposure to radiation (von Sonntag, 1988).

3. Reaction of HOCl with $O_2\cdot^-$ (Candeias et al., 1993).

4. Possibly, during the decomposition of $ONOO^-$ (Beckman et al., 1994; van der Vliet et al., 1994a).

I. Reactions of Hydroxyl Radical

The hydroxyl radical attacks almost every molecule found in vivo, with rate constants of $\geq 10^9\ M^{-1}\ sec^{-1}$ (von Sonntag, 1988). Thus, almost everything in a cell is an OH˙ scavenger. Examples are glucose (rate constant $\sim 10^9\ M^{-1}\ sec^{-1}$, present at millimolar concentrations in body fluids) and albumin [rate constant $> 10^{10}\ M^{-1}\ sec^{-1}$ (Smith et al., 1992)]. Hence, suggestions that natural or synthetic "antioxidants" act by scavenging OH˙ in vivo are chemically unlikely.

An "antioxidant" that affects OH˙-dependent damage in vivo is more likely to act by blocking OH˙ formation, e.g., by removing its precursors ($O_2\cdot^-$, H_2O_2, $ONOO^-$, HOCl), or by chelating transition metal ions. Because there is some confusion in the literature about metal-chelating agents, it is worth listing some basic principles as to how they can work.

1. The binding of a metal ion to the chelator could alter its redox potential and/or accessibility (e.g., to $O_2\cdot^-$ or H_2O_2) so as to stop it catalyzing

redox reactions. An example is the binding of iron to transferrin or lactoferrin (reviewed in Halliwell and Gutteridge, 1989, 1990).

2. Their binding to the "chelator" does not stop $OH^{\cdot}$ formation, but because $OH^{\cdot}$ is formed at the binding site, the "chelator" absorbs it and "spares" a more important target. For example, copper ions bound to albumin can still form $OH^{\cdot}$, and the protein is damaged (Marx and Chevion, 1986). The albumin, by binding copper and targeting damage to itself, is acting as a "sacrificial antioxidant" that protects more important targets, such as plasma lipoproteins and cell membranes (Halliwell, 1988). The binding of copper ions to the amino acid histidine in plasma might also be a protective mechanism: formation of $OH^{\cdot}$ radicals detectable in free solution is suppressed (Rowley and Halliwell, 1983), but the histidine is destroyed (Uchida and Kawakishi, 1986). Both albumin and histidine might thus represent safe temporary transport forms for plasma copper ions absorbed from the gut, until they are cleared from the circulation by the liver. Histidine-containing dipeptides, found in many mammalian tissues, may also act as antioxidants by copper ion chelation (Kohen et al., 1988).

2. Measuring Hydroxyl Radical Scavenging

The definitive technique for investigating reactions of "antioxidants" with $OH^{\cdot}$ is pulse radiolysis (Bielski, 1985; Butler et al., 1988), but approximate rate constants can be obtained more easily. For example, the spin-trap 5,5-dimethyl-1-pyrroline-N-oxide (DMPO) reacts fast with $OH^{\cdot}$ (rate constant $> 10^9 \ M^{-1} \ sec^{-1}$). An added $OH^{\cdot}$ scavenger will compete for $OH^{\cdot}$ and decrease the DMPO–OH electron spin resonance (ESR) signal, and its reaction rate constant can be obtained from a competition plot (Finkelstein et al., 1980). DMPO also reacts with $O_2^{\cdot -}$, although much more slowly [rate constant $\sim 10 \ M^{-1} \ sec^{-1}$ (Finkelstein et al., 1980)] to give a different adduct, and $O_2^{\cdot -}$ scavenging can similarly be measured by examining the effects on the ESR signal of the DMPO–OOH adduct.

Another example of a detector for $OH^{\cdot}$ is the sugar 2-deoxy-D-ribose (Halliwell et al., 1987). In the "deoxyribose method" for studying reactions with $OH^{\cdot}$, the $OH^{\cdot}$ is generated by a mixture of ascorbic acid, H_2O_2, and Fe^{3+}–EDTA. It attacks deoxyribose, degrading it into fragments that give a chromogen on heating with thiobarbituric acid (TBA) at low pH. If an $OH^{\cdot}$ scavenger is added, it competes with deoxyribose for $OH^{\cdot}$, inhibits chromogen formation, and competition plots allow the calculation of rate constants.

Competition methods are widely used, but key controls are often not reported in the literature. It is essential to show that

1. The "antioxidant" does not interfere with $OH^{\cdot}$ generation, e.g., by reacting with $OH^{\cdot}$ precursors in the reaction mixture, such as H_2O_2 or, if metal ion-dependent systems are used to make $OH^{\cdot}$, by chelating them (the deoxyribose assay avoids this by using iron ions chelated to EDTA).

2. The "antioxidant" does not interfere with product measurement. It should not inhibit when added to the reaction mixture at the end of the incubation, e.g., with the TBA reagents during the deoxyribose assay, or after the DMPO–OH ESR spectrum has developed. Many compounds, e.g., ascorbate and certain thiols, such as captopril (Bartosz et al., 1996), reduce DMPO–OH to an ESR silent species.

3. In chromogenic assays such as the deoxyribose assay, the attack of OH· on the "antioxidant" does not generate a false chromogen, e.g., omitting deoxyribose from the reaction mixture should eliminate color development.

3. Inhibition of Metal Ion-Dependent Hydroxyl Radical Formation

Some "antioxidants" may block OH· formation by chelating metal ions. The deoxyribose method can also be used to test this possibility. When iron ions are added to the reaction mixture as $FeCl_3$ (not EDTA chelated), some of them bind to deoxyribose. They still appear to catalyze OH· formation, but because the OH· immediately attacks deoxyribose, OH· scavengers (at moderate concentrations) do not inhibit chromogen formation (Gutteridge, 1984). However, an "antioxidant" can inhibit in this version of the deoxyribose assay if it chelates iron ions away from the deoxyribose and renders them inactive or poorly active in generating OH·. For example, citrate, a poor OH· scavenger but a good iron chelator, is a very effective inhibitor in this version of the deoxyribose assay but not when EDTA is present (Aruoma and Halliwell, 1988). Hence, this version of the deoxyribose assay (subject to the controls described earlier) indicates the potential ability of a compound to interfere with the "site-specific" generation of OH· (Gutteridge, 1984; Aruoma and Halliwell, 1988).

If an "antioxidant" binds metal ions and decreases the amount of OH· detected, two possibilities exist:

1. The "antioxidant"–metal ion complex cannot catalyze OH· formation.

2. OH· is still made but is largely intercepted by the antioxidant. To distinguish between these mechanisms, the fate of the antioxidant in the reaction mixture can be examined; it will be chemically modified if it reacts with OH·.

C. Peroxyl Radicals

The formation of peroxyl radicals ($RO_2·$) is a key step in lipid peroxidation (Halliwell and Gutteridge, 1989), but they can also be formed from DNA and proteins and when thiyl (RS·) radicals combine with oxygen (Dean et al., 1993; Sevilla et al., 1989; Willson, 1985). Peroxyl radical scavengers may be water soluble (e.g., dealing with radicals from DNA, thiols, proteins) or lipid soluble (e.g., the chain-breaking antioxidant inhibitors of lipid peroxidation).

1. Measuring Peroxyl Radical Scavenging

Peroxyl radicals can be generated by pulse radiolysis (Willson, 1985) and their reactions with "antioxidants" studied. ESR spin-trapping studies of peroxyl radical reactions can also be carried out (e.g., Davies *et al.*, 1993).

Another approach is the total (peroxyl) radical-trapping antioxidant parameter (TRAP) assay (Wayner *et al.*, 1987), which is much used (in various versions) to study antioxidants in biological fluids. Peroxyl radicals are generated at a controlled rate by the thermal decomposition of a water-soluble "azo initiator," such as 2,2'-azobis(2-amidinopropane)hydrochloride (AAPH). Decomposition produces carbon-centered radicals, which react fast with O_2 to give peroxyl radicals that then attack a lipid to cause peroxidation. By analyzing the effect of an "antioxidant" on the lag time to onset and the rate of peroxidation, information about its mechanism of action (Wayner *et al.*, 1987) and a relative rate for its reaction with RO_2˙ (Darley-Usmar *et al.*, 1989) can be obtained. Lipid "targets" can be endogenous lipids in biological fluids or added lipids, often linoleic acid/ester. Studies of the ability to protect against AAPH-induced peroxidation have been used to show, for example, that ascorbate is an excellent scavenger of water-soluble RO_2˙ (Wayner *et al.*, 1987; Darley-Usmar *et al.*, 1989), whereas desferrioxamine is not as good (Darley-Usmar *et al.*, 1989). AAPH-derived radicals also inactivate the enzyme lysozyme, providing a protein target for studies of protection by "antioxidants" (Lissi and Clavero, 1990; Paya *et al.*, 1992).

The carbon-centered radicals produced by AAPH decomposition can do direct damage [e.g., to DNA (Hiramoto *et al.*, 1993)] and can deplete antioxidants (Soriani *et al.*, 1994). Thus, one must ensure that reaction mixtures contain enough O_2 to convert them completely into peroxyl radicals.

2. Lipid-Soluble Peroxyl Radicals

It is difficult to generate "clean" lipophilic peroxyl radicals *in vitro*. One exception is trichloromethylperoxyl, formed by exposing a mixture of CCl_4, propan-2-ol, and buffer to ionizing radiation (Alfassi *et al.*, 1993). Rate constants for the reaction of several "antioxidants" with CCl_3O_2˙ have been determined (reviewed in Aruoma *et al.*, 1995). However, because CCl_3O_2˙ is more reactive than nonhalogenated peroxyl radicals, the results should be taken only as approximations of relative reactivity with the peroxyl radicals formed during lipid peroxidation.

D. Studies of Lipid Peroxidation

To test lipid antioxidant activity directly, the ability of an "antioxidant" to inhibit the peroxidation of lipoproteins, tissue homogenates, fatty acid/

ester emulsions, liposomes, or membranes (e.g., erythrocytes, liposomes, microsomes) can be examined. Such studies are widespread and cannot be discussed in detail here, but a few points can be made.

1. The lipid systems are usually kept under ambient pO_2, but some antioxidants [e.g., β-carotene (Burton and Ingold, 1984)] work better at lower pO_2. Variable results can arise if rapid peroxidation depletes O_2 during the reaction. This can be a special problem in ESR studies using narrow-bore ESR tubes.

2. Accurate measurement of peroxidation is not easy (reviewed in Packer, 1994). One *must* ensure that an apparent "antioxidant" action is not caused by interference with the assay. For example, much of the "lipid-peroxidation inhibitory effect" of carnosine and anserine in microsomes is due to interference with the TBA test (Aruoma *et al.*, 1989).

3. How lipid peroxidation is started. If azo initiators (e.g., AAPH) are used, it can be difficult to distinguish whether an antioxidant is acting by direct scavenging of the initiator-derived peroxyl radicals or by scavenging the chain-propagating peroxyl radicals generated from the lipid substrate. Another problem is that lipophilic antioxidants added to reaction mixtures seldom partition completely into membranes or lipoproteins (e.g., α-tocopherol added to low density lipoprotein (LDL) suspensions enters the LDL very inefficiently).

Lipid peroxidation is often started by adding metal ions, e.g., as $CuSO_4$ (for LDL), $FeSO_4$, $FeCL_3$ plus ascorbate, or $FeCL_3$-ADP plus NADPH (for microsomes). In these cases, an "antioxidant" effect could occur not only by peroxyl radical scavenging, but also by metal ion chelation. However, these two possibilities can be distinguished. If the antioxidant is acting only by chelation, it will not be consumed during the reaction, as shown by, for example, HPLC analysis. A chain-breaking antioxidant is consumed as it scavenges peroxyl radicals.

1. Microsomal Peroxidation Assays

These are extremely popular, and some authors appear to think that "microsomes" are subcellular organelles! In fact, "microsomes" prepared by differential centrifugation of tissue homogenates are a complex mixture of vesicles from endoplasmic reticulum, plasma membrane, and other cell membranes. They contain variable amounts of endogenous antioxidants, such as α-tocopherol. Hence an "antioxidant" inhibiting microsomal lipid peroxidation could (in addition to the mechanisms discussed earlier) be acting by "recycling" endogenous antioxidants. For example, dihydrolipoate does not inhibit iron/ascorbate-dependent peroxidation in liposomes (Scott *et al.*, 1994), but it recycles the vitamin E radical in microsomes to inhibit peroxidation (Scholich *et al.*, 1989). If microsomal lipid peroxidation is started by adding NADPH plus Fe^{3+}-ADP, a control (usually measuring NADPH consumption) is needed to check that the "antioxidant" does not inhibit the enzymic reduction of Fe^{3+}-ADP. Because the addition of NADPH to microsomes activates cytochromes P450, added "antioxidants" could be metabolized to products more (or less) active in inhibiting peroxidation.

Microsomal lipid peroxidation is often started by adding iron/ascorbate. This avoids problems with P450 activation. However, ascorbate may be capable of reducing "antioxidant" radicals (generated as they scavenge peroxyl radicals) back to the antioxidant, thus enhancing their action. Such a reaction will only occur if the antioxidant-derived radicals become accessible for reduction at the membrane surface. The classic example is the recycling of the α-tocopheryl radical by ascorbate, but the same effect may occur with other antioxidants, such as certain flavonoids. Hence, the antioxidant activity of some lipid-soluble, chain-breaking antioxidants may appear to be greater if ascorbate is present.

Overall, it is wise to evaluate anti-lipid peroxidation ability using several different lipid substrates, with peroxidation started by different mechanisms.

E. Hypochlorous Acid

Neutrophils contain myeloperoxidase, which uses H_2O_2 to oxidize Cl^- into HOCl (Weiss, 1989). Eosinophils contain a similar enzyme, which prefers to oxidize bromide (Br^-) ions and presumably produces HOBr (Mayeno et al., 1989). Hypohalous acids contribute not only to phagocyte killing of foreign organisms (although the extent of the contribution is uncertain), but also to tissue damage. For example, HOCl inactivates α_1-antiproteinase, an important inhibitor of certain serpins, such as elastase (Weiss, 1989). HOCl is a powerful oxidizer of -SH groups on cell surfaces and can inhibit membrane transport systems, as well as leading to chlorination of tyrosine residues (Weiss, 1989; Kettle, 1996).

1. Hypochlorous Acid Scavenging Assays

"Antioxidants" that prevent HOCl-mediated damage could scavenge HOCl directly and/or inhibit its production by myeloperoxidase. Myeloperoxidase can be assayed by standard tests of peroxidase activity (Halliwell and Gutteridge, 1989) (e.g., oxidation of guaiacol to a chromogen in the presence of H_2O_2) or by measuring its production of HOCl (Kettle and Winterbourn, 1988). Often the former type of assay is easier when looking for inhibitors because the latter type can give confusing results if compounds that also scavenge HOCl are tested. If an apparent inhibition of myeloperoxidase is found, it should be checked whether the "antioxidant" is really inhibiting myeloperoxidase or is simply acting as a competing substrate, e.g., several thiols are not only HOCl scavengers but also substrates for myeloperoxidase (Svensson and Lindvall, 1988). The plant phenol 4-hydroxy-3-methoxyacetophenone (apocynin) inhibits neutrophil $O_2^{\cdot-}$ release *in vitro*, apparently because it is oxidized by myeloperoxidase to generate the "real" inhibitor (Hart *et al.*, 1990).

Once established that an "antioxidant" does not inhibit myeloperoxidase, HOCl scavenging can then be examined using myeloperoxidase/H_2O_2/Cl^- as an HOCl source. More simply, HOCl can be made by acidifying sodium hypochlorite (Green *et al.*, 1985). If a physiologically relevant concentration of an "antioxidant" is mixed with α_1-antiproteinase, a good HOCl scavenger should protect the α_1-antiproteinase against inactivation when HOCl is added (Wasil *et al.*, 1987). Controls are needed to show that the "antioxidant" does not interfere with this assay by:

1. inactivating elastase directly,
2. stopping α_1-antiproteinase from inhibiting elastase, or
3. reactivating α_1-antiproteinase after inactivation by HOCl.

If an "antioxidant" fails to protect α_1-antiproteinase against HOCl, it may be that

1. It reacts too slowly (if at all) with HOCl.
2. It reacts with HOCl to form a "long-lived" oxidant that is itself capable of inactivating α_1-antiproteinase (Weiss, 1989). Such products are formed, for example, when taurine reacts with HOCl. An alternative screening assay for HOCl scavenging involves testing to see if the "antioxidant" can prevent HOCl-dependent oxidation of 5-thio-2-nitrobenzoic acid (Ching *et al.*, 1994).

Even if an "antioxidant" could act by scavenging HOCl *in vivo*, the possibility of forming toxic reaction products should be considered (Uetrecht, 1983).

F. Heme Proteins/Peroxides

Mixtures of H_2O_2 with heme proteins (including hemoglobin and myoglobin) oxidize many substrates and catalyze lipid peroxidation, apparently by generating amino acid radicals and heme-associated oxo-iron species (Rao *et al.*, 1994; Kelman *et al.*, 1994; Evans *et al.*, 1994; Giulivi and Cadenas, 1994). Cytochrome c can also accelerate oxidative damage in the presence of peroxides (Evans *et al.*, 1994). Such reactions may contribute to ischemia–reperfusion injury, atherosclerosis, neurodegenerative disease, muscle injury, and chronic inflammation (Kelman *et al.*, 1994; Evans *et al.*, 1994; Giulivi and Cadenas, 1994; Rice-Evans *et al.*, 1989). The ability of a substance to react with activated heme proteins can be examined spectrophotometrically by looking for loss of the ferryl myoglobin (or hemoglobin) spectrum as the compound reduces it to the ferrous or ferric state (Giulivi and Cadenas, 1994; Rice-Evans *et al.*, 1989). "Antioxidants" can also be tested for the ability to inhibit oxidative protein damage or lipid peroxidation by heme protein/peroxide mixtures (Kelman *et al.*, 1994; Evans *et al.*, 1994; Giulivi and Cadenas, 1994; Rice-Evans *et al.*, 1989).

Exposure of heme proteins to excess H_2O_2 causes heme breakdown and iron ion release (Gutteridge, 1986). Some antioxidants, such as ascorbate, prevent this (Rice-Evans *et al.*, 1989), providing an additional assay method for testing the effects of "antioxidants." Another application of myoglobin/ H_2O_2 systems is the development of a colorimetric assay for "total antioxidant status." Myoglobin/H_2O_2 is used to oxidize 2,2'-azinobis(3-ethylbenzothiazoline-6-sulfonate) (ABTS) into a colored radical cation, $ABTS^{·+}$, which reacts with several antioxidants (Rice-Evans and Miller, 1994).

G. Peroxynitrite

Peroxynitrite is easily prepared (Beckman *et al.*, 1994), allowing its reactions with "antioxidants" to be investigated. Careful control of pH is essential; peroxynitrite solutions are highly alkaline and can be contaminated with H_2O_2. The bicarbonate content of the reaction mixture affects peroxynitrite reactivity (Beckman *et al.*, 1994; van der Vliet *et al.*, 1994b). The rate constants for the reaction of several scavengers with $ONOO^-$ have been determined (Beckman *et al.*, 1994; Pryor and Squadrito, 1995). Methods for testing the ability of antioxidants to protect against $ONOO^-$-mediated damage include studies of their ability to protect against inactivation of α_1-antiproteinase and nitration of tyrosine by $ONOO^-$ *in vitro* (Beckman *et al.*, 1994; Pryor and Squadrito, 1995; Whiteman *et al.*, 1996).

H. Singlet Oxygen

Oxygen has two singlet states, but the $^1\Delta g$ state is probably the most important. Singlet $O_2{}^1\Delta g$, although not a free radical, is a powerful oxidizing agent, able to attack several molecules, including polyunsaturated fatty acids. Singlet oxygen can be produced by photosensitization reactions (Sies, 1991; Halliwell and Gutteridge, 1989). It is also formed when O_3 reacts with human body fluids (Kanofsky and Sima, 1993), when $ONOO^-$ reacts with H_2O_2 (Di Mascio *et al.*, 1994), and by self-reaction of peroxyl radicals during lipid peroxidation (Wefers, 1987).

I. Assessing Singlet Molecular Oxygen Quenching

Singlet O_2 ($^1\Delta g$) can be generated by photosensitization reactions, but one must ensure that damage to a target molecule is due to singlet O_2 rather than to direct interactions with the excited state of the sensitizer or by reactions involving other ROS, such as $O_2{}^{·-}$ and $OH^·$, that are often generated in light/pigment systems (Halliwell and Gutteridge, 1989). A technique has been described (Midden and Wang, 1983) in which $^1O_2$ generated by an immobilized sensitizer is allowed to diffuse a short distance to reach the target molecule. Singlet O_2 can also be generated by the thermal decomposition of endoperoxides, such as 3,3'-(1,4-naphthylidene)dipropanoate (Wagner *et al.*, 1993). Quenching of O_2 ($^1\Delta g$) can often also be measured as a decrease in the lifetime of $^1O_2$ photoemission at 1270 nm using an infrared detector (Sundquist *et al.*, 1994).

III. Proving That a Putative Antioxidant Is Important *in Vivo*

The tests outlined in this chapter enable one to examine the possibility that a given compound acts directly as an antioxidant *in vivo*. The tests may clearly show that a direct antioxidant role is unlikely. Alternatively, they could show that an antioxidant action is feasible, in that the compound shows protective action at concentrations within the range present *in vivo*. How then does one prove that the compound actually does act as an antioxidant *in vivo*?

For some naturally occurring antioxidants, it has been possible to remove them and observe increased oxidative damage, e.g., mutants of *Escherichia coli* lacking both MnSOD and FeSOD show severe damage when grown aerobically (Touati, 1989). For dietary antioxidants, the effect of depletion can be studied, e.g., prolonged vitamin E deficiency in patients with disorders of intestinal fat absorption produces neurodegeneration (Harding *et al.*, 1985).

These approaches are not feasible for most putative "antioxidants." Evidence supporting their antioxidant role *in vivo* can be provided by at least two approaches:

1. Is the compound depleted under conditions of oxidative stress? For example, ascorbate is rapidly lost at sites of oxidative stress (Buettner and Juriewicz, 1993). However, antioxidant action *in vivo* need not result in antioxidant depletion; vitamin E and perhaps certain flavonoids can be "recycled" by ascorbate.

2. If an "antioxidant" scavenges radicals, is it degraded into products whose concentrations can be measured and shown to increase during oxidative stress? Thus, ascorbate produces ascorbate radical (Buettner and Juriewicz, 1993), and attack of ROS on urate gives allantoin, cyanuric acid, and parabanic acid (Kaur and Halliwell, 1990).

As far as the ability of nutrients and drugs to act as antioxidants *in vivo* is concerned, specific assays are being developed to measure rates of oxidative damage to protein, DNA, and lipid (reviewed in Halliwell, 1996). Steady-state and total body oxidative damage to these targets can now be approximated, providing a tool to examine the effects of "antioxidants" *in vivo*.

References

Alfassi, Z. B., Huie, R. E., and Neta, P. (1993). Rate constants for reaction of perhaloalkyl peroxyl radicals with alkanes. *J. Phys. Chem.* **97**, 6835–6838.

Aruoma, O. I., and Halliwell, B. (1988). The iron binding and hydroxyl radical scavenging action of anti-inflammatory drugs. *Xenobiotica* **18**, 459–470.

Aruoma, O. I., Laughton, M. J., and Halliwell, B. (1989). Carnosine, homocarnosine and anserine: Could they act as antioxidants *in vivo*? *Biochem. J.* **264**, 863–869.

Aruoma, O. I., Spencer, J. P. E., Butler, J., and Halliwell, B. (1995). Reaction of plant-derived and synthetic antioxidants with trichloromethylperoxyl radicals. *Free Rad. Res.* **22**, 187–190.

Bartosz, M., Kedziora, J., and Bartosz, G. (1996). The copper complex of captopril is not a superoxide dismutase mimic. *Free Rad. Res.* **24**, 391–396.

Beckman, J. S., Chen, J., Ischiropoulos, H., and Crow, J. P. (1994). Oxidative chemistry of peroxynitrite. *Methods Enzymol.* **233**, 229–240.

Bielski, B. H. J. (1985). Reactivity of HO_2/O_2^- radicals in aqueous soution. *J. Phys. Chem. Ref. Data* **14**, 1041–1100.

Buettner, G. R., and Jurkiewicz, B. A. (1993). Ascorbate free radical as a marker of oxidative stress: An EPR study. *Free Rad. Biol. Med.* **14**, 49–55.

Bull, C., McClune, G. J., and Fee, J. A. (1983). The mechanism of Fe-EDTA catalyzed superoxide dismutation. *J. Am. Chem. Soc.* **105**, 5290–5300.

Burton, G. W., and Ingold, K. U. (1984). β-Carotene, an unusual type of lipid antioxidant. *Science* **224**, 569–573.

Butler, J., Hoey, B. M., and Lea, J. S. (1988). The measurement of radicals by pulse radiolysis. *In* "Free Radicals, Methodology and Concepts" (C. Rice-Evans and B. Halliwell, eds.), pp. 457–479. Richelieu Press, London.

Candeias, L. P., Patel, K. B., Stratford, M. R. L., and Wardman, P. (1993). Free hydroxyl radicals are formed on reaction between the neutrophil-derived species superoxide anion and hypochlorous acid. *FEBS Lett.* **333**, 151–153.

Ching, T. L., de Jong, J., and Bast, A. (1994). A method for screening hypochlorous acid scavengers by inhibition of the oxidation of 5-thio-2-nitrobenzoic acid: Application to anti-asthmatic drugs. *Anal. Biochem.* **218**, 377–381.

Cochrane, C. G. (1991). Cellular injury by oxidants. *Am. J. Med.* **92**, (Suppl. 3C), 23S–30S.

Corbett, J. T. (1989). The scopoletin assay for hydrogen peroxide: A review and a better method. *J. Biochem. Biophys. Methods* **18**, 297–308.

Daniels, V. (1989). Oxidative damage and the preservation of organic artefacts. *Free Rad. Res. Commun.* **5**, 213–220.

Darley-Usmar, V. M., Hersey, A., and Garland, L. G. (1989). A method for the comparative assessment of antioxidants as peroxyl radical scavengers. *Biochem. Pharmacol.* **38**, 1465–1469.

Davies, M. J., Gilbert, B. C., and Haywood, R. M. (1993). Radical-induced damage to bovine serum albumin: Role of the cysteine residue. *Free Rad. Res. Commun.* **18**, 353–367.

Dean, R. T., Gieseg, S., and Davies, M. J. (1993). Reactive species and their accumulation on radical damaged proteins. *Trends Biochem. Sci.* **18**, 437–441.

Di Mascio, P., Bechara, E. J. H., Medeiros, M. H. G., Briviba, K., and Sies, H. (1994). Singlet molecular oxygen production in the reaction of peroxynitrite with hydrogen peroxide. *FEBS Lett.* **355**, 287–289.

Evans, P. J., Akanmu, D., and Halliwell, B. (1994). Promotion of oxidative damage to arachidonic acid and α_1-antiproteinase by anti-inflammatory drugs in the presence of the haem proteins myoglobin and cytochrome c. *Biochem. Pharmacol.* 48, 2173–2179.

Finkelstein, E., Rosen, G. M., and Rauckman, E. J. (1980). Spin trapping: Kinetics of the reaction of superoxide and hydroxyl radicals with nitrones. *J. Am. Chem. Soc.* 102, 4994–4999.

Flint, D. H., Tuminello, J. F., and Emptage, M. H. (1993). The inactivation of Fe-S cluster containing hydro-lyases by superoxide. *J. Biol. Chem.* 268, 22369–22376.

Frei, B., Forte, T. M., Ames, B. N., and Cross, C. E. (1991). Gas phase oxidants of cigarette smoke induce lipid peroxidation and changes in lipoprotein properties in human blood plasma. *Biochem. J.* 277, 133–138.

Fridovich, I. (1989). Superoxide dismutases: An adaptation to a paramagnetic gas. *J. Biol. Chem.* 264, 7761–7764.

Giulivi, C., and Cadenas, E. (1994). Ferrylmyoglobin: Formation and chemical reactivity toward electron-donating compounds. *Methods Enzymol.* 233, 189–202.

Green, T. R., Fellman, J. H., and Eicher, A. L. (1985). Myeloperoxidase oxidation of sulfur-centered and benzoic acid hydroxyl radical scavengers. *FEBS Lett.* 192, 33–36.

Gutteridge, J. M. C. (1984). Reactivity of hydroxyl and hydroxyl-like radical discriminated by release of thiobarbituric-acid-reactive material from deoxyribose, nucleosides and benzoate. *Biochem. J.* 224, 761–767.

Gutteridge, J. M. C. (1986). Iron promoters of the Fenton reaction and lipid peroxidation can be released from haemoglobin by peroxides. *FEBS Lett.* 201, 291–295.

Gutteridge, J. M. C., and Quinlan, G. J. (1992). Antioxidant protection against organic and inorganic oxygen radicals by normal human plasma: The important primary role for iron-binding and iron-oxidizing proteins. *Biochim. Biophys. Acta* 1159, 248–254.

Halliwell, B. (1985). Use of desferrioxamine as a probe for iron-dependent formation of hydroxyl radicals. *Biochem. Pharmacol.* 34, 229–233.

Halliwell, B. (1988). Albumin: An important extracellular antioxidant? *Biochem. Pharmacol.* 37, 569–571.

Halliwell, B. (1990). How to characterize a biological antioxidant. *Free Rad. Res. Commun.* 9, 1–32.

Halliwell, B. (1991). Drug antioxidant effects: A basis for drug selection. *Drugs* 42, 569–605.

Halliwell, B. (1996). Oxidative stress, nutrition and health: Experimental strategies for optimization of nutritional antioxidant intake in humans. *Free Rad. Res.,* in press.

Halliwell, B., and Gutteridge, J. M. C. (1989). "Free Radicals in Biology and Medicine," 2nd Ed. Clarendon Press, Oxford.

Halliwell, B., and Gutteridge, J. M. C. (1990). Role of free radicals and catalytic metal ions in human disease: An overview. *Methods Enzymol.* 186, 1–85.

Halliwell, B., Gutteridge, J. M. C., and Aruoma, O. I. (1987). The deoxyribose method: A simple "test tube" assay for determination of rate constants for reactions of hydroxyl radicals. *Anal. Biochem.* 165, 215–219.

Halliwell, B., Hu, M. L., Louie, S., Duvall, T. R., Tarkington, B. R., Motchnik, P., and Cross, C. E. (1992). Interaction of nitrogen dioxide with human plasma. *FEBS Lett.* 313, 62–66.

Harding, A. E., Matthews, S., Jones, S., Ellis, C. J. K., Booth, I. W., and Muller, D. P. R. (1985). Spinocerebellar degeneration associated with a selective defect of vitamin E absorption. *N. Engl. J. Med.* 313, 32–35.

Hart, B. A. T., Simons, J. M., Knaan-Shanzer, S., Bakker, N. P. M., and Labadie, R. P. (1990). Antiarthritic activity of the newly developed neutrophil oxidative burst antagonist apocynin. *Free Rad. Biol. Med.* 9, 127–131.

Henry, L. E. A., Halliwell, B., and Hall, D. O. (1976). The superoxide dismutase activity of various photosynthetic organisms measured by a new and rapid assay technique. *FEBS Lett.* 66, 303–306.

Hiramoto, K., Johkoh, H., Sako, K., and Kikugawa, K. (1993). DNA breaking activity of the carbon-centered radical generated from 2,2'-azobis(2-amidinopropane)hydrochloride (AAPH). *Free Rad. Res. Commun.* 19, 323–332.

Hu, M. L., Louie, S., Cross, C. E., Motchnik, P., and Halliwell, B. (1992). Antioxidant protection against hypochlorous acid in human plasma. *J. Lab. Clin. Med.* **121**, 257–262.

Huie, R. E., and Padmaja, S. (1993). The reaction of NO with superoxide. *Free Rad. Res. Commun.* **18**, 195–199.

Kanofsky, J. R., and Sima, P. D. (1993). Singlet-oxygen generation at gas-liquid interfaces: A significant artifact in the measurement of singlet-oxygen yields from ozone-biomolecule reactions. *Photochem. Photobiol.* **58**, 335–340.

Kaur, H., and Halliwell, B. (1990). Action of biologically-relevant oxidizing species upon uric acid. *Chem. Biol. Interact.* **73**, 235–247, 1990.

Kelman, D. J., De Gray, J. A., and Mason, R. P. (1994). Reaction of myoglobin with hydrogen peroxide forms a peroxyl radical which oxidizes substrates. *J. Biol. Chem.* **269**, 7458–7463.

Kettle, A. J. (1996). Neutrophils convert tyrosyl residues in albumin to chlorotyrosine. *FEBS Lett.* **379**, 103–106.

Kettle, A. J., and Winterbourn, C. C. (1988). The mechanism of myeloperoxidase-dependent chlorination of monochlordimedon. *Biochim. Biophys. Acta* **957**, 185–191.

Kettle, A. J., Carr, A. C., and Winterbourn, C. C. (1994). Assays using horseradish peroxidase and phenolic substrates require superoxide dismutase for accurate determination of hydrogen peroxide production by neutrophils. *Free Rad. Biol. Med.* **17**, 161–164.

Kohen, R., Yamamoto, Y., Cundy, K. C., and Ames, B. N. (1988). Antioxidant activity of carnosine, homocarnosine and anserine present in muscle and brain. *Proc. Natl. Acad. Sci. USA* **85**, 3175–3179.

Lissi, E. A., and Clavero, N. (1990). Inactivation of lysozyme by alkylperoxyl radicals. *Free Rad. Res. Commun.* **10**, 177–184.

Marshall, K. A., Reiter, R. J., Poeggeler, B., Aruoma, O. I., and Halliwell, B. (1996). Evaluation of the antioxidant activity of melatonin *in vitro. Free Rad. Biol. Med.,* in press.

Marx, G., and Chevion, M. (1986). Site-specific modification of albumin by free radicals: Reaction with copper(II) and ascorbate. *Biochem. J.* **236**, 397–400.

Mayeno, A. N., Curran, A. J., Roberts, R. L., and Foote, C. S. (1989). Eosinophils preferentially use bromide to generate halogenating agents. *J. Biol. Chem.* **264**, 5660–5668.

Midden, W. R., and Wang, S. Y. (1983). Singlet oxygen generation for solution kinetics; clean and simple. *J. Am. Chem. Soc.* **105**, 4129–4135.

Packer, L. (ed.) (1994). Oxygen radicals in biological systems. *Methods Enzymol.* **233**, 1–711.

Paya, M., Halliwell, B., and Hoult, J. R. S. (1992). Peroxyl radical scavenging by a series of coumarins. *Free Rad. Res. Commun.* **17**, 293–298.

Pryor, W. A., and Squadrito, G. L. (1995). The chemistry of peroxynitrite: A product from the reaction of nitric oxide with superoxide. *Am. J. Physiol.* **268**, L699–L722.

Rao, S. I., Wilks, A., Hamberg, M., and Ortiz de Montellano, P. (1994). The lipoxygenase activity of myoglobin: Oxidation of linoleic acid by the ferryl oxygen rather than protein radical. *J. Biol. Chem.* **269**, 7210–7216.

Reznick, A. Z., Cross, C. E., Hu, M., Suzuki, Y. J., Khwaja, S., Safadi, A., Motchnik, P. A., Packer, L., and Halliwell, B. (1992). Modification of plasma proteins by cigarette smoke as measured by protein carbonyl formation. *Biochem. J.* **286**, 607–611.

Rice-Evans, C., and Miller, M. J. (1994). Total antioxidant status in plasma and body fluids. *Methods Enzymol.* **234**, 279–293.

Rice-Evans, C., Okunade, G., and Khan, R. (1989). The suppression of iron release from activated myoglobin by physiological electron donors and by desferrioxamine. *Free Rad. Res. Commun.* **7**, 45–54.

Rowley, D. A., and Halliwell, B. (1983). Superoxide-dependent and ascorbate-dependent formation of hydroxyl radicals in the presence of copper salts: A physiologically significant reaction? *Arch. Biochem. Biophys.* **225**, 279–284.

Roy, D., and Liehr, J. G. (1991). Elevated 8-hydroxydeoxy-guanosine levels in DNA of diethylstilboestrol-treated syrian hamsters: Covalent DNA damage by free radicals generated by redox cycling of diethylstilboestrol. *Cancer Res.* **51**, 3882–3885.

Scholich, H., Murphy, M. E., and Sies, H. (1989). Antioxidant activity of dihydrolipoate against microsomal lipid peroxidation and its dependence on α-tocopherol. *Biochim. Biophys. Acta* **1001**, 256–261.

Scot:, B. C., Aruoma, O. I., Evans, P. J., O'Neill, C., van der Vliet, A., Cross, C. E., Tritschler, H., and Halliwell, B. (1994). Lipoic and dihydrolipoic acids as antioxidants: A critical evaluation. *Free Rad. Res.* **20**, 119–133.

Sevilla, M. D., Yan, M., Becker, D., and Gillich, S. (1989). ESR investigations of the reactions of radiation-produced thiyl and DNA peroxyl radicals; formation of sulfoxyl radicals. *Free Rad. Res. Commun.* **6**, 21–24.

Sies, H. (ed.) (1991). "Oxidative Stress: Oxidants and Antioxidants." Academic Press, London.

Smith, C. A., Halliwell, B., and Aruoma O. I. (1992). Protection by albumin against the pro-oxidant actions of phenolic dietary components. *Food Chem. Toxicol.* **30**, 483–489.

Soriani, M., Pietraforte, D., and Minetti, M. (1994). Antioxidant potential of anaerobic human plasma: Role of serum albumin and thiols as scavengers of carbon radicals. *Arch. Biochem. Biophys.* **312**, 180–188.

Sundquist, A. R., Briviba, K., and Sies, H. (1994). Singlet oxygen quenching by carotenoids. *Methods Enzymol.* **234**, 384–388.

Svensson, B. E., and Lindvall, S. (1988). Myeloperoxidase-oxidase oxidation of cysteamine. *Biochem. J.* **249**, 521–530.

Touati, D. (1989). The molecular genetics of superoxide dismutase in *E. coli. Free Rad. Res. Commun.* **8**, 1–9.

Uchida, K., and Kawakishi, S. (1986). Selective oxidation of imidazole ring in histidine residues by the ascorbic acid-copper ion system. *Biochem. Biophys. Res. Commun.* **138**, 659–665.

Uetrecht, J. P. (1983). Idiosyncratic drug reactions: Possible role of reactive metabolites generated by leukocytes. *Pharmacol. Rev.* **6**, 265–273.

van der Vliet, A., O'Neill, C. A., Halliwell, B., Cross, C. E., and Kaur, H. (1994a). Aromatic hydroxylation and nitration of phenylalanine and tyrosine by peroxynitrite. *FEBS Lett.* **339**, 89–92.

van der Vliet, A., Smith, D., O'Neill, C. A., Kaur, H., Darley-Usmar, V., Cross, C. E., and Halliwell, B. (1994b). Interactions of peroxynitrite with human plasma and its constituents: Oxidative damage and antioxidant depletion. *Biochem. J.* **303**, 295–301.

Varma, S. D. (1989). Radio-isotopic determination of subnanomolar amounts of peroxide. *Free Rad. Res. Commun.* **5**, 359–368, 1989.

von Sonntag, C. (1988). "The Chemical Basis of Radiation Biology." Taylor & Francis, London.

Wagner, J. R., Motchnik, P. A., Stocker, R., Sies, H., and Ames, B. N. (1993). The oxidation of blood plasma and low density lipoprotein components by chemically generated singlet O_2. *J. Biol. Chem.* **268**, 18502–18506.

Wasil, M., Halliwell, B., Moorhouse, C. P., Hutchison, D. C. S., and Baum, H. (1987). Biologically-significant scavenging of the myeloperoxidase-derived oxidant hypochlorous acid by some anti-inflammatory drugs. *Biochem. Pharmacol.* **36**, 3847–3850.

Wayner, D. D. M., Burton, G. W., Ingold, K. U., Barclay, L. R. C., and Locke, S. J. (1987). The relative contributions of vitamin E, urate, ascorbate and proteins to the total peroxyl radical-trapping antioxidant activity of human blood plasma. *Biochim. Biophys. Acta* **924**, 408–419.

Wefers, H. (1987). Singlet oxygen in biological systems. *Bioelectrochem. Bioenerg.* **18**, 91–104.

Weiss, S. J. (1989). Tissue destruction by neutrophils. *N. Engl. J. Med.* **320**, 365–376.

Whiteman, M., Tritschler, H., and Halliwell, B. (1996). Protection against peroxynitrite-dependent tyrosine nitration and α_1-antiproteinase inactivation by oxidized and reduced lipoic acid. *FEBS Lett.* **379**, 74–76.

Willson, R. L. (1985). Organic peroxy free radicals as ultimate agents in oxygen toxicity. *In* "Oxidative Stress" (H. Sies, ed.), pp. 41–72. Academic Press, London.

Wiseman, H., and Halliwell, B. (1993). Carcinogenic antioxidants: Diethylstilboestrol, hexoestrol and 17α-ethynyloestradiol. *FEBS Lett.* **332**, 159–163.

Ann M. Bode

Physiology Department
School of Medicine and Health Sciences
University of North Dakota
Grand Forks, North Dakota 58201

Metabolism of Vitamin C in Health and Disease

I. Introduction and Overview

The amount of literature and research on vitamin C (ascorbic acid) and its influence in a variety of disease processes is overwhelming. A search of Medline from 1990 until the present will result in more than 4500 citations that include the key words vitamin C or ascorbic acid. In addition to original research articles, a number of review articles regarding vitamin C and its association with a variety of disease conditions also have been published since 1994. Diseases reviewed most recently appear to include vitamin C and diabetes (Crawford, 1995; Thompson and Godin, 1995); vitamin C and cardiovascular disease (Bendich and Langseth, 1995; Crawford, 1995; Defeudis, 1995; Gey, 1995); vitamin C and cancer (Bendich and Langseth, 1995; Crawford, 1995; Flagg *et al.*, 1995; Kodama *et al.*, 1995); vitamin C and asthma (Bielory and Gandhi, 1994); vitamin C and the common cold (Hemila and Herman,

1995); and vitamin C and cataract (Bendich and Langseth, 1995; Thompson and Godin, 1995). As is apparent from this list, a vast amount of literature exists regarding the association of vitamin C with the risk and/or occurrence of a variety of diseases.

Therefore, in an attempt not to be redundant, this chapter focuses on three aspects that together have not been extensively reviewed: (1) the most recent experimental studies where a *change* in the status of vitamin C is associated with a disease state; (2) the most recent evidence regarding the proposed mechanism(s) for the observed changes in vitamin C status in disease states; and (3) the effectiveness of vitamin C as a therapeutic or preventative agent in certain disease states.

II. Changes in Vitamin C Status Associated with Disease

A. Introduction

Ascorbic acid functions in a variety of roles, including (1) prevention of scurvy (Ghorbani and Eichler, 1994); (2) acceleration of hydroxylation reactions in the synthesis of collagen (Sauberlich, 1994), carnitine (Ha *et al.*, 1994), and norepinephrine (Friedman and Kaufman, 1965); (3) amidation of peptide hormones (Glembotski, 1984); (4) regeneration of vitamin E (Chan, 1993); and (5) protection against photooxidative damage in the (Organisciak *et al.*, 1985). Its effectiveness as a free radical scavenger has been reviewed by Rose and Bode (1993). Vitamin C is a very good antioxidant (Frei *et al.*, 1989) found at high levels in a variety of tissues in comparison to plasma levels. It is involved in many physiological chemical reactions in which it is subsequently oxidized to a compound known as dehydroascorbic acid. Accumulation of dehydroascorbic acid at micromolar (Rose *et al.*, 1992) concentrations has been suggested to disrupt cell membranes and act as a neurotoxin (Hisanaga *et al.*, 1992) and may cause islet cell dysfunction (Merlini and Caramia, 1965). In addition, injections of dehydroascorbic acid have for many years been known to be diabetogenic (Patterson, 1950).

In normal, healthy tissues, dehydroascorbic acid generally appears to be recycled immediately back to ascorbic acid. The suggestion has been made that the maintenance of a high ratio of ascorbic acid to dehydroascorbic acid in a particular tissue may be an indicator of the health of that tissue (Rose, 1989). A normal recycling between the oxidized and the reduced forms of the vitamin appears to be absolutely necessary for maintaining a normally high ratio. However, the physiological mechanisms for recycling and the consequences of impaired recycling are not clearly understood and may be much more important than has been generally believed.

Accumulated evidence suggests that a dysfunction in ascorbic acid metabolism might have a role in aging (Bates and Cowen, 1988) and in certain diseases, including diabetes, leukemia, ocular disease, and chronic inflammatory diseases (Rose, 1989). These types of studies suggest that alterations in vitamin C levels may be due either to a disease-related depletion or to an impairment of the recycling mechanisms for ascorbic acid. The implications of a dysfunction in recycling are that cells and tissues are persistently exposed to low levels of ascorbic acid and elevated levels of dehydroascorbic acid, resulting in increased susceptibility of the cell or tissue to oxidative damage.

B. Compromised Vitamin C Metabolism in Diabetes

Over the years, researchers in diabetes have found decreased leukocyte (Cunningham *et al.,* 1991), plasma, and selected tissue concentrations of ascorbic acid (Rikans, 1981; Yew, 1983) and persistently increased concentrations of dehydroascorbic acid (Yew, 1983). In the author's laboratory, severely depressed levels of ascorbic acid (50% of control) and increased dehydroascorbic acid levels in streptozotocin-induced diabetic rat liver and kidney compared to control rats have been found (Bode, Yavarow, *et al.,* 1993; Schell and Bode, 1993). At least one group has shown an inverse correlation between plasma and urinary ascorbic acid levels and glycosylated hemoglobin (Yue *et al.,* 1990) in diabetic patients. In some cases, supplementation with ascorbic acid has been shown to reduce the occurrence of cataracts and the γ-crystalline leakage from lenses in diabetic rats (Linklater *et al.,* 1990). However, in a study in which diabetic patients with microangiopathy received supplementation, Sinclair and colleagues (1991) concluded that a metabolic impairment exists in ascorbate metabolism in diabetes that was not correctable with supplementation. In addition to the streptozotocin-induced, diabetic rat model, which is used extensively, changes in ascorbic acid and dehydroascorbic acid metabolism have been observed in tissues and plasma of spontaneously diabetic BioBreeding rats during development and before the onset of diabetes (Behrens and Madera, 1991). The BioBreeding rat is considered by some (Marliss *et al.,* 1982) to be a better model for diabetes than the streptozotocin rat in that it resembles type I insulin-dependent diabetes more closely. The resemblance includes a spontaneous development of diabetes, including symptoms of weight loss, hypoinsulinemia, hyperglycemia, and ketoacidosis, and, unlike the streptozotocin model, will die if not maintained with insulin (Rodrigues and McNeill, 1992).

The relationship between glucose and ascorbic acid in diabetes is an interesting one. Some researchers have suggested that glucose and ascorbic acid may interfere with one another in gaining access to the cell. For example, in human fibroblast cultures, ascorbic acid stimulated collagen and proteoglycan production, but the stimulation was inhibited by 25 mM glucose (Fisher *et al.,* 1991), suggesting that high glucose levels in diabetes can interfere with the action of ascorbic acid and may explain partially the low levels of ascorbate observed in certain tissues.

In other studies, oxidative stress has been implicated in the complications of diabetes where changes in vitamin C levels were observed. Markers of oxidative stress, including increased levels of conjugated dienes and malondialdehyde in plasma, red cell membranes, and urine, were observed simultaneously with decreased levels of vitamin C, α-tocopherol, and retinol in diabetic rats (Young *et al.*, 1995). In the same study, a combination of ascorbate and desferrioxamine treatment returned all parameters to normal, leading the authors to conclude that treatment with antioxidants can reduce both oxidative stress and protein glycation and may help to reduce the risk of developing diabetic complications (Young *et al.*, 1995). However, based on studies with erythrocytes, in which glucose and ascorbic acid were both shown to undergo transition metal-catalyzed oxidation with hydrogen peroxide production, Ou and Wolff (1994) concluded that ascorbic acid oxidation was more likely to be a source of oxidative stress in diabetes than was hyperglycemia.

C. Compromised Vitamin C Metabolism in Other Diseases

Reduced and total ascorbic acid levels were observed to be significantly decreased in inflamed mucosa from Crohn's disease patients and in mucosa from ulcerative colitis patients (Buffinton and Doe, 1995). At the same time, reduction of dehydroascorbic acid by dehydroascorbate reductase was observed to be significantly decreased, leading Buffinton and Doe (1995) to suggest that the capacity of the inflamed mucosa to maintain the concentration of reduced ascorbic acid also was diminished. Depletion of ascorbic acid has been observed in acute pancreatitis (Braganza *et al.*, 1995), in weight-losing gastrointestinal cancer patients (Georgiannos *et al.*, 1993), in glutathione depletion by treatment of lung tumors with buthionine sulfoximine (BSO) (Thanislass *et al.*, 1995), in iron overload (Dabbagh *et al.*, 1994), in preeclampsia (Mikhail *et al.*, 1994), in hypertension (Tse *et al.*, 1994), in infection (Tanzer and Ozalp, 1993), in smoking (Schectman, 1993), in acute myocardial infarction (Singh *et al.*, 1994), 48 hr following cerebrovascular accident (Sharpe *et al.*, 1994), and following cardiopulmonary bypass surgery (Ballmer *et al.*, 1994).

The question is whether these types of alterations in vitamin C levels are due to a disease-related depletion, to an impairment of the recycling mechanisms for ascorbic acid, or to a combination of both.

III. Mechanism of Changes in Vitamin C Status Associated with Disease

Two questions must be answered to explain the differences observed in vitamin C levels in disease: (1) How does vitamin C get into the cell/tissue to perform its many functions; and (2) how are sufficient levels of vitamin C maintained in cells and tissues?

A. Transport of Vitamin C

At least two reviews have presented the latest thoughts on the transport and metabolism of ascorbic acid. One group presents an overview of transport in the eye (Rose and Bode, 1991) and the other (Goldenberg and Schweinzer, 1994) gives a general overview of transport systems in animal and human tissues, highlighting the similarities of ascorbate transporters and glucose transporters and the influence of sodium on ascorbate transport in many tissues.

Transport systems appear to be necessary for a sufficient supply of vitamin C, even though most mammals have the capacity to synthesize vitamin C from glucose. However, humans, primates, and guinea pigs lack a functional gulonolactone oxidase and therefore must obtain an adequate supply of vitamin C in their diet. In comparison to plasma, many tissues appear to have an increased need to conserve or sequester the reduced form of vitamin C. Plasma vitamin C levels are about 40 μM whereas cellular levels as high as 14 mM have been observed under certain conditions (Washko et $al.$, 1993) (i.e., in activated neutrophils). Conversely, as stated earlier, the oxidized form, dehydroascorbic acid, is normally found at very low levels in plasma and tissues. For example, in the various ocular tissues, ascorbic acid levels range from 400 μM to about 2 mM and dehydroascorbic acid levels are very low (Fig. 1). Tissues therefore are obviously capable of transporting vitamin C into cells against a concentration gradient.

The mechanisms of vitamin C transport and accumulation are a topic of intense investigation. Research has shown that both the reduced and the oxidized forms of vitamin C are transported. Two mechanisms have been suggested for the ability of a cell to accumulate ascorbic acid: (1) active transport systems driven by ATP hydrolysis directly or indirectly via a sodium gradient; and (2) preferential uptake of dehydroascorbic acid and subsequent intracellular reduction.

Ascorbic acid has been shown to accumulate in normal (Washko et $al.$, 1990) neutrophils and in activated human neutrophils to a level as high as 10-fold above the millimolar concentrations present in the normal neutrophils (Washko et $al.$, 1993). Earlier, isolated human mononuclear leukocytes were found by this same group to maintain a large concentration gradient of ascorbic acid across the plasma membrane (Bergsten et $al.$, 1990). The suggestion has been made that neutrophils accumulate high levels of ascorbic acid to protect themselves from oxidative stress (Wolf, 1993).

Substantial evidence has been accumulated suggesting that the reduced form, ascorbic acid, is transported into many tissues by a two-component system: a high-affinity transporter and a low-affinity transporter. These activities have been studied in normal human fibroblasts (Butler et $al.$, 1991; Welch et $al.$, 1993) and in human B lymphocytes (Bergsten et $al.$, 1995).

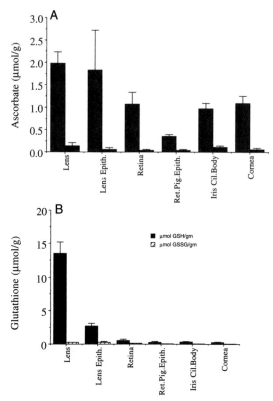

FIGURE I   (A) Ascorbate and dehydroascorbate concentration in various bovine ocular tissues. Compounds were measured according to Schell and Bode (1993). ☐, μmol ascorbic acid/g tissues; ▨, μmol dehydroascorbic acid/g tissue. (B) Glutathione concentration in various bovine ocular tissues. Glutathione was measured according to Griffith (1986). ■, μmol glutathione/g tissue; ▨, μmol glutathione disulfide/g tissue.

The high-affinity transporters have reported K_m values within the physiological range (5–150 μM) of plasma levels whereas the low-affinity transporters are reported to have apparent K_m values of 1 to 5 mM. The mechanism of transport was earlier suggested to involve the oxidation of the external ascorbic acid to dehydroascorbic acid, preferential transmembrane translocation of dehydroascorbic acid, and intracellular reduction to ascorbic acid within minutes (Washko *et al.*, 1993). A similar suggestion was made by another researcher at approximately the same time (Wolf, 1993). The former group presented evidence that ascorbate and dehydroascorbic acid are transported by two distinct mechanisms and that dehydroascorbic acid transport and accumulation as ascorbic acid occurred faster than ascorbate transport (Welch *et al.*, 1995) in human neutrophils and fibroblasts. In addition, they showed that ascorbate transport was sodium dependent and that dehydroascorbic acid transport was sodium independent (Welch *et al.*, 1995).

Glucose has been shown to inhibit the uptake and accumulation of ascorbic acid by both transport activities in human neutrophils in a concentration-dependent manner (Washko et al., 1991), and Washko and Levine (1992) reported later that the inhibition of the high-affinity transport activity by glucose was noncompetitive, whereas inhibition of the low-affinity activity by glucose was competitive. Ascorbic acid has been shown to inhibit glucose-induced insulin release from single pancreatic islets in a dose-dependent and completely reversible manner (Bergsten et al., 1994). These types of studies suggest that glucose and ascorbate may share the same transport system and are key to our understanding of ascorbic acid metabolism.

In an important study by Vera and colleagues (1993), transport and accumulation of vitamin C was investigated in an *Xenopus laevis* oocyte system in which the GLUT1 transporter was overexpressed. This group also studied the transport and accumulation of vitamin C in human neutrophils (Vera et al., 1993) and in a human myeloid leukemia cell line (HL-60 cells) (Vera et al., 1994). These authors identified two functional activities for the transport of dehydroascorbic acid, a high- and low-affinity transport activity (Vera et al., 1993, 1994), confirming the work of others discussed earlier. Based on these studies, Vera and colleagues (1993, 1994) concluded that mammalian facilitative hexose transporters are efficient transporters of dehydroascorbic acid in the systems studied.

In almost every cell type studied where transport of dehydroascorbic acid occurred, dehydroascorbic acid accumulated as the reduced form intracellularly. The mechanism or the site for the reduction of dehydroascorbic acid to ascorbic acid is not known; however, data suggest that the transport of extracellular dehydroascorbic acid and the intracellular reduction and accumulation of ascorbic acid are separate processes, at least in HL-60 cells (Vera et al., 1995). Significantly, Vera and colleagues (1995) were able to show that the transport of dehydroascorbic acid by HL-60 cells is sodium independent and probably occurs via glucose transporters whereas the subsequent intracellular reduction of dehydroascorbic acid and the accumulation of ascorbic acid are sodium sensitive and occur independently of the transport process.

Rat cerebral astrocytes also have been shown to accumulate millimolar concentrations of ascorbate against a concentration gradient (Siushansian and Wilson, 1995) by a sodium-dependent process. In addition, when dibutyryl cyclic AMP was added, uptake and efflux of ascorbic acid were increased, leading Siushansian and Wilson (1995) to conclude that dibutyryl cyclic AMP was responsible for the regulation of the cytosolic concentration of ascorbate in these cells. Transforming growth factor-β was shown to regulate ascorbate transport and concentration in UMR-106 rat osteosarcoma cells (Wilson and Dixon, 1995). These cells also accumulate and reduce vitamin C from culture medium by a sodium-dependent process

(Wilson and Dixon, 1995). Other studies of cells and tissues that preferentially take up, reduce dehydroascorbic acid, and accumulate it as ascorbic acid include pig lacrimal gland (Dreyer and Rose, 1993) and cerebrum (Rose, 1993) and bovine corneal endothelial cells (Bode *et al.*, 1991).

Ascorbate transport and metabolism in response to an oxidant stress have been investigated in human erythrocytes (May *et al.*, 1995a,b). These studies confirmed that erythrocytes transport both the reduced and the oxidized forms of ascorbate and accumulate dehydroascorbic acid as ascorbic acid at rates greater than they do the reduced form (May *et al.*, 1995a). These researchers suggest further that erythrocytes may contribute to the antioxidant capacity of blood by regenerating ascorbate and possibly by exporting ascorbate-derived reducing equivalents through a transmembrane oxidoreductase (May *et al.*, 1995b).

B. Recycling/Regeneration of Vitamin C

Despite the progress being made regarding the transport mechanisms for ascorbic acid and dehydroascorbic acid, many aspects remain unclear. One of the most confusing issues surrounds the mechanism(s) and/or sites for the reduction of dehydroascorbic acid and the maintenance of the reduced form both intracellularly and extracellularly. The fact that extracellular dehydroascorbic acid accumulates intracellularly as ascorbic acid is indisputable; however, the exact mechanism or site for the reduction is not known. One may ask why the recycling of vitamin C is important, especially when ascorbic acid can be obtained in abundant amounts in the diet. However, controversy exists as to what amount is actually needed to maintain optimal health. Current RDA levels are based on the minimum amount needed to prevent scurvy; whether this amount is adequate under all conditions is not known. Because many tissues concentrate ascorbate at levels greater than that found in the plasma, the suggestion is that they have an increased need for the vitamin. Overwhelming amounts of evidence suggest that ascorbic acid plays a major role in protecting tissues from oxidative stress and free radical damage. Ascorbic acid and dehydroascorbic acid levels have been shown to be altered with diabetes, cataract, heart disease, cancer, and aging where researchers have found decreased plasma, leukocyte, and selected tissue concentrations of ascorbic acid and persistently increased concentrations of the oxidized form, dehydroascorbic acid. As stated earlier, research suggests that the alteration in vitamin C levels may be due either to a disease-related depletion or to an impairment of the recycling mechanisms for ascorbic acid.

In humans, some suggestion has been made that reduced ascorbic acid levels in diabetes may be due to an increased excretion of the vitamin or to an inadequate dietary intake. However, at least one study has confirmed that reduced ascorbic acid levels in insulin-dependent (type I) diabetes mellitus are not due to an increased urinary excretion of ascorbic acid (Seghieri *et al.*, 1994), and results of a vitamin C questionnaire and a 4-day food diary in 30 diabetic patients led Sinclair and colleagues (1994) to conclude that the low ascorbic acid levels are a consequence of the disease and are not due to an inadequate intake. In addition, differences observed between women and men and the young and old in the plasma vitamin C levels could not be explained by differences in renal handling of the vitamin (Oreopoulos *et al.*, 1993). Other explanations must be sought.

The implications of a dysfunction in recycling are that cells and tissues are persistently exposed to low levels of ascorbic acid and elevated levels of dehydroascorbic acid, resulting in an increased susceptibility to oxidative damage leading to diabetic complications. Based on these types of studies, one may conclude that because recycling of vitamin C is extremely important in maintaining optimal health, determination of the mechanism(s) for maintaining ascorbic acid in its reduced form is essential. This topic has been reviewed by Rose and Bode (1992).

Some controversy exists as to whether the recycling of vitamin C is an enzymatic or nonenzymatic process. The primary nonenzymatic mechanism is related to glutathione. Specifically, glutathione is a nonprotein thiol that has been shown to maintain ascorbic acid in its reduced form both *in vitro* and *in vivo*. However, the extent of its role in the recycling *in vivo* is not known but most certainly occurs. Glutathione is a ubiquitous compound found in virtually every tissue, ranging from micromolar to millimolar concentrations. For example, in ocular tissues, reduced glutathione levels range from less than 0.01 to 15 mM (Fig. 1B) and correlate fairly well with ascorbic acid levels (0.3 to 2 mM, Fig. 1A). Winkler and associates (1994) have maintained that the *in vivo* reduction of dehydroascorbic acid occurs primarily by a nonenzymatic interaction with glutathione and have presented convincing evidence that the reduced and oxidized forms of glutathione and ascorbic acid have important nonenzymatic interactions both *in vitro* and *in vivo*. Even so, enzymatic reduction of dehydroascorbic acid to ascorbic acid has been convincingly reported in various mammalian tissues.

The idea of a protein-mediated reduction of dehydroascorbic acid is not new. Laboratories have been attempting to purify a specific enzyme for many years. However, several enzymes have been implicated in the reduction process. These proteins include dehydroascorbate reductase, thioltransferase (also called glutaredoxin), protein disulfide isomerase, and semidehydroascorbate reductase (also called ascorbyl-free radical reductase or monodehydroreductase).

The purification and full characterization of a specific dehydroascorbate reductase in mammalian tissues have been challenging due possibly to (1) the instability of substrates and products (specifically ascorbic acid and dehydroascorbic acid); (2) the lack of a fast, efficient, and reliable assay;

the best way to measure ascorbic acid is by HPLC; spectrophotometric assays are generally considered to be inconclusive and nonspecific; (3) the occurrence of a significant chemical reduction by glutathione and/or other sulfhydrals that interferes with the catalyzed reaction; and (4) the involvement of more than one enzyme.

I. Evidence for the Existence of a Specific Dehydroascorbate Reductase

Historically, Hughes (1964) was the first to present data proposing the existence of dehydroascorbate reductase in mammalian tissues. More recently, studies have been conducted in this laboratory and with collaborators to determine if a specific dehydroascorbic acid-reducing enzyme exists, and results indicate that a variety of tissues appear to have the ability to enzymatically reduce dehydroascorbic acid. A factor has been identified and partially characterized in a variety of tissues, including human placenta (Choi and Rose, 1989), rat colon (Rose et al., 1988), rat kidney (Rose, 1988), rat and calf cerebral tissue (Rose, 1993), and bovine iris ciliary body (Bode et al., 1993). The enzyme seems to vary in hydrogen donor requirement by tissue and species, but is acid and temperature labile, is inhibited by trypsin, is pH dependent, and shows increasing activity with increasing protein concentration. The question of a cofactor requirement is an interesting one. The cytosolic dehydroascorbic acid-reducing activity detected in human placenta, ocular tissues, and rat kidney and liver is maximal in the presence of both reduced glutathione (GSH) and reduced nicotinamide adenine dinucleotide phosphate (NADPH). However, the dehydroascorbic acid-reducing activity in rat heart and rat colon appears to require NADPH only as a cofactor and not GSH. The evidence strongly suggests that a group or family of enzymes with dehydroascorbic acid-reducing activity exists, but their exact identity is still unknown.

Dehydroascorbate reduction varies from tissue to tissue. In iris ciliary body (Bode et al., 1993), enzymatic activity has been well characterized. Apparent K_m values were observed as follows: K_{mDHAA} was 0.24 mM and K_{mGSH} was 0.5 mM. V_{max} values observed were 15.6 and 16.1 nmol/min · mg for dehydroascorbic acid and GSH, respectively. These values suggest that the reduction may be catalyzed by an enzyme other than thioltransferase (K_{mDHAA} between 0.6 and 2 mM and K_{mGSH} between 2 and 9 mM) or protein disulfide isomerase (K_{mDHAA} of 1 mM and K_{mGSH} of 4 mM) (Bode et al., 1993). Interestingly, in many tissues (i.e., neuroendocrine), ascorbic acid levels appear to correlate well with enzymatic dehydroascorbate-reducing activity in the same tissues (Fig. 2).

One of only a very few reports of dehydroascorbate reductase activity associated with a membrane fraction was noted in which a dehydroascorbate reductase activity was detected in a mitochondrial fraction isolated from the cortex-epithelial fraction of bovine lens tissue (Rose et al., 1995).

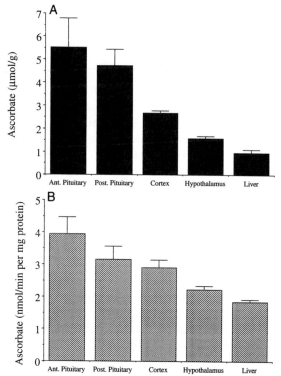

FIGURE 2   Correlation of ascorbate levels (A) and dehydroascorbate reduction (B) in various rat neuroendocrine and endocrine tissues.

The most convincing evidence of the existence of proteins with dehydroascorbate reductase activity was presented by Wells and associates (1990). Dehydroascorbate reductase activity was recognized as an intrinsic activity of thioltransferase and protein disulfide isomerase *in vitro* (Wells *et al.*, 1990). For thioltransferase, apparent K_m and V_{max} for dehydroascorbate were reported as being between 0.2 and 2.2 mM and 6–27 nmol min^{-1}, respectively, and for glutathione between 1.6 and 8.7 mM and 11–30 nmol min^{-1}, respectively, both within physiological range. For protein disulfide isomerase, K_m for dehydroascorbic acid was 1.0 mM and the V_{max} was 8 nmol min^{-1}, and for GSH were 3.9 mM and 14 nmol min^{-1}, respectively. Cellular roles for these enzymes have been proposed and reviewed by Wells and Xu (1994). However, conclusive evidence as to their importance, in dehydroascorbate reduction is still needed.

Evidence for the existence of a specific dehydroascorbate reductase is stronger with the reported purification of a cytosolic rat liver protein with glutathione-dependent dehydroascorbate reductase activity (Maellaro *et al.*, 1994). The protein was described as having a molecular weight between

30,000 and 48,000 and an apparent K_m for dehydroascorbate of 0.245 mM and a V_{max} of 1.9 μmol/min per mg of protein; for GSH, the apparent K_m was reported to be 2.8 mM and a V_{max} of 4.5 μmol/min per mg of protein. These values are well within the physiological range. The protein had no significant sequence similarity to any known protein sequences and is yet to be completely characterized. In addition, the same group (Delbello *et al.*, 1994) purified an NADPH-dependent dehydroascorbate reductase with an apparent K_m for dehydroascorbic acid of 4.6 mM and a V_{max} of 1.55 units/mg of protein; for NADPH, the K_m and V_{max} were 4.3 μM and 1.10 units/mg of protein, respectively. The apparent K_m value for dehydroascorbic acid is well beyond known physiological concentrations for this compound, making it an unlikely candidate for *in vivo* dehydroascorbic acid reduction. The protein was subsequently identified as 3-α-hydroxysteroid dehydrogenase. Whether either of these proteins functions physiologically in the reduction of dehydroascorbic acid remains to be seen. The three reports cited may explain partially why a specific dehydroascorbate reductase had not been purified in mammalian tissues (e.g., more than one enzyme has the reported activity) before now.

It has been suggested that the *in vivo* regeneration of ascorbic acid involves a complex system of events as illustrated in Fig. 3. Reaction 1 may include glucose-6-phosphate dehydrogenase, isocitrate dehydrogenase, or malic enzyme, any of which could provide NADPH for the reduction of

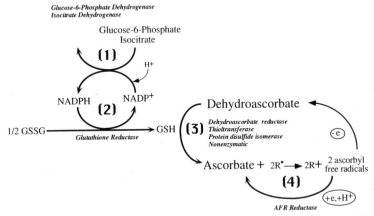

FIGURE 3   Ascorbate regenerating pathways. (1) NADPH-generating system: NADPH is obtained from reactions catalyzed by glucose-6-phosphate dehydrogenase and/or isocitrate dehydrogenase. (2) Glutathione-generating system: Glutathione is required for the reduction of dehydroascorbate to ascorbate and is obtained from the reaction catalyzed by glutathione reductase. (3) Ascorbate recycling systems: Dehydroascorbate is reduced to ascorbate in a reaction believed to be catalyzed by enzymes having dehydroascorbate reductase activity and, in addition, a chemical reduction also most certainly occurs *in vivo*. (4) Ascorbate may also be formed in a reaction catalyzed by ascorbyl-free radical reductase.

glutathione disulfide in a reaction (reaction 2) catalyzed by glutathione reductase. The reduced glutathione then could either react chemically with dehydroascorbic acid directly or through a reaction catalyzed by a dehydroascorbate-reducing enzyme (a specific dehydroascorbic acid-reductase, thioltransferase, or protein disulfide isomerase) (reaction 3). In the scavenging or oxidation process, evidence suggests that ascorbic acid may pass through an intermediate form, the ascorbyl-free radical, which is relatively nonreactive but highly unstable. Ascorbyl-free radicals may spontaneously decay (half-life in microseconds) via the loss of an unpaired electron (e) with the net production of DHAA. In some tissues, ascorbyl-free radicals may be reduced to ascorbic acid by a membrane-associated protein referred to as ascorbyl-free radical reductase or semidehydroascorbate reductase (reaction 4). Ascorbic acid production via the coupling of the hexose monophosphate shunt through NADPH to glutathione reductase and subsequent dehydroascorbic acid reduction was demonstrated and shown to be significantly decreased in diabetic liver correlating with the decreased levels of vitamin C observed (Bode *et al.*, 1993). The impaired ascorbic acid production and altered levels of vitamin C were shown to be related directly to a decrease in glucose-6-phosphate dehydrogenase activity in diabetic liver (Bode *et al.*, 1993). Interestingly, in an unrelated study (Disimplicio *et al.*, 1995), decreased vitamin C levels in insulin-dependent diabetics were significantly correlated with a decreased activity of platelet thioltransferase, an enzyme with intrinsic dehydroascorbate-reducing activity.

2. Evidence for the Existence of a Specific Semidehydroascorbate Reductase (Ascorbyl-Free Radical Reductase)

Semidehydroascorbate reductase or ascorbyl free radical reductase activity is associated with membranes including microsomes, outer mitochondrial membrane, and plasma membrane. This enzyme is believed to catalyze the reduction of the ascorbyl-free radicals to ascorbate, and its activity has been studied in a variety of tissues over the years (Diliberto *et al.*, 1982; Khatami *et al.*, 1986; Nishino and Ito, 1986). It has been purified and characterized in plants (Sano *et al.*, 1995), but less is known about its function in mammalian tissues. Ascorbyl-free radical reductase is thought to be associated with the stabilization of extracellular ascorbate (Schweinzer and Goldenberg, 1993; Goldenberg and Schweinzer, 1995). The evidence for this is related to the fact that when ascorbate is incubated in the presence and absence of cells, autooxidation of ascorbate in the media is lessened when cells are present. Reduced nicotinamide adenine dinucleotide (NADH) appears to be the electron donor for the process and the stabilization is time and temperature dependent.

An interesting series of studies has appeared in which Schweinzer and colleagues (1993) earlier concluded that the stabilization of ascorbic acid by K-562 cells was not due to a membrane redox system but probably to a nonspecific chelation of metals. However, they do suggest that these cells may prevent the oxidation of ascorbic acid in the surrounding medium by three mechanisms, including a semidehydroascorbic acid reductase, a slow conversion of dehydroascorbic acid to ascorbic acid by a mechanism in which dehydroascorbic acid may be taken up, reduced, and released into the medium, or by the action of peptides released by the cells into the medium, which can chelate transition metal ions (Goldenberg and Schweinzer, 1995). The same group detected and partially characterized an ascorbyl-free radical reductase activity at the surface of erythroleukemic cells (Schweinzer and Goldenberg, 1993). The activity was shown to be inhibited by thenoyltrifluoroacetone but not by ouabain, monensin, or tetraethylammonium. Significantly, the cells were able to generate ascorbate from dehydroascorbate and that activity was not inhibited by thenoyltrifluoroacetone, suggesting that the processes are separable. These data support an earlier study (Coassin et al., 1991) in which the two activities were characterized as occurring separately by a two-electron reduction pathway (GSH-dehydroascorbate reductase) and by a one-electron reduction pathway (NADH-semidehydroascorbate reductase) in pig tissues. Others have also suggested that a plasma membrane enzymatic redox system is responsible for the stabilization of extracellular ascorbic acid in the presence of cells (Rodriguezaguilera and Navas, 1994).

Other studies on the detection and function of this enzyme in mammals include a report of activity in human lens (Bando and Obazawa, 1991). An inverse relationship was observed between enzyme activity and insoluble protein content in lens (Bando and Obazawa, 1991), supporting an earlier report by the same group in which decreased ascorbyl-free radical reductase activity was observed in cataractous human lens and was associated with increases in insoluble lens protein (Bando and Obazawa, 1990). A more recent report by the same group presented the separation and partial characterization of what they referred to as a major and minor ascorbyl-free radical reductase from human lens cortex (Bando and Obazawa, 1994). Significantly, this appears to be the verification of an ascorbate-free radical reductase from human lens or other tissue and supports further the existence of enzymatic ascorbic acid recycling mechanisms in mammals. In addition, an NADH-dependent ascorbyl free radical reductase activity has been studied in purified plasma membranes from rat liver (Villalba et al., 1993a,b). The activity was inhibited by glutathione and by reagents that block thiol groups (Villaba et al., 1993a). The NADH-dependent ascorbyl-free radical reductase activity did not survive detergent solubilization, suggesting that more than one protein may be involved in the reduction of the ascorbyl-free radical (Villalba et al., 1993b).

IV. Evidence for Therapeutic Effectiveness of Vitamin C _____

In recent years, oxidative stress has become implicated as a key player in the etiology of a variety of diseases, including diabetes, cancer, cataract, and cardiovascular disease. Due to this connection, much debate surrounds the effectiveness or lack of effectiveness of antioxidants, including vitamin C, in the treatment of a variety of ailments. In two separate reviews (Gershoff, 1993; Sauberlich, 1994), the point was made that even though nutritionists are trumpeting the health effects of antioxidant vitamins, including vitamin C, research studies have not provided an unequivocal answer as to how much of which antioxidant is necessary to achieve the preventative effects in humans such as those observed in animal, tissue, and cell studies. Obviously more precise work is needed before a recommendation may be made. In addition, a lack of knowledge exists regarding the function(s), mechanism(s) of action, and interactions of vitamin C with other antioxidants, and the serious functional consequences of a deficiency of ascorbic acid in certain diseases are just beginning to be realized.

Nonetheless, a fair amount of evidence suggests that vitamin C may be effective in the treatment of a variety of conditions, especially those in which its metabolism has been shown to be affected. For those interested, a review of the clinical use of vitamin C, problems surrounding its use, and an overview of the debate surrounding its use can be found in Kodama and Kodama (1994). The remainder of this chapter focuses on research regarding the effectiveness of vitamin C in the treatment of variety of experimentally induced and clinically relevant conditions.

A. Vitamin C and Diabetes

The majority of human and animal studies are consistent in that vitamin C metabolism is observed to be deficient in diabetes (type I and type II). Whether the low levels of ascorbic acid are a result of the disease or a cause of the disease is unclear. Regardless, the low levels would suggest that diabetic tissues are not well protected against oxidative stress and may therefore be more susceptible to oxidative damage. The suggestion has been made that supplementation with ascorbic acid (1 g or less/day) might be a safe and effective means of alleviating the chronic complications of diabetes (Wang et al., 1994).

Vitamin C infusion in healthy and diabetic humans was shown to raise plasma levels of the vitamin and to improve whole body glucose disposal by mediating an improvement in nonoxidative glucose metabolism (Paolisso et al., 1994). Paolisso and colleagues (1994) concluded that plasma vitamin C levels seemed to play a role in modulating insulin action in older normal humans and in diabetic humans. In another clinical study, infusion of vitamin C combined with insulin injection was used to treat three diabetic patients and resulted in the patients being able to maintain higher plasma levels of vitamin C and in all three patients showing clinical improvements in their condition (Kodama and Kodama, 1993), suggesting that the vitamin C treatment augmented the action of insulin.

Ascorbic acid supplementation was shown to improve glycemic control and decrease fasting blood glucose, triglycerides, and cholesterol levels in noninsulin-dependent diabetes mellitus (NIDDM) subjects (Eriksson and Kohvakka, 1995). Sorbitol is known to accumulate in diabetes through the polyol pathway due to increased aldose reductase activity and has been implicated in diabetic complications. Supplementation with ascorbic acid in diabetic patients for 12–30 days resulted in decreased erythrocyte sorbitol with no effect on glucose levels (Cunningham et al., 1994; Wang et al., 1994). One group showed that the mechanism for the decrease was an inhibition of aldose reductase activity by ascorbic acid and suggested that vitamin C would be superior to aldose reductase inhibitors normally used (Cunningham et al., 1994).

However, treatment with high doses (6 g/day for 4 weeks) of ascorbic acid produced no side effects but failed to normalize glomerular hyperfiltration in insulin-dependent diabetes mellitus patients (Klein et al., 1995). Authors concluded that low ascorbic acid levels in kidney were either not corrected or were not responsible for glomerular hyperfiltration observed in diabetes. In addition, another study indicated that 2-g/day ascorbic acid supplementation delayed the insulin response to a glucose challenge in normal subjects (Johnston and Yen, 1994), resulting in the prolonged elevation of glucose levels and depressed insulin levels.

B. Vitamin C and Cancer

A great deal of interest exists in the use of vitamins in experimental, clinical, and epidemiological studies for cancer chemoprevention and treatment. However, little is actually known about the anticancer mechanisms by which vitamins may act. In a comprehensive review, data are presented regarding the role and rationale for the use of vitamins A, E, and C in cancer chemoprevention and treatment (Lupulescu, 1994). A variety of studies exist showing that vitamins have a variety of effects on cell function and that they can exert both cytotoxic and proliferative consequences. Vitamin C has been suggested to have a potential anticancer effect, particularly for cancer of the stomach, rectum, and cervix (Schwartz et al., 1994), and low levels of vitamin C have been significantly correlated to an increased risk of gastric carcinogenesis (Hansson et al., 1994; Zhang et al., 1994).

Free radicals are believed to play a part in tumor development and growth (Landolt et al., 1994) and, because of this, interest in the usefulness of antioxidant nutrients in the treatment of cancer has increased. Riordan and colleagues (1995) presented data suggesting that high doses of vitamin C are effective in selectively killing tumor cells and suggest that ascorbic acid may be useful as a cytotoxic chemotherapeutic agent. Some researchers suggest that the antitumor activity of ascorbic acid is due to its oxidation and subsequent metabolites and not to its action as a vitamin (Leung *et al.*, 1993), a hypothesis shared by others (Tsao *et al.*, 1995).

Data suggest that neuroblastoma cells have high levels of iron-rich ferritin and therefore could be especially vulnerable to the pro-oxidant effects of vitamin C (Lode *et al.*, 1994). This hypothesis was confirmed by others who observed that ascorbic acid was cytotoxic to neuroblastoma cell and the mechanism was ascribed to ascorbic acid's action as a prooxidant (Baader *et al.*, 1994). Sodium nitroprusside, a nitrogen oxide-generating agent, combined with ascorbate treatment is suggested to an effective means of treating brain tumors (Lee and Wurster, 1994). In a study using 3T6 fibroblasts, data suggested that the cytotoxicity of ascorbate was induced by hydrogen peroxide because the effects were inhibited by the addition of catalase (Arakawa *et al.*, 1994).

Garland and associates (1993) reviewed epidemiological evidence on the relationship between vitamins A, C, E, and selenium and breast cancer risk. Vitamin A was the only vitamin with a protective effect. This conclusion was supported by the results of a comprehensive analysis of vitamin C, E, and A intake in 89,494 women, which revealed no significant effect of large doses of vitamins C or E on protecting women from breast cancer, although low intakes of vitamin A were associated with an increased risk of the disease (Hunter *et al.*, 1993). At the same time, in a survey of data compiled on cancer patients from three hospitals in Scotland, ascorbate-supplemented cancer patients were shown to have a median overall survival time almost double that of controls (Cameron, 1991; Cameron and Campbell, 1991).

C. Vitamin C and the Common Cold

No consistent unequivocal evidence has been found confirming that vitamin C supplementation prevents the common cold. However, a review of 21 studies revealed that vitamin C supplementation apparently is associated with a reduction in the duration and severity of the symptoms, suggesting that treatment with vitamin C may be helpful in that respect (Hemila, 1994).

D. Vitamin C and Cardiovascular Disease

Evidence from a variety of research studies supports the hypothesis that oxidized low-density lipoprotein (LDL) has a role in atherosclerosis. Supplementation with vitamin C increases the oxidation resistance of atherogenic lipoproteins in human plasma (Nyyssonen *et al.*, 1994), and a substantial amount of evidence suggests that dietary antioxidants, including vitamin C, decrease the risk of atherosclerosis by protecting low-density lipoproteins from oxidation (Retsky *et al.*, 1993; Tonstad, 1995). Vitamin C has been shown to be effective in preventing oxidized LDL-induced leukocyte adhesion in vascular endothelium and leukocyte-platelet aggregate formation in the blood stream (Lehr *et al.*, 1995). The same group (Lehr *et al.*, 1994) reported previously that vitamin C effectively prevented leukocyte adhesion induced by cigarette smoke.

Ascorbic acid has been used to treat systolic and essential hypertension (Ghosh *et al.*, 1994) with no difference observed between treatment or placebo but with decreases observed in both.

Gaziano (1994) discussed and reviewed coronary artery disease and antioxidant supplementation and suggested that, in general, supplementation is associated with a decreased risk of coronary artery disease. An overview of ongoing clinical trials has also been presented (Gaziano, 1994). Another review summarizes the evidence for the association of antioxidants and coronary heart disease (van Poppel *et al.*, 1994). At least two other studies support the hypothesis that an increased antioxidant vitamin intake protects against coronary heart disease in men (Sardesai, 1995; Todd *et al.*, 1995). However, the results of most epidemiological studies suggest an association but have not produced unequivocal evidence that high intakes of specific vitamins prevent heart disease (van Poppel *et al.*, 1994).

E. Vitamin C and Ischemia–Reperfusion Injury

Free radicals are suggested to be a cause of ischemia–reperfusion injury in tissue, hence leading to an increased interest in the use of antioxidants in the prevention of this type of injury. Vitamin C has been shown to reduce or limit reperfusion injury after ischemia (Zaccaria *et al.*, 1994). High-dose vitamin C supplementation decreased lipid peroxidation and protected the myocardium from ischemia–reperfusion injury during and after open-heart surgery (Dingchao *et al.*, 1994). The use of ascorbic acid in preventing free radical damage in reoxygenated liver indicates that ascorbic acid is effective in preserving cellular integrity and ATP levels in postischemic tissue (Ozaki *et al.*, 1995).

F. Vitamin C and Other Diseases

In addition to the brief summaries just outlined, vitamin C has been used in treating a variety of conditions from schizophrenia to wound healing. A summary of some of these results are outlined in the following paragraphs.

Evidence suggests that vitamin C may have a role in augmenting the behavioral and therapeutic effects of neuroleptic drugs (Pierce *et al.*, 1994).

Vitamin C has been suggested to be useful in treating chronic schizophrenia (Sandyk and Kanofsky, 1993) and as being useful in conjunction with other compounds for slowing the progression of Parkinson's disease (Pardo et al., 1993). However, at least one study suggests that serum vitamin C concentrations are unrelated to the risk of developing Parkinson's disease (Fernandez Calle et al., 1993).

Vitamin C intake has been associated with a protective effect on pulmonary function (forced expiratory volume in 1 sec) (Schwartz and Weiss, 1994); however, high vitamin C levels in premature infants have been associated with an inhibition of the antioxidant capacity of ceruloplasmin (Silvers et al., 1994). The suggestion is that high levels of vitamin C may compromise antioxidant mechanisms and make the premature child more susceptible to oxidative damage (Powers et al., 1995).

Ascorbate in combination with N-acetylcysteine has been used successfully to increase glutathione levels, leading to a decrease in erythrocyte turnover in patients with hereditary glutathione deficiency (Jain et al., 1994). It is well known that ascorbate spares glutathione in glutathione deficiency (Martenson et al., 1993; Meister, 1995) and that glutathione esters delay the onset of scurvy in ascorbic acid deficiency in guinea pigs (Martensson et al., 1993; Meister, 1995).

Some research has been done in the effect of ascorbic acid on wound healing. Supplementation with vitamin C increases the collagen content in wounded and intact skin, but ascorbic acid levels were lower and dehydroascorbic acid levels were higher in wounded skin (Kim et al., 1994), perhaps suggesting an increased turnover of the vitamin in tissue regeneration. Studies in rabbits have suggested that ascorbic acid may have a therapeutic role in the repair of corneal alkali burns (Saika, 1993). In dogs, high-dose vitamin C treatment has been shown to diminish postburn lipid peroxidation and microvascular leakage of fluid and protein (Matsuda et al., 1993).

V. Significance and Conclusions

The recent death of Linus Pauling led an editor to comment "One pity of Linus Pauling's death is that even vitamin C could not keep him living longer." Evidence is accumulating supporting vitamin C as a key protective nutrient in mammals under normal and stressed conditions. Maintaining a high ascorbic acid/dehydroascorbic acid ratio appears to be imperative for healthy tissues. The mechanism underlying normal ascorbic acid metabolism in tissues is as yet unclear as are the implications and complications of impaired metabolism such as is observed in certain disease states, and the jury is still out regarding the effectiveness of vitamin C as a therapeutic agent.

Despite these issues, some conclusions may be formulated: (1) Adequate vitamin C levels are essential in maintaining optimal health. The research evidence is overwhelming that vitamin C may have a role in preventing or at least impeding the progress of diabetes, cataract, heart disease, cancer, aging, and a variety of other disease states. Therefore the process of ascorbic acid conservation is of increasing interest. (2) The protective action of vitamin C appears to be at least partially related to its role as an effective antioxidant. (3) Research has shown that the transport and metabolism of the reduced and/or oxidized forms are influenced by environmental and metabolic demands. (4) An efficient recycling between the oxidized and the reduced forms of vitamin C is extremely important. (5) Vitamin C levels may be regulated and controlled tightly by both nonenzymatic and enzymatic recycling mechanisms that may become dysfunctional in certain disease states. (6) Although promising, more research is needed to determine the therapeutic value of vitamin C.

References

Arakawa, N., Nemoto, S., et al. (1994). Role of hydrogen peroxide in the inhibitory effect of ascorbate on cell growth. *J. Nutr. Sci. Vitaminol. Tokyo* **40**(3), 219–227.

Baader, S. L., Bruchelt, G., et al. (1994). Uptake and cytotoxicity of ascorbic acid and dehydroascorbic acid in neuroblastoma (SK-N-SH) and neuroectodermal (SK-N-LO) cells. *Anticancer Res.* **14**(1a), 221–227.

Ballmer, P. E., Reinhart, W. H., et al., (1994). Depletion of plasma vitamin C but not of vitamin E in response to cardiac operations. *J. Thorac. Cardiovasc. Surg.* **108**(2), 311–320.

Bando, M., and Obazawa, H. (1990). Activities of ascorbate free radical reductase and H_2O_2-dependent NADH oxidation in senile cataractous human lenses. *Exp. Eye Res.* **50**(6), 779–786.

Bando, M., and Obazawa, H. (1991). Regional and subcellular distribution of ascorbate free radical reductase activity in the human lens. *Tokai. J. Exp. Clin. Med.* **16**(5–6), 217–222.

Bando, M., and Obazawa, H. (1994). Soluble ascorbate free radical reductase in the human lens. *Jpn. J. Ophthalmol.* **38**(1), 1–9.

Bates, C. J., and Cowen, T. D. (1988). Effects of age and dietary vitamin C on the contents of ascorbic acid and acid-soluble thiol in lens and aqueous humour of guinea pigs. *Exp. Eye Res.* **46**, 937–945.

Behrens, W. A., and Madera, R. (1991). Vitamin C and vitamin E status in the spontaneously diabetic BB rat before the onset of diabetes. *Metabolism* **40**(1), 72–76.

Bendich, A., and Langseth, L. (1995). The health effects of vitamin c supplementation: A review. *J. Am. Coll. Nutr.* **14**(2), 124–136.

Bergsten, P., Amitai, G., et al. (1990). Millimolar concentrations of ascorbic acid in purified human mononuclear leukocytes: Depletion and reaccumulation. *J. Biol. Chem* **265**(5), 2584–2587.

Bergsten, P., Moura, A. S., et al. (1994). Ascorbic acid and insulin secretion in pancreatic islets. *J. Biol. Chem.* **269**(2), 1041–1045.

Bergsten, P., Yu, R., et al. (1995). Ascorbic acid transport and distribution in human b lymphocytes. *Arch. Biochem. Biophys.* **317**(1), 208–214.

Bielory, L., and Gandhi, R. (1994). Asthma and vitamin C. *Ann. Allergy* **73**(2), 89–96.

Bode, A. M., Green, E., *et al.* (1993). Ascorbic acid regeneration by bovine iris-caliary body. *Curr. Eye Res.* 12(7), 593–601.

Bode, A. M., Vanderpool, S. S., *et al.* (1991). Ascorbic acid uptake and metabolism by corneal endothelium. *Invest. Opththal. Vis. Sci.* 32(8), 2266–2271.

Bode, A. M., Yavarow, C. R., *et al.* (1993). Enzymatic basis for altered ascorbic acid and dehydroascorbic acid levels in diabetes. *Biochem. Biophys. Res. Commun.* 191(3), 1347–1353.

Braganza, J. M., Scott, P., *et al.* (1995). Evidence for early oxidative stress in acute pancreatitis. Clues for correction. *Int. J. Pancreatol* 17(1), 69–81.

Buffinton, G. D., and Doe, W. F. (1995). Altered ascorbic acid status in the mucosa from inflammatory bowel disease patients. *Free Rad. Res.* 22(2), 131–143.

Butler, J. D., Bergsten, P., *et al.* (1991). Ascorbic acid accumulation in human skin fibroblasts. *Am. J. Clin. Nutr.* 54(Suppl. 6), 1144s–1146s.

Cameron, E. (1991). Protocol for the use of vitamin C in the treatment of cancer. *Med. Hypotheses* 36(3), 190–194.

Cameron, E., and Campbell, A. (1991). Innovation vs. quality control: An 'unpublishable' clinical trial of supplemental ascorbate in incurable cancer. *Med. Hypotheses* 36(3), 185–189.

Chan, A. C. (1993). Partners in defense, vitamin E and vitamin C. *Can. J. Physiol. Pharmacol.* 71(9), 725–731.

Choi, J. L., and Rose, R. C. (1989). Transport and metabolism of ascorbic acid in human placenta. *Am. J. Physiol.* 257(Cell. Physiol. 26), C110–C113.

Coassin, M., Tomasi, A., *et al.* (1991). Enzymatic recycling of oxidized ascorbate in pig heart: One electron vs two-electron pathway. *Arch. Biochem. Biophys.* 290(2), 458–462.

Crawford, R. D. (1995). Proposed role for a combination of citric acid and ascorbic acid in the production of dietary iron overload: A fundamental cause of disease. *Biochem. Mol. Med.* 54(1), 1–11.

Cunningham, J. J., Ellis, S. L., *et al.* (1991). Reduced mononuclear leukocyte ascorbic acid content in adults with insulin-dependent diabetes mellitus consuming adequate dietary vitamin C. *Metabolism* 40(2), 146–149.

Cunningham, J. J., Mearkle, P. L., *et al.* (1994). Vitamin C: An aldose reductase inhibitor that normalizes erythrocyte sorbitol in insulin-dependent diabetes mellitus. *J. Am. Coll. Nutr.* 13(4), 344–350.

Dabbagh, A. J., Mannion, T., *et al.* (1994). The effect of iron overload on rat plasma and liver oxidant status in vivo. *Biochem. J.* 300(Part 3), 799–803.

Defeudis, F. V. (1995). Excess EDRF/NO, a potentially deleterious condition that may be involved in accelerated atherogenesis and other chronic disease states. *Gen. Pharmacol.* 26(4), 667–680.

Delbello, B., Maellaro, E., *et al.* (1994). Purification of NADPH-dependent dehydroascorbate reductase from rat liver and its identification with 3 alpha-hydroxysteroid dehydrogenase. *Biochem. J.* 304(Part 2), 385–390.

Diliberto, E. J., Dean, G., *et al.* (1982). Tissue, subcellular, and submitochondrial distributions of semidehydroascorbate reductase; possible role of semidehydroascorbate reductase in cofactor regeneration. *J. Neurochem.* 39, 563–568.

Dingchao, H., Zhiduan, Q., *et al.* (1994). The protective effects of high-dose ascorbic acid on myocardium against reperfusion injury during and after cardiopulmonary bypass. *Thorac. Cardiovasc. Surg.* 42(5), 276–278.

Disimplicio, P., Degiorgio, L. A., *et al.* (1995). Glutathione, glutathione utilizing enzymes and thioltransferase in platelets of insulin-dependent diabetic patients: Relation with platelet aggregation and with microangiopathic complications. *Eur. J. Clin. Invest.* 25(9), 665–669.

Dreyer, R., and Rose, R. C. (1993). Lacrimal gland uptake and metabolism of ascorbic acid. *Proc. Soc. Exp. Biol. Med.* 202, 212–216.

Eriksson, J., and Kohvakka, A. (1995). Magnesium and ascorbic acid supplementation in diabetes mellitus. *Ann. Nutr. Metab.* **39**(4), 217–223.

Fernandez Calle, P., Jimenez Jimenez, F. J., *et al.* (1993). Serum levels of ascorbic acid (vitamin C) in patients with Parkinson's disease. *J. Neurol. Sci.* **118**(1), 25–28.

Fisher, E., McLennan, S. V., *et al.* (1991). Interaction of ascorbic acid and glucose on production of collagen and proteoglycan by fibroblasts. *Diabetes* **40**, 371–376.

Flagg, E. W., Coates, R. J., *et al.* (1995). Epidemiologic studies of antioxidants and cancer in humans. *J. Am. Coll. Nutr.* **14**(5), 419–427.

Frei, B., England, L., *et al.* (1989). Ascorbate is an outstanding antioxidant in human blood plasma. *Proc. Natl. Acad. Sci. USA* **86**, 6377–6381.

Friedman, S., and Kaufman, S. (1965). 3,4-Dihydroxyphenylethylamine β-hydroxylase. *J. Biol. Chem.* **240**, 4763.

Garland, M., Willett, W. C., *et al.* (1993). Antioxidant micronutrients and breast cancer. *J. Am. Coll. Nutr.* **12**(4), 400–411.

Gaziano, J. M. (1994). Antioxidant vitamins and coronary artery disease risk. *Am. J. Med.* **97**(3a), 18S–21S; discussion 22S–28S.

Georgiannos, S. N., Weston, P. M. T., *et al.* (1993). Micronutrients in gastrointestinal cancer. *Br. J. Can.* **68**(6), 1195–1198.

Gershoff, S. N. (1993). Vitamin C (ascorbic acid): New roles, new requirements?" *Nutr. Rev.* **51**(11), 313–326.

Gey, K. F. (1995). Ten-year retrospective on the antioxidant hypothesis of arteriosclerosis: Threshold plasma levels of antioxidant micronutrients related to minimum cardiovascular risk. *J. Nutr. Biochem.* **6**(4), 206–236.

Ghorbani, A. J., and Eichler, C. (1994). Scurvy. *J. Am. Acad. Dermatol.* **30**(5pt2), 881–883.

Ghosh, S. K., Ekpo, E. B., *et al.* (1994). A double-blind, placebo-controlled parallel trial of vitamin C treatment in elderly patients with hypertension. *Gerontology* **40**(5), 268–272.

Glembotski, C. C. (1984). The α-amidation of α-melanocyte stimulating hormone in intermediate pituitary requires ascorbic acid. *J. Biol. Chem.* **259**, 13041–13048.

Goldenberg, H., and Schweinzer, E. (1994). Transport of vitamin C in animal and human cells. *J-Bioenerg-Biomembr* **26**(4), 359–367.

Goldenberg, H., and Schweinzer, E. (1995). Mechanisms of vitamin c stabilization by k562 erythroleukemic cells. *Protoplasma* **184**(1–4), 220–228.

Griffith, O. W. (1986). Glutathione and glutathione disulphide. In "Methods of Enzymology," 3rd Ed., Vol. VIII, pp. 521–529. VCH Publishers, Deerfield Beach, FL.

Ha, T. Y., Otsuka, M., *et al.* (1994). Ascorbate indirectly stimulates fatty acid utilization in primary cultured guinea pig hepatocytes by enhancing carnitine synthesis. *J. Nutr.* **124**(5), 732–737.

Hansson, L. E., Nyren, O., *et al.* (1994). Nutrients and gastric cancer risk: A population-based case-control study in Sweden. *Int. J. Cancer* **57**(5), 638–644.

Hemila, H. (1994). Does vitamin C alleviate the symptoms of the common cold? A review of current evidence. *Scand. J. Infect. Dis.* **26**(1), 1–6.

Hemila, H., and Herman, Z. S. (1995). Vitamin C and the common cold: A retrospective analysis of Chalmers' review. *J. Am. Coll. Nutr.* **14**(2), 116–123.

Hisanaga, K., Sagar, S. M., *et al.* (1992). Ascorbate neurotoxicity in cortical cell culture. *Ann. Neurol.* **31**, 562.

Hughes, R. E. (1964). Reduction of dehydroascorbic acid by animal tissues. *Nature* **4949**, 1068–1069.

Hunter, D. J., Manson, J. E., *et al.* (1993). A prospective study of the intake of vitamins C, E, and A and the risk of breast cancer [see comments]. *N. Engl. J. Med* **329**(4), 234–240.

Jain, A., Buist, N. R. M., *et al.* (1994). Effect of ascorbate or *n*-acetylcysteine treatment in a patient with hereditary glutathione synthetase deficiency. *J. Pediatr.* **124**(2), 229–233.

Johnston, C. S., and Yen, M. F. (1994). Megadose of vitamin c delays insulin response to a glucose challenge in normoglycemic adults. *Am. J. Clin. Nutr.* **60**(5), 735–738.

Khatami, M., Roel, L. E., *et al.* (1986). Ascorbate regeneration in bovine ocular tissues by NADH-dependent semidehydroascorbate reductase. *Exp. Eye Res.* **43**, 167–175.

Kim, M., Otsuka, M., *et al.,* (1994). The distribution of ascorbic acid and dehydroascorbic acid during tissue regeneration in wounded dorsal skin of guinea pigs. *Int. J. Vitam. Nutr. Res.* **64**(1), 56–59.

Klein, F., Juhl, B., *et al.* (1995). Unchanged renal haemodynamics following high dose ascorbic acid administration in normoalbuminuric IDDM patients. *Scand. J. Clin. Lab Invest.* **55**(1), 53–59.

Kodama, M., and Kodama, T. (1994). Is Linus Pauling, a vitamin C advocate, just making much ado about nothing? *In Vivo* **8**(3), 391–400.

Kodama, M., Kodama, T., *et al.* (1993). Diabetes mellitus is controlled by vitamin C treatment. *In Vivo* **7**(6a), 535–542.

Kodama, M., Kodama, T., *et al.* (1995). Vitamin C, steroid and environmental carcinogenesis. *Int. J. Oncol.* **6**(4), 797–815.

Landolt, H., Langemann, H., *et al.* (1994). Levels of water-soluble antioxidants in astrocytoma and in adjacent tumor-free tissue. *J. Neurooncol.* **21**(2), 127–133.

Lee, Y. S., and Wurster, R. D. (1994). Potentiation of anti-proliferative effect of nitroprusside by ascorbate in human brain tumor cells. *Cancer Lett.* **78**(1–3), 19–23.

Lehr, H. A., Frei, B., *et al.* (1994). Vitamin C prevents cigarette smoke-induced leukocyte aggregation and adhesion to endothelium in vivo. *Proc. Natl. Acad. Sci. U.S.A.* **91**(16), 7688–7692.

Lehr, H. A., Frei, B., *et al.* (1995). Protection from oxidized LDL-induced leukocyte adhesion to microvascular and macrovascular endothelium in vivo by vitamin C but not by vitamin E. *Circulation* **91**(5), 1525–1532.

Leung, P. Y., Miyashita, K. *et al.* (1993). Cytotoxic effect of ascorbate and its derivatives on cultured malignant and nonmalignant cell lines. *Anticancer Res.* **13**(2), 475–480.

Linklater, H. A., Dzialoszynski, T., *et al.* (1990). Modelling cortical cataractogenesis. XI. Vitamin C reduces γ-crystallin leakage from lenses in diabetic rats. *Exp. Eye Res.* **51**, 241–247.

Lode, H. N., Bruchelt, G., *et al.* (1994). Ascorbic acid induces lipid peroxidation on neuroectodermal SK-N-LO cells with high endogenous ferritin content and loaded with Mab-ferritin immunoconjugates. *Anticancer. Res.* **14**(5a), 1903–1906.

Lupulescu, A. (1994). The role of vitamins a, beta-carotene, e and c in cancer cell biology. *Int. J. Vitam. Nutr. Res.* **64**(1), 3–14.

Maellaro, E., Delbello, B., *et al.* (1994). Purification and characterization of glutathione-dependent dehydroascorbate reductase from rat liver. *Biochem. J.* **301**(Part 2), 471–476.

Marliss, E. B., Nakhooda, A. F., *et al.* (1982). The diabetic syndrome of the "BB" Wistar rat: Possible relevance to Type I (insulin-dependent) diabetes in man. *Diabetologia* **22**, 225–232.

Martensson, J., Han, J., *et al.* (1993). Glutathione ester delays the onset of scurvey in ascorbate-deficient guinea pigs. *Proc. Natl. Acad. Sci. U.S.A.* **90**(1), 317–321.

Matsuda, T., Tanaka, H., *et al.* (1993). The effects of high-dose vitamin C therapy on postburn lipid peroxidation. *J. Burn. Care Rehabil.* **14**(6), 624–629.

May, J. M., Qu, Z. C., *et al.* (1995a). Ascorbate is the major electron donor for a transmembrane oxidoreductase of human erythrocytes. *Biochim. Biophys. Acta* **1238**(2), 127–136.

May, J. M., Qu, Z. C., *et al.* (1995b). Ascorbic acid recycling enhances the antioxidant reserve of human erythrocytes. *Biochemistry* **34**(39), 12721–12728.

Meister, A. (1995). Mitochondrial changes associated with glutathione deficiency. *Biochim. Biophys. Acta* **1271**(1), 35–42.

Merlini, D., and Caramia, F. J. C. (1965). Effect of dehydroascorbic acid on the islets of Langerhans of the rat pancreas. *J. Cell Biol.* **26**, 245–261.

Mikhail, M. S., Anyaegbunam, A., *et al.* (1994). Preeclampsia and antioxidant nutrients: Decreased plasma levels of reduced ascorbic acid, alpha-tocopherol, and beta-carotene in women with preeclampsia. *Am. J. Obstet. Gynecol.* **171**(1), 150–157.

Nishino, H., and Ito, A. (1986). Subcellular distribution of OM cytochrome b-mediated NADH-semidehydroascorbate reductase activity in rat liver. *J. Biochem.* **100**, 1523–1531.

Nyyssonen, K., Porkkala, E., *et al.* (1994). Increase in oxidation resistance of atherogenic serum lipoproteins following antioxidant supplementation: A randomized double-blind placebo-controlled clinical trial. *Eur. J. Clin. Nutr.* **48**(9), 633–642.

Oreopoulos, D. G., Lindeman, R. D., *et al.* (1993). Renal excretion of ascorbic acid: Effect of age and sex. *J. Am. Coll. Nutr.* **12**(5), 537–542.

Organisciak, D. T., Wang, H.-M., *et al.* (1985). The protective effect of ascorbate in retinal light damage of rats. *Invest. Ophthalmol. Vis. Sci.* **26**, 1580.

Ou, P. M., and Wolff, S. P. (1994). Erythrocyte catalase inactivation (H2O2 production) by ascorbic acid and glucose in the presence of aminotriazole: Role of transition metals and relevance to diabetes. *Biochem. J.* **303**(Part 3), 935–939.

Ozaki, M., Fuchinoue, S., *et al.* (1995). The in vivo cytoprotection of ascorbic acid against ischemia reoxygenation injury of rat liver. *Arch. Biochem. Biophys.* **318**(2), 439–445.

Paolisso, G., Damore, A., *et al.* (1994). Plasma vitamin c affects glucose homeostasis in healthy subjects and in non-insulin-dependent diabetics. *Am. J. Physiol.* **266**(2 Part 1), E261–E268.

Pardo, B., Mena, M. A., *et al.* (1993). Ascorbic acid protects against levodopa-induced neurotoxicity on a catecholamine-rich human neuroblastoma cell line. *Mov. Disord.* **8**(3), 278–284.

Patterson, J. W. (1950). The diabetogenic effect of dehydroascorbic and dehydroisoascorbic acids. *J. Biol. Chem.* **183**, 81.

Pierce, R. C., Clemens, A. J., *et al.* (1994). Repeated treatment with ascorbate or haloperidol, but not clozapine, elevates extracellular ascorbate in the neostriatum of freely moving rats. *Psychopharmacol. Berl.* **116**(1), 103–109.

Powers, H. J., Loban, A., *et al.* (1995). Vitamin C at concentrations observed in premature babies inhibits the ferroxidase activity of caeruloplasmin. *Free Radic. Res.* **22**(1), 57–65.

Retsky, K. L., Freeman, M. W., *et al.* (1993). Ascorbic acid oxidation product(s) protect human low density lipoprotein against atherogenic modification: Anti- rather than prooxidant activity of vitamin C in the presence of transition metal ions. *J. Biol. Chem.* **268**(2), 1304–1309.

Rikans, L. E. (1981). Effect of alloxan diabetes on rat liver ascorbic acid. *Horm. Metab. Res.* **13**, 123.

Riordan, N. H., Riordan, H. D., *et al.* (1995). Intravenous ascorbate as a tumor cytotoxic chemotherapeutic agent. *Med. Hypotheses* **44**(3), 207–213.

Rodrigues, B., and McNeill, J. H. (1992). The diabetic heart: Metabolic causes for the development of a cardiomyopathy. *Cardiovas. Res.* **26**, 913–922.

Rodriguezaguilera, J. C., and Navas, P. (1994). Extracellular ascorbate stabilization: Enzymatic or chemical process? *J. Bioenerg. Biomemb.* **26**(4), 379–384.

Rose, C., Devamanoharan, P. S., *et al.* (1995). Dehydroascorbate reductase activity in bovine lens. *Int. J. Vitam. Nutr. Res.* **65**(1), 40–44.

Rose, R. C. (1988). Renal metabolism of the oxidized form of ascorbic acid (DHAA). *Am. J. Physiol.* **256**(Renal Fluid Electrolyte Physiol. 25, F267.

Rose, R. C. (1989). The ascorbate redox potential of tissues: A determinant or indicator of disease? *NIPS* **4**, 190–195.

Rose, R. C. (1993). Cerebral metabolism of oxidized ascorbate. *Brain Res.* **708**, 49–55.

Rose, R. C., and Bode, A. M. (1991). Ocular ascorbate transport and metabolism. *Comp. Biochem. Physiol. A* **100**(2), 273–285.

Rose, R. C., and Bode, A. M. (1992). Tissue-mediated regeneration of ascorbic acid: Is the process enzymatic? *Enzyme* **46**(4–5), 196–203.

Rose, R. C., and Bode, A. M. (1993). Biology of free radical scavengers: An evaluation of ascorbate. *FASEB J.* **7**(12), 1135–1142.

Rose, R. C., Choi, J. L., *et al.* (1988). Intestinal transport and metabolism of oxidized ascorbic acid (DHAA). *Am. J. Physiol.* **254**(Gastrointest. Liver Physiol. 17, G824–G828.

Rose, R. C., Choi, J. L., *et al.* (1992). Short term effects of oxidized ascorbic acid on bovine corneal endothelium and human placenta. *Life Sci.* **50**, 1543–1549.

Saika, S. (1993). Ascorbic acid and proliferation of cultured rabbit keratocytes. *Cornea* **12**(3), 191–198.

Sandyk, R., and Kanofsky, J. D. (1993). Vitamin C in the treatment of schizophrenia. *Int. J. Neurosci.* **68**(1–2), 67–71.

Sano, S., Miyake, C., *et al.* (1995). Molecular characterization of monodehydroascorbate radical reductase from cucumber highly expressed in *Escherichia coli*. *J. Biol. Chem.* **270**(36), 21354–21361.

Sardesai, V. M. (1995). Role of antioxidants in health maintenance. *Nutr. Clin. Pract.* **10**(1), 19–25.

Sauberlich, H. E. (1994). Pharmacology of vitamin C. *Annu. Rev. Nutr.* **14**, 371–391.

Schectman, G. (1993). Estimating ascorbic acid requirements for cigarette smokers. *Ann. N.Y. Acad. Sci.* **686**, 335–345.

Schell, D. A., and Bode, A. M. (1993). Measurement of ascorbic acid and dehydroascorbic acid in mammalian tissue utilizing HPLC and electrochemical detection. *Biomed. Chrom.* **7**, 267–272.

Schwartz, J., and Weiss, S. T. (1994). Relationship between dietary vitamin C intake and pulmonary function in the First National Health and Nutrition Examination Survey (NHANES I). *Am. J. Clin. Nutr.* **59**(1), 110–114.

Schwartz, L. H., Urban, T., *et al.* (1994). Anticancer effect of antioxidant vitamin and minerals. *Presse Med.* **23**(39), 1826–1830.

Schweinzer, E., and Goldenberg, H. (1993). Monodehydroascorbate reductase activity in the surface membrane of leukemic cells: Characterization by a ferricyanide-driven redox cycle. *Eur. J. Biochem.* **218**(3), 1057–1062.

Schweinzer, E., Waeg, G., *et al.* (1993). No enzymatic activities are necessary for the stabilization of ascorbic acid by K-562 cells. *FEBS Lett.* **334**(1), 106–108.

Seghieri, G., Martinoli, L., *et al.* (1994). Renal excretion of ascorbic acid in insulin dependent diabetes mellitus. *Int. J. Vitam. Nutr. Res.* **64**(2), 119–124.

Sharpe, P. C., Mulholland, C., *et al.* (1994). Ascorbate and malondialdehyde in stroke patients. *Ir. J. Med. Sci.* **163**(11), 488–491.

Silvers, K. M., Gibson, A. T., *et al.* (1994). High plasma vitamin C concentrations at birth associated with low antioxidant status and poor outcome in premature infants. *Arch. Dis. Child* **71**(1 Spec No), F40–F44.

Sinclair, A. J., Girling, A. J., *et al.* (1991). Disturbed handling of ascorbic acid in diabetic patients with and without microangiopathy during high dose ascorbate supplementation. *Diabetologia* **34**, 171–175.

Sinclair, A. J., Taylor, P. B., *et al.* (1994). Low plasma ascorbate levels in patients with type 2 diabetes mellitus consuming adequate dietary vitamin C. *Diabet. Med.* **11**(9), 893–898.

Singh, R. B., Niaz, M. A., *et al.* (1994). Plasma levels of antioxidant vitamins and oxidative stress in patients with acute myocardial infarction. *Acta Cardiol.* **49**(5), 441–452.

Siushansian, R., and Wilson, J. X. (1995). Ascorbate transport and intracellular concentration in cerebral astrocytes. *J. Neurochem.* **65**(1), 41–49.

Tanzer, F., and Ozalp, I. (1993). Leucocyte ascorbic acid concentration and plasma ascorbic acid levels in children with various infections. *Mater. Med. Pol.* **25**(1), 5–8.

Thanislass, J., Raveendran, M., *et al.* (1995). Buthionine sulfoximine-induced glutathione depletion: Its effect on antioxidants, lipid peroxidation and calcium homeostasis in the lung. *Biochem. Pharmacol.* **50**(2), 229–234.

Thompson, K. H., and Godin, D. V. (1995). Micronutrients and antioxidants in the progression of diabetes. *Nutr. Res.* **15**(9), 1377–1410.

Todd, S., Woodward, M., *et al.* (1995). An investigation of the relationship between antioxidant vitamin intake and coronary heart disease in men and women using logistic regression analysis. *J. Clin. Epidemiol.* 48(2), 307–316.

Tonstad, S. (1995). Antioxidants and cardiovascular disorders—epidemiologic aspects. Should high risk patients receive supplementation? *Tidsskr-Nor-Laegeforen* 115(2), 227–229.

Tsao, C. S., Dunham, W. B., *et al.* (1995). Growth control of human colon tumor xenografts by ascorbic acid, copper, and iron. *Cancer J.* 8(3), 157–163.

Tse, W. Y., Maxwell, S. R., *et al.* (1994). Antioxidant status in controlled and uncontrolled hypertension and its relationship to endothelial damage. *J. Hum. Hypertens.* 8(11), 843–849.

van Poppel, G., Kardinaal, A., *et al.* (1994). Antioxidants and coronary heart disease. *Ann. Med.* 26(6), 429–434.

Vera, J. C., Rivas, C. I., *et al.* (1994). Human HL-60 myeloid leukemia cells transport dehydroascorbic acid via the glucose transporters and accumulate reduced ascorbic acid. *Blood* 84(5), 1628–1634.

Vera, J. C., Rivas, C. I., *et al.* (1995). Resolution of the facilitated transport of dehydroascorbic acid from its intracellular accumulation as ascorbic acid. *J. Biol. Chem.* 270(40), 23706–23712.

Vera, J. C., Rivas, C. I., Fischbarg, J., and Golde, D. W. (1993). Mammalian facilitative hexose transporters mediate the transport of dehydroascorbic acid. *Nature* 364, 79–82.

Villalba, J. M., Canalejo, A., *et al.* (1993a). Thiol groups are involved in NADH-ascorbate free radical reductase activity of rat liver plasma membrane. *Biochem. Biophys. Res. Commun.* 192(2), 707–713.

Villalba, J. M., Canalejo, A., *et al.* (1993b). NADH-ascorbate free radical and -ferricyanide reductase activities represent different levels of plasma membrane electron transport. *J. Bioenerg. Biomembr.* 25(4), 411–417.

Wang, H., Zhang, Z. B., *et al.* (1994). Reduction of erythrocyte sorbitol by ascorbic acid in patients with diabetes mellitus. *Chung-Hua-I-Hsueh-Tsa-Chih* 74(9), 548–551, 583.

Washko, P., and Levine, M. (1992). Inhibition of ascorbic acid transport in human neutrophils by glucose. *J. Biol. Chem.* 267(33), 23568–23574.

Washko, P., Rotrosen, D., *et al.* (1990). Ascorbic acid accumulation in plated human neutrophils. *FEBS Lett.* 260(1), 101–104.

Washko, P., Rotrosen, D., *et al.* (1991). Ascorbic acid in human neutrophils. *Am. J. Clin. Nutr.* 54(Suppl. 6), 1221s–1227s.

Washko, P. W., Wang, Y., *et al.* (1993). Ascorbic acid recycling in human neutrophils. *J. Biol. Chem.* 268(21), 15531–15535.

Welch, R. W., Bergsten, P., *et al.* (1993). Ascorbic acid accumulation and transport in human fibroblasts. *Biochem. J.* 294(Pt 2), 505–510.

Welch, R. W., Wang, Y. H., *et al.* (1995). Accumulation of vitamin C (ascorbate) and its oxidized metabolite dehydroascorbic acid occurs by separate mechanisms. *J. Biol. Chem.* 270(21), 12584–12592.

Wells, W. W., and Xu, D. P. (1994). Dehydroascorbate reduction. *J. Bioenerg. Biomemb.* 26(4), 369–377.

Wells, W. W., Xu, D. P., *et al.* (1990). Mammalian thioltransferase (glutaredoxin) and protein disulfide isomerase have dehydroascorbate reductase activity. *J. Biol. Chem.* 265(26), 15361–15364.

Wilson, J. X., and Dixon, S. J. (1995). Ascorbate concentration in osteoblastic cells is elevated by transforming growth factor-beta. *Am. J. Physiol. Endocrinol. Metab.* 31(4), E565–E571.

Winkler, B. S., Orselli, S. M., *et al.* (1994). The redox couple between glutathione and ascorbic acid: A chemical and physiological perspective. *Free Rad. Biol. Med.* 17(4), 333–349.

Wolf, G. (1993). Uptake of ascorbic acid by human neutrophils. *Nutr. Rev.* 51(11), 337–338.

Yew, M. S. (1983). Effect of streptozotocin diabetes on tissue ascorbic acid and dehydroascorbic acid. *Horm. Metabol. Res.* 15, 158.

Young, I. S., Tate, S., *et al.* (1995). The effects of desferrioxamine and ascorbate on oxidative stress in the streptozotocin diabetic rat. *Free Rad. Biol. Med.* **18**(5), 833–840.

Yue, D. K., McLennan, S., *et al.* (1990). Abnormalities of ascorbic acid metabolism and diabetic control: Differences between diabetic patients and diabetic rats. *Diab. Res. Clin. Pract.* **9**, 239–244.

Zaccaria, A., Weinzweig, N., *et al.* (1994). Vitamin C reduces ischemia-reperfusion injury in a rat epigastric island skin flap model. *Ann. Plast. Surg.* **33**(6), 620–623.

Zhang, L., Blot, W. J., *et al.* (1994). Serum micronutrients in relation to pre-cancerous gastric lesions. *Int. J. Cancer* **56**(5), 650–654.

Maret G. Traber

Department of Molecular and Cell Biology
University of California
Berkeley, California 94720

Regulation of Human Plasma Vitamin E

I. Introduction

There is renewed interest in the role of vitamin E in human nutrition because of the demonstration that vitamin E deficiency occurs in humans consuming apparently adequate diets. New techniques in understanding the basic biology of vitamin E has led to the revelation of some startling interactions—a neurologic disorder is caused by genetic defects in a liver protein necessary for the incorporation of vitamin E into plasma lipoproteins!

The studies leading to this conclusion are highlighted in this chapter. Information about the plasma transport of vitamin E was gleaned using stable isotopes and gas chromatography/mass spectrometry to analyze lipoprotein fractions. Techniques of molecular biology were instrumental in the isolation of the hepatic tocopherol transport protein and in the identification of genetic defects in this protein, which ultimately causes vitamin E deficiency

Advances in Pharmacology, Volume 38

49

in humans. These studies now open the way for new approaches in investigating the function of vitamin E, which may include more than just antioxidant function (Traber and Packer, 1995).

II. Plasma Kinetics

A. Lipoprotein Transport

Vitamin E, because it is fat soluble, was recognized long ago to be transported in the circulation in plasma lipoproteins (Lewis et al., 1954; McCormick et al., 1960). Because vitamin E readily exchanges and equilibrates between lipoproteins (Bjornson et al., 1975), it was not obvious that a specific protein was required for the incorporation of vitamin E into lipoproteins. Studies using various deuterated tocopherols were instrumental in demonstrating the critical role of the liver and the tocopherol transfer protein in this process, as summarized below.

1. Absorption, Chylomicron Secretion, and Catabolism

Vitamin E (Gallo-Torres, 1970; Traber et al., 1986), along with dietary fats (Cohn et al., 1988a), is absorbed from the intestine and secreted into the circulation in chylomicrons. Vitamin E is poorly absorbed in the absence of bile acid secretion, as occurs in patients with cholestatic liver disease or other causes of fat malabsorption (Sokol et al., 1983), or when there is an inability to secrete apolipoprotein B, as occurs in patients with abetalipoproteinemia (Traber et al., 1994a).

Although the human diet contains a variety of forms of vitamin E (Sheppard et al., 1993), the plasma is preferentially enriched in α-tocopherol. Furthermore, administration of supplemental vitamin E (when given as α-tocopherol or α-tocopheryl acetate) not only increases α-tocopherol concentrations, but it also decreases plasma γ-tocopherol concentrations (Handelman et al., 1985; Baker et al., 1986). This process is not dependent on differential absorption of the various forms of vitamin E. During absorption and chylomicron secretion in humans following the administration of deuterium-labeled tocopherols (Traber et al., 1988, 1990a,c, 1992a), discrimination did not occur between forms of vitamin E, such as α- and γ-tocopherols or between RRR- and SRR-α-tocopherols. Thus, the various forms of vitamin E are absorbed in proportion to their abundance in the diet. Similar results have been obtained using unlabeled tocopherols (Meydani et al., 1989; Traber and Kayden, 1989).

During chylomicron catabolism, vitamin E is distributed to all of the circulating lipoproteins (Traber et al., 1988, 1990a,c, 1992a). The various vitamin E forms are probably transferred to high-density lipoproteins (HDL), which then transfer these forms to other lipoproteins (Kayden and Bjornson,

1972; Bjornson *et al.*, 1975; Massey, 1984; Granot *et al.*, 1988; Traber *et al.*, 1992b). This transfer process does not discriminate between vitamin E forms; following an equimolar dose of *RRR*- and *SRR*-α-tocopherols and γ-tocopherol, all these forms are present in similar concentrations in all the lipoproteins at early time points (up to about 12 hr) (Traber *et al.*, 1988, 1990a,c, 1992a). Furthermore, the phospholipid transfer protein (PLTP) may aid in the redistribution of HDL vitamin E to other lipoproteins (Kostner *et al.*, 1995).

2. Very Low Density Lipoprotein Secretion and Catabolism

The liver is the key regulator of the forms and amounts of vitamin E that are secreted into the plasma (Traber, 1994; Traber and Sies, 1996). Vitamin E along with dietary lipids is delivered to the liver by chylomicron remnants (Gotto *et al.*, 1986). These lipids are then repackaged by the liver and secreted into the plasma in very low density lipoproteins (VLDL), which have a triglyceride-rich core and apolipoprotein B-100 as a major apolipoprotein (Gotto *et al.*, 1986). Although vitamin E is secreted in VLDL (Bjørneboe *et al.*, 1987; Cohn *et al.*, 1988b), it is only *RRR*-α-tocopherol that is preferentially incorporated into VLDL (Traber *et al.*, 1990b). Studies of nascent VLDL isolated from perfusates of livers from cynomolgus monkeys fed 24 hr previously with various deuterated tocopherols have demonstrated that the liver, not the intestine, discriminates between tocopherols (Traber *et al.*, 1990b).

The plasma becomes preferentially enriched in *RRR*-α-tocopherol as a result of VLDL catabolism (Traber *et al.*, 1988, 1990a,c, 1992a). During the conversion of VLDL to low density lipoproteins (LDL) in the circulation, a portion of the *RRR*-α-tocopherol remains with the core lipids, whereas some is transferred to HDL. Equilibration of *RRR*-α-tocopherol between LDL and HDL occurs because these two lipoproteins readily exchange tocopherol without the assistance of any transfer proteins (Kayden and Bjornson, 1972; Bjornson *et al.*, 1975; Massey, 1984; Granot *et al.*, 1988; Traber *et al.*, 1992b). However, the distribution of α-tocopherol between LDL and HDL is dependent on plasma concentrations of these two lipoproteins; HDL contains a higher α-tocopherol to protein ratio when plasma HDL levels are elevated (Traber *et al.*, 1992b).

B. Turnover Rates

A kinetic model of plasma vitamin E transport was developed from the results of studies in humans using deuterium-labeled stereoisomers of α-tocopherol (*RRR*- and *SRR*-, natural and synthetic forms of vitamin E, respectively) (Traber *et al.*, 1994b). In normal subjects given equimolar amounts of these two labeled stereoisomers, plasma was preferentially enriched in *RRR*-α-tocopherol within 24 hr. Fractional disappearance rates of deuterium-labeled RRR-α-tocopherol (0.4 ± 0.1 pools per day) were significantly ($P < 0.01$) slower than for SRR- (1.2 ± 0.6). Thus, labeled RRR-α-tocopherol apparently leaves the plasma compartment slowly. However, the opposite is true. Although plasma-labeled RRR-α-tocopherol concentrations appear to change slowly, both RRR- and SRR-α-tocopherols leave the plasma similarly, but only RRR-α-tocopherol is returned to the plasma. RRR-α-tocopherol rapidly leaves the plasma, but is incorporated into VLDL by the liver and is resecreted back into the plasma (Traber *et al.*, 1990a,b, 1992a). The differences (0.8 ± 0.6 pools/day) between the rates of disappearance for RRR- and SRR-α-tocopherols estimate the rate that RRR-α-tocopherol is returned to the plasma. This recycling of RRR-α-tocopherol accounts for nearly 1 pool of α-tocopherol per day. Based on an estimated plasma α-tocopherol concentration of 25 μM and a 4-liter plasma volume, the plasma pool size can be calculated to equal 100 μmol. Thus, approximately 74 μmol/day or 3.1 μmol/hr of RRR-α-tocopherol is reincorporated into the plasma. The net effect of this process is the apparent slow disappearance of RRR-α-tocopherol from the plasma.

III. Discrimination between Tocopherols

Vitamin E, a potent peroxyl radical scavenger, is a chain-breaking antioxidant, which prevents the propagation of free radical damage in biological membranes (Burton *et al.*, 1983b; Ingold *et al.*, 1987b). The term vitamin E includes eight naturally occurring molecules: four tocopherols and four tocotrienols. Tocopherols have a phytyl tail, whereas tocotrienols have an isoprenoid tail; the four forms of tocopherols and tocotrienols differ in the number of methyl groups on the chromanol nucleus (α has three, β and γ have two, and δ has one). The biological activity of the various vitamin E forms roughly correlate with their antioxidant activities; the order of relative peroxyl radical scavenging reactivities of α-, β-, γ-, and δ-tocopherols (100, 60, 25, and 27, respectively) (Burton and Ingold, 1981) is the same as the relative order of their biological activities (1.5, 0.75, 0.15, and 0.05 mg/IU, respectively) (Bunyan *et al.*, 1961). However, on closer examination, the relationship between antioxidant and biological activities breaks down. α-Tocotrienol has only one third of the biological activity of α-tocopherol (Bunyan *et al.*, 1961; Weimann and Weiser, 1991), yet it has higher (Serbinova *et al.*, 1993) or equivalent (Suarna *et al.*, 1993) antioxidant activity. A vitamin E analog [2,4,6,7-tetramethyl-2-(4', 8', 12'-trimethyltridecyl)-5-hydroxy-3,4-dihydrobenzofuran] with equivalent biological activity to RRR-α-tocopherol (Ingold *et al.*, 1990) has 1.5 times the antioxidant activity (Burton *et al.*, 1983a). Furthermore, synthetic vitamin E (*all rac*-α-tocopherol) contains equal amounts of eight different stereoisomers of α-tocopherol, which have equivalent antioxidant activity, but each of which has a different biological activity (Weiser *et al.*, 1986)—the biologic activities of the 2-*S* forms are generally lower than the 2-*R* forms (Weiser *et al.*, 1986). Overall, the highest biological activity is found in molecules with three methyl groups and a free hydroxyl on the chromane ring with the phytyl tail meeting the ring in the *R* orientation. This specific structural requirement suggests specific interactions of vitamin E with proteins.

Discrimination between forms of vitamin E has been studied using deuterated tocopherols in rats (Cheng *et al.*, 1987; Ingold *et al.*, 1987a; Burton *et al.*, 1988), guinea pigs (Burton *et al.*, 1990), monkeys (Traber *et al.*, 1990b), and in humans, both in normal subjects (Traber *et al.*, 1990a, 1992a, 1993) and in patients with genetic abnormalities of lipoprotein metabolism (Traber *et al.*, 1992a). Preferential secretion of *RRR*-α-tocopherol in VLDL has been observed consistently. γ-Tocopherol, which has a phytyl tail in the natural conformation, and *SRR*-α-tocopherol, which has the opposite conformation at the 2-position, had similar plasma disappearance kinetics and disappeared from the plasma more rapidly than *RRR*-α-tocopherol (Traber *et al.*, 1992a). Apparently, γ- and *SRR*-α-tocopherols are transported in lipoproteins in a nonspecific manner.

A. Tocopherol Transfer Protein

Given the hydrophobic nature of vitamin E, tocopherol transfer proteins may be required for intracellular trafficking. Perhaps the best studied example is the hepatic α-tocopherol transfer protein, first partially purified from rat liver cytosol by Catignani and Bieri (1977) and shown by Murphy and Mavis (1981) to have transfer activity. This hepatic α-tocopherol transfer protein (30–35 kDa) has now been purified to homogeneity from rat (Sato *et al.*, 1991; Yoshida *et al.*, 1992) and human liver (Kuhlenkamp *et al.*, 1993; Arita *et al.*, 1995), and the amino acid sequence of both the rat (Sato *et al.*, 1993) and the human (Arita *et al.*, 1995) has been reported. This protein recognizes the following features of vitamin E: (1) the fully methylated, intact chromane ring with the free 6-OH group; (2) the presence of the phytyl side chain; and (3) the stereochemical configuration 2R position.

The α-tocopherol transfer protein is present only in hepatocytes, not in any other liver cells, other tissues, or in plasma (Yoshida *et al.*, 1992). The purified protein transfers α-tocopherol between liposomes and microsomes (Sato *et al.*, 1991; Arita *et al.*, 1995). Both α- and β-tocopherols are effective competitors of this transfer activity, γ-tocopherol is about one-half as effective, and δ-tocopherol is about one-third as effective; α-tocopheryl acetate, tocopherol quinone, and cholesterol are ineffective competitors (Sato *et al.*, 1991). This ability to transfer tocopherol is likely responsible for the incorporation of *RRR*-α-tocopherol into VLDL, as discussed earlier.

Tissues other than the liver may also have distinct binding proteins necessary for the regulation and delivery of *RRR*-α-tocopherol to specific intracellular sites. Prasad *et al.* (1981) described various cytosolic *RRR*-α-tocopherol-binding proteins in neuroblastoma and glioma cells in culture. Dutta-Roy *et al.* (1993a,b) found that both liver and heart contain a 14.2-kDa α-tocopherol-binding protein, which transfers α-tocopherol in preference to γ- or δ-tocopherol. Nalecz *et al.* (1992) isolated multiple α-tocopherol-binding proteins (81, 58, and 31 kDa) from cytosol of cultured smooth muscle cells using an α-tocopherol affinity column and reported that the 58-kDa polypeptide could also be eluted from the column by chromanol and the 81-kDa polypeptide by phytol.

Membrane-binding proteins for *RRR*-α-tocopherol (major component kDa 65 and a minor component kDa 125) also have been described by Kitabchi and Wimalasena (1982) and Wimalasena *et al.* (1982) on human erythrocytes. Specific binding of α-tocopherol to erythrocyte membranes may be involved in the prevention of hemolysis because α-tocopherol is more effective than other tocopherols in preserving erythrocytes (Urano *et al.*, 1992). Specific binding sites for *RRR*-α-tocopherol were found by Kunisaki *et al.* (1992, 1993) on cultured bovine aortic endothelial cells. These binding sites exhibited time- and temperature-dependent saturation, and were suggested to be involved in the regulation of prostacyclin production (Kunisaki *et al.*, 1992). Furthermore, the specific binding affinity was decreased when the endothelial cells were cultured with high glucose concentrations (Kunisaki *et al.*, 1993), suggesting that glycation or oxidative modification of membrane proteins can prevent vitamin E binding. Taken together, these studies suggest that cell membranes from various tissues have vitamin E-binding sites.

IV. Genetic Studies

The central role of the liver in maintaining plasma vitamin E concentrations and in discriminating between various forms of vitamin E has been demonstrated using deuterated tocopherols (Traber *et al.*, 1990c, 1993). Genetic defects in the hepatic α-tocopherol transfer protein have now been reported in patients with isolated vitamin E deficiency (Ouahchi *et al.*, 1995). Arita *et al.* (1995) isolated the α-tocopherol transfer protein from human liver cytosol and reported its cDNA sequence. The human protein has 97% homology to the rat protein (Sato *et al.*, 1993) and some homology both to the retinaldehyde-binding protein in the retina and to sec14, a phospholipid transfer protein (Arita *et al.*, 1995). The human gene has been localized to the 8q13.1–13.3 region of chromosome 8 (Arita *et al.*, 1995; Doerflinger *et al.*, 1995).

This genetic defect in the α-tocopherol transfer protein is associated with a characteristic syndrome. Patients have deficient plasma vitamin E concentrations, but if they are given vitamin E supplements (about 1 g per day), then plasma concentrations rapidly reach normal concentrations (Sokol *et al.*, 1988). However, if supplementation is halted, then plasma vitamin E concentrations fall rapidly to as low as 1/100 of normal. These patients are characterized by a progressive peripheral neuropathy with a specific "dying back" of the large caliber axons of the sensory neurons, which results in ataxia (Sokol *et al.*, 1988). Vitamin E supplements halt the progression of the neurologic disorder and, in some cases, improvement has been noted (Sokol, 1993).

Studies in patients with Friedreich's ataxia led to the characterization of genetic defects in the α-tocopherol transfer protein. The most common autosomal recessive ataxia is Friedreich's ataxia, which is linked to a defect on chromosome 9 and is characterized by absent tendon reflexes, deep sensory loss, and cerebellar and Babinski signs (Belal *et al.*, 1995). From a large number of inbred Tunisian families, Ben Hamida *et al.* (1993a,b) identified a small group of patients with ataxia who had defects on chromosome 8, not 9. Subsequent investigation demonstrated that patients from six families who had neurologic abnormalities also had extraordinarily low plasma vitamin E concentrations (Ben Hamida *et al.*, 1993a). Their genetic defect was mapped to the same location (chromosome 8) as the α-tocopherol transfer protein (Ouahchi *et al.*, 1995). The neurologic syndrome in these patients is similar to Friedreich's ataxia (Ben Hamida *et al.*, 1993a,b), but now their disorder is termed ataxia with vitamin E deficiency (AVED). The similarity of the clinical description of Friedreich's ataxia and AVED emphasizes the necessity for measuring the vitamin E concentrations in patients with peripheral neurologic disorders because vitamin E supplementation can reverse neurologic symptoms in early stages of AVED and prevent further progression of the neurologic disorder due to vitamin E deficiency (Sokol *et al.*, 1985; Brin *et al.*, 1986; Sokol, 1993).

The biochemical defect in patients with AVED [previously called familial isolated vitamin E (FIVE) deficiency] was described using deuterated tocopherols (Traber, Sokol, *et al.*, 1990; Traber *et al.*, 1993). Following oral administration of the labeled α-tocopherol stereoisomers to patients and control subjects, both groups similarly absorbed and secreted the two stereoisomers in chylomicrons. Both groups also had similar plasma concentrations of d_3-SRR-α-tocopherol throughout the study. However, patients and controls differed in their abilities to maintain plasma d_6-RRR-α-tocopherol concentrations. Controls maintained plasma d_6-RRR-α-tocopherol concentrations by preferentially secreting it in VLDL. Three of seven patients were "nondiscriminators"—they did not discriminate between the two stereoisomers. Their plasma d_6-RRR-α-tocopherol concentrations declined rapidly and identically to d_3-SRR-α-tocopherol. The remaining patients were intermediate between nondiscriminators and controls in their ability to discriminate and maintain plasma d_6-RRR-α-tocopherol concentrations. Based on these kinetic data, Traber *et al.* (1993) suggested (1) that a hepatic α-tocopherol-binding protein, which preferentially incorporates *RRR*-α-tocopherol into VLDL, is required to maintain plasma *RRR*-α-tocopherol concentrations; (2) that nondiscriminators are lacking this protein or have a marked defect in the *RRR*-α-tocopherol-binding region of the protein; and (3) that patients who discriminate, but have difficulty maintaining plasma *RRR*-α-tocopherol concentrations have a less severe defect, or perhaps a defect in the transfer function of the protein.

The genetic defect in the α-tocopherol transfer protein has been identified in some of the same patients who have been analyzed using deuterium-labeled tocopherols and these are shown in Table I. The gene for the protein contains five exons; these have been defined and are labeled "a" through "e" from the proximal to the distal portion of the gene (Gotoda *et al.*, 1995). A family with three affected siblings (Traber *et al.*, 1987, 1990c; Sokol *et al.*, 1988) has an impaired incorporation of α-tocopherol into VLDL (Traber *et al.*, 1990c), but discriminates between stereoisomers of α-tocopherol (Traber *et al.*, 1993). These patients are heterozygous for two mutations in the tocopherol transfer protein (Hentati *et al.*, 1995; Ouahchi *et al.*, 1995). One mutation changes the arginine at position 192 to a histidine; the other mutation results in a truncation of the terminal 38% of the protein (Hentati *et al.*, 1995; Ouahchi *et al.*, 1995). Another patient (No. 4 in Table I) with severe vitamin E deficiency symptoms (Stumpf *et al.*, 1987; Traber *et al.*, 1987, 1990c; Sokol *et al.*, 1988) has been described in a preliminary report to be homozygous for this latter mutation (Hentati *et al.*, 1995). This patient does not discriminate between stereoisomers of α-tocopherol and has impaired incorporation of α-tocopherol into VLDL (Traber *et al.*, 1990c, 1993).

Truncations of the terminal portion of the α-tocopherol transfer protein result in severe vitamin E deficiency; this is clearly demonstrated in patients 4 and 6, and in the patients from the 17 Tunisian families (Table I). All these patients have a truncation of the terminal portion of the protein, extremely low plasma vitamin E concentrations, and severe neurologic abnormalities.

Interestingly, one patient who discriminates between stereoisomers of α-tocopherol (Table I, patient 7) did not experience neurologic difficulties until after he reached 50 years of age or so (Yokota *et al.*, 1987). The observations in this patient prompted Gotoda *et al.* (1995) to initiate a large population study near the patient's home on an isolated Japanese island, where 801 inhabitants were screened. The patient was found to be homozygous for a thymine to guanine transversion that causes the histidine at position 101 to be replaced with glutamine; 21 other island dwellers were heterozygous for this mutation. These heterozygotes significantly lower serum vitamin E concentrations than a control population.

TABLE I Ability of Patients with Ataxia with Vitamin E Deficiency to Discriminate between Stereoisomers of α-Tocopherol and Their Genetic Defects in the α-Tocopherol Transfer Protein

Patient (references to clinical descriptions)	Response to natural and synthetic stereoisomers of α-tocopherol[a]	Location of the genetic defect[a]	Mutation	Defect	Resultant protein defect	Ref.
1, 2 (Traber et al., 1987, 1990; Sokol et al., 1988)	Discriminator (Traber et al., 1993)	Nucleotide 514 Exon c	Heterozygote insertion [513 insTT(G)][b]	Frameshift and stop codon	Truncation by 38%	Hentati et al. (1995); Ouahchi et al. (1995)
		Nucleotide 576 Exon d	A to G	192 arginine → histidine		
4 (Stumpf et al., 1987; Traber et al., 1987, 1990; Sokol et al., 1988)	Nondiscriminator (Traber et al., 1993)	Nucleotide 514 Exon c	Homozygote (513 insTT) insertion	Frameshift and stop codon	Truncation by 38%	Hentati et al. (1995)
6 (Burck et al., 1981; Kohlschütter et al., 1988)	Nondiscriminator (Traber et al., 1993)	Nucleotide 530 and 532 Exon c	Homozygote 4-bp insertion (530AG → GTAAGT)	Frameshift and stop codon	Truncation by >30%	Ouahchi et al. (1995)
7 (Yokota et al., 1987; Gotoda et al., 1995)	Discriminator (Traber et al., 1993)	Nucleotide 303 Exon b	T to G	101 histidine → glutamine	Transfer function decreased to 11.4 ± 2.2% of normal	Gotoda et al. (1995)
17 families (Doerflinger et al., 1995; Ouahchi et al., 1995)	Unknown	Nucleotide 744 Exon e	Delete A	Replacement of last 30 amino acids with an aberrant 14 amino acid peptide	Truncation by ~10%	Ouahchi et al. (1995)

[a] Exon mapping was from Gotoda et al. (1995).
[b] The 513 insertion TT was reported in Hentati et al. (1995) and insertion of TG was reported in Ouahchi et al. (1995).

The identification of the location of specific mutations in the α-tocopherol transfer protein now allows identification of functional portions of the protein. Genetic defects that result in truncations of the terminal portion of the α-tocopherol transfer protein cause the most severe cases of human vitamin E deficiency. Patients incapable of discriminating between *RRR*- and *SRR*-α-tocopherol also have truncation defects (Table I). It is notable that the patients who are heterozygous for a truncation mutation in the α-tocopherol transfer protein are able to discriminate between stereoisomers (Hentati *et al.*, 1995; Ouahchi *et al.*, 1995), suggesting that the 192 arginine to histidine mutation results in a protein that can bind vitamin E. Taken together, these data suggest that the terminal portion of the protein is essential for binding vitamin E and that a dysfunctional α-tocopherol transfer protein, which is incapable of binding vitamin E, is incapable of enriching lipoproteins with vitamin E, even if its transfer capability is not impaired.

Defective transfer function of the α-tocopherol transfer protein has also been demonstrated. A discriminator patient preferentially incorporated natural *RRR*-α-tocopherol into lipoproteins, but at a reduced rate (Traber *et al.*, 1993). This patient's mutation, when expressed in COS-7 cells, produced a functionally defective transfer protein with approximately 11% of the activity of the wild-type protein (Gotoda *et al.*, 1995). Thus, the region around histidine 101 must be important for transfer activity. This region has a high degree of homology, both with the retinaldehyde-binding protein present in retina and with the yeast SEC14 protein (Arita *et al.*, 1995). These latter two proteins are both lipid-binding/transfer proteins, again suggesting that this region is involved in transfer function.

V. Conclusions

The year 1995 has been an important year for advances in the understanding of human vitamin E nutrition. It is now clear that the liver controls plasma vitamin E concentrations through the incorporation in plasma VLDL by the α-tocopherol transfer protein and that vitamin E deficiency in humans results from genetic defects in this protein. Vitamin E deficiency in humans results in a peripheral neuropathy that arises from the resultant impaired lipoprotein delivery of vitamin E to nervous tissues; α-tocopherol concentrations of peripheral nerves are especially sensitive to variations in plasma vitamin E (Pillai *et al.*, 1993a,b).

It is also clear that vitamin E trafficking is mediated by the hepatic α-tocopherol transfer protein, if not additionally by other cellular tocopherol transfer proteins. Defects in the gene for the hepatic α-tocopherol transfer protein have now begun to yield information about the structure/function relationships of this protein. There are at least two functional domains: an α-tocopherol-binding region and a region involved in transfer. Studies using site-directed mutagenesis will be helpful to delineate specific functional regions.

Regulation of the hepatic α-tocopherol transfer protein remains to be elucidated. Is the protein regulated by the redox status of the liver, the rate of lipoprotein synthesis, the availability of substrate α-tocopherol, or by a combination of these? Identification, isolation, and demonstration of the function of other tissue tocopherol transfer proteins are also important areas for future studies, especially since α-tocopherol has a putative regulatory role in cell proliferation and signal transduction (Boscoboinik *et al.*, 1991; Stauble *et al.*, 1994).

References

Arita, M., Sato, Y., Miyata, A., Tanabe, T., Takahashi, E., Kayden, H., Arai, H., and Inoue, K. (1995). Human alpha-tocopherol transfer protein: cDNA cloning, expression and chromosomal localization. *Biochem. J.* **306**, 437–443.

Baker, H., Handelman, G. J., Short, S., Machlin, L. J., Bhagavan, H. N., Dratz, E. A., and Frank, O. (1986). Comparison of plasma α- and γ-tocopherol levels following chronic oral administration of either all-rac-α-tocopheryl acetate or RRR-α-tocopheryl acetate in normal adult male subjects. *Am. J. Clin. Nutr.* **43**, 382–387.

Belal, S., Hentati, F., Ben Hamida, C., and Ben Hamida, M. (1995). Friedreich's ataxia-vitamin E responsive type: The chromosome 8 locus. *Clin. Neurosci.* **3**, 39–42.

Ben Hamida, C., Doerflinger, N., Belal, S., Linder, C., Reutenauer, L., Dib, C., Gyapay, G., Bignal, A., Le Paslier, D., Cohen, D., Pandolfo, M., Mokini, V., Novelli, G., Hentati, F., Ben Hamida, M., Mandel, J. L., and Koenig, M. (1993a). Localization of Friedreich ataxia phenotype with selective vitamin E deficiency to chromosome 8q by homozygosity mapping. *Nature Genet.* **5**, 195–200.

Ben Hamida, M., Belal, S., Sirugo, G., Ben Hamida, C., Panayides, K., Ionannou, P., Beckmann, J., Mandel, J. L., Hentati, F., Koenig, M., and Middleton, L. (1993b). Friedreich's ataxia phenotype not linked to chromosome 9 and associated with selective autosomal recessive vitamin E deficiency in two inbred Tunisian families. *Neurology* **43**, 2179–2183.

Bjørneboe, A., Bjørneboe, G.-E. A., Hagen, B. F., Nossen, J. O., and Drevon, C. A. (1987). Secretion of α-tocopherol from cultured rat hepatocytes. *Biochim. Biophys. Acta* **922**, 199–205.

Bjornson, L. K., Gniewkowski, C., and Kayden, H. J. (1975). Comparison of exchange of α-tocopherol and free cholesterol between rat plasma lipoprotein and erythrocytes. *J. Lipid Res.* **16**, 39–53.

Boscoboinik, D., Szewczyk, A., Hensey, C., and Azzi, A. (1991). Inhibition of cell proliferation by alpha-tocopherol: Role of protein kinase C. *J. Biol. Chem.* **266**, 6188–6194.

Brin, M. F., Pedley, T. A., Emerson, R. G., Lovelace, R. E., Gouras, P., MacKay, C., Kayden, H. J., Levy, J., and Baker, H. (1986). Electrophysiological features of abetalipoproteinemia: Functional consequences of vitamin E deficiency. *Neurology* **36**, 669–673.

Bunyan, J., McHale, D., Green, J., and Marcinkiewicz, S. (1961). Biological potencies of ε- and ζ₁-tocopherol and 5-methyltocol. *Brit. J. Nutr.* **15**, 253–257.

Burck, U., Goebel, H. H., Kuhlendahl, H. D., Meier, C., and Goebel, K. M. (1981). Neuromyopathy and vitamin E deficiency in man. *Neuropediatrics* **12**, 267–278.

Burton, G. W., Hughes, L., and Ingold, K. U. (1983a). Antioxidant activity of phenols related to vitamin E: Are there chain-breaking antioxidants better than α-tocopherol? *J. Am. Chem. Soc.* **105**, 5950–5951.

Burton, G. W., and Ingold, K. U. (1981). Autoxidation of biological molecules. I. The antioxidant activity of vitamin E and related chain-breaking phenolic antioxidants in vitro. *J. Amer. Chem. Soc.* **103**, 6472–6477.

Burton, G. W., Ingold, K. U., Foster, D. O., Cheng, S. C., Webb, A., Hughes, L., and Lusztyk, E. (1988). Comparison of free α-tocopherol and α-tocopheryl acetate as sources of vitamin E in rats and humans. *Lipids* **23**, 834–840.

Burton, G. W., Joyce, A., and Ingold, K. U. (1983b). Is vitamin E the only lipid-soluble, chain-breaking antioxidant in human blood plasma and erythrocyte membranes? *Arch. Biochem. Biophys.* **221**, 281–290.

Burton, G. W., Wronska, U., Stone, L., Foster, D. O., and Ingold, K. U. (1990). Biokinetics of dietary RRR-α-tocopherol in the male guinea pig at three dietary levels of vitamin C and two levels of vitamin E: Evidence that vitamin C does not "spare" vitamin E in vivo. *Lipids* **25**, 199–210.

Catignani, G. L., and Bieri, J. G. (1977). Rat liver α-tocopherol binding protein. *Biochim. Biophys. Acta* **497**, 349–357.

Cheng, S. C., Burton, G. W., Ingold, K. U., and Foster, D. O. (1987). Chiral discrimination in the exchange of α-tocopherol stereoisomers between plasma and red blood cells. *Lipids* **22**, 469–473.

Cohn, J. S., McNamara, J. R., Cohn, S. D., Ordovas, J. M., and Schaefer, E. J. (1988a). Plasma apolipoprotein changes in the triglyceride-rich lipoprotein fraction of human subjects fed a fat-rich meal. *J. Lipid Res.* **29**, 925–936.

Cohn, W., Loechleiter, F., and Weber, F. (1988b). α-Tocopherol is secreted from rat liver in very low density lipoproteins. *J. Lipid Res.* **29**, 1359–1366.

Doerflinger, N., Linder, C., Ouahchi, K., Gyapay, G., Weissenbach, J., Le Paslier, D., Rigault, P., Belal, S., Ben Hamida, C., Hentati, F., Hamida, M. B., Pandoifo, M., Di Donato, S., Sokol, R., Kayden, H., Landrieu, P., Durr, A., Brice, A., Goutières, F., Kohlschütter, A., Sabouraud, P., Benemar, A., Yahyaovi, M., Mandel, J.-L., and Koenig, M. (1995). Ataxia with vitamin E deficiency: Refinement of genetic localization and analysis of linkage disequilibrium by using new markers in 14 families. *Am. J. Hum. Genet.* **56**, 1116–1124.

Dutta-Roy, A., Gordon, M., Leishman, D., Paterson, B. J., Duthie, G. G., and James, W. P. T. (1993a). Purification and partial characterisation of an alpha-tocopherol-binding protein from rabbit heart cytosol. *Mol. Cell. Biochem* **123**, 139–144.

Dutta-Roy, A. K., Leishman, D. J., Gordon, M. J., Campbell, F. M., and Duthie, G. G. (1993b). Identification of a low molecular mass (14.2 kDa) alpha-tocopherol-binding protein in the cytosol of rat liver and heart. *Biochem. Biophys. Res. Commun.* **196**, 1108–1112.

Gallo-Torres, H. (1970). Obligatory role of bile for the intestinal absorption of vitamin E. *Lipids* **5**, 379–384.

Gotoda, T., Arita, M., Arai, H., Inoue, K., Yokota, T., Fukuo, Y., Yazaki, Y., and Yamada, N. (1995). Adult-onset spinocerebellar dysfunction caused by a mutation in the gene for the alpha-tocopherol-transfer protein. *N. Engl. J. Med.* **333**, 1313–1318.

Gotto, A. M., Pownall, H. J., and Havel, R. J. (1986). Introduction to the plasma lipoproteins. *Methods Enzymol.* **128**, 3–41.

Granot, E., Tamir, I., and Deckelbaum, R. J. (1988). Neutral lipid transfer protein does not regulate α-tocopherol transfer between human plasma lipoproteins. *Lipids* **23**, 17–21.

Handelman, G. J., Machlin, L. J., Fitch, K., Weiter, J. J., and Dratz, E. A. (1985). Oral α-tocopherol supplements decrease plasma γ-tocopherol levels in humans. *J. Nutr.* **115**, 807–813.

Hentati, A., Deng, H.-X., Hung, W.-Y., Ahmed, S., He, X., Stumpf, D. A., and Siddique, T. (1995). Human α-tocopherol transfer protein: Gene organization and identification of mutations in patients with familial vitamin E deficiency. *Am. J. Hum. Genet.* **57**(Suppl. 4), A214.

Ingold, K. U., Burton, G. W., Foster, D. O., and Hughes, L. (1990). Further studies of a new vitamin E analogue more active than alpha-tocopherol in the rat curative myopathy bioassay. *FEBS Lett.* **267**, 63–65.

Ingold, K. U., Burton, G. W., Foster, D. O., Hughes, L., Lindsay, D. A., and Webb, A. (1987a). Biokinetics of and discrimination between dietary *RRR*- and *SRR*-α-tocopherols in the male rat. *Lipids* **22**, 163–172.

Ingold, K. U., Webb, A. C., Witter, D., Burton, G. W., Metcalfe, T. A., and Muller, D. P. R. (1987b). Vitamin E remains the major lipid-soluble, chain-breaking antioxidant in human plasma even in individuals suffering severe vitamin E deficiency. *Arch. Biochem. Biophys.* **259**, 224–225.

Kayden, H. J., and Bjornson, L. K. (1972). The dynamics of vitamin E transport in the human erythrocyte. *Ann. N.Y. Acad. Sci.* **203**, 127–140.

Kitabchi, A. E., and Wimalasena, J. (1982). Specific binding sites for d-alpha-tocopherol on human erythrocytes. *Biochim. Biophys. Acta* **684**, 200–206.

Kohlshütter, A., Hubner, C., Jansen, W., and Lindner, S. G. (1988). A treatable familial neuromyopathy with vitamin E deficiency, normal absorption, and evidence of increased consumption of vitamin E. *J. Inher. Metab. Dis.* **11**, 149–152.

Kostner, G. M., Oettl, K., Jauhianinen, M., Ehnholm, C., Esterbauer, H., and Dieplinger, H. (1995). Human plasma phospholipid transfer protein accelerates exchange/transfer of alpha-tocopherol between lipoproteins and cells. *Biochem. J.* **305**, 659–667.

Kuhlenkamp, J., Ronk, M., Yusin, M., Stolz, A., and Kaplowitz, N. (1993). Identification and purification of a human liver cytosolic tocopherol binding protein. *Prot. Exp. Purific.* **4**, 382–389.

Kunisaki, M., Umeda, F., Inoguchi, T., and Nawata, H. (1992). Vitamin E binds to specific binding sites and enhances prostacyclin production by cultured aortic endothelial cells. *Thromb. Haemo.* **68**, 744–751.

Kunisaki, M., Umeda, F., Yamauchi, T., Masakado, M., and Nawata, H. (1993). High glucose reduces specific binding for d-alpha-tocopherol in cultured aortic endothelial cells. *Diabetes* **42**, 1138–1146.

Lewis, L. A., Quaife, M. L., and Page, I. H. (1954). Lipoproteins of serum, carriers of tocopherol. *Am. J. Physiol.* **178**, 221–222.

Massey, J. B. (1984). Kinetics of transfer of α-tocopherol between model and native plasma lipoproteins. *Biochim. Biophys. Acta* **793**, 387–392.

McCormick, E. C., Cornwell, D. G., and Brown, J. B. (1960). Studies on the distribution of tocopherol in human serum lipoproteins. *J. Lipid Res.* **1**, 221–228.

Meydani, M., Cohn, J. S., Macauley, J. B., McNamara, J. R., Blumberg, J. B., and Schaefer, E. J. (1989). Postprandial changes in the plasma concentration of α- and γ-tocopherol in human subjects fed a fat-rich meal supplemented with fat-soluble vitamins. *J. Nutr.* **119**, 1252–1258.

Murphy, D. J., and Mavis, R. D. (1981). Membrane transfer of α-tocopherol. *J. Biol. Chem.* **256**, 10464–10468.

Nalecz, K., Nalecz, M., and Azzi, A. (1992). Isolation of tocopherol-binding proteins from the cytosol of smooth muscle A7R5 cells. *Eur. J. Biochem.* **209**, 37–42.

Ouahchi, K., Arita, M., Kayden, H., Hentati, F., Ben Hamida, M., Sokol, R., Arai, H., Inoue, K., Mandel, J.-L., and Koenig, M. (1995). Ataxia with isolated vitamin E deficiency is caused by mutations in the α-tocopherol transfer protein. *Nature Genet.* **9**, 141–145.

Pillai, S. R., Traber, M. G., Steiss, J. E., and Kayden, H. J. (1993a). Depletion of adipose tissue and peripheral nerve α-tocopherol in adult dogs. *Lipids* **28**, 1095–1099.

Pillai, S. R., Traber, M. G., Steiss, J. E., Kayden, H. J., and Cox, N. R. (1993b). α-Tocopherol concentrations of the nervous system and selected tissues of dogs fed three levels of vitamin E. *Lipids* **28**, 1101–1105.

Prasad, K. N., Gaudreau, D., and Brown, J. (1981). Binding of vitamin E in mammalian tumor cells in culture. *Proc. Soc. Exp. Biol. Med.* **166**, 167–174.

Sato, Y., Arai, H., Miyata, A., Tokita, S., Yamamoto, K., Tanabe, T., and Inoue, K. (1993). Primary structure of alpha-tocopherol transfer protein from rat liver: Homology with cellular retinaldehyde-binding protein. *J. Biol. Chem.* **268**, 17705–17710.

Sato, Y., Hagiwara, K., Arai, H., and Inoue, K. (1991). Purification and characterization of the α-tocopherol transfer protein from rat liver. *FEBS Lett.* **288**, 41–45.

Serbinova, E. A., Tsuchiya, M., Goth, S., Kagan, V. E., and Packer, L. (1993). Antioxidant action of α-tocopherol and α-tocotrienol in membranes. *In* "Vitamin E in Health and Disease" (L. Packer and J. Fuches, eds.), pp. 235–243. Dekker, New York.

Sheppard, A. J., Pennington, J. A. T., and Weihrauch, J. L. (1993). Analysis and distribution of vitamin E in vegetable oils and foods. *In* "Vitamin E in Health and Disease" (L. Packer and J. Fuchs, eds.), pp. 9–31. Dekker, New York.

Sokol, R. J. (1993). Vitamin E deficiency and neurological disorders. *In* "Vitamin E in Health and Disease" (L. Packer and J. Fuchs, eds.), pp. 815–849. Dekker, New York.

Sokol, R. J., Guggenheim, M., Iannaccone, S. T., Barkhaus, P. E., Miller, C., Silverman, A., Balistreri, W. F., and Heubi, J. E. (1985). Improved neurologic function after long-term correction of vitamin E deficiency in children with chronic cholestasis. *N. Engl. J. Med.* **313**, 1580–1586.

Sokol, R. J., Heubi, J. E., Iannaccone, S., Bove, K. E., Harris, R. E., and Balistreri, W. F. (1983). The mechanism causing vitamin E deficiency during chronic childhood cholestasis. *Gastroenterology* **85**, 1172–1182.

Sokol, R. J., Kayden, H. J., Bettis, D. B., Traber, M. G., Neville, H., Ringel, S., Wilson, W. B., and Stumpf, D. A. (1988). Isolated vitamin E deficiency in the absence of fat malabsorption-familial and sporadic cases: Characterization and investigation of causes. *J. Lab. Clin. Med.* **111**, 548–559.

Stauble, B., Boscoboinik, D., Tasinato, A., and Azzi, A. (1994). Modulation of activator protein-1 (AP-1) transcription factor and protein kinase C by hydrogen peroxide and d-alpha-tocopherol in vascular smooth muscle cells. *Eur. J. Biochem.* **226**, 393–402.

Stumpf, D. A., Sokol, R., Bettis, D., Neville, H., Ringel, S., Angelini, C., and Bell, R. (1987). Friedreich's disease. V. Variant form with vitamin E deficiency and normal fat absorption. *Neurology* **37**, 68–74.

Suarna, C., Food, R. L., Dean, R. T., and Stocker, R. (1993). Comparative antioxidant activity of tocotrienols and other natural lipid-soluble antioxidants in a homogeneous system, and in rat and human lipoproteins. *Biochim. Biophys. Acta* **1166**, 163–170.

Traber, M. G. (1994). Determinants of plasma vitamin E concentrations. *Free Rad. Biol. Med.* **16**, 229–239.

Traber, M. G., Burton, G. W., Hughes, L., Ingold, K. U., Hidaka, H., Malloy, M., Kane, J., Hyams, J., and Kayden, H. J. (1992a). Discrimination between forms of vitamin E by humans with and without genetic abnormalities of lipoprotein metabolism. *J. Lipid Res.* **33**, 1171–1182.

Traber, M. G., Burton, G. W., Ingold, K. U., and Kayden, H. J. (1990a). *RRR*- and *SRR*-α-tocopherols are secreted without discrimination in human chylomicrons, but *RRR*-α-tocopherol is preferentially secreted in very low density lipoproteins. *J. Lipid Res.* **31**, 675–685.

Traber, M. G., Ingold, K. U., Burton, G. W., and Kayden, H. J. (1988). Absorption and transport of deuterium-substituted $2R,4'R,8'R$-α-tocopherol in human lipoproteins. *Lipids* **23**, 791–797.

Traber, M. G., and Kayden, H. J. (1989). Preferential incorporation of α-tocopherol vs γ-tocopherol in human lipoproteins. *Am. J. Clin. Nutr.* **49**, 517–526.

Traber, M. G., Kayden, H. J., Green, J. B., and Green, M. H. (1986). Absorption of water miscible forms of vitamin E in a patient with cholestasis and in rats. *Am. J. Clin. Nutr.* **44**, 914–923.

Traber, M. G., Lane, J. C., Lagmay, N., and Kayden, H. J. (1992b). Studies on the transfer of tocopherol between lipoproteins. *Lipids* **27**, 657–663.

Traber, M. G., and Packer, L. (1995). Vitamin E: Beyond antioxidant function. *Am. J. Clin. Nutr.* **62**(Suppl.), 1501S–1509S.

Traber, M. G., Rader, D., Acuff, R., Brewer, H. B., and Kayden, H. J. (1994a). Discrimination between *RRR*- and *all rac*-α-tocopherols labeled with deuterium by patients with abetalipoproteinemia. *Atherosclerosis* **108**, 27–37.

Traber, M. G., Ramakrishnan, R., and Kayden, H. J. (1994b). Human plasma vitamin E kinetics demonstrate rapid recycling of plasma *RRR*-α-tocopherol. *Proc. Natl. Acad. Sci. USA* **91**, 10005–10008.

Traber, M. G., Rudel, L. L., Burton, G. W., Hughes, L., Ingold, K. U., and Kayden, H. J. (1990b). Nascent VLDL from liver perfusions of cynomolgus monkeys are preferentially enriched in *RRR*-compared with *SRR*-α tocopherol: Studies using deuterated tocopherols. *J. Lipid Res.* **31**, 687–694.

Traber, M. G., and Sies, H. (1996). Vitamin E in humans: Demand and delivery. *Annu. Rev. Nutr.* **16**, 321–347.

Traber, M. G., Sokol, R. J., Burton, G. W., Ingold, K. U., Papas, A. M., Huffaker, J. E., and Kayden, H. J. (1990c). Impaired ability of patients with familial isolated vitamin E deficiency to incorporate α-tocopherol into lipoproteins secreted by the liver. *J. Clin. Invest.* **85**, 397–407.

Traber, M. G., Sokol, R. J., Kohlschütter, A., Yokota, T., Muller, D. P. R., Dufour, R., and Kayden, H. J. (1993). Impaired discrimination between stereoisomers of α-tocopherol in patients with familial isolated vitamin E deficiency. *J. Lipid Res.* **34**, 201–210.

Traber, M. G., Sokol, R. J., Ringel, S. P., Neville, H. E., Thellman, C. A., and Kayden, H. J. (1987). Lack of tocopherol in peripheral nerves of vitamin E-deficient patients with peripheral neuropathy. *N. Engl. J. Med.* **317**, 262–265.

Urano, S., Inomori, Y., Sugawara, T., Kato, Y., Kitahara, M., Hasegawa, Y., Matsuo, M., and Mukai, K. (1992). Vitamin E: Inhibition of retinol-induced hemolysis and membrane-stabilizing behavior. *J. Biol. Chem.* **267**, 18365–18370.

Weimann, B. J., and Weiser, H. (1991). Functions of vitamin E in reproduction and in prostacyclin and immunoglobulin synthesis in rats. *Am. J. Clin. Nutr.* **53**, 1056S–1060S.

Weiser, H., Vecchi, M., and Schlachter, M. (1986). Stereoisomers of α-tocopheryl acetate. IV. USP units and α-tocopherol equivalents of *all-rac*-, 2-*ambo*- and *RRR*-α-tocopherol evaluated by simultaneous determination of resorption-gestation, myopathy and liver storage capacity in rats. *Int. J. Vit. Nutr. Res.* **56**, 45–56.

Wimalasena, J., Davis, M., and Kitabchi, A. E. (1982). Characterization and solubilization of the specific binding sites for d-alpha-tocopherol from human erythrocyte membranes. *Biochem. Pharmacol.* **31**, 3455–3461.

Yokota, T., Wada, Y., Furukawa, T., Tsukagoshi, H., Uchihara, T., and Watabiki, S. (1987). Adult-onset spinocerebellar syndrome with idiopathic vitamin E deficiency. *Ann. Neurol* **22**, 84–87.

Yoshida, H., Yusin, M., Ren, I., Kuhlenkamp, J., Hirano, T., Stolz, A., and Kaplowitz, N. (1992). Identification, purification and immunochemical characterization of a tocopherol-binding protein in rat liver cytosol. *J. Lipid Res.* **33**, 343–350.

Mary E. Anderson

Department of Biochemistry
Cornell University Medical College
New York, New York 10021

Glutathione and Glutathione Delivery Compounds

I. Glutathione

A. Overview of Functions

This chapter provides a brief background in glutathione function and metabolism and discusses currently available methods for increasing cellular glutathione levels.

Glutathione (γ-glutamylcysteinylglycine; GSH) is the major cellular antioxidant and is found in high concentrations in most mammalian cells (1 to 10 mM) [for reviews on GSH metabolism, see Meister (1989, 1991, 1995) and Meister and Anderson (1983)].[1] Intracellular GSH is maintained in its thiol form by glutathione disulfide (GSSG) reductase, which requires

---

[1] Due to space limitations and the large number of papers on GSH and GSH delivery compounds, all references in the literature could not be included.

Advances in Pharmacology, Volume 38

NADPH. GSH has several functions (Fig. 1), including roles in metabolism, transport, catalysis (coenzyme), and maintenance of the thiol moieties of proteins and the reduced form of other molecules such as cysteine, coenzyme A, and antioxidants such as ascorbic acid; it is also used in the formation of deoxyribonucleic acids. GSH participates nonenzymatically and enzymatically (GSH S-transferases) in the protection against toxic compounds. Perhaps one of its most important functions is in the protection against oxidative damage caused by reactive oxygen species (ROS), many of which are generated during normal metabolism. GSH can react nonenzymatically with ROS, and GSH peroxidase (and non-Se peroxidase) catalyzes the destruction of hydrogen peroxide and hydroperoxides.

GSH is synthesized intracellularly by the consecutive actions of γ-glutamylcysteine (1) and GSH (2) synthetases:

(1) L-Glutamate + L-cysteine + ATP $\Leftrightarrow$ L-γ-glutamyl-L-cysteine + ADP + P_i
(2) L-γ-Glutamyl-L-cysteine + glycine + ATP $\Leftrightarrow$ GSH + ADP + P_i

The synthesis of GSH is limited by the availability of substrates; cysteine is usually the limiting substrate. γ-Glutamylcysteine synthetase is nonallosterically feedback inhibited by GSH (K_i about 1.5 mM) (Richman and Meister, 1973). Thus, under physiological conditions, γ-glutamylcysteine synthetase is probably not operating at its maximal rate.

The degradation of the γ-glutamyl moiety of GSH (or GSH S-conjugates) is catalyzed by γ-glutamyl transpeptidase, a membrane-bound enzyme whose active site is on the external surface of certain cells. GSH is normally trans-

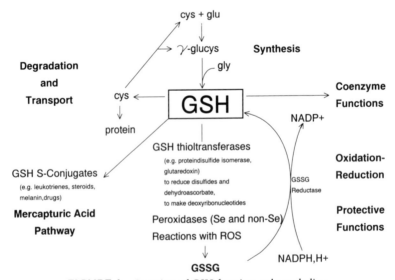

FIGURE I   Overview of GSH function and metabolism.

ported out of cells where transpeptidation occurs in the presence of amino acids, and a γ-glutamyl amino acid is formed. This can then be transported into cells where it, but not GSH, is a substrate for γ-glutamyl cyclotransferase and forms amino acid and 5-oxoproline. 5-Oxoproline is ring opened by 5-oxoprolinase to form glutamate. Most L-amino acids participate in transpeptidation, but cystine is one of the best acceptor amino acids. When cystine participates, γ-glutamylcystine is formed, transported, and reduced to cysteine and γ-glutamylcysteine. Cysteine can be used for the two-step pathway of GSH biosynthesis (1 and 2), and γ-glutamylcysteine can be used directly by GSH synthetase (2) to form GSH (Anderson and Meister, 1983). This alternative pathway of GSH synthesis serves to conserve cysteine moieties; however, the extent to which it functions physiologically is not yet known.

B. GSH Deficiencies

I. Inborn Errors of Metabolism

Although relatively rare, there are patients with inborn errors of GSH metabolism (Meister and Larsson, 1994). A deficiency of GSSG reductase is associated with erythrocyte hemolysis and early cataract formation. Deficiency of glucose-6-phosphate dehydrogenase deficiency leads to decreased levels of NADPH and is thus associated with hemolytic anemia in response to oxidative stress. There are also deficiencies of γ-glutamylcysteine synthetase and GSH synthetase; these are also associated with hemolytic anemia and impaired central nervous system function.

2. L-Buthionine-SR-Sulfoximine (BSO)-Induced GSH Deficiency

Although cultured cells from individuals with inborn errors of GSH metabolism provide some useful information, *in vivo* studies are necessary to study the effects of GSH deficiency, gain information for therapy, and understand more clearly the functions of GSH. Various nonspecific oxidizing compounds (e.g., hydroperoxides, diethylmaleate, phorone, diamide), increased hyperoxia, and radiation have been used to produce oxidative stress and they also usually deplete GSH; however, these compounds may affect other cellular targets and are not specific only for GSH. Amino acid sulfoximines (Griffith *et al.,* 1979; Griffith and Meister, 1979) were developed as specific inhibitors of γ-glutamylcysteine synthetase. The most commonly employed sulfoximine is BSO. The administration of BSO to mice leads to large decreases (10% to 20% of control) in GSH levels of most tissues. Instead of depleting GSH, BSO inhibits the resynthesis of GSH, which replaces the GSH that is normally exported from cells. Treatment of rodents with BSO also sensitizes them to the toxic effects of mercuric ions, cadmium ions, melphalan, cisplatin, and radiation, and it is currently in clinical trials in combination anticancer therapy. Treatment of human lymphocytes and T cells with BSO blocks their activation (Suthanthiran *et al.*, 1990). The remaining 10% to 20% of cellular GSH found after BSO treatment is in the mitochondria pool, which turns over very slowly. Mitochondria do not synthesis GSH, but effectively transport it from the cytosol (Griffith and Meister, 1985; Meister, 1995). Because mitochondria lack catalase, GSH peroxidase, which is found in mitochondria, protects against the normally produced mitochondrial ROS.

The long-term administration of BSO to adult mice leads to severe GSH deficiency and is characterized by mitochondria swelling and vacuolization (Martensson and Meister, 1989). Damage has been observed in skeletal muscle, lung type 2 lamellar bodies, jejunum, and colon (reviewed in Meister, 1991). The administration of BSO to rodents provides a model for endogenously produced oxidative stress (Martensson *et al.*, 1991). Interestingly, in this model, plasma triglycerides and cholesterol levels are substantially increased, whereas tissue ascorbate levels are substantially decreased (Martensson and Meister, 1991). The short-term administration of BSO to neonatal rodents leads to cataracts (Calvin *et al.*, 1986). An oxidative stress model using newborn rodents was also developed (Martensson *et al.*, 1991). Newborn rats are treated twice a day on days 2 to 5 after birth with BSO (3 mmol/kg). Cerebral cortex, lung, liver, and proximal renal tubule are also damaged in the newborn rodent model, and the animals die in less than a week (Martensson *et al.*, 1991). Guinea pigs, like humans and newborn rats, do not synthesize ascorbate, and BSO administration caused oxidative damage similar to that found in the newborn rodent model, except in the lens and brain where the blood–brain and blood–lens barrier prevented the transport of BSO (Martensson *et al.*, 1993; Meister, 1995). No damage to heart or stomach was observed in either model of oxidative stress, suggesting that further study is required. The oxidative damage just described can be largely prevented by treatment with GSH monoesters (see Section II,D) and by ascorbate (Meister, 1992, 1994), but not by GSH.

Elegant studies (Martensson *et al.*, 1991; Martensson and Meister, 1991; Meister, 1992, 1994, 1995) have described a physiological interrelationship between GSH and ascorbate. In the newborn rat–BSO model of oxidative stress, administration of either GSH monoester (see Section III,D) or ascorbate prevented death and protected against some of the decreases in mitochondrial GSH levels. Thus, ascorbate spares GSH. [Interestingly, very high doses of ascorbate (5 mmol/kg per day) were fatal to newborn rats (Martensson *et al.*, 1991).] A guinea pig model of oxidative stress was developed by feeding an ascorbate-deficient diet (Martensson *et al.*, 1993). After 21 days on the scorbutic diet, the tissue levels of ascorbate decreased and the levels of GSH decreased in mitochondria, but not cytosol. The animals showed signs of scurvy (hematomas, swelling of joints, loss of osteoid material from bones, and fractures). The administration of GSH monoester to increase cellular GSH levels (see Section III,D) prevented or significantly delayed (40 days was the longest observation time) the development of scurvy. These results suggest that GSH, supplied as GSH monoester, spares ascorbate. This "sparing" of ascorbate by GSH probably occurs because of the reduction of dehydroascorbate to ascorbate, which is normally carried out by protein disulfide isomerase (Wells *et al.*, 1990, 1992), and by GSH doing some of the functions of ascorbate. Thus, there is apparently some overlap in functions between ascorbate and GSH.

3. Disease Associated with Oxidative Damage

GSH deficiency, as discussed in Section I,B,1, occurs in inborn errors of GSH metabolism and in the BSO model of oxidative stress. However, GSH deficiency has been reported in a number of diseases. Low thiol levels have been reported in the peripheral blood mononuclear cells (PBMC) (Eck *et al.*, 1989) and in the epithelial lining fluid (ELF) (Buhl *et al.*, 1989) of HIV-infected patients. Similarly, low PBMC levels of GSH have been reported in chronically infected hepatitis C patients (Suarez *et al.*, 1993). Patients with type II diabetes are reported to have decreased GSH and increased GSSG levels in whole blood (Forrester *et al.*, 1990). GSH levels are also lower in blood and colonic tissue of patients with ulcerative colitis (Fields *et al.*, 1994). In rabbit and rat skin burn models, tissue and mitochondria GSH levels decrease significantly (Martensson *et al.*, 1992; Sabeh *et al.*, 1995). GSH levels in ELF are decreased in patients with idopathetic pulmonary fibrosis (Cantin *et al.*, 1989) and also in patients with adult respiratory distress syndrome (ARDS) (Pacht *et al.*, 1991).

ROS have been widely implicated in the pathology of numerous diseases such as atherosclerosis, ARDS, Parkinson's, ischemia–reperfusion injury, rheumatoid arthritis, cancer, and AIDS, to mention only a few (see, for example, Cutler *et al.*, 1995; Pace and Leaf, 1995; Packer and Cadenas, 1995; Martinez-Cayuela, 1995), and amelioration is clearly desirable. Antioxidants such as GSH, ascorbate, α-tocopherol, lipoic acid, β-carotene, and superoxide dismutase have been suggested as potential therapies. Perhaps the best evidence for a protective role of ascorbic acid and glutathione in atherosclerosis comes from early studies of Willis (1953) and confirmed by Ginter (1978). Willis (1953) fed cholesterol to scorbutic guinea pigs and found that atherosclerosis was inhibited in those given ascorbate. Ascorbate and GSH may decrease the formation of oxidized LDL, which is the form implicated in the development of atherosclerosis. Studies discussed in Section I,B,2 have shown that GSH and ascorbate "spare" each other. Thus, it is likely that GSH and ascorbate, and possibly other antioxidants, have some overlapping and some nonoverlapping protective functions. This chapter focuses on therapies based on GSH delivery compounds.

II. Glutathione Delivery Compounds

A. Introduction

Increasing cellular GSH levels may have therapeutic effects. Although the administration of cysteine, GSH, or N-acetylcysteine (NAC) might raise cellular GSH levels, this may not always occur and would not be useful in the BSO model of oxidative stress (Section I,B,3). Cysteine is reported to be toxic to cultured cells (Nishiuch et al., 1976) and to newborn mice (Olney et al., 1971) and rats (Karlsen et al., 1981). The mechanism of the toxicity is not clear, but several possibilities have been considered (Anderson and Meister, 1987; Cooper et al., 1982). Administered GSH is not effectively transported into cells (Meister, 1991), but rather it is degraded extracellularly, and the products are transported into cells where they may be used for GSH resynthesis. Thus, while administered GSH may supply substrates for its resynthesis, its administration is not a very effective way to increase cellular GSH levels. NAC administration can increase cellular GSH levels, but first it must be deacetylated to cysteine, which then may be used for GSH synthesis. Methionine can serve as a source of cellular cysteine via the cystathionase pathway; however, this pathway may not function in neonates and certain other patients such as those with liver disease.

B. L-2-Oxothiazolidine-4-Carboxylic Acid: Cysteine Delivery Compound

Studies on the substrate specificity of 5-oxoprolinase (Section I,A) found that an analog of 5-oxoproline, L-2-oxothiazolidine-4-carboxylate (OTC), is an effective substrate (Williamson and Meister, 1981, 1982). The enzymatic product is thought to be S-carboxy cysteine, which rapidly hydrolyzes to give cysteine; D-OTC is neither a substrate nor an inhibitor of the enzyme. 5-Oxoprolinase is found in many tissues, except the lens and erythrocyte. Treatment with BSO, as expected, prevents GSH levels from increasing after OTC administration because BSO blocks the use of cysteine by γ-glutamylcysteine synthetase. The administration of OTC to mice treated with acetaminophen (2.5 mmol/kg) produced larger increases in hepatic GSH levels than did NAC, and OTC also protects against lethal doses of acetaminophen (Williamson et al., 1982). Experiments in which ^{35}S-labeled OTC was administered to mice showed that label was found in all organs and fluids examined (Meister et al., 1986). The specific radioactivity of cysteine was much greater than of GSH, except in kidney and liver where the rate of GSH synthesis is very high. Most of the ^{35}S found in the urine was unmetabolized OTC. The administration of OTC to mice leads to increased tissue levels of cysteine skeletal muscle, liver, pancreas, and brain. GSH levels were increased in liver, pancreas, and, to a lesser extent, in skeletal muscle; little or no increases were found in spleen, brain, kidney, and heart. The administration of OTC increased cysteine levels significantly in various regions of rat brain and were especially pronounced in the cerebellum (Anderson and Meister, 1989a).

Studies using cultured human peritoneal mesothelial cells showed that OTC is effective in stimulating proliferation and decreasing cell death, suggesting that OTC may be useful in peritoneal dialysis (Breborowicz et al., 1993). OTC raised GSH levels and protected Chinese hamster ovary cells from oxygen radical stress (Weitberg, 1987). OTC was also reported to be effective in protecting isolated perfused rat hearts from ischemic damage (Shug and Madsen, 1994). OTC was effective in promoting rat growth and increasing tissue GSH in rats fed a sulfur amino acid-deficient diet (Jain et al., 1995).

Interesting preliminary reports have appeared (Wilson et al., 1993; Josephs et al., 1993) that suggest that OTC administration reduces AZT-induced bone marrow hypoplasia in mice and enhances the antiviral activity of AZT in cultured lymphocytes. In vitro studies (Lederman et al., 1995) on human peripheral blood mononuclear cells showed that OTC treatment inhibited HIV expression; in transient expression experiments, OTC also inhibited HIV-1 promoter activity and NF-κB binding activity in activated Jurkat cells.

In a human clinical trial (Porta et al., 1991), OTC was administered orally at 0.15 or 0.45 mmol/kg to healthy subjects. Plasma OTC levels peaked at about 45 to 60 min. Plasma cysteine levels peaked at about 2 hr, and plasma GSH levels were unchanged during the 8-hr study. Lymphocyte levels of cysteine and GSH increased about two- to threefold after 2 to 3 hr. Thus, OTC is an effective cysteine delivery agent and is converted into GSH in humans. A phase I/II trial (Kalayjian et al., 1994) in asymptomatic HIV-infected patients showed that GSH levels in whole blood GSH increased after 6 weeks of treatment with OTC (100 mg/kg, twice weekly) and there was a decrease in β_2-microglobulin levels; these results suggest that OTC is metabolized to GSH in these patients and that further clinical studies are indicated.

C. γ-Glutamylcyst(e)ine

Compounds that supply cellular cysteine increase cellular GSH levels, but such increases are controlled by the feedback inhibition of γ-glutamylcysteine synthetase. γ-Glutamyl amino acids are transported into kidney (Griffith et al., 1979) and possibly other tissues. We administered γ-glutamylcysteine, γ-glutamylcysteine disulfide, or γ-glutamylcystine to mice and found significantly increased kidney levels of GSH; γ-glutamylcystine was the most effective (Anderson and Meister, 1983). Later studies (Meister et al., 1986) using differentially ^{35}S-labeled γ-glutamylcystine (Anderson, 1985) showed that when the label was in the γ-glutamylcysteine moiety, it was more readily converted to GSH than when the ^{35}S was in the cysteine moiety of γ-glutamylcystine. As discussed earlier (Section I,A,1), these results support the theory of the alternative pathway of GSH synthesis. Pileblad and Magnusson (1992) showed that the intercerebroventricular administration of γ-glutamylcysteine increases rat brain GSH levels. Further studies on the effectiveness of these compounds to increase GSH levels in other tissues are needed.

D. Glutathione Monoesters

A very effective way to increase cellular GSH levels is to supply an analog of GSH that is readily transported into many tissues and converted into GSH. GSH (glycyl) monoester was synthesized and administered to mice; the monoester was found to increase GSH levels in liver and kidney as well as protect against acetaminophen toxicity (Puri and Meister, 1983); GSH levels were raised by the monoester even after BSO pretreatment. Various monoesters of GSH, such as methyl, ethyl, isopropyl, n-propyl, and higher, have been prepared (Anderson et al., 1994; Anderson and Meister, 1989b, Puri and Meister, 1983); we prefer GSH monoethylester (GEE) or GSH monoisopropylester (GiPE) because of the low toxicity of the alcohols produced from the hydrolysis of the ester. GSH esters are transported into human lymphoid cells and fibroblasts and elevate cellular GSH levels (Wellner et al., 1984). These studies also showed that CEM lymphoid cells pretreated with BSO and then treated with GSH monoester were protected cells against radiation; partial protection was observed when GSH was added before or GSH monoester was added after irradiation, suggesting that GSH may participate in repair.

The administration of ^{35}S-labeled GEE to mice led to the appearance of ^{35}S in kidney, liver, spleen, lung, and heart (Anderson et al., 1985). GEE administration led to increased GSH levels in kidney, liver, spleen, pancreas, and heart; plasma and urinary GSH levels were elevated only slightly in comparison with the very large increases produced by GSH administration. GSH levels are increased whether GEE is administered intraperitoneally, subcutaneously, or orally. GEE is also transported into erythrocytes and slowly converted into GSH. GEE is also transported into rat cerebrospinal fluid (Anderson et al., 1989) and into the lens and brain of newborn rats where it protects against oxidative damage in the BSO newborn rat model (Martensson et al., 1989). GSH monoesters protect against the toxicity of cadmium ion, mercuric ions, cisplatin, acetaminophen, melphalan, cyclophosphamide, radiation, and the BSO model of oxidative stress (Meister, 1991, 1995). GiPE is reported to protect against ischemia damage in rat brain (Gotoh et al., 1994) and may protect against the effects of hypoxia/hyperglycemia in rat brain hippocampus (Shibata et al., 1995). GEE, NAC, and GSH decrease HIV replication in chronically infected monocyte cells (Kalebic *et al.*, 1991).

There have been occasional reports (Vos and Roos-Verhey, 1988; Tsan *et al.*, 1989; Scadutto *et al.*, 1988) of toxicity of GSH monoesters that have been attributed to various impurities, especially metal ions (Levy *et al.*, 1993).

E. GSH Diester

GEE preparations occasionally contain slight (1% to 15%) GSH di(gly-cyl and α-carboxyl)ester impurities, and we found that these preparations consistently increase erythrocyte thiol levels more than GEE preparations without GSH diester "impurities." GSH diethyl ester (GDE) was synthesized and found to be more effectively transported than GEE into human erythrocytes, lymphocytes, ovarian tumor cells, and fibroblasts than GEE (Levy *et al.*, 1993); this result was also found in studies using a murine macrophage cell line (P388D$_1$) (Minhas and Thornalley, 1995). GDE may be more effectively transported into cells than GEE because of its increased hydrophobicity and/or because of its net charge in solution (a solution of GEE free base is slightly acidic, whereas one of GDE is almost neutral).

When human erythrocytes were incubated with GDE, there was a rapid increase in cellular thiols (GDE and GEE), then there was a steady increase in GEE levels. When human erythrocytes were rapidly preloaded with GDE, washed, and resuspended in phosphate buffer saline and the extracellular medium analyzed with time, GDE was the predominate species found. These studies (Levy *et al.*, 1993) suggest that GDE is rapidly transported into (and out of) cells and is hydrolyzed to GEE, which is slowly exported and thus is "trapped" within cells. GDE can also be considered a GEE delivery system (Fig. 2).

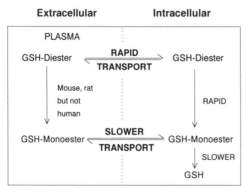

FIGURE 2   Transport and cleavage of GSH esters (from Levy *et al.*, 1993).

GDE is rapidly hydrolyzed to GEE by the plasma of rats and mice, but not that of humans, guinea pigs, hamsters, rabbits, or sheep (Levy *et al.,* 1993). Thus, our preliminary *in vivo* studies were carried out in hamsters; the administration of GDE (5 mmol/kg) increased hamster liver GSH levels 3-fold and this increase is much greater increase than that found after administration of an equivalent dose of GEE which only increased GSH levels about 0.7-fold. These results are especially encouraging because the conversion of 1 mol of GEE to GSH also releases 1 mol of ethanol, whereas the conversion of 1 mol of GDE is expected to produce 2 mol of ethanol and might cause toxicity. It should be noted that the interaction of GDE with metal ions does not apparently produce cellular toxicity like that observed with GEE. We have seen no evidence of toxicity of GDE in our preliminary studies. Further *in vivo* studies of GDE are needed and will be facilitated by a rapid, high yield synthetic route to the preparation of GDE.

III. Summary

Deficiency of GSH (BSO model) demonstrates the need for cellular protection from endogenous ROS. The pathology of various diseases is often associated with ROS and oxidative stress. The several methods for modulating cellular GSH levels presented in this chapter offer selective tools to study mechanisms and offer potential therapy of human diseases associated with GSH deficiency and oxidative stress.

Acknowledgments

This research was partially supported by a grant (AI 31804) from the National Institutes of Health and by a Fellowship from the Danish Research Academy.

This chapter is dedicated to Alton Meister (1922–1995), a pioneer in glutathione, amino acid, and enzyme biochemistry.

References

Anderson, M. E. (1985). Preparation of γ-glutamyl amino acids by chemical and enzymatic methods. *Methods Enzymol.* 113, 555–564.

Anderson, M. E., Levy, E. J., and Meister, A. (1994). Preparation and use of glutathione monoesters. *Methods Enzymol.* 234, 492–499.

Anderson, M. E., and Meister, A. (1983). Transport and direct utilization of γ-glutamylcyst(e)ine for glutathione synthesis. *Proc. Natl. Acad. Sci. U.S.A.* 80, 707–711.

Anderson, M. E., and Meister, A. (1987). Intracellular delivery of cysteine. *Methods Enzymol.* 143, 313–325.

Anderson, M. E., and Meister, A. (1989a). Marked increase of cysteine levels in many regions of the brain after administration of 2-oxothiazolidine-4-carboxylate. *FASEB J.* 3, 1632–1636.

Anderson, M. E., and Meister, A. (1989b). Glutathione monoesters. *Anal. Biochem.* **183**, 16–20.

Anderson, M. E., Powrie, F., Puri, R. N., and Meister, A. (1985). Glutathione monoethyl ester: Preparation, uptake by tissues, and conversion to glutathione. *Arch. Biochem. Biophys.* **239**, 538–548.

Anderson, M. E., Underwood, M., Bridges, R. J., and Meister, A. (1989). Glutathione metabolism at the blood-cerebrospinal fluid barrier. *FASEB J.* **3**, 2527–2531.

Breborowicz, A., Witowski, J., Martis, L., and Oreopoulos, D. G. (1993). Enhancement of viability of human peritoneal mesothelial cells with glutathione precursor: L-2-Oxothiazolidine-4-carboxylate. *Adv. Periton. Dialy.* **9**, 21–24.

Buhl, R., Holroyd, K. J., Mastrangeli, A., Cantin, A. M., Jaffe, H. A., Wells, F. B., Saltini, C., and Crystal, R. G. (1989). Systemic glutathione deficiency in symptom-free HIV-seropositive individuals. *Lancet* ii, 1294–1298.

Calvin, H. L., Medvedovsky, C., and Worgul, B. V. (1986). Near-total glutathione depletion and age-specific cataracts induced by buthionine sulfoximine in mice. *Science* **233**, 553–555.

Cantin, A. M., Hubbard, R. C., and Crystal, R. G. (1989). Glutathione deficiency in the epithelial lining fluid of the lower respiratory tract in idiopathic pulmonary fibrosis. *Am. Rev. Respir. Dis.* **139**, 370–372.

Cooper, A. J. L., Haber, M. T., and Meister, A. (1982). On the chemistry and biochemistry of 3-mercaptopyruvate acid, the α-keto acid analog of cysteine. *J. Biol. Chem.* **257**, 816–826.

Cutler, R. G., Packer, L., Bertram, and Mori, A., eds. (1995). "Oxidative Stress and Aging." Birkhauser Verlag, Boston.

Eck, H.-P., Gmuender, H., Hartmann, M., Petzoldt, D., Daniel, V., and Droege, W. (1989). Low concentrations of acid soluble thiol (cysteine) in the blood plasma of HIV-1 infected patients. *Biol. Chem. Hoppe Seyler* **370**, 101–108.

Fields, J. A., Keshavarzian, A., Eiznhamer, D., Frommel, T., Winship, D., and Holmes, E. W. (1994). Low levels of blood and colonic glutathione in ulcerative colitis. *Gastroenterology* **106**, A680.

Forrester, T. E., Badaloo, V., Bennett, F. I., and Jackson, A. A. (1990). Excessive excretion of 5-oxoproline and decreased levels of blood glutathione in type II diabetes mellitus. *Eur. J. Clin. Nutr.* **44**, 847–850.

Ginter, E. (1978). Marginal vitamin C deficiency, lipid metabolism, and atherogenesis. *Adv. Lipid Res.* **16**, 167–220.

Gotoh, O., Yamamoto, M., Tamura, A., and Sano, K. (1994). Effect of YM737, a new glutathione analog, on ischemic brain edema. *Acta Neurochir.* **60**(Suppl.), 318–320.

Griffith, O. W., Anderson, M. E., and Meister, A. (1979). Inhibition of glutathione biosynthesis by prothionine (S-n-propyl homocysteine sulfoximine), a selective inhibitor of γ-glutamylcysteine synthetase. *J. Biol. Chem.* **254**, 1205–1210.

Griffith, O. W., and Meister, A. (1979). Potent and specific inhibition of glutathione synthesis by buthionine sulfoximine (S-n-butyl homocysteine sulfoximine). *J. Biol. Chem.* **254**, 7558–7560.

Jain, A., Madsen, D. C., Auld, P. A. M., Frayer, W. W., Schwartz, M. K., Meister, A., and Martensson, J. (1995). L-2-Oxothiazolidine-4-carboxylate, a cysteine precursor, stimulates growth and normalizes tissue glutathione concentrations in rats fed a sulfur amino acid-deficient diet. *J. Nutr.* **125**, 851–856.

Josephs, S., Asuncion, C., Sun, C., Jacobson, P., and Webb, L. (1993). Procysteine (L-2-oxothiazolidine-4-carboxylic acid) reduces the toxicity of AZT (zidovudine) and enhances the antiviral activity of AZT in cultured peripheral blood mononuclear cells (PBMC). *In* "IXth International Conference on AIDS (June 6–11, 1993, Berlin), 1. [Abstract 230].

Kalayjian, R. C., Skowron, G., Emgushov, R.-T., Chance, M., Spell, S. A., Borum, P. R., Webb, L. S., Mayer, K. H., Jackson, J. B., Yen-Lieberman, B., Story, K. O., Rowe, W. B., Thompson, K., Goldberg, D., Trimbo, S., and Lederman, M. M. (1994). A phase I/II trial of intravenous L-2-oxothiazolidine-4-carboxylic acid (procysteine) in asymptomatic HIV-infected subjects. *J. Acq. Immune Def. Syndr.* **7**, 369–374.

Kalebic, T., Kinter, A., Poli, G., Anderson, M. E., Meister, A., and Fauci, A. S. (1991). Suppression of HIV expression in chronically infected monocytic cells by glutathione, glutathione ester, and N-acetyl cysteine, *Proc. Natl. Acad. Sci. U.S.A.* **88**, 896–990.

Karlsen, R. L., Grofava, I., Malthe-Sorensen, D., and Fornum, E. (1981). Morphological changes in rat brain induced by L-cysteine injection in newborn animals. *Exp. Brain Res.* **208**, 167–180.

Lederman, M. M., Georger, D., Dando, S., Schmelzer, R., Averill, L., and Goldberg, D. (1995). L-2-Oxothiazolidine-4-carboxylic acid (procysteine) inhibits expression of the human immunodeficiency virus and expression of the interleukin-2 receptor alpha chain. *J. Acq. Immune Def. Syndr.* **8**, 107–115.

Levy, E. J., Anderson, M. E., and Meister, A. (1993). Transport of glutathione diethyl ester into human cells. *Proc. Natl. Acad. Sci. U.S.A.* **90**, 9171–9175.

Levy, E. J., Anderson, M. E., and Meister, A. (1994). Preparation and properties of glutathione diethyl ester and related derivatives. *Methods Enzymol.* **234**, 499–505.

Martensson, A., Goodwin, C. W., and Blake, R. (1992). Mitochondrial glutathione in hypermetabolic rats following burn injury and thyroid hormone administration: Evidence of a selective effect on brains glutathione by burn injury. *Metabolism* **41**, 273–277.

Martensson, J., Han, J., Griffith, O. W., and Meister, A. (1993). Glutathione ester delays the onset of scurvy in ascorbate-deficient guinea pigs. *Proc. Natl. Acad. Sci. U.S.A.* **90**, 317–321.

Martensson, J., Jain, A., Stole, E., Frayer, W., Auld, P. A. M., and Meister, A. (1991). Inhibition of glutathione synthesis in the newborn rat: A model for endogenously produced oxidative stress. *Proc. Natl. Acad. Sci. U.S.A.* **88**, 9360–9364.

Martensson, J., and Meister, A. (1989). Mitochondrial damage in muscle occurs after marked depletion of glutathione and is prevented by giving glutathione monoester. *Proc. Natl. Acad. Sci. U.S.A.* **86**, 471–475.

Martensson, J., and Meister, A. (1991). Glutathione deficiency decreases tissue ascorbate levels in newborn rats: Ascorbate spares glutathione and protects. *Proc. Natl. Acad. Sci. U.S.A.* **88**, 4656–4660.

Martinez-Cayuela, M. (1995). Oxygen free radicals and human disease. *Biochimie* **77**, 147–161.

Meister, A. (1989). Metabolism and function of glutathione. *In* "Glutathione: Chemical, Biochemical and Medical Aspects" (D. Dolphin, R. Poulson, and O. Avramovic, eds.), pp. 367–474. Wiley, New York.

Meister, A. (1991). Glutathione deficiency produced by inhibition of its synthesis, and its reversal; applications in research and therapy. *Pharmacol. Therap.* **51**, 155–194.

Meister, A. (1992). On the antioxidant effects of ascorbic acid and glutathione. *Biochem. Pharmacol.* **44**, 1905–1915.

Meister, A. (1994). Glutathione-ascorbate acid antioxidant system in animals. *J. Biol. Chem.* **269**, 9397–9400.

Meister, A. (1995). Strategies for increasing cellular glutathione. *In* "Biothiols in Health and Disease" (L. Packer and E. Cadenas, eds.), pp. 165–188. Dekker, New York.

Meister, A., and Anderson, M. E. (1983). Glutathione. *Annu. Rev. Biochem.* **52**, 711–760.

Meister, A., Anderson, M. E., and Hwang, O. (1986). Intracellular delivery of cysteine and glutathione delivery systems. *J. Am. Coll. Nutr.* **5**, 137–151.

Meister, A., and Larsson, A. (1994). Glutathione synthetase deficiency and other disorders of the γ-glutamyl cycle. *In* "The Metabolic Basis of Inherited Disease" (C. R. Scriver, A. L. Beaudet, W. S. Sly, and D. Valle, eds.), 7th Ed., pp. 1461–1477. McGraw Hill, New York.

Minhas, H., and Thornalley, P. J. (1995). Comparison of the delivery of reduced glutathione into P388D$_1$ cells by reduced glutathione and its mono- and diethyl ester derivatives. *Biochem. Pharmacol.* **49**, 1475–1482.

Nishiuch, Y., Sasaki, M., Nakayasu, M., and Oikawa, A. (1976). Cytotoxicity of cysteine in culture media. *In Vitro* **12**, 635.

Olney, J. W., Ho, O.-L., and Rhee, V. (1971). Cytotoxic effect of acid and sulphur containing amino acids on the infant mouse central nervous system. *Brain Res.* **14**, 61–76.

Pace, G. W., and Leaf, C. D. (1995). The role of oxidative stress in HIV disease. *Free Rad. Biol. Med.* **19**, 523–528.

Pacht, E. R., Timerman, A. P., Lykens, M. G., and Merola, A. J. (1991). Deficiency of alveolar fluid glutathione in patients with sepsis and the adult respiratory distress syndrome. *Chest* **100**, 1397–1403.

Packer, L., and Cadenas, E., eds. (1995). "Biothiols in Health and Disease." Dekker, New York.

Pileblad, E., and Magnusson, T. (1992). Increase in rat brain glutathione following intracerebro-ventricular administration of γ-glutamylcysteine. *Biochem. Pharmacol.* **44**, 895–903.

Porta, P., Aebi, S., Summer, K., and Lauterburg, B. H. (1991). L-2-Oxothiazolidine-4-carboxylic acid, a cysteine prodrug: Pharmacokinetics and effects on thiols in plasma and lymphocytes in human. *J. Pharmacol. Exp. Ther.* **257**, 331–334.

Puri, R. N., and Meister, A. (1983). Transport of glutathione, as γ-glutamylcysteinylglycyl ester, into liver and kidney. *Proc. Natl. Acad. Sci. U.S.A.* **80**, 5258–5260.

Richman, P. G., and Meister, A. (1973). Regulation of γ-glutamylcysteine synthetase by nonallosteric feedback inhibition by glutathione. *J. Biol. Chem.* **250**, 1422–1426.

Sabeh, F., Baxter, C. R., and Norton, S. J. (1995). Skin burn injury and oxidative stress in liver and lung tissues of rabbit models. *Eur. J. Clin. Chem. Clin. Biochem.* **33**, 323–328.

Scadutto, R. C., Jr., Gattone, V. H., II, Grotyohann, L. W., Wertz, J., and Martin, L. F. (1988). Effect of an altered glutathione content on renal ischemic injury. *Am. J. Physiol.* **255**, F911–F921.

Shibata, S., Tominaga, K., and Watanabe, S. (1995). Glutathione protects against hypoxic/hypoglycemic decreases in 2-deoxyglucose uptake and presynaptic spikes in hippocampal slices. *Eur. J. Pharmacol.* **273**, 191–195.

Shug, A. L., and Madsen, D. C. (1994). Protection of the ischemic rat heart by procysteine and amino acids. *J. Nutr. Biochem.* **5**, 356–359.

Suarez, M., Beloqui, O., Ferrer, J. V., Fil, B., Qian, C., Garcia, N., Civeira, P., and Prieto, J. (1993). Glutathione depletion in chronic hepatitis C. *Int. Hepatol. Commun.* **1**, 215–221.

Suthanthiran, M., Anderson, M. E., Sharma, V. K., and Meister, A. (1990). Glutathione regulates activation-dependent DNA synthesis in highly purified T lymphocytes stimulated via CD2 and CD3 Antigens. *Proc. Natl. Acad. Sci. U.S.A.* **87**, 3343–3347.

Tsan, M.-F., White, J. E., and Rosano, C. L. (1989). Modulation of endothelial GSH concentrations: Effect of exogenous GSH and GSH monoethyl ester. *J. Appl. Physiol.* **66**, 1029–1034.

Vos, O., and Roos-Verhey, W. S. D. (1988). Endogenous versus exogenous thiols in radioprotection. *Pharmacol. Ther.* **39**, 169–177.

Weitberg, A. B. (1987). The effect of L-2-oxothiazolidine on glutathione levels in cultured mammalian cells. *Mutat. Res.* **191**, 189–191.

Wellner, V. P., Anderson, M. E., Puri, R. N., Jensen, G. L., and Meister, A. (1984). Radioprotection by glutathione ester: Transport of glutathione ester into human lymphoid cells and fibroblasts. *Proc. Natl. Acad. Sci. U.S.A.* **81**, 4732–4735.

Wells, W. W., Xu, D. P., Yang, Y., and Rocque, P. A. (1990). Mammalian thioltransferase (glutaredoxin) and protein disulfide isomerase have dehydroascorbate reductase activity. *J. Biol. Chem.* **265**, 15361–15364.

Wells, W. W., Yang, Y., Deits, T. L., and Gan, Z.-R. (1992). Thioltransferases. *Adv. Enzymol.* **66**, 149–201.

Williamson, J. M., Boettcher, B., and Meister, A. (1982). Intracellular cysteine delivery system that protects against toxicity by promoting glutathione synthesis. *Proc. Natl. Acad. Sci. U.S.A.* **79**, 6246–6249.

Williamson, J. M., and Meister, A. (1981). Stimulation of hepatic glutathione formation by administration of L-2-oxothiazolidine-4-carboxylate, a 5-oxo-L-prolinase substrate. *Proc. Natl. Acad. Sci. U.S.A.* **78**, 936–939.

Williamson, J. M., and Meister, A. (1982). New substrates of 5-oxo-L-prolinase. *J. Biol. Chem.* **257**, 12039–12042.

Willis, G. C. (1953). An experimental study of the intimal ground substance in atherosclerosis. *Can. Med. Assoc. J.* **69**, 17–22.

Wilson, D. M., White, R. D., Webb, L. E., Bender, J. G., Pippin, L. L., and Goldberg, D. L. (1993). Amelioration of AZT-induced bone marrow hypoplasia in mice cotreated with glutathione prodrug, procysteine. *In* "IXth International Conference on AIDS (June 7–11, 1993, Berlin). [Abstract]

Lester Packer
Sashwati Roy
Chandan K. Sen

Department of Molecular and Cell Biology
University of California at Berkeley
Berkeley, California 94720

α-Lipoic Acid: A Metabolic Antioxidant and Potential Redox Modulator of Transcription

I. Introduction

There has been a great deal of interest in the antioxidant properties of α-lipoate (Fig. 1), which has long been known as an essential cofactor in oxidative metabolism. This chapter discusses the metabolic role as well as the antioxidant properties of α-lipoate and its reduced form, dihydrolipoate. In addition, the effects of this antioxidant in modulating the redox-sensitive transcription factor nuclear factor κB (NF-κB) are evaluated. Because NF-κB is involved in a wide variety of acute inflammatory responses, as well as many other aspects of rapid responses in cells, we have chosen this system to explore the action of α-lipoate and dihydrolipoate on transcription factors.

Advances in Pharmacology, Volume 38

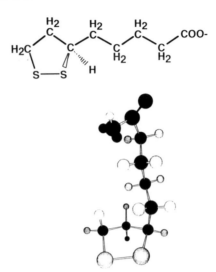

FIGURE 1   Structure of *R*-lipoate.

II. Role of Lipoate in Metabolism

A. History

α-Lipoate has been known to be an essential cofactor in oxidative metabolism since the 1950s. As early as 1937, there were reports of compounds exhibiting the same biological activity as α-lipoate. These compounds were variously called "potato growth factor" (Snell *et al.*, 1937), "acetate replacing factor" (Guirard *et al.*, 1946), "pyruvate oxidation factor" (O'Kane and Gunsalus, 1952), and "protogen" (Stokstad *et al.*, 1949). In 1951, Reed and colleagues reported the purification of α-lipoic acid. It was recognized that "potato growth factor," "acetate replacing factor," "pyruvate oxidation factor," and "protogen" were probably all the same compound, α-lipoic acid. α-Lipoic acid is also known as 1,2-dithiolane-3-pentanoic acid, 1,2-dithiolane-3-valeric acid, and 6,8-thioctic acid.

α-Lipoic acid is found as lipoamide in five proteins in eukaryotes, where it is covalently attached to a lysyl residue. Four of these proteins are found in the three α-keto acid dehydrogenase complexes, the pyruvate dehydrogenase complex (PDC), the branched chain keto acid dehydrogenase complex, and the α-ketoglutarate dehydrogenase complex. Three are in the E2 enzyme, dihydrolipoyl acyltransferase, which is different in each of the complexes and specific for the substrate of the complex. One is found in protein X, which is the same in each complex. The fifth lipoamide moiety is found in the glycine cleavage system (Fig. 2).

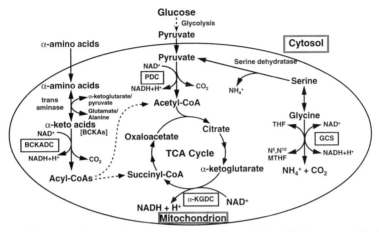

FIGURE 2   Sites of action of α-lipoic acid in energy metabolism. PDC, pyruvate dehydrogenase complex; α-KGDC, α-ketoglutarate dehydrogenase complex; BCKADC, branched chain
α-keto acid dehydrogenase complex; GCS, glycine cleavage system; BCKAs, branched chain α-
keto acids; THF, tetrahydrofolate; N^5N^{10}MTHF, N^5N^{10}-methylene tetrahydrofolate. (Adapted
from Patel and Smith, 1994, p. 69).

B. α-Keto Acid Dehydrogenases

The α-keto dehydrogenase complexes are composed of multiple copies
of three enzymes: the α-keto acid (pyruvate, branched chain α-keto acid,
or α-ketoglutarate) dehydrogenase or E1; the dihydrolipoyl acyltransferase
or E2; and the dihydrolipoyl dehydrogenase or E3. These three enzymes
catalyze five reactions that oxidatively decarboxylate their substrates. Lipoamide is involved as a moeity to which an acyl group is attached, transferring
it from thiamine pyrophosphate on E1 to coenzyme A to produce acyl-CoA.
In the process the lipoamide is reduced, with its disulfide linkage broken.
E3 is the enzyme that reoxidizes the lipoamide for another round of catalysis
using NAD$^+$, which is converted to NADH.

Dihydrolipoyl acyltransferase consists of three regions: a catalytic domain, a subunit binding domain, and 1–3 lipoyl domains, the number depending on the complex and the species. These domains are linked by highly
mobile sequences rich in alanine, proline, and charged amino acids. The
purpose of the repeating lipoyl domains, to which lipoate can attach, is not
known at present. The *Escherichia coli* pyruvate dehydrogenase complex,
which contains three lipoyl domains, undergoes no adverse change in function when the number of repeating domains is reduced to one (Perham,
1991); in contrast, when the number is increased to between four and
nine, there is a marked reduction in the catalytic activity of the E2 enzyme
(Machado *et al.*, 1992). It is also not known to what extent repeating lipoyl
domains are lipoylated *in vivo* (Patel and Smith, 1994). For example, the

human pyruvate dehydrogenase complex contains two lipoyl domains and it is not known whether both are lipoylated or just one. Studies on *E. coli* dihydrolipoyl acyltransferase (which contain three lipoyl domains) suggest that one is all that is necessary. This may have some bearing on the effects of supplemental α-lipoic acid. At present it is assumed that the effects of this compound in improving pathological conditions (e.g., diabetic polyneuropathy) are due to its antioxidant effects (discussed later), but it should not be overlooked that α-lipoate also plays a key role in metabolism and is intimately linked to intermediary metabolism. It is possible that there is additional lipoylation of lipoyl domains when exogenous α-lipoate is supplied to the organism.

C. Protein X

Each PDC contains about six copies of protein X, which contains a lipoyl domain to which α-lipoate is bound and which can accept an acetyl group (Patel and Roche, 1990; Lawson et al., 1991a,b). However, protein X has no catalytic activity, and site-directed mutagenesis reveals that removal of its lipoyl group does not affect the catalytic activity of the PDC (Lawson et al., 1991a,b). Protein X appears to play a structural role aiding in the binding of E3, which is essential for PDC activity. Limited proteolysis of protein X decreases E3 binding ability.

D. Glycine Cleavage System

The glycine cleavage system catalyzes the oxidation of glycine to CO_2, ammonia, and forming NADH and 5,10-methylenetetrahydrofolate. It contains four proteins, the P-, H-, T-, and L-proteins; the α-lipoate is covalently attached to a lysine in the H-protein and is involved in the transfer of the methylamine moiety formed after oxidative decarboxylation of the glycine. There are four highly conserved residues around the lipoylation site (Lys59) in the H-protein: Gly43, Gly70, Glu56, and Glu63. Site-directed mutagenesis was used to study these residues in bovine H-protein (Fujiwara et al., 1991), and it was found that replacement of the Glu56 and Gly70 decreased the efficiency of lipoylation, whereas replacement of Gly43 had no effect, despite the fact that it is completely conserved in all sequences examined. Replacement of Glu63 with aspartic acid did not change the efficiency of lipoylation, whereas replacement with glutamine lowered the efficiency of lipoylation to 26% that of the wild type, indicating that the anionic charge on this residue is important in lipoylation. These results indicate that Glu56 and Gly70, and the amionic nature of Glu63, are important in maintaining the proper conformation and microenvironment around the lipoic acid attachment site.

E. Does Dihydrolipoamide Dehydrogenase Serve a Purpose Outside of Multienzyme Complexes?

The dihydrolipoamide dehydrogenase of α-keto acid dehydrogenase complexes and the glycine cleavage system is the component that oxidizes dihydrolipoamide to lipoamide. The reaction can also go in the reverse direction at the expense of NADH. There are seveal reports of dihydrolipoamide activity that are not associated with the α-keto acid dehydrogenase complexes or the glycine cleavage systems. Archaebacteria contain neither of these complexes, yet dihydrolipoamide dehydrogenase has been found in halophilic (Danson *et al.*, 1984), thermoacidophilic (Smith *et al.*, 1987), and methanophilic archaebacteria. In the thermoacidophilic archaebacterium *Thermaplasma acidophilum,* the enzyme appears to be associated with the plasma membrane. Dihydolipoamide dehydrogenase has also been purified from *E. coli* (Richarme, 1989), where it is found in mutants that do not contain the dihydrolipoamide dehydrogenase normally associated with the pyruvate and α-ketoglutarate acid dehydrogenase complexes. This enzyme, also localized to the plasma membrane, has been suggested to be involved in binding protein-dependent transport. In addition, eukaryotic cells contain dihydrolipoamide not associated with the multienzyme complexes. In its bloodstream form, the parasite that causes sleeping sickness *Trypanosoma brucei,* contains a single mitochondrion that does not contain the multienzyme complexes. It has been found, however, to contain dihydrolipoamide dehydrogenase associated with the plasma membrane and, in addition, to contain lipoic acid (presumably the substrate of the enzyme) at a level of 8.3 nmol/g protein. It is also possible that rat adipocytes contain a plasma membrane-associated dihydrolipoamide dehydrogenase (Danson, 1987).

The function of these plasma membrane-associated dihydrolipoamide dehydrogenases is not known. In *E. coli* it appears to be associated with sugar transport. It has been speculated that in eukaryotes it may be connected with insulin-stimulated hexose transport (Danson, 1987). This is based on the observations that an arsenical (sequesters lipoamide) blocks insulin-stimulated hexose transport in 3T3-L1 adipocytes (Frost and Lane, 1985) and that thiol–disulfide interchange may be involved downstream from the insulin receptor in the mechanism of its action (Bernier *et al.*, 1987). Although a function for dihydrolipoamide dehydrogenase in hexose transport remains speculative, it is interesting to note that several groups have reported increased glucose uptake *in vitro* (Haugaard and Haugaard, 1970; Singh and Bowman, 1970; Bashan *et al.*, 1993) and *in vivo* in animals (Henriksen *et al.*, 1994) and humans (Jacob *et al.*, 1994) with type II (noninsulin-dependent) diabetes treated with α-lipoic acid, a substrate for dihydrolipoamide dehydrogenase that could be involved in such thiol–disulfide exchange.

F. Reduction of Exogenous Lipoate

Much of the current interest in α-lipoate is centered on the effects of the free, nonprotein-bound compound, which is supplied exogenously to cell systems, animals, and humans. It has been shown that such exogenously supplied α-lipoate is absorbed, transported to tissues, and reduced to dihydrolipoate, part of which is released into the extracellular space. Hence, it is of interest to investigate the mechanisms whereby exogenously supplied α-lipoate is reduced to dihydrolipoate.

The mitochondrial E3 enzyme, dihydrolipoamide dehydrogenase (EC 1.8.1.4), of mitochondrial α-keto acid dehydrogenases is capable of reducing α-lipoate to dihydrolipoate, at the expense of NADH. This reaction is the reverse of the one normally catalyzed by the enzyme in its multi-enzyme complex. The enzyme shows a marked preference for the natural, R-enantiomer of α-lipoate (Pick *et al.*, 1995). Glutathione reductase, found mainly in the cytosol, catalyzes the reduction of glutathione disulfide, at the expense of NADPH. This enzyme also catalyzes the reduction of α-lipoate, although at a much lower rate than with its natural substrate, with a preference for the nonnatural S-enantiomer of α-lipoate (Pick *et al.*, 1995; Haramaki *et al.*, 1996). The ability of this enzyme to reduce α-lipoate is not surprising in that its tertiary structure is very similar to dihydrolipoamide dehydrogenases (Jentoft *et al.*, 1992); however, in light of this similarity in tertiary structure, the difference in stereospecificity of the enzymes is unexpected. Tissues vary in their ratios of NADH-dependent to NADPH-dependent reduction of α-lipoate, and the variation corresponds with the mitochondrial content. Tissues with greater mitochondrial density have a higher ratio of NADH-dependent reduction, presumably because of a greater dihydrolipoamide dehydrogenase content (Haramaki *et al.*, 1996).

G. Other Metabolic Effects of Exogenous α-Lipoate

It has been shown that exogeneous lipoamide dramatically improves functional recovery in ischemic rat heart. This effect is probably due to its effects on the altered metabolism in the ischemic tissue (Sumegi *et al.*, 1994); namely, it activates the pyruvate dehydrogenase complex, increases pyruvate oxidation, and activates the tricarboxylic cycle, probably by decreasing the acetyl-CoA/CoA ratio in mitochondria by acting as an acetyl group acceptor. α-Lipoate also improves recovery in ischemic-reperfused hearts (Serbinova *et al.*, 1994; Haramaki *et al.*, 1993), and it is possible that at least part of the mechanism is through a lipoamide-like metabolic action, in addition to its antioxidant effects.

III. Antioxidant Properties

A. Lipoate

α-Lipoic acid scavenges hydroxyl radicals, hypochlorous acid, singlet oxygen, and nitric oxide. In addition, it chelates a number of transition metals.

α-Lipoic acid is a potent hydroxyl radical scavenger. α-Lipoic acid (1 mM) completely eliminated the Fenton (2 mM H_2O_2 + 0.2 mM $FeSO_4$)-derived DMPO-OH (DMPO, 5,5-dimethylpyrroline-N-oxide) adduct signal (Suzuki et al., 1991). Another study (Scott et al., 1994), using a similar hydroxyl radical-generating system (2.8 mM H_2O_2, 0.05 mM $FeCl_3$, 0.1 mM EDTA, and 0.1 mM ascorbate) but a different assay for the radical (deoxyribose degradation), also found α-lipoic acid to be a hydroxyl radical scavenger. In this study, a rate constant of 4.7×10^{10} M^{-1} sec^{-1} for the scavenging reaction was determined.

Studies from our laboratory with a novel hydroxyl radical-producing compound, N,N'-bis(2-hydroperoxy-2-methoxyethyl)-1,4,5,8-naphthalene-tetracarboxylic diimide (NP-III), confirm that α-lipoate is a hydroxyl radical scavenger and eliminate the possibility that the effect is simply due to metal chelation (Matsugo et al., 1995a). NP-III produces hydroxyl radicals on illumination with UVA light. The production of the hydroxyl radical was followed by measuring the oxidation of bovine serum albumin (BSA) and apolipoprotein B (apoB) of human low-density lipoprotein, as well as by using electron spin resonance (ESR) to detect DMPO-OH spin adducts. At a concentration of 1–2 mM, α-lipoate completely eliminated hydroxyl radical production as assessed by DMPO adducts and protein carbonyl formation in apoB and BSA; the rate constant for reaction with the hydroxyl radical was found to be 1.92×10^{10} M^{-1} sec^{-1}, comparable with that found by Scott et al. (1994), essentially a diffusion-limited rate.

There is similar agreement about the ability of α-lipoic acid to scavenge hypochlorous acid. Haenen and Bast (1991) and Scott et al. (1994) independently found that 50 μM α-lipoic acid almost completely abolished the inactivation of α_1-antiproteinase by 50 μM HOCl. This is in contrast to the glutathione redox couple. Although reduced glutathione is a potent scavenger of HOCl, the oxidized form glutathione disulfide is not (Haenen and Bast, 1991). The authors of this study speculate that the greater reactivity of α-lipoic acid compared to glutathione disulfide may be due to the somewhat strained conformation of the five-membered ring in the intramolecular disulfide form of α-lipoic acid. There is no such strain on the intermolecular disulfide of glutathione disulfide, perhaps explaining its lack of reactivity in this system.

α-Lipoic acid has been reported to scavenge singlet oxygen in at least four different systems. Two early studies showed that α-lipoic acid reacted with singlet oxygen generated by rubrene autoxidation (Stevens et al., 1974) or by photosensitized oxidation of methylene (Stary et al., 1975). These experiments were carried out in organic solvents. Later studies conducted under more physiological conditions also showed that α-lipoic acid is an effective scavenger of singlet oxygen. Kaiser et al. (1989) generated singlet oxygen by thermolysis of an endoperoxide and detected it by chemiluminescence. In this system, α-lipoic acid reacted with singlet oxygen with a rate constant of 1.38×10^8 M^{-1} sec^{-1}. Also, in experiments in which singlet oxygen was generated by thermolysis of endoperoxide and detected by single-strand DNA breaks, α-lipoic acid was confirmed to be a scavenger of singlet oxygen (Devasagayam et al., 1993).

In addition, it has been observed that α-lipoate decreases nitric oxide production in macrophages (H. Kobuchi, personal communication, 1995). RAW 264.7 macrophages were coincubated with lipopolysaccharide (LPS) and various doses of α-lipoate for 24 hr. Nitrite in the culture medium was measured using the Griess reagent. At doses of α-lipoate above 200 μM there was a significant effect on NO production, with more than 50% inhibition at 0.5 mM. It is not possible in this system to determine whether the inhibition was due to direct interactions with NO, to decreased induction of inducible nitric oxide synthase, or to some other effect of α-lipoate. However, in experiments in which NO was produced directly by sodium nitroprusside, α-lipoate decreased NO generation.

α-Lipoic acid may also exert an antioxidant effect in biological systems through transition metal chelation. It has been found to decrease Cd^{2+}-induced toxicity in isolated hepatocytes, although the authors speculate that the effect was due to the conversion of α-lipoic acid to dihydrolipoate (DHLA), which was the true chelating agent (Müller and Menzel, 1990). Two studies indicate that α-lipoic acid may chelate iron, whereas one indicates that it does not. Devasagayam et al. (1993) observed that α-lipoic acid is effective against singlet oxygen-induced DNA single strand breaks. They also reported that the efficacy of α-lipoate was lower when EDTA was added to the system, indicating that part of the effect was due to iron chelation. Scott et al. (1994) found that α-lipoic acid inhibited the site-specific degradation of deoxyribose by a $FeCl_3/H_2O_2$/ascorbate system, indicating that it was able to remove iron ions bound to deoxyribose. In contrast, in a rat liver microsomal system in which lipid peroxidation was induced by Fe_3SO_4 (Bast and Haenen, 1988), α-lipoic acid did not inhibit peroxidation. α-Lipoate forms stable complexes with Mn^{2+}, Cu^{2+}, and Zn^{2+}, with the complex being almost entirely with the carboxylate group (Sigel et al., 1978). These investigators also examined the complexes formed with bisnorlipoate and tetranorlipoate, two and four carbon shorter homologs of α-lipoic acid, respectively. These were found to be more stable than those with α-lipoic acid, presumably because the shorter chain allows the dithiolane ring to also participate in chelation. The effects of α-lipoic acid on copper have been extended to oxidation situations. α-Lipoic acid was effective in preventing Cu^{2+}-catalyzed ascorbic acid oxidation, increased the partitioning of Cu^{2+} into *n*-octanol from aqueous solution, and inhibited Cu^{2+}-catalyzed liposomal peroxidation (Ou *et al.*, 1995). These observations indicate that LA is a copper chelator.

B. Dihydrolipoate

With a redox potential of -0.32 V (Searls and Sanadi, 1960) for the DHLA/LA redox couple, dihydrolipoic acid is a potent redundant; for comparison, the redox potential of the glutathione/glutathione disulfide (GSH/GSSG) couple is -0.24 V (Scott *et al.*, 1963). DHLA reduces GSSG to GSH, but GSH is incapable of reducing α-lipoic acid to DHLA (Jocelyn, 1967). DHLA scavenges hypochlorous acid and peroxyl, superoxide, hydroxyl, and nitric oxide radicals.

In the systems in which α-lipoic acid was tested for reaction with hypochlorous ion, DHLA was also found to be effective, with about the same scavenging ability (Haenen and Bast, 1991). DHLA is an effective peroxyl radical scavenger in several different systems. Kagan *et al.* (1992) used 2,2′-azobis (2-amidinopropane) hydrochloride (AAPH) or 2,2′-azobis (2,4-dimethylvaleronitrile) (AMVN) to generate peroxyl radicals in aqueous or lipid systems and found that DHLA scavenged peroxyl radicals in these systems. It was also found to scavenge peroxyl radicals (Suzuki *et al.*, 1993) and $CCl_3O_2\cdot$ radicals generated in aqueous solution (Scott *et al.*, 1994). In the latter system a rate constant of $2.7 \times 10^7 \; M^{-1} \, sec^{-1}$ was calculated.

Using an NP-III system similar to that used with α-lipoate (Matsugo *et al.*, 1995b), we found that DHLA was also effective in scavenging the hydroxyl radical completely, eliminating the ESR DMPO-OH signal produced on illumination of NP-III with UV light at a dose of $500 \; \mu M$.

In a study in which $O_2\cdot^-$ was generated by xanthine–xanthine oxidase and detected by ESR using a DMPO spin trap (Suzuki *et al.*, 1991). DHLA was found to react with the superoxide radical with a rate constant of $3.3 \times 10^5 \; M^{-1} \, sec^{-1}$. The reaction was confirmed by assaying the decreasing sulfhydryl content of DHLA and by detecting an increase in H_2O_2. These results were confirmed by this group in a later study (Suzuki *et al.*, 1993) using the same system for generating superoxide radical and competition with epinephrine oxidation to assess the scavenging ability of DHLA. In this study, a rate constant of $7.3 \times 10^5 \; M^{-1} \, sec^{-1}$ was found in good agreement with the previous work. In contrast to these studies, Scott *et al.* (1994), using a similar superoxide generating system (hypoxanthine–xanthine oxidase) and detecting superoxide by $O_2\cdot^-$-dependent nitro blue tetrazolium reduction, found no scavenging effect of DHLA. It is difficult to explain the discrepancy in the results of the two groups, and further work is clearly indicated.

Similar to α-lipoate, it has been observed that DHLA reacts with nitric oxide (unpublished finding).

C. Redox Interactions with Other Antioxidants

DHLA appears to be able to regenerate other antioxidants, such as ascorbate and (indirectly) vitamin E, from their radical forms. Vitamin E is the major chain-breaking antioxidant that protects membranes from lipid peroxidation (Burton and Ingold, 1981). Vitamin E exists in biological membranes in a low molar ratio to unsaturated phospholipids, usually less than 0.1 nmol per milligram of membrane protein or, in other words, one molecule per 1000 to 2000 membrane phospholipid molecules, which are the main target of oxidation in membranes. Lipid peroxyl radicals can be generated in membranes at the rate of 1–5 nmol per milligram of membrane protein per minute, yet destructive oxidation of membrane lipids does not normally occur, nor is vitamin E rapidly depleted. Furthermore, deficiency states for vitamin E are remarkably difficult to induce in adult animals. These apparent paradoxes can be explained by "vitamin E recycling" in which the antioxidant ability of vitamin E is continuously restored by other antioxidants. Those antioxidants that recycle vitamin E are vitamin C, ubiquinols, and thiols (Packer, 1992; Sies, 1993).

Evidence for vitamin E recycling by DHLA has come from a number of studies. DHLA protects against microsomal lipid peroxidation, but only in the presence of vitamin E (Scholich *et al.*, 1989). DHLA prolonged the lag phase prior to the onset of low-level chemiluminescence, loss of vitamin E, and the rapid accumulation of thiobarbituric acid reactive substances in normal, but not in vitamin E-deficient, microsomes. α-Lipoic acid was not effective in either system. DHLA could have exerted its effect in this system by directly reducing tocopheroxyl radical or by reducing other antioxidants (such as ascorbate), which then regenerated vitamin E. There may be a weak direct interaction between DHLA and the tocopheroxyl radical. The presence of DHLA reduces the tocopheroxyl radical ESR signal in liposomes exposed to UV light (unpublished data). The use of UV light, which directly produces tocopheroxyl radicals, eliminates the possibility of iron chelation or other antioxidant effects of DHLA, and the use of liposomes eliminates the possibility of DHLA acting through other antioxidants. Therefore, in this system it seems likely that DHLA, which partitions mainly in the aqueous phase, directly reduces tocopheroxyl radicals at the membrane/water interface. However, the effect is weak, and the major recycling of vitamin E by DHLA in biological systems probably occurs through other antioxidants.

Bast and Haenen (1990) have proposed that DHLA prevents lipid peroxidation by reducing glutathione disulfide, which in turn recycles vitamin E. This proposal is based on the observation that the combination of DHLA and glutathione disulfide, but not DHLA alone, prevented lipid peroxidation induced by Fe^{2+} ascorbate (Bast and Haenen, 1988). Alternatively, Kagan *et al.* (1992) proposed that DHLA protects membranes against oxidation by recycling ascorbate, which in turn recycles vitamin E. Their ESR studies indicated a DHLA-mediated reduction of ascorbyl radicals generated in the course of ascorbate oxidation by chromanoxyl (vitamin E or short-chain vitamin E homologs) radicals in DOPC liposomes and also showed that DHLA interacts with NADPH- or NADH-dependent electron transport chains to recycle vitamin E. Ascorbate-dependent recycling of vitamin E by DHLA has also been observed in human low-density lipoproteins (Kagan *et al.*, 1992) and erythrocyte membranes (Constantinescu *et al.*, 1993). In addition, evidence shows that α-lipoic acid administration *in vivo* can increase the level of ubiquinol in the face of oxidative stress (Gotz *et al.*, 1994) and ubiquinol is known to recycle vitamin E (Kagan *et al.*, 1990). Hence, current evidence points to the contention that DHLA can recycle vitamin E via glutathione, vitamin C, or ubiquinol (Fig. 3), but the relative contributions of each avenue are not well defined.

In fact, protection of other antioxidants by α-lipoic was suggested as early as 1959, by Rosenberg and Culik (1959) who stated, with startling prescience, "*α-Lipoic acid, and even more so its dihydro derivative into which it is converted rapidly after entering cellular metabolism, might act as an antioxidant for ascorbic acid and tocopherols* (p. 92)." In the studies of Rosenberg and Culik, α-lipoic acid was found to prevent symptoms of both vitamin E and vitamin C deficiency. We have found similar protective effects of α-lipoic acid administration in tocopherol-deficient hairless mice (Podda *et al.*, 1994b). Such results are consistent with a recycling of tocopherol and/or ascorbate by α-lipoic acid, but could also be explained by the ability of α-lipoic acid to spare ascorbate and vitamin E through its separate but overlapping radical-scavenging effects.

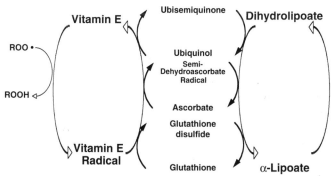

FIGURE 3 Pathways by which dihydrolipoate recycles vitamin E and other antioxidants. Lipoate is reduced to dihydrolipoate by the NADH-dependent dihydrolipoamide dehydrogenase (the E3 enzyme of α-ketoacid dehydrogenase complexes) and also by NADPH-dependent glutathione reductase.

α-Lipoic acid also causes an increase in intracellular glutathione. Peina-dao *et al.* (1989) showed that thiols appeared in the perfusate when α-lipoate was infused into the liver, but the exact nature of the thiols was not identified. Busse *et al.* (1992) added α-lipoic acid to murine neuroblastoma and melanoma cell lines, and observed a dose-dependent increase in GSH content of 30–70% compared to untreated controls. These investigators found similar evaluations in lung, liver, and kidney cells of mice injected (ip) daily with doses of 4, 8, or 16 mg/kg α-lipoic acid for 11 days. These results have been confirmed in human Jurkat cell lines, in which intracellular concentrations of GSH increase approximately 50% in 5 hr after the addition of α-lipoic acid to the culture medium (Han *et al.*, 1995). Such elevations in GSH cannot be explained by the reduction of GSSG because GSSG is normally present at less than 10% the concentration of GSH (Halliwell and Gutteridge, 1989). Following addition to the cell culture medium, α-lipoate is taken up by the cells and reduced to dihydrolipoate (Handelman *et al.*, 1994). Intracellular dihydrolipoate is rapidly released to the culture medium rich in cystine where this amino acid is reduced to the precursor amino acid for glutathione synthesis, cysteine. The extracellular reduction of cystine to cysteine and the rapid entry of the latter to the cell to stimulate glutathione synthesis may be one possible way of intracellular glutathione increase in response to α-lipoate treatment.

D. Lipoic Acid as a Metabolic Antioxidant

There are two ways in which α-lipoate is intimately connected to cell metabolism and redox state. As discussed earlier, α-lipoate has long been known for its role in oxidative metabolism. In addition, because the reducing power for reduction of exogenous α-lipoate to DHLA comes from both NADH and NADPH (Pick *et al.*, 1995), α-lipoate may modulate NADH/NAD$^+$ and NADPH/NADP$^+$ ratios, thus affecting numerous aspects of cell metabolism.

α-Lipoate has several features that, in combination, make it a unique "ideal" antioxidant. It is soluble in both aqueous and lipid environments and is readily absorbed from the diet, transported, taken up by cells, and reduced to dihydrolipoate in various tissues (Handelman *et al.*, 1994; Podda *et al.*, 1994a,b). The dihydrolipoate, thus formed, is exported and can provide antioxidant protection to the extracellular compartment and to nearby cells. DHLA is capable of affecting a wide range of other antioxidants, such as glutathione, vitamin C, and indirectly vitamin E. This combination of traits makes α-lipoate an intriguing possibility for the treatment of a host of conditions related to oxidative stress. Some of the most exciting possibilities stem from its effects on the transcription factor NF-κB.

IV. NF-κB Role in Normal and Abnormal Cell Function

A. Overview

In response to stresses such as injury and infection, organisms must rapidly marshal a host of responses. Various genes must be activated, and a central, coordinating factor in this activation of rapid response gene is NF-κB. In addition, NF-κB plays a crucial role in modulating gene expression during growth and development. In pathological conditions such as viral infection, NF-κB has far-reaching significance for a variety of pathological conditions in which inflammation, growth, or viral activation occur, such as atherosclerosis and HIV infection (AIDS).

NF-κB is actually a family of Rel transcription factors that share common characteristics. They consist of hetero- or homodimeric proteins in association with an inhibitory protein family, IκB. Phosphorylation and dissociation of IκB from Rel protein dimers, followed by its degradation, allow the dimeric DNA-binding NF-κB to enter the nucleus and regulate genes (Fig. 4) (Brown *et al.*, 1993; Siebenlist *et al.*, 1994).

A host of genes have been shown to be modulated by NF-κB, including genes for cytokines and growth factors, immunoreceptors, adhesion molecules, acute-phase proteins, transcription factors and regulators, NO-synthase, and viral genes (Siebenlist *et al.*, 1994).

B. Role of Redox Balance in Modulating NF-κB Activation

Several lines of evidence show that reactive oxygen species (ROS) may be the final common signal for a number of stimuli that activate NF-κB (Brown *et al.*, 1993; Sen and Packer, 1996).

First, many of the stimuli that activate NF-κB also are known to cause intracellular increases in ROS. These include tumor necrosis-factor (TNF)-α, IL-1, phorbol 12-myristate 13-acetate (PMA), LPS, UV light, and gamma irradiation (Schreck *et al.*, 1992a,b; Geng *et al.*, 1993; Schieven *et al.*, 1993).

Second, administration of hydrogen peroxide can directly stimulate NF-κB activation (Schreck *et al.*, 1991, 1992a). However, several other ROS (hydroxyl radical, singlet oxygen, etc.) have not been effective (Siebenlist *et al.*, 1994).

Third, a diverse array of antioxidants can block NF-κB activation. These include N-acetyl-L-cysteine (Meyer *et al.*, 1992), dithiocarbamates (Meyer *et al.*, 1992; Schreck *et al.*, 1992b), vitamin E derivatives (Suzuki and Packer, 1993, 1994a,b,c), catechol derivatives (Suzuki and Packer, 1994c), dithiolthione (Sen *et al.*, 1996), and α-lipoate (Suzuki *et al.*, 1992, 1995;

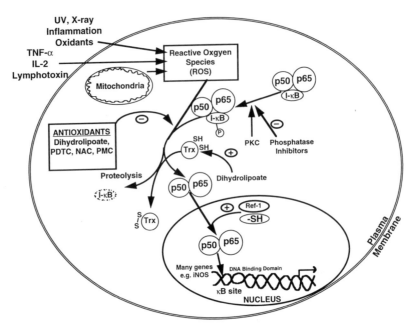

FIGURE 4 Modulation of NF-kB activation by α-lipoate. The exact loci of antioxidant action in the inhibition of NF-κB activation are unclear. Antioxidants, however, inhibit the dissociation of IκB from the Rel dimer. TNFα, tumor necrosis factor α; IL-2, interleukin 2; PDTC, pyrrolidine dithiocarbamate; NAC, N-acetylcysteine; Trx, thioredoxin; Ref-1 (nuclear) redox factor 1; PKC, protein kinase C; iNOS, inducible nitric oxide synthase; PMC, 2,2,5,7,8-pentamethyl-6-hydroxychromanol.

Packer and Suzuki, 1993; Bessho *et al.*, 1994; Suzuki and Packer, 1994a). Conversely, in certain instances agents which deplete cellular thiol antioxidants have been observed to induce NF-κB activation (Staal *et al.*, 1990).

C. Effects of Lipoate and Dihydrolipoate on NF-κB

A number of researchers have reported antioxidants, including ebselen, 2-mercaptoethanol, glutathione, pyrrolidinedithiocarbamate (PDTC), diethyldithiocarbamate, disulfiram, orthophenanthroline, anetholdithiolthione, desferrioxamine, vitamin C, L-cysteine, and butyllated hydroxyanisole (BHA), to inhibit NF-κB activation. We have shown that vitamin E derivates are potent suppressors of NF-κB activation (Suzuki and Packer, 1993, 1994a,b,c). Incubation of human Jurkat T cells with vitamin E acetate or α-tocopheryl succinate exhibited a concentration-dependent inhibition of NF-κB activation. α-Tocopherol or succinate at these concentrations had no apparent side effects. Also, 2,2,5,7,8-pentamethyl-6-hydroxychromanol was extremely effective, causing complete inhibition of NF-κB activation at 10 μM. Therefore, it was of interest to us to examine the effects of α-lipoate on NF-κB activation.

Suzuki *et al.* (1992) observed that the incubation of Jurkat T cells with a millimolar concentration of lipoate prior to the stimulation with TNF-α or PMA inhibited NF-κB activation in a dose-dependent manner, with 4 mM required for complete inhibition. α-Lipoate at these concentrations did not affect either cell viability or oct-1 DNA-binding activity. Direct supplementation of DHLA in the incubation medium also inhibited the NF-κB activation induced by TNF-α (Suzuki and Packer, 1994a). Although the potency of DHLA on NF-κB inhibition was comparable to that of α-lipoic acid, millimolar concentrations of DHLA affected cell viability. Both *R*- and *S*-enantiomers of lipoate are effective in inhibiting NF-κB activation simulated by TNF-α treatment in Jurkat T cells after 2 hr of preincubation (Suzuki and Packer, 1994a). DNA binding of NF-κB occurs after the activation and involves phosphorylation, dissociation, and degradation of the IκB subunit. DHLA enhanced DNA binding of NF-κB, whereas α-lipoate was inhibited (Suzuki *et al.*, 1995). Hence, there is an apparent paradox in the redox regulation of NF-κB. The observations from this study suggest that the α-lipoate action preventing the NF-κB DNA-binding activity is, at least in part, the underlying mechanism for the observed inhibitory action against NF-κB activity. A dual action of dihydrolipoate, by inhibiting the TNFα-induced NF-κB activation and potentiating the DNA binding of activated NF-κB, suggests that two modes of redox regulation indeed exist in cell signaling for NF-κB: (1) a requirement of oxidative processes in activating NF-κB, and (2) a requirement of reductive processes in binding to DNA.

D. Possible Therapeutic Roles for Lipoate Involving Modification of NF-κB Activity

I. HIV Infection and AIDS

Acquired immunodeficiency syndrome results from infection with a human immunodeficiency virus (HIV-1 or HIV-2), which eventually destroys a specific subset (CD4$^+$) of helper T lymphocytes, so that the patient ultimately yields to opportunistic infection and certain neoplasms. The long terminal repeat (LTR) region of HIV-1 proviral DNA contains two binding sites for NF-κB, which activate transcription by binding to the sequence 5'-GGGACTTTCC-3' in the κ enhancer where it interacts with the transcription apparatus. In this case, and for other viruses as well, the virus has usurped normal cellular machinery in order to transcribe its own DNA. Hence, strategies that decrease this transcription offer possible therapies for HIV infection, to delay the onset of AIDS symptoms, and are actively being pursued.

Herzenberg and co-workers observed that NAC inhibited HIV LTR-directed expression of the β-galactosidase gene in response to TNFα and phorbol ester (Roederer *et al.*, 1990) and that NAC inhibited and diamide stimulated NF-κB activations (Staal *et al.*, 1990). NAC is known to increase the intracellular level of reduced glutathione, whereas diamide increases the amount of oxidized glutathione and depletes reduced glutathione. The finding that NAC blocks TNFα-induced NF-κB responses was also reported by Mihm *et al.* (1991). In our laboratory, NAC was observed to be effective in inhibiting okadaic acid-induced activation of NF-κB in Wurzburg T cells (Sen and Packer, 1996).

Baur *et al.* (1991) found that α-lipoate inhibited the replication of HIV-1 in cultured lymphoid T cells. Jurkat, SupT1, and Molt-4 cells were infected with HTLV IIB and HIV-1 Wal, and α-lipoate was added 16 hr after infection (Baur *et al.*, 1991). A dose-dependent inhibition of reverse transcriptase activity and plaque formation was observed. At a concentration of 70 μg/ml (around 0.35 mM), reverse transcriptase activity was decreased by 90%, and 35 μg/ml (around 0.17 mM) completely eliminated plaque-forming units.

Although the exact physiological activator of HIV-1 is not yet determined, TNFα appears to play an important role. Considerable evidence indicates generation of ROS in TNFα responses (Matthews *et al.*, 1987; Yamauchi *et al.*, 1989; Zimmerman *et al.*, 1989). Furthermore, TNFα has been found to induce manganous superoxide dismutase in some cells, and its implications as a possible mechanism of cellular resistance to cytoxicity have been suggested (Wong and Goeddel, 1988; Wong *et al.*, 1989). These finding suggest a signal transduction cascade as follows:

$$TNF\alpha \rightarrow ROS \rightarrow NF\text{-}\kappa B \rightarrow \text{expression of HIV-1}$$

Agents currently used in the therapy of AIDS include reverse transcriptase inhibitors, e.g., AZT, DDC, DDI, which act at the final step in this sequence. However, these agents exhibit toxicity (Schinazi *et al.*, 1992), and the use of high concentrations is not desirable. Natural compounds that can affect the HIV life cycle are attractive in that they may support the actions of more aggressive antiviral agents in combination therapy without a risk of toxicity. α-Lipoate is a natural compound physiologically involved as a cofactor for metabolism and has been used clinically in the treatment of various diseases. Our studies and those of Baur *et al.* (1991) have shown that α-lipoic acid is potent in its inhibitory action against NF-κB and against HIV-1 replication. It is more effective, at lower doses, than other suggested therapeutic thiol-enhancing substances. In HIV-infected patients and in SIV-infected rhesus macaques, there are decreased cysteine and intracellular GSH levels, which have been hypothesized to play a role in progression of the disease (Droge *et al.*, 1994); α-lipoate has been shown to enhance intracellular glutathione levels both *in vitro* (Busse *et al.*, 1992; Han *et al.*, 1995) and *in vivo* (Busse *et al.*, 1992). α-Lipoate therapy has few, if any, side effects (Packer *et al.*, 1995). Therefore the use of α-lipoate as an AIDS therapeutic agent appears logical and desirable.

2. Atherosclerosis

Oxidants appear to be involved in two steps of the atherosclerotic process (Andalibi *et al.*, 1993). LDL particles enter the artery wall and some remain; in this environment the LDL becomes mildly oxidized, possibly by ROS released by vascular cells. The mildly oxidized LDL contains a component or components, hypothesized to be oxidized phospholipid, which causes the endothelium to secrete inflammatory molecules that result in the recruitment of monocytes and their differentiation to macrophages. Further oxidation of the LDL results in its uptake by macrophages, their conversion to foam cells, and deposition in the vascular wall as part of the fatty-streak lesions.

NE-κB appears to be a central transcription factor in the expression of a number of the genes induced by the early oxidation step in atherosclerosis, including serum amyloid A and macrophage colony-stimulating factor (Visvanathan and Goodbourn, 1989; Li and Liao, 1991). When C57BL/6J mice and C3H/HeJ mice, which are susceptible and not susceptible to formation of aortic fatty streaks, respectively, are fed an atherogenic diet, only the C57BL/6J mice show activation of NF-κB (Liao *et al.*, 1993).

3. Diabetes

There is a further acceleration of atherosclerosis in diabetes, which is hypothesized to be due to advanced glycosylation end products (AGE) interacting with their endothelial receptor to induce the expression of vascular cell adhesion molecule-1 (VCAM-1) (Schmidt *et al.*, 1995), an early feature in the pathogenesis of atherosclerosis (Marui *et al.*, 1993). In cultured human endothelial cells, exposure to AGEs induced expression of VCAM-1, associated with increased levels of VCAM-1 transcripts (Schmidt *et al.*, 1995); electrophoretic mobility shift assays indicated that there was also induction of specific DNA-binding activity for NF-κB in the VCAM-1 promoter, which was blocked by NAC. Other studies also indicate that the induction of expression of VCAM-1 in response to a variety of signals is mediated by an NF-κB-like DNA-binding protein and is blocked by a variety of antioxidants, including pyrrolidine dithiocarbamate and NAC (Marui *et al.*, 1993). The authors hypothesize that accelerated atherosclerosis in diabetes may be due to the presence of larger amounts of AGE, a greater degree of interaction of these with the receptor for AGE (AGE-RAGE interaction), activation of NF-κB, expression of VCAM-1, and priming of the diabetic vasculature for enhanced interaction with circulating monocytes.

In nuclear magnetic resonance studies, we have found that α-lipoate can prevent early glycation reactions in albumin, probably through nonspecific interactions with hydrophobic-binding sites on the albumin molecule (Kawabata and Packer, 1994; Schepkin *et al.*, 1994). Hence, this nonantioxidant effect of α-lipoate may be effective in decreasing early reactions in the production of AGE. This, combined with the attenuating effect of α-lipoate on NF-κB activation, suggests a role for α-lipoate in dampening the acceleration of atherosclerosis seen in diabetes.

Thus, especially in diabetes, there appears to be three sites for α-lipoate to be effective in preventing or delaying the development of atherosclerotic lesions. It may help delay or prevent the initial formation of mildly oxidized LDL, thus decreasing the activation of NF-κB and thus decreasing the likelihood of recruitment of monocytes and their differentiation to macrophages. Later, more severe oxidation of LDL may also be decreased or prevented by α-lipoate, thus decreasing the uptake of LDL by macrophages and the consequent formation of foam cells. In diabetes, the ability of α-lipoate to modulate glycation reactions and its likely ability to decrease NF-κB activation in response to AGE-RAGE interactions may play a role in decreasing atherosclerotic pathogenesis associated with diabetes.

References

Andalibi, A., Liao, F., Imes, S., Fogelman, A. M., and Lusis, A. J. (1993). Oxidized lipoproteins influence gene expression by causing oxidative stress and activating the transcription factor NF-κB. *Biochem. Soc. Transactions* **21**(Pt 3), 651–655.

Bashan, N., Burdett, E., Klip, A. (1993). Effect of thioctic acid on glucose transport. *In* "Third International Thioctic Acid Workshop" Vol. 3, pp. 218–223. Frankfurt, Germany, Universimed Verlag.

Bast, A., and Haenen, G. R. M. M. (1988). Interplay between lipoic acid and glutathione in the protection against microsomal lipid peroxidation. *Biochim. Biophys. Acta* **963**, 558–561.

Bast, A., and Haenen, G. R. M. M. (1990). Regulation of lipid peroxidation of glutathione and lipoic acid: involvement of liver microsomal vitamin E free radical reductase. *In* "Antioxidants in Therapy and Preventive Medicine" (I. Emerit, L. Packer, and C. Auclair, eds.), pp. 111–116. Plenum Press, New York.

Baur, A., Harrer, T., Peukert, M., Jahn, G., Kalden, J. R., and Fleckenstein, B. (1991). Alpha-lipoic acid is an effective inhibitor of human immuno-deficiency virus (HIV-1) replication. *Klin. Wochenschr.* **69**, 722–724.

Bernier, M., Laird, D. M., and Lane, M. D. (1987). Insulin-activated tyrosine phosphorylation of a 15-kilodalton protein in intact 3T3-L1 adipocytes. *Proc. Natl. Acad. Sci. USA* **84**, 1844–1848.

Bessho, R., Matsubara, K., Kubota, M., Kuwakado, K., Hirota, H., Wakazono, Y., Lin, Y. W., Okuda, A., Kawai, M., and Nishikomori, R. (1994). Pyrrolidine dithiocarbamate, a potent inhibitor of nuclear factor kappa (NF-kappa B) activation, prevents apoptosis in human promyelocytic leukemia HL-60 cells and thymocytes. *Biochem. Pharmacol.* **48**, 1883–1889.

Brown, K., Park, S., Kanno, T., Franzoso, G., and Siebenlist, U. (1993). Mutual regulation of the transcriptional activator NF-κB and its inhibitor, I kappa B-a. *Proc. Natl. Acad. Sci. USA* **90**, 2532–2536.

Burton, G. W., and Ingold, K. U. (1981). Autoxidation of biological molecules. I. The antioxidant activity of vitamin E and related chain-breaking phenolic antioxidant in vitro. *J. Am. Chem. Soc.* **103**, 6472–6477.

Busse, E., Zimmer, G., Schopohl, B., and Kornhuber, B. (1992). Influence of alpha-lipoic acid on intracellular glutathione in vitro and in vivo. *Arzneimittel-Forschung* **42**, 829–831.

Constantinescu, D., Han, D., and Packer, L. (1993). Vitamin E recycling in human erythrocyte membranes. *J. Biol. Chem.* **268**, 10906–10913.

Danson, M. J. (1987). Dihydrolipoamide dehydrogenase: A 'new' function for an old enzyme? *Biochem. Soc. Trans.* **16**, 87–89.

Danson, M. J., Eisenthal, R., Hall, S., Kessell, S. R., and Williams, D. L. (1984). Dihydrolipoamide dehydrogenase from halophilic archaebacteria. *Biochem. J.* **218**, 811–818.

Devasagayam, T. P. A., Subramanian, M., Pradhan, D. S., and Sies, H. (1993). Prevention of singlet oxygen-induced DNA damage by lipoate. *Chem.-Biol. Interact.* **86**, 79–92.

Droge, W., Schulze-Osthoff, K., Mihm, S., Galter, D., Schenk, H., Eck, H. P., Roth, S., and Gmunder, H. (1994). Functions of glutathione and glutathione disulfide in immunology and immunopathology. *FASEB J.* **8**, 1131–1138.

Frost, S. C., and Lane, M. D. (1985). Evidence for the involvement of vicinal sulfhydryl groups in insulin-activated hexose transport by 3T3-L1 adipocytes. *J. Biol. Chem.* **260**, 2646–2652.

Fujiwara, K., Okamura-Ikeda, K., and Motokawa, Y. (1991). Lipoylation of H-protein of the glycine cleavage system. *FEBS Lett.* **293**, 115–118.

Geng, Y., Zhang, B., and Lotz, M. (1993). Protein tyrosine kinase activation is required for lipopolysaccharide induction of cytokines in human blood monocytes. *J. Immunol.* **151**, 6692–6700.

Gotz, M. E., Dirr, A., Burger, R., Wanetzky, B., Weinmuller, M., Chan, W. W., Chen, S. C., Reichmann, H., Rausch, W. D., and Riederer, P. (1994). Effect of lipoic acid on redox state of coenzyme Q in mice treated with 1-methyl-4-phenyl-1,2,3,6-tetrahydropyridine and diethyldithiocarbamate. *Eur. J. Pharmacol.* **266**, 291–300.

Guirard, B. M., Snell, E. E., and Williams, R. J. (1946). The nutritional role of acetate for lactic acid bacteria. I. The response to substances related to acetate. *Arch. Biochem. Biophys.* **9**, 361–379.

Haenen, G. R., and Bast, A. (1991). Scavenging of hypochlorous acid by lipoic acid. *Biochem. Pharmacol.* **42**, 2244–2246.

Halliwell, B., and Gutteridge, J. M. C. (1989). "Free Radicals in Biology and Medicine," 2nd Ed. Clarendon Press, Oxford, England.

Han, D., Tritschler, H. J., and Packer, L. (1995). Alpha-lipoic acid increases intracellular glutathione in a human T-lymphocyte Jurkat cell line. *Biochem. Biophys. Res.* **207**, 258–264.

Handelman, G. J., Han, D., Tritschler, H., and Packer, L. (1994). Alpha-lipoic acid reduction by mammalian cells to the dithiol form, and release into the culture medium. *Biochem. Pharmacol.* **47**, 1725–1730.

Haramaki, N., Han, D., Handelman, G. J., and Packer, L. (1996). Cytosolic and mitochondrial systems for cellular reduction of lipoic acid. Submitted for publication.

Haramaki, N., Packer, L., Assadnazari, H., and Zimmer, G. (1993). Cardiac recovery during post-ischemic reperfusion is improved by combination of vitamin E with dihydrolipoic acid. *Biochem. Biophys. Res. Commun.* **196**, 1101–1107.

Haugaard, N., and Haugaard, E. S. (1970). Stimulation of glucose utilization by thioctic acid in rat diaphragm incubated in vitro. *Biochim. Biophys. Acta* **222**, 583–586.

Henriksen, E. J., Jacob, S., Tritschler, H., Wellel, K., Augustin, H. J., and Dietze, G. J. (1994). Chronic thioctic acid treatment increases insulin-stimulated glucose transport activity in skeletal muscle of obese Zucker rats. *Diabetes* (Suppl.) **1**, 122A.

Jacob, S., Henriksen, E. J., Tritschler, H. J., Clancy, D. E., Simon, I., Schiemann, A. L., Ulrich, H., Jung, I., Dietze, G. J., and Augustin, H. J. (1994). Thioctic acid enhances glucose-disposal in patients with Type 2 diabetes. *Horm. Metab.,* in press.

Jentoft, J. E., Shoham, M., Jurst, D., and Patel, M. S. (1992). A structural model for human dihydrolipoamide dehydrogenase. *Proteins* **14**, 88–101.

Jocelyn, P. C. (1967). The standard redox potential of cysteine-cystine from the thiol-disulphide exchange reaction with glutathione and lipoic acid. *Eur. J. Biochem.* **2**, 327–331.

Kagan, V., Serbinova, E., and Packer, L. (1990). Antioxidant effects of ubiquinones in microsomes and mitochondria are mediated by tocopherol recycling. *Biochem. Biophys. Res. Commun.* **169**, 851–857.

Kagan, V. E., Serbinova, E. A., Forte, T., Scita, G., and Packer, L. (1992). Recycling of vitamin E in human low density lipoproteins. *J. Lipid Res.* **33**, 385–397.

Kagan, V. E., Shvedova, A., Serbinova, E., Khan, S., Swanson, C., Powell, R., and Packer, L. (1992). Dihydrolipoic acid: A universal antioxidant both in the membrane and in the aqueous phase. *Biochem. Pharmacol.* **44**, 1637–1649.

Kaiser, S., Di Mascio, P., and Sies, H. (1989). Lipoat und Singulettsauerstoff. "Thioctsäure (H. O. Borbe and H. Ulrich, eds.), pp. 69–76. Frankfurt, pmi Verlag GmbH.

Kawabata, T., and Packer, L. (1994). Alpha-lipoate can protect against glycation of serum albumin, but not low density lipoprotein. *Biochem. Biophys. Res. Commun.* **203**, 99–104.

Lawson, J. E., Behal, R. H., Reed, L. J. (1991). Disruption and mutagenesis of the *Saccharomyces cervisiae* PDX1 gene encoding the protein X component of the pyruvate dehydrogenase complex. *Biochemistry* **30**, 2834–2839.

Lawson, J. E., Niu, X.-D., Reed, L. J. (1991). Functional analysis of the domains of dihydrolipoamide acetyltransferase from *Saccharamyces cerevisiae*. *Biochemistry* **30**, 11249–11254.

Li, X. X., and Liao, W. S. (1991). Expression of rat serum amyloid A1 gene involves both C/EBP-like and NF kappa B-like transcription factors. *J. Biol. Chem.* **266**, 15192–15201.

Liao, F., Andalibi, A., deBeer, F. C., Fogelman, A. M., and Lusis, A. J. (1993). Genetic control of inflammatory gene induction and NF-kappa B-like transcription factor activation in response to an atherogenic diet in mice. *J. Clin. Invest.* **91**, 2572–2579.

Machado, R. S., Clark, D. P., and Guest, J. R. (1992). Construction and properties of pyruvate dehydrogenase complexes with up to nine lipoyl domains per lipoate acetyltransferase chain. *FEMS Microbiol. Lett.* **100**, 243–248.

Marui, N., Offermann, M. K., Swerlick, R., Kunsch, C., Rosen, C. A., Ahmad, M., Alexander, R. W., and Medford, R. M. (1993). Vascular cell adhesion molecule-1 (VCAM-1) gene transcription and expression are regulated through an antioxidant-sensitive mechanism in human vascular endothelial cells. *J. Clin. Invest.* **92**, 1866–1874.

Matsugo, S., Yan, L.-J., Han, D., Tritschler, H. J., and Packer, L. (1995a). Elucidation of antioxidant activity of alpha-lipoic acid toward hydroxyl radical. *Biochem. Biophys. Res. Comun.* **208**, 161–167.

Matsugo, S., Yan, L.-J., Han, D., Tritschler, H. J., and Packer, L. (1995b). Elucidation of antioxidant activity of dihydrolipoic acid toward hydroxyl radical using a novel hydroxyl radical generator NP-III. *Biochem. Mol. Biol. Int.* **37**, 375–383.

Matthews, N., Neale, M. L., Jackson, S. K., and Stark, J. M. (1987). Tumour cell killing by tumour necrosis factor: Inhibition by anaerobic conditions, free-radical scavengers and inhibitors of arachidonate metabolism. *Immunology* **52**, 153–155.

Meyer, R., Caselmann, W. H., Schluter, V., Schreck, R., Hofschneider, P. H., and Baeuerle, P. A. (1992). Hepatitis B virus transactivator MHBst: Activation of NF-kappa B, selective inhibition by antioxidants and integral membrane localization. *EMBO J.* **11**, 2992–3001.

Mihm, S., Ennen, J., Pessara, U. K., and Droge, W. (1991). Inhibition of HIV-1 replication and NF-kappa B activity by cysteine and cysteine derivatives. *AIDS* **5**, 497–503.

Müller, L., and Menzel, H. (1990). Studies on the efficacy of lipoate and dihydrolipoate in the alteration of cadmium^{2+} toxicity in isolated hepatocytes. *Biochim. Biophys. Acta* **1052**, 386–391.

O'Kane, D. J., and Gunsalus, I. C. (1952). Accessory factor requirement for pyruvate oxidation. *J. Bacteriol.* **54**, 20–21.

Ou, P., Tritschler, H., and Wolff, S. P. (1995). Thioctic (lipoic) acid: A therapeutic metal-chelating antioxidant? *Biochem. Pharmacol.* **50**, 123–126.

Packer, L. (1992). New horizons in vitamin E research—the vitamin E cycle, biochemistry and clinical applications. *In* "Lipid-Soluble Antioxidants: Biochemistry and Clinical Applications" (A. S. H. Ong and L. Packer, eds.), pp. 1–16. Birkhauser Verlag, Boston.

Packer, L., Suzuki, Y. J. (1993). Vitamin E and alpha-lipoate: Role in antioxidant recycling and activation of the NF-κB transcription factor. *Mol. Aspects Med.* **14**, 229–239.

Packer, L., Witt, E. H., and Tritschler, H. J. (1995). Alpha-lipoic acid as a biological antioxidant. *Free Rad. Biol. Med.* **19**, 227–250.

Patel, M. S., and Roche, T. E. (1990). Molecular biology and biochemistry of pyruvate dehydrogenase complexes. *FASEB J.* **4**, 3224–3233.

Patel, M. S., and Smith, R. L. (1994). Biochemistry of lipoic acid containing proteins: Past and present. *In* "The Evolution of Antioxidants in Modern Medicine" (K. Schmidt, A. T. Diplock, and H. Ulrich, eds.), pp. 65–77. Hippocrates Verlag, Stuttgart.

Peinado, J., Sies, H., and Akerboom, T. P. M. (1989). Hepatic lipoate uptake. *Arch. Biochem. Biophys.* **273**, 389–395.

Perham, R. N. (1991). Domains, motifs, and linkers in 2-oxo acid dehydrogenase multienzyme complexes: A paradigm in the design of a multifunctional protein. *Biochemistry* **30**, 8501–8512.

Pick, U., Haramaki, N., Constantinescu, A., Handelman, G. J., Tritschler, H. J., and Packer, L. (1995). Glutathione reductase and lipoamide dehydrogenase have opposite stereospecificities for alpha-lipoic acid enantiomers. *Biochem. Biophys. Res. Commun.* **206**, 724–730.

Podda, M., Han, D., Koh, B., Fuchs, J., and Packer, L. (1994). Conversion of lipoic acid to dihydrolipoic acid in human keratinocytes. *Clin. Res.* **42**, 41a.

Podda, M., Tritschler, H. J., Ulrich, H., and Packer, L. (1994). α-Lipoic acid supplementation prevents symptoms of vitamin E deficiency. *Biochem. Biophys. Res. Commun.* **204**, 98–104.

Reed, L. J., DeBusk, B. G., Gunsalus, I. C., and Hornberger, C. S. (1951). Crystalline α-lipoic acid: A catalytic agent associated with pyruvate dehydrogenase. *Science* **114**, 93–94.

Richarme, G. (1989). Purification of a new dihydrolipoamide dehydrogenase from *Escherichia coli. J. Bacteriol.* **171**, 6580–6585.

Roederer, M., Staal, F. J. T., Faju, P. A., Ela, S. W., Herzenberg, L. A., and Herzenberg, L. A. (1990). Cytokine-stimulated human immunodeficiency virus replication is inhibited by N-acetyl-L-cysteine. *Proc. Natl. Acad. Sci. USA* **87**, 4884–4888.

Rosenberg, H. R., and Culik, R. (1959). Effect of α-lipoic acid on vitamin C and vitamin E deficiencies. *Arch. Biochem. Biophys.* **80**, 86–93.

Schepkin, V., Kawabata, T., and Packer, L. (1994). NMR study of liopoic acid binding to bovine serum albumin. *Biochem. Mol. Biol. Intl.* **33**, 879–886.

Schieven, G. L., Kirihara, J. M., Myers, D. E., Ledbetter, J. A., and Uckun, F. M. (1993). Reactive oxygen intermediates activate NF-κB in a tyrosine-kinase dependent mechanism and in combination with vanadate activate the p56 lck and p59 fyn tyrosine kinase in human lymphocytes. *Blood* **82**, 1212–1220.

Schinazi, R. F., Mead, J. R., and Feorino, P. M. (1992). Insights into HIV chemotherapy. *AIDS Res. Hum. Retrovir.* **8**, 963–990.

Schmidt, A. M., Hori, O., Chen, J. X., Li, J. F., Crandall, J., Ahzng, J., Cao, R., Yan, S. D., Brett, J., and Stern, D. (1995). Advanced glycation endproducts interacting with their endothelial receptor induce expression of vascular cell adhesion molecule-1 (VCAM-1) in cultured human endothelial cells and in mice: A potential mechanism for the accelerated vasculopathy of diabetes. *J. Clin. Invest.* **96**, 1395–1403.

Scholich, H., Murphy, M. E., and Sies, H. (1989). Antioxidant activity of dihydrolipoate against microsomal lipid peroxidation and its dependence on α-tocopherol. *Biochim. Biophys. Acta* **1001**, 256–261.

Schreck, R., Albermann, K., and Baeuerle, P. A. (1992). NF-κB: An oxidative stress-responsive transcription factor of eukaryotic cells (a review). *Free Rad. Res. Commun.* **17**, 221–237.

Schreck, R., Meier, B., Maennel, D. N., Droge, W., and Baeuerle, A. (1992). Dithiocarbamates as potent inhibitors of nuclear factor κB activation in intact cells. *J. Exp. Med.* **175**, 1181–1194.

Schreck, R., Rieber, P., and Baeuerle, P. (1991). Reactive oxygen intermediates are apparently widely used messengers in the activation of NF-κB transcription factor and HIV-1. *EMBO J* **10**, 2247–2258.

Scott, B. C., Aruoma, O. I., Evans, P. J., O'Neill, C., van der Vliet, A., Cross, C. E., Tritschler, H., and Halliwell, B. (1994). Lipoic and dihydrolipoic acids as antioxidants: A critical evaluation. *Free Rad. Res.* **20**, 119–133.

Scott, I., Duncan, W., and Ekstrand, V. (1963). Purification and properties of glutathione reductase of human erythrocytes. *J. Biol. Chem.* **238**, 3928–3939.

Searls, R. L., and Sanadi, D. R. (1960). α-Ketoglutaric dehydrogenase. 8. Isolation and some properties of a flavoprotein component. *J. Biol. Chem.* **235**, 2485–2491.

Sen, C. K., and Packer, L. (1996). Antioxidant and redox regulation of gene transcription. *FASEB J.*, **10**, 709–720.

Sen, C. K., Traber, K., and Packer, L. (1996). Inhibition of NF-kB activation in human T-cell lines by anetholdithiolthione. *Biochem. Biophys. Res. Commun.* **218**, 148–153.

Serbinova, E., Khwaja, S., Reznick, A. Z., and Packer, L. (1994). Thioctic acid protects against ischemia-reperfusion injury in the isolated perfused Langendorff heart. *Free Rad. Res. Commun.* **17**, 49–58.

Siebenlist, U., Franzoso, G., and Brown, K. (1994). Structure, regulation and function of NF-kB. *Annu. Rev. Cell Biol.* **10**, 405–455.

Sies, H. (1993). Strategies of antioxidant defense. *Eur. J. Biochem.* **215**, 213–219.

Sigel, H., Prijs, B., McCormick, D. B., and Shih, J. C. H. (1978). Stability of binary and ternary complexes of a-lipoate and lipoate derivatives with Mn^{2+}, Cu^{2+}, and Zn^{2+} in solution. *Arch. Biochem. Biophys.* **187**, 208–214.

Singh, H. P. P., and Bowman, R. H. (1970). Effect of D,L-alpha lipoic acid on the citrate concentration and phosphofructokinase activity of perfused hearts from normal and diabetic rats. *Biochem. Biophys. Res. Commun.* **41**, 555–561.

Smith, L. C., Bungard, S. J., Danson, M. J., and Hough, D. W. (1987). "Purification and characterization of glucose dehydrogenase from the thermoacidophilic archaebacterium Thermoplasma acidophilum. *Biochem. Soc. Trans.* **15**, 1097.

Snell, E. E., Strong, F. M., and Peterson, W. H. (1937). Growth factors for bacteria. VI. Fractionation and properties of an accessory factor for lactic acid bacteria. *Biochem. J.* **31**, 1789–1799.

Staal, F. J., Roederer, M., Herzenberg, L. A., and Herzenberg, L. A. (1990a). Intracellular thiols regulate activation of nuclear factor kappa B and transcription of human immunodeficiency virus. *Proc. Natl. Acad. Sci. USA* **87**(24), 9943–9947.

Stary, F. E., Jindal, S. J., and Murray, R. W. (1975). Oxidation of α-lipoic acid. *J. Org. Chem.* **40**, 58–62.

Stevens, B., Perez, S. R., and Small, R. D. (1974). The photoperoxidation of unsaturated organic molecules. IX. Lipoic acid inhibition of rubrene autoperoxidation. *Photochem. Photobiol.* **19**, 315–316.

Stokstad, E. L. R., Hoffman, C. E., Regan, M. A., Fordham, D., and Jukes, T. H. (1949). Observations on an unknown growth factor for *T. geleii*. *Arch. Biochem. Biophys.* **20**, 75–82.

Sumegi, B., Butwell, N. B., Malloy, C. R., and Sherry, A. D. (1994). Lipoamide influences substrate selection in post-ischemic perfused rat hearts. *Biochem. J.* **297**, 109–113.

Suzuki, Y. J., Aggarwal, B. B., and Packer, L. (1992). Alpha-lipoic acid is a potent inhibitor of NF-kB activation in human T cells. *Biochem. Biophys. Res. Commun.* **189**, 1709–1715.

Suzuki, Y. J., Mizuno, M., Tritschler, H. J., and Packer, L. (1995). Redox regulation of NF-kB DNA binding activity by dihydrolipoate. *Biochem. Mol. Biol. Int.* **36**, 241–246.

Suzuki, Y. J., and Packer, L. (1993). Inhibition of NF-kB activation by vitamin E derivatives. *Biochem. Biophys. Res. Commun.* **193**, 277–283.

Suzuki, Y. J., and Packer, L. (1994a). Alpha-lipoic acid is a potent inhibitor of NF-kappa B activation in human T cells: Does the mechanism involve antioxidant activities. *In* "Biological Oxidants and Antioxidants" (L. Packer and E. Cadenas, eds.), Hippokrates Verlag, Stuttgart.

Suzuki, Y. J., and Packer, L. (1994b). Inhibition of NF-kB DNA binding activity by alpha-tocopheryl succinate. *Biochem. Mol. Biol. Intl.* **31**, 693–700.

Suzuki, Y. J., and Packer, L. (1994c). Inhibition of NF-kB transcription factor by catechol derivatives. *Biochem. Mol. Bio. Intl.* **32**, 299–305.

Suzuki, Y. J., Tsuchiya, M., and Packer, L. (1991). Thioctic acid and dihydrolipoic acid are novel antioxidants which interact with reactive oxygen species. *Free Rad. Res. Commun.* **15**, 255–263.

Suzuki, Y. J., Tsuchiya, M., and Packer, L. (1993). Antioxidant activities of dihydrolipoic acid and its structural homologues. *Free Rad. Res. Commun.* **18**, 115–122.

Visvanathan, K. V., and Goodburn, S. (1989). Double-stranded RNA activities binding of NF-kappa B to an inducible element in the human beta-interferon promoter. *EMBO J.* **8**, 1129–1138.

Wong, G. H., Elwell, J. H., Oberley, L. W., and Goeddel, D. V. (1989). Manganous superoxide dismutase is essential for cellular resistance to cytotoxicity of tumor necrosis factor. *Cell* **58**, 923–931.

Wong, G. H. W., and Goeddel D. V. (1988). Induction of manganous superoxide dismutase by tumor necrosis factor: Possible protective mechanism. *Science* **242**, 941–944.

Yamauchi, N., Kuriyama, H., Watanabe, N., Neda, H. M., and Nitsu, Y. (1989). Intracellular hydroxyl radical production induced by recombinant human tumor necrosis factor and its implication in the killing of tumor cells in vitro. *Cancer Res.* **49**, 1671–1675.

Zimmerman, R. J., Chan, A., and Leadon, S. A. (1989). Oxidative damage in murine tumor cells treated in vitro by recombinant human tumor necrosis factor. *Cancer Res.* **49**, 1644–1648.

Russel J. Reiter

Department of Cellular and Structural Biology
The University of Texas Health Science Center at San Antonio
San Antonio, Texas 78284

Antioxidant Actions of Melatonin

I. Introduction

N-Acetyl-5-methoxytryptamine (melatonin) was isolated from the mammalian pineal gland and chemically characterized by Lerner and colleagues (1959). Its production in the pineal of all mammalian species, including humans, is regulated by the photoperiodic environment with the daily period of light being associated with minimal melatonin production, whereas during darkness the synthesis and secretion of the molecule increase markedly. Because of this light-dependent regulatory mechanism, melatonin is produced and released in a circadian rhythm in all vertebrates studied (Reiter, 1991).

Because of its marked circadian rhythm of production, the physiological levels of melatonin in bodily fluids and tissues vary over the light:dark cycle. In most species where it has been measured, levels of melatonin in the blood during the day are less than 15 pg/ml. At night, these levels increase markedly, sometimes reaching values of 150 pg/ml. Similar rhythms, but generally of a lower amplitude (50–75% less), in melatonin levels are found in a variety of other fluids, including saliva, cerebrospinal fluid, ovarian follicular fluid, and anterior eye chamber fluid. When exogenously administered, blood melatonin levels can increase several thousandfold (Menendez-Pelaez et al., 1993).

Relatively little is known about tissue levels of melatonin and its subcellular distribution. Being both highly lipid as well as relatively water soluble (Shida et al., 1994), melatonin is readily taken up by cells. Furthermore, it easily traverses the blood–brain barrier as well as other morphophysiological barriers and, therefore, it is believed to have ready access to every cell in the organism (Menendez-Pelaez and Reiter, 1993). In the nuclei of cerebro-cortical cells of rats, nighttime levels of melatonin are about 30 pg/mg DNA, three- to fourfold higher than daytime values (Menendez-Palaez et al., 1993). This day:night difference coincides with the circadian variation in blood levels of melatonin and the nighttime rise in tissue concentrations of the indole is prevented by pinealectomy. When melatonin is exogenously administered, nuclear concentrations in cerebrocortical cells rose to 150 pg/mg DNA. There is a paucity of information concerning the concentration of melatonin in other subcellular compartments.

II. Melatonin as a Free Radical Scavenger

Although well known for years to have hormonal (Bartness et al., 1993), sleep-inducing (Dawson and Encel, 1993), and chronobiological (Cassone, 1990) effects, melatonin has also been shown to have antioxidant properties. This feature of melatonin, first suggested by Ianas et al. (1991), may prove to be the phylogenetically oldest function of this ubiquitously acting molecule (Hardeland et al., 1995). It has been theorized that melatonin may have evolved coincident with oxygen-based metabolism as a protector against free radicals generated from the diradical. This would mean that the other actions of melatonin, which are in general better known, came about somewhat later in evolution.

In the initial report, published in a relatively inaccessible journal, Ianas and colleagues (1991) claimed that melatonin has both antioxidant and prooxidant actions. They generated free radicals using a combination of luminol and H_2O_2 and used chemiluminescence as an index of free radical production. In this system, melatonin reportedly served to quench free radical generation when its concentration exceeded 0.25 mM; at lower concentrations melatonin was reported to be prooxidative.

In 1993, the free radical scavenging role of melatonin was confirmed using another model system (Tan et al., 1993a). In this case, hydroxyl radicals ($\cdot$OH) were generated during the exposure of H_2O_2 to 254 nm ultraviolet light (UV), with the radicals being trapped using 5,5-dimethyl-pyrroline-N-oxide (DMPO); the resultant ·OH-DMPO adducts were identified by high-performance liquid chromatography with electrochemical detection and electron spin resonance spectroscopy. Using this system the efficacy of melatonin as a ·OH quencher was compared with that of the known scavengers glutathione (GSH) and mannitol. This test did not uncover any prooxidant activity of melatonin but did show that it was an efficient ·OH scavenger. Dose–response studies revealed that the concentrations required to neutralize 50% (IC_{50}) of the radicals generated was 21 μM for melatonin, 123 μM for GSH, and 283 μM for mannitol (Tan et al., 1993a). These findings were interpreted to mean that melatonin, at least under the conditions of this study, was a more efficient ·OH scavenger than was either GSH or mannitol. In the same report, melatonin was compared with several chemically related molecules in terms of their ·OH quenching activity; serotonin, N-acetylserotonin, or 5-methoxytryptamine did not compare favorably to melatonin as a neutralizer of ·OH generated by the photolysis of H_2O_2. These structure–activity studies suggested that the methoxy group at position 5 of the indole nucleus and the N-acetyl group on the side chain are both necessary for the efficient scavenging activity of melatonin (Fig. 1). The importance of the methoxy group to the free radical scavenging activity of melatonin has been confirmed (Scaiano, 1995).

Follow-up studies by Poeggeler et al. (1994) also showed that melatonin, based on its loss of fluorescence, is quickly oxidized in the presence of ·OH. In a system wherein $FeSO_4$ and H_2O_2 were used to generate ·OH, melatonin was quickly oxidized; by comparison, Fe^{2+} was incapable of efficiently oxidizing melatonin and likewise the indoleamine was not oxidized by its exposure to H_2O_2 alone. Like Tan and co-workers (1993a), Poeggeler et al. (1994) also observed that melatonin fluorescence was quickly lost in a solution where H_2O_2 was exposed to UV light. Finally, when the photosensitizer riboflavin was coincubated with H_2O_2, the addition of melatonin led to a rapid oxidative degradation of melatonin. These findings were interpre-

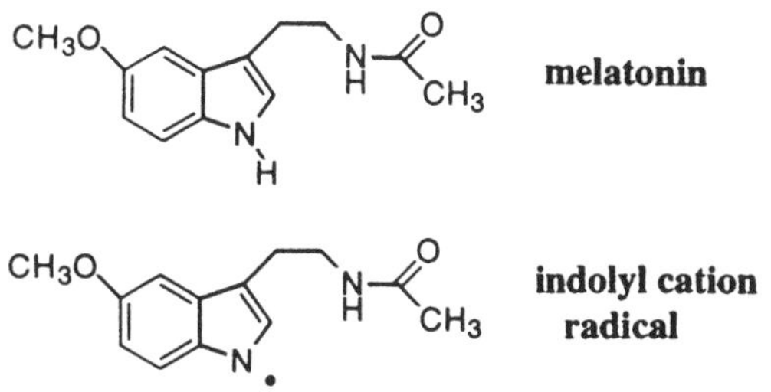

FIGURE I   Melatonin, N-acetyl-5-methoxytryptamine, is produced in and released from the pineal gland of all mammals, including humans. Although its endocrine, soporific, and chronologic properties are well known, it is also an efficient free radical scavenger and antioxidant. Upon electron donation, melatonin gives rise to the indolyl cation radical.

ted to mean that at least under *in vitro* conditions, melatonin efficiently detoxifies highly reactive oxidants by electron donation. The loss of an electron by melatonin causes melatonin itself to become a radical, namely, the indolyl cation radical (Fig. 1), which then is believed to scavenge a superoxide anion radical ($O_2\cdot^-$) (Hardeland *et al.*, 1993; Poeggeler *et al.*, 1994). Whether the indolyl cation radical can be reduced back to melatonin is currently being investigated.

A year after the first reports claiming that melatonin was a scavenger of the highly toxic ·OH, Pieri *et al.* (1994) compared the peroxyl radical (ROO·) scavenging activity of melatonin with that of trolox (water-soluble vitamin E), ascorbic acid (vitamin C), and GSH. Using β-phycoerythrin as a fluorescent indicator and 2-2′-azo-bis(2-amidopropane)dihydrochloride as a ROO· initiator, Pieri and co-workers found that melatonin > trolox > ascorbic acid > GSH in scavenging the ROO·; indeed, melatonin was about twice as effective as the water-soluble vitamin E. The following year this group used another assay system to confirm the ROO· scavenging activity of the four molecules with the same results, i.e., among the four agents tested, melatonin was the most effective in detoxifying the ROO· (Fig. 2) (Pieri *et al.*, 1995). On the basis of their results, they deduced that because melatonin was a twofold better ROO· scavenger than trolox, the indole scavenges four ROO· per molecule.

Marshall *et al.* (1996) used a battery of *in vitro* tests to examine the scavenging activity of melatonin. Their findings, consistent with the observations of Pieri *et al.* (1994, 1995), also established that melatonin is an excellent trichloromethylperoxyl radical ($CCl_3O_2\cdot$) scavenger. Furthermore, they found that although melatonin also reacts with hypochlorous acid (HOCl), the reaction is slow. Also, melatonin does not directly scavenge the $O_2\cdot^-$. More important, Marshall and co-workers (1996) reported that

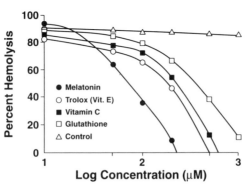

FIGURE 2   Inhibition of ROO·-induced lysis of erythrocytes by four different antioxidants. This and other work suggests that melatonin is an efficient scavenger of the ROO·. Adapted from the work of Pieri and colleagues (1995).

melatonin is devoid of prooxidative actions, at least in the series of tests they performed.

Scaiano (1995) used laser excitation of di-*tert*-butyl peroxide to initiate cleavage of the molecule into two *tert*-butoxyl radicals ($Bu^tO\cdot$). Melatonin was then shown to interact with the radical, thereby neutralizing it and generating what Scaiano referred to as an indolyl-like radical, whose proposed structure is identical to that believed to be generated when melatonin contributes an electron to $\cdot OH$ (Hardeland *et al.*, 1993). Scaiano (1995) also found that melatonin scavenges the cumyloxyl radical. The rate constants for the interaction of melatonin with the *tert*-butoxyl and cumyloxyl radical were calculated to be 3.4 and $6.7 \times 10^7\ M^{-1}\ sec^{-1}$, respectively. This high reactivity of melatonin with free radicals supports the idea that melatonin is likely an excellent antioxidant *in vivo* (Reiter *et al.*, 1995).

Poeggeler and co-workers (1995) investigated the antioxidative properties of melatonin, trolox, GSH, and ascorbate using the radical trapping agent, 2,2′-azinobis-3-ethylbenzothiazoline-6-sulfuric acid (ABTS); the aim of these studies was to determine if melatonin works synergistically with the phenolic and thiolic antioxidants in scavenging the $ABTS^{\cdot+}$ when ABTS is coincubated with H_2O_2 and $FeSO_4$. The results showed in fact that melatonin, which works by electron donation, and the chain-breaking antioxidants act in concert to detoxify the $ABTS^{\cdot+}$. The implication of these findings is that *in vivo*, where melatonin coexists with the common chain-breaking antioxidants, melatonin may cooperatively interact in the antioxidative defense of the cell.

III. Melatonin as an Antioxidant

A. Antioxidative Protection against DNA Damage *in Vivo*

Nuclear DNA is readily oxidized by toxic-free radicals generated intracellularly by a variety of processes. After Tan *et al.* (1993a) described the efficient radical scavenging activity of melatonin in a cell-free system, they immediately undertook studies to test its effectiveness as an antioxidant *in vivo*. When rats are treated with the carcinogen safrole, the resulting damage to nuclear DNA is believed to be, at least in part, a result of the generation of free radicals by the xenobiotic. This being the case, Tan and colleagues (1993b) proposed that the coadministration of melatonin with safrole would protect the DNA against radical-related damage. Rats treated with 300 mg/ kg safrole exhibited extensive hepatic DNA adduct products 24 hr later. When melatonin was given in doses 1500-fold (0.2 mg/kg) and 750-fold (0.4 mg/kg) less than that of the safrole, it diminished DNA adduct formation by >50 and >90%, respectively (Tan *et al.*, 1993b). Thus, at least at pharmacological levels, albeit at much lower doses than that of the carcino-gen, melatonin was effective in protecting nuclear DNA from the damaging effects of safrole. Accepting that this protection was a consequence of the detoxification of free radicals, melatonin obviously gets into the nucleus in sufficiently high concentrations to provide on-site protection and to reduce radical damage to DNA (Menendez-Pelaez and Reiter, 1993).

As a follow-up to these pharmacological investigations, Tan *et al.* (1994) also performed tests that took advantage of the circadian rhythm of melato-nin production. Because blood and tissue melatonin levels are normally low during the day and high at night (Reiter, 1991), rats were injected either early in the light phase or early in the dark phase with 100 mg/kg safrole; 8 hr later, hepatic DNA adduct formation was estimated. The animals treated with safrole at night, when physiological levels of melatonin were elevated, had less DNA damage than rats given safrole during the day, when tissue melatonin levels are reduced. Likewise, they reported that rats that were pinealectomized, a procedure that greatly reduces their physiological melatonin levels, exhibited the highest hepatic DNA damage after safrole administration. Finally, giving supplemental melatonin greatly reduced he-patic DNA adduct formation that resulted from safrole injection. These findings led Tan and co-workers (1994) to conclude that melatonin, even at amounts normally produced in the organism, is sufficient to provide significant antioxidative protection.

B. Antioxidative Protection against DNA Damage
in Vitro

Ionizing radiation is known to damage DNA due to the generation of free radicals as a consequence of the homolytic scission of H_2O. Vijayalaxmi *et al.* (1995a,b) took advantage of this information and surmised that melato-nin, because of its free radical-neutralizing capacity, may protect against the damage resulting from high energy radiation. Human peripheral blood lymphocytes were isolated and some were incubated for 20 min with concen-trations of melatonin ranging from 0.5 to 2 mM. Thereafter, melatonin-treated and nonmelatonin-treated cells were exposed to 150 cGy γ-radiation and damage to nuclear DNA was estimated using a variety of cytogenetic parameters, i.e., exchange aberrations, acentric fragments, and micronuclei. For each index measured, melatonin, in a dose–response manner, diminished DNA damage, which was a consequence of free radicals generated by the radiation (Vijayalaxmi *et al.*, 1995a,b). At a concentration of 2 mM, melato-nin diminished nuclear damage by roughly 70%; an equivalent protection against ionizing radiation-induced damage in this study was achieved by a 500 times greater dose (1 M) of dimethyl sulfoxide (DMSO), a known radioprotective agent.

The results suggest either that melatonin is a much better radioprotective agent or that it gets into the nucleus 500 times more efficiently than does DMSO. In either case, considering the lack of toxicity of melatonin and the ease with which it is absorbed and enters cells (Menendez-Pelaez and Reiter, 1993), its potential utility as a radioprotector should not be overlooked. This also is emphasized by the observations of Blinkenstaff and co-workers (1994) who reported that mice exposed to 950 cGy ionizing radiation had a much better 30-day survival rate when they were pretreated with melatonin or melatonin homologs compared to untreated animals. Also, it has been found that sufficiently high blood and tissue levels of melatonin can be achieved by its oral administration such that lymphocytes collected from these individuals are protected ($>60\%$) from the cytogenetic damage induced by 150 cGy ionizing radiation (Vijayalaxmi *et al.*, unpublished results).

C. Antioxidative Protection against Lipid Peroxidation *in Vivo*

In light of the findings of Pieri and colleagues (1994, 1995) and Marshall *et al.* (1996) that melatonin is a potent ROO· scavenger, it would be predicted that the indole may afford protection against free radical-induced lipid peroxidation. Studies to test this possibility *in vivo* have been carried out and the findings uniformly suggest an antioxidative action of melatonin in lipid-rich environments. Melchiorri *et al.* (1995a) used the highly toxic herbicide paraquat to induce oxidative damage to the lungs and liver of rats. The end points used in these studies included tissue levels of malondialdehyde (MDA) and 4-hydroxyalkenals (4HDA) and total and oxidized glutathione (GSSG) levels. Paraquat, at doses of either 20 or 70 mg/kg, induced large rises in tissue MDA + 4HDA levels as well as a drop in total glutathione and a rise in GSSG concentrations within 2 hr after its administration; these are all changes indicative of increased oxidative stress. Melatonin (10 mg/kg), given prior to paraquat and at 6-hr intervals thereafter, counteracted the oxidative changes measured as a consequence of paraquat toxicity. Subsequent studies by the same group (Melchiorri *et al.*, 1996) also showed that the 24-hr LD_{50} for paraquat was increased threefold in rats cotreated with melatonin.

The endotoxin lipopolysaccharide (LPS), which is produced by gram-negative bacteria, induces lipid peroxidation in a variety of tissues via free radical mechanisms; these effects are typically counteracted by conventional antioxidants such as coenzyme Q_{10} and α-tocopherol. Considering the reported antioxidative potential of melatonin, Sewerynek and co-workers (1995a,b) tested the ability of this indole to combat LPS-induced oxidative changes *in vivo*. Oxidative damage to the liver of rats was evaluated by measuring hepatic MDA + 4HDA levels, changes in total glutathione and

GSSG, and cellular morphology. Bacterial LPS (10 mg/kg) significantly increased hepatic MDA + 4HDA levels and GSSG concentrations at 6 hr after its administration with the magnitude of the changes either being reduced or totally inhibited by melatonin treatment (Sewerynek *et al.*, 1995b). Morphologically, the liver of rats treated with LPS exhibited cellular damage and a marked infiltration of the tissue by inflammatory leukocytes, a response typical of LPS-damaged tissue. Again, melatonin attentuated these responses. In general, the protective effects of melatonin against LPS-induced oxidative changes *in vivo* seem somewhat less pronounced than those observed in rats treated with paraquat (Melchiorri *et al.* 1995a).

Carbon tetrachloride (CCl_4) is a prooxidative agent commonly used to experimentally induce oxidative damage. The liver is particularly sensitive to CCl_4 toxicity with a cascade of free radical events being mediated by the metabolic products of CCl_4, the trichloromethyl and the trichloromethylperoxyl radical. Lipid peroxidation in a number of organs is a consequence of CCl_4 administration. When rats were given an intraperitoneal injection of CCl_4 (5 ml/kg), melatonin was not effective in preventing the rise in lipid peroxidation products in the liver, although it did so in the kidneys (Daniels *et al.*, 1995). The authors assumed that the amount of melatonin administered (10 mg/kg) was insufficient in overcoming hepatic lipid peroxidation because CCl_4 is so readily taken up by the liver while in the kidneys; where CCl_4 is not so heavily concentrated, melatonin was effective.

D. Antioxidative Protection against Lipid Peroxidation
in Vitro

Pierrefiche and colleagues (1993) compared the ability of several indoles as inhibitors of lipid peroxidation in mouse brain homogenates. They found that melatonin was only a weak antioxidant under these conditions and that the chief hepatic metabolite of melatonin, 6-hydroxymelatonin, was in fact more effective in limiting the peroxidation of lipids. Although this observation requires confirmation, the findings do suggest that part of the antioxidative activity of melatonin *in vivo* may be due to its conversion to this hepatic metabolite.

Subsequent studies have shown that melatonin markedly limits oxidative damage to lipids under *in vitro* conditions. Melchiorri *et al.* (1995b) found melatonin to diminish, in a dose–response manner, the accumulation of MDA + 4HDA in rat brain homogenates treated with kainic acid. Kainic acid is a nondegradable analog of the excitatory neurotransmitter, which is highly neurotoxic. When injected into animals, kainic acid typically induces seizures and causes extensive neuronal damage. In the study of Melchiorri and colleagues (1995b), homogenates of cerebral cortex, cerebellum, hippocampus, hypothalamus, and striatum collected from both Wistar and Sprague–Dawley rats exhibited marked lipid peroxidation when incubated with 11.7 mM kainic acid; melatonin over a range of doses (0.1–4 mM) suppressed the accumulation of the products of lipid peroxidation. In some cases the highest concentration of melatonin reduced MDA + 4HDA levels in kainic acid-treated homogenates below those in control samples not incubated with kainic acid. Although the findings do not prove that melatonin's protection was a result of its scavenging of free radicals, it is likely that a portion of its beneficial effects were a consequence of this property of the indole (Reiter, 1995a).

Studies similar to those Melchiorri *et al.* (1995b), described earlier, were carried out by Sewerynek and colleagues (1995c) and Chen *et al.* (1995) using a different free radical generating scheme. The reactive oxygen intermediate H_2O_2, although not generally considered to be highly toxic, does in the presence of transition metals, e.g., Fe^{2+} or Cu^{1+}, generate the devastatingly damaging ·OH. Thus, H_2O_2 is often used as a means to experimentally induce oxidative damage, and it was utilized by Sewerynek *et al.* (1995c) on rat brain homogenates to increase lipid peroxidation as indicated by MDA + 4HDA levels. Incubating brain homogenates with 5 mM H_2O_2 led to large increases in MDA + 4HDA concentrations presumably related in part to the ability of melatonin to directly scavenge the ·OH (Tan *et al.*, 1993a); melatonin in a dose–reponse manner suppressed the production of free radical-damaged lipid products. Studies using rat retinal homogenates resulted in a similar protective action of melatonin against free radical attack (Chen *et al.*, 1995). When retinal homogenates were incubated with a combination of H_2O_2 and $FeSO_4$, oxidatively damaged lipid products dramatically increased in a time-dependent manner with melatonin, depending on concentration, either totally or partially inhibiting the oxidative process.

CCl_4, as noted earlier, is also a powerful generator of oxidative damage. When hepatic homogenates or microsomes were incubated with 0.5 mM CCl_4, lipid peroxidation was elevated in these preparations; the addition of melatonin prevented CCl_4 from inducing rises in MDA + 4HDA (Daniels *et al.*, 1995). The authors believed that the protective effect of melatonin in these studies was a consequence of its free radical scavenging capacity.

E. Other Evidence Supporting the Antioxidative Function of Melatonin

Giusti *et al.* (1995) tested the ability of melatonin to protect cultured cerebellar neurons from the excitotoxic effects of glutamate and kainic acid. Cellular viability, as determined by quantitative staining with 3-(4,5-dimethylthiazol-2-yl)-2,5-diphenyl tetrazolium bromide, was increased as a consequence of the cotreatment of cells with the neurotoxic agents and melatonin. Interestingly, however, whereas melatonin protected cerebellar neurons from kainic acid and glutatmate-induced damage, it was not effective in protecting the cells from N-methyl-D-aspartate excitotoxicity. Also, melatonin was only effective as a protective agent against kainic acid when the compounds were simultaneously added to the culture mediate, but not if the cells were pretreated or posttreated with melatonin. With specific tests, the authors eliminated several mechanisms by which the protective actions of melatonin could have been mediated and they concluded the indole was most likely acting as an antioxidant in its protective action on cerebellar neurons (Giusti *et al.*, 1995).

The same workers followed their *in vitro* studies with an *in vivo* experiment where rats were treated with kainic acid alone on in combination with melatonin. Treatment of rats with kainic acid alone was followed by neurobehavioral changes as well as morphological and biochemical alterations in the brain (Giusti *et al.*, 1996). The masticatory movements, head nodding, and "wet-dog" shakes normally associated with kainic acid treatment were prevented when melatonin was administered. Also, the kainic acid-induced pyramidal cell loss in the hippocampus was eliminated by melatonin as was the associated drop in brain norepinephrine and the rise in 5-hydroxylindoleacetic acid levels. Using brain synaptosomes, these workers also fluorimetrically measured the formation of reactive oxygen species and found that they were diminished by melatonin. Thus, Giusti and colleagues (1996) concluded that melatonin protects the brain from kianic acid, at least in part due to its direct free radical scavenging ability.

Singlet oxygen ($^1O_2$) is a reactive molecule that is produced in cells upon the electronic excitation of O_2. It is generally accepted the $^1O_2$ is cytotoxic. Cagnoli *et al.* (1995) incubated cerebellar granule cells with a photosensitive dye, rose bengal, and then exposed the cultures to light, a process that generates $^1O_2$. Using this system, these workers found that melatonin inhibited DNA fragmentation, a marker of apoptosis, and cell death, which were induced by $^1O_2$. The findings, according to the authors (Cagnoli *et al.*, 1995), are consistent with the ability of melatonin to directly neutralize $^1O_2$ toxicity.

Sainz and colleagues (1995) have also found melatonin to be antipoptotic. In both *in vivo* and *in vitro* studies, apoptosis was induced in the thymus by exposing the cells to the glucocorticoid dexamethasone. The end point measured included increased levels of DNA fragmentation and increased numbers of apoptotic cells, both of which were in part inhibited when melatonin was also given. One of several explanations given for this effect of melatonin was its ability to scavenge free radicals that initiate apoptotic events.

In another indirect assessment of the potential antioxidative action of melatonin, Pentney and Bubenik (1995) induced colitis in mice by providing dextran sodium sulfate (DSS) in the drinking water. The severity of the resulting colitis, evaluated by diarrhea and occult blood in the feces as well as morphological lesions of the gastrointestinal mucosa, was significantly reduced by the daily intraperitoneal injection of melatonin. Whereas melatonin may have protected the mice from colitis via several mechanisms, it is possible that the ability of melatonin to scavenge toxic radicals may have been involved. Feeding mice DDS induces the overgrowth of *Escherichia coli* in the intestine; these bacteria are known to produce the endotoxin, LPS, which generates free radicals that in turn cause damage to the mucosal lining of the gut. Melatonin, as shown by Sewerynek *et al.* (1995b,d) is known to reduce LPS-induced oxidative damage, and the findings of Pentney and Bubenik (1995) may also be, in part, explainable on the basis of the direct scavenging action of melatonin.

At the molecular level, melatonin has been shown to inhibit the activation of the transcriptional regulator nuclear factor (NF)-κB, whose activation involves free radicals as second messengers. Using HeLa S3 cells, Mohan and colleagues (1995) stimulated NF-κB binding to DNA by exposing the cells to tumor necrosis factor-α, phorbol 12-myristate 13-acetate, or ionizing radiation, all of which generate intracellular free radicals. Exogenously added melatonin, at a concentration as low as 10 μM, inhibited the activation of NF-κB by each of these agents, with the mechanism of the inhibitory effect of melatonin probably involving the ability of indole to neutralize free radicals.

IV. Effect of Melatonin on Enzymes Related to the Antioxidative Defense System

Exogenously administered melatonin stimulated the activity of an important antioxidative enzyme in the brain, glutathione peroxidase (GPx) (Barlow-Walden *et al.*, 1995). In the study in question, pharmacological levels of melatonin (500 mg/kg) increased whole brain GPx activity at 30 and 180 min after its intraperitoneal administration; the rise in GPx activity was correlated with a similar increase in brain concentrations of melatonin. The action of melatonin may have been via the recently discovered nuclear melatonin receptors (Acuña-Castroviejo *et al.*, 1994). The effect of melatonin on GPx activity does not seem to be exclusively a pharmacological phenomenon because Pablos and colleagues (1995) have also found, in chicks, a rise in nighttime levels of brain GPx activity with the nocturnal increase in enzyme activity disappearing when the birds are kept in light at night; light during the normal dark period prevents the usual increase in melatonin production and secretion (Reiter, 1991). By stimulating GPx activity, melatonin would promote the metabolism of H_2O_2 and hydroperoxides, thereby reducing the generation of toxic-free radicals.

According to Pierrefiche and Laborit (1995), melatonin also stimulates hepatic and cerebral glucose-6-phosphate dehydrogenase (G6PDH) activity in mice, thereby increasing NADPH levels; in so doing, melatonin promotes the enzymatic conversion, via glutathione reductase, of GSSG to glutathione. Glutathione is a necessary cofactor for GPx.

Melatonin also reportedly regulates the activity of 5-lipoxygenase, an enzyme that is restricted to granulocytes, monocytes/macrophages, mast cells, and B lymphocytes (Carlberg and Weisenberg, 1995). This action of melatonin was shown to involve a nuclear melatonin receptor. The action of melatonin in reference to 5-lipoxygenase activity is inhibitory. 5-Lipoxygenase catalyzes the conversion of arachadonic acid to leukotrienes that are involved in a variety of pathophysiological conditions, including inflammatory bowel disease. The suppressive effect of melatonin on colitis, described earlier (Pentney and Bubenik, 1995), could in part be via a modulation of 5-lipoxygenace activity as well as via a direct free radical scavenging activity. The decrease in 5-lipoxygenase expression in B lymphocytes induced by melatonin would also suggest a potentially important role for this indole in inflammatory and immune processes.

Finally, melatonin has been found to suppress the activity of nitric oxide synthase (NOS) probably via a mechanism involving its binding to calmodulin (CaM) (Pozo $et\ al.$, 1994). Considering that nitric oxide interacts with O_2 to generate peroxynitrite anion, which can degrade into the $\cdot OH$, the inhibitory effect of melatonin on NOS may also contribute to its antioxidative capability.

V. Concluding Remarks

Only a few years ago the role of melatonin as an antioxidant was unknown. Collectively, data indicate a role for melatonin in the antioxidative defense system of the organism. Whether the actions of melatonin reported herein are merely pharmacological or whether they have physiological relevance remains to be clarified (Reiter $et\ al.$, 1995). The preliminary findings of Tan $et\ al.$ (1994) and Pablos $et\ al.$ (1995) suggest that melatonin is protective against free radical damage even at physiological concentrations. Even if only pharmacologically relevant, however, the findings have important implications in that the molecule has no known toxicity and is readily absorbed when administered via any route.

The possibility that melatonin may be involved in age-related free radical diseases and aging itself has received some support (Reiter, 1995b,c). Of particular interest is the possible use of melatonin as a protective agent against free radical-related neurodegenerative diseases. This specifically relates to the fact that melatonin, when exogenously administered, rapidly gets into the brain (Menendez-Pelaez $et\ al.$, 1993) where it acts as both a direct and an indirect scavenger (Tan $et\ al.$, 1993a; Pieri $et\ al.$, 1994; Barlow-Walden $et\ al.$, 1995). Also, because of the marked protective action of melatonin against environmental toxins such as paraquat (Melchiorri $et\ al.$,

1995a, 1996) and CCl$_4$ (Daniels *et al.*, 1995), it may have utility in the treatment after exposure to these agents. Finally, melatonin has demonstrated an ability to greatly diminish free radical damage that is a consequence of ionizing radiation (Vijayalaxmi *et al.*, 1995a,b) and, considering its inhibitory effects on NF-κB (Mohan *et al.*, 1995), it could be considered as a potential treatment for individuals infected with human immunodeficiency virus. These potential uses are also suggested by the virtual absence of toxicity of melatonin.

One consistent feature of endogenous melatonin production in the pineal gland is that it drops steadily throughout life such that, in most old animals (including human), blood and tissue levels are low. Considering the potential role of free radicals in the processes of aging, the loss of melatonin could be a factor contributing to the declining functional integrity of cellular organelles, cells, organs, and organ systems as organisms age. Whether supplemental melatonin would avert age-related degenerative processes is currently being investigated.

References

Acuña-Castroviejo, D., Reiter, R. J., Menendez-Pelaez, A., Pablos, M. I., and Burgos, A. (1994). Characterization of high-affinity melatonin binding sites in purified cell nuclei of rat liver. *J. Pineal Res.* **16**, 100–112.

Barlow-Walden, L. R., Reiter, R. J., Abe, M., Pablos, M., Menendez-Pelaez, A., Chen, L. D., and Poeggeler, B. (1995). Melatonin stimulates brain glutathione peroxidase activity. *Neurochem. Int.* **26**, 497–502.

Bartness, T. J., Powers, J. B., Hastings, M. H., Bittman, E. L., and Goldman, B. D. (1993). The timed infusion paradigm for melatonin delivery: What has it taught us about the melatonin signal, its reception, and the photoperiodic control of seasonal responses? *J. Pineal Res.* **15**, 151–190.

Blinkenstaff, R. T., Brandstadter, S. M., Reddy, S., and Witt, R. (1994). Potential radioprotective agents. I. Homologs of melatonin. *J. Pharm. Sci.* **83**, 493–498.

Cagnoli, C. M., Atabay, C., Kharlamova, E., and Manev, H. (1995). Melatonin protects neurons from singlet oxygen-induced apoptosis. *J. Pineal Res.* **18**, 222–226.

Carlberg, C., and Weisenberg, I. (1995). The orphan receptor family RZR/ROR, melatonin and 5-lipoxygenase: An unexpected relationship. *J. Pineal Res.* **18**, 171–178.

Cassone, V. M. (1990). Effect of melatonin on vertebrate circadian systems. *Trends Neurosci.* **13**, 457–464.

Chen, L. D., Melchiorri, D., Sewerynek, E., and Reiter, R. J. (1995). Retinal lipid peroxidation *in vitro* in inhibited by melatonin. *Neurosci. Res. Commun.* **17**, 151–158.

Daniels, W. M. W., Reiter, R. J., Melchiorri, D., Sewerynek, E., Pablos, M. I., and Ortiz, G. G. (1995). Melatonin counteracts lipid peroxidation induced by carbon tetrachloride but does not restore glucose-6-phosphatase activity. *J. Pineal Res.* **19**, 1–6.

Dawson, D., and Encel, N. (1993). Melatonin and sleep in humans. *J. Pineal Res.* **15**, 1–12.

Giusti, P., Gusella, M., Lipartiti, M., Milani, D., Zhu, W., Vicini, S., and Manev, H. (1995). Melatonin protects primary cultures of cerebellar granule neurons from kainate but not from N-methyl-D-aspartate excitatoxicity. *Exp. Neurol.* **131**, 39–46.

Giusti, P., Lipartiti, M., Franceschini, D., Schiavo, N., Floreani, M., and Manev, H. (1996). Neuroprotection by melatonin from kainate-induced excitotoxicity in rats. *FASEB. J.* 10, 891–896.

Hardeland, R., Reiter, R. J., Poeggeler, B., and Tan, D. X. (1993). The significance of the metabolism of the neurohormone melatonin: Antioxidative protection and formation of bioactive substances. *Neurosci. Biobehav. Rev.* 17, 347–357.

Hardeland, R., Balzer, I., Poeggeler B, Fuhrberg, B., Uria, H., Behrmann, G., Wolf, R., Meyer, T. J., and Reiter, R. J. (1995). On the primary functions of melatonin in evolution: Mediation of photoperiodic signals in a unicell, photoxidation and scavenging of free radicals. *J. Pineal Res.* 18, 104–111.

Ianas, O., Olnescu, R., and Badescu, I. (1991). Melatonin involvement in oxidative stress. *Rom. J. Endocrinol.* 29, 147–153.

Lerner, A. B., Case, J. B., and Heinzelman, R. V. (1959). Structure of melatonin. *J. Am. Chem. Soc.* 81, 6084–6085.

Marshall, K. A., Reiter, R. J., Poeggeler, B., Aruoma, O. I., and Halliwell, B. (1996). Evaluation of the antioxidant activity of melatonin *in vitro. Free Rad. Biol. Med.,* in press.

Melchiorri, D., Reiter, R. J., Attia, A. M., Hara, M., Burgos, A., and Nistico, G. (1995a). Potent protective effect of melatonin on *in vivo* paraquat-induced oxidative damage in rats. *Life Sci.* 56, 83–89.

Melchiorri, D., Reiter, R. J., Sewerynek, E., Chen, L. D., and Nistico, G. (1995b). Melatonin reduces kainate-induced lipid peroxidation in homogenates of different brain regions. *FASEB J.* 9, 1205–1210.

Melchiorri, D., Reiter, R. J., Sewerynek, E., Hara, M., Chen, L. D., and Nistico, G. (1996). Paraquat toxicity and oxidative damage: Reduction by melatonin. *Biochem. Pharmacol.* 51, 1095–1099.

Menendez-Palaez, A., and Reiter, R. J. (1993). The distribution of melatonin in mammalian tissues: The relative importance of nuclear versus cytosolic localization. *J. Pineal Res.* 15, 59–69.

Menendez-Pelaez, A., Poeggeler, B., Reiter, R. J., Barlow-Walden, L. R., Pablos, M. I., and Tan, D. X. (1993). Nuclear localization of melatonin in different mammalian tissues; immunological and radioimmuno-assay evidence. *J. Cell. Biochem.* 53, 373–382.

Mohan, N., Sadeghi, K., Reiter, R. J., and Meltz, M. (1995). The neurohormone melatonin inhibits cytokine, mitogen and ionizing radiation induced NF-κB. *Biochem. Mol. Biol. Int.* 37, 1063–1070.

Pablos, M. I., Agapito, M. T., Gutierrez, R., Recio, J. M., Reiter, R. J., Barlow-Walden, L. R., Acuña-Castroviejo, D., and Menendez-Pelaez, A. (1995). Melatonin stimulates the activity of the detoxifying enzyme glutathione peroxidase in several tissues of chicks. *J. Pineal Res.,* 19, 111–115.

Pentney, P. T., and Bubenik, G. A. (1995). Melatonin reduces the severity of dextran-induced colitis. *J. Pineal Res.* 19, 31–39.

Pieri, C., Marra, M., Moroni, F., Recchioni, R., and Marcheselli, F. (1994). Melatonin: A peroxyl radical scavenger more effective than vitamin E. *Life Sci.* 55, PL271–PL276.

Pieri, C., Moroni, F., Marra, M., Marcheselli, F., and Ricchioni, R. (1995). Melatonin as an efficient antioxidant. *Arch. Gerontol. Geriatr.* 20, 159–165.

Pierrefiche, G., and Laborit, H. (1995). Oxygen radicals, melatonin, and aging. *Exp. Gerontol.* 30, 213–227.

Pierrefiche, G., Topall, G., Courbain, I., and Henriet, H. (1993). Antioxidant activity of melatonin in mice. *Res. Commun. Chem. Pathol. Pharmacol.* 80, 211–223.

Poeggeler, B., Reiter, R. J., Hardeland, R., Sewerynek, E., Melchiorri, D., and Barlow-Walden, L. R. (1995). Melatonin, a mediator of electron transfer and repair reactions, acts synergistically with the chain-breaking antioxidants ascorbate, trolox, and glutathione. *Neuroendocrinol. Lett.* 17, 87–92.

Poeggeler, B., Saarela, S., Reiter, R. J., Tan, D. X., Chen, L. D., Manchester, L. C., and Barlow-Walden, L. R. (1994). Melatonin—a highly potent endogenous radical scavenger and electron donor: New aspects of the oxidation chemistry of this indole assessed *in vitro*. *Ann. N.Y. Acad. Sci.* **738**, 419–420.

Pozo, D., Reiter, R. J., Calvo, J. A., and Guerrero, J. M. (1994). Physiological concentrations of melatonin inhibit nitric oxide synthase in rat cerebellum. *Life Sci.* **55**, PL455–PL460.

Reiter, R. J. (1991). Pineal melatonin: Cell biology of its synthesis and of its physiological interactions. *Endocr. Rev.* **12**, 151–180.

Reiter, R. J. (1995a). Oxidative processes and antioxidative defense mechanisms in the aging brain. *FASEB J.* **9**, 526–533.

Reiter, R. J. (1995b). Functional pleiotropy of the neurohormone melatonin: Antioxidant protection and neuroendocrine regulation. *Front. Neuroendocrinol.* **16**, 383–415.

Reiter, R. J. (1995c). The pineal gland and melatonin in relation to aging: A summary of the theories and of the data. *Exp. Gerontol.* **30**, 199–212.

Reiter, R. J., Melchiorri, D., Sewerynek, E., Poeggeler, B., Barlow-Walden, L. R., Chuang, J. I., Ortiz, G. G., and Acuña-Castroviejo, D. (1995). A review of the evidence supporting melatonin's role as an antioxidant. *J. Pineal Res.* **18**, 1–11.

Sainz, R. M., Mayo, J. C., Uria, H., Kolter, M., Antolin, I., Rodriquez, C., and Menendez-Pelaez, A. (1995). The pineal neurohormone melatonin prevents in vivo and in vitro apoptosis in thymocytes. *J. Pineal Res.* **19**, 178–188.

Scaiano, J. C. (1995). Exploratory laser flash photolysis study of free radical reactions and magnetic field effects in melatonin chemistry. *J. Pineal Res.* **19**, 189–195.

Sewerynek, E., Abe, M., Reiter, R. J., Barlow-Walden, L. R., Chen, L. D., McCabe, T. J., Roman, L. J., and Diaz-Lopez, B. (1995a). Melatonin administration prevents lipopolysaccharide-induced oxidative damage in phenobarbital-treated animals. *J. Cell. Biochem.* **58**, 436–444.

Sewerynek, E., Melchiorri, D., Chen, L. D., and Reiter, R. J. (1995b). Melatonin reduces both basal and bacterial lipopolysaccharide-induced lipid peroxidation *in vitro*. *Free Rad. Biol. Med.* **19**, 903–909.

Sewerynek, E., Melchiorri, D., Ortiz, G. G., Peoggeler, B., and Reiter, R. J. (1995c). Melatonin reduces H_2O_2 induced lipid peroxidation in homogenates of different rat brain regions. *J. Pineal Res.* **19**, 51–56.

Sewerynek, E., Melchiorri, D., Reiter, R. J., Ortiz, G. G., and Lewinski, A. J. (1995d). Lipopolysaccharide-induced hepatotoxicity is inhibited by the antioxidant melatonin. *Eur. J. Pharmacol.* **293**, 327–334.

Shida, C. S., Castrucci, A. M. L., and Lamy-Freund, M. T. (1994). High melatonin solubility in aqueous medium. *J. Pineal Res.* **16**, 198–201.

Tan, D. X., Chen, L. D., Poeggeler, B., Manchester, L. C., and Reiter, R. J. (1993a). Melatonin: A potent, endogenous hydroxyl radical scavenger. *Endocr. J.* **1**, 57–60.

Tan, D. X., Poeggeler, B., Reiter, R. J., Chen, L. D., Chen, S., Manchester, L. C., and Barlow-Walden, L. R. (1993b). The pineal hormone melatonin inhibits DNA adduct formation induced by the chemical carcinogen safrole *in vivo*. *Cancer Lett.* **70**, 65–71.

Tan, D. X., Reiter, R. F., Chen, L. D., Poeggeler, B., Manchester, L. C., and Barlow-Walden, L. R. (1994). Both physiological and pharmacological levels of melatonin reduce DNA adduct formation induced by the carcinogen safrole. *Carcinogenesis* **15**, 215–218.

Vijayalaxmi, Reiter, R. J., and Meltz, M. (1995a). Melatonin protects human blood lymphocytes from radiation-induced chromosome damage. *Mut. Res.* **346**, 23–31.

Vijayalaxmi, Reiter, R. J., Sewerynek, E., Poeggeler, B., Leal, B. Z., and Meltz, M. (1995b). Marked reduction of radiation-induced micronuclei in human blood lymphocytes pre-treated with melatonin. *Radiat. Res.* **143**, 102–106.

Erik Løvaas

The Norwegian College of Fishery Science
University of Tromsø
9037 Tromsø, Norway

Antioxidative and Metal-Chelating Effects of Polyamines[1]

I. Introduction

During the past decade evidence has accumulated indicating that poly-amines are potent antioxidants and anti-inflammatory agents, protecting indispensable cellular components such as nucleic acids and polyunsaturated fatty acids from oxidative damage. The purpose of this chapter is to summa-rize present knowledge on the antioxidative effects of polyamines and to draw attention toward possible pharmacological applications.

Polyamines are essential for life and participate in a bewildering number of seemingly unrelated processes. Because of their cationic nature they tend to interact electrostatically with anions: In some situations they mimic the effects of simple cations like Ca^{2+} and Mg^{2+} by competing for the same binding sites on receptors, enzymes, and membranes. Poly-

---

[1] Dedicated to the memory of David E. Green: Long gone but not forgotten.

Advances in Pharmacology, Volume 38

amines may thus modulate signal transduction and enzyme activity. For example, polyamines compete with Ca^{2+} for the anionic sites on phospholipid membranes, thereby modulating the transduction and transmission of receptor-mediated signals (Koenig *et al.*, 1983). Polyamines also modulate the activities of key regulatory enzymes of signal transduction and of the cell cycle by lowering the K_m value for phosphate substrates (Nordlie *et al.*, 1979). Furthermore, polyamines change membrane fluidity by electrostatically cross-linking membrane components and thus indirectly modify a number of membrane activities simultaneously. Polyamines, and in particular spermine, may thus have gross regulatory functions on cell and organelle metabolism.

Polyamines also bind cations like Cu and Fe, and by this act as cellular protectors. The inherent attribute of polyamines as metal chelators has generally been overlooked by biologists. It is, however, well known that other nitrogen-containing compounds such as porphyrins and chlorophylls form strong complexes with Co, Ni, Fe, and Cu. The chelate effect results from the presence of a number of complexing groups in the same molecule, and polyamines with a high number of nitrogen groups will thus act as a stronger chelator.

The main thesis of this chapter is that the antioxidative effect of polyamines is due to a combination of their anion- and cation-binding properties. The binding of polyamines to anions (phospholipid membranes, nucleic acids) contributes to a high local concentration at cellular sites particularly prone to oxidations, whereas the binding to cations efficiently prevents the site-specific generation of "active oxygen" (i.e., hydroxyl radicals and singlet oxygen).

For the purpose of this chapter, antioxidants are defined as molecules which intercept or prevent free radical chain reactions. This definition includes metal-chelating molecules such as ceruloplasmine, ferritin, desferoxamin, and polyamines, but excludes regulatory and repair functions of polyamines which counteract oxidative stress. For example, polyamines are known to modulate the activity of NADPH oxidase and thus reduce the oxidative burst of polymorphonuclear macrophages. This regulatory function cannot be classified as an antioxidative effect by the common usage of the word, but the function certainly counteracts oxidative stress. Similarly, polyamines are known to stimulate DNA repair enzymes as well as to stabilize DNA, and polyamines may thus protect against mutagenesis. Also, this regulatory function protects against oxidative stress, but not by a mechanism which can be classified as "antioxidative." However, because the intention of this contribution is to summarize the effects of polyamines with respect to protection against oxidative stress, some reference will be made to the regulatory functions of polyamines as well.

II. Chemistry and Type Reactions of Polyamines _____

Polyamines are naturally occurring unbranched aliphatic amines, with molecular weights ranging from 88 (putrescine) to 202 (spermine).

Putrescine: $NH_2CH_2CH_2CH_2CH_2NH_2$
Spermidine: $NH_2CH_2CH_2CH_2NHCH_2CH_2CH_2CH_2NH_2$
Spermine: $NH_2CH_2CH_2CH_2NHCH_2CH_2CH_2CH_2NHCH_2CH_2CH_2CH_2NH_2$

At physiological pH, putrescine, spermidine, and spermine are protonated and possess two, three, and four charges, respectively. There is free rotation around each bond, giving rise to a great number of conformations. Polyamines are soluble both in water and in a number of organic solvents. In aqueous solution they are highly solvated and exist in an extended conformation. In the extended state, spermine has a length of approximately 16 Å, whereas in the compact "curled" state the diameter is 6.5 Å with a central core of about 4.1 Å. Exogeneously administered polyamines readily enter a diversity of cell types, often by active uptake. Because of their high solubility in both lipid and water they probably distribute to all cellular compartments. This lack of compartmentalization makes polyamines unique among antioxidants; they are confined neither to lipophilic environments (like vitamin E) nor to aqueous environments (like vitamin C).

A. Interaction with Anions

Polyamines interact electrostatically with a number of organic anions modulating both the tertiary structure and the biological activity of proteins and nucleic acids (Vertino et al., 1987). They have been shown to stabilize and modulate membrane functions (Bratton, 1994; Besford et al., 1993; Khan et al., 1991; Tkachenko et al., 1991; Schuber, 1989) and to protect DNA against both thermal and alkaline denaturation and radiation-induced damage and mutagenesis (Muscari et al., 1995; Shigematsu et al., 1994; Khan et al., 1992a,b; Cozzi et al., 1991). Polyamines also induce the condensation of DNA and facilitate the transition of DNA from a B-to-Z conformation [see references in Vertino et al. (1987)].

The profound effects of polyamines on membrane fusions were among the earliest documented in the polyamine field (Tabor and Tabor, 1964, 1972; Bachrach, 1973; Cohen, 1971). A biological membrane containing 20% acidic phospholipids has an electrostatic potential of -40 mV in the aqueous diffuse double layer close to the surface (Chung et al., 1985). The addition of polyamines changes the polarity of the membrane. Repulsive forces between membranes are thus diminished, leading to aggregation of phospholipids and to fusion of cells and organelles. Accordingly, it has been suggested that polyamines modulate phenomena which require proximity of cell membranes (Hong et al., 1983).

In vitro studies have shown that polyamines can bind to phospholipid vesicles in order of effectiveness: spermine > spermidine > putrescine (Tadolini et al., 1984). Binding depends on the electrical charge of the phospholipid (Tadolini et al., 1985a,b; Chung et al., 1985), decreasing in the order of phosphatidic acid (PA, -2) > phosphatidylserine (PS, -1) > phosphatidylinositol (PI, -1) > phosphatidyletanolamine (PE, 0) > phosphatidylcholine (PC, 0) (charges in parentheses) (Tadolini et al., 1985b; Igarashi et al., 1982). Cardiolipin seems to be an exception from this trend. Despite having a charge of -2, its binding strength is reported to be less than that of PI, which is peculiar and hard to accept.

The binding strength of polyamines to an anionic target molecule can be expressed by an equilibrium constant which quantifies the stability of the polyamine–anion complex. Stability constants basically reflect the interaction between polyamines and phosphate head groups on target molecules. However, steric hindrances, proximity of phosphate groups, and redistribution of electrons by neighboring groups on the target molecule also modulate the binding strength. Some stability constants for spermine with organic anions are shown in Table I. The interaction of polyamines with DNA has also been investigated by other workers, e.g., Morgan et al. (1986) and Igarashi et al. (1982), whereas Tadolini and Hakim (1988) have measured the stability constants of spermine to liposomes with a varied phospholipid composition.

When assessing the effect of polyamines on cells, the relative amounts of different cell components such as DNA, RNA, and phospholipids must be taken into consideration together with their stability constants. By such assessment it has been found that lymphocytes and liver cells contain virtually no free spermine (see Table I), as most is bound to DNA and RNA (Watanabe et al., 1991). Only a small fraction (less than 1%) is bound to

TABLE I Percentage of Total Spermine Bound to Cell Components (Rat Liver)

Molecule	Log K	% bound	N^a
DNA	2.52	8.7	0.27
RNA	3.18	82.0	0.97
Phospholipid	2.65	6.8	0.08
ATP	2.95	2.0	0.72
Free		1.9	0.72

^{a} N is mol spermine bound to 100 mol of phosphate. Ionic conditions: 2 mM Mg^{2+}, 100 mM K$^+$. Data compiled and recalculated from Watanabe et al. (1991).

phospholipids, and the amount of spermine available in the cell is too low to provide a complete "coverage" of the membrane. However, biological membranes differ greatly in composition: Membranes that are particularly prone to oxidative attack, such as mitochondria and chloroplasts, also have a particularly high affinity for polyamines (see Section III,A). Polyamines thus tend to accumulate at specific sites within the cell, and local concentrations may be much higher than what should be expected from data on whole cells.

B. Interaction with Cations

Polyamines form complexes with several metal ions, including Ni^{2+}, Co^{2+}, Cu^{2+}, and Zn^{2+} (Hares et al., 1956; Bertsch et al., 1958). Table II summarizes data on the interaction of polyamines with Cu(II) ions. The stability constants of the chelates increase with the number of N-groups and their positions within the molecule as well as with the chain length of the polyamine. There is a correlation between the stability constants listed in Table II and the antioxidative efficacy of polyamines (Løvaas, 1991a,b).

From Table II it can be noted that a synthetic macrocyclic derivative of spermine has a far higher stability constant than the open chain homolog. The reason for this is that formation of a chelate from an open chain structure requires an input of dehydration energy as well as changes in conformation from an extended to a more compact and rigid arrangement in the metal chelate. These changes are associated with significant decreases

TABLE II Logarithms of Binding Constants for Polyamine/Cu(II)

	$Log\ K_i$	Reference
Di-amines		
1,2-Ethanediamine	10.36	Oldham et al. (1988)
1,3-Propanediamine	9.45	Oldham et al. (1988)
1,4-Butanediamine (PUT)		
1,5-Pentandiamine (CAD)		
Tri-amines		
4-Azaoctane-1,8-diamine (SPD)	11.61	Palmer and Powell (1974a,b)
4-Azaheptane-1,7-diamine	14.20	Palmer and Powell (1974a,b)
3-Azaheptane-1,7-diamine	13.44	Palmer and Powell (1974a,b)
3-Azahexane-1,6-diamine	16.60	Palmer and Powell (1974a,b)
3-Azapentane-1,5-diamine	15.80	Palmer and Powell (1974a,b)
Tetra-amines		
4,9-Diazadodecane-1,12-diamine (SPM)	14.70	Palmer and Powell (1974a,b)
Macrocyclic analog of spermine	24.80	Martell (1981)
4,8-Diazaundecane-1,11-diamine	17.00	Palmer and Powell (1974a,b)
3,7-Diazanonane-1,9-diamine	23.90	Palmer and Powell (1974a,b)
3,6-Diazaoctane-1,8-diamine	20.80	Palmer and Powell (1974a,b)

in entropy due to the restriction on the rotational and vibrational motion of the ligand (Martell, 1981). In free solution these factors tend to reduce the stability constants for the metal–ligand complex so that a weaker complex is formed with the open chain structure.

One might then wonder why macrocyclic polyamines do not occur in nature since they are so much more efficient as metal chelators. A simple answer could be that the disadvantage of a low stability constant by far is compensated for by the advantages of an open chain structure. The open chain structure permits close interaction between polyamines and the anionic sites on membranes and DNA/RNA, thus increasing the local concentration of polyamines at the sites of oxidative attack. It also appears that positively charged polyamines interact more easily with positively charged metals when the Coulombic repulsion between the metal and the polyamine is reduced by charge neutralization. Such charge neutralization takes place when polyamines are adsorbed on negatively charged biomolecules such as phospholipids, DNA, and RNA. Thus, the conformational change from an open chain to a closed metal–ligand complex would be facilitated in the presence of negatively charged counterions. This view is supported by Tadolini (1988a) who showed that the antioxidative effect of spermine is enhanced in the presence of phosphate, ADP, ATP, etc. In any case, polyamines have high affinity for both anions (Table I) and cations (Table II), thus offering unique possibilities for the prevention of oxidative damage.

The structure of a Cu(II)–spermine complex, as determined by X-ray crystallography (Boggs and Donohue, 1975), shows the nitrogen donor atoms arranged in a square plane around a central metal ion, indicating binding through charge transfer (Fig. 1). This arrangement has a resemblance to metal binding in porphyrins and phtalocyanins, which has rigid macrocylic structures containing four completely resonance-linked aromatic nitrogen donor atoms arranged in a square plane around a central metal ion.

C. Type Reactions of Polyamines: Are They Antioxidants?

The first indication that polyamines could act as antioxidants dates back to 1979, when polyamines was found to inhibit lipid peroxidation in rat liver microsomes (Kitada et al., 1979). In 1984, spermine was found to offer unique stability to marine oils far exceeding that of traditional antioxidants such as BHT, ethoxyquin, and vitamin E. Due to pending patent application (Løvaas, 1991a), these data were first published in 1991 (Løvaas, 1991b).

The lack of literature on polyamines as antioxidants was striking at the time. Even today, relatively few papers have focused on the matter. However, a survey of available publications reveals several interesting points supporting the hypothesis of polyamines as important biological antioxidants:

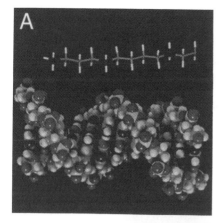
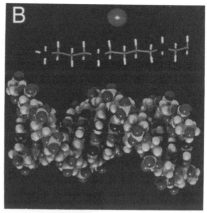
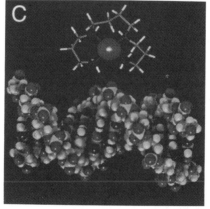

FIGURE I The dynamic interaction of spermine with anions (DNA) and cations (iron). (A) Spermine is adsorbed to DNA by inonic interactions. (B) Intermediate situation, where iron is approaching the complex. (C) Spermine/Fe complex is formed, with a simultaneous change in the conformation of spermine. The soluble complex then diffuses away from DNA, thus preventing a site-specific oxidation. Spermine structures are based on crystallographic data (Boggs and Donohue, 1975; Iitaka and Huse, 1965).

- the level of polyamines is high in organs under oxidative stress (summary data in Løvaas, 1995)
- the biosynthesis of polyamines is induced by oxidative stress induced by UVB irradiation (Verma *et al.*, 1979; Hillebrand *et al.*, 1990), $O_2\cdot^-$ (Fisher *et al.*, 1988), O_3 (Langebartels *et al.*, 1991), and peroxides (Binder *et al.*, 1989; Harari *et al.*, 1989)
- there is an active uptake of polyamines in cells and organelles subject to oxidative stress (Seiler and Dezeure, 1990; Byers and Pegg, 1989; McCormack and Johnson, 1991; Toninello *et al.*, 1992)

- cells defective in polyamine synthesis are exceptionally sensitive to oxygen (Minton *et al.*, 1990; Balasundaram *et al.*, 1993), to paraquat (Minton *et al.*, 1990), and to radiation damage (see later)
- polyamines protect isolated DNA against oxidative attack (Khan *et al.*, 1992a,b; Muscari *et al.*, 1995)

Only a few publications have questioned the biological importance of polyamines as antioxidants, e.g., Kogure *et al.* (1993) found that spermine accelerated iron-induced lipid peroxidation in mitochondria. It is well known that the level of polyamines is increased in cancer, indicating an adverse function of polyamines in cell biology. However, polyamines are neither associated with carcinogenic promotion (Peterson *et al.*, 1980) nor are they specific to the process (Black *et al.*, 1989). Løvaas (1995) advanced the hypothesis that polyamines prevent cancer in healthy cells and protect cancer cells against oxidative attack by leukocytes. However, the mechanism by which polyamines acts as antioxidants has been a matter of controversy. It has been suggested that polyamines scavenge one or more of the following oxygen species: $O_2^{\cdot-}$, $OH^{\cdot}$, $RO^{\cdot}$, $ROO^{\cdot}$, $^1O_2^*$ (Khan *et al.*, 1992a,b), and O_3 (Bors *et al.*, 1989; Langebartels *et al.*, 1991) (see Table IV for more references). It has also been proposed that polyamines inhibit metal-catalyzed oxidations (Tadolini, 1988b; Muscari *et al.*, 1995; Løvaas, 1995; Løvaas and Carlin, 1991).

For polyamines to act as scavengers they should have rate constants comparable to the half-life of the active oxygen species they are supposed to act on. Estimated half-lives in seconds for some radicals are $OH^{\cdot}$ (10^{-9}), $RO^{\cdot}$ (10^{-6}), $^1O_2^*$ (10^{-5}), $ROO^{\cdot}$ (7), and semiquinone radical (days) (Sies, 1993). Rate data for the reaction between oxygen radicals and different types of antioxidants reveal that polyamines are inefficient scavengers of $OH^{\cdot}$, $RO^{\cdot}$, and $O_2^{\cdot-}$ radicals, and that far better scavengers are available for these radicals (Table III). Current evidence, particularly based on the observations of Bors *et al.* (1989), Tadolini *et al.* (1992), and Løvaas (1995), favors the notion that polyamines prevent metal-catalyzed oxidations or, alternatively, quench $^1O_2^*$ (Khan *et al.*, 1992a,b). Some comments on the reported antioxidative effects of polyamines seem appropriate:

1. Superoxide Radicals ($O_2^{\cdot-}$)

The low rate constants for the reaction between $O_2^{\cdot-}$ and polyamines rule out the possibility that polyamines, at physiological concentrations, can act as efficient superoxide scavengers (Table III). The possibility that polyamine–metal complexes act as superoxide dismutase mimetics has been excluded by Løvaas and Carlin (1991) and by Kimura *et al.* (1981). However, because $O_2^{\cdot-}$ causes the liberation of iron from storage proteins like ferritin (Biemond *et al.*, 1988) it is possible that polyamines prevent radical formation subsequent to $O_2^{\cdot-}$ formation by chelating iron. This would prevent the $O_2^{\cdot-}$-driven Fen-

TABLE III   Antioxidant Activities Expressed as Bimolecular Reaction Rate Constants (M^{-1} sec^{-1}) (k $\times$ 10^{-6})

Antioxidant	OH·	t-BuO·	$O_2\cdot^-$	$^1O_2{}^*$	Reference
Ascorbate	11,000	2200	0.0575		Bors et al. (1989)
Propyl gallate	12,000	200	0.26		Bors et al. (1989)
Quercetin	4300	2500	0.089		Bors et al. (1989)
Trolox	34,500	690	0.20		Bors et al. (1989)
β-Carotene				30,000	Bellus (1979)
Superoxide dismutase				2600	Bellus (1979)
α-Tocopherol				250	Bellus (1979)
Histidine				170	Bellus (1979)
Tryptophan				90	Bellus (1979)
Methionine				30	Bellus (1979)
Putrescine	113	<0.15[a]	0.00011		Bors et al. (1989)
Spermidine	125	<0.15[a]	0.00029	10	Bors et al. (1989); Khan et al. (1992a)
Spermine	130	4.2	0.00029	12	Bors et al. (1989); Khan et al. (1992a)

[a] Not observable (value is detection limit of system).

ton reaction so that destructive OH· radicals are not formed. Complexation of iron could also prevent formation of $^1O_2^*$, as described next.

2. Singlet Oxygen ($^1O_2^*$)

Tertiary and secondary amines are known to be physical quenchers of singlet oxygen (Quannès and Wilson, 1968; Monroe, 1985). They quench $^1O_2^*$ by a charge-transfer process (Quannès and Wilson, 1968). Chemical reaction is minimal, and the amines are not destroyed in the process. However, polyamines are not very efficient quenchers of $^1O_2^*$ compared to other antioxidants available (e.g., β-carotene, Table III). However, it is possible that polyamines prevent $^1O_2^*$ formation by intercepting electron transfer from metals to $O_2\cdot^-$ (Cabrini et al., 1989; Khan, 1991). It is also possible that polyamine–metal chelates are more efficient scavenges of $^1O_2^*$ than "naked" polyamines since it has been known for a long time that metal chelates are good quenchers of $^1O_2^*$ (with rate constants in the order of 3–30 $\times$ 10^9 M^{-1} sec^{-1}) (Bellus, 1979).

3. Ozone (O_3)

Ozone is a remarkably reactive compound that combines with almost any type of organic molecule, including uncharged amines (Keinan and Masur, 1977). Ozone also reacts with lipids by a process that involves an unstable carbonyloxide which can decompose in a variety of ways. One prominent product is lipid hydroperoxides, subsequently giving rise to chain propagating alkoxyl radicals by the catalytic action of iron. In principle, polyamines can prevent O_3 damages by a number of mechanisms, including direct scavenging of O_3, or by preventing metal-catalyzed degradation of lipid hydroperoxides. Because conjugated polyamines, such as p-coumarolyputrescine, caffeoylputrescine, and feruloylputrescine, are efficient scavengers of OH· radicals, such compounds have been suggested to prevent O_3 damages in plants (Ormrod and Beckerson, 1986).

4. Hydroxyl Radical (OH·), Alkoxyl Radicals (RO·), and Peroxyl Radicals (ROO·)

Aliphatic amines react only slowly with hydroxyl radicals (Getoff and Schwörer, 1970), thus rendering suggested radical-scavenging properties of polyamines rather questionable. Rate constants for the reaction of polyamines with OH· are reported to be approximately $0.1 \times 10^9 \ M^{-1} \ sec^{-1}$ (Bors et al., 1989). These values are somewhat lower than that of the related but simpler compound ethyleneamine, ranging from 0.5 (Spinks and Woods, 1976) to $5 \times 10^9 \ M^{-1} \ sec^{-1}$ (Buxton et al., 1988). Polyamines are also poor scavengers of RO· (Bors et al., 1989) and ROO· radicals (DeLange and Glazer, 1989), and compounds such as ascorbate and urate (DeLange and Glazer, 1989) are much more efficient antioxidants. For example, ROO· radicals generated by the thermal decomposition of AAPH[2] was only 50% inhibited by 3 mM spermine (DeLange and Glazer, 1989). In comparison, 1 μM urate gave 100% inhibition, clearly indicating that polyamines were inferior as chain breakers.

III. Antioxidative Effects of Polyamines _____

The proposition of polyamines as important biological antioxidants has emerged from independent lines of experimental approaches by groups working in different research areas, such as food processing, physical chemistry, and cell biology/medical research. This chapter describes in some detail data originating from areas relating to:

- prevention of lipid peroxidation in oils, liposomes, microsomes, cell cultures, and whole organs,
- prevention of radiation damage (γ-rays, X-rays, UV irradiation, PUVA, visible light) and $^1O_2^*$ damage on DNA and cell cultures,
- investigations on anti-inflammatory effects in leukocytes and intact animals,
- investigations on the toxic effect of oxygen in microorganisms and defective in polyamine synthesis, and
- prevention of ozone damage and physical stress damage of plants.

[2] AAPH [2,2'-azobis(2-amidinopropane)dihydrochloride] gives rise to ROO· radicals by a reaction between a carbon radical (R·) and molecular oxygen (Niki, 1990).

Some events on the acceptance of polyamines as antioxidants are summarized in Table IV.

A. Prevention of Lipid Peroxidation

Kitada *et al.* (1979) were probably the first to suggest that polyamines act as antioxidants, although earlier work on the effects of polyamines on fatty acid oxidation and on radiation protection now can be interpreted as antioxidative effects. Kitada *et al.* (1979) observed that polyamines inhibited NADPH- and ascorbic acid-dependent lipid peroxidation in rat liver microsomes, and it was suggested that the effect was related to the binding of polyamines to phospholipids. A subsequent paper noted that spermine also prevented the loss of the catalyzing activity of cytochrome P450 caused by lipid peroxidation (Kitada *et al.*, 1980). Furthermore, it was found that spermine neither acted as a chain breaker nor reduced the generation of superoxide radicals (Kitada *et al.*, 1981).

Shortly after it was demonstrated that polyamines prevented the oxidation of marine oils (Løvaas, 1991b). It was noted that the efficacy increased with increasing number of amine groups, so that spermine was more efficient than spermidine and putrescine. Spermine also had a stabilizing effect on α-tocopherol and carotenoid pigments. This was expected on the basis of the antioxidative effect on polyunsaturated fatty acids, but the effect was surprisingly high compared to other antioxidants tested. Based on observations that acid stimulates the free radical decomposition of peroxides, which act as initiation centers for further propagation reactions, it was speculated that the antioxidative effect of polyamines was due to neutralization of acid or, alternatively, that acid promoted an increased solubility of trace amounts of iron, thus stimulating the Fenton reaction. In brief, it was demonstrated that polyamines had an unsurpassed antioxidative effect on fish oil, but a fully satisfactory explanation for the effect could not be given.

In 1984, Tadolini *et al.* observed that polyamines inhibited lipid peroxidation in liposomes, but only when bound to the negative charges on the surface. Binding of spermine to liposomes highly decreased the reactivity of both Fe^{2+} and Fe^{3+} versus superoxide, whereas free spermine had no effect. Later the inhibitory effect of spermine was found to be related to the relative binding affinity of spermine and Fe to the phospholipid vesicles. In membranes with high affinity for Fe and low affinity for spermine, more polyamine was needed to competitively displace iron (Tadolini *et al.*, 1985a). In accordance with these observations, Tadolini *et al.* (1985a) suggested that spermine mobilizes iron from the anionic sites of the phospholipids by a cation-exchange process. The autoxidation of Fe^{2+} was later observed to be inhibited by spermine, but only in the presence of phosphorous-containing compounds mimicking phosphatidic acid (AMP, ADP, ATP) (Tadolini, 1988a). Compounds mimicking phosphatidylcolin (CDP-choline) or phos-

TABLE IV Experimental Systems Used and Suggested Antioxidative Mechanism for Polyamines

Year	Experimental system	Oxidation induced by	Function	Mechanism	Reference
1979	Microsomes	NADPH/ascorbate	Antioxidant		Kitada et al. (1979)
1980	Cytochrome c	HX/XO	Antioxidant	$O_2^{\cdot-}$	Vanella et al. (1980)
1983	Edema	Carrageenan	Anti-inflamm.		Bird et al. (1983)
1984	Edema	Carrageenan	Anti-inflamm.	$O_2^{\cdot-}$ (?)	Oyanagui (1984)
1984	Fish oil	Temperature/O_2	Antioxidant		Løvaas (1991b)
1984	Liposomes	Fe^{2+}	Antioxidant	Fe	Tadolini et al. (1984)
1986	ESR	HX/XO	Antioxidant	$O_2^{\cdot-}$	Drolet et al. (1986)
1986	Leukocytes	PMA	Anti-inflamm.	$O_2^{\cdot-}$	Kafy et al. (1986)
1986	Plant (tomato)	O_3	Antisenescence		Ormrod and Beckerson (1986)
1988	Lysosomes	HX/XO	Antioxidant	$O_2^{\cdot-}$	Kafy and Lewis (1988)
1989	Phycoerythrin	AAPH decomp.	Antioxidant	$ROO^{\cdot}$ (?)	DeLange and Glazer (1989)
1989	Plant (tobacco)	O_3	Antioxidant	Fe (?)	Bors et al. (1989)
1989	Plant (barley)	O_3	Antioxidant?		Rowland-Bamford et al. (1989)
1989	Plant (zucchini)	Chilling	Antioxidant		Kramer and Wang (1989)
1990	E. coli	Paraquat	Antioxidant?	$O_2^{\cdot-}$ (?)	Minton et al. (1990)
1991	Plant (cucumber)	UVB	Antioxidant		Kramer et al. (1991)
1991	Cell culture (CHO)	PUVA	Radioprotector		Cozzi et al. (1991)
1991	DNA	X-rays	Radioprotector	$OH^{\cdot}$	Held and Awad (1991)
1991	Leukocytes/HA	HX/XO-PMA	Antioxidant	Fe	Løvaas and Carlin (1991)
1992	DNA	NDPO$_2$ decomp.	Antioxidant	$^1O_2^*$	Khan et al. (1992a)
1992	Liposomes	Fe^{2+}	Antioxidant (?)	$O_2^{\cdot-}$	Pavlovic et al. (1992)
1993	S. cerevisiae	O_2	Antioxidant (?)	$O_2^{\cdot-}$	Balasundaram et al. (1993)
1993	Homogenates	Fe^{2+}/H_2O_2	Antioxidant		Matkovics et al. (1993)
1993	Cell culture (CHO)	γ-radiation	Radioprotector		Prager et al. (1993)
1994	Cell culture (V79)	PUVA	Radioprotector		Williams et al. (1994)
1995	DNA	H_2O_2/ascorbate	Antioxidant	Fe (?)	Muscari et al. (1995)

Abbreviations: AAPH, 2,2'-azobis(2-amidinopropane); Antiinflamm., antiinflammatory agent; HA, hyaluronic acid; HX/XO, hypoxanthine/xanthine oxidase; NL²O$_2$, endoperoxide of 3,3'-(1,4-naphtylidene)dipropionate; PMA, phorbol myristate acetate.

phatidylinositol (glycerophosphoinositol) had no effect on the autoxidation of Fe^{2+}. On the basis of these findings, inhibition of lipoperoxidation by polyamines was suggested to be due to their ability to form a ternary complex with the phospholipid polar head and the Fe^{2+} catalyst. Comparing the antioxidative effect of spermine on microsomal and mitochondrial membranes from sea bass, it was found that lipid peroxidation in mitochondria was nearly abolished by the addition of 100 μM spermine, whereas no effect was found for microsomes (Cabrini et al., 1989). Since it was established that spermine neither binds to nor protects vesicles containing high levels of phosphatidylinositol from peroxidation (Tadolini et al., 1985a), it was concluded that mitochondrial membranes are protected due to their low level of phosphatidylinositol. The most striking difference between mitochondrial and microsomal membranes is, however, the high level of cardiolipin in the former (Table V). Tadolini and colleagues did not consider this remarkable difference in phospholipid composition between mitochondria and microsomes, probably because they had earlier observed the binding strength of spermine to decrease in order of effectiveness: phosphatidic acid > phosphatidylserine > phosphatidylinositol > cardiolipin (Tadolini et al., 1985b; Meers et al., 1986). If cardiolipin is placed at the other end of the series, which seems quite reasonable due to the double negative charge on cardiolipin, the results become easier to interpret. The antioxidative effect of polyamines would then correlate with the negative charge density on the membranes. These experiments should be repeated.

Pavlovic et al. (1992) have reported that polyamines inhibit Fe^{2+}-catalyzed lipoperoxidation in liposomes and rat liver homogenates. They suggested that the antioxidative effect was due to an interaction of polyamines with $O_2 \cdot^-$, but it now seems much more reasonable to ascribe the antioxidative effect to iron chelation. Matkovics et al. (1993) showed that polyamines prevent oxidative damages in erythrocytes, brain homogenates,

TABLE V Phospholipids in Mitochondrial and Microsomal Membranes

Phospholipid	Charge	Microsomes	Mitochondria
Phosphatidylcholine	0	62.0	46.5
Phosphatidylethanolamine	0	19.0	25.2
Cardiolipin	−2		12.8
Phosphatidylinositol	−1	6.1	2.2
Phosphatidylserine	−1	3.2	7.2
Sphingomyelin	0	4.5	
Lysophosphatidylcholine	0	2.2	1.5
Lysophosphatidylethanolamine	0	1.0	1.3
Unidentified		2.0	2.5
"Charge density" (calculated)		−0.093	−0.350

Note. Data based on Cabrini et al. (1989).

and guinea pig hearts. Oxidation of the brain homogenates was induced by the addition of H_2O_2 and Fe^{2+}. Polyamines completely inhibited the oxidation. However, because lipid peroxidation was measured by the malonedialdehyde (MDA) assay, which is notoriously erroneous (Halliwell and Chirico, 1993; Janero, 1990), some caution must be exercised in the interpretation of these results.

B. Polyamines and Inflammation

The previous sections have focused on the metal-chelating properties of polyamines and how binding to anionic sites can protect against metal-catalyzed site-specific oxidations. This section will examine the possible functions of polyamines in inflammations. An inflammation is characterized by a respiratory burst of activated neutrophils and macrophages, leading to the destruction of invading microorganisms. In this respect an inflammation performs a useful function. However, inflammations may also be detrimental to "self" as the respiratory burst is nonspecific. Consequently, it is imperative to limit the extent of damage to healthy tissue. This can be accomplished by the way of "salvage molecules" protecting healthy cells against destruction. The identity of such molecules is at present unknown. However, it has been suggested that polyamines act as such cellular protectors in inflammatory reactions. This suggestion is based on the facts that harmful metals are liberated during an inflammation, that polyamine synthesis is induced by inflammation, that polyamines chelate metals, and that polyamines protect biomolecules against oxidative attack. Some of these aspects have been reviewed (Løvaas, 1995). In addition, polyamines are recognized as anti-inflammatory agents, i.e., they inhibit the respiratory burst of neutrophils as well as lymphocyte proliferation. These anti-inflammatory effects are mediated by the binding of polyamines to anionic sites (phospholipid membranes) or to anionic signal molecules (PIP_2). The anti-inflammatory effects of polyamines must be clearly distinguished from the salvaging (or antioxidative) effect, which is based on cation binding.

The following sections will first summarize some aspects of inflammation, then go on to the protective effects of polyamines in inflammations, and finally turn to the regulatory effects of polyamines on inflammation.

I. Iron, Tissue Damage, and Polyamines

It is a paradox of inflammation biology that while neutrophils are essential for host defense, they are also involved in the pathology of various inflammatory conditions (Weiss, 1989; Smith, 1994), in the propagation of ischemia–reperfusion injury (Ricevuti et al., 1991), and in the initiation of atherosclerosis by the oxidation of low-density lipoproteins (Scaccini and Jialal, 1994; Abdalla et al., 1992). Reactive oxygen species generated by neutrophils may also initiate tumour development due to damage to DNA

(Weitzman and Gordon, 1990; Trush and Kensler, 1991). These events are believed to be related to iron pathology, as described later.

Nearly all the body iron is located in iron-binding proteins such as transferrin, lactoferrin, ferritin, cytochromes, and hemoglobin. Many studies have shown that protein-bound iron does not catalyze $OH^{\cdot}$ formation (Gutteridge *et al.*, 1981; Halliwell, 1978; Winterbourn, 1981). However, during inflammation, catalytic iron is mobilized by the following events:

- Plasma proteins, including ferritin, leak from the arterioles into the extravascular space. For example, in UVB-irradiated skin the local inflammatory response results in a 10-fold increase of ferritin in epidermis (Bissett *et al.*, 1991).
- Superoxide radicals generated by the oxidative burst promote iron release from ferritin (Biemond *et al.*, 1988; Ryan and Aust, 1992; Reif, 1992). Liberated iron further amplifies the inflammatory reaction (Blake *et al.*, 1981), and a vicious cycle, in which lymphocytes and macrophages are attracted to the inflamed site, is thus established. Also, $O_2 \cdot^-$ radicals act as chemotactic factors (Hurst and Barrette, 1989), contributing to a positive feedback situation that can propagate the inflammatory reaction indefinitely.
- The inflammation causes a drop in pH to less than 5 (Etherington *et al.*, 1981), and the solubility of uncomplexed Fe^{3+} is increased several million times. For example, a drop in pH from 7.4 to 4.8 will increase the solubility of Fe^{3+} from 10^{-22} to 10^{-14} M, as given by the equation $[Fe^{3+}] = 10 \times [H^+]^3$.

The combination of these events causes a "reshuffling" of iron from storage proteins to anionic sites such as phospholipid membranes and DNA. The scene is thus set for site-specific oxidations. This is clearly destructive and it seems reasonable to assume that protective mechanisms have developed to counteract such damage caused by redistributed iron.

In accordance with this view, several low molecular weight chelating agents have been evaluated for possible protective effects. For example, it is known that chelating agents like diethylenetriaminepentaacetic acid, bathophenantroline, uric acid (Ames *et al.*, 1981; Hochstein *et al.*, 1984; Sevanian *et al.*, 1985; Rowley and Halliwell, 1985), and desferrioxamine (Carlin, 1985) prevent tissue damage by activated phagocytes. Because polyamines have high affinities for iron, it is tempting to suggest that they protect against inflammation-mediated destructions of tissue *in vivo*. In this respect it is interesting to note that there is an accumulation of polyamines in inflammatory exudates (Bird and Lewis, 1981) and that patients with persisting inflammations, such as rheumatoid arthritis, have an increased level of polyamines in the synovial fluid (Yukioka *et al.*, 1992). It is also suggestive that stimulated lymphocytes display a steady increase in ornithine decarboxylase (ODC) (Hunt and Fragonas, 1992) and that the polyamine level increases rapidly in activated lymphocytes (Kay and Lindsay, 1973; Otani *et al.*, 1982). It is also interesting to note that metal chelators such as desferrioxamine inhibit the increase in ODC activity in stimulated lymphocytes (Hunt and Fragonas, 1992). This observation falls in line with previously mentioned results on a correlation between the biosynthesis of polyamines with oxidative stress, suggesting that synthesis is induced at times of need.

In 1977 Byrd *et al.* found that micromolar quantities of spermine and spermidine reversibly inhibited inflammatory responses of murine spleen cells. Similar effects were later observed by Bird *et al.* (1983) and by Oyanagui (1984). In an effort to pinpoint the anti-inflammatory mechanism, the radical scavenging effect of polyamines was studied by Kafy *et al.* (1986). They found that polyamines scavenge $O_2 \cdot^-$ generated both by xanthine oxidase and by stimulated polymorphonuclear leukocytes (PMNLs), but only at unphysiologically high concentrations (50 mM). Similar results were obtained by Mizui *et al.* (1987), who also observed that lipid peroxidation induced by Fe^{2+} was partially prevented by physiological levels of polyamines, suggesting a role as metal chelators. In 1988, Kafy and Lewis found that $O_2 \cdot^-$ was extremely lytic to lysosomes and that polyamines protected against both $O_2 \cdot^-$-induced lysis and lipid peroxidation. It is now clear (see Table III) that polyamines do not scavenge $O_2 \cdot^-$ radicals at physiological concentrations and that the protective effects must be related to metal binding or perhaps to scavenging of $^1O_2^*$. The finding that spermine prevents Fe^{2+}-catalyzed/$O_2 \cdot^-$-mediated degradation of hyaluronic acid (Løvaas and Carlin, 1991) supports the suggestion that polyamines act as metal-chelating agents.

2. Polyamines as Anti-inflammatory Agents

The previously described discussion of polyamine pharmacology puts a main emphasis on the cation-binding, or antioxidative, effects of polyamines. However, polyamines also modulate inflammatory processes by their ability to interact with anions. This function is mediated by the inhibition of both the locomotion and the respiratory burst of phagocytes (Løvaas and Carlin, 1991; Ferrante *et al.*, 1986).

The NADPH-oxidase of human phagocytes is a powerful generator of superoxide radicals, achieving rates of 10^{10} molecules $O_2 \cdot^-$/cell/min. The production of superoxide is normally highly regulated, meaning that in the resting state there is no contact between the intracellular NADPH and the extracellular oxygen. The activation process requires the assembly of large macromolecular complexes of proteins from different cellular compartments after stimulation of the cell (Jesaitis *et al.*, 1991). Key elements in this process are (Fig. 2):

- Activation of phospholipase C (PLC), which subsequently catalyzes the production of a second messenger [diacylglycerol - (DAG)] by acting on phosphatidyl inositol diphosphate (PIP_2).

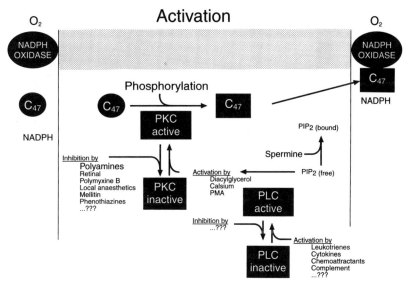

FIGURE 2 Events in the activation of NADP-oxidase. See text for details.

- DAG then activates free phosphokinase C (PKC), which then attaches to the membrane and at the same time is activated.
- PKC then phosphorylates soluble proteins (e.g., C_{47}) which then attach to the membrane and become part of an activated NADPH-oxidase complex.

The current notion is that polyamines inhibit this complex activation process by two different mechanisms: Spermine may bind to PIP_2 (Chung et al., 1985; Meers et al., 1986; Reaven et al., 1993; Parchment et al., 1990), thus making this substrate inaccessible for PLC, or it may inhibit the binding of PKC to the membrane. Polyamine inhibition of PLC has indeed been demonstrated (Eichberg et al., 1981; Nahas and Graff, 1982; Das et al., 1987).

PKC is known to fluctuate rapidly between a soluble inactive form and a membrane-bound active form. The enzyme binds at a specific membrane site containing phosphatidylserine (PS) and Ca^{2+}. The transition from the free to the membrane-bound state is brought about by DAG, liberated by the action of PLC. PKC is also activated by phorbol 12-myristate 13-acetate (PMA), also known as tetradecanoylphorbolacetate (TPA). The enzyme is inhibited noncompetitively by spermine with respect to PS and Ca^{2+} (Ciaccio et al., 1993) and also when the stimulation is brought about by PMA (Liu and Mori, 1993). The inhibitory effect of spermine is partly due to its interaction with Ca^{2+}-binding sites on the membrane (PS) (Meers et al., 1986; Afanasev et al., 1993; Linseman et al., 1993; Moruzzi et al., 1990) and partly to its direct interaction with the catalytic site of the enzyme

(Muntané et al., 1993). A full understanding of the role of polyamines as regulators of the respiratory burst has not yet emerged.

In addition to the modulation of the respiratory burst, polyamines also regulate the multiplication of lymphocytes. Data on this matter are conflicting: Some workers find that polyamines prevent lymphocyte proliferation by downregulating IL-2 production[3] in monocytes (Flescher et al., 1989). This action seems to require the presence of polyamine oxidase, which converts polyamines to aldehydes, ammonia, and H_2O_2 (Flescher et al., 1994). A similar anti-inflammatory effect has been observed by Colombatto et al. (1988), who suggest that spermine, or its oxidation products, inhibits the release of inflammatory mediators or that polyamines inhibit the synthesis of these compounds. However, polyamines have also been observed to stimulate lymphocyte proliferation, and an inhibitor of polyamine synthesis difluoromethylornithine (DFMO) prevents proliferation (Yukioka et al., 1987; Bowlin et al., 1987).

In summary, polyamines modulate inflammatory processes and possibly also protect against tissue damages caused by iron released during the inflammation. The rather complex picture that has emerged is due to the fact that polyamines are multifunctional molecules. Some effects are explained by their ability to chelate cations, whereas other effects are due to their ability to bind to phospholipid membranes and other anionic compounds.

C. Radiation Protection and Protection against Mutagenesis

Polyamines may prevent photooxidative damages in animal as well as plant cells. This suggestion is based on the following observations:

- In skin, moderate exposure to UV irradiation promotes an inflammatory state [references in Kligman et al. (1985), review in Søndergaard et al. (1985)] characterized by an activation of neutrophils, generation of $O_2\cdot^-$ radicals, and liberation of iron, as described in the previous section.
- At the same time there is an induction of ODC, and the level of polyamines in the skin increases dramatically in a dose-dependent manner (Verma et al., 1979; Hillebrand et al., 1990). Other antioxidants in skin are either unaffected by irradiation or destroyed (Løvaas, 1995).
- Metal chelators like desferroxamine significantly reduce free radical generation in skin (Jurkiewicz and Buettner, 1994). It seems probable that polyamines have a similar effect.

---

[3] Interleukins are naturally occurring proteins that are important for the activation of lymphocytes and for the regulation of their activities. IL-2 stimulates the proliferation of T lymphocytes as well as lymphocytes that kill infected body cells.

- Polyamines dramatically reduce radiation damages on DNA (Cozzi *et al.*, 1991) and mutation frequencies (Shigematsu *et al.*, 1994).
- Polyamines protect DNA against $^1O_2^*$-mediated oxidation of guanine residues and prevent native supercoiled plasmid DNA from transforming into the open circular form (Khan *et al.*, 1992a,b).
- In plants, polyamine biosynthesis is induced by UV radiation (Hagege *et al.*, 1990; Kramer *et al.*, 1991). Polyamines are also enriched in photosystem II, suggesting a protective effect on this oxidation-susceptible structure (Kotzabasis *et al.*, 1993).

Some of these aspects have been reviewed (Løvaas, 1995). The present discussion is confined to studies on polyamine-manipulated cells, mutation studies, and studies on isolated DNA. Polyamine antimutagenesis might in many instances operate via a stimulation of the error-correcting function of DNA polymerase (Chiu and Sung, 1972), although some effects are certainly due to the chelation of metals or to scavenging of $^1O_2^*$.

The antimutagenic effects of polyamines were initially observed in 1962, when polyamines were discovered to prevent the spontaneous resistance to antibiotics in wild-type bacteria (Sevag and Drabble, 1962). During the next 13 years (reviewed in Clarke and Shankel, 1975) the antimutagenic effect of polyamines was described for many bacterial strains. The mechanism of action remained elusive, and it was cautioned that polyamines display such a diversity of biochemical effects that it could be extremely difficult to identify the sites of polyamine antimutagenesis. New experimental tools have emerged since 1975, and recent data support the idea that the antioxidative effects of polyamines are of importance for the protection of DNA.

I. Studies on Polyamine-Manipulated Cells

In 1978 a specific inhibitor of polyamine synthesis [2-difluoromethylornithine (DFMO)] became available (Metcalf *et al.*, 1978), and this led to an explosion of studies on the biological functions of polyamines. The use of DFMO as a tool has been complicated by the fact that DFMO only acts on the initial step of the biosynthesis of polyamines. In practice it has thus proved difficult to deplete cells for the higher homolog of the polyamine series (spermine), while the levels of putrescine and spermidine are more easily manipulated (Gerner *et al.*, 1988; Prager *et al.*, 1993). DFMO treatment, therefore, does not deplete cells for polyamines, although the total pool is reduced. Caution must thus be exercised when interpreting these data. The current notion is that polyamines are involved in the postradiation repair process, but protective effects are not excluded.

By investigating the effect of X-rays on human tumor cells, Courdi *et al.* (1986) found that polyamine depletion by DFMO made the cells more sensitive to radiation. The protective effect was suggested to be due to the stabilization of DNA by polyamines or to stimulation of postradiation repair processes. A similar conclusion was obtained by Gerner *et al.* (1988), who found that CHO cells treated with DFMO were more sensitive to repeated X-ray irradiations than the controls. The addition of 1 mM of putrescine *after* the irradiation significantly increased cell survival. It was thus obvious that putrescine was involved in the postirradiation repair process. However, data did not exclude the possibility that spermine prevented the initial damage since the level of spermine was unaffected by DFMO treatment. DFMO also sensitized Chinese hamster ovary (CHO) cells to aerobic γ-radiation (Prager *et al.*, 1993). Williams *et al.* (1994) investigated the effect of polyamine depletion on the cytotoxic response to psoralen + UVA radiation (PUVA), γ-rays, and UVC on V79 cells *in vitro*. Polyamine depletion was found to sensitize the cells for DNA damages caused by PUVA and γ-rays, but not to UVC irradiation. The addition of exogenous putrescine prior to irradiation restored the normal polyamine content of the cells and also reversed the cytotoxicity. This effect could be due to prevention of initial damage or to modulation of postdamage DNA repair. Similar results were obtained by Cozzi *et al.* (1991), who found that low levels of polyamines (0.1–1.0 μM) dramatically reduced damage to DNA. It was suggested that the protective effects could be due to competition of polyamines and psoralen for the same binding sites on DNA or that polyamines stabilized the DNA to render it less accessible to psoralen.

2. Mutation Studies

Studies on bacteria with a deficiency in the polyamine synthesis have shown that the supplementation of spermine dramatically reduces the spontaneous or induced mutation frequency (Nestmann, 1977; Clarke and Shankel, 1988; Shankel and Clarke, 1990). Spermine has been shown to protect against γ-ray-induced mutations at the *hprt* locus in Chinese hamster AA8 cells (Shigematsu *et al.*, 1994). Spermine (10 or 1 μM) administered 30 min before or 3 hr after irradiation reduced mutation frequencies by factors of 2.2, 1.2, 1.9, and 2.2, respectively. These data suggest an involvement of polyamines in repair processes, although it also was argued that the antioxidative (Løvaas, 1991b) and DNA-producing effects (Tabor and Tabor, 1976; Held and Awad, 1991) of polyamines in part could explain the enhanced tolerance against γ-irradiation.

3. Studies on Isolated DNA

The protective effects of polyamines have also been investigated on isolated DNA (Held and Awad, 1991; Muscari *et al.*, 1995; Khan *et al.*, 1992a,b). Such studies have the obvious merit of separating the antioxidative effects of polyamines from the repair or error-correcting effects. Held and Awad (1991) have investigated the effects of polyamines and thiols on the radiation sensitivity of bacterial DNA. It was demonstrated that polyamines protected DNA against radiation damages by X-rays in both O_2-saturated and N_2-saturated DNA solutions. The protection was much better under aerobic conditions, and the effect was suggested to be due to OH· scavenging. This explanation is not supported by kinetic data on the reactivity between polyamines and OH· radicals (see Table III). Polyamines have been shown to efficiently protect native supercoiled plasmid DNA from transformations into the open circular form by $^1O_2^*$ (Khan et al., 1992a,b). Such transformations are brought about by the reaction of $^1O_2^*$ with deoxyguanosine; oxidized guanine residues lose base-pairing specificity [see Piette (1991) for review and discussion]. Oxidation of guanine in DNA causes mutagenesis and loss of transforming activity, as well as some single strand breaks. Because $^1O_2^*$ is known to be generated by various types of irradiation, including UVA radiation by a type II photodynamic process (Gange et al., 1986; Pathak, 1986), this observation is relevant for the function of polyamines as protectors of radiation-induced damage.

Muscari et al. (1995) have described the protective effect of spermine on DNA exposed to oxidative stress. It was found that isolated DNA was oxidized by H_2O_2/ascorbic acid and that the oxidation could be prevented by the addition of 100 μM spermine. Spermine did not prevent the oxidation of deoxyguanosine alone, suggesting that the polyamine should be bound to the DNA strands to exert its antioxidative effect. Based on previous work by Tadolini (1988b) and Løvaas and Carlin (1991), the antioxidative effect of spermine was suggested to be due to metal binding.

D. Other Possible Antioxidative Effects

I. Polyamines and Neurodegenerative Diseases

During the last decade there has been an increasing interest in the protective effects of polyamines in neurodegenerative diseases such as Alzheimer's and Parkinson's disease. Since the brain accounts for 2% of the body weight only, but consumes 20% of the total inspired oxygen, some neurodegenerative diseases may be due to a lack of antioxidative protection. The fact that the brain contains high levels of polyunsaturated fatty acids (Tappel, 1973; Rice-Evans and Burdon, 1993) and nonheme iron (Connor and Fine, 1986; Roskams and Connor, 1994) opens up for extensive neuronal damage due to free radicals. Metal-chelating antioxidants are known to preserve brain morphology and function (Reiter, 1995), but information on natural occurring scavengers is more sparse. Gilad has found that polyamines can protect against ischemia-induced nerve cell death in gerbil forebrain (Gilad and Gilad, 1991, 1992), whereas other workers have observed that polyamines enhance recovery after sciatic nerve trauma in the rat (Kauppila, 1992; Kauppila et al., 1988) and that spermine promotes the survival of primary cultured brain neurons (Abe et al., 1993). Whether these effects are related to the antioxidative action of polyamines remains to be investigated.

2. Paraquat

Paraquat is a bipyridinium herbicide which after ingestion may cause acute respiratory failure. Substantial evidence shows that the toxicity of paraquat (PQ^{2+}) is due to the generation of oxygen radicals involving the oxidation of NADPH:

$$NADPH + PQ^{2+} \rightarrow NADP^+ + PQ^+$$
$$PQ^+ + O_2 \rightarrow PQ^{2+} + O_2{\cdot}^-$$

The paraquat radical (PQ^+) is rapidly reoxidized by oxygen with the concomitant production of superoxide ($O_2{\cdot}^-$). Superoxide in turn is reduced to H_2O_2. Hydrogen peroxide may then be reduced to $OH{\cdot}$ in an iron-catalyzed Fenton-type reaction.

It is well known that polyamines prevent paraquat toxicity (van der Wal et al., 1990). The proposed mechanism is that paraquat, which structurally is related to polyamines, is taken up in cells by the polyamine transporter and that high levels of polyamines thus competitively block the paraquat uptake. However, in light of the antioxidative effects of polyamines, there is a possibility that polyamines prevent the toxicity of paraquat by inhibiting the Fenton reaction. Other metal chelators, such as desferroxamine, protect against paraquat toxicity in this manner (van der Wal et al., 1990). It has been shown that paraquat toxicity is increased in Escherichia coli cells defective in the synthesis of polyamines (Minton et al., 1990). The increased sensitivity of these polyamine-deficient mutants is eliminated by growth in a medium containing spermidine or by endogenous supplementation by the use of a $speE^+D^+$ plasmid. No paraquat toxicity was seen in the absence of oxygen. A number of possible explanations were suggested, among those the antioxidative effect of spermine.

3. Prevention of Ozone Damages on Plants

Ozone gives rise to the oxidation of polyunsaturated fatty acids (Pryor et al., 1982) and is very destructive to biological membranes. The rising level of ozone in the atmosphere is presently causing grave damage on crops and forest, and is of serious concern for plant breeders and for the human society.

It is known that the synthesis of polyamines is induced by O_3 in plants (Langebartels et al., 1991), and it is proposed that polyamines act as a defense against oxidative stress. In principle, this can be effectuated in a number of ways, including direct scavenging of O_3, or by preventing radical chain reactions initiated by the metal-catalyzed degradation of lipid hydroperoxides.

IV. Conclusions

This chapter has advanced the view that polyamines act as cellular protectors by their capacity to bind cations and as cellular regulators by their ability to bind anions. These dual properties form a basis for the bewildering number of effects ascribed to polyamines.

The ability of polyamines to bind cations adds new and important dimensions to their biological functions. Thus new directions will certainly take this property into account, and future research will deal with how polyamine synthesis is regulated by oxidative stress, how polyamines prevent tissue and cell damage caused by iron liberation, and what happens to polyamine–metal complexes. Are such complexes excreted or do they participate in the iron loading of storage proteins like ferritin? What are the redox potentials of polyamine complexes? At the present time it is premature to say anything about the pharmacological potential of polyamines. It is, however, tempting to suggest applications in dermatology, in the prevention of atherosclerosis and reperfusion injuries, and in neurological disorders. It is also possible that polyamines can be applied for the retardation of aging processes.

Experiments designed to confirm these potentially important actions of polyamines should be given high priority.

Acknowledgments

This work was supported by a grant from the Norwegian Research Council. I thank J. E. Olsen, R. E. Olsen, and T. Strøm for helpful discussions and advice.

References

Abe, K., Chida, N., Nishiyama, N., and Saito, H. (1993). Spermine promotes the survival of primary cultured brain neurons. *Brain Res.* **605**, 322–326.

Abdalla, D. S. P., Campa, A., and Monteiro, H. P. (1992). Low density lipoprotein oxidation by stimulated neutrophils and ferritin. *Atherosclerosis* **97**, 149–159.

Afanasev, I. B., Dorozhko, A. I., Polozova, N. I., Kuprianova, N. S., Brodskii, A. V., Ostrachovitch, E. A., and Korkina, L. G. (1993). Is superoxide an initiator of microsomal lipid peroxidation. *Arch. Biochem. Biophys.* **302**, 200–205.

Ames, B. N., Cathcart, R., Schwiers, E., and Hochstein, P. (1981). Uric acid provides an antioxidant defense in humans against oxidant- and radical-caused ageing and cancer: A hypothesis. *Proc. Natl. Acad. Sci. USA* **78**, 6858–6862.

Bachrach, U. (1973). *In* "Function of Natural Occurring Polyamines" (U. Bachrach, ed.). Academic Press, New York.

Balasundaram, D., Tabor, C. W., and Tabor, H. (1993). Oxygen toxicity in a polyamine-depleted *spe2*Delta mutant of *Saccharomyces cerevisiae. Proc. Natl. Acad. Sci. USA* **90**, 4693–4697.

Bellus, D. (1979). Physical quenchers of singlet molecular oxygen. *Adv. Photochem.* **11**, 105–205.

Bertsch, C. R., Fernelius, W. C., and Block, B. P. (1958). A thermodynamic study of some complexes of metal ions with polyamines. *J. Phys. Chem.* **62**, 444–450.

Besford, R. T., Richardson, C. M., Campos, J. L., and Tiburcio, A. F. (1993). Effect of polyamines on stabilization of molecular complexes in thylakoid membranes of osmotically stressed oat leaves. *Planta* **189**, 201–206.

Biemond, P., Swaak, A. J., Van Eijk, H. G., and Koster, J. F. (1988). Superoxide dependent iron release from ferritin in inflammatory diseases. *Free Rad. Biol. Med.* **4**, 185–198.

Binder, R. L., Volpenhein, M. E., and Motz, A. A. (1989). Characterization of the induction of ornithine decarboxylase activity by benzoyl peroxide in SENCAR mouse epidermis. *Carcinogenesis* **10**, 2351–2357.

Bird, J., and Lewis, D. A. (1981). A relationship between putrescine and the anti-inflammatory activity of sponge exudate. *Br. J. Pharmacol.* **74**, 204–205.

Bird, J., Mohd Hidir, S., and Lewis, D. A. (1983). Putrescine: A potent endogenous anti-inflammatory substance in inflammatory exudates. *Agents Actions* **13**, 342–347.

Bissett, D. L., Chatterjee, R., and Hannon, D. P. (1991). Chronic ultraviolet radiation-induced increase in skin iron and the photoprotective effect of topically applied iron chelators. *Photochem. Photobiol.* **54**, 215–223.

Black, H. S., Young, A. R., and Gibbs, N. K. (1989). Effects of butylated hydroxytoluene upon PUVA-tumorigenesis and induction of ornithine decarboxylase activity in the mouse. *J. Photochem. Photobiol. B* **3**, 91–100.

Blake, D. R., Hall, N. D., Bacon, P. A., Dieppe, P. A., Halliwell, B., and Gutteridge, J. M. C. (1981). The importance of iron in rheumatoid disease. *Lancet* **2**, 1142–1144.

Boggs, R., and Donohue, J. (1975). Spermine copper(II) perchlorate. *Acta Cryst.* **B31**, 320–322.

Bors, W., Langebartels, C., Michel, C., and Sandermann, J. H. (1989). Polyamines as radical scavengers and protectants against ozone damage. *Phytochemistry* **28**, 1589–1595.

Bors, W., Michel, C., Erben-Russ, M., Kreileder, B., Tait, D., and Saran, M. (1984). *in* "Oxygen Radicals in Chemistry and Biology," pp. 95–99. de Gruyter, Berlin.

Bowlin, T. L., McKnown, B. J., Babcock, G. F., and Sunkara, P. S. (1987). Intracellular polyamine biosynthesis is required for interleukin 2 responsiveness during lymphocyte mitogenesis. *Cell. Immunol.* **106**, 420–427.

Bratton, D. L. (1994). Polyamine inhibition of transbilayer movement of plasma membrane phospholipids in the erythrocyte ghost. *J. Biol. Chem.* **269**, 22517–22523.

Buxton, G. V., Greenstock, C. L., Helman, W. P., and Ross, A. B. (1988). Critical review of rate constants for reactions of hydrated electrons, hydrogen atoms and hydroxyl radicals ($\cdot$OH/$\cdot$O) in aqueous solution. *J. Phys. Chem. Ref. Data* **17**, 513–886.

Byers, T. L., and Pegg, A. E. (1989). Properties and physiological function of the polyamine transport system. *Am. J. Physiol.* **257**, C545–C553.

Byrd, W. J., Jacobs, D. M., and Amoss, M. S. (1977). Synthetic polyamines added to cultures containing bovine sera reversibly inhibit *in vitro* parameters of immunity. *Nature* **267**, 621–623.

Cabrini, L., Tadolini, B., Landi, L., Fiorentini, D., and Sechi, A. M. (1989). The influence of polyunsaturated fatty acids on spermine inhibition of lipoxidation: Studies on liposomes prepared with microsomal and mitochondrial phospholipids of sea bass (*Dicentrarchus labrax* L.) and rat liver. *Comp. Biochem. Physiol. B* **93**, 647–651.

Carlin, G. (1985). Peroxidation of linolenic acid promoted by human polymorphonuclear leukocytes. *Free Rad. Biol. Med.* **1**, 255–261.

Chiu, J. F., and Sung, S. C. (1972). Effect of spermidine on the activity of DNA polymerases. *Biochim. Biophys. Acta* **281**, 535–542.

Chung, L., Kaloyanides, G., McDaniel, R., McLaughlin, A., and McLaughlin, S. (1985). Interaction of gentamicin and spermine with bilayer membranes containing negatively charged phospholipids. *Biochemistry* **24**, 442–452.

Ciaccio, M., Valenza, M., Tesoriere, L., Bongiorno, A., Albiero, R., and Livrea, M. A. (1993). Vitamin A inhibits doxorubicin-induced membrane lipid peroxidation in rat tissues *in vivo*. *Arch. Biochem. Biophys.* **302**, 103–108.

Clarke, C. H., and Shankel, D. M. (1975). Antimutagenesis in microbial systems. *Bacteriol. Rev.* **39**, 33–53.

Clarke, C. H., and Shankel, D. M. (1988). Antimutagens against spontaneous and induced reversion of a *lacZ* frameshift mutation in *E. coli* K-12 strain ND-160. *Mutat. Res.* **202**, 19–23.

Cohen, S. S. (1971). *In* "Introduction to the Polyamines" (S. S. Cohen, ed.). Prentice-Hall, Englewood Cliffs, NJ.

Colombatto, S., Fasulo, L., and Grillo, M. A. (1988). Polyamines in rat liver during experimental inflammation. *Agents Actions* 24, 326–330.

Connor, J. R., and Fine, R. E. (1986). The distribution of transferrin immunoreactivity in the rat central nervous system. *Brain Res.* 368, 319–328.

Courdi, A., Milano, G., Bouclier, M., and Lalanne, C. M. (1986). Radiosensitization of human tumor cells by alpha-difluoromethylornithine. *Int. J. Cancer* 38, 103–107.

Cozzi, R., Perticone, P., Bona, R., and Polani, S. (1991). Antimutagenic activities of naturally occurring polyamines in Chinese hamster ovary cells *in vitro*. *Environ. Mol. Mutagen.* 18, 207–211.

Das, I., de Belleroche, J., and Hirsch, S. (1987). Inhibitory action of spermidine on formyl-methionyl-leucyl-phenylalanine stimulated inositol phosphate production in human neutrophils. *Life Sci.* 41, 1037–1041.

DeLange, R. J., and Glazer, A. N. (1989). Phycoerythrin fluorescens-based assay for peroxy radicals: A screen for biologically relevant protective agents. *Anal. Biochem.* 177, 300–306.

Drolet, G., Dumbroff, E. B., Legge, R. L., and Thompson, J. E. (1986). Radical scavenging properties of polyamines. *Phytochemistry* 25, 367–371.

Eichberg, J., Zetusky, W. J., Bell, M. E., and Cavanagh, E. (1981). Effects of polyamines on calcium-dependent rat brain phosphatidylinositol-phosphodiesterase. *J. Neurochem.* 36, 1868–1871.

Etherington, D. J., Pugh, D., and Silver, I. A. (1981). Collagen degradation in an experimental inflammatory lesion: Studies on the role of the macrophage. *Acta Biol. Med. Germ.* 40, 1625–1636.

Ferrante, A., Maxwell, G. M., Rencis, V. O., Allison, A. C., and Morgan, D. M. L. (1986). Inhibition of the respiratory burst of human neutrophils by the polyamine oxidase-polyamine system. *Int. J. Immunopharmacol.* 8, 411–417.

Fisher, S. M., Cameron, G. S., Baldwin, J. K., Jasheway, D. W., and Patrick, K. E. (1988). Reactive oxygen in the tumor promotion stage of skin carcinogenesis. *Lipids* 23, 592–597.

Flescher, E., Bowlin, T. L., Ballester, A., Houk, R., and Talal, N. (1989). Increased polyamines may downregulate interleukin 2 production in rheumatoid arthritis. *J. Clin. Invest.* 83, 1356–1362.

Flescher, E., Ledbetter, J. A., Schieven, G. L., Vela Roch, N., Fossum, D., Dang, H., Ogawa, N., and Talal, N. (1994). Longitudinal exposure of human T lymphocytes to weak oxidative stress suppresses transmembrane and nuclear signal transduction. *J. Immunol.* 153, 4880–4889.

Gange, R. W., Park, Y. K., Auletta, M., Kagetsu, N., Blackett, A. D., and Parrish, J. A. (1986). *In* "The Biologic Effects of UVA Radiation" (F. Urbach and R. W. Gange, eds.), pp. 57–67. Praeger, New York.

Gerner, E. W., Tome, M. E., Fry, S. E., and Bowden, G. T. (1988). Inhibition of ionizing radiation recovery process in polyamine-depleted chinese hamster cells. *Cancer Res.* 48, 4881–4885.

Getoff, N., and Schwörer, F. (1970). Pulsradiolytische Bestimmung von Geschwindigkeitskonstanten der Reaktionen einiger Amine mit OH und e_{aq}^-. *Int. J. Radiat. Phys. Chem.* 2, 81.

Gilad, G. M., and Gilad, V. H. (1991). Polyamines can protect against ischemia-induced nerve cell death in gerbil forebrain. *Exp. Neurol.* 111, 349–355.

Gilad, G. M., and Gilad, V. H. (1992). Polyamines in neurotrauma: Ubiquitous molecules in search of a function. *Biochem. Pharmacol.* 44, 401–407.

Guarnieri, C., Georgountzos, A., Caldarera, I., Flamigni, F., and Ligabue, A. (1987). Polyamines stimulate superoxide production in human neutrophils activated by N-fMet-Leu-Phe but not by phorbol myristate acetate. *Biochim. Biophys. Acta* 930, 135–139.

Gutteridge, J. M. C., Rowley, D. A., and Halliwell, B. (1981). Superoxide-dependent formation of hydroxyl radicals in the presence of iron salts. *Biochem. J.* **199**, 263–265.

Hagege, D., Kevers, C., Boucaud, J., Duyme, M., and Gaspar, T. (1990). Polyamines, phospholipids, and peroxides in normal and habituated sugar beet calli. *J. Plant Physiol.* **136**, 641–645.

Halliwell, B. (1978). Superoxide-dependent formation of hydroxyl radicals in the presence of iron salts. *FEBS Lett.* **96**, 238–242.

Halliwell, B., and Chirico, S. (1993). Lipid peroxidation: Its mechanism, measurement, and significance. *Am. J. Clin. Nutr.* **57**, Suppl. 715S–725S.

Harari, P. M., Tome, M. E., Fuller, D. J. M., Carper, S. W., and Gerner, E. W. (1989). Effects of diethyldithiocarbamate and endogenous polyamine content on cellular responses to hydrogen peroxide cytotoxicity. *Biochem. J.* **260**, 487–490.

Hares, G. B., Fernelius, W. C., and Douglas, B. E. (1956). Equilibrium constants for the formation of complexes between metal ions and polyamines. *J. Am. Chem. Soc.* **78**, 1816–1818.

Held, K. D., and Awad, S. (1991). Effects of polyamines and thiols on the radiation sensitivity of bacterial transforming DNA. *Int. J. Radiat. Biol.* **59**, 699–710.

Hillebrand, G. G., Winslow, M. S., Benzinger, M. J., Heitmeyer, D. A., and Bissett, D. L. (1990). Acute and chronic ultraviolet radiation induction of epidermal ornithine decarboxylase activity in hairless mice. *Cancer Res.* **50**, 1580–1584.

Hochstein, P., Hatch, L., and Sevanian, A. (1984). *Method Enzymol.* **105**, 162–166.

Hong, K., Schuber, F., and Papahadjopoulos, D. (1983). Polyamines: Biological modulators of membrane function. *Biochim. Biophys. Acta* **732**, 469–472.

Hunt, N. H., and Fragonas, J.-C. (1992). Effects of anti-oxidants on ornithine decarboxylase in mitogenically activated T lymphocytes. *Biochim. Biophys. Acta Mol. Cell Res.* **1133**, 261–267.

Hurst, J. K., and Barrette, W. C. J. (1989). Leukocytic oxygen activation and microbicidal oxidative toxins. *Crit. Rev. Biochem. Mol. Biol.* **24**, 271–328.

Igarashi, K., Sakamoto, I., Goto, N., Kashiwagi, K., Honma, R., and Hirose, S. (1982). Interaction between polyamines and nucleic acids or phospholipids. *Arch. Biochem. Biophys.* **219**, 438–443.

Iitaka, Y., and Huse, Y. (1965). The crystal structure of spermine phosphate hexahydrate. *Acta Cryst.* **18**, 110–121.

Janero, D. R. (1990). Malondialdehyde and thiobarbituric acid-reactivity as diagnostic indices of lipid peroxidation and peroxidative tissue injury. *Free Rad. Biol. Med.* **9**, 515–540.

Jesaitis, A. J., Quinn, M. T., Mukherjee, G., Ward, P. A., and Dratz, E. A. (1991). Death by oxygen: Radical views. *New Biol.* **3**, 651–655.

Jurkiewicz, B. A., Bissett, D. L., and Buettner, G. R. (1995). Effect of topically applied tocopherol on ultraviolet radiation-mediated free radical damage in skin. *J. Invest. Dermatol.* **104**, 484–488.

Jurkiewicz, B. A., and Buettner, G. R. (1994). Ultraviolet light-induced free radical formation in skin: An electron paramagnetic resonance study. *Photochem. Photobiol.* **59**, 1–4.

Kafy, A. M. L., Haigh, C. G., and Lewis, D. A. (1986). In vitro interactions between endogenous polyamines and superoxide anion. *Agents Actions* **18**, 555–559.

Kafy, A. M. L., and Lewis, D. A. (1988). Antioxidant effects of exogenous polyamines in damage of lysosomes inflicted by xanthine oxidase or stimulated polymorphonuclear leucocytes. *Agents Actions* **24**, 145–151.

Kauppila, T., Stenberg, D., and Porkka-Heiskanen, T. (1988). Putative stimulants for functional recovery after neural trauma: Only spermine was effective. *Exp. Neurol.* **99**, 50–58.

Kauppila, T. (1992). Polyamines enhance recovery after sciatic nerve trauma in the rat. *Brain Res.* **575**, 299–303.

Kay, J. E., and Lindsay, V. J. (1973). Polyamine synthesis during lymphocyte activation: Induction of ornithine decarboxylase and S-adenosyl methionine decarboxylase. *Exp. Cell Res.* **77**, 428–436.

Keinan, E., and Masur, Y. (1977). *J. Org. Chem.* **42**, 844.

Khan, A. U. (1991). The discovery of the chemical evolution of singlet oxygen: Some current chemical, photochemical, and biological applications. *Int. J. Quantum Chem.* **39**, 251–267.

Khan, A. U., Mei, Y.-H., and Wilson, T. (1992a). A proposed function for spermine and spermidine: Protection of replicating DNA against damage by singlet oxygen. *Proc. Natl. Acad. Sci. USA* **89**, 11426–11427.

Khan, A. U., Di Mascio, P., Medeiros, M. H. G., and Wilson, T. (1992b). Spermine and spermidine protection of plasmid DNA against single-strand breaks induced by singlet oxygen. *Proc. Natl. Acad. Sci. USA* **89**, 11428–11430.

Khan, N. A., Quemener, V., and Moulinoux, J.-P. (1991). Polyamine membrane transport regulation. *Cell Biol. Int. Rep.* **15**, 9–24.

Kimura, E., Sakonaka, A., and Nakamoto, M. (1981). Superoxide dismutase activity of macrocyclic polyamine complexes. *Biochim. Biophys. Acta* **678**, 172–179.

Kitada, M., Igarashi, K., Hirose, S., and Kitagawa, H. (1979). Inhibition by polyamines of lipid peroxide formation in rat liver microsomes. *Biochem. Biophys. Res. Commun.* **87**(2), 388–394.

Kitada, M., Naito, Y., Igarashi, K., Hirose, S., Kanakubo, Y., and Kitagawa, H. (1981). Possible mechanism of inhibition by polyamines of lipid peroxidation in rat liver microsomes. *Res. Commun. Chem. Pathol. Pharmacol.* **33**(3), 487–497.

Kitada, M., Yamaguchi, N., Igarashi, K., Hirose, S., and Kitagawa, H. (1980). Effects of polyamines on ethylmorphine N-demethylation in rat liver microsomes. *Japan J. Pharmacol.* **30**, 579–586.

Kligman, L. H., Akin, F. J., and Kligman, A. M. (1985). The contribution of UVA and UVB to connective tissue damage in hairless mice. *J. Invest. Dermatol.* **84**, 272–276.

Koenig, H., Goldstone, A., and Lu, C. Y. (1983). Polyamines regulate calcium fluxes in a rapid plasma membrane response. *Nature* **305**, 530–534.

Kogure, K., Fukuzawa, K., Kawano, H., and Terada, H. (1993). Spermine accelerates iron-induced lipid peroxidation in mitochondria by modification of membrane surface charge. *Free Rad. Biol. Med.* **14**, 501–507.

Kotzabasis, K., Fotinou, C., Roubelakis-Angelakis, K. A., and Ghanotakis, D. (1993). Polyamines in the photosynthetic apparatus: Photosystem II highly resolved subcomplexes are enriched in spermine. *Photosynth. Res.* **38**, 83–88.

Kramer, G. F., Norman, H. A., Krizek, D. T., and Mirecki, R. M. (1991). Influence of UV-B radiation on polyamines, lipid peroxidation and membrane lipids in cucumber. *Phytochemistry* **30**, 2101–2108.

Kramer, G. F., and Wang, C. Y. (1989). Correlation of reduced chilling injury with increased spermine and spermidine levels in zucchini squash. *Physiol. Plant* **76**, 479–484.

Krupa, S. V., and Kickert, R. N. (1989). *Environ. Pollut.* **61**, 263.

Langebartels, C., Kerner, K., Leonardi, S., Schraudner, M., Trost, M., Heller, W., and Sandermann, J. H. (1991). Biochemical plant responses to ozone. *Plant Physiol.* **95**, 882–889.

Linseman, K. L., Larson, P., Braughler, J. M., and McCall, J. M. (1993). Iron-initiated tissue oxidation: Lipid peroxidation, vitamin E destruction and protein thiol oxidation. Inhibition by a novel antioxidant, U-78517F. *Biochem. Pharmacol.* **45**, 1477–1482.

Liu, J., and Mori, A. (1993). Monoamine metabolism provides an antioxidant defense in the brain against oxidant- and free radical-induced damage. *Arch. Biochem. Biophys.* **302**, 118–127.

Løvaas, E. (1991a). Antioxidative effects of polyamines. *J. Am. Oil. Chem. Soc.* **68**, 353–358.

Løvaas, E. (1991b). Antioxidants for the use in the preparation of easily oxidizable compounds, and a method for the protection of easily oxidizable compounds and application of the antioxidants. *EP Patent 209 509 B1.*

Løvaas, E. (1994). *In* "Polyamines: Biological and Clinical Aspects" (C. M. Caldarera, C. Clô, and M. S. Moruzzi, eds.), pp. 161–167. Editrice Clueb, Bologna.

Løvaas, E. (1995). Hypothesis: Spermine may be an important epidermal antioxidant. *Med. Hypoth.* **45**, 59–67.

Løvaas, E., and Carlin, G. (1991). Spermine: An anti-oxidant and anti-inflammatory agent. *Free Rad. Biol. Med.* **11**, 455–461.

Martell, A. E. (1981). In "Development of Iron Chelators for Clinical Use" (A. E. Martell, W. French Anderson, and D. G. Badman, eds.), pp. 67–104. Elsevier North-Holland, New York.

Matkovics, B., Kecskemeti, V., Varga, S. I., Novak, Z., and Kertesz, Z. (1993). Antioxidant properties of di- and polyamines. *Comp. Biochem. Physiol. B* **104**, 475–479.

McCormack, S. A., and Johnson, L. R. (1991). Role of polyamines in gastrointestinal mucosal growth. *Am. J. Physiol. Gastrointest. Liver Physiol.* **260**, G795–G806.

Meers, P., Hong, K., Bentz, J., and Papahadjopoulos, D. (1986). Spermine as a modulator of membrane fusion: Interactions with acidic phospholipids. *Biochemistry* **25**, 3109–3118.

Metcalf, B. W., Bey, P., Danzin, C., Jung, M. J., Casara, J. P., and Vevert, J. P. (1978). *J. Am. Chem. Soc.* **100**, 2551.

Minton, K. W., Tabor, H., and Tabor, C. W. (1990). Paraquat toxicity is increased in *Escherichia coli* defective in the synthesis of polyamines. *Proc. Natl. Acad. Sci. USA* **87**, 2851–2855.

Mizui, T., Shimono, N., and Doteuchi, M. (1987). A possible mechanism of protection by polyamines against gastric damage induced by acidified ethanol in rats: Polyamine protection may depend on its antiperoxidative properties. *Jpn. J. Pharmacol.* **44**, 43–50.

Monroe, B. M. (1985). In "Singlet O_2" (A. A. Frimer, ed.), pp. 177–224. CRC, Boca Raton, FL.

Morgan, J. E., Blankenship, J. W., and Mathews, H. R. (1986). Association constants for the interaction of double-stranded and single-stranded DNA with spermine, putrescine, diaminopropane, N^1- and N^8-acetyl-spermidine, and magnesium: Determination from analysis of the broadening of thermal denaturation curves. *Arch. Biochem. Biophys.* **246**, 225–232.

Moruzzi, M. S., Monti, M. G., Piccinini, G., Marverti, G., and Tadolini, B. (1990). Effect of spermine on association of protein kinase C with phospholipid vesicles. *Life Sci.* **47**, 1475–1482.

Muntané, J., Puig-Parellada, P., Fernandez, Y., Mitjavila, S., and Mitjavila, M. T. (1993). Antioxidant defenses and its modulation by iron in carrageenan-induced inflammation in rats. *Clin. Chim. Acta* **214**, 185–193.

Muscari, C., Guarnieri, C., Stefanelli, C., Giaccari, A., and Caldarera, C. M. (1995). Protective effect of spermine on DNA exposed to oxidative stress. *Mol. Cell. Biochem.* **144**, 125–129.

Nahas, N., and Graff, G. (1982). Inhibitory activity of polyamines on phospholipase C from human platelets. *Biochem. Biophys. Res. Commun.* **109**(3), 1035–1040.

Nestmann, E. R. (1977). Antimutagenic effects of spermine and guanosine in continuous cultures of *Escherichia coli* mutator strain *mutH⁻*. *Mol. Gen. Genet.* **152**, 109–110.

Niki, E. (1990). Free radical initiators as source of water- or lipid-soluble peroxyl radicals. *Methods Enzymol.* **186**, 100–107.

Nordlie, R. C., Johnson, W. T., Cornatzer, H. W. E., and Twedell, G. W. (1979). Stimulation by polyamines of carbamylphosphate : glucose phosphotransferase and glucose-6-phosphate phosphohydrolase activities of multifunctional glucose-6-phosphatase. *Biochim. Biophys. Acta* **585**, 12–23.

Oldham, K. T., Guice, K. S., Ward, P. A., and Johnson, K. J. (1988). The role of oxygen radicals in immune complex injury. *Free Rad. Biol. Med.* **4**, 387–397.

Ormrod, D. P., and Beckerson, D. W. (1986). Polyamines as antiozonants for tomato. *HortScience* **21**, 1070–1071.

Otani, S., Kuramoto, A., Matsui, I., and Morisawa, S. (1982). Induction of ornithine decarboxylase in guinea pig lymphocytes by the divalent cation ionophore A 23187: Effect of dibutyryladenosine 3′,5′-monophosphate. *Eur. J. Biochem.* **125**, 35–40.

Oyanagui, Y. (1984). Anti-inflammatory effects of polyamines in serotonin and carrageenan paw edemata: Possible mechanism to increase vascular permeability inhibitory protein level which is regulated by glucocorticoids and superoxide radical. *Agents Actions* **14**, 228–237.

Palmer, B. N., and Powell, H. K. J. (1974a). Complex formation between 4,9-diazadodecane-1, 12-diamine (spermine) and copper(II) ions and protons in aqueous solution. *J. C. S. Dalton* 2086–2089.

Palmer, B. N., and Powell, H. K. J. (1974b). Polyamine complexes with seven-membered chelate rings: Complex formation of 3-azaheptane-1,7-dimaine, 4-azaoctane-1,8-diamine (spermidine), and 4,9-diazadodecane-1,12-diamine (spermine) with copper(II) and hydrogen ions in aqueous solution. *J.C.S. Dalton* 2089–2092.

Parchment, R. E., Lewellyn, A., Swartzendruber, D., and Pierce, G. B. (1990). Serum amine oxidase activity contributes to crisis in mouse embryo cell lines. *Proc. Natl. Acad. Sci. USA* **87**, 4340–4344.

Pathak, M. A. (1986). *In* "The Biologic Effects of UVA Radiation" (F. Urbach and R. W. Gange, eds.), pp. 156–167. Praeger, New York.

Pavlovic, D. D., Uzunova, P., Galabova, T., Peneva, V., Sokolova, Z., Bjelakovic, G., and Ribarov, S. (1992). Polyamines as modulators of lipoperoxidation. *Gen. Physiol. Biophys.* **11**, 203–211.

Peterson, A. O., McCann, V., and Black, H. S. (1980). Dietary modification of UV-induced epidermal ornithine decarboxylase. *J. Invest. Dermatol.* **75**, 408–410.

Piette, J. (1991). Biological consequences associated with DNA oxidation mediated by singlet oxygen. *J. Photochem. Photobiol. B* **11**, 241–260.

Prager, A., Terry, N. H. A., and Murray, D. (1993). Influence of intracellular thiol and polyamine levels on radioprotection by aminothiols. *Int. J. Radiat. Biol.* **64**, 71–81.

Pryor, W. A., Lightsey, J. W., and Prier, D. G. (1982). *In* "Lipid Peroxides in Biology and Medicine" (K. Yagi, ed.), pp. 1–22. Academic Press, New York.

Quannès, C., and Wilson, T. (1968). Quenching of singlet oxygen by tertiary aliphatic amines: Effect of DABCO. *J. Am. Chem. Soc.* **90**, 6527–6528.

Reaven, P. D., Khouw, A., Beltz, W. F., Parthasarathy, S., and Witztum, J. L. (1993). Effect of dietary antioxidant combinations in humans: Protection of LDL by vitamin E but not by β-carotene. *Arterioscler. Thromb.* **13**, 590–600.

Reif, D. W. (1992). Ferritin as a source of iron for oxidative damage. *Free Rad. Biol. Med.* **12**, 417–427.

Reiter, R. J. (1995). Oxidative processes and antioxidative defence mechanisms in the aging brain. *FASEB J.* **9**, 526–533.

Rice-Evans, C., and Burdon, R. (1993). Free radical-lipid interactions and their pathological consequences. *Prog. Lipid Res.* **32**, 71–110.

Ricevuti, G., Mazzone, A., Pasotti, D., de Servi, S., and Specchia, G. (1991). Role of granulocytes in endothelial injury in coronary heart disease in humans. *Atherosclerosis* **91**, 1–14.

Roskams, A. J. I., and Connor, J. R. (1994). Iron, transferrin, and ferritin in the rat brain during development and aging. *J. Neurochem.* **63**, 709–716.

Rowland-Bamford, A. J., Borland, A. M., Lea, P. J., and Mansfield, T. A. (1989). The role of arginine decarboxylase in modulating the sensitivity of barley to ozone. *Environ. Pollut.* **61**, 95–106.

Rowley, D. A., and Halliwell, B. (1985). Formation of hydroxyl radicals from NADH and NADPH in the presence of copper salts. *J. Inorganic Biochem.* **23**, 103–108.

Ryan, T. P., and Aust, S. D. (1992). The role of iron in oxygen-mediated toxicities. *Crit. Rev. Toxicol.* **22**, 119–141.

Scaccini, C., and Jialal, I. (1994). LDL modification by activated polymorphonuclear leukocytes: A cellular model of mild oxidative stress. *Free Rad. Biol. Med.* **16**, 49–55.

Schuber, F. (1989). Influence of polyamines on membrane function. *Biochem. J.* **260**, 1–10.

Seiler, N., and Dezeure, F. (1990). Polyamine transport in mammalian cells. *Int. J. Biochem.* **22**, 211–218.

Sevag, M. G., and Drabble, W. T. (1962). Prevention of the emergence of drug-resistant bacteria by polyamines. *Biochem. Biophys. Res. Commun.* **8**, 446–452.

Sevanian, A., Davies, K. J. A., and Hochstein, P. (1985). Conservation of vitamin C by uric acid in blood. *Free Rad. Biol. Med.* **1**, 117–124.

Shankel, D. M., and Clarke, C. H. (1990). Specificity of antimutagens against chemical mutagens in microbial systems. *Basic Life Sci.* **52**, 457–460.

Shigematsu, N., Schwartz, J. L., and Grdina, D. J. (1994). Protection against radiation-induced mutagenesis at the *hprt* locus by spermine and N,N″-(dithiodi-2,1-ethanediyl)bis-1,3-propanediamine (WR-33278). *Mutagenesis* **9**, 355–360.

Sies, H. (1993). Strategies for antioxidative defense. *Eur. J. Biochem.* **215**, 213–219.

Smith, J. A. (1994). Neutrophils, host defense, and inflammation: A double-edged sword. *J. Leukocyte Biol.* **56**, 672–686.

Spinks, J. W. T., and Woods, R. J. (1976). *In* "An Introduction to Radiation Chemistry" (J. W. T. Spinks and R. J. Woods, eds.), pp. 330–331. Wiley, New York.

Søndergaard, J., Bisgaard, H., and Thorsen, S. (1985). Eicosanoids in skin UV inflammation. *Photodermatology* **2**, 359–366.

Tabor, C. W., and Tabor, H. (1976). 1,4-Diaminobutane (putrescine), spermidine, and spermine. *Annu. Rev. Biochem.* **45**, 285–306.

Tabor, H., and Tabor, C. W. (1972). Biosynthesis and metabolism of 1,4-diaminobutane, spermidine, spermine, and related amines. *Adv. Enzymol.* **36**, 203–268.

Tabor, H., and Tabor, C. W. (1964). Spermidine, spermine, and related amines. *Pharmacol. Rev.* **16**, 245–300.

Tadolini, B. (1988a). Polyamine inhibition of lipoperoxidation. *Biochem. J.* **249**, 33–36.

Tadolini, B. (1988b). The influence of polyamine-nucleic acid complexes on Fe^{2+} autoxidation. *Mol. Cell Biochem.* **83**, 179–185.

Tadolini, B., Cabrini, L., Landi, L., Varani, E., and Pasquali, P. (1984). Polyamine binding to phospholipid vesicles and inhibition of lipid peroxidation. *Biochem. Biophys. Res. Commun.* **122**, 550–555.

Tadolini, B., Cabrini, L., Landi, L., Varani, E., and Pasquali, P. (1985a). Inhibition of lipid peroxidation by spermine bound to phospholipid vesicles. *Biogenic Amines* **3**, 97–106.

Tadolini, B., Cabrini, L., Varani, E., and Sechi, A. M. (1985b). Spermine binding and aggregation of vesicles of different phospholipid composition. *Biogenic Amines* **3**, 87–96.

Tadolini, B., and Hakim, G. (1988). Interaction of polyamines with phospholipids: Spermine and Ca^{2+} competition for phosphatidylserine containing liposomes. *Adv. Exp. Med. Biol.* **250**, 481–490.

Tadolini, B., Motta, P., and Sechi, A. M. (1992). Phospholipid polar heads affect the generation of oxygen active species by Fe^{2+} autoxidation. *Biochem. Int.* **26**, 987–994.

Tappel, A. L. (1973). Lipid peroxidation damage to cell components. *Fed. Proc.* **32**, 1870–1881.

Tkachenko, A. G., Rosenblat, G. F., Chudinov, A. A., and Raev, M. B. (1991). The role of the cell energetic status and polyamines in phospholipid content of membranes in *Escherichia coli* in the course of aerobic-anaerobic transitions. *Curr. Microbiol.* **22**, 151–153.

Toninello, A., Via, L. D., Siliprandi, D., and Garlid, K. D. (1992). Evidence that spermine, spermidine, and putrescine are transported electrophoretically in mitochondria by a specific polyamine uniporter. *J. Biol. Chem.* **267**, 18393–18397.

Trush, M. A., and Kensler, T. W. (1991). An overview of the relationship between oxidative stress and chemical carcinogenesis. *Free Rad. Biol. Med.* **10**, 201–209.

van der Wal, N. A., van Oirschot, J. F., van Dijk, A., Verhoef, J., and van Asbeck, B. S. (1990). Mechanism of protection of alveolar type II cells against paraquat-induced cytotoxicity by deferoxamine. *Biochem. Pharmacol.* **39**, 1665–1671.

Vanella, A., Rapisarda, A., Pinturo, R., and Rizza, V. (1980). Inhibitor effect of polyamines on reduction of cytochrome c by superoxide anion. *Biochem. Exp. Biol.* **16**, 165–170.

Verma, A. K., Lowe, N. J., and Boutwell, R. K. (1979). Induction of mouse epidermal ornithine decarboxylase activity and DNA synthesis by ultraviolet light. *Cancer Res.* **39**, 1035–1040.

Vertino, P. M., Bergeron, R. J., Cavanaugh, P. F., and Porter, C. V. (1987). Structural determinants of spermidine–DNA interactions. *Biopolymers* **26**, 691–703.

Walters, J. D., Sorboro, D. M., and Chapman, K. J. (1992). Polyamines enhance calcium mobilization in fMet-Leu-Phe-stimulated phagocytes. *FEBS Lett.* **304**, 37–40.

Watanabe, S., Kusama Eguchi, K., Kobayashi, H., and Igarashi, K. (1991). Estimation of polyamine binding to macromolecules and ATP in bovine lymphocytes and rat liver. *J. Biol. Chem.* **266**, 20803–20809.

Weiss, S. J. (1989). Tissue destruction by neutrophils. *N. Engl. J. Med.* **320**, 365–376.

Weitzman, S. A., and Gordon, L. I. (1990). Inflammation and cancer: Role of phagocyte-generated oxidants in carcinogenesis. *Blood* **76**, 655–663.

Williams, J. R., Casero, R. A., and Dillehay, L. E. (1994). The effect of polyamine depletion on the cytotoxic response to PUVA, gamma rays and UVC in V79 cells in vitro. *Biochem. Biophys. Res. Commun.* **201**, 1–7.

Winterbourn, C. C. (1981). Hydroxyl radical production in body fluids: Roles of metal ions, ascorbate and superoxide. *Biochem. J.* **198**, 125–131.

Yukioka, K., Otani, S., Matsui Yuasa, I., Shibata, T., Nishizawa, Y., Morii, H., and Morisawa, S. (1987). Polyamine biosynthesis is necessary for interleukin-2-dependent proliferation but not for interleukin-2 production or high-affinity interleukin-2 receptor expression. *J. Biochem. Tokyo.* **102**, 1469–1476.

Yukioka, K., Wakitani, S., Yukioka, M., Furumitsu, Y., Shichikawa, K., Ochi, T., Goto, H., Matsui-Yuasa, I., Otani, S., Nishizawa, Y., and Morii, H. (1992). Polyamine levels in synovial tissues and synovial fluids of patients with rheumatoid arthritis. *J. Rheumatol.* **19**, 689–692.

Ludmila G. Korkina*
Igor B. Afanas'ev[†]

*Institute of Pediatric Hematology
Moscow 117513, Russia

[†] Vitamin Research Institute
Moscow 177820, Russia

Antioxidant and Chelating Properties of Flavonoids

I. Introduction

Flavonoids are a group of naturally occurring, low molecular weight polyphenols of plant origin, which formally should be considered as benzo-γ-pyrone derivatives. This chapter discusses several groups of flavonoids: flavones (I), flavonols (Ia), flavanones (II), and flavanols (IIa) (Fig. 1). The members of these groups differ by the number and the positions of hydroxyl substituents in rings A and B. In addition, there are flavonoids that cannot be included in these groups but that are important biological and pharmaceutic agents. Among them are catechin (III), (+)-cianidanol (IV), and others.

Flavonoids have been reported to exert multiple biological effects (Havsteen, 1983) and to exhibit anti-inflammatory, antiallergic, antiviral, and anticancer activities. It has been suggested that flavonoid activities depend heavily on their antioxidant (Larson, 1988) and chelating properties (Cavallini et al., 1978; Afanas'ev et al., 1989, 1995; Wu et al., 1995). Therefore,

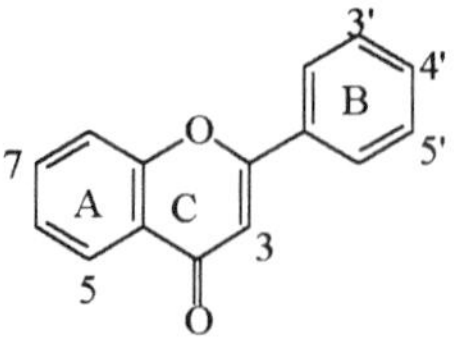

Flavones (I)
Flavonols (Ia, 3-OH)

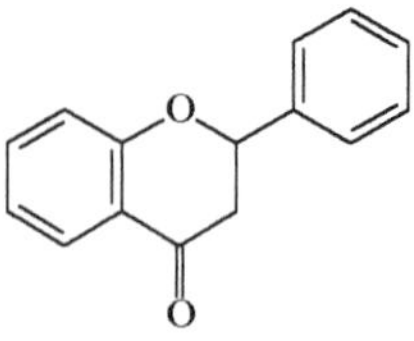

Flavanones (II)
Flavanols (IIa, 3-OH)

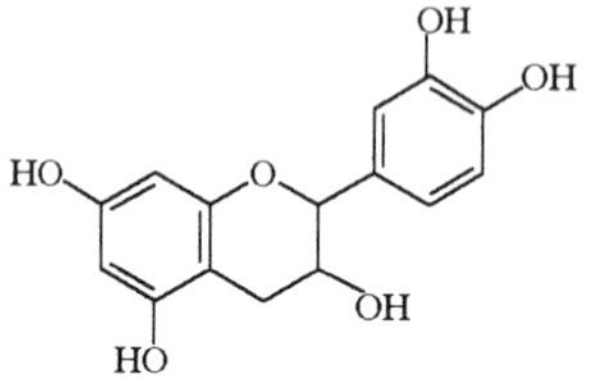

Catechin (III)

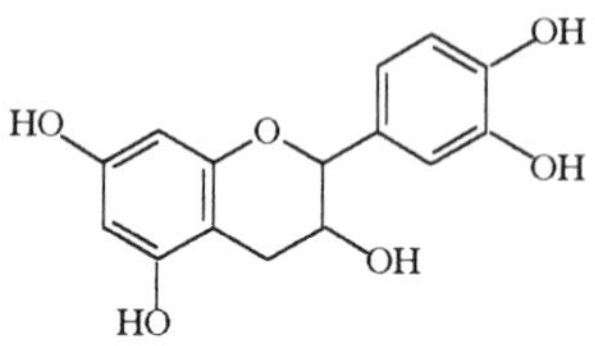

(+)-Cyanidanol (IV)

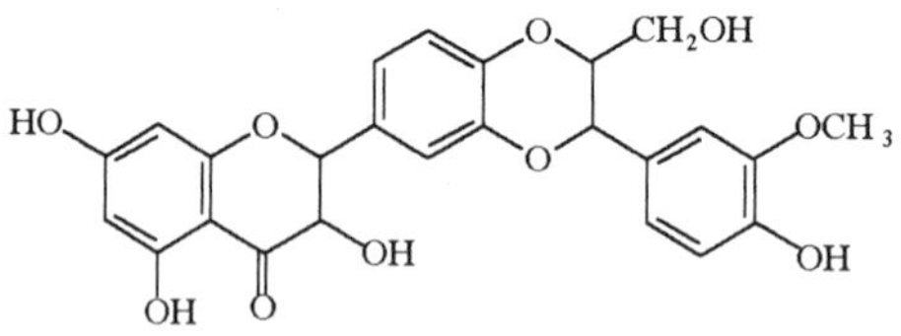

Silybin (V)

FIGURE I   Flavonoids.

numerous studies of the effects of flavonoids on the *in vitro* and *in vivo* free radical-mediated processes have been carried out (see later).

This chapter considers the mechanisms of free radical-scavenging, antioxidant, and chelating activities of flavonoids. Furthermore, the effects of flavonoids on free radical-mediated nonenzymatic and enzymatic processes and free radical production by cells will be discussed. The final sections will regard the protective effects of flavonoids against cellular and tissue damage, their toxic effects on tumor cells, and their mutagenic and antimutagenic properties.

II. Free Radical-Scavenging, Antioxidant, and Chelatory Activities of Flavonoids

A. Interaction of Flavonoids with Free Radicals and Metal Ions

Being polyphenols, flavonoids are the good scavengers of free radicals due to high reactivities of their hydroxyl substituents in a hydrogen atom abstraction reaction:

$$Fl(OH) + R\cdot \rightarrow Fl(O\cdot) + RH \qquad (1)$$

The rate constants of Eq. (1) depend on the dissociation energy of the O-H bond $D(O\text{-}H)$ and the one-electron reduction potential of the ($FlOH/FlO\cdot$) pair. Unfortunately, the values of $D(O\text{-}H)$ for flavonoids are unknown; therefore, in several studies the reactivities of flavonoids as free radical scavengers were estimated on the basis of their reduction potentials. The values of reduction potentials of flavonoids are cited in Table I. Although there is a significant difference between the results obtained by different authors, it seems that the reactivities of flavonoids increase with increasing a number of hydroxyl substituents in ring B.

The effects of flavonoids on various superoxide-generating systems have been studied. It has been shown that quercetin, myricetin, quercitrin, and rutin inhibit superoxide production by xanthine/xanthine oxidase measured as the formation of DMPO-OOH spin adducts (Ueno *et al.*, 1984a,b), by the reduction of nitro blue tetrazolium (Robak and Gryglewski 1988; Hanasaki *et al.*, 1994), or by the oxidation of hydroxylamine (Negre-Salvayre *et al.*, 1991). Yuting *et al.* (1990) have studied the inhibitory effect of flavonoids on superoxide production by the phenazine methosulfate-NADH system. Pincemail *et al.* (1989) have shown that flavonoids from the Ginkgo biloba extract possess superoxide dismuting activity. Flavonoids suppressed the release of superoxide by zymosan-, FMLP-, IgG-, asbestos-, zeolite-, and PMA-stimulated PMNs (Blackburn *et al.*, 1987; Lonchampt *et al.*, 1989; Korkina *et al.*, 1992a; Ursini *et al.*, 1994). Rutin was found to be an especially effective inhibitor of oxygen radical production by leukocytes of Fanconi anemia patients (Korkina *et al.*, 1992b).

TABLE I Values of Reduction Potentials and Rate Constants for the Reaction of Flavonoids with Superoxide Ion (E_7 at pH 7.0)

Flavonoid	$E_7(V)$	$k_2{}^a 10^4\ M^{-1}\ sec^{-1}$
Catechin (III)	0.57^a	1.8
Fisetin (Ia, 3,3',4',7-(OH)$_4$)	0.214^b	1.3
Quercetin (Ia, 3,3',4',5,7-(OH)$_5$)	$0.60^a\ 0.398^b$	4.7
Rutin (Ia, 3-rutinose,3',4',5,7-(OH)$_4$)	$0.60^a\ 0.295^b$	5.1
Hesperetin (II, 5,7,3'-(OH)$_3$,4-OMe)		0.59
Hesperidin (II,3-rutinose,5,7,3'(OH)$_3$, 4-OMe)	0.72^a	2.8
Kaempferol (Ia, 3,4',5,7-(OH)$_4$)	ca. $0.95^a\ 0.209^b$	0.24
Galangin (Ia, 3,5,7-(OH)$_3$)		0.088
Silybin (V)	0.62 or 0.76^c	
Dihydroquercetin (IIa, 3,3',4',5,7-(OH)$_5$)	0.083^b	
Luteolin (I, 5,7,3',4'-(OH)$_4$)	0.299^b	

[a] Jovanovic *et al.* (1994).
[b] Bors *et al.* (1995).
[c] Gyorgy *et al.* (1992).

Flavonoids apparently exhibit a double effect on superoxide-generating systems: via the inhibition of the enzymes responsible for superoxide production such as xanthine oxidase (Robak and Gryglewski, 1988; Hanasaki *et al.*, 1994) and protein kinase C (Blackburn *et al.*, 1987; Ursini *et al.*, 1994) or by the direct interaction with superoxide ion (Eq. 2).

$$Fl(OH) + O_2 \cdot^- \rightarrow Fl(O \cdot) + HOO^- \qquad (2)$$

The importance of the latter mechanism was proven by the observation of a direct reaction between rutin and superoxide ion (Afanas'ev *et al.*, 1989) and by the measurement of the rate constants of similar reactions (Jovanovic *et al.*, 1994).

One should expect a high reactivity of flavonoids in the reaction with extremely reactive hydroxyl radicals. Indeed, many flavonoids effectively scavenge hydroxyl radicals produced by the photolysis of hydrogen peroxide (Husain *et al.*, 1987) or in the Fenton reaction (Hanasaki *et al.*, 1994; Sanz *et al.*, 1994). For example, rutin is a 100-fold superior hydroxyl scavenger to mannitol (Hanasaki *et al.*, 1994). However, these authors also found that morin and some other oxidizible flavonoids are able to enhance hydroxyl radical production in the Fenton reaction (see later).

The ability of flavonoids to form complexes with iron ions (the catalysts of the Fenton reaction) has been shown spectrophotometrically (Afanas'ev *et al.*, 1989; Wu *et al.*, 1995). These complexes are apparently unable to catalyze the Fenton reaction, making the chelating activity of flavonoids an important factor of their inhibitory action on free radical-mediated processes.

B. Effects of Flavonoids on Lipid Peroxidation

The effects of flavonoids on the *in vitro* and *in vivo* lipid peroxidation have been studied extensively. Flavonoids inhibit the *in vitro* peroxidative processes such as the autoxidation of linoleic acid (Torel *et al.*, 1986), the oxidation of low-density lipoproteins (Mangiapane *et al.*, 1992), the peroxidation of phospholipid membranes (Terao *et al.*, 1994), microsomal and mitochondrial lipid peroxidation (Bindoli *et al.*, 1977; Das and Ray, 1988; Afanas'ev *et al.*, 1989; Cholbi *et al.*, 1991), lysis and peroxidation in human erythrocytes (Sorata *et al.*, 1984; Maridonneau-Parini *et al.*, 1986), lipid peroxidation in erythrocyte ghosts (Negre-Salvayre *et al.*, 1991), the autoxidation of rat brain homogenates (Kozlov *et al.*, 1994), and photooxidation and lipid peroxidation in chloroplasts (Wagner *et al.*, 1988). Flavonoids also inhibit lipid peroxidation and the deterioration of the lisosomal membrane in the light mitochondrial fraction of rat liver (Decharneux *et al.*, 1992). However, some flavonoids stimulate lipid peroxidation. Thus, Maridonneau-Parini *et al.* (1986) showed that kaempferol inhibits lipid peroxidation in human erythrocytes, rutin has no effect, and myricetin enhances it. The effect of myricetin, which has three hydroxyl substituents in the B ring, is most probably a consequence of its oxidizability.

Both free radical scavenging and chelating properties are apparently responsible for the inhibitory effect of flavonoids on lipid peroxidation. Thus it was found (Afanas'ev *et al.*, 1989) that rutin is a much more stronger inhibitor in the case of NADPH-dependent microsomal lipid peroxidation, which is catalyzed by iron ions, than in the case of lipid peroxidation initiated by CCl₄, which is independent of iron ions catalysis. The role of the chelating activity of rutin in lipid peroxidation may be explained by the formation of inactive iron–rutin complexes [forming at the oxidation of ferrous into ferric ions inside of the complex (Kozlov *et al.*, 1994)], which are unable to catalyze the Fenton reaction. It has also been shown that flavonoids are capable of inhibiting lipid peroxidation by the extraction of iron ions from iron-loaded hepatocytes (Morel *et al.*, 1993).

These data make rutin a potential pharmaceutic agent, especially against the pathological states characterized by an enhanced level of iron. Animal studies with iron-overloading rats support this proposal. It was found (Afanas'ev *et al.*, 1995) that rutin inhibited *in vivo* lipid peroxidation in liver microsomes and the oxygen radical production by neutrophils and macrophages, but only slightly affected these processes in normal animals. From comparison of lucigenin- and luminol-amplified chemiluminescence produced by phagocytes, it was also shown that rutin sharply decreased the conversion of innocuous superoxide ion into harmful hydroxyl radicals via the superoxide-driven Fenton reaction.

C. Effects of Flavonoids on Enzymatic Activity

Flavonoids have been shown to inhibit a wide range of enzymes, including ATPases, aldole reductase, phosphodiesterases, protein tyrosine kinases, and so on. The inhibitory activity of flavonoids may depend on their antioxidant and chelatory properties. For example, the ability of flavonoids to affect free radical-mediated processes seems to be based on the inhibition of cyclooxygenase (Gryglewski *et al.*, 1987), lipoxygenase (Yoshimoto *et al.*, 1983; Ratty *et al.*, 1988), microsomal monooxygenase (Beyeler *et al.*, 1988), and glutathione *S*-transferase (Merlos *et al.*, 1991). Hodnick *et al.* (1986, 1987, 1994) has shown that many flavonoids having a catechol moiety in the B ring inhibit the mitochondrial succinoxidase and NADH oxidase activities. It was supposed that the primary inhibition site was complex I (the NADH-ubiquinone reductase) and the second one was complex II (the succinate-ubiquinone reductase). It was also shown that flavonols quercetin, kaempferol, morin, and the flavone luteolin are effective inhibitors of rat liver cytosolic GSH *S*-transferase activity (Merlos *et al.*, 1991; Zhang and Das, 1994), whereas quercetin and tannin inhibit neuronal constitutive endothelial NO synthase (Chiesi and Schwaller, 1995). Beyeler *et al.* (1988)

supposed that the inhibitory effect of flavonones and (+)-cyanidanol on microsomal monooxygenase activity is a consequence of their chelating activity, namely, the formation of a complex with cytochrome P450 via ligand binding.

D. Suppression by Flavonoids of Free Radical-Induced Cellular and Tissue Damage

Antioxidant and chelatory activities of flavonoids are probably the most important factors of their protective action against free radical-mediated damage in cells and tissue, although some other mechanisms cannot be excluded. Flavonoids (quercetin, kaempferol, catechin, and taxifolin) suppress the cytotoxicity of superoxide ion and hydrogen peroxide on Chinese hamster V79 cells (Nakayama *et al.*, 1993; Nakayama, 1994). The flavonoid morin was shown to be an effective protector of human cells of the cardiovascular system against oxygen radical-mediated damage (Wu *et al.*, 1994b). Myricetin and quercetin, the constituents of Ginkgo biloba extract (EGb), suppress oxidative processes in brain neurons, possibly explaining the beneficial action of these flavonoids on brain neurons subjected to ischemia (Oyama *et al.*, 1994). The flavonoid purpurogallin (which is extractable from nutgalls and chemically identical to an oxidation product of pyrogallol) markedly protects a variety of cells such as erythrocytes, hepatocytes, kidney cells, and cardiocytes in cultures against hypoxic injury. It is also quite effective as a protector of the rabbit heart damage induced by ischemia/reperfusion (Wu *et al.*, 1994a). Administration of (+)-catechin to rats injected with bromotrichloromethane decreases oxidative damage to blood and tissue caused by this toxic compound (Chien and Tappel, 1995). Many flavonoids are apparently protective against free radical-mediated injury by ionizing radiation. Thus, it has been found that 10 structurally different flavonoids exhibit anticlastogenic effects in the whole body γ-irradiated mice (Shimoi *et al.*, 1994). A good correlation between antioxidative and radioprotective activities of flavonoids indicates that the protection against the clastogenic action of irradiation may be attributed to the hydroxyl radical scavenging potency or/and the inhibition of oxygen radical-generated enzymes.

III. Prooxidant Activity of Flavonoids ─────────────────────

In previous sections the inhibitory effects of flavonoids on free radical-mediated processes based on their free radical scavenging and chelatory properties have been discussed. However, it should be remembered that flavonoids are not a homogenous group of compounds with similar chemical properties and that some of them and their complexes with transition metals are readily oxidized by molecular oxygen to form new active oxygen species. For example, Laughton *et al.* (1989) measured hydroxyl radical formation during the autoxidation of quercetin and myricetin using a deoxyribose assay. It was concluded that the reaction was catalyzed by iron ions because it was accelerated by the addition of Fe-EDTA. Canada *et al.* (1990) confirmed the formation of hydroxyl radicals in this process using DMPO as a spin-trapping agent. Thus the autoxidation of flavonoids may be responsible for the cytotoxic activity of these compounds.

IV. Cytotoxic Effects of Flavonoids against Tumor Cells

In vitro inhibition of cell proliferation by flavonoids in several tumor cell lines has been reported in several studies. Thus, Ramanathan *et al.* (1994) demonstrated the cytotoxic effects of quercetin, luteolin, and buteinon but not rutin on Raji lymphoma cells. These authors concluded that flavonoids do not exert cytotoxicity on Raji lymphoma cells through an oxidative mechanism because, in contrast to γ-linolenic acid, they did not affect the basal cellular peroxidation level.

Hirano *et al.* (1994) have regarded 28 naturally occurring and synthetic flavonoids as novel antileukemic compounds with potent cytostatic activity and low cytotoxicity against normal cells. They found that flavonoids significantly inhibit the growth of the human promyelocytic leukemia cell line HL-60, with their antiproliferative efficacy being either equivalent or even higher than that for traditional anticancer agents such as etoposide, doxorubicin, vincristine, and methotrexate. Furthermore, the antiproliferative action of flavonoids appears to be specific to leukemia cells rather than normal lymphocytes because they are less suppressive on the normal lymphocyte blastogenesis. It was supposed that flavonoids inhibit the HL-60 cell growth by a nontoxic mechanism, possibly via the cessation of DNA, RNA, and protein synthesis of leukemic cells.

Quercetin turned out to be a powerful *in vitro* antiproliferative agent against other human cancer cell lines such as primary colorectal, ovarian, lymphoblastoid, and breast cancer cells [Scambia *et al.*, (1994b) and references therein]. In all these cases quercetin modulated the growth of cancer cells acting through the type II estrogen-binding sites. However, it was proposed (Scambia *et al.*, 1994b) that quercetin may inhibit the proliferation of human ovarian cancer cells by the enhancement of transforming growth factor β_1 (TGFβ_1) secretion at posttranscriptional levels (Scambia *et al.*, 1994a).

In addition to direct anticancer activity, quercetin exhibits synergistic antiproliferative effects with some chemotherapeutic agents such as cisplatin (Scambia *et al.*, 1992), cytosine arabinoside (Teofili *et al.*, 1992), and adriamycin (Scambia *et al.*, 1994b). In addition, quercetin is apparently able to suppress the drug resistance of tumor cells, for example, via the inhibition of the overexpression of P-glycoprotein on the surface of human breast cancer line MCF-7 cells (Scambia *et al.*, 1994b). In contrast, Avila *et al.* (1994) failed to find the changes in P-glycoprotein expression on the surface of the human breast cancer line MDA-MB468, although they confirmed that quercetin affects the growth and cell cycle progression of tumor cells, arresting them at the G_2–M phase. Quercetin is capable of inhibiting heat shock protein synthesis and inducing apoptosis in several lines of human tumor cells, including chronic myeloid and acute T-lymphocytic leukemias, Burkitt lymphoma, ovarian adenocarcinoma, and human erythroleukemia (Wei *et al.*, 1994; Elia and Santoro, 1994). The induction of apoptosis can be a plausible explanation for the pronounced antitumor activity of quercetin. Apoptosis, which is characterized by the condensation of nuclear chromatin, DNA fragmentation, and the loss of membrane integrity, is now recognized as a free radical-stimulated and free radical-regulated process (Wood and Youle, 1994). Therefore, the induction of apoptosis by flavonoids may be mediated by free radicals, especially if one takes into account the fact that flavonoids are potent inducers of DNA breakage and protein fragmentation (Ahmed *et al.*, 1992, 1994).

Further evidence of the antitumor effects of flavonoids has been obtained in animal experiments with chemically induced carcinogenesis. Thus, quercetin applied topically is active against the 12-O-tetradecanoylphorbol-13-acetate (TPA)-promoted dimethylbenz[*a*]anthracene-induced skin tumor (Kato *et al.*, 1983), and, when provided with a diet, quercetin is an inhibitor of the rat mammary cancer induced by the derivatives of benzanthracene and nitrosourea (Verma *et al.*, 1988). Dietary rutin and quercetin significantly reduce the tumor incidence and multiplicity in azoxymethanol-induced colonic neoplasia, being effective on the stage of tumor promotion (Deschner *et al.*, 1991). Skin application of silymarin leads to suppression of the activity of ornithine decarboxylase, a well-known biochemical marker of tumor promotion, and mRNA overexpressed after TPA induction (Agarval *et al.*, 1994).

The exact mechanism responsible for the antitumor effect of flavonoids is not yet thoroughly understood. It is possible that flavonoids (mainly quercetin) inhibit the growth of malignant cells, affecting various metabolic pathways such as the activation of glycolytic enzymes or protein synthesis, heat shock protein induction, freezing the cell cycle, interaction with estrogen type II binding sites, induction of protein and DNA fragmentation, and induction of apoptosis. As practically all of these processes may be modulated by oxygen-free radicals, it is possible to assume that flavonoids affect the oxidant/antioxidant balance in malignant cells, leading to irreversible cell damage and death. It is important that in contrast with their effects on tumor cells, flavonoids appear to be protective against the DNA and chromosome damage induced by a variety of clastogenic factors. For example, the *in vivo* and *in vitro* treatment of mouse spleen lymphocytes with galangin suppresses the induction of bleomycin-induced chromosome aberrations (Heo *et al.*, 1994).

V. Mutagenic and Antimutagenic Properties of Flavonoids

Flavonoids can apparently manifest both mutagenic and antimutagenic effects. Present data seem to indicate that the interaction of flavonoids with active oxygen species is an important factor determining their mutagenic/ antimutagenic activity. Thus it has been shown (Ueno *et al.*, 1984b) that quercetin is mutagenic under *in vitro* (but not *in vivo*) conditions. Because SOD enhanced the quercetin mutagenicity, and superoxide ion diminished it, it was concluded that the interaction of quercetin with superoxide led to quercetin degradation and, correspondingly, to the loss of its mutagenicity. It was suggested that flavonoids have to be metabolically activated to acquire mutagenic activity, but it is not always an obligatory condition. For example, quercetin possesses a weak genuine mutagenic activity for *Salmonella typhimurium* strains whereas kaempherol requires metabolic activation (Brown and Dietrich, 1979; Vrijsen *et al.*, 1990).

The antimutagenic activity of flavonoids is most probably a consequence of their antioxidant and chelatory properties. Thus, it has been shown (Korkina *et al.*, 1992) that rutin is an effective inhibitor of oxygen radical-mediated mutagenic effects of mineral fibers and dusts on human lymphocytes.

References

Afanas'ev, I. B., Dorozhko, A. I., Brodskii, A. V., Kostyuk, V. A., and Potapovitch, A. I. (1989). Chelating and free radical scavenging mechanisms of inhibitory action of rutin and quercetin in lipid peroxidation. *Biochem. Pharmacol.* 38, 1763–1769.

Afanas'ev, I. B., Ostrachovitch, E. A., Abramova, N. E., and Korkina, L. G. (1995). Different antioxidant activities of bioflavonoid rutin in normal and iron-overloading rats. *Biochem. Pharmacol.* 50, 627–637.

Agarwal, R., Kativar, S. K., Lundgren, D. W., and Mukhtar. (1994). Inhibitory effect of silymarin, an anti-hepatotoxic flavonoid, on 12-O-tetradecanoylphorbol-13-acetate-induced epidermal ornithine decarboxylase activity and mRNA in SENCAR mice. *Carcinogenesis* 15, 1099–1103.

Agullo, G., Gamet, L., Besson, C., Demigne, C., and Remesy, C. (1994). Quercetin exerts a preferential cytotoxic effect on active dividing colon carcinoma HT29 and Caco-2 cells. *Cancer Lett.* 87, 55–63.

Ahmed, M. S., Ainley, K., Parish, J. H., and Hadi, S. M. (1994). Free radical-induced fragmentation of proteins by quercetin. *Carcinogenesis* 15, 1627–1630.

Ahmed, M. S., Fasal, F., Rahman, A., Hadi, S. M., and Parish, J. H. (1992). Activities of flavonoids for the cleavage of DNA in the presence of Cu(II): Correlation with generation of active oxygen species. *Carcinogenesis* **13**, 605–608.

Avila, M. A., Velasco, J. A., Cansado, J., and Notario, V. (1994). Quercetin mediates the down-regulation of mutant p53 in the human breast cancer cell line MDA-MB468. *Cancer Res.* **54**, 2424–2428.

Beyeler, S., Testa, B., and Perrissoud, D. (1988). Flavonoids as inhibitors of rat liver monooxygenase activities. *Biochem. Pharmacol.* **37**, 1971–1979.

Bindoli, A., Cavallini, L., and Silipandri, N. (1977). Inhibitory action of silymarin on lipid peroxidation formation in rat liver mitochondria and microsomes. *Biochem. Pharmacol.* **26**, 2405–2409.

Blackburn, W. D., Jr., Heck, L. W., and Wallace, R. W. (1987). The bioflavonoid quercetin inhibits degranulation, superoxide production, and the phosphorylation of specific neutrophil proteins. *Biochem. Biophys. Res. Commun.* **144**, 1229–1236.

Bors, W., Michel, C., and Schikora, S. (1995). Interaction of flavonoids with ascorbate and determination of their univalent redox potentials: A pulse radiolysis study. *Free Rad. Biol. Med.* **19**, 45–52.

Brown, J. P., and Dietrich, P. S. (1979). Mutagenicity of plant flavonoids in the Salmonella/mammalian microsome test: Activation of flavonol glycosides by mixed glycosidases from rat cecal bacteria and other sources. *Mutat. Res.* **66**, 223–240.

Canada, A. T., Giannella, E., Nguyen, T. D., and Mason, R. P. (1990). The production of reactive oxygen species by dietary flavonoids. *Free Rad. Biol. Med.* **9**, 441–449.

Cavallini, L., Bindoli, A., and Silipandri, N. (1978). Comparative evaluation of antiperoxidative action of silymarin and other flavonoids. *Pharm. Res. Commun.* **10**, 133–136.

Chien, H., and Tappel, A. L. (1995). Protection of vitamin E, selenium, trolox C, ascorbic acid palmitate, acetylcystein, coenzyme Qo, coenzyme Q10, beta-carotene, canthaxanthin, and (+)-catechin against oxidative damage to rat blood and tissue in vivo. *Free Rad. Biol. Med.* **18**, 949–953.

Chiesi, M., and Schwaller, P. (1995). Inhibition of constitutive endothelial NO-synthase activity by tannin and quercetin. *Biochem. Pharmacol.* **49**, 495–501.

Chimoi, K., Masuda, S., Furugori, M., Esaki, S., and Kinae, N. (1994). Radioprotective effect of antioxidative flavonoids in γ-ray irradiated mice. *Carcinogenesis* **15**, 2669–2672.

Cholbi, M. R., Paya, M., and Alcaraz, M. J. (1991). Inhibitory effects of phenolic compounds on CCl$_4$-induced microsomal lipid peroxidation. *Experientia* **47**, 195–199.

Das, M., and Ray, P. K. (1988). Lipid antioxidant properties of quercetin in vitro. *Biochem. Int.* **17**, 203–209.

Deschner, E. E., Ruperto, J., Wong, G., and Newmark, H. L. (1991). Quercetin and rutin as inhibitors of azoxymethanol-induced colonic neoplasia. *Carcinogenesis* **12**, 1193–1196.

Decharneux, T., Dubois, F., Beauloye, C., Wattiaux-De Conick, S., and Wattiaux, R. (1992). Effect of various flavonoids on lysosomes subjected to an oxidative or an osmotic stress. *Biochem. Pharmacol.* **44**, 1243–1248.

Elangovan, V., Balasubramanian, S., Sekar, N., and Govindasamy, N. (1994). Studies on the chemopreventive potential of some naturally-occurring bioflavonoids in 7,12-dimethylbenz(a)anthracene-induced carcinogenesis in mouse skin. *J. Clin. Biochem. Nutr.* **17**, 153–160.

Elia, G., and Santoro, M. G. (1994). Regulation of heat shock protein synthesis by quercetin in human erythroleukaemia cells. *Biochem. J.* **300**, 201–209.

Grinberg, L. N., Rachmilewitz, E. A., and Newmark, H. (1994). Protective effects of rutin against hemoglobin oxidation. *Biochem. Pharmacol.* **48**, 643–649.

Gryglewski, R. J., Korbut, R., Robak, J., and Swies, J. (1987). On the mechanism of antithrombotic action of flavonoids. *Biochem. Pharmacol.* **36**, 317–322.

Gyorgy, I., Antus, S., Glazovics, A., and Foldiak, G. (1992). Substituent effects in the free radical reactions of silybin: Radiation-induced oxidation of the flavonoid at neutral pH. *Int. J. Radiat. Biol.* **61**, 603–609.

Hanasaki, Y., Ogawa, S., and Fukui, S. (1994). The correlation between active oxygens scavenging and antioxidative effects of flavonoids. *Free Rad. Biol. Med.* **16**, 845–850.

Havsteen, B. (1983). Flavonoids, a class of natural products of high pharmacological potency. *Biochem. Pharmacol.* **32**, 1141–1148.

Heo, M. Y., Lee, S. J., Kwon, C. H., Kim, S. W., Sohn, D. H., and Au, W. W. (1994). Anticlastogenic effects of galangin against bleomycin-induced chromosomal aberrations in mouse spleen lymphocytes. *Mutat. Res.* **311**, 225–229.

Hirano, T., Gotoh, M., and Oka, K. (1994). Natural flavonoids and lignans are potent cytostatic agents against human leukemic HL-60 cells. *Life Sci.* **55**, 1061–1069.

Hodnick, W. F., Kung, F. S., Roettger, W. J., Bohmont, C. W., and Pardini, R. S. (1986). Inhibition of mitochondrial respiration and production of toxic oxygen radicals by flavonoids: A structure–activity study. *Biochem. Pharmacol.* **35**, 2345–2357.

Hodnick, W. F., Bohmont, C. W., Capps, C., and Pardini, R. S. (1987). Inhibition of the mitochondrial NADH-oxidase (NADH-coenzyme Q oxidoreductase) enzyme system by flavonoids: A structure–activity study. *Biochem. Pharmacol.* **36**, 2873–2874.

Hodnick, W. F., Duval, D. L., and Pardini, R. S. (1994). Inhibition of mitochondrial respiration and cyanide-stimulated generation of reactive oxygen species by selected flavonoids. *Biochem. Pharmacol.* **47**, 573–580.

Huang, H. C., Wang, H. R., and Hsieh, L. M. (1994). Antiproliferative effect of baicalein, a flavonoid from a Chinese herb, on vascular smooth muscle cell. *Eur. J. Pharmacol.* **251**, 91–93.

Husain, S. R., Cillard, J., and Cillard, P. (1987). Hydroxyl radical scavenging activity of flavonoids. *Phytochemistry* **26**, 2489–2491.

Jovanovic, S., Steenken, S., Tosic, M., Marjanovic, B., and Simic, M. G. J. (1994). Flavonoids as antioxidants. *J. Am. Chem. Soc.* **116**, 4846–4851.

Kato, R., Nakadate, T., Yamamoto, S., and Sugimura, T. (1983). Inhibition of 12-O-tetradecanoylphorbol-13-acetate induced tumor promotion and ornithine decarboxylase activity by quercetin: Possible involvement by lipoxygenase inhibition. *Carcinogenesis* **4**, 1301–1305.

Korkina, L. G., Durnev, A. D., Suslova, T. B., Cheremisina, Z. P., Daugel-Dauge, N. O., and Afansa'ev, I. B. (1992a). Oxygen radical-mediated mutagenic effect of asbestos on human lymphocytes: Suppression by oxygen radical scavengers. *Mutat. Res.* **265**, 245–253.

Korkina, L. G., Samochatova, E. V., Maschan, A. A., Suslova, T. B., Cheremisina, Z. P., and Afanas'ev, I. B. (1992b). Release of active oxygen radicals by leukocytes of Fanconi anemia patients. *J. Leukocyte Biol.* **52**, 357–362.

Kozlov, A. V., Ostrachovitch, E. A., and Afanas'ev, I. B. (1994). Mechanism of inhibitory effects of chelating drugs on lipid peroxidation in rat brain homogenates. *Biochem. Pharmacol.* **47**, 795–799.

Larson, L. (1988). The antioxidants of higher plants. *Phytochemistry* **27**, 959–978.

Laughton, M. J., Halliwell, B., Evans, P. J., and Hoult, J. R. S. (1989). Antioxidant and prooxidant action of the plant phenolics quercetin, gossipol and myricetin. *Biochem. Pharmacol.* **38**, 2859–2865.

Lonchampt, M., Guardiola, B., Sicot, N., Bertrand, M., Pedrix, L., and Duhault, J. (1989). Protective effect of a purified flavonoid fraction against reactive oxygen radicals. *Arzneim.-Forsch.* **39**, 882–885.

Mangiapane, H., Thomson, J., Salter, A., Brown, S., Bell, G. D., and White, D. A. (1992). The inhibition of the oxidation of low density lipoprotein by (+)-catechin, a naturally occurring flavonoid. *Biochem. Pharmacol.* **43**, 445–451.

Maridonneau-Parini, I., Braquet, P., and Garay, R. P. (1986). Heterogenous effect of flavonoids on K^+ loss and lipid peroxidation induced by oxygen-free radicals in human red cells. *Pharm. Res. Commun.* **18**, 61–73.

Merlos, M., Sanchez, R. M., Camarasa, J., and Adzet, T. (1991). Flavonoids as inhibitors of rat liver cytosolic glutathione S-transferase. *Experientia* **47**, 616–619.

Morel, I., Lescoat, G., Cogrel, P., Sergent, O., Pasdeloup, N., Brissot, P., Cillard, P., and Cillard, J. (1993). Antioxidant and iron-chelating activities of the flavonoids catechin, quercetin and diosmetin on iron-loaded rat hepatocyte cultures. *Biochem. Pharmacol.* **45**, 13–19.

Nakayama, T. (1994). Suppression of hydroperoxide-induced cytotoxicity by polyphenols. *Cancer Res.* **54**, 1991s–1993s.

Nakayama, T., Yamada, M., Osawa, T., and Kawakishi, S. (1993). Suppression of active oxygen-induced cytotoxicity by flavonoids. *Biochem. Pharmacol.* **45**, 265–267.

Namgoong, S. Y., Son, K. H., Chang, H. W., Kang, S. S., and Kim, H. P. (1994). Effects of naturally occurring flavonoids on mitogen-induced lymphocyte proliferation and mixed lymphocyte culture. *Life Sci.* **54**, 313–320.

Negre-Salvayre, A., Affany, A., Hariton, C., and Salvayre, R. (1991). Additional antilipoperoxidant activities of alpha-tocopherol and ascorbic acid on membrane-like systems are potentiated by rutin. *Pharmacology* **42**, 262–272.

Oyama, Y., Fuchs, P. A., Katayama, N., and Noda, K. (1994). Myricetin and quercetin, the flavonoids constituents of Ginkgo bilola extract, greatly reduce oxidative metabolism of both resting and Ca^{2+}-loaded brain neurons. *Brain Res.* **635**, 125–129.

Perez-Guerrero, C., Martin, M. J., and Marhuenda, E. (1994). Prevention by rutin of gastric lesions induced by ethanol in rats: Role of endogenous prostaglandins. *Gen. Pharmacol.* **25**, 575–580.

Pincemail, J., Dupuis, M., Nasr, C., Haus, P., Haag-Berrurier, M., Anton, R., and Deby, C. (1989). Superoxide anion scavenging effect and superoxide dismutase activity of Ginkgo biloba extract. *Experientia* **45**, 708–712.

Ramanathan, R., Das, N. P., and Tan, C. H. (1994). Effects of γ-linolenic acid, flavonoids, and vitamins on cytotoxicity and lipid peroxidation. *Free Rad. Biol. Med.* **16**, 43–48.

Ratty, A. K., Sunamoto, J., and Das, N. P. (1988). Interaction of flavonoids with 1,1-diphenyl-2-picrylhydrazyl free radical, liposomal membranes and soybean lipoxygenase-1. *Biochem. Pharmacol.* **37**, 989–995.

Robak, J., and Gryglewski, R. J. (1988). Flavonoids are scavengers of superoxide anions. *Biochem. Pharmacol.* **37**, 837–841.

Sanz, M. J., Ferrandiz, M. L., Cejudo, M., Terencio, M. C., Gil, B., Bustos, G., Ubeda, A., Gunasegaran, R., and Alcaraz, M. J. (1994). Influence of a series of natural flavonoids on free radical generating systems and oxidative stress. *Xenobiotica* **24**, 689–699.

Scambia, G., Panici, P. B., Ranelletti, F. O., Ferrandina, G., De Vincenzo, R., Piantelli, M., Masciullo, V., Bonanno, G., Isola, G., and Mancuso, S. (1994a). Quercetin enhances transforming growth factor beta 1 secretion by human ovarian cancer cells. *Int. J. Cancer* **57**, 211–215.

Scambia, G., Ranelletti, F. O., Benedetti Panici, P., Piantelli, M., Bonanno, G., De Vincenzo, R., Ferrandina, G., Maggiano, N., Capelli, A., and Mancuso, S. (1992). Inhibitory effect of quercetin on primary ovarian and endometrial cancers and synergistic antiproliferative activity with cis-diamminedichloroplatinum(II). *Gynecol. Oncol.* **45**, 13–19.

Scambia, G., Ranelletti, F. O., Panici, P. B., De Vincenzo, R., Bonanno, G., Ferrandina, G., Piantelli, M., Bussa, S., Rumi, C., and Cianfriglia, M. (1994b). Quercetin potentiates the effect of adriamycin in a multidrug-resistant MCF-7 human breast-cancer cell line: P-glycoprotein as a possible target. *Cancer Chemother. Pharmacol.* **34**, 459–464.

Shimoi, K., Masuda, S., Furugori, M., Esaki, S., and Kinae, N. (1994). Radioprotective effect of antioxidative flavonoids in γ-ray irradiated mice. *Carcinogenesis* **15**, 2669–2672.

Sorata, Y., Takahama, U., and Kimura, M. (1984). Protective effect of guercetin and rutin on photosensitized lysis of human erythrocytes in the presence of hematoporphyrin. *Biochem. Biophys. Acta* **799**, 313–317.

Teofili, L., Pierelli, L., Lovino, M. S., Leone, G., Scambia, G., De Vincenzo, R., Benedetti, F. O., and Larocca, L. M. (1992). The combination of quercetin and cytosine arabinoside synergistically inhibits leukemic growth. *Leukocyte Res.*, **16**, 497–502.

Terao, J., Piskula, M., and Yao, Q. (1994). Protective effect of epicatechin, epicatechin gallate, and quercetin on lipid peroxidation in phospholipid bilayers. *Arch. Biochem. Biophys.* **308**, 278–284.

Torel, J., Cillard, J., and Cillard, P. (1986). Antioxidant activity of flavonoids and reactivity with peroxy radical. *Phytochemistry* **25**, 383–387.

Ueno, I., Kohno, M., Haraikawa, K., and Hirono, I. (1984a). Interaction between quercetin and superoxide radicals: Reduction of the quercetin mutagenicity. *J. Pharm. Dyn.* **7**, 798–803.

Ueno, I., Kohno, M., Yoshihira, K., and Hirono, I. (1984b). Quantitative determination of the superoxide radicals in the xanthine oxidase reaction by measurement of the electron spin resonance signal of the superoxide radical spin adduct of 5,5-dimethyl-1-pyrroline-1-oxide. *J. Pharm. Dyn.* **7**, 563–569.

Ursini, F., Maiorino, M., Morazzoni, P., Roveri, A., and Pifferi, G. (1994). A novel antioxidant flavonoid (IdB 1031) affecting molecular mechanisms of cellular activation. *Free Rad. Biol. Med.* **16**, 547–553.

Verma, A. K., Johnson, J. A., Gould, M. N., and Tanner, M. A. (1988). Inhibition of 7,12-dimethylbenz(a)anthracene and N-nitrosomethylurea-induced rat mammary cancer by dietary flavonol quercetin. *Cancer Res.* **48**, 5754–5758.

Vrijsen, R., Mishotte, Y., and Boeye, A. (1990). Metabolic activation of quercetin mutagenicity. *Mutat. Res.* **232**, 243–248.

Wagner, G. R., Youngman, R. J., and Elstner, E. F. (1988). Inhibition of chloroplast photooxidation by flavonoids and mechanisms of the antioxidative action. *J. Photochem. Photobiol.* **B1**, 451–460.

Wei, Y. Q., Zhao, X., Kariya, Y., Fukata, H., Teshigawara, K., and Uchida, A. (1994). Induction of apoptosis by quercetin: Involvement of heat shock protein. *Cancer Res.* **54**, 4952–4957.

Wood, K. A., and Youle, R. J. (1994). Apoptosis and free radicals. *Ann. N.Y. Acad. Sci.* **738**, 400–407.

Wu, T.-W., Fung, K. P., Zeng, L.-H., Wu, J., Hempel, A., Grey, A. A., and Camerman, N. (1995). Molecular properties and myocardial salvage effects of morin hydrate. *Biochem. Pharmacol.* **49**, 537–543.

Wu, T.-W., Wu, J., Zeng, L.-H., Au, J.-X., Carey, D., and Fung, K. P. (1994a). Purpurogallin: In vivo evidence of a novel and effective cardioprotector. *Life Sci.* **54**, 23–28.

Wu, T.-W., Zeng, L. H., Wu, J., and Fung, K. P. (1994b). Morin: A wood pigment that protects three types of human cells in the cardiovascular system against oxyradical damage. *Biochem. Pharmacol.* **47**, 1099–1103.

Yoshimoto, T., Furukawa, M., Yamamoto, S., Horie, T., and Watanabe-Kohno, S. (1983). Flavonoids: Potent inhibitors of arachidonate 5-lipoxygenase. *Biochem. Biophys. Res. Commun.* **116**, 612–618.

Yuting, C., Rongliang, Z., Zhongjian, J., and Yong, J. (1990). Flavonoids as superoxide scavengers and antioxidants. *Free Rad. Biol. Med.* **9**, 19–21.

Zhang, K., and Das, N. P. (1994). Inhibitory effects of plant polyphenols on rat liver glutathione S-transferases. *Biochem. Pharmacol.* **47**, 2063–2068.

PART II

Synthetic Antioxidants and Enzyme Mimics

Jean-Baptiste Galey

Department of Chemistry
L'OREAL Research Center
93600 Aulnay sous bois, France

Potential Use of Iron Chelators against Oxidative Damage

I. Introduction

Iron is an essential element for all living organisms. It is critically involved in dioxygen transport and in a wide variety of cellular events ranging
from respiration to DNA synthesis. The central position in life processes
comes from its flexible coordination chemistry and redox potential, which
can be finely tuned by coordinating ligands. However, these physical properties which enable iron to be an essential factor for a wide range of proteins
involved in controlled redox reactions also allow iron to be toxic when not
carefully handled by proteins and shielded from surrounding media (Ryan
and Aust, 1992). Indeed, free or weakly bound ferrous iron is very sensitive
to oxygen and leads to the formation of ferric iron and superoxide which
rapidly dismutates to hydrogen peroxide. Ferrous iron can also react with
hydrogen peroxide, leading to the hydroxyl radical by Fenton chemistry.
Moreover, in the presence of physiological reductants, iron can redox cycle

Advances in Pharmacology, Volume 38

167

between the two oxidation states, thereby generating the production of highly reactive oxygen species continuously (Fig. 1) (see Halliwell and Gutteridge, 1989). In addition, it is essential to recall that although triplet dioxygen cannot directly react with biomolecules in the ground state, iron, as well as other transition metals, can relieve the spin restriction of oxygen and dramatically enhances the rates of oxidation (Miller *et al.*, 1990). Therefore, cells have to maintain the concentration of free iron as low as possible so as to avoid its interaction with oxygen and reduced oxygen species.

Hence, iron is not "free" in biological systems but is carefully handled by transport and storage proteins. In plasma, iron is transported by transferrin which maintains the concentration of ferric iron sufficiently low to avoid significant hydroxyl radical formation under nonpathological situations. Within the cytosol, iron is incorporated into ferritin, in the form of inorganic ferric iron for which interaction with oxygen is minimal.

Finally, iron is transported to specific cell compartments where it is incorporated into apoproteins or cofactors. In summary, under nonpathological conditions, iron levels are tightly controlled, and iron-catalyzed free radical reactions are kept minimal. However, in some situations the iron balance can be disturbed either locally or systemically. In systemic iron overload, excess iron saturates the binding sites of transferrin, allowing free iron to circulate and oxidize heart muscle cells ultimately leading to heart failure (Hider and Singh, 1992; Nathan, 1995). In other respects, in oxidative stress situations, traces of redox-active metal ions, especially iron, are locally released from their normal sites and then participate in Fenton chemistry (Ryan and Aust, 1992; Reif, 1992).

Systemic iron overload is associated with several diseases such as hereditary hemochromatosis and thalassemia major, which is characterized by a defective production of the globin chain of hemoglobin. This leads to anemia and consequently is treated by repetitive blood transfusions. At each transfusion, about 250 mg of iron is injected, which inevitably leads to an excess of iron in the body. In most of these systemic iron overload situations, treatment by iron chelators such as desferrioxamine (Fig. 2) is the only effective way to remove excess iron (see Martell *et al.*, 1981; Porter *et al.*, 1989).

This chapter focuses on the potential use of iron chelators for the treatment of conditions involving oxidative damage or local disturbance of iron

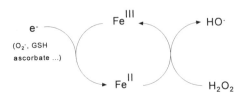

FIGURE I   Reductant-driven redox cycling of iron.

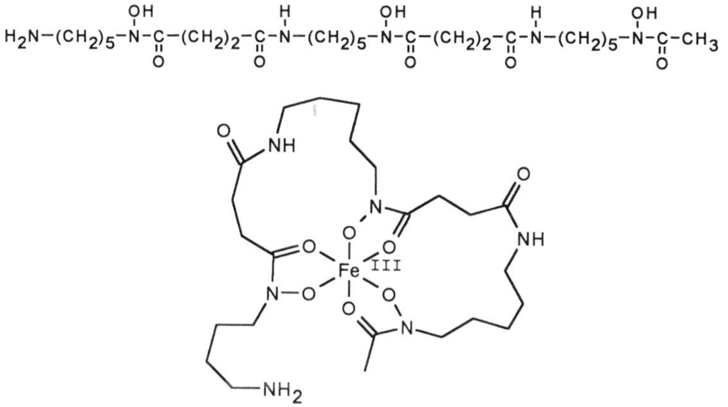

FIGURE 2 Desferrioxamine (DFO) and ferrioxamine.

homeostasis, with a special emphasis on the possible side effects related to interactions of the chelators with normal iron metabolism.

Iron chelators could also be designed for the treatment of pathological conditions which do not involve oxidative damage or systemic iron overload. For instance, drugs able to specifically chelate iron from ribonucleotide reductase could find important applications as antineoplastic or antiviral agents if they could be targeted to specific cells or tissues. Selective inhibitors of lipoxygenases could also have some relevance for the treatment of psoriasis, asthma, and atherosclerosis. The antimalarial activity of some iron chelators such as desferrioxamine has also triggered intensive research to find new compounds active in this field (Gordeuk *et al.*, 1994). However, this chapter does not deal with the potential utilization of iron chelators for the treatment of such diseases that do not primarily involve oxidative stress.

II. Iron Homeostasis

Normal human adults maintain a level of approximately 4–5 g of iron, mostly in the form of hemoglobin within red cells. About 10% is contained in myoglobin, cytochromes, and other iron-containing enzymes. The remaining 30% are distributed between storage proteins ferritin and its lysosomal degradation product hemosiderin. Transferrin only accounts for 0.1–0.2% of the total body iron (Table I). In humans, the average daily absorption of iron is 1–3 mg/day. About the same amount is lost by desquamation of cells, mostly from the gut but also from skin and via urine and bile. Thus, under normal circumstances, iron is used by the body in an almost closed circuit (Crichton, 1991; Fontecave and Pierre, 1993a). Iron is absorbed

TABLE I   Iron-Containing Proteins in the Human Body and Their Function

Protein	Type of iron	Function	Total body iron (%)
Hemoglobin	Heme	Oxygen transport	55–60
Ferritin/hemosiderin	Nonheme	Iron storage	30
Transferrin	Nonheme	Iron transport	0.1
Myoglobin	Heme	Oxygen transport	10
Cytochromes a, b, and c	Heme	Electron transport	
Ferredoxin	Nonheme	Electron transport	
NADH dehydrogenase	Nonheme	Electron transport	
Cytochrome P450	Heme	Drug detoxification	
Ribonucleotide reductase	Nonheme	DNA synthesis	
Proline hydroxylase	Nonheme	Collagen synthesis	
Phenylalanine hydroxylase	Nonheme	Amino acid metabolism	
Tyrosine hydroxylase	Nonheme	Catecholamine synthesis	
Peroxidases	Heme	H_2O_2 catabolism	
Catalase	Heme	H_2O_2 dismutation	
Lipoxygenase	Nonheme	Eicosanoid synthesis	
Cyclooxygenase	Heme	Eicosanoid synthesis	
Aconitase	Nonheme	Citric acid cycle	
Succinate dehydrogenase	Nonheme	Citric acid cycle	
Lactoferrin	Nonheme	Iron sequestration	

through the gut and is transported to transferrin, a glycoprotein with a molecular weight of 80,000 with two high affinity-binding sites for ferric iron ($K = 10^{21}$) (Fig. 3). Plasma transferrin (about 30 μM) is only 20–30%

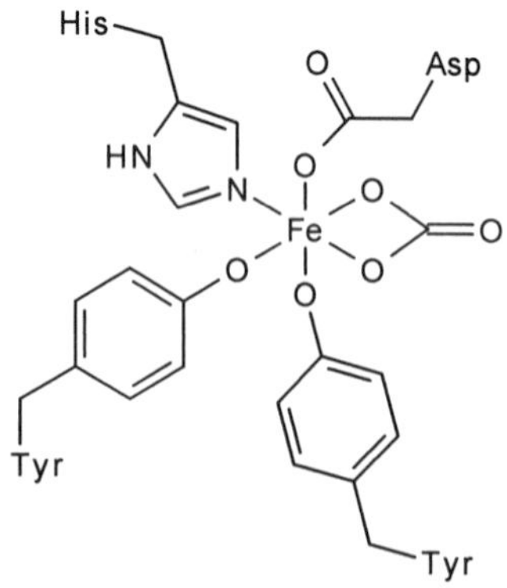

FIGURE 3   Binding sites of lactotransferrin.

saturated normally and the concentration of free ferric iron in the plasma is thus maintained very low (below $10^{-12}\,M$). This extremely low concentration of iron also accounts for much of the bacteriostatic effect of human plasma by depriving microorganisms of iron which is required for their growth.

Iron is specifically directed from the blood to tissue where it is required, mainly erythroid marrow. Cells that require iron for maturation or proliferation express high densities of transferrin receptors on the surface of their plasma membrane. Cellular uptake of iron transferrin then involves internalization of the transferrin receptor complex into an acidic endosome. The low pH facilitates iron mobilization from transferrin and perhaps also helps solubilize released ferric iron. Once iron has been released from transferrin, apotransferrin is secreted back into the extracellular space. The endosome contains a specific NADH-dependent reductase to reduce iron to the ferrous form, which is apparently transported to the cytosol. Within the cytosol, iron is incorporated into ferritin probably as Fe(II), which becomes oxidized within the protein core. Ferritin has a molecular weight of about 450,000 and is composed of 24 equivalent subunits arranged such that six transprotein channels are formed leading to an inner core. One molecule of ferritin can store up to 4500 iron atoms in the form of a regular lattice of ferric–hydroxyde–phosphate inside the core. It is normally only 20% saturated. To be utilized by the cell, ferritin iron has to be reduced by a not yet identified physiological reducing system (Crichton, 1991).

In systemic iron overload, the proportion of storage iron present as hemosiderin increases. Hemosiderin is found as water-insoluble brown aggregates, especially in areas of old hemorrhage or blood extravasation. In hemosiderin, iron is present similar to iron in ferritin, but is also associated with carbohydrates, lipids and heme. It does not catalyze Fenton chemistry.

Inside the cell, a low molecular mass iron pool is also present, although it has not been characterized in detail. It is supposed to be present at very low concentration (around 1 μM), mainly in a reduced form and to serve as a transit pool with a high turnover. This low molecular mass iron is likely to be able to catalyze oxidative damage from which cells are protected by enzymatic and nonenzymatic antioxidants. It must be pointed out that iron within ferritin is regularly turned over, with a half-life of about 72 hr and will pass through this low molecular mass iron pool as a consequence (Porter *et al.*, 1989).

The cellular homeostasis of iron is maintained by the coordinate regulation of the expression of the transferrin receptor and ferritin at a posttranscriptional level (Klausner *et al.*, 1993). It is now accepted that the expression of these proteins is regulated by intracellular iron levels. The way by which cytoplasmic iron regulates the expression of ferritin and transferrin receptor involves its interaction with a protein called iron responsive element-binding protein (IRE-BP). This protein has a binding site for iron. The iron-free form of IRE-BP inhibits ferritin mRNA translation by occupying a specific portion of ferritin mRNA called the iron responsive element (IRE). The iron-free IRE-BP also regulates the expression of transferrin receptor by stabilizing its mRNA (Klausner et al., 1993). Thus, when the intracellular iron pool is low, IRE-BP is iron free so that ferritin synthesis is repressed and transferrin receptor expression is upregulated. Conversely, when intracellular iron is high, the IRE-BP loses its affinity for IRE and thus ferritin synthesis is derepressed whereas transferrin receptor mRNA is degraded more rapidly so that the transferrin receptor is not expressed at the cell surface. Thus, IRE-BP is literally "sensing" intracellular iron in order to strictly regulate its level. The IRE-BP sequence has been found to be similar to that of mitochondrial aconitase which is a 83-kDa protein catalyzing the conversion of citrate to isocitrate. It contains an iron sulfur cluster, and the active enzyme requires the fully assembled [4Fe-4S] cluster. It is proposed that the lability of one iron atom is critically involved in reversible iron sensing by modifying the IRE-BP mRNA binding affinity. Loss of iron from aconitase indeed seems to induce a conformational change resulting in an increase in binding affinity for IRE. Conversely, fully assembled cluster aconitase has low affinity for IRE (Basilion et al., 1994; Gardner et al., 1995). It has also been proposed that nitric oxide could elicit the activation of IRE binding by triggering the disassembly of the iron sulfur cluster independently of iron deprivation (Pantopoulos and Hentze, 1995).

III. Iron Release during Oxidative Stress

Despite the remarkable efficacy of the iron transport and storage systems, various conditions associated with oxidative stress have been shown to induce a local release of iron from normal sites. Most of the work on this topic has been performed in in vitro systems. Detection of an increase of free iron concentrations in vivo in some pathological situations has also been reported, but mostly from indirect methods. It is indeed extremely difficult to monitor the concentration of "free" iron. Most of the reported methods necessitate homogenization of tissues, which may alter the existing equilibrium between free and bound iron as well as its oxidation state. Such methods include atomic absorption spectrometry or colorimetric assay in the presence of a suitable iron chelator such as ferrozine. Another important method is based on the measurement of bleomycin-detectable iron in tissue samples (Gutteridge et al., 1981). Bleomycin is indeed able to intercalate DNA and to bind iron in such a way that in the presence of a reducing agent and dioxygen, highly oxidizing species are formed which induce DNA strand breaks (Stubbe and Kozarich, 1987). It is thus possible to detect with great sensitivity "free" iron in biological systems using bleomycin-dependent degradation of DNA. Ferrali et al. (1989) have also reported an HPLC determination of "free iron" (desferrioxamine-chelatable iron) which involves several steps of purification. The electron paramagnetic resonance determination of intracellular iron has also been proposed, using desferrioxamine or sodium nitrite-treated samples (Kozlov *et al.*, 1992). It is noteworthy that the direct fluorescent monitoring of free iron concentration in intact cells is difficult because, contrary to calcium (Tsien *et al.*, 1982), iron is paramagnetic and therefore the use of specific fluorescent chelators as iron probes is compromised due to fluorescence quenching by iron. Nevertheless, two original fluorimetric approaches have been reported. Lytton *et al.* (1992) described the use of a membrane-permeable fluorescent derivative of desferrioxamine allowing an indirect quantitation of intracellular iron after cell lysis and dissociation of the nonfluorescent complex. However, the cytoplasmic pool of chelatable iron was monitored by Breuer *et al.* (1995) by studying intracellular fluorescence quenching of the metal-sensitive probe calcein.

A. Iron Release from Ferritin

The main source of intracellular iron release obviously is ferritin. Release of iron from ferritin requires reduction in the presence of a ferrous iron acceptor (Watt *et al.*, 1985; Thomas and Aust, 1986; Jacobs *et al.*, 1989). Physiological reductants such as ascorbate and glutathione do not release iron from ferritin at significant rates. As early as 1955, Mazur and colleagues demonstrated that xanthine oxidase was able to mobilize iron from ferritin in the presence of xanthine, either in the presence or in the absence of oxygen. It was further shown that iron release was mediated, at least in part, by superoxide (Biemond *et al.*, 1986). Reductants other than superoxide have also been shown to release iron from ferritin. Reduced flavins, for instance, have been reported to do so (Funk *et al.*, 1985). This is also the case for many redox-cycling xenobiotics whose redox potential lies below that of ferritin, e.g., $E'^{o} < -0.23$ V/NHE (normal hydrogen electrode). This includes toxic herbicide paraquat ($E'^{o} = -0.45$ V/NHE), the antineoplastic drug adriamycin ($E'^{o} = -0.33$ V/NHE), and the diabetes-inducing chemical alloxan ($E'^{o} = -0.44$ V/NHE). These compounds undergo facile one-electron reduction by NADPH cytochrome P450 reductase to a free radical that can either autoxidize yielding superoxide or directly induce reductive iron release from ferritin. However, redox-cycling drugs whose redox potential is higher, such as menadione ($E'^{o} = -0.20$ V/NHE), do not directly release iron from ferritin, although they can generate superoxide (Miller *et al.*, 1990). The standard potential for superoxide is $E'^{o} = -0.33$ V/NHE. However, this value is taken for 1 atmosphere oxygen and a more useful value of -0.16 V/NHE can be taken referring to 1 M dissolved oxygen. Therefore, from a thermodynamic point of view, the ability of superoxide to release iron from ferritin is dubious. However, it should be kept in mind that such a consideration refers to equilibrium conditions which is generally not the case of biological systems. Moreover, the presence of ferrous iron acceptors which can continuously displace the equilibrium is likely to allow ferric iron reduction even when thermodynamic conditions are not maintained. In biological medium, various ferrous iron acceptors such as porphyrins or apoproteins could be available (Fontecave and Pierre, 1993b). Therefore, it is generally accepted that superoxide can release iron from ferritin. Moreover, it is now believed by some authors that one of the main ways by which superoxide induces molecular damage is through its capacity to release iron from various iron proteins. Nitric oxide has also been reported to be able to induce iron release from ferritin (Reif and Simmons, 1990), although these results have been contradicted (Laulhère and Fontecave, 1995). UVA irradiation has been proposed to induce reductive iron release from ferritin *in vitro* (Aubailly *et al.*, 1991), although these observations also have been discussed by others (Vile *et al.*, 1995) as resulting from loosely bound iron outside the core of unpurified ferritin. However, it has also been shown that oxidative stress resulting from UVA irradiation leads to a heme oxygenase increase in ferritin in human skin fibroblasts (Vile and Tyrrell, 1993), which suggests the involvement of free iron in some of the effects of UVA irradiation. In any case, the early degradation of ferritin during oxidative stress is likely to expand the intracellular free iron pool that subsequently activates the synthesis of ferritin, thus limiting the prooxidant challenge (Cairo *et al.*, 1995).

B. Iron Release from Other Sites

Apart from ferritin, another potentially significant source of free iron may be other iron-containing proteins. Superoxide has been shown to rapidly oxidize enzymes that contain [4Fe-4S] clusters at their active sites with a concomitant loss of ferrous iron from the cluster. Among these enzymes, aconitase and 6-phosphogluconate dehydratase have been especially studied (Fridovich, 1986; Gardner *et al.*, 1995). Chronic exposure to ultraviolet light has been shown to result in an increased skin level of nonheme iron. This iron accumulation is likely to be the consequence of a UVB radiation-induced capillary damage and leakage of protein-bound iron (Bissett *et al.*, 1991). Inside red cells, which are rich in antioxidant enzymes, iron is tightly bound to hemoglobin, but injured or lysed red cells can release their iron into the surrounding medium. Indeed, hemoglobin has been shown to release iron ions on exposure to excess hydrogen peroxide. In addition, a protein-bound oxidizing species, probably a heme ferryl species, is formed and is able to stimulate lipid peroxidation. Moreover, glutathione has been shown to induce heme degradation in the presence of oxygen with a concomitant release of iron (Atamna and Ginsburg, 1995). Considering the amount of iron in red cells (approximately 60% of total body iron), red cell injury by various mechanisms has therefore the potential to release important amounts of iron at the site of injury and thus to induce oxidative damage. In particular, bleeding as a result of injury could lead to hemoglobin liberation with a possible compounding effect (Trenam *et al.*, 1992).

C. Consequences

Once released, free iron will catalyze the formation of oxidative damage as mentioned earlier, mostly through its ability to undergo redox cycling between the two oxidation sites and to generate a hydroxyl radical able to oxidize the surrounding molecules. Although it is generally far less monitored than lipid peroxidation as a marker of oxidative damage, oxidative damage to proteins and DNA is also very critical. Both can nonspecifically bind iron either as a ferrous or as a ferric form and therefore undergo site-specific damage. It is generally believed that because of the presence of physiological reductants inside cytosol, the ferrous form predominates. Once Fe(II) is bound to a target molecule, hydroxyl radicals produced by a reaction with hydrogen peroxide will react very closely to the metal-binding site according to a so-called site-specific Fenton reaction (Halliwell *et al.*, 1992). This type of damage is insensitive to inhibition by hydroxyl radical scavengers because the hydroxyl radical reacts immediately at the place where it is formed. In other words, any hydroxyl radical scavenger could not compete for such a reaction except at excessively high concentrations.

Because of its polyanionic nature, DNA is known to bind various metal ions (Eisinger *et al.*, 1962) and is therefore especially prone to iron-dependent site-specific oxidative damage. Single strand breaks are indeed easily formed during the incubation of DNA with micromolar concentrations of iron and a reductant such as ascorbate in aerated aqueous medium. DNA can also be oxidatively damaged by a variety of other reactions, leading to abasic sites and modifications of the bases or the sugar. It has been suggested that double strand breaks result from a metal-catalyzed, site-specific Fenton reaction because of a "multihit" effect resulting from possible multiple hydroxyl radical formation at the same site (Chevion, 1988).

The metal ion-catalyzed oxidation of proteins by such mechanisms has been extensively studied (Stadtman, 1990). It is noteworthy that only a few amino acid residues are modified with relatively little peptide bond cleavage. These features clearly distinguish site-specific damage from random hits that occur during the exposure to ionizing radiations where global modifications of many different amino acid residues and extensive fragmentation occur. Main modifications lead to the formation of protein carbonyls. In the case of lysine residues, a mechanism has been proposed in which hydrogen abstraction at the α-amino carbon occurs with a subsequent single electron transfer from cation radical to ferric iron, leading to an imminium cation and ferrous iron. Then, an aldehyde derivative is formed on spontaneous hydrolysis. It must be pointed out that such a mechanism is catalytic and thus can be repeated on other target molecules or on other sites of the same macromolecule. The iron-catalyzed inactivation of glutamine synthetase has been associated with the conversion of a single histidine to an asparagine and of a single arginine to a glutamic semialdehyde residue (Farber and Levine, 1986). Oxidatively modified proteins could mark them for preferential protease degradation but could also contribute to the accumulation of inactive proteins observed during aging and in pathological states associated with oxidative stress (Stadtman, 1992).

IV. Iron Chelators

A. Design

The design of clinically useful iron chelators has been extensively described in excellent reviews (see Porter *et al.,* 1989; Martell *et al.,* 1981). Most of the reported approaches are currently focused on the treatment of systemic iron overload by well-designed chelators. Moreover, all these strategies are oriented on ferric iron chelators, considering that Fe(III) is more stable and is thus a better target under aerobic conditions. It is indeed noteworthy that all natural siderophores so far reported are powerful ferric iron chelators. However, more or less specific Fe(II) chelators also exist, although their clinical usefulness has not been fully investigated.

However, there are few reports on the development of iron chelators specifically designed for the treatment of oxidative stress-associated situations, especially in the case of long-term treatment. The main reason is probably that such a design is a real nightmare for medicinal chemists in terms of safety margins due to the possible interaction of chelators with normal iron metabolism. Indeed, if some side effects can be supported in the case of thalassemia or hemochromatosis treatment because it is a matter of life or death, it is not necessarily the case for other situations. However, when speaking of iron chelation therapy against oxidative stress, it is important to distinguish between short-term and long-term treatment. Indeed, as pointed out by Halliwell *et al.* (1992), the administration of strong iron chelators for long periods to patients who do not have iron overload is dangerous. The side effects of DFO, including neurotoxicity, unfortunately suggest that DFO or other powerful iron chelators are unlikely to be useful for the treatment of chronic diseases involving oxidative stress. However, strong iron chelators might be useful in acute situations, e.g., for short periods of treatment during which temporary iron deprivation and increase in iron excretion could be tolerable. Ischemia/reperfusion injury or redoxcycling xenobiotics poisoning for instance may be managed by strong iron chelators (Halliwell *et al.*, 1992). The possible use of iron chelators for long-term treatment is specifically discussed in Section C.

The properties required for an iron chelator oriented against oxidative stress are close to those already reported for an ideal iron chelator designed for the treatment of systemic iron overload (Porter *et al.*, 1989):

1. Specificity versus other metal ions
2. Acceptable tissue distribution and metabolism
3. Oral bioavailability

However, some points have to be especially emphasized: (1) the interaction of the chelator with normal iron metabolism, and (2) the capacity of the corresponding iron chelate to catalyze Fenton chemistry. Concerning the side effects anticipated from the clinical use of an iron chelator, apart from those related to a direct toxicity of the chelator and the corresponding iron chelate, careful attention has to be paid to risks of inhibition of iron-containing enzymes, redistribution of iron to more susceptible sites of the body, encouragement of bacterial growth, and iron depletion.

I. Interaction with Normal Iron Metabolism

The capacity of a given chelator to interact with normal iron metabolism can be anticipated roughly by its affinity for iron and its bioavailability. By definition, chelation requires the presence of two or more atoms on the same molecule capable of metal binding, i.e., forming a coordinate bond, the interaction between an electron donor (chelating ligand) and an electron acceptor (metal). Depending on the number of covalently linked donor groups associated with the chelating agent, varying stoichiometry of metal : ligand can be found in order to satisfy the coordination requirement of the metal ion. Both ferrous and ferric ions have a coordination number of six, i.e., most, but not all, of $Fe(II)$ and $Fe(III)$ complexes are octahedral. Bidentate ligands therefore form 3 : 1 iron complexes, whereas hexadentate ligands form 1 : 1 complexes. However, some ligands may also form polynuclear chelates. Examples of biologically relevant donor groups for the coordination of iron are given in Table II. The stability of a metal chelate in solution is influenced by the number of donor groups present on the same molecule according to the so-called chelate effect, i.e., hexadentate ligands form more stable complexes than the corresponding bidentate or tridentate ligands. Hexadentate ligands also have greater scavenging ability at low concentrations ($< 20 \ \mu M$) and are less likely to dissociate (Hider and Singh, 1992). Hard metals ion acceptors such as $Fe(III)$ will interact more strongly with rather small negatively charged atoms. However, softer metals ions, such as $Fe(II)$, tend to form strong bonds with neutral ligands or ligands with low negative charge and high polarizability.

The affinities between metal ion and ligand are usually expressed as the equilibrium constants for the formation of a complex from hydrated metal

TABLE II   Examples of Biologically Relevant Donor Groups

Negative monodentate donors

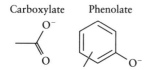

Carboxylate   Phenolate

Neutral monodentate donors

Tertiary amine   Imidazole   Pyridyl

Bidentate combinations

Catecholate   Hydroxamate   Aminocarboxylate   Amino phenolate

ion and the fully dissociated form of the ligand. Thus for the combination of bidentate ligand with ferric iron, the equilibria may be expressed by:

$$Fe^{3+} + L \rightleftarrows FeL^{3+} \qquad K_1 = [FeL^{3+}]/[Fe^{3+}][L] \qquad (1)$$

$$FeL^{3+} + L \rightleftarrows FeL_2^{3+} \qquad K_2 = [FeL_2^{3+}]/[FeL^{3+}][L] \qquad (2)$$

$$FeL_2^{3+} + L \rightleftarrows FeL_3^{3+} \qquad K_3 = [FeL_3^{3+}]/[FeL_2^{3+}][L] \qquad (3)$$

(stepwise formation constants)

Another way of expressing the relation is

$$Fe^{3+} + \ L \rightleftarrows FeL^{3+} \qquad \beta_1 = [FeL^{3+}]/[Fe^{3+}][L] \qquad (4)$$

$$Fe^{3+} + 2L \rightleftarrows FeL_2^{3+} \qquad \beta_2 = [FeL_2^{3+}]/[Fe^{3+}][L]^2 \qquad (5)$$

$$Fe^{3+} + 3L \rightleftarrows FeL_3^{3+} \qquad \beta_3 = [FeL_3^{3+}]/[Fe^{3+}][L]^3 \qquad (6)$$

$$\text{where } \beta_3 = K_1 K_2 K_3 = [FeL_3^{3+}]/[Fe^{3+}][L]^3 \qquad (7)$$

(overall formation constant)

For hexadentate, the expression simplifies as

$$Fe^{3+} + L \rightleftarrows FeL^{3+} \qquad K = [FeL^{3+}]/[Fe^{3+}][L] \qquad (8)$$

Stability constants for various metal ions of some metal chelators are given in Table III.

The simple comparison of overall formation constants of iron chelators is not necessarily a good criterion of their relative effectiveness. Indeed, competition between the ferric iron and the hydrogen ion for the ligand donor group, measured by ligand protonation constants, greatly influences chelating efficiency in an aqueous solution. For this reason, affinity for iron may be better represented by the free iron concentration in equilibrium at pH 7.4 in the presence of excess ligand expressed in pM (see Martell *et al.*, 1987; Harris *et al.*, 1979). The higher the pM, the more effective the ligand.

TABLE III Stability Constants of Some Ligands for Various Metal Ions

Name	Structure	Type of constant	Fe(III)	Fe(II)	Cu(II)	Zn(II)	Ca(II)
NTA		$\log K_1$	15.9[a]	8.8[c]	13.1[a]	10.4[a]	6.5[a]
		$\log \beta_2$	24.3[b]				
Salicylic acid		$\log K_1$		6.6[a]	10.6[a]		0.5[a]
		$\log \beta_3$	35.3[b]				
1,10-Phenanthroline		$\log K_1$		5.8[a]	6.3[a]	6.4[a]	0.5[a]
		$\log \beta_3$		16.5[a]	18[a]	17[a]	
8-Hydroxy-quinoline		$\log \beta_2$		15[a]	23.4[a]	17.6[a]	3.3[a]
		$\log \beta_3$	36.9[b]				

[a] From Dawson *et al.* (1986).
[b] From Martell *et al.* (1981).
[c] From Singh and Hider (1988).

To summarize, although the stability constant for iron is certainly not the only parameter that can predict the capacity of a chelator to mobilize "safe" iron, it can be assumed that if the constant is higher than the affinity constant of iron for most metalloproteins and especially for transferrin, i.e., 10^{21}, the chelator is thermodynamically able to compete for iron of these proteins, and therefore side effects can be anticipated.

The ability of a given chelator to remove iron from iron proteins such as transferrin is not only based on thermodynamic but also on kinetic considerations. Indeed, rates of removal of iron from transferrin may vary dramatically according to the chelator (Kretchmar Nguyen *et al.*, 1993). Finally, iron mobilization from iron-containing proteins is also obviously dependent on tissue distribution, especially the membrane permeability of the chelator. The partition coefficient of the chelator between octanol and water at pH 7.4, measured as log P, gives an estimation of the facility to penetrate the cell membrane by simple diffusion as well as of potential oral activity. A log P in the range of 0.2 to 1 has been proposed to be a good compromise for allowing oral activity and hepatocyte penetration, yet lacking acute toxicity (Porter *et al.*, 1989). Very hydrophilic iron chelators are likely to have only access to extracellular iron. However, some hydrophilic iron chelators might also access the cytosol using specific transport systems (Molenda *et al.*, 1994). Conversely, relatively lipophilic chelators are more likely to have access to the intracellular low molecular mass iron pool but also to intracellular iron-containing proteins such as IRE-BP, with more potential side effects. Moreover, small lipophilic compounds are also more likely to penetrate the blood–brain barrier, with possible side effects in the central nervous system (CNS). It must be pointed out that desferrioxamine, although very hydrophilic, is known to cause CNS side effects (vide infra), and therefore it can be supposed that it can cross the blood–brain barrier, possibly according to nonfacilitated reptation processes.

2. Catalysis of Fenton Chemistry

The capacity of a chelator to protect against the formation of damaging oxidizing species is obviously critical. This capacity is influenced by several thermodynamic and kinetic factors. The spin state of an iron complex conditions its possible interaction with oxygen and therefore its capacity to relieve the spin restriction for the autoxidation of various ground state molecules. Fe(II) complexes can be high or low spin, depending on the ligand field of the coordinating molecule. Strong ligands such as those containing imine nitrogen donor atoms form low spin chelates, i.e., the six electrons from the 3d orbital, which are split in energy because of electrostatic interaction with the donor groups, are paired, thus leading to a diamagnetic species. When the ligand field is not strong enough, as in the case of EDTA, electrons are unpaired and the corresponding chelate is paramagnetic. Fe(III) is high spin in most of its complexes except those with the strongest ligands such as 1,10-phenanthroline or bipyridyl.

In addition, the capacity of an iron chelate to catalyze Fenton chemistry is related to its redox potential. Generally speaking, chelators in which oxygen atoms ligate the metal will tend to prefer the oxidized form. Thus, they decrease the redox potential of Fe(II)/Fe(III). Conversely, chelators in which nitrogen atoms primarily bind the metal prefer the ferrous form and tend to increase the redox potential of iron. It is noteworthy that the stability constant of a given ligand for both forms of iron gives an estimation of the redox potential of chelated iron, provided the same complexes are formed by both ferric and ferrous forms (Miller *et al.*, 1990). Redox potential and affinity constants for both ferric and ferrous iron of some well-known chelators are given in Table IV.

To be able to catalyze the formation of hydroxyl radicals in the presence of hydrogen peroxide, an iron complex has to fit two simultaneous conditions (Fontecave and Pierre, 1993b):

1. The ferric chelate has to be reducible by physiological reductants [e.g., its standard redox potential must be higher than -0.324 V/ NHE (NADPH/NADP$^+$ redox couple)].
2. Single electron transfer from ferrous chelate to hydrogen peroxide must be possible [e.g., its redox potential must lie below $+0.46$ V/ NHE (H$_2$O$_2$/HO$\cdot$, OH$^-$)].

Iron-catalyzed hydroxyl radical formation also requires at least one iron coordination site occupied by water or a readily dissociable group allowing access to superoxide and/or hydrogen peroxide (Graf *et al.*, 1984; Gutteridge, 1990). This explains the ability of certain well-known iron chelates to be powerful catalysts of Fenton chemistry. Indeed, although EDTA is a potential hexadentate ligand, it is too small to reach all octahedric

TABLE IV Stability Constants for Fe(III) and Fe(II) and Standard Reduction Potential at pH 7 versus NHE (E'°) for Some Iron Chelates

	log K Fe(III)	*log K Fe(II)*	E'°/V^a
DFO	31[a]	9[b]	-0.45
Transferrin	21[a]	2[a]	-0.40
DTPA	27[c]	16[c]	$+0.03$
EDTA	25[c]	14[c]	$+0.12$
o-Phenanthroline		16[b,d]	$+1.15$

[a] From Fontecave and Pierre (1993b).
[b] From Dawson *et al.* (1986).
[c] From Singh and Hider (1988).
[d] log β_3.

coordination sites. The structure of the ferric iron EDTA complex includes a seventh coordination site occupied by water (Lind *et al.*, 1964). Moreover, the redox potential of iron EDTA complex is $E'^\circ = +0.12$ V/NHE, which means that it can alternatively be reduced by physiological reductants and oxidized by hydrogen peroxide, thus generating highly oxidizing species that are probably hydroxyl radicals despite some controversy regarding a possible formation of ferryl-type species (Rush and Koppenol, 1986; Yamazaki and Piette, 1991; Burkitt, 1993). Chemically related structures such as EDDA and NTA also form iron complexes that are powerful catalysts of Fenton chemistry. These properties probably account for their renal toxicity (Li *et al.*, 1987; Liu and Okada, 1994). Studies on the DTPA–iron complex ($E'^\circ = +0.03$ V/NHE) provided conflicting evidence concerning its capacity to catalyze the cycling Fenton reaction (Egan *et al.*, 1992; Burkitt and Gilbert, 1990), although it is generally accepted that the DTPA–ferric chelate is not reducible because it lacks a free coordination site. However, the triphenanthroline–iron complex ($E'^\circ = +1.15$ V/NHE) will not react with hydrogen peroxide, and the desferrioxamine–iron complex ($E'^\circ = -0.40$V/NHE) is not reducible by physiological reductants. However, as already mentioned, such thermodynamic considerations refer to standard conditions at equilibrium, and it is likely that some reductase might be able to reduce iron even from ferrisiderophores such as ferrioxamine by continuously displacing the equilibrium in the presence of ferrous iron acceptors. Such a mechanism is probably involved in iron mobilization by various microorganisms. Some iron chelates may also be able to generate hydroxyl radicals in such a way that damage is directed to the chelator, thereby sparing more important targets which bind iron such as DNA.

B. Main Classes of Iron Chelators

I. Siderophores

Desferrioxamine belongs to a class of natural molecules called siderophores which have been evolved by microorganisms in order to provide iron for their growth. Siderophores are very specific low molecular weight iron chelators which solubilize and transport ferric iron in aqueous media. They are synthesized and secreted when microorganisms are grown under iron-deficient conditions. Iron siderophore complexes then interact with membrane receptors, and iron is released in the cell by various reductive, proteolytic, and hydrolytic processes. Most siderophores utilize three secondary hydroxamate or catecholate groups to provide very stable and specific octahedral coordination of ferric iron. Their affinity constants are extremely high, up to 10^{52} (Martell *et al.*, 1981). Major siderophores containing catecholate groups are enterobactin and agrobactin, whereas ferrichrome, desferrioxamine, rhodotorulic acid, pseudobactin, and mycobactins represent the major hydroxamate-containing class.

The potential use of siderophores and analogs for the treatment of systemic iron overload was recognized early (see Bergeron *et al.,* 1991; Miller, 1989). However, these agents, with the exception of desferrioxamine, have limitations for use *in vivo* because of their restricted bioavailability and their propensity to remove ferric iron from both ferritin and transferrin. Another drawback for the clinical use of natural siderophores is that they can deliver iron to some bacteria and fungi, thereby enhancing their pathogenicity. DFO (Desferal), originally isolated from *Streptomyces pilosus,* is currently the only accepted drug for iron chelation therapy and has been used for more than 20 years for the treatment of systemic iron overload. It has a high selectivity for iron and therefore avoids the elevated excretion of other metals such as zinc. The affinity constants for various metals of DFO are: Fe^{3+}, 10^{31}; Al^{3+}, 10^{25}; Cu^{2+}, 10^{14}; Zn^{2+}, 10^{11}; and Ca^{2+}, 10^3 (Porter *et al.,* 1989). However, the drug is not active when given orally and therefore requires subcutaneous or intravenous administration. The daily required dose is about 50 mg/kg when administered in an overnight subcutaneous infusion for the treatment of transfusion-induced iron overload. Typical plasma concentrations of 20–50 μM have been reported in humans, with a short half life of 5–10 min. Part of the success of DFO is likely because of its relatively low capacity of iron scavenging from transferrin, probably because DFO does not trigger the conformational change of the protein that is required for iron removal (Kretchmar Nguyen *et al.,* 1993). Moreover, DFO has a poor ability to cross biological membranes (Lloyd *et al.,* 1991) and therefore its main effect is probably related to iron scavenging in extracellular fluids instead of in intracellular fluids.

DFO has a low general acute toxicity, likely to be linked to its low lipid solubility (Porter and Huens, 1989). However, severe toxic effects on vision, hearing, and renal function have been reported. Moreover, therapy with DFO has been associated with an increased susceptibility to acute infections, thrombocytopenia, and pulmonary complications. Maternal toxicity of DFO has been assessed (Bosque *et al.,* 1995). DFO also inhibits cell proliferation, presumably by the inhibition of ribonucleotide reductase (Bomford *et al.,* 1986).

DFO might also have therapeutic uses in protecting against oxidative damage (Gutteridge *et al.,* 1979). Because large doses can be injected into animals and humans, DFO was proposed for use as a "probe" to study the importance of iron-dependent free radical reactions as mediators of tissue injury (Halliwell, 1985). It has since been shown in a variety of experimental conditions that DFO efficiently inhibits lipid peroxidation and protects cells or animals against oxidative damage (vide infra) in various conditions. The protective effect of DFO comes from its high affinity for ferric iron (K = 10^{31}) and the low redox potential of the corresponding ferrioxamine complex (E'° = -0.45 V/NHE) which avoids iron reduction. DFO has also been reported to affect eicosanoid synthesis at millimolar concentrations

(Laughton *et al.*, 1989) and to react with superoxide and hydroxyl radicals to form a reactive nitroxide free radical *in vitro* which could inactive enzymes (Davies *et al.*, 1987). However, the significance of these reactions *in vivo* has been controversial considering that the concentrations of DFO that can be achieved *in vivo* in biological fluids are generally less than 50 μM, ruling out other mechanism than iron chelation (Halliwell, 1989).

2. Hydroxypyridinones

Hydroxypyridinone ligands have been widely investigated for ferric iron chelation. Their success is mainly due to their relative selectivity for ferric iron and to their unique ability to be uncharged under physiological conditions both as free ligand and as iron complex (Porter *et al.*, 1989). This allows most of these compounds to be orally active and to cross cell membranes. However, their lipophilicity also enables them to penetrate numerous cells where they can exert undesired toxic effects. Hydroxypyridinones can apparently efficiently remove iron from transferrin. The most extensively studied compound of this series is bidentate 1,2-dimethyl-3-hydroxypyrid-4-one (CP20, L1) (Fig. 4). L1 forms a very stable iron complex (log β_3 = 35.7) which apparently does not catalyze hydroxyl radicals formation as 3 : 1 chelate (Singh and Hider, 1988). However, 2 : 1 and 1 : 1 hydroxypyridinone–iron chelates may be able to catalyze hydroxyl radicals formation because of free coordination sites (Kontoghiorghes, 1995). A prospective clinical trial of L1 (deferiprone) was conducted on patients with thalassemia

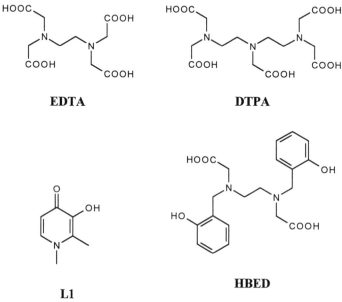

FIGURE 4   EDTA, DTPA, L1, and HBED.

major (Olivieri *et al.*, 1995). Although L1 is orally active and induces sustained decreases in body iron, it has a much lower therapeutic ratio than DFO (Nathan, 1995). Indeed, being a bidentate ligand, L1 must be present at a high concentration to effectively scavenge iron, e.g., close to toxic levels. Required doses represent 75 mg/kg three times a day. At such a regimen, L1 was shown to cause agranulocytosis and arthralgia or arthritis during clinical studies. Second-generation hydroxypyridinone compounds are now being developed with a more favorable safety margin. It is in particular believed that small structural changes will have a great influence on the metabolism of the chelator and therefore on its efficacy and toxicity (Porter *et al.*, 1994). Incorporation of a single negative charge on 3-hydroxy-4-pyridinones has been proposed to enhance the organ specificity of the chelator by exploiting monoanionic transport systems in the liver and kidney (Molenda *et al.*, 1994). It is hoped that such charged molecules may be less toxic, as their penetration through the membrane should be restricted to organs with suitable anionic transport systems.

3. Aminocarboxylates

Amino polycarboxylates such as EDTA and DTPA are well-known metal chelators with relatively high affinities for iron but also for other transitions metals. Their use for iron chelation therapy was proposed early. DTPA has been used in patients with iron overload who could not use DFO. However, it is not orally active and causes zinc depletion. To increase the selectivity for iron, various ligands were synthetized by replacing carboxylates with hard phenolate donors. Martell and co-workers have synthetized HBED (Fig. 4), in which two carboxylate groups of EDTA are replaced by two phenolate groups with an increase of affinity constant ($K = 10^{39}$) of 14 log units compared to EDTA (L'Eplattenier *et al.*, 1967). Although when given orally, it loses part of its activity, HBED is a promising compound for the treatment of iron overload (Grady and Hershko, 1990). Moreover, the dimethyl ester of HBED is orally active (Pitt *et al.*, 1986). Results from a phase I clinical trial of HBED have confirmed the potential usefulness of this chelator (Grady *et al.*, 1994). A large number of chemically related structures have been synthetized and tested as potential drugs for iron chelation therapy (Martell *et al.*, 1987). For instance, HBED analogs containing phenolic groups substituted on pyridine rings have been carefully investigated (Taliaferro *et al.*, 1984). Apart from iron chelation therapy, these compounds are also studied for potential application as radiopharmaceuticals involving Ga(III) and In(III).

The ability of EDTA and DTPA iron chelate to catalyze Fenton chemistry has been discussed in the previous section. However, although it is incomparably used less than DFO *in vitro* to assess the participation of iron to oxidative damage, HBED efficiently inhibits the formation of hydroxyl radicals catalyzed by iron (vide infra), presumably because it decreases its redox potential and fully occupies the six coordination sites (Dean and Nicholson, 1994).

ICRF-187 (dexrazoxane) is a drug in phase III clinical trials which shows great promise against doxorubicin-induced cardiotoxicity. This drug acts presumably via its hydrolysis product ADR-925, which has a structure similar to that of EDTA (Fig. 5) and is therefore able to chelate iron which is believed to be involved in anthraquinone toxicity. Indeed, ICRF-187 is uncharged at neutral pH and therefore can enter cells where it can be cleaved by nonspecific hydrolytic enzymes to generate ADR-925 able to chelate intracellular iron as well as other metals (Hasinoff, 1989). However, like iron EDTA, the iron ADR-925 chelate is a good catalyst of formation of hydroxyl radicals, and therefore ADR-925 generated *in vivo* from ICRF-187 would probably not avoid intracellular hydroxyl radical production (Thomas *et al.*, 1993; Malisza and Hasinoff, 1995). Nevertheless, ADR-925 efficiently removes iron from the iron–anthracycline complex and therefore could also avoid site-specific damage on DNA by orienting damage in the bulk or toward less sensitive targets.

4. Miscellaneous

The 21-aminosteroids such as U-74500A (Fig. 6) were designed as membrane-specific antioxidants by attaching a membrane-localizing steroid to an antioxidant amine which could act either as lipid peroxyl radical scavengers or as iron chelators (Braughler *et al.*, 1987). Some of these compounds were found to be potent inhibitors of iron-catalyzed lipid peroxidation *in vitro*, as well as protecting drugs against cerebral ischemia and central nervous system trauma *in vivo*. U-74500A has been shown to chelate Fe(II) with substantial, although not quantified, affinity. Moreover, it was shown to be a ferric iron reductant (Ryan and Petry, 1993). However, although it is suspected that the U-74500A inhibition of lipid peroxidation is partially related to a modulation of iron redox chemistry, additional mechanisms are also probable.

Pyridoxal isonicotinoyl hydrazone (PIH, Fig. 6) is a strong iron chelator able to remove iron from various sites (Ponka *et al.*, 1979). Its use has been

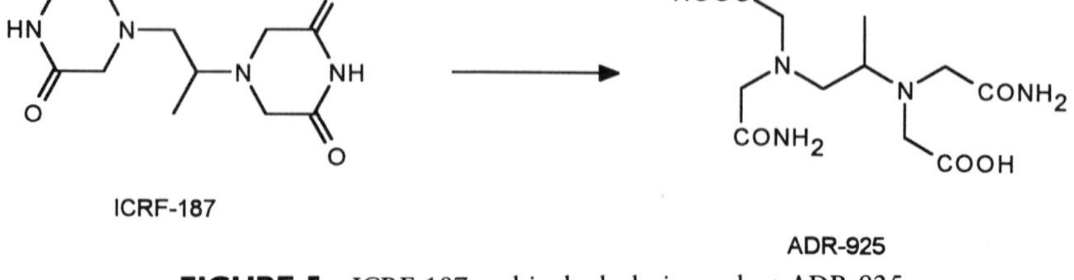

ICRF-187

ADR-925

FIGURE 5   ICRF-187 and its hydrolysis product ADR-925.

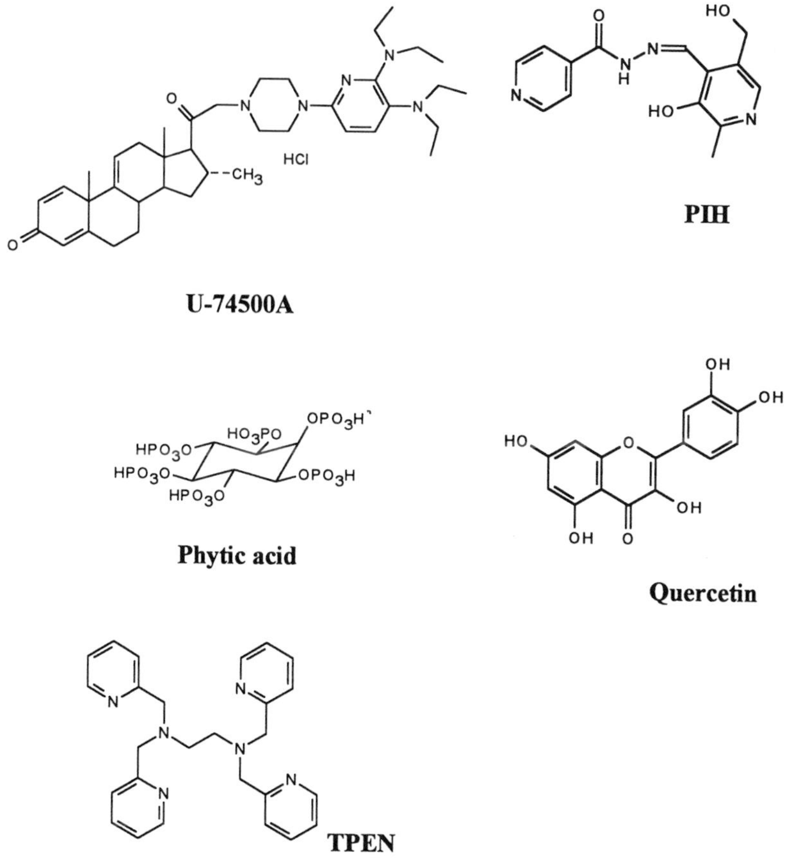

U-74500A

PIH

Phytic acid

Quercetin

TPEN

FIGURE 6 Miscellaneous chelators.

proposed for oral therapy in transfusional iron overload (Williams *et al.*, 1982). PIH acts as a tridentate neutral ligand and the X-ray structure of its iron complex has been described (Murphy *et al.*, 1985). However, PIH appears to be a potent inhibitor of DNA synthesis. Although the use of PIH and chemically related structures as protecting compounds against oxidative damage could have some relevance, they do not seem to have been fully investigated.

Phytic acid (Fig. 6) has been claimed as a potent natural antioxidant for plants acting both as an iron chelator and as an inhibitor of polyphenol oxidase. It is proposed that phytic acid inhibits the Fenton reaction and lipid peroxidation by accelerating ferrous iron-mediated oxygen reduction (Graf *et al.*, 1987). Burkitt and Gilbert (1990) have confirmed the inactivation efficiency of phytic acid but suggested that phytate acts by preventing ferric chelate reduction. Antioxidant properties of uric acid were interpreted in terms of iron chelation in a form which does not promote hydroxyl radical formation (Davies *et al.*, 1986). Indeed, it was proposed that urate can form a 2 : 1 complex with ferric iron with an overall affinity constant of $\beta_2 = 10^{11}$.

The antioxidant properties of flavonoids are generally interpreted by favorable reduction potentials of the phenoxyl radicals which allow an efficient scavenging of alkylperoxyl and superoxide radicals (Jovanovic *et al.*, 1994; Bors and Saran, 1987). Nevertheless, the iron-chelating activities of catechol-containing flavonoids such as quercetin (Fig. 6) could probably account for part of their antioxidant activity (Morel *et al.*, 1993). Careful chemical and physical characterization of iron chelates of such polyphenolic compounds is still lacking. O-Phenanthroline and other 2,2'-bipyridyl compounds are known as strong lipophilic ferrous iron chelators which inhibit the formation of hydroxyl radicals by the Fenton reaction because the corresponding iron chelate has a high redox potential preventing single-electron exchange with hydrogen peroxide. However, all the aromatic amine-containing ligands do not behave in the same way. N,N,N',N'-Tetrakis(2-pyridylmethyl)ethylenediamine (TPEN, Fig. 6) is a powerful transition metal chelator, although nonspecific for iron, which readily penetrates into living cells. Moreover, the ferrous chelate of TPEN has been shown to possess superoxide dismutase activity *in vitro* and *in vivo* (Nagano *et al.*, 1989). However, the complex generates hydroxyl radicals in a Fenton-type reaction in the presence of hydrogen peroxide and physiological reductants (Juliano *et al.*, 1992).

C. Oxidative Stress-Activatable Iron Chelators

From thermodynamic reasons, there seems to be a physical impossibility in designing a clinically efficient iron chelator with a large safety margin for long-term use. As a matter of fact, the affinity constant for iron of such a compound should be relatively low in order to avoid side effects related to iron mobilization from iron proteins. At the same time, to compete for iron traces released during oxidative stress, the association constant must be high enough compared to those of endogenous, nonspecific metal-binding molecules. Indeed, intracellular medium is rich in various small molecules able to nonspecifically bind iron, e.g., amino acids, nucleotides, or citrate. It is noteworthy that the affinity constant for citrate of ferric iron is $K_1 = 10^{11.4}$, whereas for ferrous iron it is $K_1 = 10^{4.4}$ (Singh and Hider, 1988). Moreover, macromolecules such as DNA also nonspecifically bind metal ions (Eisinger *et al.*, 1962).

A possible approach to this drawback is to design a chelator with an affinity for iron lower than 10^{21} (affinity for iron of transferrin) which undergoes a large increase when (and only when) free catalytic iron is released. One way to do so is to take benefit of the oxidizing species formed

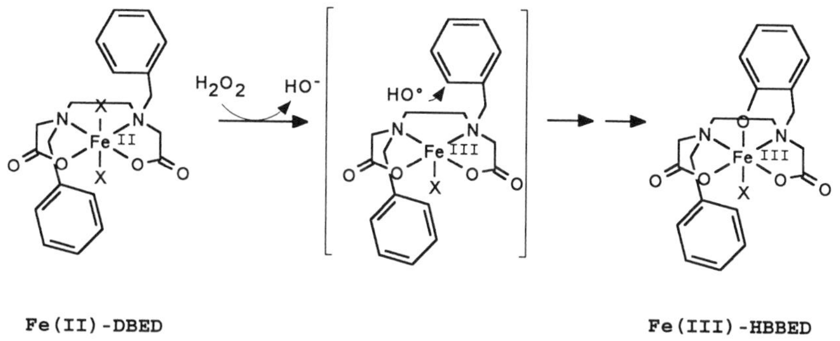

Fe(II)-DBED   Fe(III)-HBBED

FIGURE 7 Monohydroxylation of iron DBED chelate yielding HBBED iron chelate (X is axial ligand).

during oxidative stress by favoring a chemical oxidation of the chelator into an "activated" oxidized form with stronger binding capacity. A particular interest of such a strategy would be its localization in time and space. Indeed, the "activated" chelating entity would be generated only at sites where and when iron and hydrogen peroxide are available. This could be especially useful for diseases that require preventive long-term treatment.

We have looked for such a strategy using ligands bearing phenyl moieties which could be oxidized in phenol, e.g., a hard donor group with a good affinity for iron. This concept has been illustrated with a new series of substituted N,N'-dibenzylethylenediamine-N,N'-diacetic acid (DBED) called "oxidative stress-activatable iron chelators" (Galey *et al.*, 1995). Indeed, in the presence of hydrogen peroxide, the DBED ferrous iron chelate undergoes a site-specific intramolecular hydroxylation leading to N-(2-hydroxybenzyl)-N'-benzylethylenediamine-N,N'-diacetic acid (HBBED) as one of the products (Fig. 7). Aromatic hydroxylation in position 2 of the phenyl moiety is very favored for steric reasons. The affinity for iron of HBBED is dramatically increased ($\Delta \log K > 13$) compared to DBED because it bears a phenolate moiety able to occupy a supplementary coordinating site (Table V). Moreover, although it still has a free coordination site, the

TABLE V Ligand pK_a Values and Stability Constants for Fe(III) (log K) of DBED, HBED, HBBED, and EDDA

	EDDA	*DBED*	*HBBED*	*HBED*
pK_1	6.4	4.8	4.7	4.6
pK_2	9.5	9.6	9.2	8.3
pK_3			11.0	11.0
pK_4				12.5
log K	17	<15	28	39.7

Fe(III) HBBED chelate was not found to be a catalyst of Fenton chemistry, probably because of a sufficient decrease of redox potential of iron. It is noteworthy that HBED is not a product of the reaction, although it could be expected to be formed from HBBED chelate according to a similar pathway. However, the yield of HBBED is low, and various other oxidation side products are formed. In order to increase electron density so as to favor the electrophilic addition of HO· to the aromatic ring, a trimethoxy derivative OR10141 (Fig. 8) was investigated. It was shown that the yield of monohydroxylation of OR10141 to OR10163 (>80%) is much higher than in the case of DBED. Moreover, in pure *in vitro* systems, OR10141 is able to protect biological molecules, proteins, lipids, and DNA against iron-catalyzed oxidative damage (Galey *et al.*, 1996). Therefore, it suggests that the affinity for iron of OR10141 is high enough to partially withdraw nonspecifically bound iron from target molecules. One important question concerns the subsequent possible metabolism of the ferric OR10163 chelate that could be formed intracellularly following the protective action of OR10141 against oxidative challenge. In any case, the amount of OR10163 generated is proportional to that of released iron, and therefore iron deprivation of the cell is not a likely consequence. Moreover, it can be speculated

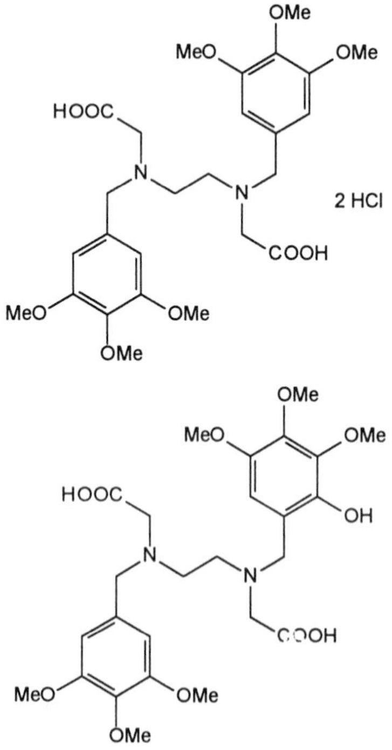

FIGURE 8   OR10141 and its monohydroxylation product OR10163.

that such a transient strong iron chelation by OR10163 could give the opportunity to the cell to activate multiple mechanisms to adapt to the prooxidant challenge even if iron is further released by a reductive mechanism from the OR10163 chelate. Because of the relatively low affinity of OR10141 for iron (log K $<$ 15), it is hoped that the side effects related to "safe" iron chelation of such compounds will be low. Therefore, it is believed that the use of OR10141 in chronic situations in which treatment for prolonged periods of time are required could be more feasible than the use of strong iron-binding chelators. However, OR10141 is a relatively hydrophilic compound, and its membrane permeability is poor. It now remains to achieve sufficient bioavailability either by synthesizing prodrugs or by vectorization of the chelator in order to favor membrane penetration and therefore to allow the chelator to reach a sufficient intracellular concentration. Chemical modifications of the chelator could also include the addition of a supplementary coordination site on the molecule, e.g., starting from a pentadentate ligand so as to increase the initial affinity for iron and to favor the competition with endogenous ligands. The corresponding hydroxylated ligand would be hexadentate, i.e., more likely to inhibit Fenton chemistry.

V. Protective Effect of Iron Chelators against Oxidative Damage

A. Pure *in Vitro* Systems

A number of biochemical assays are available for evaluating the ability of iron complexes to catalyze the formation of hydroxyl radicals. They are all based on the use of a detector molecule for hydroxyl radicals using various analytical methods such as gas chromatography, spectrophotometry, or spectrofluorimetry (Kontoghiorghes *et al.*, 1986; Singh and Hider, 1988; Smith *et al.*, 1990, Dean and Nicholson, 1994). HPLC determination of salicylate hydroxylation products and electron spin resonance detection of the spin adduct of spin-trap DMPO are among the most widely used methods (see Burkitt and Gilbert, 1991). Experiments based on relaxation and linearization of supercoiled DNA using electrophoresis also allow a sensitive detection of single strand breaks generated by hydroxyl radicals (Toyokuni *et al.*, 1992). The formation of thiobarbituric acid-reactive substances from deoxyribose has been extensively used for screening iron-chelating agents (Gutteridge, 1990). From these studies, it generally appears that the only uniformly protective iron chelators are strong hexadentate chelators such as DFO and HBED (Dean and Nicholson, 1994).

B. Cellular Systems

Protection of mammalian cells by *o*-phenanthroline against the deleterious effects of activated oxygen species was first demonstrated by Mello

Filho *et al.* (1984). In that work, it was shown that 100 μM o-phenanthroline completely protected mouse fibroblasts against cell killing by 15–35 μM hydrogen peroxide as measured both by colony-forming ability assays and by DNA single strand breaks. It was proposed that o-phenanthroline prevents the formation of DNA strand breaks by chelating intracellular iron in a nonreactive form. Hydrophilic chelators such as DFO or DTPA afforded no protection, suggesting that the iron chelator has to enter the cell to exert its protective effect (Mello Filho and Meneghini, 1985). Much *in vitro* work on cultured cells has confirmed the ability of some membrane-permeant iron chelators to protect cells against oxidative injury. Most of the studies have been performed using hydrogen peroxide (Tachon, 1990), but results with other sources of oxidative challenge were also obtained (Keyse and Tyrrell, 1990; Adamson and Harman, 1993). For instance, a protective effect of DFO was observed on alveolar type II cells treated with paraquat (Van der Wal *et al.*, 1990). Although these chelators protect cells against a short-term toxic exposure to oxidative stress at apparently nontoxic concentrations, they can become toxic themselves when applied chronically (Gille *et al.*, 1992, Ganeshaguru *et al.*, 1992). Moreover, o-phenanthroline, the most extensively used lipophilic chelator, is not specific for iron and, although it chelates iron in a nonreactive form, it also forms a DNA damaging complex with copper (Birnboim, 1992) and is sometimes used as a zinc chelator (Kiss, 1994). However, DFO is much more specific for iron but, being hydrophilic, its rate of penetration into cells is generally very slow, and therefore it has to be used at very high (often 10 mM or more) nonphysiologic concentrations (Starke and Farber, 1985). The intracellular DFO concentration in the erythrocyte has been evaluated in the range of 30–40 μM after a 90-min incubation with 10 mM of the drug at 37°C (Ferrali *et al.*, 1992). Extracellular iron chelators, such as hydroxyethyl starch-conjugated desferrioxamine, have also been reported to protect cells from hypoxia/reoxygenation (Paller and Hedlund, 1994).

It has been argued that hydrogen peroxide was unlikely to reach cellular DNA since it should be very rapidly consumed by detoxifying enzymes catalase and glutathione peroxidase. Moreover, since a protection by the calcium chelator Quin 2 was also observed, it was proposed that some of the toxic effects of hydrogen peroxide were linked to a rise in intracellular calcium generated by oxidative stress, which may stimulate endonuclease-dependent DNA damage (Cantoni *et al.*, 1989). It is noteworthy that cell protection by Quin 2 was obtained with low concentrations (1 μM) of the membrane-permeant prodrug acetoxymethyl ester (Quin 2 AM). However, Quin 2 is also able to bind iron and copper with respective affinity constants of 10^8 and 10^{12}. Hence Quin 2 effects should not be uncritically linked to effects on calcium homeostasis (Sandström *et al.*, 1994). Although binding of Quin 2 to iron does not prevent hydroxyl radical formation, it inhibits DNA damage, presumably by preventing iron binding to DNA and thereby directing the hydroxyl radical to other less sensitive targets (Burkitt *et al.*, 1994).

C. In Vivo Situations Unrelated to Systemic Iron Overload

I. Inflammatory Diseases

Infiltration by neutrophils and macrophages into an inflamed area will produce high local concentrations of superoxide. This respiratory burst is likely to induce a local iron release from ferritin. Degradation of ferritin by the autophagic vacuolar apparatus in macrophages during inflammation is also considered to be responsible for iron release (Sakaida *et al.*, 1990). The possible interest of iron chelation therapy against inflammatory skin disorders has been suggested (Trenam *et al.*, 1992), but no clinical studies have yet been reported. Synovial fluid of patients with rheumatoid arthritis contains increased amounts of both low molecular weight iron species and ferritin-bound iron. The presence of iron associated to hydrogen peroxide generated by activated inflammatory cells is believed to participate in the degradation of the involved tissues. Therefore, although iron is certainly not the only factor involved in the pathogenesis of rheumatoid arthritis, a good rationale exists in favor of the use of iron chelators to prevent damage. Studies of experimental arthritis in rats showed that the chronic inflammatory phase was reduced by DFO, although it was unable to prevent the acute phase (Blake *et al.*, 1983). So far, only limited clinical studies have been conducted mainly with DFO and, to a lesser extent, hydroxypyridinone L1 in small groups of patients for relatively short periods of time. In DFO-treated patients, serious side effects were observed, including anemia and cerebral toxicity with relatively disappointing results (Polson *et al.*, 1985). Larger trials, especially with a longer duration of treatment, are now needed to establish the potential usefulness of iron chelators for the treatment of rheumatoid arthritis (Voest *et al.*, 1994).

2. Postischemic/Reperfusion Injury

Intracellular free iron increases dramatically during ischemia, probably because of the accumulation of reducing equivalents which arise during ischemia and also because of ischemia-induced acidosis (Voogt *et al.*, 1992). Such free iron is believed to catalyze the production of a pulse of hydroxyl radicals when oxygen tension is suddenly restored during reperfusion. Therefore, the use of iron chelators has been proposed to reduce postischemic tissue damage. A number of studies have shown that DFO improved survival and physiological function in various models of cerebral or cardiac ischemia/reperfusion (Reddy *et al.*, 1989; Komara *et al.*, 1986). A significant protection was also obtained with L1 on isolated rat hearts as measured by contractility after reperfusion (Van der Kraaij *et al.*, 1989). These results suggest that iron chelators could be useful in the treatment of stroke and heart attack. Lazaroids of U-74500A series, for instance, have provided interesting results in animal model systems of ischemia/reperfusion injury to the brain or spinal cord (Braughler *et al.*, 1989). Iron chelation could also be beneficial during cardiopulmonary bypass surgery for which a transient ischemia occurs as well as in organ transplantation by increasing the viability of transplanted tissues [see Herschko (1992) and reference cited therein].

3. Intoxication by Xenobiotics

Several xenobiotics which can undergo redox cycling have the capacity to generate superoxide and to release iron from ferritin inducing tissue damage. The toxicity of the herbicide paraquat has been extensively studied. Paraquat is selectively accumulated by lung tissue via the polyamine transport system and induces a progressive respiratory failure after accidental ingestion or inhalation. Studies on mice have shown that DFO treatment reduces the mortality and prolongs the survival of paraquat-treated animals (Kohen and Chevion, 1985). However, these results were not confirmed by others (Osheroff *et al.*, 1985). Moreover, it has been suggested in some circumstances, that the paraquat cation radical is able to reduce ferrioxamine and thus allows the formation of hydroxyl radicals (Borg and Schaich, 1986). This could explain the reported potentiation of the toxicity of paraquat in rats by high doses of DFO.

The cardiotoxicity of anthracyclines such as doxorubicin unfortunately limits their clinical usefulness, although they are very effective anticancer drugs. Doxorubicin toxicity is believed to be mediated by redox cycling of the iron–anthracycline complex; the drug is accumulated in cardiac tissue. Therefore iron chelators such as DFO have been studied for possible protection against cardiotoxicity. Treatment with ICRF-187 (dexrazoxane) has been shown to protect against doxorubicin cardiotoxicity without disturbing the antitumor effect. It is currently in phase III clinical trials (Malisza and Hasinoff, 1995).

4. Atherosclerosis

Oxidized low-density lipoprotein plays an important role in atherogenesis. The importance of copper and iron in oxidative modifications of LDL increasing its affinity for subendothelial macrophages and its atherogenicity is also well known. Gruel samples from advanced atherosclerotic lesions were found to contain significant amounts of bleomycin-detectable iron (Smith *et al.*, 1992). However, direct experimental proof of the actual participation of catalytic iron in the generation of atherosclerotic lesions is still lacking, as is the origin and form of such iron. Studies were performed in the 1950s and 1960s using EDTA as therapy for atherosclerosis. However, the rationale for these trials seems originally to have been calcium chelation from atherosclerotic plaques rather than iron chelation. No significant improvement was observed for EDTA-treated patients (Olszewer *et al.*, 1990).

EDTA is certainly not a good candidate for assessing the participation of iron-dependent oxidative damage in the development of atherosclerosis because EDTA is an excellent catalyst of Fenton chemistry for the reasons discussed earlier. Clinical studies using more relevant iron chelators for the prevention of atherosclerotic vascular disease are needed.

5. Neurodegenerative Diseases

Several areas of the human brain are rich in iron in a protein-bound form, although cerebrospinal fluid has a low content in transferrin and only moderate amounts of SOD and glutathione peroxidase (Halliwell *et al.,* 1992). Thus, being very rich in polyunsaturated fatty acids, brain may be especially prone to iron-dependent oxidative damage. Ischemic or traumatic brain injury, by causing a partial homogenization, is thought to be able to increase catalytic iron availability and to induce lipid peroxidation. Dopamine-rich regions of the brain may be especially sensitive to altered iron metabolism, first because monoamine oxidase produces hydrogen peroxide, and second because dopamine firmly binds iron and as a catechol can undergo redox cycling, thereby generating reduced oxygen species. Although direct demonstration is lacking, evidence suggests that oxidative stress is involved in Parkinson's disease and is associated with the loss of dopaminergic neurons. An increased iron content have been found in substantia nigra of postmortem parkinsonian brain. The cause of this perturbation of iron homeostasis and the physical nature of the iron (which could be complexed with neuromelanin) are not known in detail (Gerlach *et al.,* 1994). Altered brain metabolism has also been proposed to be linked to Alzheimer's diseases (Olanow and Arendash, 1994), but it is not known whether this alteration is an early or a late event in the disease process. It has been suggested that the toxicity of β-amyloid protein toward some neural cells is mediated by hydrogen peroxide generated by activation of an NADPH oxidase-like enzyme (Behl *et al.,* 1994). A potential use of iron chelators which could cross the blood–brain barrier has been proposed. Intramuscular injection of DFO into patients with Alzheimer's disease was found to have a significant effect on their rate of decline (McLachlan *et al.,* 1991). However, DFO also binds aluminium, which is also suspected to play a role in the development of the disorder. Some iron chelators have been studied for their ability to decrease the iron content of brain in a model of iron-loaded rats (Ward *et al.,* 1995). DFO and hydroxypyridinones (30–100 mg/kg) were found to significantly reduce the brain iron content; however, they also induced a decrease in the level of dopamine, suggesting a removal of iron from tyrosine hydroxylase.

VI. Concluding Remarks

Although considerable work has been performed on iron chelation therapy for treatment of systemic iron overload, there are relatively few reports on chelators specifically focused on the treatment of pathological conditions associated with oxidative stress, with the exception of ICRF-187 and lazaroids.

Some encouraging results have emerged from animal studies as well as from limited trials. However, there is a need for larger randomized clinical studies to evaluate the clinical usefulness of iron chelators in the various oxidative stress-associated pathological situations unrelated to systemic iron overload. Because of the ubiquitous role of iron in biological processes, special attention will have to be paid to side effects related to "safe" iron chelation, including the inhibition of iron-containing enzymes such as ribonucleotide reductase and tyrosine hydroxylase as well as iron depletion. Indeed, iron deficiency may result in anemia, but increasing evidence shows that it may also affect work performance, neurological function, immune response, and epithelial tissues [see Crichton (1991) and reference cited therein].

DFO and hydroxypyridinones such as L1 are still good candidates for future investigations of acute oxidative stress situations, especially because they can be used in humans. Nevertheless, they also show side effects when used for prolonged periods of time, and DFO is not orally active. Covalent attachment of DFO to biocompatible polymers may be a promising way to diminish toxic effects (Hallaway *et al.*, 1989). However, there is also a need for new orally active iron chelators with less toxicity, especially in the prospect of long-term use. The potential usefulness of such an iron chelator may be extremely broad. The "oxidative stress activation" concept described in this chapter could provide a rational approach to minimize side effects and therefore to allow chelation therapy for long periods of time. Results obtained *in vitro* with some compounds of the DBED series as well as toxicology data are encouraging.

References

Adamson, G. M., and Harman, A. W. (1993). Oxidative stress in cultured hepatocytes exposed to acetaminophen. *Biochem. Pharmacol.* **45**, 2289–2294.

Atamna, H., and Ginsburg, H. (1995). Heme degradation in the presence of glutathione. *J. Biol. Chem.* **270**, 24876–24883.

Aubailly, M., Santus, R., and Salmon, S. (1991). Ferrous iron release from ferritin by ultraviolet-A radiations. *Photochem. Photobiol.* **54**, 769–773.

Basilion, J. P., Rouault, T. A., Massinople, C. M., Klausner, R. D., and Burgess, W. H. (1994). The iron-responsive element-binding protein: Localization of the RNA-binding site to the aconitase active-site cleft. *Proc. Natl. Acad. Sci. U.S.A.* **91**, 574–578.

Behl, C., Davis, J. B., Lesley, R., and Schubert, D. (1994). Hydrogen peroxide mediates amyloid b protein toxicity. *Cell* **77**, 817–827.

Bergeron, R. J., Wiegand, J., Dionis, J. B., Egli-Karmakka, M., Frei, J., Huxley-Tencer, A., and Peter, H. H. (1991). Evaluation of desferrithiocin and its synthetic analogues as orally effective iron chelators. *J. Med. Chem.* **34**, 2072–2078.

Biemond, P., Swaak, A. J. G., Beindorf, C. M., and Koster, J. F. (1986). Superoxide-dependent and independent mechanisms of iron mobilization from ferritin by xanthine oxidase. *Biochem. J.* **239**, 169–173.

Birnboim, H. C. (1992). Effect of lipophilic chelators on oxyradical-induced DNA strand breaks in human granulocytes: paradoxical effect of 1,10-phenanthroline. *Arch. Biochem. Biophys.* **294**, 17–21.

Bissett, D. L., Chatterjee, R., and Hannon, D. P. (1991). Chronic ultraviolet radiation-induced increase in skin iron and the photoprotective effect of topically applied iron chelators. *Photochem. Photobiol.* **54**, 215–223.

Blake, R. R., Hall, N. D., Bacon, P. A., Dieppe, P. A., Halliwell, B., and Gutteridge, J. M. G. (1983). Effect of a specific iron chelating agent on animal models of inflammation. *Ann. Rheum. Dis.* **42**, 89–93.

Bomford, A., Isaac, J., Roberts, S., Edwards, A., Young, S., and Williams, R. (1986). The effect of desferrioxamine on transferrin receptors, the cell cycle and growth rates of human leukaemic cells. *Biochem. J.* **236**, 243–249.

Borg, D. C., and Schaich, K. (1986). Prooxidant action of desferrioxamine: Fenton-like production of hydroxyl radicals by reduced ferrioxamine. *J. Free Rad. Biol. Med.* **2**, 237–243.

Bors, W., and Saran, M. (1987). Radical scavenging by flavonoïd antioxidants. *Free Rad. Res. Commun.* **2**, 289–294.

Bosque, M. A., Domingo, J. L., and Corbella, J. (1995). Assessment of the developmental toxicity of deferoxamine. *Arch. Toxicol.* **469**, 467–471.

Braughler, J. M., Hall, E. D., Jacobsen, E. J., McCall, J. M., and Means, E. D. (1989). The 21-aminosteroids: Potent inhibitors of lipid peroxidation for the treatment of central nervous system trauma and ischemia. *Drugs Future* **14**, 143–152.

Braughler, J. M., Pregenzer, J. F., Chase, R. L., Duncan, L. A., Jacobsen, E. J., and McCall, J. M. (1987). Novel 21-amino steroids as potent inhibitors of iron dependent lipid peroxidation. *J. Biol. Chem.* **262**, 10438–10440.

Breuer, W., Epsztejn, S., and Cabantchik, Z. I. (1995). Iron acquired from transferrin by K562 cells is delivered into a cytoplasmic pool of chelatable iron(II). *J. Biol. Chem.* **270**, 24209–24215.

Burkitt, M. J. (1993). ESR spin trapping studies into the nature of the oxidizing species formed in the Fenton reaction: Pitfalls associated with the use of 5,5-dimethyl-1-pyrroline-N-oxide in the detection of the hydroxyl radical. *Free Rad. Res. Commun.* **18**, 43–57.

Burkitt, M. J., and Gilbert, B. C. (1990). Model studies of the iron-catalysed Haber-Weiss cycle and the ascorbate-driven Fenton reaction. *Free Rad. Res. Commun.* **10**, 265–280.

Burkitt, M. J., and Gilbert, B. C. (1991). The autoxidation of iron(II) in aqueous systems: The effects of iron chelation by physiological, nonphysiological and therapeutic chelators on the generation of reactive oxygen species and the inducement of biomolecular damage. *Free Rad. Res. Commun.* **14**, 107–123.

Burkitt, M. J., Milne, L., Tsang, S. Y., and Tam, S. C. (1994). Calcium indicator dye Quin2 inhibits hydrogen peroxide-induced DNA strand break formation via chelation of iron. *Arch. Biochem. Biophys.* **311**, 321–328.

Cairo, G., Tacchini, L., Pogliaghi, G., Anzon, E., Tomasi, A., and Bernelli-Zazzera, A. (1995). Induction of ferritin synthesis by oxidative stress, transcriptional and post-transcriptional regulation by expansion of the "free" iron pool. *J. Biol. Chem.* **270**, 700–703.

Cantoni, O., Sestili, P., Cattabeni, F., Bellomo, G., Pou, S., Cohen, M., and Cerutti, P. (1989). Calcium chelator Quin 2 prevents hydrogen peroxide-induced DNA breakage and cytotoxicity. *Eur. J. Biochem.* **182**, 209–212.

Chevion, M. (1988). A site-specific mechanism for free radical induced biological damage: The essential role of redox-active transition metals. *Free Rad. Biol. Med.* **5**, 27–37.

Crichton, R. R. (1991). "Inorganic Biochemistry of Iron Metabolism." Elis Horwood, New York/London.

Davies, K. J. A., Sevanian, A., Muakkassah-Kelly, S. F., and Hochstein, P. (1986). Uric acid-iron ion complexes: A new aspect of the antioxidant functions of uric acid. *Biochem. J.* **235**, 747–754.

Davies, M. J., Donkor, R., Dunster, C. A., Gee, C. A., Jonas, S., and Willson, R. L. (1987). Desferrioxamine (desferal) and superoxide free radicals: Formation of an enzyme-damaging nitroxide. *Biochem. J.* **246**, 725–729.

Dawson, R. M. C., Elliott, D. C., Elliott, W. H., and Jones, K. M. (1986). *In* "Data for Biochemical Research," 3rd Ed. Clarendon Press, Oxford.

Dean, R. T., and Nicholson, P. (1994). The action of nine chelators on iron-dependent radical damage. *Free Rad. Res.* **20**, 83–101.

Egan, T. J., Barthakur, S. R., and Aisen, P. (1992). Catalysis of the Haber-Weiss reaction by iron-diethylenetriaminepentaacetate. *J. Inorg. Biochem.* **48**, 241–249.

Eisinger, J., Shulman, R. G., and Szymanski, B. M. (1962). Transition metal binding in DNA solutions. *J. Chem. Phys.* **36**, 1721–1729.

Farber, J. M., and Levine, R. L. (1986). Sequence of a peptide susceptible to mixed-function oxidation: Probable cation binding site in glutamine synthetase. *J. Biol. Chem.* **261**, 4574–4578.

Ferrali, M., Ciccoli, L., and Comporti, M. (1989). Allyl alcohol-induced hemolysis and its relation to iron release and lipid peroxidation. *Biochem. Pharmacol.* **38**, 1819–1825.

Ferrali, M., Signorini, C., Ciccoli, L., and Comporti, M. (1992). Iron release and membrane damage in erythrocytes exposed to oxidizing agents, phenylhydrazine, divicine and isouramil. *Biochem. J.* **285**, 295–301.

Fontecave, M., and Pierre, J. L. (1993a). Iron: Metabolism, toxicity and therapy. *Biochimie* **75**, 767–773.

Fontecave, M., and Pierre, J. L. (1993b). Fer et peroxyde d'hydrogène: Aspects chimiques d'un problème biologique fondamental. *Bull. Soc. Chim. Fr.* **130**, 77–85.

Fridovich, I. (1986). Biological effects of the superoxide radical. *Arch. Biochem. Biophys.* **274**, 1–11.

Funk, F., Lenders, J. P., Crichton, R. R., and Schneider, W. (1985). Reductive mobilization of ferritin iron. *Eur. J. Biochem.* **152**, 167–172.

Galey, J.-B., Dumats, J., Beck, I., Fernandez, B., and Hocquaux, M. (1995). N,N'-Bis-dibenzyl ethylenediaminediacetic acid (DBED): A site specific hydroxyl radical scavenger acting as an "oxidative stress activatable iron chelator" in vitro. *Free Rad. Res.* **22**, 67–86.

Galey, J.-B., Dumats, J., Genard, S., Destrée, O., Pichaud, P., Catroux, P., Marrot, L., Beck, I., Fernandez, B., Barre, G., Seite, M., Hussler, G., and Hocquaux, M. (1996). N,N'-Bis-(3,4,5-trimethoxybenzyl) ethylenediamine N,N'-diacetic acid as a new iron chelator with potential medicinal applications against oxidative stress. *Biochem. Pharmacol.* **51**, 103–115.

Ganeshaguru, K., Lally, J. M., Piga, A., Hoffbrand, A. V., and Kontoghiorghes, G. J. (1992). Cytotoxic mechanisms of iron chelators. *Drugs Today* **28** (Suppl. A), 29–34.

Gardner, P. R., Raineri, I., Epstein, L. B., and White, C. W. (1995). Superoxide radical and iron modulate aconitase activity in mammalian cells. *J. Biol. Chem.* **270**, 13399–13405.

Gerlach, M., Ben-Schachar, D., Riederer, P., and Youdim, M. B. H. (1994). Altered brain metabolism of iron as a cause of neurodegenerative diseases. *J. Neurochem.* **63**, 793–807.

Gille, J. J. P., Van berkel, C. G. M., and Joenje, H. (1992). Effect of iron chelators on the cytotoxic and genotoxic action of hyperoxia in chinese hamster ovary cells. *Mutat. Res.* **275**, 31–39.

Gordeuk, V. R., Thuma, P. E., and Brittenham, G. M. (1994). Iron chelation therapy for malaria. *In* "Progress in Iron Research" (C. Hershko *et al.*, eds.), pp. 371–383. Plenum Press, New York.

Grady, R. W., and Hershko, C. (1990). HBED: A potential oral iron chelator. *Ann. N.Y. Acad. Sci.* **612**, 361–368.

Grady, R. W., Salbe, A. D., Hilgartner, M. W., and Giardina, P. J. (1994). Results from a phase I clinical trial of HBED. *In* "Progress in Iron Research" (C. Hershko *et al.,* eds.), pp. 351–359. Plenum Press, New York.

Graf, E., Empson, K. L., and Eaton, J. W. (1987). Phytic acid: A natural antioxydant. *J. Biol. Chem.* **262**, 11647–11650.

Graf, E., Mahoney, J. R., Bryant, R., and Eaton, J. W. (1984). Iron-catalysed hydroxyl radical formation: Stringent requirement for free iron coordination site. *J. Biol. Chem.* **259**, 3620–3624.

Gutteridge, J. M. C. (1990). Superoxide-dependent formation of hydroxyl radicals from ferric-complexes and hydrogen peroxide: An evaluation of fourteen iron chelators. *Free Rad. Res. Commun.* **9**, 119–125.

Gutteridge, J. M. C., Richmond, R., and Halliwell, B. (1979). Inhibition of the iron-catalysed formation of hydroxyl radicals from superoxide and of lipid peroxidation by desferrioxamine. *Biochem. J.* **184**, 469–472.

Gutteridge, J. M. C., Rowley, D. A., and Halliwell, B. (1981). Superoxide-dependent formation of hydroxyl radicals in the presence of iron salts: Detection of "free" iron in biological systems by using bleomycin-dependent degradation of DNA. *Biochem. J.* **199**, 263–265.

Hallaway, P. E., Eaton, J. W., Panter, S. S., and Hedlund, B. E. (1989). Modulation of deferoxamine toxicity and clearance by covalent attachment to biocompatible polymers. *Proc. Natl. Acad. Sci. U.S.A.* **86**, 10108–10112.

Halliwell, B. (1985). Use of desferrioxamine as a "probe" for iron-dependent formation of hydroxyl radicals. *Biochem. Pharmacol.* **34**, 229–233.

Halliwell, B. (1989). Protection against tissue damage in vivo by desferrioxamine: What is its mechanism of action? *J. Free Rad. Biol. Med.* **7**, 645–651.

Halliwell, B. (1990). How to characterize a biological antioxidant. *Free Rad. Res. Commun.* **9**, 1–32.

Halliwell, B. (1992). Reactive oxygen species and the central nervous system. *J. Neurochem.* **59**, 1609–1623.

Halliwell, B., and Gutteridge, J. M. C. (1989). "Free Radicals in Biology and Medicine," 2nd Ed. Clarendon Press, Oxford.

Halliwell, B., Gutteridge, J. M. C., and Cross, C. E. (1992). Free radicals, antioxidants, and human disease: Where are we now? *J. Lab. Clin. Med.* **119**, 598–620.

Harris, W. R., Carrano, C. J., and Raymond, K. N. (1979). Coordination chemistry of microbial iron transport compounds. 16. Isolation, characterisation, and formation constants of ferric aerobactin. *J. Am. Chem. Soc.* **101**, 2722–2727.

Hasinoff, B. B. (1994). Pharmacodynamics of the hydrolysis-activation of the cardioprotective agent (+)-1,2-bis(3,5-dioxopiperazin-1-yl) propane. *J. Pharm. Sci.* **83**, 64–67.

Herschko, C. (1992). Iron chelators in medicine. *Mol. Aspects Med.* **13**, 113–165.

Hider, R. C., and Singh, S. (1992). Iron chelating agents with clinical potential. *Procs. Roy. Soc. Ed.* **99B**, 137–168.

Iuliano, L., Pedersen, J. Z., Ghiselli, A., Pratico, D., Rotilio, G., and Violi, F. (1992). Mechanism of reaction of a suggested superoxide-dismutase mimic, Fe(II)-N,N,N',N'-tetrakis(2-pyridylmethyl)ethylenediamine. *Arch. Biochem. Biophys.* **293**, 153–157.

Jacobs, D. J., Watt, G. D., Frankel, R. B., and Papaefthymiou, G. C. (1989). Redox reactions associated with iron release from mammalian ferritin. *Biochemistry* **28**, 1650–1655.

Jovanovic, S. V., Steenken, S., Tosic, M., Marjanovic, M., and Simic, M. G. (1994). Flavonoïds as antioxidants. *J. Am. Chem. Soc.* **116**, 4846–4851.

Keyse, S. M., and Tyrrell, R. M. (1990). Induction of the heme oxygenase gene in human skin fibroblasts by hydrogen peroxide and UV-A 365 nm radiation. *Carcinogenesis* **11**, 787–791.

Kiss, Z. (1994). The zinc chelator, 1,10-phenanthroline enhances the stimulatory effects of protein kinase C activators and staurosporine, but not sphingosine and H2O2, on phopholipase D activity in NIH 3T3 fibroblasts. *Biochem. J.* **298**, 93–98.

Klausner, R. D., Rouault, T. A., and Harford, J. B. (1993). Regulating the fate of mRNA: The control of cellular iron metabolism. *Cell* **72**, 19–28.

Kohen, R., and Chevion, M. (1985). Paraquat toxicity is enhanced by iron and reduced by desferrioxamine in laboratory mice. *Biochem. Pharmacol.* **34**, 1841–1843.

Komara, J. S., Nayini, N. R., Bialik, H. A., Indrien, R. J., Evans, A. T., Garritano, A. M., Hoehner, T. J., Jacobs, W. A., Huang, R. R., Krause, G. S., White, B. C., and Aust, S. D. (1986). Brain iron delocalization and lipid peroxidation following cardiac arrest. *Ann. Emerg. Med.* **15**, 384–389.

Kontoghiorghes, G. J. (1995). Comparative efficacy and toxicity of desferrioxamine, deferiprone and other iron and aluminium chelating drugs. *Toxicol. Lett.* **80**, 1–18.

Kontoghiorghes, G. J., Jackson, M. J., and Lunec, J. (1986). *In vitro* screening of iron chelators using models of free radical damage. *Free Rad. Res. Commun.* **2**, 115–124.

Kozlov, A. V., Yegorov, D. Y., Vladimirov, Y. A., and Azizova, O. A. (1992). Intracellular free iron in liver tissue and liver homogenate: Studies with electron paramagnetic resonance on the formation of paramagnetic complexes with desferal with nitric oxide. *Free Rad. Biol. Med.* **13**, 9–16.

Kretchmar Nguyen, S. A., Craig, A., and Raymond, K. N. (1993). Transferrin: The role of conformational changes in iron removal by chelators. *J. Am. Chem. Soc.* **115**, 6758–6764.

Laughton, M. J., Moroney, M. A., Hoult, J. R. S., and Halliwell, B. (1989). Effects of desferrioxamine on eicosanoid production in two intact cell systems. *Biochem. Pharmacol.* **38**, 189–193.

Laulhère, J.-P., and Fontecave, M. (1996). Nitric oxide does not promote iron release from ferritin. *BioMetals* **9**, 10–14.

L'Eplattenier, F., Murase, I., and Martell, A. E. (1967). New multidentate ligands. VI. Chelating tendencies of N,N'-Di(2-hydroxybenzyl)ethylenediamine-N,N'-diacetic acid. *J. Am. Chem. Soc.* **89**, 837–843.

Li, J. L., Okada, S., Hamazaki, S., Ebina, Y., and Midorikawa, O. (1987). Subacute nephrotoxicity and induction of renal cell carcinoma in mice with ferric nitrilotriacetate. *Cancer Res.* **47**, 1867–1869.

Lind, M. D., Hamor, M. J., Hamor, T. A., and Hoard, J. L. (1964). Stereochemistry of ethylenediamine-tetraacetato complexes. *Inorg. Chem.* **3**, 34–43.

Liu, M., and Okada, S. (1994). Induction of free radicals and tumors in the kidney of Wistar rats by ferric ethylenediamine-N,N'-diacetic acid. *Carcinogenesis* **15**, 2817–2821.

Lloyd, J. B., Cable, H., and Rice-Evans, C. (1991). Evidence that desferrioxamine cannot enter cells by passive diffusion. *Biochem. Pharmacol.* **41**, 1361–1363.

Lytton, S. D., Mester, B., Libman, J., Shanzer, A., and Cabantchik, Z. I. (1992). Monitoring of iron(III) removal from biological sources using a fluorescent siderophore. *Anal. Biochem.* **205**, 326–333.

Malisza, K. L., and Hasinoff, B. B. (1995). Doxorubicin reduces the iron(III) complexes of the hydrolysis products of the antioxidant cardioprotective agent dexrazoxane (ICRF-187) and produces hydroxyl radicals. *Arch. Biochem. Biophys.* **316**, 680–688.

Martell, A. E., Anderson, W. F., and Badman, D. G., eds. (1981). "Development of Iron Chelators for Clinical Use." Elsevier North Holland, New York.

Martell, A. E., Motekaitis, R. J., Murase, I., Sala, L. F., Stoldt, R., and Ng, C. Y. (1987). Development of iron chelators for Cooley's anemia. *Inorgan. Chim. Acta.* **138**, 215–230.

McLachlan, D. R. C., Dalton, A. J., Kruck, T. P. A., Bell, M. Y., Smith, W. L., Kallow, W., and Andrews, D. F. (1991). Intramuscular desferrioxamine in patients with Alzheimer's disease. *Lancet* **337**, 1304–1308.

Mazur, A., Baez, S., and Schorr, E. (1955). The mechanism of iron release from ferritin as related to its biological properties. *J. Biol. Chem.* **213**, 147–160.

Mello Filho, A. C., and Meneghini, R. (1985). Protection of mammalian cells by o-phenanthroline from lethal and DNA-damaging effects produced by active oxygen species. *Biochim. Biophys. Acta* **847**, 82–89.

Mello Filho, A. C., Hoffmann, M. E., and Meneghini, R. (1984). Cell killing and DNA damage by hydrogen peroxide are mediated by intracellular iron. *Biochem. J.* **218**, 273–275.

Miller, D. M., Buettner, G. R., and Aust, S. D. (1990). Transition metals as catalysts of "autoxidation" reactions. *Free Rad. Biol. Med.* **8**, 95–108.

Miller, M. J. (1989). Synthesis and therapeutic potential of hydroxamic base siderophores and analogues. *Chem. Rev.* **89**, 1563–1579.

Molenda, J. J., Jones, M. J., Cecil, K. M., and Basinger, M. A. (1994). Structure/activity relationships afffecting the ability of monoanionic 3-hydroxypyrid-4-ones to mobilize iron. *Chem. Res. Toxicol.* **7**, 815–822.

Morel, I., Lescoat, G., Cogrel, P., Sergent, O., Pasdeloup, N., Brissot, P., Cillard, P., and Cillard, J. (1993). Antioxidant and iron-chelating activities of the flavonoïds catechin, quercetin, and diosmetin on iron-loaded rat hepatocyte culture. *Biochem. Pharmacol.* **45**, 13–19.

Murphy, T. B., Rose, N. J., Schomaker, V., and Aruffo, A. (1985). Synthesis of iron(III) aroyl hydrazones containing pyridoxal and salicyladehyde: The crystal and molecular structure of two iron(III)-pyridoxal isonicotinoyl hydrazone complexes. *Inorgan. Chim. Acta* **108**, 183–194.

Nagano, T., Hirano, T., and Hirobe, M. (1989). Superoxide dismutase mimics based on iron in vivo. *J. Biol. Chem.* **264**, 9243–9249.

Nathan, D. G. (1995). An orally active iron chelator. *N. Engl. J. Med.* **332**, 953–954.

Olanow, C. W., and Arendash, G. W. (1994). Metals and free radicals in neurodegeneration. *Curr. Opin. Neurol.* **7**, 548–558.

Olivieri, N. F., Brittenham, G. M., Matsui, D., Berkovitch, M., Blendis, L. M., Cameron, R. G., McClelland, R. A., Liu, P. P., Templeton, D. M., and Koren, G. (1995). Iron-chelation therapy with oral deferiprone in patients with thalassemia major. *N. Engl. J. Med.* **332**, 918–922.

Olszewer, E., Sabbag, F. C., and Carter, J. P. (1990). A pilot double-blind study of sodium-magnesium EDTA in peripheral vascular disease. *J. Natl. Med. Assoc.* **82**, 173–177.

Osheroff, K. M., Schaich, K. M., Drew, R. T., and Borg, D. C. (1985). Failure of desferrioxamine to modify the toxicity of paraquat in rats. *J. Free Rad. Biol. Med.* **1**, 71–82.

Paller, M. S., and Hedlund, B. E. (1994). Extracellular iron chelators protect kidney cells from hypoxia/reoxygenation. *Free Rad. Biol. Med.* **17**, 597–603.

Pantopoulos, K., and Hentze, M. W. (1995). Nitric oxide signaling to iron regulatory protein: Direct control of ferritin mRNA translation and transferrin receptor mRNA stability in transfected fibroblasts. *Proc. Natl. Acad. Sci. U.S.A.* **92**, 1267–1271.

Pitt, C. G., Bao, Y., Thompson, J., Wani, M. C., Rosenkrantz, H., and Metterville, J. (1986). Esters and lactones of phenolic amino carboxylic acids: Prodrugs for iron chelation. *J. Med. Chem.* **29**, 1231–1237.

Polson, R. J., Jawed, A., Bomford, A., Berry, H., and Williams, R. (1985). Treatment of rheumatoid arthritis with desferrioxamine: Relation between stores of iron before treatment and side effects. *Br. Med. J. (Clin. Res. Ed.)* **291**, 448.

Ponka, P., Borova, J., Neuwirt, J., and Fuchs, O. (1979). Mobilization of iron from reticulocytes; identification of pyridoxal isonicotinoyl hydrazones as a new iron chelating agent. *FEBS Lett.* **97**, 317–321.

Porter, J. B., and Huens, E. R. (1989). The toxic effects of desferrioxamine. *Bailliere's Clin. Haematol.* **2**, 459–474.

Porter, J. B., Huens, E. R., and Hider, R. C. (1989). The development of iron chelating drugs. *Bailliere's Clin. Haematol.* **2**, 257–292.

Porter, J. B., Singh, S., Hoyes, K. P., Epemolu, O., Abeysinghe, R. D., and Hider, R. C. (1994). Lessons from preclinical and clinical studies with 1,2-diethyl-3-hydroxypyrid-4-one, CP94 and related compounds. *In* "Progress in Iron Research" (C. Hershko *et al.*, eds.), pp. 361–370. Plenum Press, New York.

Reddy, B. R., Kloner, R. A., and Przyklenk, K. (1989). Early treatment with deferoxamine limits myocardial ischemic/reperfusion injury. *Free Rad. Biol. Med.* **7**, 45–52.

Reif, D. W. (1992). Ferritin as a source of iron for oxidative damage. *Free Rad. Biol. Med.* **12**, 417–427.

Reif, D. W., and Simmons, R. D. (1990). Nitric oxide mediates iron release from ferritin. *Arch. Biochem. Biophys.* **283**, 537–541.

Rush, J. D., and Koppenol, W. H. (1986). Oxidizing intermediates in the reaction of ferrous EDTA with hydrogen peroxide. *J. Biol. Chem.* **261**, 6730–6733.

Ryan, T. P., and Aust, S. D. (1992). The role of iron in oxygen-mediated toxicities. *Crit. Rev. Toxicol.* **22**, 119–141.

Ryan, T. P., and Petry, T. W. (1993). The effects of 21-aminosteroids on the redox status of iron in solution. *Arch. Biochem. Biophys.* **300**, 699–704.

Sakaida, S., Kyle, M. E., and Farber, J. L. (1990). Autophagic degradation of protein generates a pool of ferric iron required for the killing of cultured hepatocytes by an oxidative stress. *Mol. Pharmacol.* **37**, 435–442.

Samuni, A., Aranovich, J., Godinger, D., Chevion, M., and Czapski, G. (1983). On the cytotoxicity of vitC and metal ions: A site specific mechanism. *Eur. J. Biochem.* **137**, 119–124.

Sandström, B. E., Granström, M., and Marklund, S. L. (1994). New roles for quin2: Powerful transition metal ion chelator that inhibits copper but potentiates iron-driven, Fenton-type reactions. *Free Rad. Biol. Med.* **16**, 177–185.

Singh, S., and Hider, R. C. (1988). Colorimetric detection of the hydroxyl radical: Comparison of the hydroxyl-radical-generating ability of various iron complexes. *Anal. Biochem.* **171**, 47–54.

Smith, C., Mitchinson, M. J., Aruoma, O. I., and Halliwell, B. (1992). Stimulation of lipid peroxidation and hydroxyl radical generation by the contents of human artherosclerotic lesions. *Biochem. J.* **286**, 901–905.

Smith, J. B., Cusumano, J. C., and Babbs, C. F. (1990). Quantitative effects of iron chelators on hydroxyl radical production by the superoxide-driven Fenton reaction. *Free Rad. Res. Commun.* **8**, 101–106.

Stadtman, E. R. (1990). Metal ion-catalysed oxidation of proteins: Biochemical mechanism and biological consequences. *Free Rad. Biol. Med.* **9**, 315–325.

Stadtman, E. R. (1992). Protein oxidation and aging. *Science* **257**, 1220–1224.

Starke, P. E., and Farber, J. L. (1985). Ferric iron and superoxide ions are required for the killing of cultured hepatocytes by hydrogen peroxide. *J. Biol. Chem.* **260**, 10099–10104.

Stubbe, J., and Kozarich, J. W. (1987). Mechanisms of bleomycin-induced DNA degradation. *Chem. Rev.* **87**, 1107–1136.

Tachon, P. (1989). Ferric and cupric ions requirement for DNA single-strand breakage by H_2O_2. *Free Rad. Res. Commun.* **7**, 1–10.

Taliaferro, C. H., Motekaitis, R. J., and Martell, A. E. (1984). New multidentates ligands. 22. N,N'-Dipyridoxylethylenediamine-N,N'-diacetic acid: A new chelating ligand for trivalent metal ions. *Inorg. Chem.* **23**, 1188–1192.

Thomas, C., Vile, G. F., and Winterbourn, C. C. (1993). The hydrolysis product of ICRF-187 promotes iron-catalysed hydroxyl radical production via the Fenton reaction. *Biochem. Pharmacol.* **45**, 1967–1972.

Thomas, C. G., and Aust, S. D. (1986). Reductive release of iron from ferritin by cation free radicals of paraquat and other bipyridyls. *J. Biol. Chem.* **261**, 13064–13070.

Toyokuni, S., and Sagripanti, J. L. (1992). Iron-mediated DNA damage: Sensitive detection of DNA strand breakage catalysed by iron. *J. Inorg. Biochem.* **47**, 241–248.

Trenam, C. W., Blake, D. R., and Morris, C. J. (1992). Skin inflammation: Reactive oxygen species and the role of iron. *J. Invest. Dermatol.* **99**, 675–682.

Tsien, R., Pozzan, T., and Rink, T. J. (1982). Calcium homeostasis in intact lymphocytes: Cytoplasmic free calcium monitored with a new, intracellularly trapped fluorescent indicator. *J. Cell Biol.* **94**, 325–334.

Van der Kraaij, A. M. M., Van Eijk, H. G., and Koster, J. F. (1989). Prevention of post ischemic cardiac injury by the orally active iron chelator 1,2-dimethyl-3-hydroxy-4-pyridone (L1) and the antioxidant (+)-cyanidanol-3. *Circulation* **80**, 158–164.

Van der Wal, N. A. A., Van Oirschot, J. F. L. M., Van Dijk, A., Verhoef, J., and Van Asbeck, B. S. (1990). Mechanism of protection of alveolar type II cells against paraquat-induced cytotoxicity by deferoxamine. *Biochem. Pharmacol.* **39**, 1665–1671.

Vile, G. F., and Tyrrell, R. M. (1993). Oxidative stress resulting from ultraviolet A irradiation of human skin fibroblasts leads to a heme oxygenase-dependent increase in ferritin. *J. Biol. Chem.* **268**, 14678–14681.

Voest, E. E., Vreugdenhil, G., and Marx, J. J. M. (1994). Iron chelating agents in non-iron overload conditions. *Ann. Intern. Med.* **120**, 490–499.

Voogt, A., Sluiter, W., van Eijk, H. G., and Koster, J. F. (1992). Low molecular weight iron and the oxygen paradox in isolated rat hearts. *J. Clin. Invest.* **90**, 2050–2055.

Ward, R. J., Dexter, D., Florence, A., Aouad, F., Hider, R., Jenner, P., and Crichton, R. R. (1995). Brain iron in the ferrocene-loaded rat: Its chelation and influence on dopamine metabolism. *Biochem. Pharmacol.* **49**, 1821–1826.

Watt, G. D., Frankel, R. B., and Papaefthymiou, G. C. (1985). Reduction of mammalian ferritin. *Proc. Natl. Acad. Sci. U.S.A.* **82**, 3640–3643.

Williams, A., Hoy, T., Pugh, A., and Jacobs, A. (1982). Pyridoxal complexes as potential chelating agents for oral therapy in transfusion iron overload. *J. Pharm. Pharmacol.* **34**, 730–732.

Yamazaki, I., and Piette, L. H. (1991). EPR spin-trapping study on the oxidizing species formed in the reaction of the ferrous ion with hydrogen peroxide. *J. Am. Chem. Soc.* **113**, 7588–7593.

Ian A. Cotgreave

Biochemical Toxicology Unit
Institute of Environmental Medicine
Karolinska Institute
S-17177 Stockholm, Sweden

N-Acetylcysteine: Pharmacological Considerations and Experimental and Clinical Applications

I. Introduction

N-Acetylcysteine (NAC) is a thiol-containing compound that has been used in clinical practice since the mid-1950s. Historically, NAC was introduced for the treatment of congestive and obstructive lung diseases, primarily those associated with hypersecretion of mucus, e.g., chronic bronchitis and cystic fibrosis. NAC has also been used since the mid-1970s as the drug of choice in the treatment of paracetamol intoxication. In more recent times, the drug has enjoyed a renaissance, and its applications now include attempts to use it in the treatment of pulmonary oxygen toxicity and adult respiratory distress syndrome (ARDS), with exploratory studies being performed in the treatment of a variety of acute and chronic inflammations and a number of disorders involving the immune system, such as human immunodeficiency virus (HIV-1) infections. In addition to its clinical uses, the compound has, in recent years, been used increasingly as a tool for basic research, particu-

Advances in Pharmacology, Volume 38

larly into basic molecular and cellular aspects of the regulation of apoptosis and gene transcription. Here, this research has been particularly conducted at the level of the regulation of transcription factors.

The apparent diversity in the pharmacological use of NAC is rather unique and is due to the multifaceted chemical properties of the cysteinyl thiol of the molecule. These include its nucleophilicity and redox interactions, particularly with other members of the group XIV elements (O and Se), providing "scavenger" and antioxidant properties, as well as the ability to undergo transhydrogenation or thiol-disulfide exchange (TDE) reactions with other thiol redox couples.

This chapter covers various pharmacological properties of NAC, particularly focusing on the human pharmacokinetic data, the metabolism of the compound in biological systems, and the pharmacodynamic properties of the parent thiol drug and various metabolites emanating from it. The chapter will then detail some aspects of the biochemical and clinical applications of NAC ranging from the redox modulation of biochemical components of cells, through effects of the compound on intact cells and tissues, extending to therapeutic effects in human disease states.

II. Biopharmaceutical and Pharmacological Aspects _____

A. Analysis of NAC in Biological Systems

The assay of NAC in biological systems presents several problems. First, the compound has few physical properties, apart from the redox activity and nucleophilicity of its thiol, which allow its direct detection. Second, as a typical thiol, NAC may oxidize to disulfide species or undergo transhydrogenation reactions with other thiol redox couples, resulting in the potential introduction of artifacts during the manipulation of biological samples. Finally, biological systems contain low molecular weight thiols, such as cysteine and glutathione, which possess similar physical and chemical properties to NAC and are its primary metabolites. Thus, the ideal analytical approach for NAC in complex biological systems must "trap" the reduced form of NAC as quickly as possible.

This has been accomplished by the quantitative chemical derivatization of NAC with a variety of electrophilic agents resulting in stable adducts. These adducts are often more readily available to chromatographic separation than the parent compound, mostly on reversed-phase high-performance liquid chromatography (HPLC) columns, and exhibit physical properties, such as fluorescence, which facilitate detection and quantitation. Several suitable analytical procedures have been published based on the use of N-(1-pyrene)maleimide and N-(7-dimethylamino-4-methylcoumarinyl) maleimide (Kågedahl and Källberg, 1982), 4-(aminosulfonyl)-7-fluoro-

2,1,3-benzoxadiazole (Toyoóka and Imai, 1983), eosine-5-maleimide (Majima *et al.*, 1995), ammonium 7-fluoro-2,1,3-benzoxadiazole (Imai and Toyoóka, 1987), 2,4-dinitro-1-fluorobenzene (Lewis *et al.*, 1985), *o*-phthaladehyde (Gabard and Masher, 1991), 2-chloro-4,5-bis (*p*-N,N-dimethylamino sulfonylphenyl)oxazole, 2-fluoro-4,5-diphenyloxazole (Yoyoóka *et al.*, 1983), and monobromobimane (Cotgreave and Moldéus, 1987).

The use of these reagents has been, in most cases, optimized to account for a number of restrictions. First, because most of these reagents react with the thiolate anion of NAC and the derivatizations are conducted at alkaline pH, and as the rate of oxidation of NAC rapidly increases with increasing pH, the derivatizing agent must react quantitatively and rapidly with the available NAC in the sample, preferably *in situ*, without prior extraction of the thiol. Second, due to the presence of endogenous low molecular weight and protein thiols in biological samples, enough of the reagent must be added to the sample to allow quantitative recovery of the NAC adduct in the presence of these other thiols. Finally, the physical properties of the adducts (e.g., fluorescence) should impart enough sensitivity to allow the accurate quantitation of NAC in body fluids during its pharmacological usage. The agents detailed earlier have generally fulfilled these analytical criteria and been successfully applied to the analysis of reduced NAC in body fluids such as blood, plasma, and urine. However, each reagent presents different practical restrictions for its usage. It should be remembered that any complete assay protocol for NAC in biological systems must include the entire redox status of the thiol, i.e., the oxidized NAC in homo- and heterologous low molecular weight disulfides and NAC in mixed disulfides with protein. This has generally been accomplished by separating the soluble and proteinaceous portions of the sample, by acid precipitation, and by subsequent reduction of these components with reducing agents such as dithiothreitol, followed by derivatization of the "extra" NAC released from these oxidized forms (Cotgreave and Moldéus, 1987).

B. Basic Pharmacokinetics of NAC in Humans

The use of reagents such as monobromobimane provided some of the first human pharmacokinetic data for NAC in the mid-1980s (Cotgreave and Moldéus, 1986, 1987; Moldéus *et al.*, 1986; Cotgreave *et al.*, 1986). In one typical experiment following a single iv infusion of 200 mg NAC, peak plasma levels (ca. 200 μM) declined rapidly and biphasically ($\alpha T_{1/2}$ and $\beta T_{1/2}$ = 6 and 40 min, respectively) (Cotgreave and Moldéus, 1987). Infused NAC also rapidly forms disulfides in plasma, which prolongs the existence of the drug in the plasma to up to 6 hr (Cotgreave and Moldéus, 1987). However, following oral ingestion of 200 mg NAC, the free thiol is largely undetectable, with low levels of oxidized NAC detected for several hours after the dose (Cotgreave and Moldéus, 1987). These early data also indicated that the drug was less than 5% bioavailable from the oral formulation. This was also confirmed in larger populations (Borgström *et al.*, 1986; Olsson *et al.*, 1988). The poor oral bioavailability of NAC probably underlies the observation that free NAC is still undetectable and total NAC does not accumulate in plasma following repeated oral doses of 200 mg NAC to healthy volunteers (Cotgreave *et al.*, 1987b). Further pharmacokinetic data were made available from different oral formulations of NAC and different dose regimes for the compound (De Bernardi di Valserra *et al.*, 1989; De Carro *et al.*, 1989; Borgström and Kågedahl, 1990; Watson and McKinney, 1991; Holdiness, 1991), generally suggesting that the drug itself does not accumulate in the body, rather its oxidized forms and reduced and oxidized metabolites.

In addition to pharmacokinetic studies dealing with NAC at doses indicated in the treatment of various human pulmonary diseases, attempts have also been made to define the pharmacokinetics of the drug following high dose administration, as indicated during its use as a chemical adjunct in chemotherapy or as a general chemoprotective agent (Pendyala and Creaven, 1995).

C. Metabolism of *N*-Acetylcysteine in Animals and Humans

One factor behind the low oral bioavailability of NAC may be extensive first-pass metabolism in the body. Early work had demonstrated that the liver might be involved as NAC was effectively deacetylated by isolated hepatocytes and supported GSH synthesis in these cells (Thor *et al.*, 1979). Additionally, it was demonstrated that intestinal deacetylation of NAC was extensive in rats and that the major metabolites entering the hepatic portal circulation were cysteine, cystine, and sulfite (Cotgreave *et al.*, 1987a). This was subsequently confirmed in homogenates of rat intestine and by *in situ* perfusion of the rat small intestine by Sjödin *et al.* (1989). Homogenates of rat lung and liver also deacetylate NAC, as do human liver homogenates. The deacetylation of NAC is regio-specific with the D-isomer being poorly metabolized in rat and human tissues. Additionally, the homodisulfide of L-NAC is poorly deacetylated in rat and human tissues (Sjödin *et al.*, 1989). Human endothelial cells are also able to deacetylate NAC and utilize the liberated cysteine for the support of intracellular GSH biosynthesis. This indicates that the human endothelium may contribute to the metabolic clearance of NAC administered intravenously and may supply NAC-derived cysteine equivalents to other cell types in the underlying tissue or in the circulation (Cotgreave *et al.*, 1991). In addition, it was noted that the deacetylation of NAC during first-pass metabolism resulted in an elevation of circulatory levels of GSH in the rat, probably as a result of stimulated efflux from the liver (Cotgreave *et al.*, 1987a). This phenomenon was confirmed in humans by several groups following repeated oral administration of the drug (Cotgreave *et al.*, 1987b; Burgunder *et al.*, 1989; Bridgeman *et al.*, 1991). More recent experiments, however, have indicated different origins for this extra plasma GSH in different species. Thus, Poulsen *et al.* (1993) have measured the net splanchnic efflux of GSH from the liver of healthy volunteers and cirrhotics and found this to be unaffected by NAC infusion at a level indicated in paracetamol overdose. This indicates an extrahepatic source for plasma GSH emanating from NAC.

Other systemic metabolic phenomena which are of relevance to the therapeutic use of NAC include indirect effects on general thiol homeostasis in extracellular body fluids. Thus, the infusion of reduced NAC directly into the circulation must be envisaged to disturb the redox regulation of endogenous plasma thiols (Cotgreave and Moldéus, 1986). An example of this comes from work by Hultberg *et al.* (1994) in which they clearly demonstrate that the oral administration of NAC to human volunteers depresses plasma levels of total homocysteine, whereas reduced homocysteine levels rose significantly, suggesting displacement of the reduced amino thiol from plasma protein mixed disulfides. Similarly, in the intracellular setting, NAC has been shown to readily displace GSH from cellular mixed disulfides in human red cells, thus elevating the effect levels of this tripeptide without stimulation of its synthesis (Russel *et al.*, 1994).

Additionally, because NAC has been indicated in treatment of a variety of pulmonary diseases (see later), efforts have been made to determine if systemic metabolites of the drug or the drug itself can accumulate, either within the lung tissue or within the alveolar space. Thus, although early work in humans failed to detect NAC or its primary endogenous thiol metabolites in alveolar lavage fluid or cells (Cotgreave *et al.*, 1987b) using standard oral dosing regimes, recent work in rats indicated that NAC can elevate cysteine levels in lavage fluid following the intraperitoneal administration of the drug to levels which alleviate toxic insult in the lung (Lailey and Upshall, 1994).

III. General Biochemical and Cellular Effects

A. N-Acetylcysteine as a Chemical Antioxidant

Increasing evidence shows that oxidants play a major role in the development of a variety of human disease states. The sources of these oxidants vary and include activated inflammatory cells, cells undergoing redox cycling of xenobiotics, and exposure to environmental media such as cigarette smoke. Because NAC is used therapeutically in several disorders related to these sources, it has been suggested to function as an antioxidant. Thus, NAC has been demonstrated to effectively reduce free radical species and other oxidants. The interaction with free radical species results in the intermediate formation of NAC thiyl radicals, with NAC disulfide as the major end product (Moldéus et al., 1986). Of special interest is the interaction of NAC with reactive oxygen intermediates such as the superoxide anion ($\cdot O_2^-$), hydrogen peroxide (H_2O_2), hydroxyl radical ($\cdot OH$), and hypochlorous acid (HOCl), all of which are released by inflammatory cells. N-Acetylcysteine has been shown to directly reduce $\cdot OH$, H_2O_2, and HOCl (Moldéus et al., 1986; Aruoma et al., 1989). NAC does not, however, appear to interact with $\cdot O_2^-$ to any major extent (Aruoma et al., 1989). The interaction of NAC with H_2O_2 is fairly slow, whereas the interaction between $\cdot OH$ and NAC is extremely rapid, with a calculated rate constant of $1.36 \times 10^{10}\ M^{-1}\ sec^{-1}$. NAC is also a powerful scavenger of HOCl, and low concentrations of NAC have been demonstrated to protect cells from HOCl-induced toxicity. NAC has also been shown to protect against the HOCl-induced contraction of guinea pig tracheal smooth muscle (Bast et al., 1991) and to protect against the HOCl-induced inactivation of α_1-antiproteinase (Aruoma et al., 1989). Additionally, and interestingly, NAC at the cellular level has been shown to suppress the production of reactive oxygen metabolites from human neutrophils during their oxidative burst in vitro (Gressier et al., 1993). However, similar experiments demonstrated that NAC actually enhances receptor-mediated phagocytosis in these cells (Ohman et al., 1992). Both effects occur at levels of the thiol achievable during intravenous infusion of the drug. However, all of the studies in which NAC has been shown to interact directly with oxidants and to protect against oxidant-induced macromolecular damage and cytotoxicity have been performed in vitro, and the significance of a direct antioxidative effect of NAC in vivo is questionable, particularly in view of the pharmacokinetics of the reduced form of the drug. NAC may also exert its antioxidant effect indirectly by facilitating GSH biosynthesis and supplying GSH for GSH peroxidase-catalyzed reactions. Apart from the demonstration that NAC indeed supports GSH biosynthesis in biological systems, there are little data describing effects of this on GSH peroxidase-dependent detoxication, either in vitro or in vivo. An interesting possibility here is to use NAC together with a synthetic GSH peroxidase mimetic. For instance, NAC serves as a thiol substrate for the GSH peroxidase-mimetic Ebselen, and this combination protects isolated hepatocytes from oxidative damage (Cotgreave et al., 1987c). Whether or not such a mechanism may occur in vivo remains to be established.

B. Cell and Gene Modulatory Effects of NAC

A number of both fundamental and specialized cellular functions have been shown to be sensitive to NAC. Many of these may be directly related to the mechanisms of action of the compound in vivo. Emerging data on the ability of NAC to affect gene regulation will be presented, particularly at the level of redox-sensitive transcription factors. Such alterations have direct ramifications in the mechanism of cellular adaptation to altered redox environment, as well as to processes controlling cell differentiation and cell deletion by apoptosis. It should be kept in mind, however, that most of the effects elicited by NAC are not specific, as other thiol-containing compounds elicit similar responses.

A growing number of redox-sensitive transcription factors are being identified, which can potentially relate alterations to intracellular redox status with adaptive alterations to gene expression. Such factors include nuclear factor κB (NF-κB) and activator protein 1 (AP-1). For a more extensive review of this area, see Chapter 15 in this volume. Early work by Schreck *et al.* (1991) has shown that an increase in the intracellular flux of oxidants activates NF-κB by initiating dissociation of an inhibitory subunit, I-κB, and the subsequent translocation of the active cytosolic protein complex to the nucleus, with consequent binding to a consensus binding motif in the promoter of various genes. In this early work, it was quickly discovered that thiol-containing reducing agents, such as NAC, effectively inhibit this activation in intact cells.

One group of cellular genes which are known to be controlled, at least in part, by the action of NF-κB are those for intercellular adhesion molecules. The controlled expression of these molecules is fundamental to the process of both acute and chronic inflammation, as well as to atherosclerosis. N-Acetylcysteine has been shown to inhibit the expression of vascular cell adhesion molecule-1 (VCAM-1) in human endothelial cells (Marui *et al.,* 1993; Weber *et al.,* 1994), resulting in a blockade of monocyte adhesion in response to stimuli such as tumor necrosis factor-α (TNF-α) or interleukin-1 (IL-1) (Weber *et al.,* 1994). This was clearly shown to result from the suppression of NF-κB mobilization (Weber *et al.,* 1994). The incubation of human endothelial cells with NAC was also shown to suppress the expression of endothelial–leukocyte adhesion molecule-1 (ELAM-1), in association with decreased monocyte adhesion (Faruqi *et al.,* 1994). It is interesting to note, however, that vitamin E was far more effective than NAC and that vitamin E-dependent inhibition was not associated with an inhibition of NF-κB translocation, despite downregulation of the expression of the ELAM-1 gene. This example illustrates that there is no simple relationship between transcription factor sensitivity to redox changes and alterations to gene expression. Indeed, the regulation of expression of individual genes relies on the concerted action of many factors, whose correct orchestration dictates the response to a particular stimulus. Consequently, it is of interest to note that NF-κB itself is not always activated under conditions of oxidative stress in cells (Brennan and O'Neill, 1995).

In addition to NF-κB the DNA-binding activities of other transcription factors are thought to be sensitive to redox changes to their respective proteins, as revealed by responses to agents such as NAC. An example of this is the zinc finger glucocorticoid receptor (GR) protein, which binds to the consensus glucocorticoid response element and is a close relative of another redox-sensitive factor, SP-1. The GR has been shown to be activated under oxidative stress, in a manner counteracted by NAC (Esposito *et al.*, 1995).

The expression of the human heme oxygenase-1 (HO-1) gene has been shown to be associated with many situations of increased oxidative stress. Rizzardini *et al.* (1994) have demonstrated that NAC suppresses the expression of murine hepatic HO-1 in response to endotoxin stress. Other stress genes, the expression of which is inhibited by NAC, include a mammalian heat-shock protein-27 (HSP-27) kinase (Huot *et al.*, 1995), a human urokinase-type plasminogen activator (Egawa *et al.*, 1994), the human protooncogenes c-*jun* (Collart *et al.*, 1995) and c-*fos* (Bergelson *et al.*, 1994), glutathione transferase Ya and NAD(P)H : quinone reductase (Bergelson *et al.*, 1994), and cyclin D_1 (Huang *et al.*, 1995). In most of these cases, however, the exact molecular mechanism of the effect, in terms of events at the level of the individual transcription factors, was not revealed. However, two important features of the use of NAC seem to be common to most studies. First, the inhibitory effects are generally achieved at extremely high concentrations of NAC, often >5 mM, which are not achieved in body fluids given the known pharmacokinetics of the compound. Second, the response of the individual transcription factors to NAC in the cell may be complex, resulting from both direct, NAC-dependent reduction of the proteins and indirect effects via intracellular GSH. Indeed, the absolute levels of cellular GSH seem to be of paramount importance in the regulation of many of these transcription factors in intact cells.

C. Effects of NAC on Cellular Proliferation and Apoptosis

Many studies have shown a correlation between alterations in the intracellular redox status and the rates of proliferation of cells (see Chapter 15 in this volume). In an elegant series of studies, Burdon *et al.* (1994) have demonstrated that the maintenance of a low level flux of reactive oxygen metabolites seems important to the maintenance of normal cell proliferation. These authors have shown that many antioxidants, including NAC, have powerful anti-proliferative effects in normal cells (Burdon *et al.*, 1994). N-Acetylcysteine has also been shown to inhibit pro-oxidant-induced increases in cellular proliferation and transformation (Parfett and Pilon, 1995). It should be remembered, however, that oxidative stress-related changes to cellular proliferation may only superimpose on more basal growth factor-dependent mechanisms. This will complicate eventual interpretations of the effects of compounds such as NAC. Indeed, Das *et al.* (1992) have shown that NAC reverses the growth-inhibitory action of transforming growth factor-β (TGF-β) in endothelial cells.

More and more attention has been paid to the process of cell deletion termed apoptosis. Following the initial observations by Hockenbury *et al.* (1993), in which an antioxidant function was ascribed to Bcl-2, a protein shown to protect cells from oxidant-induced apoptosis, a situation of intracellular oxidative stress has been implicated in the molecular mechanism of apoptosis. It was then demonstrated that thiol-containing agents, such as NAC, effectively suppressed apoptosis which could be associated with a situation of intracellular stress, particularly following oxidant insult, incubation of cells with stimuli such as TNF-α, or the depletion of intracellular antioxidants, particularly GSH and enzymes such as superoxide dismutase. Such effects have been demonstrated in T-cell hybridomas (Sandström *et al.*, 1994), embryonic cortical neurons (Ratan *et al.*, 1994), spinal chord neurons (Rothstein *et al.*, 1994), neuroblastoma cells (Talley *et al.*, 1995), endothelial cells (Abello *et al.*, 1994), peripheral lymphocytes (Zamzami *et al.*, 1995), and WEHI 231 B lymphocytes (Fang *et al.*, 1995). Once again, the anti-apoptotic effect of NAC may be due to both direct action of the parent compound and/or the stimulated synthesis of GSH, which is thought to play a central role in mediating some of the apoptotic observations described earlier. Indeed, an interesting report from Ferrari *et al.* (1995) clearly demonstrated that both L-NAC (metabolically deacetylatable) and D-NAC (metabolically inert) effectively prevented apoptosis in PC12 neuronal cells. N-Acetylcysteine has also been shown to affect apoptotic processes in HIV-1-infected cells (see later).

IV. Specific Therapeutic Effects of NAC in Animals and Humans

N-Acetylcysteine has been traditionally utilized in a number of diseases, which will be discussed first. However, because NAC is well tolerated and elicits few side effects in humans, coupled with the emerging molecular and cell biological evidence of the effects of this thiol drug, attempts are being made to test the efficacy of the compound in a growing variety of disease models in animals and in their corresponding human syndromes. These more exploratory aspects of the use of the compound will then be reviewed.

A. NAC as a Paracetamol Antidote

One of the best studied examples of the clinical use of NAC is provided by its activity as an antidote against paracetamol (acetaminophen) toxicity, through supporting hepatic GSH biosynthesis. Acetaminophen induces hepatic necrosis when taken in overdose. This toxicity is due to a reactive metabolite of acetaminophen, N-acetyl-p-benzoquinone imine (NAPQI). NAPQI is normally detoxified by conjugation with GSH but, in cases of overdose, GSH becomes depleted and there is a lack of cysteine for resynthesis which is compensated for by NAC (Moldéus, 1981). Appropriate dosage regimes and ample clinical evidence for the efficacy of NAC have been previously reviewed (Prescott, 1983).

B. NAC as a Modulator of Pulmonary Injury

I. Diseases Associated with Chronic Pulmonary Inflammation

During the 1980s, attempts were made to use orally administered NAC clinically in the treatment of a variety of lung diseases; the pathophysiological mechanisms of which are thought to include elements of oxidative stress, particularly in association with the induction and maintenance of chronic inflammation. From early work it was known that when NAC was inhaled in an aerosol form, the fluidity of airway mucus was considerably increased (Sheffner, 1963). This provided therapeutic advantages to patients developing copious mucus in the airways, particularly during chronic bronchitis and cystic fibrosis. When attempts were made to use orally administered NAC in chronic bronchitis, limited therapeutic advantages were noted in the form of decreased numbers of exacerbations experienced by the patients (Boman et al., 1983). Despite these observations, the mechanism of this therapeutic activity is uncertain, as NAC itself could not be detected in association with bronchoalveolar fluid or cells from humans receiving therapeutic doses of NAC (Cotgreave et al., 1987b). Some insight into the mechanism of the therapeutic effect of NAC may be provided by work reported by Riise et al. (1994). N-Acetylcysteine treatment of smokers with nonobstructive chronic bronchitis, smokers with chronic bronchitis and chronic obstructive pulmonary disease (COPD), and nonsymptomatic smokers was shown to decrease the numbers of bacteria recovered from their lungs by lavage. The reasons for this might involve an increased host defense against these pathogens. Thus, it is interesting to note that Oddera et al. (1994) have shown that NAC is able to considerably enhance the bactericidal activity of human alveolar macrophages and blood neutrophils in vitro in association with a protective effect on the viability of these phagocytes. In a similar study using alveolar macrophages and peripheral blood monocytes (PBM) from COPD patients, NAC was shown to increase the anti-fungal activity of PBM in vitro, but not in vivo (Hansen et al., 1994). Another interesting therapeutic aspect of NAC in the treatment of chronic obstructive lung diseases was revealed in a study involving mild bronchitis, in which the subjective well-being of the patients was significantly improved by the drug (Vecchiarelli et al., 1994).

Several attempts have been made to apply orally administered NAC to other chronic inflammatory diseases in the lung. For instance, Meyer et al.

(1994) demonstrated that NAC significantly elevated GSH levels in the alveolar lavage fluid of patients with idiopathic pulmonary fibrosis. This may provide remedial effects on the rate and extent of the development of fibrotic lesions in these patients, although such clinical improvements await confirmation.

2. Diseases Associated with Acute Pulmonary Inflammations

The adult respiratory distress syndrome (ARDS) is a life-threatening disease associated with the induction of a severe, acute pulmonary inflammation. A large part of the pathophysiology associated with ARDS is dependent on massive neutrophil infiltration into the lung and the release of cytotoxic mediators, including oxidants. In animal models, endotoxin-induced ARDS is clearly ameliorated by intraperitoneal NAC, which has been shown to improve survival, reduce structural damage and edema in the lung, and lower the systemic release of pro-inflammatory arachidonic acid metabolites (Peddersen et al., 1993). Interestingly, a single oral dose of NAC has also been shown to effectively diminish the symptoms of immunologically induced acute alveolitis in the rat (Scala et al., 1993). In ARDS patients, however, despite the demonstration that indirect symptoms of ARDS, such as the systemic depletion of blood GSH levels, can be compensated for by the use of NAC, particularly by infusion (Bernard, 1991), evidence for clear clinical benefits of the treatment is still debatable. For instance, a study reported by Suter et al. (1994) failed to demonstrate a significant protective effect of intravenous NAC on the development of the disease, or of mortality, despite a clearly improved systemic oxygenation of the patients and a lowered requirement for forced ventilation.

3. Diseases Associated with the Inhalation of Airborne Oxidants

The inhalation of oxygen at high partial pressures for extended periods has been shown to induce considerable structural and functional damage in the lung by the direct cytotoxicity of the gas, by its reactive oxygen metabolites, and by the induction of inflammation. Evidence is emerging that NAC can effectively diminish hyperoxic injury in animals. For instance, Langely and Kelly (1993) demonstrated beneficial effects of intraperitoneally administered NAC on hyperoxic injury in preterm guinea pigs. NAC prevented fluid and protein infiltration into the alveolar space, but had no effect on the movement of neutrophils into the lung lumen. Lung GSH levels were also unaltered in the animals. The authors thus conclude that the beneficial effects were provided by NAC itself. Support for this suggestion comes from a report by Särnstrand et al. (1995) who demonstrated that both L-NAC and D-NAC effectively protect rodent lungs from hyperoxic injury. Again, clinical evidence for the efficacy of NAC in hyperoxia is not as convincing. In a study in which the effect of NAC on hyperoxic ventilation during hypoxemia was tested, intravenous infusion of NAC before and during ventilation only caused a small, but significant, increase in the cardiac output of the patients, in association with decreased systemic vascular resistance (Reinhart *et al.*, 1995).

Therapeutic effects of NAC have been investigated for the inhalation of a variety of other inhaled oxidant toxins. Indeed, experiments in rodents have revealed protective effects of NAC on the effects of cigarette smoke (Sohn *et al.*, 1993), whereas pulmonary toxicity induced in rats by the inhalation of nitrogen dioxide was clearly not affected by the administration of NAC to the animals, even by the relatively large infusion doses used (Meulenbelt *et al.*, 1994).

C. Other Anti-inflammatory Actions of NAC and Miscellaneous Actions Related to Antioxidant Function

I. Ischemia–Reperfusion Injury

Experiments largely performed in rodent models have suggested that NAC may be beneficial in the treatment of oxidant-induced injury incurred during ischemia–reperfusion in a variety of tissues. Thus, NAC has been shown to protect the functionality of the myocardium during periods of ischemia *in vitro* (Tang *et al.*, 1991). Similarly, Knuckey *et al.* (1995) have demonstrated a neuroprotective capacity for NAC during periods of transient forebrain ischemia in rats. Thus, NAC partially improved hippocampal neuron survival after transient ischemia, but the efficacy of the drug was decreased with increasing length of the ischemic period.

2. Sepsis

Another inflammatory condition in which NAC has received attention as a possible therapeutic agent is that of septic shock. Bakker *et al.* (1994) have demonstrated that the intravenous administration of NAC to dogs subjected to endotoxin-induced septic shock enhanced cardiac output and the oxygenation of peripheral tissues, as well as diminished the systemic release of the potent pro-inflammatory mediator TNF-α.

Several human studies on the value of NAC in sepsis have been performed but remain inconclusive as to the true value of the treatment (Henderson and Hayes, 1994). However, a clinical study by Spies *et al.* (1994) has shown transient improvements in tissue oxygenation in circa half of the septic shock patients treated. This was correlated with a higher survival rate in these NAC responders.

3. Miscellaneous Conditions

Attempts have been made to apply NAC to a variety of other conditions, some of which may be dependent on inflammatory events and the induction of oxidative stress in tissues. Thus, Jones *et al.* (1994) failed to indicate therapeutic effects of NAC in the treatment of stable cirrhosis, despite increased oxygen delivery to the tissues, whereas orally administered NAC was shown to clearly protect the rat stomach against acute, ethanol-induced toxicity (Barreto *et al.,* 1993). Orally administered NAC has also been shown to significantly suppress enhanced TNF-α production and release into the circulation in streptozotocin-induced diabetes in rats (Sagara *et al.,* 1994).

Several groups have convincingly demonstrated that NAC may have beneficial clinical effects on human syndromes associated with muscle fatigue. Thus, Khawli and Reid (1994) have shown that NAC depresses contractile function and inhibits fatigue in rat diaphragm bundles *in vitro,* whereas the same group demonstrated that intravenous NAC had some beneficial effect on human limb muscle during fatiguing exercise (Reid *et al.,* 1994). Sen *et al.* (1994) have provided further evidence for an effect of NAC on some oxidative stress-related biochemical parameters in rats and humans subjected to physical exercise, including blood GSH redox status and circulatory products of lipid peroxidation.

Finally, NAC has been shown to decrease erythrocyte turnover and to elevate lymphocyte and plasma levels of GSH in 5-oxoprolinuria, a hereditary disease in which the synthesis of cellular GSH is deficient (Jain *et al.,* 1994). N-Acetylcysteine has also been shown to interfere with the biochemical aspects of cellular ageing. In a lifetime study in rats, Miquel *et al.* (1995) have remarkably demonstrated that a continuous dietary supplementation of NAC is clearly protective against an age-related decline in the oxidative phosphorylation capacity of liver mitochondria.

D. NAC in Treatment of Hypertension

Hypertensive patients treated with organic nitrates frequently develop tolerance to the effects of these drugs. This tolerance has been suggested to be associated with, among other factors, the depletion of intracellular sulfhydryl groups in vascular smooth muscle. There are numerous investigations where NAC, as well as other thiols, have been used in order to reverse this tolerance. Even though the results are somewhat contradictory, it is evident that NAC prevents the development of tolerance or at least partially restores the effect of the organic nitrates (Abrams, 1991; Horowitz, 1991). The mechanism by which NAC reverses the tolerance is not known. It is of relevance, however, that only the L-isomer of NAC is active. Because L-NAC is further metabolized to cysteine and GSH, whereas D-NAC is not (Newman *et al.,* 1990), an enzymatic step is required for activity. The mechanism may also involve the intra- or extracellular formation of S-nitrosothiol and/or NO (Fung *et al.,* 1988). It has also been shown that NAC may ellicit some of its effects via inhibition of the renin–angiotensin system and modulation of the counterregulatory vasoconstriction which often presides. Indeed, NAC was shown to reduce the pressor response of angiotensin I in rats, as well as lower angiotensin-converting enzyme (ACE) levels in plasma. In the same study, a beneficial effect of NAC on the renin–angiotensin systems was also noted in healthy humans receiving isosorbide treatment (Bosegaard *et al.*, 1993).

E. NAC as an Antimutagen and Anticarcinogen

N-Acetylcysteine is an antimutagen and anticarcinogen both *in vitro* and *in vivo*. The drug has, thus, been demonstrated to inhibit the mutagenicity of both directly acting carcinogens and procarcinogens in *in vitro* mutagenicity tests (De Flora *et al.*, 1991). *N*-Acetylcysteine has also been demonstrated to inhibit DNA adduct formation *in vivo* after the administration of acetylaminofluorene (AAF) or benzo[*a*]pyrene to rats (De Flora *et al.*, 1991). In a study by Izzotti *et al.* (1995), orally administered NAC was shown to depress the numbers of DNA adducts formed in rat tracheal epithelial cells after extended periods of exposure to mainstream tobacco smoke *in vivo*. NAC is also able to protect the function of certain enzymes involved in DNA replication and repair (Cesarone *et al.*, 1992). In addition to protective effects at the initiation stage of chemical carcinogenesis, Albini *et al.* (1995) have shown that NAC may also provide chemoprotective effects by inhibiting the invasiveness of malignant cells. These authors showed that NAC inhibited the colonization of gelatin *in vitro* by a variety of human and murine tumour cells in association with a potent inhibition of gelatinase activity in the cells.

Given these chemical and biochemical effects, it is not surprising that NAC has anticarcinogenic effects in various animal models and has been shown to decrease the formation of lung tumours in urethane-treated mice (De Flora *et al.*, 1986), prevent the formation of AAF (Cesarone *et al.*, 1987)- and hydrazine (Wilpart *et al.*, 1986)-induced sebaceous squamocellular carcinomas of the zymbal glands of rats, and inhibit azoxymethane (Reddy *et al.*, 1993)-induced colon cancer in rats, although effects were only marginal in the latter study. As a result of these encouraging results in animal models, a large multicenter clinical trial has been initiated to study the chemoprotective potential of vitamin A and NAC in the development of second primary tumors in patients treated for lung, larynx, and oral cancer. The outcome of the study is, at present, unknown (Cianfriglia *et al.*, 1994).

F. The Immunomodulatory Potential of NAC and Its Use in HIV-I Therapy

The potential of *N*-acetylcysteine as a modulator of immune system functions and responses to physiological and toxicological stimuli is receiving increasing attention, particularly in view of the emerging data on its ability to interfere with gene regulation. Senaldi *et al.* (1994) have demonstrated a protective effect for topically applied NAC in a model of hapten-induced irritant and contact hypersensitivity in mice. Here, NAC was proposed to exhibit an anti-allergenic effect by preventing a NF-κB-induced increase in the transcription of TNF-α in the epidermis. This protective effect is, however, probably due to a local anti-inflammatory effect in the skin and not due to direct interaction with immunologically competent cells. Several groups are also investigating the ability of NAC to prevent the immunosuppressive effects of UV irradiation (Van den Broek and Beijersbergen van Hanegouen, 1995) and its potential as a protective adjunct in the clinical use of radiosensitizers in photodynamic therapy (Baas *et al.*, 1994). Again, general antioxidative mechanisms for NAC were discussed and direct effects on lymphocyte function were not studied.

Perhaps of more interest to immunobiology has been the demonstration of direct modulatory effects of NAC on T-cell function, especially during cases of HIV infection in humans. A number of studies have demonstrated that NAC can modulate functions in normal human T lymphocytes. For instance, Eylar *et al.* (1993) have shown that NAC enhances T-cell mitogenesis in the presence of accessory cells, IL-2 secretion, and, remarkably, cell growth in peripheral T cells *in vitro*. The conjugation of natural killer (NK) cells to target tumor cells has also been shown to be increased by NAC, resulting in increased NK-dependent cytotoxicity (Malorna *et al.*, 1994). In an interesting study by Flescher *et al.* (1994), NAC was clearly shown to preserve T-cell function (IL-2 release) in polyamine oxidase-treated cells. Once again, however, the effect appears to be due to a general antioxidative action of NAC, perhaps related to alterations to the redox status and activities of transcription factors such as NF-κB and AP-1 (Flescher *et al.*, 1994). Other groups have noted inhibitory effects of NAC, and the naturally occurring thiol antioxidant lipoate, an NF-κB activation in Jurkat T lymphocytes, but it should be remembered that these inhibitory effects, as well as most of the previously mentioned effects on normal T-cell function, occur at relatively high concentrations (mM) of NAC.

Of perhaps the most interest at present is that NAC has been shown to have interesting modulatory effects on T cells from individuals infected with HIV-1. In 1989 a link was provided by Eck *et al.* (1989) between plasma thiol and T-cell thiol homeostasis and the HIV infection. Thus, it was shown that the cysteine concentration in plasma, as well as the intracellular GSH concentrations in peripheral blood mononuclear cells and monocytes, is decreased in patients infected with HIV-1 and in individuals with immunodeficiency (Eck *et al.*, 1989). It was also shown that T lymphocytes are dependent on the supply of extracellular cysteine for the maintenance of intracellular cysteine and for GSH biosynthesis, as these cells exhibit poor uptake of cystine (Dröge *et al.*, 1991). It was soon discovered that the long terminal repeat region of the proviral DNA contains a number of consensus-binding sites for NF-κB and that the activation of this transcription factor, for instance during oxidative stress initiated by the depletion of intracellular cysteine and/or GSH, stimulates the transcription of viral genes and leads to viral replication (Roederer *et al.*, 1992). Thus, a link was found between the aberrant T-cell thiol homeostasis in and around circulatory T cells and the control of HIV gene expression. Attempts were then made to modulate the expression of NF-κB-dependent genes in T cells from HIV patients using thiol agents. Thus, cysteine and NAC have been found to inhibit the expression of NF-κB-dependent genes *in vitro* (Mihm *et al.*, 1991; Schreck *et al.*, 1991), and both cysteine and NAC were shown to inhibit HIV-1 replication in infected T cells, as well as in normal peripheral blood mononuclear cells (Roederer *et al.*, 1991; Raju *et al.*, 1994). In the latter case, the authors showed that NAC was able to replenish the intracellular GSH levels of human peripheral blood mononuclear cells, in association with an inhibition of viral replication. Similarly, NAC has been shown to inhibit viral replication in human macrophages (Newman *et al.*, 1994) and monoblastoid U937 cells (Malorini *et al.*, 1993), but again at relatively high concentrations of the thiol in both cases. Interestingly, in the latter study it was also shown that NAC also inhibited apoptosis in the virally infected cells, an effect that was not reproduced by Sandström *et al.* (1993) in a human T-cell line constitutively expressing HIV-1.

Taken together, these data, mainly obtained in *in vitro* systems, indicate that a cysteine deficiency in HIV-infected T lymphocytes may lead to the overexpression of the NF-κB-dependent genes and to enhanced HIV replication. Chronic treatment of patients infected with the HIV-1 virus with NAC has, thus, been indicated (Roederer *et al.*, 1992), and a number of clinical trials have been initiated in order to test the efficacy of NAC. In a small-scale study, Witschi *et al.* (1995) demonstrated that although circulatory cysteine levels were considerably elevated in patients treated chronically with NAC, there was no significant increase in the GSH levels of either plasma or peripheral blood mononuclear cells. This suggested that the low levels of GSH present in the cells of HIV patients result from impaired synthesis of the tripeptide and not as a consequence of oxidative stress. Similarly, in a very limited study, Järnstrand and Åkerlund (1994) reported that, although plasma cysteine levels were elevated in the patients, only some beneficial therapeutic effects were noted. The results of more extensive clinical trials with NAC in patients at various stages of the HIV infection are awaited. Of particular interest is if NAC therapy can actually give therapeutic improvements that can be correlated to discrete molecular events such as cellular GSH levels, the expression of HIV-related genes, and the response of virally infected cells to normal immunological surveillance, including the induction of apoptosis.

V. Summary

The diversity of application of the thiol drug NAC in both the experimental setting, as a tool for the study of the mechanisms and consequences of oxidative stress, and the clinical setting, as a therapeutic agent, clearly reflects the central role played by the redox chemistries of the group XVI elements, oxygen and sulfur, in biology. As our understanding of such redox processes increases, particularly their roles in specific pathophysiological processes, new avenues will open for the use of NAC in the clinical setting. As a drug, NAC represents perhaps the ideal xenobiotic, capable of directly entering endogenous biochemical processes as a result of its own metabolism. Thus, it is hoped that the experience gained with this unique agent will help in future efforts to design antioxidants and chemoprotective principles which are able to more accurately utilize endogenous biochemical processes for cell- or tissue-specific therapy.

References

Abello, P. A., Fidler, S., and Buchman, T. G. (1994). Thiol reducing agents modulate TNF-α induced apoptosis in porcine endothelial cells. *Shock* **2**, 79–83.

Abrams, J. (1991). Interactions between organic nitrates and thiol groups. *Am. J. Med.* **91**, 106–112.

Albini, A., Dágostini, F., Giunciuglio, D., Pagliari, I., Balansky, R., and De Flora, S. (1995). Inhibition of invasion, gelatinase activity, tumour take and metastasis of malignant cells by N-acetylcysteine. *Int. J. Cancer* **61**, 121–129.

Aruoma, O. K., Halliwell, B., Hoey, B. M., and Butler, J. (1989). The antioxidant action of N-acetylcysteine: Its reaction with hydrogen peroxide, hydroxyl radical, superoxide and hypochlorous acid. *Free Rad. Biol. Med.* **6**, 593–597.

Baas, P., Oppelaar, H., van der Valk, M. A., van Zandwijk, N., and Stewart, F. A. (1994). Partial protection of photodynamic-induced skin reactions in mice by N-acetylcysteine: A preclinical study. *Photochem. Photobiol.* **59**, 448–454.

Bakker, J., Zhang, H., Depierreux, M., van Asbeck, S., and Vincent, J. L. (1994). Effects of N-acetyl cysteine in endotoxic shock. *J. Crit. Care* **9**, 236–243.

Barreto, J. C., Smith, G. S., Tornwell, M. S., and Miller, T. A. (1993). Protective action of oral N-acetylcysteine against gastric injury: Role of hypertonic sodium. *Am. J. Physiol.* **264**, G422–G426.

Bast, A., Haenen, G. R. M. M., and Doelman, C. J. A. (1991). Oxidants and antioxidants: State of art. *Am. J. Med.* **91**, 2–13.

Bergelson, S., Pinkus, R., and Daniel, V. (1994). Intracellular glutathione levels regulate Fos/ Jun induction and activation of glutathione S-transferase gene expression. *Cancer Res.* **54**, 36–40.

Bernard, G. R. (1991). N-acetylcysteine in experimental and clinical acute lung injury. *Am. J. Med.* **91**, 54–59.

Boman, G., Bäcker, U., Larsson, S., Melander, B., and Wåhlander, L. (1983). Oral acetylcysteine reduces exacerbation rate in chronic bronchitis: Report of a trial organized by the Swedish Society for Pulmonary Diseases. *Eur. J. Respir. Dis.* **64**, 405–415.

Borgström, L., and Kågedahl, B. (1990). Dose-dependent pharmacokinetics of N-acetylcysteine after oral dosing to man. *Biopharm. Drug Dispos.* **11**, 131–136.

Borgström, L., Kågedahl, B., and Paulsen, O. (1986). Pharmacokinetics of N-acetylcysteine in man. *Eur. J. Clin. Pharmacol.* **31**, 217–222.

Bosegaard, S., Aldershvile, J., Poulsen, H. E., Christensen, S., Dige-Petersen, H., and Giese, J. (1993). N-acetylcysteine inhibits angiotensin converting enzyme *in vivo. J. Pharmacol. Exp. Ther.* **265**, 1239–1244.

Brennan, P., and O'Neill, L. A. (1995). Effect of oxidants and antioxidants on nuclear factor kappa b activation in three different cell lines: Evidence against a universal hypothesis involving oxygen radicals. *Biochem. Biophys. Acta* **1260**, 167–175.

Bridgeman, M. M. E., Marsden, M., MacNee, W., Flenley, D. C., and Ryle, A. P. (1991). Cysteine and glutathione concentrations in plasma and bronchoalveolar lavage fluid after treatment with N-acetylcysteine. *Thorax* **46**, 39–42.

Burdon, R. H., Alliangana, D., and Gill, V. (1994). Endogenously generated active oxygen species and cellular glutathione levels in relation to BHK-21 cell proliferation. *Free Rad. Res.* **21**, 121–133.

Burgunder, J. M., Varriale, A., and Lauterberg, B. (1989). Effect of N-acetylcysteine on plasma cysteine and glutathione following paracetamol administration. *Eur. J. Clin. Pharmacol.* **36**, 127–131.

Cesarone, C. F., Scarabelli, L., Giannoni, P., and Orunesu, M. (1992). Differential assay and biological significance of poly(ADP-ribose)-polymerase activity and DNA synthesis in isolated liver nucleii. *Mutat. Res.* **245**, 157–163.

Cesarone, C. F., Scarabelli, L., Orunesu, M., Bagnasco, M., and De Flora, S. (1987). Effects of aminothiols in 2-acetylaminofluorene-treated rats. I. Damage and repair of liver DNA, hyperplastic foci and zymbal gland tumours. *In Vivo* **1**, 85–91.

Cianfriglia, F., Iofrida, R. V., Calpicchio, A., and Manieri, A. (1994). The chemoprevention of oral carcinoma with vitamin A and/or N-acetylcysteine. *Minerva Stomatol.* **43**, 255–261.

Collart, F. R., Horio, M., and Huberman, E. (1995). Heterogeneity in c-jun expression in normal and malignant cells exposed to either ionizing radiation or hydrogen peroxide. *Radiat. Res.* **142**, 188–196.

Cotgreave, I. A., Berggren, M., Jones, T. J., Dawson, J., and Moldéus, P. (1987a). Gastrointestinal metabolism of N-acetylcysteine in the rat, including an assay for sulfite in biological systems. *Biopharm. Drug Dispos.* **8**, 377–385.

Cotgreave, I. A., Eklund, A., Larsson, K., and Moldéus, P. (1987b). No penetration of orally administered N-acetylcysteine into bronchalveolar lavage fluid. *Eur. J. Respir. Dis.* **70**, 73–77.

Cotgreave, I. A., Grafström, R., and Moldéus, P. (1986). Modulation of pneumotoxicity by cellular glutathione and precursors. *Bull. Eur. Physiopathol. Respir.* **22**, 263–266.

Cotgreave, I. A., and Moldéus, P. (1986). Methodologies for the application of monobromobimane to the simultaneous analysis of reduced and oxidised soluble and protein thiol components of biological systems. *Biochem. Biophys. Methods* **13**, 231–249.

Cotgreave, I. A., and Moldéus, P. (1987). Methodologies for the analysis of reduced and oxidised N-acetylcysteine in biological systems. *Biopharm. Drug Dispos.* **8**, 365–375.

Cotgreave, I. A., Moldéus, P., and Schuppe, I. (1991). The metabolism of N-acetylcysteine by human endothelial cells. *Biochem. Pharmacol.,* **42**, 13–15.

Cotgreave, I. A., Sandy, M. S., Berggren, M., Moldéus, P. W., and Smith, M. T. (1987c). N-Acetylcysteine and glutathione dependent protective effect of PZ51 (Ebselen) against diquat-induced cytotoxicity in isolated rat hepatocytes. *Biochem. Pharmacol.* **36**, 2899–2904.

Das, S. K., White, A. C., and Fanburg, B. L. (1992). Modulation of transforming growth factor beta-1 antiproliferative effects on endothelial cells by cysteine, cystine and N-acetylcysteine. *J. Clin. Invest.* **90**, 1649–1656.

De Bernardi di Valserra, M., Mautone, G., Barindelli, E., Lualdi, P., Feletti, F., and Galmozzi, M. R. (1989). Bioavailability of suckable tablets of N-acetylcysteine in man. *Eur. J. Clin. Pharmacol.* 37, 419–421.

De Carro, L., Ghizzi, A., Costa, R., Longo, A., Ventresca, G. P., and Lodola, E. (1989). Pharmacokinetics and bioavailability of oral N-acetylcysteine in healthy volunteers. *Arzneimettelforschung* 39, 382–386.

De Flora, S., Astengo, M., Serra, D., and Bennicelli, C. (1986). Inhibition of urethane-induced lung tumours in mice by N-acetylcysteine. *Cancer Lett.* 32, 235–241.

De Flora, S., Izzotti, A., D'Agostini, F., and Cesarone, C. F. (1991). Antioxidant activity and other mechanisms of thiols involved in the chemoprevention of mutation and cancer. *Am. J. Med.* 91, 122–130.

Dröge, W., Eck, H. P., Gmünder, H., and Mihm, S. (1991). Modulation of lymphocyte functions and immune responses by cysteine and cysteine derivatives. *Am. J. Med.* 91, 140–143.

Eck, H. P., Gmünder, H., Hartmann, M., Ptzoldt, D., Daniel, V., and Dröge, W. (1989). Low concentrations of acid-soluble thiol (cysteine) in the blood plasma of HIV-infected patients. *Biol. Chem. Hoppe-Seyler* 370, 101–108.

Egawa, K., Yoshiwara, M., and Nose, K. (1994). Effect of radical scavengers on TNF alpha-mediated activation of the uPA in cultured cells. *Experimentia* 50, 958–962.

Esposito, F., Cuccovillio, F., Morra, F., Russo, T., and Cimino, F. (1995). DNA binding activity of the glucocorticoid receptor is sensitive to redox changes in cells. *Biochim. Biophys. Acta* 1260, 308–314.

Eylar, E., Rivera-Quinones, C., Molina, C., Baez, I., Molina, F., and Mercado, C. M. (1993). N-actetylcysteine enhances T cell functions and T cell growth in culture. *Int. Immunol.* 5, 97–101.

Fang, W., Rivard, J. J., Ganser, J. A., Lebein, T. W., Nath, K. A., Mueller, D. L., and Behrens, T. W. (1995). Bcl-xl rescues WEHI 231 B lymphocytes from oxidant-mediated death following diverse apoptotic stimuli. *J. Immunol.* 155, 66–75.

Faruqi, R., de la Motte, C., and Dicorleto, P. E. (1994). Alpha-tocopherol inhibits agonist-induced monocytic cell adhesion to cultured human endothelial cells. *J. Clin. Invest.* 94, 592–600.

Ferrari, G., Yan, C. Y., and Greene, L. A. (1995). N-acetylcysteine (D- and L-isomers) prevents apoptotic death of neuronal cells. *J. Neurosci.* 15, 2857–2866.

Flescher, E., Ledbetter, J. A., Schieven, G. L., Vela-Roch, N., Fossum, D., Dang, H., Ogawa, N., and Talal, N. (1994). Longitudinal exposure of human T lymphocytes to weak oxidative stress suppresses transmembrane and nuclear signal transduction. *J. Immunol.* 153, 4880–4889.

Fung, H. L., Chong, S., Kowaluk, E., Hough, K., and Kakemi, M. (1988). Mechanisms for the phararmacological interactions of organic nitrates with thiols. *J. Pharmacol. Exp. Ther.* 245, 524–530.

Gabard, B., and Masher, H. (1991). Endogenous plasma N-acetylcysteine and single oral dose bioavailability from two different formulations as determined by a new analytical method. *Biopharm. Drug Dispos.* 12, 343–353.

Gressier, B., Cabanis, A., Lebegue, S., Brunet, C., Dine, T., Luyckx, M., Cazin, M., and Cazin, J. C. (1993). Comparison of the in vitro effects of two thiol-containing drugs on human neutrophil hydrogen peroxide production. *Meths. Find. Exp. Clin. Pharmacol.* 15, 101–105.

Hansen, N. C., Skriver, A., Brorson-Riis, L., Balslov, S., Evald, T., Maltbaek, N., Gunnersen, G., Garsdal, P., Sander, P., and Pedersen, J. Z. (1994). Orally-administered N-acetylcysteine may improve general well being in patients with mild bronchitis. *Respir. Med.* 88, 531–535.

Henderson, A., and Hayes, P. (1994). Acetylcysteine as a cytoprotective antioxidant in patients with severe sepsis: A potential new use for an old drug. *Ann. Pharmacother.* 28, 1086–1088.

Hockenbury, D. M., Oltavi, Z. N., Yin, X. M., Milliman, C. L., and Korsmeyer, S. J. (1993). Bcl 2 functions in an antioxidant pathway to prevent apoptosis. *Cell* **75**, 241–251.

Holdiness, M. R. (1991). Clinical pharmacokinetics of N-acetylcysteine. *Clin Pharmacokinet.* **20**, 123–134.

Horowitz, J. D. (1991). Thiol-containing agents in the management of unstable angina pectoris and acute myocardial infection. *Am. J. Med.* **91**, 113–117.

Huang, T. S., Duyster, J., and Wang, J. Y. (1995). Biological response to phorbol ester determined by alternative G1 pathways. *Proc. Natl. Acad. Sci. U.S.A.* **92**, 4793–4797.

Hultberg, B., Andersson, A., Masson, P., Larson, M., and Tunek, A. (1994). Plasma homocysteine and thiol compound fractions after oral administration of N-acetylcysteine. *Scand. J. Clin. Lab. Invest.* **54**, 417–422.

Huot, J., Lambert, H., Lavoie, J. N., Guidmond, A., Houle, F., and Landry, J. (1995). Characterization of a 45-kDa/54-kDa HSP27 kinase, a stress-sensitive kinase which may activate the phosphorylation-dependent protective function of mammalian 27-kDa heat shock protein HSP27. *Eur. J. Biochem.* **227**, 416–427.

Imai, K., and Toyoóka, T. (1987). Fluorimetric analysis of thiols with fluorobenzoxadiazoles. *Methods Enzymol.* **143**, 67–75.

Izzotti, A., Balansky, R., Scatolini, L., Rovida, A., and De Flora, S. (1995). Inhibition by N-acetylcysteine of carcinogen-DNA adducts in the tracheal epithelium of rats exposed to cigarette smoke. *Carcinogenesis* **16**, 669–672.

Jain, A., Buist, N. R., Kennaway, N. G., Powell, B. R., Auld, P. A., and Mårtensson, J. (1994). Effect of ascorbate or N-acetylcysteine treatment in a patient with hereditary glutathione synthetase defficiency. *J. Pediatr.* **124**, 229–233.

Järnstrand, C., and Åkerlund, B. (1994). Oxygen radical release by neutrophils of HIV infected patients. *Chem.-Biol. Interact.* **91**, 141–146.

Jones, A. L., Bangash, I. H., Bouchier, I. A., and Hayes, P. C. (1994). Portal and systemmic haemodynamic action of N-acetylcysteine in patients with stable cirrhosis. *Gut* **35**, 1290–1293.

Kågedahl, B., and Källberg, M. (1982). Determination of non-protein bound N-acetylcysteine in plasma by high performance liquid chromatography. *J. Chromatogr.* **311**, 170–175.

Khwali, F. A., and Reid, M. B. (1994). N-acetylcysteine depresses contractile function and inhibits fatigue of diaphragm *in vitro*. *J. Appl. Physiol.* **77**, 317–324.

Knuckey, N. W., Palm, D., Primiano, M., Epstein, L. G., and Johnson, C. E. (1995). N-Acetylcysteine enhances hippocampal neuronal survival after transient forebrain ischemia in rats. *Stroke* **26**, 305–310.

Lailey, A. F., and Upshall, D. G. (1994). Thiol levels in rat bronchoalveolar lavage fluid after administration of cysteine esters. *Hum. Exp. Toxicol.* **13**, 776–780.

Langley, S. C., and Kelly, F. J. (1993). N-Acetylcysteine ameliorates hyperoxic lung injury in the preterm guinea pig. *Biochem. Pharmacol.* **45**, 841–846.

Lewis, P. A., Woodward, A. J., and Maddock, J. (1985). Improved method for the determination of N-acetylcysteine in human plasma by high performance liquid chromatography. *J. Chromatogr.* **327**, 261–267.

Majima, E., Goto, S., Hori, H., Shinohara, Y., Hong, Y. M., and Terada, H. (1995). Stabilities of the fluorescent SH-reagent eosine 5-maleimide and its products with sulphydryl compounds. *Biochim. Biophys. Acta* **1243**, 336–342.

Malorini, W., Rivaben, R., Santini, M. T., and Donelli, G. (1993). N-Acetylcysteine inhibits apoptosis and decreases viral particles in HIV-chronically infected U937 cells. *FEBS Lett.* **327**, 75–78.

Malorna, W., D'Ambrosia, A., Rainaldi, G., Rivabene, R., and Viora, M. (1994). Thiol supplier N-acetylcysteine enhances conjugate formation between natural killer cells and K562 or U937 targets but increases the lytic function only against the latter. *Immunol. Lett.* **43**, 209–214.

Marui, N., Offermann, M. K., Swerlick, R., Kunsch, C., Rosen, C. A., Ahmed, M., Alexander, R. W., and Medford, R. M. (1993). Vascular cell adhesion molecule-1 (VCAM-1) gene transcription and expression are regulated through an antioxidant-sensitive mechanism in human vascular endothelial cells. *J. Clin. Invest.* **92**, 1866–1874.

Meulenbelt, J., Van Bree, L., Dormans, J., and Sangster, B. (1994). No beneficial effect of N-acetylcysteine treatment on broncho-alveolar lavage fluid variables in acute nitrogen dioxide intoxication in rats. *Hum. Exp. Toxicol.* **13**, 472–477.

Meyer, A., Buhl, R., and Magnussen, H. (1994). The effect of oral N-acetylcysteine on lung glutathione levels in ideopathic pulmonary fibrosis. *Eur. Respir. J.* **7**, 431–436.

Mihm, S., Ennen, J., Pessara, U., Kurth, R., and Dröge, W. (1991). Inhibition of HIV-1 replication and NFκB activity by cysteine and cysteine derivatives. *AIDS* **5**, 497–503.

Miquel, J., Ferrandiz, M. L., De Juan, E., Sevila, I., and Martinez, M. (1995). N-Acetylcysteine protects against age-related decline of oxidative phosphorylation in liver mitochondria. *Eur. J. Pharmacol.* **292**, 333–335.

Moldéus, P. (1981). Use of isolated hepatocytes in the study of paracetamol metabolism and toxicity. *In* "Drug Reactions and the Liver (M. Davis, J. M. Tredger, and R. Williams, eds.), pp. 114–146. Pitman Medical, London.

Moldéus, P., Cotgreave, I. A., and Berggren, M. (1986). Lung protection by a thiol containing antioxidant N-acetylcysteine. *Respiration* **50**, s31–s42.

Newman, C. M., Wassen, J. B., Taylor, G. W., Boobis, A. R., and Davies, D. S. (1990). Rapid tolerance to the hypotensive effects of glyceryl trinitrate in the rat. Prevention by N-acetyl-L-cysteine but not N-acetyl-D-cysteine. *Br. J. Pharmacol.* **99**, 825, 1990.

Newman, G. W., Balcewicz-Sablinsk, M. K., Guarnaccia, J. R., Remold, H. G., and Silberstein, D. S. (1994). Opposing regulatory effects of thioredoxin and eosinophil cytotoxicity-enhancing factor on the development of human immunodeficiency virus. *J. Exp. Med.* **180**, 359–363.

Oddera, S., Silvestri, M., Sacco, O., Eftimiadi, C., and Rossi, G. A. (1994). N-Acetylcysteine enhances *in vitro* the intracellular killing of *Staphylococcus aureus* by human alveolar macrophages and blood polymorphonuclear leukocytes and partially protects phagocytes from self-killing. *J. Lab. Clin. Med.* **124**, 293–301.

Ohman, L., Dahlgren, C., Follin, P., Lew, D., and Stendahl, O. (1992). N-Acetylcysteine enhances receptor-mediated phagocytosis by human neutrophils. *Agents Actions* **36**, 271–277.

Olsson, B., Johansson, M., Gabrielsson, J., and Bolme, P. (1988). Pharmacokinetics and bio-availability of reduced and oxidised N-acetylcysteine. *Eur. J. Clin. Pharmacol.* **34**, 77–82.

Parfett, C. L., and Pilon, R. (1995). Oxidative stress-related gene expression and promotion of morphological transformation induced in C3H/10T1/2 cells by ammonium metavanadate. *Food Chem. Toxicol.* **33**, 301–308.

Peddersen, C. O., Barth, P., Puchner, A., and von Wichert, P. (1993). N-Acetylcysteine decreases functional and structural AARDS-typical lung changes in endotoxin-treated rats. *Medizinische Klinik* **88**, 197–206.

Pendyala, L., and Creaven, P. J. (1995). Pharmacokinetic and pharmacodynamic studies of N-acetylcysteine, a potential chemoprotective agent during a phase I trial. *Cancer Epidemiol. Biomark. Prevent.* **4**, 245–251.

Poulsen, H. E., Vilstrup, H., Almdal, T., and Dalhoff, K. (1993). No net splanchnic release of glutathione in man during N-acetylcysteine infusion. *Scand. J. Gastroenterol.* **28**, 408–412.

Prescott, L. F. (1983). Paracetamol overdosage: Pharmacological considerations and clinical management. *Drugs* **25**, 290–314.

Raju, P. A., Herzenberg, L. A., Herzenberg, L. A., and Rooeder, M. (1994). Glutathione precursor and antioxidant activities of N-acetylcysteine and oxathiazolidine carboxylate compared in *in vitro* studies of HIV replication. *Aids Res. Hum. Retrovir.* **10**, 961–967.

Ratan, R. R., Murphy, T. H., and Baraban, J. M. (1994). Macromolecular synthesis inhibitors prevent oxidative stress-induced apoptosis in embryonic neurons by shunting cysteine from protein synthesis to glutathione. *J. Neurosci.* **14**, 4385–4392.

Reddy, B. S., Rao, C. V., Rivenson, A., and Kelloff, G. (1993). Chemoprevention of colon carcinogenesis by organosulfur compounds. *Cancer Res.* **53**, 3493–3498.

Reid, M. B., Stokic, D. S., Koch, S. M., Khwali, F. A., and Leis, A. A. (1994). N-Acetylcysteine inhibits muscle fatigue in humans. *J. Clin. Invest.* **94**, 2468–2474.

Reinhart, K., Spies, C. D., Meier-Hellmann, A., Bredle, D. L., Hannemann, L., Specht, M., and Schaffartzik, W. (1995). N-Acetylcysteine preserves oxygen consumption and gastric mucosal pH during hyperoxic ventilation. *Am. J. Respir. Crit. Care Med.* **151**, 773–779.

Riise, G. C., Larsson, S., Larsson, P., Jeansson, S., and Andersson, B. A. (1994). The intrabronchial microflora in chronic bronchitis patients: A target for N-acetylcysteine therapy? *Eur. Respir. J.* **7**, 94–101.

Rizzardini, M., Carelli, M., Cabello, Porras, M. R., and Cantoni, L. (1994). Mechanism of endotoxin-induced haem oxygenase mRNA accumulation in mouse liver: Synergism by glutathione depletion and protection by N-acetylcysteine. *Biochem. J.* **304**, 477–483.

Roederer, M., Ela, S. W., Staal, F. J., Herzenberg, L. A., and Herzenberg, L. A. (1992). N-Acetylcysteine: A new approach to anti-HIV therapy. *AIDS Res. Hum. Retrovir.* **8**, 209–217.

Roederer, M., Raju, P. A., Staal, F. J., Herzenberg, L. A., and Herzenberg, L. A. (1991). N-Acetylcysteine inhibits latent HIV expression in chronically-infected cells. *AIDS Res. Hum. Retrovir.* **7**, 563–567.

Rothstein, J. D., Bristol, L. A., Hosler, B., Brown, R. H., and Kuncl, R. W. (1994). Chronic inhibition of superoxide dismutase produces apoptotic death in spinal neurons. *Proc. Natl. Acad. Sci. U.S.A.* **91**, 4155–4159.

Russel, J., Spickett, C. M., Geglinski, J., Smith, W. E., McMurray, J., and Abdullah, I. B. (1994). Alterations in erythrocyte glutathione redox balance by N-acetylcysteine, captopril and exogenous glutathione. *FEBS Lett.* **347**, 215–220.

Sagara, M., Satoh, J., Zhu, X. P., Takahashi, K., Fuzukawa, M., Muto, G., Muto, Y., and Toyota, T. (1994). Inhibition with *N*-acetylcysteine of enhanced production of tumor necrosis factor in streptozotocin-induced diabetic rats. *Clin. Immunol. Immunopathol.* **71**, 333–337.

Sandström, P. A., Mannie, M. D., and Buttke, T. M. (1994). Inhibition of activation-induced death in T cell hybridomas by thiol antioxidants: Oxidative stress as a mediator of apoptosis. *J. Lekocyte Biol.* **55**, 221–226.

Sandström, P. A., Roberts, B., Folks, T. M., and Buttke, T. M. (1993). HIV gene expression enhances T cell susceptibility to hydrogen peroxide-induced apoptosis. *AIDS Res. Hum. Retrovir.* **9**, 1107–1113.

Särnstrand, B., Tunek, A., Sjödin, K., and Hallberg, A. (1995). Effects of *n*-acetylcysteine stereoisomers on oxygen-induced lung injury in rats. *Chem. Biol. Interact.* **94**, 157–164.

Scala, R., Moriggi, E., Corvasce, G., and Morelli, D. (1993). Protection by N-acetylcysteine against pulmonary endothelial cell damage by oxidant injury. *Eur. Respir. J.* **6**, 440–446.

Schreck, R., Rieber, P., and Baeuerle, P. A. (1991). Reactive oxygen intermediates as apparently widely used messengers in the activation of NF-κb transcription factor and HIV-1. *EMBO J.* **10**, 2247–2258.

Sen, C. K., Atalay, M., and Hanninen, O. (1994). Excercise-induced oxidative stress: Glutathione supplementation and deficiency. *J. Appl. Physiol.* **77**, 2177–2187.

Sen, C. K., Rankinen, T., Vaisanen, S., and Rauramää, R. (1994). Oxidative stress after human exercise: Effect of N-acetylcysteine supplementation. *J. Appl. Physiol.* **76**, 2570–2577.

Senaldi, G., Pointaire, P., Piquet, P. F., and Grau, G. E. (1994). Protective effect of N-acetylcysteine in hapten-induced irritant and contact hypersensitivity reactions. *J. Invest. Dermatol.* **102**, 934–937.

Sheffner, A. L. (1963). The reduction *in vitro* in viscocity of muoprotein solutions by a new mucolytic agent, N-acetylcysteine. *Ann. N.Y. Acad. Sci.* **106**, 298–310.

Sjödin, K., Nilsson, E., Hallberg, A., and Tunek, A. (1989). Metabolism of N-acetylcysteine: Some structural requirements for the deacetylation and consequences for the oral bioavailability. *Biochem. Pharmacol.* **38**, 3981–3985.

Sohn, H. O., Limm, H. B., Lee, Y. G., Lee, D. W., and Kim, Y. T. (1993). Effect of subchronic administration of antioxidants against cigarette smoke exposure in rats. *Arch. Toxicol.* **67**, 667–673.

Spies, C. D., Reinhart, K., Witt, I., Meier-Hellmann, A., Hanneman, L., Bredle, D. L., and Schaffartzig, W. (1994). Influence of N-acetylcysteine on direct indicators of tissue oxygenation in septic shock patients: Results from a prospective, randomized, double blind study. *Crit. Care Med.* **22**, 1738–1746.

Suter, P. M., Domenighetti, G., Schaller, M. D., Laverriere, M. C., Ritz, R., and Perret, C. (1994). N-Acetylcysteine enhances recovery from acute lung injury in man: A randomized, double-blind, placebo-controlled clinical study. *Chest* **105**, 190–194.

Talley, A. K., Dewhurst, S., Perry, S. W., Dollard, S. C., Gummulur, S., Fine, S. M., New, D., Epstein, L. G., Gendelman, H. E., and Gelbard, H. A. (1995). Tumour necrosis factor-alpha-induced apoptosis in human neuronal cells: Protection by the antioxidant N-acetylcysteine and the genes Bcl-2 and crma. *Mol. Cell. Biol.* **15**, 2359–2366.

Tang, L. D., Sun, J. Z., Wu, K., Sun, C. P., and Tang, Z. M. (1991). Beneficial effect of N-acetylcysteine and cysteine in stunned myocardium in perfused rat heart. *Br. J. Pharmacol.* **102**, 601–606.

Thor, H., Moldéus, P., and Orrenius, S. (1979). Metabolic activation and hepatotoxicity: Effects of cysteine, N-acetylcysteine, and methionine on glutathione biosynthesis and bromobenzene toxicity in isolated rat hepatocytes. *Arch. Biochem. Biophys.* **192**, 405–413.

Toyoóka, T., and Imai, K. (1983). High performance liquid chromatography and fluorimetric detection of biologically-important thiols, derivatized with ammonium 7-fluourobenzo-2-oxa-1,3-diazole-4 sulphonate (SBD-F). *J. Chromatogr.* **282**, 495–500.

Van den Broek, L. T., and Beijersbergen van Hanegouen, G. M. (1995). Topically-applied N-acetylcysteine as a protector against UVB-induced systemic immunosuppression. *J. Photochem. Photobiol. B-Biol.* **27**, 61–65.

Vecchiarelli, A., Dottorini, M., Pietrella, D., Cociani, C., Eslami, A., Todisco, T., and Bistoni, F. (1994). Macrophage activation by N-acetylcysteine in COPD patients. *Chest* **105**, 806–811.

Watson, W. A., and McKinney, P. E. (1991). Activated charcoal and N-acetylcysteine absorption: Issues in interpreting pharmacokinetic data. *D.I.C.P.* **25**, 1081–1084.

Weber, C., Erl, W., Pietsch, A., Strobel, M., Ziegler-Heibrock, H., and Weber, P. C. (1994). Antioxidants inhibit monocyte adhesion by suppressing nuclear factor kappa-b mobilization and induction of vascular adhesion molecule-1 in endothelial cells stimulated to produce radicals. *Atherosclerosis Thrombosis* **14**, 1665–1673.

Wilpart, M., Speder, A., and Roberfroid, M. (1986). Anti-initiation activity of N-acetylcysteine in experimental colonic carcinogenesis. *Cancer Lett.* **31**, 319–324.

Witschi, A., Junker, E., Schranz, C., Speck, R. F., and Lauterberg, B. H. (1995). Supplementation of N-acetylcysteine fails to increase glutathione in lymphocytes and plasma from patients with AIDS. *AIDS Res. Hum. Retrovir.* **11**, 141–143.

Yoyooka, T., Chokshi, H. P., Givens, R. S., Carlson, R. G., Lunte, S. M., and Kuwana, T. (1993). Fluorescent and chemiluminescent detection of oxazole-labelled amines and thiols. *Biomed. Chromatog.* **7**, 208–216.

Zamzami, N., Marchetti, P., Castedo, M., Zanin, C., Vassiere, J. L., Petit, P. X., and Kroemer, G. (1995). Reduction in mitochondrial potential constitutes an early irreversible step of programmed lymphocyte death *in vivo*. *J. Exp. Med.* **181**, 1661–1672.

Helmut Sies
Hiroshi Masumoto

Institut für Physiologische Chemie I
Heinrich-Heine-Universität Düsseldorf
D-40001 Düsseldorf, Germany

Ebselen as a Glutathione Peroxidase Mimic and as a Scavenger of Peroxynitrite

I. Introduction

Description of the glutathione peroxidase-like activity of the biologically active selenoorganic compound ebselen (then called PZ 51) (Müller *et al.*, 1984; Wendel *et al.*, 1984) led to extended subsequent research on this interesting molecule ranging from studies on radical reactivity through its biological properties in cells and organs to clinical settings, notably afflictions of the central nervous system. Basically, the compound is considered capable of contributing to the antioxidant defense in tissues, so that a potential pharmacological application becomes of interest.

Many of the biological properties of ebselen were explainable on the basis of its property as an enzyme mimic, carrying out the function of the selenoenzyme, GSH peroxidase (GPx), and, as found by Maiorino *et al.* (1992), of phospholipid hydroperoxide GSH peroxidase (PHGPx). In an overview (Sies, 1993), the properties of ebselen as a GSH peroxidase mimic were presented, and some of the biochemical and pharmacological reactivities were discussed; the reader is referred there for details. Sies (1993) also stated that it is obvious that the biological functions of ebselen might extend to other reactions in addition to the reduction of hydroperoxides, with the role of selenium in biology and medicine not yet being fully known. Indeed, a new aspect on biological function recently arose with the observation of an exceptionally high reactivity of ebselen with peroxynitrite,[1] an important agent active in inflammatory processes (Masumoto and Sies, 1996). At this point, it is not yet known whether this activity of ebselen can be viewed as a mimic of enzyme activity. It can be envisaged that biologically occurring selenoproteins may exhibit a novel biological role by deactivating peroxynitrite.

A further aspect, not related to the activity as an enzyme mimic as GSH peroxidase or in the deactivation of peroxynitrite, resides in the high reactivity of ebselen with thiol groups, leading to the inactivation or inhibition of enzymes. Thus, as first noted by Wendel *et al.* (1986), studying NADPH cytochrome P450 reductase in microsomes, low concentrations of ebselen are implicated. For example, a strong inhibition of lipoxygenases was observed by Schewe *et al.* (1994). This potentially interesting pharmacological aspect of ebselen reactivity has been studied with a number of other enzymes and has been put into perspective by Schewe (1995). Parnham and Graf (1991) reviewed a number of selenoorganic compounds as potential therapeutic agents.

II. Synthesis of Ebselen

The compound ebselen, 2-phenyl-1,2-benzisoselenazol-3(2*H*)-one, also called PZ 51, was synthesized at Nattermann & Cie. GmbH, Cologne, Germany, under the patents European Pat. 44,971; Ger. 3,027,073; Japan K-8256,427; U.S. 4,352,799. The synthesis is based on those described by Lesser and Weiss (1924) and Weber and Renson (1976), reacting 2-(chloroseleno)benzoyl chloride with aniline in a Schotten-Baumann reaction as described by Fischer and Dereu (1987). Several analogs and derivatives were also synthesized (Fischer and Dereu, 1987; Cantineau *et al.*, 1986; Lambert *et al.*, 1987; Engman and Hallberg, 1989; Wilson *et al.*, 1989; Cotgreave *et al.*, 1992a; Chaudière *et al.*, 1994).

III. Molecular Activities of Ebselen

A. GSH Peroxidase-like Activity

The milestones in development of current knowledge on the enzymatic reduction of hydroperoxides by GSH peroxidases go back to the mid-1950s

---

[1] The term peroxynitrite is used to refer generically to both peroxynitrite anion [oxoperoxonitrate(1-), ONOO⁻] and its conjugate acid, peroxynitrous acid (hydrogen oxoperoxonitrate, ONOOH).

[see reviews by Flohé (1989) and Sies (1993)]: The "classical" GPx of bovine erythrocytes was discovered by Mills (1957) and was shown to be a selenoprotein by Flohé *et al.* (1973) and by Rotruck *et al.* (1973). Its amino acid sequence was elucidated (Günzler *et al.*, 1984), and the three-dimensional structure was determined (Ladenstein *et al.*, 1979; Epp *et al.*, 1983).

Another GPx that acts on peroxidized phospholipids in biological membranes was isolated by Ursini *et al.* (1982, 1985). This enzyme is phospholipid hydroperoxide GSH peroxidase. It is monomeric and has been established as a selenoenzyme distinct from the classical GPx based on cDNA and amino acid sequencing (Schuckelt *et al.*, 1991). The presumed catalytic triad of selenocysteine, glutamine, and tryptophan residues in the active center has been examined by mutational analysis (Maiorino *et al.*, 1995).

GPx activity in plasma has been attributed to a protein immunologically distinct from, but homologous to, the classical GPx (Takahashi and Cohen, 1986; Takahashi *et al.*, 1990). Thus, there are at least three different glutathione peroxidases. The catalytic mechanism at the selenocysteine moiety in the active center of these different enzyme proteins seems to be similar (Flohé, 1989). The catalytic cycle is related to that deduced for ebselen.

A number of further selenopeptides and selenoproteins, many of them with a so far unknown function, have been discovered (Behne *et al.*, 1988). The presence of selenoprotein P has been associated with protection against liver damage in two oxidant injury models (Burk *et al.*, 1995a,b).

The reaction cycle of GPx for the enzymatic catalysis is thought to proceed in three main steps, involving the enzyme-bound selenocysteine, E-Cys-SeH, which is present as the selenol or more likely as the selenolate [see Flohé (1989) for a detailed discussion]. In the first step [Reaction (1)], the organic hydroperoxide, ROOH, reacts in a diffusion-controlled reaction to yield the selenenic acid, E-Cys-SeOH, and the corresponding alcohol, ROH. The following two steps consist of the sequential reduction of the selenenic acid by thiol, GSH; Reaction (2) gives the selenosulfide and water, and Reaction (3) regenerates the selenol and the disulfide, GSSG. The overall reaction is that of GPx or PHGPx [Reaction (4)].

$$\text{E-Cys-SeH} + \text{ROOH} \rightarrow \text{E-Cys-SeOH} + \text{ROH} \tag{1}$$

$$\text{E-Cys-SeOH} + \text{GSH} \rightarrow \text{E-Cys-Se-SG} + \text{H}_2\text{O} \tag{2}$$

$$\text{E-Cys-Se-SG} + \text{GSH} \rightarrow \text{E-Cys-SeH} + \text{GSSG} \tag{3}$$

$$\text{ROOH} + 2\,\text{GSH} \rightarrow \text{ROH} + \text{H}_2\text{O} + \text{GSSG} \tag{4}$$

In their chemical analysis of the GPx-like reactivity of ebselen, Fischer and Dereu (1987) proposed that formation of a diselenide, 2,2'-diselenobis-(N-phenylbenzamide), is a key step for catalytic activity, being the slowest one. According to Fischer and Dereu (1987) and to further work by Haenen *et al.* (1990), the diselenide would react with the hydroperoxide to yield the parent compound and water. It was also noted by Fischer and Dereu (1987) that the selenosulfides derived from ebselen and thiols might constitute a storage form of ebselen and eventually be responsible for transport of the drug.

However, in a kinetic study of the catalysis of the GPx reaction by ebselen, Maiorino *et al.* (1988) concluded that the mechanism appeared kinetically identical to that of the enzyme reaction, a *ter-uni-ping-pong* mechanism. Carboxymethylation of intermediates by iodoacetate to an inactive derivative suggested that a selenol moiety is involved in the mechanism. Cotgreave *et al.* (1992b) devised a method to detect 2-(phenylcarbamoyl) phenylselenol by its reaction with 1-chloro-2,4-dinitrobenzene. Morgenstern *et al.* (1992) synthesized the selenol and assayed for GPx-like activity against hydrogen peroxide. It was concluded by Morgenstern *et al.* (1992) that selenol is the predominant molecular species responsible for the GSH- or dithiothreitol-dependent peroxidase activity of ebselen.

Thus, it appears that the reaction scheme of the catalysis of the GPx reaction by ebselen occurs in analogy to the mechanism of GPx enzyme catalysis (Flohé, 1989), as was initially deduced by Wendel *et al.* (1984). The redox chemistry of selenocysteine model systems was studied by Reich and Jasperse (1987). The postulated catalytic cycle in this study also involves a selenol intermediate. It should be noted that there is an alternative pathway in which ebselen is oxidized initially by hydroperoxides to the ebselen Se-oxide (Fischer and Dereu, 1987). Subsequent reduction of the ebselen Se-oxide by thiols then regenerates the parent compound, ebselen (Fischer and Dereu, 1987; Glass *et al.*, 1989).

Unlike the enzyme-catalyzed reaction with binding sites conferring specificity for GSH, ebselen can utilize other thiols in addition to GSH, e.g., dithioerythritol (Müller *et al.*, (1985), N-acetylcysteine (Cotgreave *et al.*, 1987), or dihydrolipoate (Haenen *et al.*, 1990).

The first demonstration of the ability of ebselen to carry out the reduction of hydroperoxides at the expense of thiols as reducing equivalents, i.e., the reaction carried out by GSH peroxidase, was performed with a coupled enzymatic assay using GSH, NADPH, and glutathione disulfide reductase (H. Sies *et al.*, unpublished work, 1981); ebselen, then called PZ51, was compared to the sulfur analog, 2-phenyl-1,2-benzisothiazol-3(2*H*)-one, called PZ 25, which did not exhibit this activity. These data, together with further work on the antioxidant properties of ebselen, were published by Müller *et al.* (1984). Wendel *et al.* (1984) reported GPx-like activity by measuring the loss of GSH and extended the kinetic analysis. The activation energy for the ebselen-catalyzed reaction is 55 kJ/mol, in comparison to 36.5 kJ/mol for the enzyme-catalyzed reaction for GPx (Wendel *et al.*, 1984).

The following assay systems for the detection of GPx-like activity have been described.

1. Coupled Enzymatic Assay

As in usual coupled enzymatic assays, the test reaction (here GPx) is coupled to an indicator reaction, that of glutathione disulfide reductase, and the loss of NADPH is monitored continuously by absorbance spectrophotometry (Müller *et al.*, 1984; Maiorino *et al.*, 1988).

2. Assay of GSH Removal

This can be done by stopping the reaction and then assaying for remaining GSH, e.g., by the formation of thionitrobenzoate from Ellman's reagent (Wendel *et al.*, 1984) or of a monobromobimane adduct (Cotgreave *et al.*, 1992a).

3. Assay of Hydroperoxide Removal

A standard method using the iron–thiocyanate complex has been employed (Cotgreave *et al.*, 1992a).

B. Scavenging of Peroxynitrite

Peroxynitrite ($ONOO^-$/$ONOOH$), a strong oxidizing species, is generated in inflammatory processes from the superoxide anion radical and nitric oxide. The formation of peroxynitrite by various cell types, such as macrophages (Ischiropoulos *et al.*, 1992), Kupffer cells (Wang *et al.*, 1991), and endothelial cells (Kooy and Royall, 1994), has been detected. Because of increasing evidence for a role of peroxynitrite in biological processes (Beckman *et al.*, 1990; Koppenol *et al.*, 1992; Ramezanian *et al.*, 1996), interest has been generated in the potential defense against this reactive oxidizing species.

We have noted that the peroxynitrite-induced luminol chemiluminescence emitted from Kupffer cells (Wang *et al.*, 1991) was inhibited in the presence of low concentrations of ebselen (Wang *et al.*, 1992) and, more recently, have observed that ebselen rapidly reacts with peroxynitrite in a bimolecular fashion, yielding the selenoxide of the parent molecule, ebselen Se-oxide [2-phenyl-1,2-benzisoselenazol-3(2*H*)-one 1-oxide], as the sole selenium-containing product at 1 : 1 stoichiometry (Masumoto and Sies, 1996). The second-order rate constants for the reaction of ebselen and of 2-(methylseleno)benzanilide, an *in vivo* metabolite of ebselen, with peroxynitrite are shown in Table I and are compared with those reported for cysteine, ascorbate, and methionine. It is clear from these data (Table I) that the reaction of ebselen with peroxynitrite is the fastest reaction observed so far for a small molecule. The rate constant is about three to four orders of magnitude faster than that of biologically occurring small molecules, such as ascorbate (Bartlett *et al.*, 1995), cysteine (Radi *et al.*, 1991), and methionine (Pryor *et al.*, 1994) with peroxynitrite. The rate constant for the reaction

TABLE I   Second-Order Rate Constants for the Reaction with Peroxynitrite

Compound	$k(M^{-1}\ sec^{-1})$	T ($°C$)	Reference
Ebselen	2.0×10^6 (pH 7.2)[a]	25	H. Masumoto, R. Kissner, W. H. Koppenol, and H. Sies, unpublished results
2-(Methylseleno) benzanilide	1.2×10^4 (pH 7.2)[b]	25	H. Masumoto, R. Kissner, W. H. Koppenol, and H. Sies, unpublished results
Cysteine	5.9×10^{3c}	37	Radi *et al.* (1991)
Ascorbate	2.4×10^{2b}	25	Bartlett *et al.* (1995)
Methionine	9.5×10^{2b}	25	Pryor *et al.* (1994)
Tryptophan	1.3×10^{2b}	25	Padmaja *et al.* (1996)
Myeloperoxidase	2.0×10^{7b}	12	Floris *et al.* (1993)
Cytochrome c^{2+}	2.3×10^{5b}	25	Thomson *et al.* (1995)
Alcohol dehydrogenase	$2.6–5.2 \times 10^5$ (pH 7.4)[a]	23	Crow *et al.* (1995)

[a] Second-order rate constant for the reaction with ONOO$^-$/ONOOH.
[b] Second-order rate constant for the reaction with ONOOH.
[c] Second-order rate constant for the reaction with ONOO$^-$

of 2-(methylseleno)benzanilide is about an order of magnitude higher than the rate constant observed for methionine.

Enzymes and proteins, such as myeloperoxidase (Floris *et al.*, 1993), cytochrome c (Thomson *et al.*, 1995), and alcohol dehydrogenase (Crow *et al.*, 1995) can react substantially faster than small molecules (Table I). In particular, myeloperoxidase reacts very fast with peroxynitrite. However, ebselen reacts faster at higher pH.

Ebselen Se-oxide is readily reduced by reducing equivalents such as GSH to revert to ebselen (Fischer and Dereu, 1987; Glass *et al.*, 1989), as shown in the GPx-like catalytic cycle of ebselen. This reversible reaction would allow for a potential sustained defense line against peroxynitrite (Fig. 1).

In addition to ebselen, other reduced species derived from ebselen, such as the selenosulfides and the selenol, are expected to scavenge peroxynitrite.

One could speculate at this point that the high reactivity of selenocompounds in the scavenging of peroxynitrite is a mimic for a similar function exerted by biologically occurring selenium-containing compounds. In particular, as mentioned earlier, we are currently examining the idea whether selenoproteins or selenopeptides might, in fact, have a biological function as a defense line against peroxynitrite.

We have shown that ebselen and other selenium-containing compounds, such as selenomethionine or selenocystine, can protect DNA from single-strand break formation caused by peroxynitrite (Roussyn *et al.*, 1996).

C. Radical Scavenging

The reactivity of ebselen and related selenoorganic compounds with 1,2-dichloroethane radical cations and halogenated peroxyl radicals was

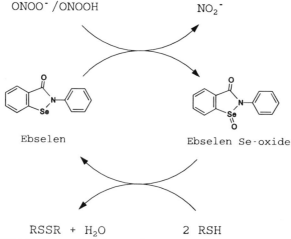

FIGURE I   Potential mechanism for peroxynitrite scavenging by ebselen.

studied by pulse radiolysis (Schöneich *et al.*, 1990). The rate constant for the reaction of ebselen with trichloromethyl peroxyl radicals was determined to be $2.9 \times 10^8 \ M^{-1} \ sec^{-1}$, whereas its sulfur analog, 2-phenyl-1,2-benzisothiazol-3-(2H)-one was oxidized at much lower rates. The rate constant observed for ebselen is comparable to that of α-tocopherol under similar conditions.

D. Singlet Oxygen Quenching

The reactivity of ebselen with singlet molecular oxygen is only sluggish; the rate constant is $2.5 \times 10^6 \ M^{-1} \ sec^{-1}$ (Scurlock *et al.*, 1991; H. Sies *et al.*, unpublished results, 1989). The sulfur analog exhibits a rate constant an order of magnitude lower than this.

E. Reactivity with Protein Thiols

As first noted by Wendel *et al.* (1986) in a study on NADPH cytochrome P450 reductase in microsomes, ebselen can inhibit enzyme activity at a low concentration. A number of enzymes have since been reported to be inhibited by ebselen. Interestingly, because many of these are of importance in inflammatory processes, Schewe (1995) concluded that it is reasonable to assume that inhibitory effects of this nature might contribute to the anti-inflammatory actions of ebselen *in vivo*.

Among the enzymes for which inhibitory effects of ebselen have been observed were purified human recombinant 5-lipoxygenase (Schewe *et al.*, 1994) and purified rabbit reticulocyte 15-lipoxygenase (Schewe *et al.*, 1994), with strong inhibition at a 0.1 μM concentration of ebselen. This is of particular interest since ebselen has been found to suppress the *in vivo* formation of proinflammatory cysteinyl leukotrienes (Wendel and Tiegs, 1986; Tabuchi *et al.*, 1995). The formation of 5-lipoxygenase products in polymorphonuclear leukocytes was inhibited at higher ebselen concentration, about 20 μM (Safayhi *et al.*, 1985).

Other enzymes for which inhibition by ebselen was reported include NADPH oxidase and protein kinase C from human granulocytes (Cotgreave *et al.*, 1989), H^+/K^+-ATPase from pig stomach (Tabuchi *et al.*, 1994), nitric oxide synthases from bovine aortic endothelium and rat peritoneal macrophages (Zembowicz *et al.*, 1993; Hattori *et al.*, 1994), and glutathione *S*-transferase from rat liver (Nikawa *et al.*, 1994a) and papain (Nikawa *et al.*, 1994b). No inhibition was found for prostaglandin H synthase from sheep vesicular glands (Schewe *et al.*, 1994), but the enzyme from human platelets had been reported to be inhibited (Kuhl *et al.*, 1985, 1986). However, Schewe (1995) argues that the latter may have been a secondary effect resulting from the suppression of an agonist-triggered rise in intracellular calcium in platelets (Brüne *et al.*, 1991; Dimmeler *et al.*, 1991). Unlike prostaglandin H synthase inhibitors, ebselen does not evoke gastric irritation in rats and, on the contrary, affords protection against gastric intolerance induced by such drugs (Leyck and Parnham, 1990).

Evidence supports the assumption that the inhibition of enzyme activity is due to a blockade of thiol groups essential for enzyme function. A protection against the inhibition by ebselen or a reversal of inhibition subsequent to the inhibition of enzyme activity by ebselen has been observed with glutathione in most of the cases mentioned in this chapter.

IV. Activity in Biological Model Systems

A. Protection against Lipid Peroxidation

The protective effect of ebselen against iron-ADP-induced lipid peroxidation in hepatic microsomal fractions using ascorbate or NADPH as a reductant has been amply documented (Müller *et al.*, 1984; Hayashi and Slater, 1986; Narayanaswami and Sies, 1990). In intact cells, it was shown that the protective effect depends on the presence of GSH. Hepatocytes depleted of GSH could not be protected by ebselen, whereas normal cells were protected (Müller *et al.*, 1985), suggesting that the GPx-like activity of ebselen is involved in the protection of intact cells against lipid peroxidation. In these experiments, several parameters of lipid peroxidation were followed, including the generation of low-level chemiluminescence and the formation of alkanes (ethane, *n*-pentane) and malondialdehyde.

Maiorino *et al.* (1992) and Noguchi *et al.* (1992) concluded that the major role of ebselen in protecting against different types of oxidative attack is to act in its capacity as a GPx mimic and, in particular, as a PHGPx mimic. Maiorino *et al.* (1988) had observed that ebselen acted at particularly high rates with the phospholipid hydroperoxides and also with cholesterol hydroperoxide and cholesterol ester hydroperoxides (Maiorino *et al.*, 1992). This conclusion came from experiments with photooxidized liposomes and peroxidized low-density lipoproteins. Miyazawa *et al.* (1993) reported that after oral administration of ebselen to rats, the levels of plasma phospholipid hydroperoxides were lowered as compared to controls.

Thomas and Jackson (1991) showed that ebselen is capable of preventing copper-dependent LDL oxidation in the presence of GSH, and Sattler *et al.* (1994) reported that ebselen plus GSH reduced HDL- and LDL-associated cholesteryl ester and phospholipid hydroperoxides. In work employing an albumin–ebselen complex, Christison *et al.* (1994) demonstrated that whole blood, but not plasma, without added GSH resulted in the reduction of LDL-associated cholesteryl linoleate hydroperoxides to the corresponding hydroxides. These data show that ebselen may be functional in controlling LDL oxidations in the circulating blood, of potential interest regarding the development of arteriosclerosis.

The radical scavenging activity of ebselen had little, if any, effect, shown by competition kinetics based on the free radical-dependent bleaching of crocin (Maiorino *et al.*, 1992), the lack of reaction with 2,2-diphenyl-1-picrylhydrazyl (DPPH), or the lack of suppression of oxidation of methyl linoleate induced by the radical generators 2,2′-azobis(amidinopropane) dihydrochloride or 2,2′-azobis(2,4-dimethylvaleronitrile) (Noguchi *et al.*, 1992). Further, Noguchi *et al.* (1992) concluded from their results that a role of ebselen in the sequestration of iron is minimal.

B. Biological Effects

Some details are mentioned in order to put the role of the enzyme mimic into perspective. Flohé (1989) in his comprehensive review has collected literature on the role of GPx in different tissues and organs and in their pathophysiology. Because GPx and PHGPx are present in most organs, an additional expression of this activity would be of interest in conditions of enhanced oxidative challenge. Further, especially in those spaces and compartments where these enzymes are not present at high activity, the low molecular weight compound, ebselen, as an enzyme mimic might have pathophysiological interest.

An overview on studies in which ebselen was employed was given by Sies (1993). In the numerous studies mentioned there and in additional studies reported since then, the beneficial (protective) effects of ebselen were observed and sometimes they were compared to the sulfur analog which was not protective. The anti-inflammatory effect was the initial property that attracted attention (Parnham *et al.*, 1985; Parnham and Graf, 1987,

1991), but the wide range of experimental conditions covered indicates a broader range of potential clinical interest.

One common basis for many of the diverse effects is that ebselen, by its property as a GPx and PHGPx mimic, can lower the peroxide tone (Hemler *et al.*, 1979), which is important in the control of lipoxygenases and cyclooxygenases.

A novel aspect of ebselen functions has been described by Hu *et al.* (1995). These authors made interesting observations on gap-junctional intercellular communication between cells, which are established by connexins. The down-regulation of intercellular communication by phorbol ester (TPA) was protected against by ebselen plus GSH, and the TPA-induced hyperphosphorylation of connexin-43 was prevented. Also, the internalization of connexin-43 from the membrane to the cell interior did not occur in the presence of ebselen and GSH. Taken together, these data were taken by Hu *et al.* (1995) to elucidate a molecular basis for the potential antitumor promotive effects of an antioxidant, ebselen.

V. Metabolism

Ebselen metabolism has been studied in isolated perfused liver (Müller *et al.*, 1988) and in intact rats, pigs, and human volunteers (Fischer *et al.*, 1988), as reviewed by Sies (1989). The metabolism after reduction to the selenol involves methylation to form 2-(methylseleno)benzanilide or glucuronidation to form 2-(glucuronylseleno)benzanilide. Whereas the latter is released into bile and finally into the urine, the former undergoes further metabolism. This includes hydroxylation at the phenyl moiety in the *para* position which, in turn, can then be glucuronidated. The microsomal metabolism of 2-(methylseleno)benzanilide was further investigated by John *et al.* (1990). Regeneration of the parent compound was observed. In pigs and in the human, the dominant metabolite in plasma and urine is the selenoglucuronide.

It is important to note that no unchanged ebselen was detectable in urine, plasma, or bile (Fischer *et al.*, 1988). the facile ring opening of the isoselenazolone ring (van Canegham 1974; Kamigata *et al.*, 1986) by rapid reaction with thiols to form a selenosulfide is a probable basis for this. Whether ebselen is attacked by sulfhydryl compounds already in the stomach or intestine before absorption or while transported through the mucosa is not yet known. In model studies using bovine serum albumin, Nomura *et al.* (1989) observed a rapid binding of ebselen to the free thiol group of albumin, forming an albumin–ebselen complex, S-[2-(phenylcarbamoyl) phenylseleno]albumin. In work on human plasma, we detected radioactivity from labeled ebselen only on albumin and negligible amounts with the globulin fractions upon gel electrophoresis (Wagner *et al.*, 1994). It is assumed that ebselen binds to the reactive thiol group at cysteine-34 in albumin, the location of bound cysteine and glutathione.

Thus, the current concept of transport of ebselen in the organism is that it is bound to proteins and that there is an interchange with low molecular weight thiols within cells and tissues. It is important to note that some of the experimental observations obtained with ebselen *in vitro* should be controlled by employing ebselen in a protein-bound form, e.g., as the albumin–ebselen complex. It is likely that the inhibitory effects of ebselen reported for isolated enzymes simply reflect the high reactivity of ebselen with protein thiols, and *in vivo* there was no detectable free ebselen (Müller *et al.*, 1988; Fischer *et al.*, 1988). One possible major intracellular binding capacity for ebselen is the class of glutathione transferases, to which ebselen was found to bind (Nikawa *et al.*, 1994a,b; Wagner *et al.*, 1994). Ziegler *et al.* (1992) showed that the selenol form of ebselen, 2-(phenylcarbamoyl)phenylselenol, can be a substrate for pig liver flavin-containing monooxygenase, as is 2-(methylseleno)benzanilide.

An early observation by Wendel *et al.* (1984) was that the selenium atom in ebselen was not bioavailable; in selenium-deficient animals the activity of the selenoprotein GPx could not be augmented by ebselen but was by selenite. The property of not entering the body pool of selenium but rather being metabolized as explained earlier may explain the lack of toxicity observed in experimental studies. In a study with a volunteer, 500 mg of [77]Se-ebselen was given orally, and urinary metabolites were followed by ^{77}Se-NMR (Dereu *et al.*, 1989). The metabolites mentioned earlier were observed, and the indicator of metabolically bioavailable selenium, trimethylselenonium as the final urinary metabolite of inorganic selenium, was blank.

VI. Clinical Assessment

The various actions of ebselen *in vitro* and in biological model systems would appear to accommodate numerous avenues of clinical assessment in a variety of potential indications. For example, a therapeutic use of ebselen as an inhibitor of restenosis after percutaneous transluminal coronary angioplasty was envisaged by Kodama *et al.* (1993).

A neuroprotective role of ebselen has been assessed clinically. Based on encouraging results obtained in animal studies (Johshita *et al.*, 1990; Dawson *et al.*, 1995), a clinical study has been carried out using ebselen in aneurysmal subarachnoid hemorrhage. Saito *et al.* (1995) reported briefly on this study, a multicenter randomized clinical trial. Eligible patients had subarachnoidal hemorrhage due to rupture of an intracranial aneurysm with Hunt & Hess grades between II and IV at admission to the hospital and who received the first drug administration within 96 hr after ictus. Ebselen was given orally at a daily dose of 300 mg per person for 2 weeks, and the study was placebo controlled. The study had 286 patients. Although the incidence of symptomatic vasospasm was 37 and 42% in the two groups, i.e., no significant difference was observed, the incidence of low-density areas in CT scans in the ebselen group (14%) was significantly less than in the placebo group (24%), $p = 0.039$. From these and other findings in this study, Saito *et al.* (1995) concluded that ebselen significantly ameliorates delayed ischemic neurological deficits and subsequent cerebral infarction in patients with severe subarachnoid hemorrhage, leading to an improvement of the overall outcome. No adverse reaction attributable to the use of ebselen was observed in this study.

VII. Other Selenoorganic Compounds

A number of attempts have been made to modify the basic structure of ebselen. Structure–activity relationships of a series of anti-inflammatory benzisoselenazolones have been reported by Parnham *et al.* (1989) and by Tarino (1986). As mentioned earlier, Wilson *et al.* (1989), Cotgreave *et al.* (1992a), and Chaudière *et al.* (1994) reported on related compounds, as did Andersson *et al.* (1994), Engman *et al.* (1994), and Galet *et al.* (1994).

The redox properties of ebselen confer activity as glutathione peroxidases to this low molecular weight compound. Thus, in the presence of thiols it acts as an enzyme mimic. The constraints of substrate specificity are much less, as the mimic is more readily accessible than active sites in proteins. This explains the relative unspecificity toward the thiol reductant: whereas the enzymes GPx and PHGPx are highly specific for GSH, the mimic also accepts other thiol compounds, some even at higher rates than GSH. Similarly, the accessibility of organic hydroperoxides as substrates is given with the relatively hydrophobic ebselen molecule, so that its range of substrates covers that of PHGPx.

There are many interesting research problems generated by these enzyme mimics. For example, this pertains to the transport pathway on proteins (Wagner *et al.*, 1994), with the albumin–ebselen complex in the plasma and its subsequent transfer to other binding sites probably being the key to the surprisingly low toxicity observed.

VIII. Concluding Remarks

Ebselen, 2-phenyl-1,2-benzisoselenazol-3(2*H*)-one, exhibits activity as an enzyme mimic, catalyzing the GSH peroxidase reaction, i.e., the reduction of a hydroperoxide at the expense of thiol. The specificity for substrates ranges from hydrogen peroxide and smaller organic hydroperoxides to membrane-bound phospholipid and cholesterylester hydroperoxides. The thiol reductant cosubstrate can be glutathione, N-acetylcysteine, dihydrolipoate, or other suitable thiol compounds.

The reaction of ebselen with peroxynitrite has a high second-order rate constant, about 1000-fold higher than that of ascorbate, cysteine, or methionine with peroxynitrite. Furthermore, ebselen inhibits enzymes such as lipoxygenases, nitric oxide synthases, NADPH oxidase, protein kinase C, and H^+/K^+-ATPase.

These molecular actions of ebselen contribute to its properties as an anti-inflammatory antioxidant. Numerous model experiments *in vitro* with isolated LDL, liposomes, microsomes, isolated cells, and organs established that ebselen protects against oxidative challenge.

Emerging results from clinical studies focusing on the indication of aneurysmal subarachnoid hemorrhage are promising.

Acknowledgments

The support of a research visit of H.M. by Drs. H. Hakusui and H. Masayasu, Daiichi Pharmaceutical Co., Tokyo, and the continuing grant support of H.S. by the National Foundation for Cancer Research are gratefully acknowledged.

References

Andersson, C. M., Hallberg, A., Linden, M., Brattsand, R., Moldeus, P., and Cotgreave, I. (1994). Antioxidant activity of some diarylselenides in biological systems. *Free. Rad. Biol. Med.* **16,** 17–28.

Bartlett, D., Church, D. F., Bounds, P. L., and Koppenol, W. H. (1995). The kinetics of the oxidation of L-ascorbic acid by peroxynitrite. *Free Rad. Biol. Med.* **18,** 85–92.

Beckman, J. S., Beckman, T. W., Chen, J., Marshall, P., and Freeman, B. A. (1990). Apparent hydroxyl radical production by peroxynitrite: Implications of endothelial injury from nitric oxide and superoxide. *Proc. Natl. Acad. Sci. USA* **87,** 1620–1624.

Behne, D., Hilmert, H., Scheid, S., Gessner, H., and Elger, W. (1988). Evidence for specific selenium target tissues and new biologically important selenoproteins. *Biochim. Biophys. Acta* **966,** 12–21.

Brüne, B., Diewald, B., and Ullrich, V. (1991). Ebselen affects calcium homeostasis in human platelets. *Biochem. Pharmacol.* **41,** 1805–1811.

Burk, R. F., Hill, K. E., Awad, J. A., Morrow, J. D., and Lyons, P. R. (1995a). Liver and kidney necrosis in selenium-deficient rats depleted of glutathione. *Lab. Invest.* **72,** 723–730.

Burk, R. F., Hill, K. E., Awad, J. A., Morrow, K. D., Kato, T., Cockwell, K. A., and Lyons, P. R. (1995b). Pathogenesis of diquat-induced liver necrosis in selenium-deficient rats: Assessment of the roles of lipid peroxidation and selenoprotein P. *Hepatology* **21,** 561–569.

Cantineau, R., Thiange, G., Plenevaux, A., Christiaens, L., and Guillaume, M. (1986). Synthesis of ^{75}Se-2-phenyl-1,2-benzisoselenazol-3(2H)-one (PZ 51, ebselen): A novel biologically active organo-selenium compound. *J. Label. Comp. Radiopharmaceut.* **23,** 59–65.

Chaudière, J., Yadan, J.-C., Erdelmeier, I., Tailhan-Lomont, C., and Moutet, M. (1994). Design of new selenium-containing mimics of glutathione peroxidase. In "Oxidative Processes and Antioxidants" (Paoletti *et al.*, eds.), pp. 165–184. Raven Press, New York.

Christison, J., Sies, H., and Stocker, R. (1994). Human blood cells support the reduction of low-density-lipoprotein-associated cholesteryl ester hydroperoxides by albumin-bound ebselen. *Biochem. J.* **304**, 341–345.

Cotgreave, I. A., Duddy, S. K., Kass, G. E. N., Thompson, D., and Moldéus, P. (1989). Studies on the anti-inflammatory activity of ebselen: Ebselen interferes with granulocyte oxidative burst by dual inhibition of NADPH oxidase and protein kinase C? *Biochem. Pharmacol.* **38**, 649–656.

Cotgreave, I. A., Moldéus, P., Brattsand, R., Hallberg, A., Andersson, C. M., and Engman, L. (1992a). Alpha-(phenylselenenyl)acetophenone derivates with glutathione peroxidase-like activity. *Biochem. Pharmacol.* **43**, 793–802.

Cotgreave, I. A., Morgenstern, R., Engman, L., and Ahokas, J. (1992b). Characterization and quantitation of a selenol intermediate in the reaction of ebselen with thiols. *Chem.-Biol. Interact.* **84**, 69–76.

Cotgreave, I. A., Sandy, M. S., Berggren, M., Moldéus, P. M., and Smith, M. T. (1987). N-Acetylcysteine and glutathione dependent protective Effect of PZ 51 (ebselen) against diquat induced cytotoxicity in isolated hepatocytes. *Biochem. Pharmacol.* **36**, 2899–2904.

Crow, J. P., Beckman, J. S., and McCord, J. M. (1995). Sensitivity of the essential zinc-thiolate moiety of yeast alcohol dehydrogenase to hypochlorite and peroxynitrite. *Biochemistry* **34**, 3544–3552.

Dawson, D. A., Masayasu, H., Graham, D. I., and Macrae, I. M. (1995). The neuroprotective efficacy of ebselen (a glutathione peroxidase mimic) on brain damage induced by transient focal cerebral ischaemia in the rat. *Neurosci. Lett.* **185**, 65–69.

Dereu, N., Fischer, H., Hilboll, G., Roemer, A., and Terlinden, R. (1989). The use of highly enriched [77]Se in metabolic studies of ebselen in man: An NMR investigation. In "Selenium in Biology and Medicine" (A. Wendel, ed.), pp. 163–168. Springer-Verlag, Heidelberg.

Dimmeler, S., Brüne, B., and Ullrich, V. (1991). Ebselen prevents inositol(1,4,5)-trisphosphate binding to its receptor. *Biochem. Pharmacol.* **42**, 1151–1153.

Engman, L., and Hallberg, A. (1989). Expedient synthesis of ebselen and related compounds. *J. Org. Chem.* **54**, 2964–2966.

Engman, L., Tunek, A., Hallberg, M., and Hallberg, A. (1994). Catalytic effects of glutathione peroxidase mimetics on the thiol reduction of cytochrome c. *Chem. Biol. Interact.* **93**, 129–137.

Epp, O., Ladenstein, R., and Wendel, A. (1983). The refined structure of the selenoenzyme glutathione peroxidase at 0.2 nm resolution. *Eur. J. Biochem.* **133**, 51–69.

Fischer, H., and Dereu, N. (1987). Mechanism of the catalytic reduction of hydroperoxides by ebselen: A selenium-77 NMR study. *Bull. Soc. Chim. Belg.* **96**, 757–768.

Fischer, H., Terlinden, R., Löhr, J. P., and Römer, A. (1988). A novel biologically active selenoorganic compound. VIII. Biotransformation of ebselen. *Xenobiotica* **18**, 1347–1359.

Flohé, L. (1989). The selenoprotein glutathione peroxidase. In "Glutathione" (D. Dolphin, R. Poulson, and O. Avramovic, eds.), pp. 643–731. Wiley, New York.

Flohé, L., Günzler, W. A., and Schock, H. H. (1973). Glutathione peroxidase: A selenoenzyme. *FEBS Lett.* **32**, 132–134.

Floris, R., Piersma, S. R., Yang, G., Jones, P., and Wever, R. (1993). Interaction of myeloperoxidase with peroxynitrite: A comparison with lactoperoxidase, horseradish peroxidase and catalase. *Eur. J. Biochem.* **215**, 767–775.

Galet, V., Bernier, J. L., Henichart, J. P., Lesieur, D., Abadie, C., Rochette, L., Lindenbaum, A., Chalas, J., Renaud-de-la-Faverie, J. F., Pfeiffer, B., *et al.* (1994). Benzoselenazolinone derivatives designed to be glutathione peroxidase mimetics feature inhibition of cyclooxygenase/5-lipoxygenase pathways and anti-inflammatory activity. *J. Med. Chem.* 37, 2903–2911.

Glass, R. S., Farooqui, R., Sabahi, M., and Ehler, K. W. (1989). Formation of thiocarbonyl compounds in the reaction of ebselen oxide with thiols. *J. Org. Chem.* 54, 1092–1097.

Günzler, W. A., Steffens, G. J., Grossmann, A., Kim, S.-M. A., Ötting, F., Wendel, A., and Flohé, L. (1984). The amino-acid sequence of bovine glutathione peroxidase. *Hoppe-Seyler's Z. Physiol. Chem.* 365, 195–212.

Haenen, G. R. M. M., de Rooij, B. M., Vermeulen, N. P. E., and Bast, A. (1990). Mechanism of the reaction of ebselen with endogenous thiols: Dihydrolipoate is a better cofactor than glutathione in the peroxidase activity of ebselen. *Mol. Pharmacol.* 37, 412–422.

Hattori, R., Inoue, R., Sase, K., Eizawa, H., Kosuga, K., Aoyama, T., Masayasu, H., Kawai, C., Sasayama, S., and Yui, Y. (1994). Preferential inhibition of inducible nitric oxide synthase by ebselen. *Eur. J. Pharmacol.* 267, R1–2.

Hayashi, M., and Slater, T. F. (1986). Inhibitory effects of ebselen on lipid peroxidation in rat liver microsomes. *Free Rad. Res. Commun.* 2, 179–185.

Hemler, M. E., Cook, H. W., and Lands, W. E. M. (1979). Prostaglandin biosynthesis can be triggered by lipid peroxides. *Arch. Biochem. Biophys.* 193, 340–345.

Hu, J., Engman, L., and Cotgreave, I. A. (1995). Redox-active chalcogen-containing glutathione peroxidase mimetics and antioxidants inhibit tumour promoter-induced downregulation of gap junctional intercellular communication between WB-F344 liver epithelial cells. *Carcinogenesis* 16, 1815–1824.

Ischiropoulos, H., Zhu, L., and Beckman, J. S. (1992). Peroxynitrite formation from macrophage-derived nitric oxide. *Arch. Biochem. Biophys.* 298, 446–451.

John, N. J., Terlinden, R., Fischer, H., Evers, M., and Sies, H. (1990). Microsomal metabolism of 2-(methylseleno)benzanilide. *Chem. Res. Toxicol.* 3, 199–203.

Johshita, H., Sasaki, T., Matsui, T., Hanamura, T., Masayasu, H., and Asano, T. (1990). Effects of ebselen(PZ51) on ischemic brain edema after focal ischemia in cats. *Acta Neurochir. Suppl.* 51, 239–241.

Kamigata, N., Takata, M., Matsuyama, H., and Kobayashi, M. (1986). Novel ring opening reaction of 2-aryl-1,2-benziselenazol-3(2H)-one with thiols. *Heterocycles* 24, 3027–3030.

Kodama, K., Hirayama, A., and Masayasu, H. (1993). Inhibitor for restenosis after percutaneous coronary arterioplasty. *PCT Int. Appl.* WO 9313, 762 (Cl. A61K31/165).

Kooy, N. W., and Royall, J. A. (1994). Agonist-induced peroxynitrite production from endothelial cells. *Arch. Biochem. Biophys.* 310, 352–359.

Koppenol, W. H., Moreno, J. J., Pryor, W. A., Ischiropoulos, H., and Beckman, J. S. (1992). Peroxynitrite, a cloaked oxidant formed by nitric oxide and superoxide. *Chem. Res. Toxicol.* 5, 834–842.

Kuhl, P., Borbe, H. O., Fischer, H., Roemer, A., and Safayhi, H. (1986). Ebselen reduces the formation of LTB4 in human and porcine leukocytes by isomerisation to its 5S, 12R-6-trans-isomer. *Prostaglandins* 31, 1029–1048.

Kuhl, P., Borbe, H. O., Römer, A., Fischer, H., and Parnham, M. J. (1985). Selective inhibition of leukotriene B4 formation by ebselen: A novel approach to anti-inflammatory therapy. *Agents Actions* 17, 366–367.

Ladenstein, R., Epp, O., Bartels, K., Jones, A., Huber, R., and Wendel, A. (1979). Structure analysis and molecular model of the selenoenzyme glutathione peroxidase at 2.8 Å resolution. *J. Mol. Biol.* 134, 199–218.

Lambert, C., Cantineau, R., Christiaens, L., Biedermann, J., and Dereu, N. (1987). Syntheses and spectral characterization of the metabolites of a new organo-selenium drug: Ebselen. *Bull. Soc. Chim. Belg.* 96, 383–389.

Lesser, R., and Weiss, R. (1924). Über selenhaltige aromatische Verbindungen (VI). *Chem. Berichte* 57, 1077–1082.

Leyck, S., and Parnham, M. J. (1990). Acute antiinflammatory and gastric effects of the selenoorganic compound ebselen. *Agents Actions* 30, 426–431.

Maiorino, M., Aumann, K.-D., Brigelius-Flohé, R., Doria, D., van den Heuvel, J., McCarthy, J., Roveri, A., Ursini, F., and Flohé, L. (1995). Probing the presumed catalytic triad of selenium-containing peroxidases by mutational analysis of phospholipid hydroperoxide glutathione peroxidase (PHGPx). *Biol. Chem. Hoppe-Seyler* **376,** 651–660.

Maiorino, M., Roveri, A., Coassin, M., and Ursini, F. (1988). Kinetic mechanism and substrate specificity of glutathione peroxidase activity of ebselen (PZ51). *Biochem. Pharmacol.* **37,** 2267–2271.

Maiorino, M., Roveri, A., and Ursini, F. (1992). Antioxidant effect of ebselen (PZ51): Peroxidase mimetic activity on phospholipid and cholesterol hydroperoxides vs free radical scanveger activity. *Arch. Biochem. Biophys.* **295,** 404–409.

Masumoto, H., and Sies, H. (1996). The reaction of ebselen with peroxynitrite. *Chem. Res. Toxicol.* **9,** 262–267.

Mills, G. C. (1957). Hemoglobin catabolism. I. Glutathione peroxidase, an erythrocyte enzyme which protects hemoglobin from oxidase breakdown. *J. Biol. Chem.* **229,** 189–197.

Miyazawa, T., Suzuki, T., Fujimoto, K., and Kinoshita, M. (1993). Elimination of plasma phosphatidylcholine hydroperoxide by a seleno-organic compound, ebselen. *J. Biochem. Tokyo* **114,** 588–591.

Morgenstern, R., Cotgreave, I. A., and Engman, L. (1992). Determination of the relative contributions of the diselenide and selenol forms of ebselen in the mechanism of its glutathione peroxidase-like activity. *Chem.-Biol. Interact.* **84,** 77–84.

Müller, A., Cadenas, E., Graf, P., and Sies, H. (1984). A novel biologically active seleno-organic compound. I. Glutathione peroxidase-like activity in-vitro and antioxidant capacity of PZ 51 (ebselen). *Biochem. Pharmacol.* **33,** 3235–3239.

Müller, A., Gabriel, H., and Sies, H. (1985). A novel biologically active selenoorganic compound. IV. Protective glutathione-dependent effect of PZ 51 (ebselen) against ADP-Fe induced lipid peroxidation in isolated hepatocytes. *Biochem. Pharmacol.* **34,** 1185–1189.

Müller, A., Gabriel, H., Sies, H., Terlinden, R., Fischer, H., and Römer, A. (1988). A novel biologically active selenoorganic compound. VII. Biotransformation of ebselen in perfused rat liver. *Biochem. Pharmacol.* **37,** 1103–1109.

Narayanaswami, V., and Sies, H. (1990). Antioxidant activity of ebselen and related selenoorganic compounds in microsomal lipid peroxidation. *Free Rad. Res. Commun.* **10,** 237–244.

Nikawa, T., Schuch, G., Wagner, G., and Sies, H. (1994a). Interaction of albumin-bound ebselen with rat liver glutathione S-transferase and microsomal proteins. *Biochem. Mol. Biol. Int.* **32,** 291–298.

Nikawa, T., Schuch, G., Wagner, G., and Sies, H. (1994b). Interaction of ebselen with glutathione S-transferase and papain in vitro. *Biochem. Pharmacol.* **47,** 1007–1012.

Noguchi, N., Gotoh, N., and Niki, E. (1994). Effects of ebselen and probucol on oxidative modifications of lipid and protein of low density lipoprotein induced by free radicals. *Biochim. Biophys. Acta* **1213,** 176–182.

Noguchi, N., Yoshida, Y., Kaneda, H., Yamamoto, Y., and Niki, E. (1992). Action of ebselen as an antioxidant against lipid peroxidation. *Biochem. Pharmacol.* **44,** 39–44.

Nomura, H., Hakusui, H., and Takegoshi, T. (1989). Binding of ebselen to plasma protein. *In* "Selenium in Biology and Medicine" (A. Wendel, ed.), pp. 189–193. Springer-Verlag, Heidelberg.

Padmaja, S., Ramezanian, M. S., Bounds, P. L., and Koppenol, W. H. (1996). Reaction of peroxynitrite with L-tryptophan. *Redox Rep.* **2,** 173–177.

Parnham, M. J., Biedermann, J., Bittner, C., Dereu, N., Leyck, S., and Wetzig, H. (1989). Structure–activity relationships of a series of anti-inflammatory benzisoselenazolones (BISAs). *Agents Actions* **27,** 306–308.

Parnham, M. J., and Graf, E. (1987). Seleno-organic compounds and the therapy of hydroperoxide-linked pathological conditions. *Biochem. Pharmacol.* **36,** 3095–3102.

Ebselen **245**

Parnham, M. J., and Graf, E. (1991). Pharmacology of synthetic organic selenium compounds. *Prog. Drug Res.* **36**, 9–47.

Parnham, M. J., Leyck, S., Dereu, N., Winkelmann, J., and Graf, E. (1985). Ebselen (PZ 51): A GSH-peroxidase-like organoselenium compound with anti-inflammatory activity. *Adv. Inflam. Res.* **10**, 397–400.

Pryor, W. A., Jin, X., and Squadrito, G. L. (1994). One- and two-electron oxidations of methionine by peroxynitrite. *Proc. Natl. Acad. Sci. USA* **91**, 11173–11177.

Radi, R., Beckman, J. S., Bush, K. M., and Freeman, B. A. (1991). Peroxynitrite oxidation of sulfhydryls. *J. Biol. Chem.* **266**, 4244–4250.

Ramezanian, M. S., Padmaja, S., and Koppenol, W. H. (1996). Nitration and hydroxylation of phenolic compounds by peroxynitrite. *Chem. Res. Toxicol.* **9**, 232–240.

Reich, H. J., and Jasperse, C. P. (1987). Organoselenium chemistry: Redox chemistry of selenocysteine model systems. *J. Am. Chem. Soc.* **109**, 5549–5551.

Rotruck, J. T., Pope, A. L., Ganther, H. E., Swanson, A. B., Hafeman, D. G., and Hoekstra, W. G. (1973). Selenium: Biochemical role as a component of glutathione peroxidase. *Science* **179**, 588.

Roussyn, I., Briviba, K., Masumoto, H., and Sies, H. (1996). Selenium-containing compounds protect DNA from damage caused by peroxynitrite. *Arch. Biochem. Biophys.* **330**, 216–218.

Safayhi, H., Tiegs, G., and Wendel, A. (1985). A novel biologically active seleno-organic compound. V. Inhibition by ebselen (PZ 51) of rat peritoneal neutrophil lipoxygenase. *Biochem. Pharmacol.* **34**, 2691–2694.

Saito, I., Abe, H., Yoshimoto, T., Asano, T., Takakura, K., Ohta, T., Kikuchi, H., and Sano, K. (1995). Multicenter randomized clinical trial of ebselen with aneurysmal subarachnoidal hemorrhage. *J. Cereb. Blood Flow Metab.* **15**, S162.

Sattler, W., Maiorino, M., and Stocker, R. (1994). Reduction of HDL- and LDL-associated cholesterylester and phospholipid hydroperoxides by phospholipid hydroperoxide glutathione peroxidase and ebselen (PZ 51). *Arch. Biochem. Biophys.* **309**, 214–221.

Schewe, C., Schewe, T., and Wendel, A. (1994). Strong inhibition of mammalian lipoxygenases by the anti-inflammatory seleno-organic compound ebselen in the absence of glutathione. *Biochem. Pharmacol.* **48**, 65–74.

Schewe, T. (1995). Molecular actions of ebselen: An anti-inflammatory antioxidant. *Gen. Pharmacol.* **26**, 1153–1169.

Schöneich, C., Narayanaswami, V., Asmus, K.-D., and Sies, H. (1990). Reactivity of ebselen and related senoorganic compounds with 1,2-dichloroethane radical cations and halogenated peroxyl radicals. *Arch. Biochem. Biophys.* **282**, 18–25.

Schuckelt, R., Brigelius-Flohé, R., Maiorino, M., Roveri, A., Reumkens, J., Strassburger, W., Ursini, F., Wolf, B., and Flohé, L. (1991). Phospholipid hydroperoxide glutathione peroxidase is a seleno-enzyme distinct from the classical glutathione peroxidase as evident from cDNA and amino acid sequencing. *Free Rad. Res. Commun.* **14**, 343–361.

Scurlock, R., Rougée, M., Bensasson, R. V., Evers, M., and Dereu, N. (1991). Deactivation of singlet molecular oxygen by organ-selenium compounds exhibiting glutathione peroxidase activity and by sulfur-containing homologs. *Photochem. Photobiol.* **54**, 733–736.

Sies, H. (1989). Metabolism and disposition of ebselen. *In* "Selenium in Biology and Medicine" (A. Wendel, ed.), pp. 153–162. Springer-Verlag, Heidelberg.

Sies, H. (1993). Ebselen, a selenoorganic compound as glutathione peroxidase mimic. *Free Rad. Biol. Med.* **14**, 313–323.

Tabuchi, Y., Ogasawara, T., and Furuhama, K. (1994). Mechanism of the inhibition of hog gastric (H$^+$,K$^+$)-ATPase by the seleno-organic compound ebselen. *Arzneimittelforschung* **44**, 51–54.

Tabuchi, Y., Sugiyama, N., Horiuchi, T., Furusawa, M., and Furuhama, K. (1995). Ebselen, a seleno-organic compound, protects against ethanol-induced murine gastric mucosal injury in both in vivo and in vitro systems. *Eur. J. Pharmacol.* **272**, 195–201.

Takahashi, K., Akasaka, M., Yamamoto, Y., Kobayashi, C., Mizoguchi, J., and Koyama, J. (1990). Primary structure of human plasma glutathione peroxidase deduced from cDNA sequences. *J. Biochem.* **108**, 145–148.

Takahashi, K., and Cohen, H. J. (1986). Selenium-dependent glutathione peroxidase protein and activity: Immunological investigations on cellular and plasma enzymes. *Blood* **68**, 640–645.

Tarino, J. Z. (1986). Evaluation of selected organic and inorganic selenium compounds for selenium-dependent glutathione peroxidase activity. *Nutr. Rep. Int.* **33**, 299–306.

Terlinden, R., Feige, M., and Römer, A. (1988). Determination of the two major metabolites of ebselen in human plasma by high-performance liquid chromatography. *J. Chromatogr.* **430**, 438–442.

Thomas, C. E., and Jackson, R. L. (1991). Lipid hydroperoxide involvement in copper-dependent and independent oxidation of low density lipoproteins. *J. Pharmacol. Exp. Ther.* **256**, 1182–1188.

Thomson, L., Trujillo, M., Telleri, R., and Radi, R. (1995). Kinetics of cytochrome c^{2+} oxidation by peroxynitrite: Implications for superoxide measurements in nitric oxide-producing biological systems. *Arch. Biochem. Biophys.* **319**, 491–497.

Ursini, F., Maiorino, M., and Gregolin, C. (1985). The selenoenzyme phospholipid hydroperoxide glutathione peroxidase. *Biochim. Biophys. Acta* **839**, 62–70.

Ursini, F., Maiorino, M., Valente, M., Ferri, L., and Gregolin, C. (1982). Purification from pig liver of a protein which protects liposomes and biomembranes from peroxidative degradation and exhibits glutathione peroxidase activity on phosphatidylcholine hydroperoxides. *Biochim. Biophys. Acta* **710**, 197–211.

van Caneghem, P. (1974). Influences comparatives de différentes substances séléniées et soufrées sur la fragilité des lysosomes et des mitochondries in vitro. *Biochem. Pharmacol.* **23**, 3491–3500.

Wagner, G., Schuch, G., Akerboom, T. P. M., and Sies, H. (1994). Transport of ebselen in plasma and its transfer to binding sites in the hepatocyte. *Biochem. Pharmacol.* **48**, 1137–1144.

Wang, J.-F., Komarov, P., Sies, H., and de Groot, H. (1991). Contribution of nitric oxide synthase to luminol-dependent chemiluminescence generated by phorbol-ester-activated Kupffer cells. *Biochem. J.* **279**, 311–314.

Wang, J.-F., Komarov, P., Sies, H., de Groot, H. (1992). Inhibition of superoxide and nitric oxide release and protection from reoxygenation injury by ebselen in rat Kupffer cells. *Hepatology* **15**, 1112–1116.

Weber, R., and Renson, M. (1976). Les chlorures de chloro-3 benzisosélénazolium-1,2: Synthèse, hydrolyse, thiolation, ammonolyse. *Bull. Soc. Chim. France* **2/8**, 1124–1126.

Wendel, A., Fausel, M., Safayhi, H., Tiegs, G., and Otter, R. (1984). A novel biologically active seleno-organic compound. II. Activity of PZ 51 in relation to glutathione peroxidase. *Biochem. Pharmacol.* **33**, 3241–3245.

Wendel, A., Otter, R., and Tiegs, G. (1986). Inhibition by ebselen of microsomal NADPH-cytochrome P 450-reductase in vitro but not in vivo. *Biochem. Pharmacol.* **35**, 2995–2997.

Wendel, A., and Tiegs, G. (1986). A novel biologically active seleno-organic compound. VI. Protection by ebselen (PZ 51) against galactosamine/endotoxin-induced hepatitis in mice. *Biochem. Pharmacol.* **35**, 2115–2118.

Wilson, S. R., Zucker, P. A., Huang, R.-R. C., and Spector, A. (1989). Development of synthetic compounds with glutathione peroxidase activity. *J. Am. Chem. Soc.* **111**, 5936–5939.

Zembowicz, A., Hatchett, R. J., Radziszewski, W., and Gryglewski, R. J. (1993). Inhibition of endothelial nitric oxide synthase by ebselen: Prevention by thiols suggests the inactivation by ebselen of a critical thiol essential for the catalytic activity of nitric oxide synthase. *J. Pharmacol. Exp. Ther.* **267**, 1112–1118.

Ziegler, D. M., Graf, P., Poulsen, L. L., Stahl, W., and Sies, H. (1992). NADPH-dependent oxidation of reduced ebselen, 2-selenylbenzanilide, and of 2-(methylseleno)benzanilide catalyzed by pig liver flavin-containing monooxygenase. *Chem. Res. Toxicol.* **5**, 163–166.

Susan R. Doctrow*
Karl Huffman*
Catherine B. Marcus*
Wael Musleh[†]
Annadora Bruce[†]
Michel Baudry[†]
Bernard Malfroy*

* Eukarion, Inc.
Bedford, Massachusetts 01730
[†] Neuroscience Program
University of Southern California
Los Angeles, California 90089

Salen–Manganese Complexes: Combined Superoxide Dismutase/ Catalase Mimics with Broad Pharmacological Efficacy

I. Introduction

Molecular oxygen, although essential for aerobic metabolism, can be converted to potentially deleterious substances, including superoxide ion, hydrogen peroxide, and hydroxyl radical, known collectively as reactive oxygen species (ROS). Increased ROS formation under pathological conditions is believed to cause cellular damage through chemical interactions with proteins, lipids, and DNA (reviewed in Halliwell and Gutteridge, 1989). In inflammatory states, ROS can be formed via a number of pathways, including activation of the NADPH oxidase system in activated polymorphonuclear neutrophils and monocyte/macrophages (Klebanoff, 1988; Anderson et al., 1991). ROS have also been implicated as being key mediators of tissue injury after ischemia and reperfusion. Under these conditions, ROS are believed to be generated initially by the enzyme xanthine oxidase acting on its endogenous substrates but, later, by infiltrating leukocytes as well (McCord, 1987; Granger, 1988). Some degree of tissue protection is provided by endogenous ROS scavengers such as superoxide dismutases (SOD) and catalase, which catalytically destroy superoxide ion and hydrogen peroxide, respectively. Also, molecules synthesized endogenously or acquired through diet, such as ascorbate (vitamin C), α-tocopherol (vitamin E), and glutathione, have antioxidant properties. Presumably, however, when ROS generation exceeds the capacity of endogenous scavengers to neutralize them, tissues become vulnerable to damage, a condition often referred to as "oxidative stress" (Sies, 1985, 1986). Consequently, antioxidant therapies are being investigated for a wide variety of disorders in which oxidative stress is believed to play a significant role in the associated pathophysiologies (reviewed in Rice-Evans and Diplock, 1993).

Various studies have investigated the clinical efficacy of SOD but, in many instances, the results of *in vivo* studies have been disappointing because, as a protein, SOD has a number of delivery and stability shortcomings. In part because of equivocal results obtained with SOD, there is an interest in developing synthetic SOD mimics that should have more favorable pharmaceutical properties than a protein such as, for example, cost-effectiveness, stability, and oral availability. The first low molecular weight SOD mimics to be described were copper complexes that showed antitumor-promoting activity in mice (Kensler *et al.*, 1983; Yamamoto *et al.*, 1990). More recently, several manganese complexes purported to mimic the active site of manganese–SOD have been described. The complex formed between manganese and the chelator desferrioxamine has been reported as having SOD activity (Darr *et al.*, 1987) and has shown some protective activity against kainate-induced neurotoxicity in the rat (Bruce *et al.*, 1992). Salen–manganese complexes, the focus of this chapter (Baudry *et al.*, 1993), and porphyrin–manganese complexes (Faulkner *et al.*, 1994) have also been reported to have SOD activity as well as more favorable stability properties than the earlier complexes. A class of manganese macrocyclic ligand complexes with SOD activity also has been described (Riley and Weiss, 1994) with one prototype reportedly showing protection in a model for myocardial ischemia–reperfusion injury (Black *et al.*, 1994).

As mentioned earlier, we have previously reported that synthetic salen–manganese have SOD activity (Baudry *et al.*, 1993) and, more recently, have found that these compounds have catalase activity as well (Malfroy-Camine and Baudry, 1995; Gonzalez *et al.*, 1995b). In this regard, the complexes may mimic properties of bacterial manganese-containing catalases, so-called "pseudocatalases" (Halliwell and Gutteridge, 1989; Dismukes, 1993). Certain other manganese complexes have been reported to possess catalase activity when assayed in organic solvents, having been designed as model compounds rather than as potential pharmaceutical agents (Mather *et al.*, 1987; Pessiki and Dismukes, 1994).

Thus, salen–manganese complexes have several characteristics that might facilitate their potential usefulness as therapeutic agents. First, as low molecular weight, synthetic molecules rather than proteinaceous antioxidant enzymes they have the potential advantages mentioned earlier that relate to cost-effectiveness and pharmaceutical formulation and delivery. Second, they act catalytically, presumably enhancing their efficiency over noncatalytic low-molecular-weight ROS scavengers such as, for example, vitamin E. Third, their ability to destroy both superoxide anion and hydrogen peroxide should enhance their protective potential in various disease states involving the production of multiple ROS species.

In this chapter, we first describe the catalytic properties of EUK-8, a prototype salen–manganese complex. We also illustrate, by the use of selected examples, the efficacy of EUK-8 in experimental models for disease. Finally, we briefly address future directions in the development of salen–manganese complexes as novel, broadly applicable potential therapeutic agents.

II. Multiple Catalytic Activities of EUK-8, a Prototype SOD/Catalase Mimic

A. EUK-8 and Other Salen–Manganese Complexes Exhibit Superoxide Scavenging Activity

EUK-8, a salen–manganese complex with the structure shown in Fig. 1, may be regarded as a prototype molecule of a new class of synthetic catalytic scavengers with combined SOD/catalase activity. The SOD activity of EUK-8 and other salen–manganese complexes (Baudry et al., 1993) has been demonstrated using a coupled "indirect" assay method (McCord et al., 1973). In this procedure, xanthine and xanthine oxidase continuously generate superoxide, which is monitored by its ability to reduce an indicator molecule to a spectrophotometrically detectable product. Addition of an agent with SOD activity results in suppression of the rate of absorbance change. As illustrated in Fig. 2a, EUK-8 suppresses the rate of reduction of the indicator molecule nitro blue tetrazolium (NBT) in such an assay system.

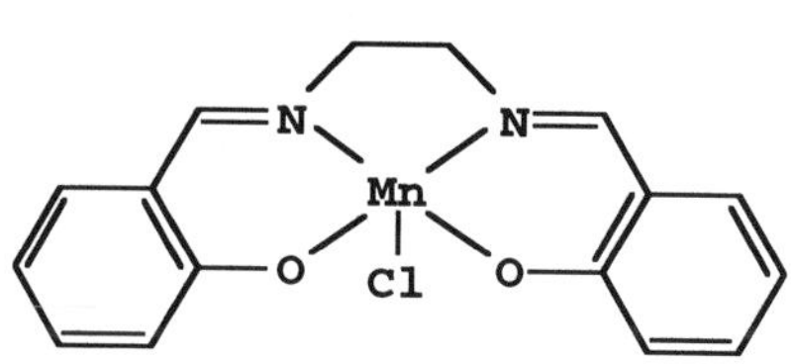

FIGURE I   Structure of the salen–manganese complex, EUK-8. [Reprinted from Gonzalez et al. (1995b). EUK-8, a synthetic superoxide dismutase and catalase mimetic, ameliorates acute lung injury in endotoxemic swine. *J. Pharmacol. Exp. Therap.* 275, 798–806.]

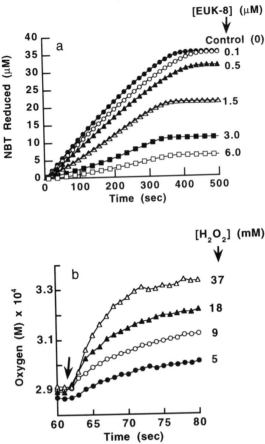

FIGURE 2   SOD and catalase activities exhibited by EUK-8. (a) EUK-8 was assayed for SOD activity as described previously (Baudry *et al.*, 1993) using xanthine and xanthine oxidase as a superoxide generating system and monitoring the superoxide-dependent reduction of NBT. The figure shows NBT reduction in the presence of various concentrations of EUK-8 or in its absence (control). EUK-8, tested at up to 12 μM, did not affect xanthine oxidase activity, based on the spectrophotometric detection of urate in equivalent reaction mixtures (data not shown). (b) EUK-8 was assayed for catalase activity as described previously (Gonzalez *et al.*, 1995b), measuring conversion of hydrogen peroxide to oxygen using a Clark-type polarographic oxygen electrode. The concentration of EUK-8 was 10 μM and the concentration of hydrogen peroxide (added at arrow) was as indicated. (c) EUK-8 was assayed for peroxidase activity in the reaction mixture employed for the catalase assay with hydrogen peroxide as indicated and with 0.5 mM 2,2'-azino-bis(3-ethylbenzthiazoline-6-sulfonic acid) (ABTS) added. The oxidation of ABTS was monitored spectrophotometrically (Childs and Bardsley, 1975).

The apparent SOD activity of EUK-8 is not due to a direct inhibition of xanthine oxidase, as EUK-8 has no effect on the activity of this enzyme, assayed by spectrophotometric detection of its other product, urate (data not shown). This observation supports the interpretation that EUK-8 exhib-

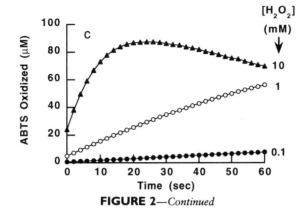

FIGURE 2—*Continued*

its true superoxide scavenging activity. The catalytic nature of this superoxide scavenging activity is supported by the observation that EUK-8 inhibits the reduction of a stoichiometric excess of NBT. (Under conditions such as those shown in Fig. 2a, the reactions cease because of xanthine depletion, resuming at similar relative rates if the substrate is replenished.) Similar results are observed if oxidized cytochrome c, rather than NBT, is used as the indicator substance.

B. Catalase Activity of EUK-8: Catalatic and Peroxidatic Reaction Pathways

EUK-8 also exhibits catalase activity, based on its ability to generate oxygen in the presence of hydrogen peroxide. As illustrated in Fig. 2b, the rate of oxygen production is dependent on hydrogen peroxide concentration and, as with the SOD reaction, a catalytic mechanism is supported by the generation of a stoichiometric excess of oxygen. Under assay conditions such as those shown in Fig. 2b, reactions cease prior to depletion of hydrogen peroxide, resuming only if more EUK-8 is added. Consistent with this observation, analysis of reaction mixtures by HPLC indicates that EUK-8 decomposes in the presence of the millimolar levels of hydrogen peroxide utilized in these *in vitro* catalase assays. Whether such decomposition would occur under conditions of *in vivo* hydrogen peroxide generation is not clear. However, salen–manganese complexes exhibiting superior *in vitro* catalase activity to EUK-8 include, not surprisingly, compounds that are resistant to such decomposition. The catalase activity of EUK-8 exhibits no apparent substrate saturation, with the initial reaction rate as well as the total number of reaction turnovers increasing as hydrogen peroxide is increased even into the molar range. Similarly, mammalian catalases have been described as exhibiting kinetics not saturable with respect to substrate (Chance *et al.*, 1979).

Consistent with its function as a catalase, EUK-8 also exhibits peroxidase activity. Catalases are well known to be capable of undergoing peroxidative as well as catalatic reactions (reviewed in Chance *et al.*, 1979). Although most such studies have focused on the heme-containing mammalian enzymes, the bacterial manganese catalases are also known to conduct peroxidative reactions (Halliwell and Gutteridge, 1989). This property is attributed to the formation of the active species compound I on reaction of the catalase with one molar equivalent of hydrogen peroxide. Spectrophotometric and kinetic evidence supports the formation of compound I, which is regarded as an active intermediate presumably formed between the active-site metal, substrate-binding moieties, and the first substrate molecule. Reaction of compound I with a second molecule of hydrogen peroxide constitutes the complete catalase reaction. Alternatively, compound I can react with certain electron donors, resulting in a peroxidative reaction whereby the donor substrate becomes oxidized. In this manner, in the presence of a suitable substrate, the catalase-driven decomposition of hydrogen peroxide proceeds via partitioning between catalatic and peroxidatic reaction pathways (Oshino *et al.*, 1973).

As shown in Fig. 2c, EUK-8 catalyzes a peroxidative reaction between hydrogen peroxide and the oxidizable substrate 2,2'-azino-bis(3-ethylbenzthiazoline-6-sulfonic acid) (ABTS). As with its catalase activity, the peroxidase activity of EUK-8 is dependent on the hydrogen peroxide concentration, with no apparent saturation reached at any concentration tested, up to 10 mM in Fig. 2c, and by up to 100 mM in other experiments. At high hydrogen peroxide concentrations, however, the kinetics of ABTS oxidation are complicated by the apparent bleaching of the oxidized product, which is known to spontaneously disproportionate (Childs and Bardsley, 1975). Under equivalent assay conditions, bovine liver catalase showed no detectable peroxidase activity toward ABTS. The catalase proteins reportedly have a limited range of suitable substrates, including short-chain alcohols which become oxidized to aldehydes (Schonbaum and Chance, 1976). Such narrow substrate specificity is likely due to steric factors, predicting that a small molecule such as EUK-8 should recognize a broader range of substrates than the proteinaceous catalases. Indeed, EUK-8 also exhibits peroxidase activity toward other donors such as, for example, reduced cytochrome c. This observation raises the concern that, in the SOD assay system described earlier, the peroxidative activity of EUK-8 toward the indicator molecule might produce an artifactual "false positive." However, the addition of bovine liver catalase to these reactions had no effect either on the rate of NBT or cytochrome c reduction or on its inhibition by EUK-8, indicating that endogenously generated hydrogen peroxide is not interfering in this manner and further supporting the conclusion that EUK-8 exhibits true SOD activity.

The relative contributions of each of these catalytic activities of EUK-8 to biological effects, such as those discussed in Sections II,C and III, remain to be determined. However, the ability of EUK-8 to conduct catalatic as well as peroxidatic reactions does support the conclusion that the molecule forms a compound I-like intermediate and, hence, functions as a true catalase mimic.

C. *In vitro* Studies Suggest that Salen–Manganese Complexes Have Advantages over Other Antioxidants

The ability of EUK-8 to suppress cellular lipid peroxidation was examined by measuring levels of malonyldialdehyde (MDA), a lipid peroxidation by-product (Esterbauer and Cheeseman, 1990; Janero, 1990; Ceconi et al., 1991). Isolated hippocampal slices were subjected to lactic acidosis, which induced a significant increase in tissue MDA, as well as MDA released into the medium (Fig. 3). This finding is consistent with those of others who have observed increased lipid peroxidation in tissues subjected to acidosis (Bralet et al., 1991; Siesjo et al., 1985). As Fig. 3 shows, this apparent increase in lipid peroxidation was prevented in slices treated with EUK-8 (Musleh et al., 1994). Lactic acidosis-induced oxidative stress is believed to result from the pH-dependent dissociation of intracellular bound iron (Bralet

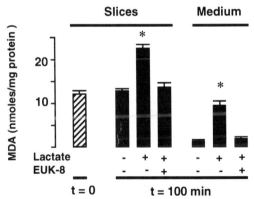

FIGURE 3   EUK-8 inhibits lactic acidosis-induced lipid peroxidation in hippocampal slices. The experiment was conducted as described (Musleh et al., 1994). Rat hippocampal slices were incubated for 100 min in a physiological buffer (pH 7.4) (control) or in a lactate buffer (pH 5, 30 mM lactic acid). EUK-8, where present, was at a concentration of 50 μM. At the end of the incubation, MDA content was determined using the thiobarbituric acid reagent as described previously (Musleh et al., 1994). The data shown represent the means ± SEM of six experiments.*$p < 0.01$, significantly different from control (Student's t test). (Reprinted from *Neuropharmacology* 33, Musleh et al., Effects of EUK-8, a synthetic catalytic superoxide scavenger, on hypoxia- and acidosis-induced damage in hippocampal slices, pp 929–934, Copyright 1994, with kind permission from Elsevier Science Ltd, The Boulevard, Langford Lane, Kidlington OX5 1GB, UK.)

et al., 1992). The resulting free iron can catalyze the generation of intracellular ROS (Halliwell and Gutteridge, 1989) and, ultimately, cellular lipid peroxidation (Braughler *et al.*, 1986). The fact that EUK-8 is a metal–ligand complex suggests the possibility that its protective effects might be related to iron chelation by a dissociated salen ligand. A number of observations are, however, inconsistent with this hypothesis. First, EUK-8 is extremely stable in solution, even under acidic conditions, e.g., having a half-life of >15 hr at pH 1.5. It is therefore unlikely to dissociate spontaneously even under conditions of lactic acidosis at pH 5. Furthermore, HPLC analysis has indicated that there is no detectable exchange of metals upon incubation of EUK-8 with iron salts. Finally, a salen–iron complex analogous to EUK-8 is unstable in aqueous solution. Thus, the antioxidant activity of EUK-8 is more likely to explain its ability to protect tissues subjected to oxidative stress than is an iron chelation effect.

EUK-8 also inhibits MDA production in brain membranes treated with iron and ascorbate in an oxygen-enriched environment. As shown in Fig. 4, EUK-8 was several orders of magnitude more potent than vitamin E in this experimental system. Again, the possibility of EUK-8 acting as an iron chelator in this assay system is not likely because of the previously mentioned considerations. In addition, the metal-free salen ligand was found to be protective in this assay system, but only at concentrations about 10-fold greater than the effective inhibitory concentrations of EUK-8 (data not shown).

The ability of EUK-8 to afford functional tissue protection was examined in an electrophysiological study employing rat hippocampal slices (Musleh

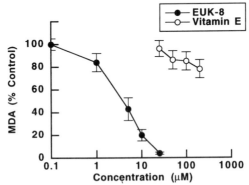

FIGURE 4   EUK-8 inhibits iron-induced lipid peroxidation in brain membranes. Rat brain membranes, prepared by homogenization of brains minus the pons and cerebellum, were incubated in an artificial cerebrospinal fluid (ASCF, Musleh *et al.*, 1994) containing 1 mM L-ascorbate and 10 μM FeCl$_2$ for 1 hr at 35°C under an atmosphere equilibrated with O$_2$:CO$_2$ (95:5). EUK-8 (●) or α-tocopherol (vitamin E, O) was also present, as indicated. Following the incubation, samples were extracted with trichloroacetic acid and assayed for MDA as described for Fig. 3.

et al., 1994). Neuronal damage was induced by subjecting the slices to periods of anoxia followed by reoxygenation. Synaptic integrity was assessed electrophysiologically by measuring, through recording electrodes placed in the CA1 striatum radiatum, excitatory postsynaptic potentials (EPSPs) generated upon stimulation of the Schaffer-commissural pathway. Both the amplitude and slope of the EPSP were compared to those obtained prior to anoxia. As shown in Fig. 5a, the anoxia–reoxygenation treatment in control medium resulted in a marked suppression of the EPSP signal, with only partial recovery of function during about 50 min of reoxygenation. Under the same conditions, slices in medium containing EUK-8 exhibited significantly greater recovery of the EPSP signal, with both amplitude and slope reaching >80% preanoxic levels during the reoxygenation period.

In contrast to the neuroprotective effect of EUK-8 in this model, the effect of SOD was deleterious. In a second set of experiments illustrated in

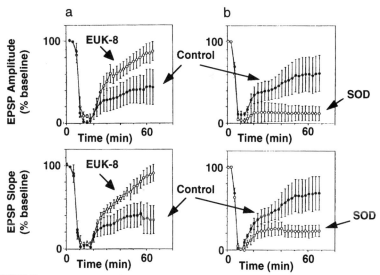

FIGURE 5 EUK-8 protects hippocampal slices from anoxia-induced synaptic dysfunction, whereas SOD exacerbates the damage. (a) The experiments were conducted as described (Musleh *et al.,* 1994). Hippocampal slices were incubated in ASCF with or without EUK-8 (50 μM), as indicated. Following a baseline recording, an anoxic episode was administered, followed by a reoxygenation period. Evoked postsynaptic potentials (EPSPs) were recorded from the slices. The figure shows both the amplitudes and the slopes of the EPSPs recorded at intervals throughout the treatment, presenting the means ± SEM for five experiments. [Reprinted from Musleh *et al.* (1994).] (b) Experiments were conducted as described for (a) except that slices were incubated in ASCF with or without bovine CuZn SOD, as indicated. The amount of SOD added was that which provided an equivalent SOD activity as did 50 μM EUK-8 (about 38 units; Baudry *et al.,* 1993). In addition, the anoxic period was shorter than that employed for (a) in order to obtain a higher level of EPSP recovery in the control slices than in (a).

Fig. 5b, slices exposed to SOD exhibited poorer recovery from anoxia–reoxygenation than those in control medium. One possible explanation for the damaging effect of SOD would be the generation of hydrogen peroxide via the superoxide dismutation reaction. Hydrogen peroxide is itself toxic and can also form the highly toxic hydroxyl radical via, for example, the Fenton reaction between hydrogen peroxide and free Fe^{2+} (Halliwell and Gutteridge, 1989). The mechanism of SOD-induced toxicity has not been investigated further with the hippocampal slice system. Studies in another experimental system have shown that acidosis induces an oxidative injury to Caco-2BBE monolayers, a model for the intestinal mucosa, and that EUK-8 and catalase are both protective (Gonzalez *et al.*, 1995a). In the same system, SOD either had no effect or severely exacerbated the acidosis-induced damage whereas combinations of SOD plus catalase were uniformly protective (Gonzalez *et al.*, 1995a; unpublished results). Taken together, these observations suggest that compounds exhibiting dual SOD/catalase activity might have distinct, qualitative advantages over those having only SOD activity. Specifically, potential toxicities due to SOD-generated hydrogen peroxide would be avoided with such compounds.

III. Efficacy of EUK-8 in Experimental Models for Disease

EUK-8 has been shown to be protective in numerous models for disease processes associated with oxidative stress, including inflammation, cardiovascular diseases, acute critical illness, autoimmune diseases, and neurological disorders. This chapter focuses on select examples from two of these broad categories, namely acute critical illness and neurological disorders.

A. EUK-8 Is Protective in a Model for Adult Respiratory Distress Syndrome

The adult respiratory distress syndrome (ARDS) is a common complication of acute medical conditions such as trauma or sepsis and has a very high mortality rate (Suchyta *et al.*, 1992; Hyers, 1993; Fowler *et al.*, 1983; Montgomery *et al.*, 1988). So far, there are no adequate therapies for ARDS, with available treatment being primarily supportive. ROS, most likely derived from inflammatory cells infiltrating the lung parenchyma, are believed to be responsible for the pathogenesis of ARDS (Simon and Ward, 1992). The involvement of ROS is supported by several observations in ARDS patients, e.g., elevated plasma levels of lipofuscin (Roumen *et al.*, 1994) and lipid peroxides (Richard *et al.*, 1990), decreased levels of endogenous ROS scavengers in the plasma (Bertrand *et al.*, 1989; Cross *et al.*, 1990; Richard *et al.*, 1990), decreased amounts of reduced glutathione in the alveolar fluid

(Pacht *et al.*, 1991), and excretion of hydrogen peroxide in exhaled breath or urine (Sznajder *et al.*, 1989; Keitzmann *et al.*, 1993; Mathru *et al.*, 1994; Baldwin *et al.*, 1986).

Gonzalez *et al.* (1995b) evaluated EUK-8 in a highly stringent porcine model for sepsis-induced ARDS. This experimental system, involving the induction of acute lung injury in pigs by endotoxin infusion, is characterized by the development of dysfunctions analogous to those seen in ARDS patients, namely severe and persistent pulmonary hypertension, decreased dynamic pulmonary compliance, pulmonary edema, and profound arterial hypoxemia (Wollert *et al.*, 1993, 1994; VanderMeer *et al.*, 1994, 1995). EUK-8 was administered intravenously using two different doses as described in the figure legends. As shown in Fig. 6, both doses of EUK-8 were protective, significantly reducing pulmonary hypertension, arterial hypoxemia, and loss of dynamic pulmonary compliance. Arterial hypoxemia, arguably among the most important indicators of pulmonary dysfunction, was abrogated by EUK-8, particularly at the higher dose. Both doses of EUK-8 were also found to abrogate pulmonary edema, as measured by the water content of lung tissue harvested at the end of the experiment (data not shown).

To address the proposed mechanism of action of EUK-8, the lung tissue was also analyzed for lipid peroxidation by measuring the content of MDA. As shown in Fig. 7, EUK-8 abrogated the increase in lung MDA, indicating that it prevented tissue lipid peroxidation. Analysis of data for individual animals further revealed that MDA levels correlated directly with lung water content ($r^2 = 0.768$, $p < 0.0005$) and inversely with arterial oxygenation at 240 min ($r^2 = -0.814$, $p < 0.0005$), lending further support to the importance of oxidative stress in the observed pulmonary pathologies.

Neither dose of EUK-8 was found to significantly affect the circulating levels of tumor necrosis factor-α, thromboxane-B_2, or the stable prostacyclin metabolite 6-ketoprostaglandin $F_{1\alpha}$, all of which become elevated by endotoxin treatment in this experimental model (Gonzalez *et al.*, 1995b). This observation supports the concept that EUK-8 protects pulmonary function by interfering with a deleterious agent(s) generated "downstream" in the inflammatory cascade, e.g., ROS, rather than with early inflammatory mediators.

This notion that the action of EUK-8 occurs relatively late in the inflammatory cascade would predict that even delayed intervention with this agent would be protective. This hypothesis is supported by a study (Gonzalez *et al.*, 1995c) showing that EUK-8 was highly effective against endotoxin-induced acute lung injury even when administered only after the endotoxin infusion had ended. The results of both studies indicate that salen–manganese complexes such as EUK-8 might offer an effective therapeutic approach to ARDS and, perhaps, to treating related multiple organ dysfunctions of similar etiology.

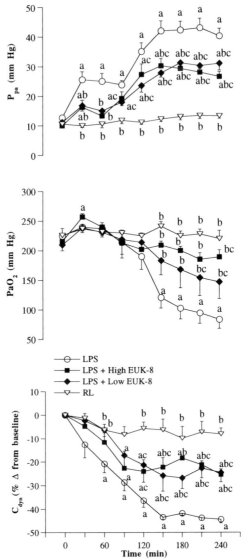

FIGURE 6 EUK-8 suppresses physiological indicators of pulmonary dysfunction in a porcine model for ARDS. The experiment was conducted as described (Gonzalez *et al.*, 1995b). Acute lung injury was induced in pigs with an infusion of endotoxin (250 μg/kg) from times 0 to 60 min (LPS group, ○). A control group received the vehicle and Ringer's lactate solution, without LPS (RL group, ▽). EUK-8-treated groups received a bolus (10 mg/kg) injection at time 0, followed by an infusion with either 1 (low EUK-8, ◆) or 3 mg/kg/hr (high EUK-8, ■) throughout the experiment. Pulmonary arterial pressure (P_{pa}), arterial oxygenation (P_{aO_2}), and dynamic pulmonary compliance (C_{dyn} as a percentage change from pretreatment baseline) were monitored as described (Gonzalez *et al.*, 1995b). Data presented are the means ± SEM for $n = 5$ (RL group) or $n = 6$ (all other groups). The results of statistical analyses, conducted as described (Gonzalez *et al.*, 1995b), were (a) $p < 0.05$ vs. baseline, (b) $p < 0.05$ vs time-matched value in the LPS group, and (c) $p < 0.05$ vs time-matched value in the RL group. [Reprinted from Gonzalez *et al.* (1995b). EUK-8, a synthetic superoxide dismutase and catalase mimetic, ameliorates acute lung injury in endotoxemic swine. *J. Pharmacol. Exp. Therap.* 275, 798–806.]

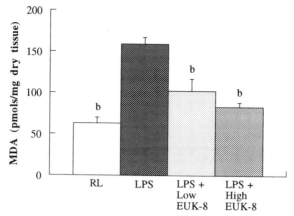

FIGURE 7   EUK-8 reduces lung tissue malonyldialdehyde (MDA) content in a porcine model for ARDS. The experimental details were as described for Fig. 6. At 240 min, lung tissue was harvested and analyzed for MDA as described (Gonzalez *et al.*, 1995b). Groups and symbols are the same as in Fig. 6, including applicable statistical tests (b and c). Note that the MDA values for the groups treated with EUK-8 did not differ significantly from the value for the RL group. [Reprinted from Gonzalez *et al.* (1995b). EUK-8, a synthetic superoxide dismutase and catalase mimetic, ameliorates acute lung injury in endotoxemic swine. *J. Pharmacol. Exp. Therap.* 275, 798–806.]

B. EUK-8 Is Protective in Models for Neurological Disorders

The brain is considered to be especially susceptible to oxidative stress, perhaps because of high oxygen consumption coupled to relatively low levels of endogenous antioxidative capacity (Jesberger and Richardson, 1990; Maestro and McDonald, 1987). ROS are thought to contribute to pathologies occurring as a result of neurological insults, such as stroke and trauma (Flamm *et al.*, 1978; Granger *et al.*, 1986; Hall and Braughler, 1989). In addition, there is increasing evidence implicating ROS in chronic neurodegenerative disorders such as Parkinson's (Dexter *et al.*, 1989; Adams and Odunze, 1991) and Alzheimer's (Behl *et al.*, 1992, 1994; Butterfield *et al.*, 1994a,b; Goodman *et al.*, 1994) diseases. Because of the susceptibility of the brain to oxidative damage, as well as the profound lack of effective treatments for many neurological disorders, several studies have focused on the potential neuroprotective effects of salen–manganese complexes such as EUK-8.

Organotypic hippocampal slice cultures, isolated from rat pups and cultured for up to 4 weeks, have been employed as experimental systems in which to model various types of disease-associated neuronal damage and to assess the effects of EUK-8. These preparations offer many of the experimental advantages of an *in vitro* culture system, such as accessibility, while preserving intact the hippocampal synaptic circuitry and anatomy that are lost in cultures of dissociated primary neurons. They have been used extensively to study neuronal death associated with excitotoxicity (Rimvall *et al.*, 1987; Vornov *et al.*, 1991), hypoxia and hypoglycemia (Newell *et al.*, 1990), and knife-cut lesions (Stoppini *et al.*, 1993).

Bruce *et al.* (1996) have investigated the ability of EUK-8 to protect organotypic hippocampal slices from neurotoxicity induced by β-amyloid peptides. Senile plaques, which are hallmark features of Alzheimer's disease, include extracellular deposits primarily composed of aggregated β-amyloid protein fragments (Masters *et al.*, 1985). The observation that these plaques are associated with areas of selective neuronal loss (Selkoe, 1991), combined with recent findings that indicate that some forms of Alzheimer's disease are related to mutations of the amyloid precursor protein (Chartier-Harlin *et al.*, 1991; Mullan and Crawford, 1994; Vanbroeckhoven, 1995), has led to extensive research into the relationships between β-amyloid protein accumulation and neuronal degeneration. The β-amyloid protein and certain of its fragments cause neurotoxicity *in vitro* and *in vivo* (Pike *et al.*, 1993; Kowall *et al.*, 1991) through mechanisms that are as yet unknown. It has been proposed, based primarily on protective effects of ROS scavengers, that ROS play a critical role in β-amyloid-mediated neuronal damage (Butterfield *et al.*, 1994a; Friedlich and Butcher, 1994).

As shown in Fig. 8, a β-amyloid peptide exhibited toxicity toward organotypic hippocampal slice cultures, based on two independent criteria for cell death. The addition of EUK-8 to the culture medium significantly reduced this toxicity. Furthermore, the β-amyloid peptide caused an increase in two indicators of oxidative stress: MDA content and dichlorofluorescin-diacetate-dependent fluorescence (Behl *et al.*, 1994; Hensley *et al.*, 1994), in untreated but not in EUK-8-treated cultures (Bruce *et al.*, 1996). These findings not only confirm that ROS contribute to β-amyloid toxicity in the organotypic hippocampal culture system but also suggest that EUK-8, by virtue of its ability to prevent oxidative stress, is protective against this form of neurotoxicity of potential relevance to Alzheimer's disease.

EUK-8 has shown efficacy in two *in vivo* models for neurodegenerative disease. In one model, mice received intraventricular injections of 6-hydroxydopamine (6-OHDA), a neurotoxin which induces selective damage to catecholaminergic neurons. This damage is believed to involve the specific uptake of the dopamine analog into the neurons, followed by a complex series of events involving 6-OHDA oxidation and leading to the acute degeneration of catecholaminergic neurons. In the experiments summarized in Fig. 9,o EUK-8 was administered intraperitoneally whereas 6-OHDA was given intracerebrally. To assess the integrity of the nigrostriatal dopaminergic pathway, the density of dopaminergic terminals was determined by the binding of [^{3}H]mazindol, a ligand for the dopamine uptake protein, to striatal membranes (Shimizu and Prasad, 1991). Based on this criterion, extensive damage was observed in the brain hemisphere ipsilateral to the

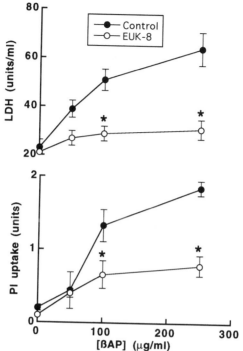

FIGURE 8 EUK-8 reduces β-amyloid peptide (βAP) toxicity in organotypic hippocampal slice cultures. The experiment was conducted as described (Bruce *et al.,* 1996). Mature organotypic hippocampal cultures were exposed to β-amyloid peptide (1–42) at the indicated concentrations for 72 hr in the presence or absence (control, ●) of 25 μM EUK-8 (○). The extent of toxicity was determined by quantifying lactate dehydrogenase (LDH) release or propidium iodide (PI) staining, as described by Bruce et al. (1996). An asterisk denotes a statistically significant difference from the control at the same β-amyloid peptide dose ($p < 0.05$). [Reprinted with kind permission of the National Academy of Sciences from Bruce *et al.* (1996).]

site of injection of 6-OHDA at a relatively high dose (Fig. 9a). Protection by EUK-8 was significant, but only partial. However, in the same mice, a lesser degree of damage was also detected in the contralateral hemisphere. On this side, full protection by EUK-8 was achieved. The contralateral damage was indicative of some leakage of 6-OHDA to the opposite side of the brain. Thus, a follow-up experiment (Fig. 9b) utilized a lower dose of 6-OHDA. With this protocol, damage was restricted to the ipsilateral side, was more modest, and was fully protected by EUK-8.

EUK-8 has also been found to be protective in mice against the neurodegenerative effects of MPTP. This toxin induces a selective damage to dopaminergic neurons in experimental animals and, in monkeys and humans, causes neurodegenerative symptoms similar to those observed in Parkinson's disease (Langston *et al.,* 1987; Snyder and D'Amato, 1986). Some evidence

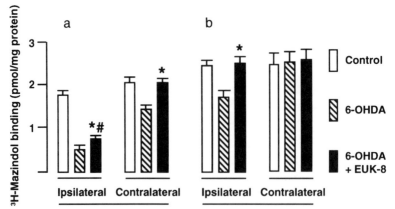

FIGURE 9 EUK-8 is protective in the 6-hydroxydopamine (6-OHDA) model for Parkinson's disease. (a) Mice received an intraventricular injection of 1 μl vehicle (control) or 6 μg 6-OHDA on day 1. Where indicated, EUK-8 (40 mg/kg) was administered intraperitoneally 24 hr prior to 6-OHDA and once on each of days 1 through 4. On day 7, striatal membranes were prepared from each brain hemisphere and analyzed for [³H]mazindol binding. Data shown are the means ± SEM (*n* = 8 to 10). An asterisk denotes EUK-8 + 6-OHDA groups showing a statistically significant difference from the hemisphere-matched 6-OHDA group (*p* < 0.05); A pound sign denotes EUK-8 + 6-OHDA groups showing a statistically significant difference from the hemisphere-matched control group (*p* < 0.05). (b) Experiment was conducted as described for (a) except that the amount of 6-OHDA administered was 3 μg.

suggests that ROS are involved in MPTP-induced neurotoxicity, including the observation that transgenic mice overexpressing SOD display resistance to the neurotoxin (Przedborski *et al.*, 1992). As shown in Fig. 10, mice

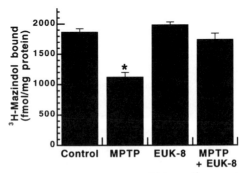

FIGURE 10 EUK-8 is protective in the MPTP model for Parkinson's disease. Mice received EUK-8, where indicated, continuously in the drinking water (5 mg/ml) starting 5 days before receiving MPTP treatment. Where indicated, mice received two intraperitoneal injections of MPTP (40 mg/kg), 24 hr apart. Six days after the second MPTP injection, mice were sacrificed and membranes were prepared from the striata and analyzed for [³H]mazindol binding. Data shown are the means ± SEM for the following group sizes: Control and EUK-8 (*n* = 4), MPTP and MPTP + EUK-8 (*n* = 6).*p* < 0.05, significantly different from the control group.

a

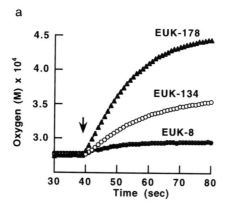

b

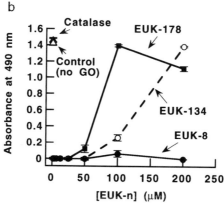

FIGURE II EUK-8 analogs exhibit enhanced catalase and cytoprotective activities.
(a) Salen–manganese complexes (each with acetate rather than chloride as the axial
ligand, see Fig. 1) were assayed for catalase activity as described in Fig. 3. Reaction
mixtures contained 10 μM salen–manganese complex and 10 mM hydrogen peroxide.
(b) Human dermal fibroblasts were treated with glucose oxidase (GO, 0.02 units/ml) in
culture medium containing 4.5 g/liter glucose. Salen–manganese complexes were present
at the indicated concentrations. Bovine liver catalase, where indicated, was present at
290 units/ml. The "control" samples contained medium with glucose but no GO. After
18 hr incubation, medium was replaced with fresh culture medium without additives
and cell viability was assessed using the XTT reagent (Boeringer Mannheim). The results
shown are the means ± SD of triplicate culture wells.

injected with MPTP exhibited a substantial loss of dopaminergic neurons,
assessed, as for the 6-OHDA model, by [³H]mazindol binding. However,
mice treated with EUK-8 were completely protected from MPTP-induced
neurotoxicity. In this experiment, EUK-8 was delivered orally, via the drinking water. In other experiments, salen–manganese complexes have also
shown efficacy in this model when delivered intraperitoneally.

As mechanisms of dopaminergic neurodegeneration in these two models
do not necessarily parallel the cascade of events occurring in Parkinson's

disease, it is difficult to make predictions from these studies regarding the potential usefulness of EUK-8 in treating Parkinson's patients. Nonetheless, the efficacy of EUK-8 in these disease models represents a crucial demonstration that the compound, administered peripherally, can cross the blood–brain barrier in pharmacologically efficacious quantities. Furthermore, the availability of compounds such as EUK-8 should enable untested hypotheses regarding the role of oxidative stress in human neurological diseases to be investigated in the clinic.

IV. Conclusions and Future Directions

EUK-8 is a combined SOD/catalase mimic and displays efficacy in numerous disease models, including several complex *in vivo* systems involving severe tissue damage. While EUK-8 has served as a prototype salen–manganese complex, additional analogs are currently being designed and evaluated. The future development plan for salen–manganese complexes will include selecting those which have been optimized with respect to catalytic activities, *in vivo* efficacy, and other key characteristics. For example, several compounds have now been identified with improved catalase activity over that of EUK-8. Figure 11 shows data obtained with two such compounds. Under equivalent assay conditions, these molecules are much more active as catalases than EUK-8 (Fig. 11a). In addition, commensurate with their enhanced catalase activity, they are considerably more effective than EUK-8 at protecting human cells from toxicity by glucose and glucose oxidase, a hydrogen peroxide-generating system (Fig. 11b). Ongoing studies are investigating whether salen–manganese complexes with improved catalytic properties over EUK-8 also show enhanced pharmacological efficacy.

In conclusion, a large body of evidence now indicates that salen–manganese complexes are a class of molecules with a novel mode of action and an extremely broad potential clinical utility. EUK-8, or a later-generation analog, will be developed as an experimental therapeutic agent for clinical indications involving oxidative stress.

Acknowledgments

We acknowledge our numerous scientific collaborators throughout the world who have evaluated and continue to evaluate our compounds in their own pharmacological model systems of interest. In particular for this review, we thank Mitchell P. Fink, Pamela K. Gonzalez, and their colleagues in Dr. Fink's laboratory at the Beth Israel Hospital, Boston (whose published work is reviewed in Section III,A), for a collaboration to evaluate our compounds in models relating to acute critical illness. In addition, we acknowledge the National Institute on Aging and the National Heart, Lung, and Blood Institute for grants awarded under the STTR program to help support follow-up research on the efficacy of salen–manganese complexes in neurodegenerative disease and in ARDS, respectively.

References

Adams, J. D., and Odunze, I. N. (1991). Oxygen radicals and Parkinson's disease. *Free Rad. Biol. Med.* **10**, 161–169.

Anderson, B. O., Brown, J. M., and Harken, A. H. (1991). Mechanisms of neutrophil-mediated tissue injury. *J. Surg. Res.* **51**, 170–179.

Baldwin, S. R., Simon, R. H., Grum, C. M., Ketai, L. H., Boxer, L. A., and Devaff, L. J. (1986). Oxidant activity in expired breath of patients with ARDS. *Lancet* **1**, 114–116.

Baudry, M., Etienne, S., Bruce, A., Palucki, M., Jacobsen, E., and Malfroy, B. (1993). Salen-manganese complexes are superoxide dismutase-mimics. *Biochem. Biophys. Res. Commun.* **192**,(2), 964–968.

Behl, C., Davis, J., Cole, G. M., and Schubert, D. (1992). Vitamin E protects nerve cells from amyloid beta protein toxicity. *Biochem. Biophys. Res. Commun.* **186**, 944–950.

Behl, C., Davis, J. B., Lesley, R., and Schubert, D. (1994). Hydrogen peroxide mediates amyloid-beta-protein toxicity. *Cell* **77**, 817–827.

Bertrand, Y., Pincemail, J., Hanique, G., Denis, B., Leenaerts, L., Vandeerberghen, L., and Deby, C. (1989). Differences in tocopherol-lipid ratios in ARDS and non-ARDS patients. *Intensive Care Med.* **15**, 87–93.

Black, S. C., Schasteen, C. S., Weiss, R. H., Riley, D. P., Driscoll, E. M., and Lucchesi, B. R. (1994). Inhibition of in vivo myocardial ischemic and reperfusion injury by a synthetic manganese-based SOD mimic. *J. Pharmacol. Exp. Ther.* **270**, 1208–1215.

Bralet, J., Bouvier, C., Schrieber, L., and Boquillon, M. (1991). Effect of acidosis on lipid peroxidation in brain slices. *Brain Res.* **539**, 175–177.

Bralet, J., Schreiber, L., and Bouvier, C. (1992). Effect of acidosis and anoxia on iron delocalization from brain homogenates. *Biochem. Pharmacol.* **43**, 979–983.

Braughler, J. M., Duncan, L. A., and Chase, R. L. (1986). The involvement of iron in lipid peroxidation. *J. Biol. Chem.* **261**, 10282–10289.

Bruce, A. J., Malfroy, B., and Baudry, M. (1996). Beta-amyloid toxicity in organotypic hippocampal cultures: Protection by EUK-8, a synthetic catalytic free radical scavenger. *Proc. Natl. Acad. Sci. USA,* **93**, 2312–2316.

Bruce, A. J., Najm, I., Malfroy, B., and Baudry, M. (1992). Effects of desferioxamine/manganese complex, a superoxide dismutase mimic, on kainate-induced pathology in rat brain. *Neurodegeneration.* **1**, 265–271.

Butterfield, D. A., Hensley, K., Harris, M., Mattson, M., and Carney, J. (1994a). Beta amyloid peptide free-radical fragments initiate synaptosomal lipoperoxidation in a sequence-specific fashion: Implications to Alzheimer's disease. *Biochem. Biophys. Res. Commun.* **200**, 710–715.

Butterfield, D. A., Hensley, K., Harris, M., Mattson, M., and Carney, J. M. (1994b). Beta-amyloid free radical production, peptide fragmentation, and neurotoxicity: Implications of molecular shrapnel to the etiology of Alzheimer's disease. *Neurobiol. Aging* **1**, 15–22.

Ceconi, C., Cargoni, A., Pasini, E., Condorelli, E., Curello, S., and Ferrari, R. (1991). Evaluation of phospholipid peroxidation as malonyldialdehyde during myocardial ischemia and reperfusion injury. *Am. J. Physiol.* **260**, H1057–H1061.

Chance, B., Sies, H., and Boveris, A. (1979). Hydroperoxide metabolism in mammalian organs. *Physiol. Rev.* **59**, 527–605.

Chartier-Harlin, M. C., Crawford, F., Houlden, H., Warren, A., Hughes, D., Fidanti, L., Goate, A., Rossor, M., Roques, P., Hardy, J., and Mullan, M. (1991). *Nature* **353**, 844–846.

Childs, R. E., and Bardsley, W. G. (1975). The steady-state kinetics of peroxidase with 2, 2′-azino-di-(3-ethylbenzthiazoline-6-sulphonic acid) as chromagen. *Biochem. J.* **145**, 93–103.

Cross, C. F., Forte, T., Stocker, R., Lowie, S., Yammamoto, Y., Ames, B. N., and Frei, B. (1990). Oxidative stress and abnormal cholesterol metabolism in patients with ARDS. *J. Lab. Clin. Med.* **115**, 396–404.

Darr, D., Zarilla, K. A., and Fridovich, I. (1987). A mimic of superoxide dismutase activity based upon desferrioxamine B and manganese (IV). *Arch. Biochem. Biophys.* **258**, 351–355.

Dexter, D. T., Carter, C. J., and Wells, F. R. (1989). Basal lipid peroxidation in substantia nigra is increased in Parkinson's disease. *J. Neurochem.* **52**, 381–389.

Dismukes, G. C. (1993). Polynuclear manganese enzymes. *In* "Bioinorganic Catalysis." (J. Reediik, ed.), Dekker, New York.

Esterbauer, H., and Cheeseman, K. H. (1990). Determination of aldehydic lipid peroxidation products: malonyldialdehyde and 4-hydroxynonenal. *Methods Enzymol.* **186**, 407–431.

Faulkner, K. M., Liochev, S. I., and Fridovich, I. (1994). Stable Mn(III) prophyrins mimic superoxide dismutase in vitro and substitute for it in vivo. *J. Biol. Chem.* **269**, 23471–23476.

Flamm, E. S., Demopoulos, H. B., Seligman, M. L., Poser, R. G., and Ransohoff, J. (1978). Free radicals in cerebral ischemia. *Stroke* **9**, 445–447.

Fowler, A. A., Hamman, R. F., Good, J. T., Benson, B. A., Baird, M., Eberle, D. J., and Petty, T. L. (1983). ARDS: Risk with common predispositions. *Ann. Intern. Med.* **98**, 593–597.

Friedlich, A. L., and Butcher, L. L. (1994). Involvement of free oxygen radicals in beta-amyloidosis: A hypothesis. *Neurobiol. Aging* **15**, 443–455.

Gonzalez, P. K., Doctrow, S. R., and Fink, M. P. (1995a). EUK-8, an SOD/Catalase mimic, protects against lactic acidosis-induced hyperpermeability and lipid peroxidation in Caco-2BBE monolayers. *FASEB J.* **9**, A953.

Gonzalez, P. K., Zhuang, J., Doctrow, S. R., Malfroy, B., Benson, P. F., Menconi, M. J., and Fink, M. P. (1995b). EUK-8, a synthetic superoxide dismutase and catalase mimetic, ameliorates acute lung injury in endotoxemic swine. *J. Pharmacol. Exp. Therap.* **275**, 798–806.

Gonzalez, P. K., Zhuang, J., Doctrow, S. R., Malfroy, B., Smith, M., Menconi, M. J., and Fink, M. P. (1995c). Delayed treatment with EUK-8, a novel synthetic superoxide dismutase (SOD) and catalase (CAT) mimetic, ameliorates acute lung injury in endotoxemic pigs. *Surg. Forum* **46**, 72–73.

Goodman, Y., Steiner, M. R., Steiner, S. M., and Mattson, M. P. (1994). Nordihydroguaiaretic acid protects hippocampal neurons against amyloid beta-peptide toxicity and attenuates free radical and calcium accumulation. *Brain Res.* **654**, 171–176.

Granger, D. N. (1988). Role of xanthine oxidase and granulocytes in ischemia-reperfusion injury. *Am. J. Physiol.* **255**, H1269–H1275.

Granger, D. N., Hollwarth, M. E., and Parks, D. A. (1986). Ischemia-reperfusion injury: Role of oxygen derived free radicals. *Acta Physiol. Scand.* **548**(Suppl), 47–63.

Hall, E. D., and Braughler, J. M. (1989). Central nervous system trauma and stroke. II Physiological and pharmacological evidence for the involvement of oxygen radicals and lipid peroxidation. *Free Rad. Biol. Med.* **6**, 303–313.

Halliwell, B., and Gutteridge, J. M. C. (1989). "Free Radicals in Biology and Medicine." Clarendon Press, Oxford.

Hensley, K., Carney, J. M., Mattson, M. P., Aksenova, M., Harris, M., Wu, J. F., Floyd, R. A., and Butterfield, D. A. (1994). A model for beta-amyloid aggregation and neurotoxicity based on free radical generation by the peptide: Relevance to Alzheimer's disease. *Proc. Natl. Acad. Sci. U.S.A.* **91**, 3270–3274.

Hyers, T. M. (1993). Prediction of survival and mortality in patients with ARDS. *New Horizons* **1**, 466–470.

Janero, D. R. (1990). Malonyldialdehyde and thiobarbituric acid-reactivity as diagnostic indices of lipid peroxidation and peroxidative tissue injury. *Free Rad. Biol. Med.* **9**, 515–540.

Jesberger, J. A., and Richardson, J. S. (1990). Oxygen free radicals and brain disfunction. *Int. J. Neurosci.* **57**, 1–17.

Keitzmann, D., Kahl, R., Muller, M., Burchardi, H., and Kettler, D. (1993). Hydrogen peroxide in expired breath condensate of patients with acute respiratory failure and with ARDS. *Intensive Care Med.* **19**, 78–81.

Kensler, T. W., Bush, D. M., and Kozumbo, W. J. (1983). Inhibition of tumor promotion by a biomimetic SOD. *Science* **221**, 75–77.

Klebanoff, S. J. (1988). Phagocytic cells: Products of oxygen metabolism. *In* "Inflammation: Basic Principles and Clinical Correlates." Raven Press, New York.

Kowall, N. W., Beal, M. F., Busciglio, J., Duffy, L. K., and Yankner, B. A. (1991). An in vivo model for the neurodegenerative effects of beta-amyloid and protection by substance P. *Proc. Natl. Acad. Sci. U.S.A.* **88**, 7247–7251.

Langston, J. W., Irwin, I., and Ricaurte, G. A. (1987). Neurotoxins, parkinsonism, and Parkinson's disease. *Pharmacol. Ther.* **32**, 19–49.

Liu, T. H., Beckman, J. S., Freeman, B. A., Hogan, E. L., and Hsu, C. Y. (1989). Polyethylene glycol conjugated superoxide dismutase and catalase reduce ischemic brain injury. *Am. J. Physiol.* **256**, 589–593.

Maestro, R. D., and McDonald, W. (1987). Distribution of superoxide dismutase, glutathione peroxidase, and catalase in developing rat brain. *Mech. Aging Dev.* **41**, 29–38.

Malfroy-Camine, B., and Baudry, M. (1995). Synthetic catalytic free radical scavengers useful as antioxidants for prevention and therapy of disease. U.S. Patent #5,403,834.

Masters, C. L., Simms, G., Weinman, N. A., Multhaup, G., McDonald, B. L., and Beyreuther, K. (1985). Amyloid plaque core protein in Alzheimer's disease and Down syndrome. *Proc. Natl. Acad. Sci. U.S.A.* **82**, 4245–4249.

Mather, P., Crowder, M., and Dismukes, G. C. (1987). Dimanganese complexes of a septadentate ligand: Functional analogs of the manganese pseudocatalase. *J. Am. Chem. Soc.* **109**, 5227–5233.

Mathru, M., Rooney, M. W., Dries, D. J., Hirsch, L. J., Barnes, L., and Tobin, M. J. (1994). Urine hydrogen peroxide during ARDS in patients with and without sepsis. *Chest* **105**, 232–236.

McCord, J. M. (1987). Oxygen-derived radicals: A link between reperfusion injury and inflammation. *Fed. Proc.* **46**, 2402–2406.

McCord, J. M., Beauchamp, C. O., Goscin, S., Misra, H. P., and Fridovich, I. (1973). Superoxide and superoxide dismutase. *In* "Oxidases and Related Redox Systems." University Press, Baltimore.

Montgomery, A. B., Stager, M. A., Carrico, C. J., and Hudson, L. D. (1988). Causes of mortality in patients with ARDS. *Am. Rev. Respir. Dis.* **132**, 485–489.

Mullan, M., and Crawford, F. (1994). The molecular genetics of Alzheimer's disease. *Mol. Neurobiol.* **9**, 1–3.

Musleh, W., Bruce, A., Malfroy, B., and Baudry, M. (1994). Effects of EUK-8, a synthetic catalytic superoxide scavenger, on hypoxia- and acidosis-induced damage in hippocampal slices. *Neuropharmacology* **33**, 929–934.

Newell, D. W., Malouf, A. T., and Franck, J. E. (1990). Glutamate-mediated selective vulnerability to ischemia is present in organotypic cultures of hippocampus. *Neurosci. Lett.* **116**, 325–330.

Oshino, N., Oshino, R., and Chance, B. (1973). The characteristic of the "peroxidatic" reactions of catalase in ethanol oxidation. *Biochem. J.* **131**, 555–567.

Pacht, E. R., Timerman, A. P., Lykens, M. G., and Merola, A. J. (1991). Deficiency of alveolar fluid GSH in patients with sepsis and ARDS. *Chest* **100**, 1397–1403.

Pessiki, P. J., and Dismukes, G. C. (1994). Structural and functional models of the dimanganese catalase enzymes. 3. Kinetics and mechanism of hydrogen peroxide dismutation. *J. Am. Chem. Soc.* **116**, 898–903.

Pike, C. J., Burdick, D., Walencewicz, A. J., Glabe, C. G., and Cotman, C. W. (1993). Neurode-generation induced by beta-amyloid peptides in vitro: The role of peptide assembly state. *J. Neurosci.* **13**, 1676–1687.

Przedborski, S., Kostic, V., Jackson-Lewis, V., Naini, A. B., Simonetti, S., Fahn, S., Carlson, E., Epstein, C., and Cadet, J. L. (1992). Transgenic mice with increased Cu/Zn SOD are resistant to MPTP-induced toxicity. *J. Neurosci.* **12**, 1658–1667.

Rice-Evans, C. A., and Diplock, A. T. (1993). Current status of antioxidant therapy. *Free Rad. Biol. Med.* **15**, 77–96.

Richard, C., Lemonnier, F., Thibault, M., Couturier, M., and Auzepy, P. (1990). Vitamin E deficiency and lipoperoxidation during ARDS. *Crit. Care Med.* **18**, 4–9.

Riley, D. P., and Weiss, R. H. (1994). Manganese macrocyclic ligand complexes as mimics of SOD. *J. Amer. Chem. Soc.* **116**, 387–388.

Rimvall, K., Keller, F., and Waser, P. G. (1987). Selective kainic acid lesions in cultured explants of rat hippocampus. *Acta Neuropathol.* **74**, 183–190.

Roda, J. M., Carceller, F., Pajares, R., and Diez, T. E. (1991). Prevention of cerebral ischemic reperfusion injury by intra-arterial administration of superoxide dismutase in the rat. *Neurol. Res.* **13**, 160–163.

Roumen, R. M. H., Hendriks, T., deMan, B. M., and Goris, R. J. A. (1994). Serum lipofuscin as a prognostic indicator of ARDS and multiple organ failure. *Br. J. Surg.* **81**, 1300–1305.

Schonbaum, G. R., and Chance, B. (1976). Catalase. *In* "The Enzymes." (P. D. Boyer, ed.), Academic Press, New York.

Selkoe, D. J. (1991). The molecular pathology of Alzheimer's disease. *Neuron* **6**, 487–498.

Shimizu, I., and Prasad, C. (1991). Relationship between ³H-mazindol binding to dopamine uptake sites and ³H-dopamine uptake in rat striatum during aging. *J. Neurochem.* **56**, 575–579.

Sies, H., ed. (1985). "Oxidative Stress." Academic Press, New York.

Sies, H. (1986). Biochemistry of oxidative stress. *Angewandte Chemie* **25**, 1058–1071.

Siesjo, B. K., Bendek, G., Koide, T., Westergang, E., and Wieloch, T. (1985). Influence of acidosis on lipid peroxidation in brain tissues in vitro. *J. Cereb. Blood Flow Metab.* **5**, 253–558.

Simon, R. H., and Ward, P. A. (1992). Adult respiratory distress syndrome. *In* "Inflammation: Basic Principles and Clinical Correlates." (J. I. Gallin, I. M. Goldman, R. Snyderman, eds.), Raven Press, New York.

Snyder, S. H., and D'Amato, R. J. (1986). MPTP, a neurotoxin relevant to the pathophysiology of Parkinson's disease. *Neurology* **36**, 250–258.

Stoppini, L., Buchs, P. A., and Muller, D. (1993). Lesion-induced sprouting and synapse formation in hippocampal organotypic cultures. *Neuroscience* **57**, 985–994.

Suchyta, M. R., Clemmer, T. P., Elliott, C. G., Orme, J. F., and Weaver, L. K. (1992). The adult respiratory distress syndrome: A report of survival and modifying factors. *Chest* **101**, 1074–1079.

Sznajder, J. I., Fraiman, A., Hall, J. B., Sanders, W., Schmidt, G., Crawford, G., Nahum, A., Factor, P., and Wood, L. D. H. (1989). Increased hydrogen peroxide in the expired breath of patients with acute hypoxemic respiratory failure. *Chest* **96**, 606–612.

Vanbroeckhoven, C. L. (1995). Molecular genetics of Alzheimer's disease: Identification of genes and gene mutations. *Eur. Neurol.* **35**, 8–19.

VanderMeer, T. J., Menconi, M. J., O'Sullivan, B. P., Larkin, V. A., Wang, H., Kradin, R. L., and Fink, M. P. (1994). Bactericidal/permeability-increasing protein ameliorates acute lung injury in porcine endotoxemia. *J. Appl. Physiol.* **76**, 2006–2014.

VanderMeer, T. J., Menconi, M. J., O'Sullivan, B. P., Larkin, V. A., Wang, H., Sofia, M., and Fink, M. P. (1995). Acute lung injury in endotoxemic pigs: Role of leukotriene B4. *J. Appl. Physiol.* **78**, 1121–1131.

Vornov, J. J., Tasker, R. C., and Coyle, J. T. (1991). Direct observation of the agonist-specific regional vulnerability to glutamate, NMDA, and kanaite neurotoxicity in organotypic hippocampal cultures. *Exp. Neurol.* **114**, 11–12.

Wollert, P. S., Menconi, M. J., O'Sullivan, B. P., Wang, H., Larkin, V., and Fink, M. P. (1993). LY255283, a novel leukotriene B4 receptor antagonist, limits activation of neutrophils and prevents acute lung injury induced by endotoxin in pigs. *Surgery* **114**, 191–98.

Wollert, P. S., Menconi, M. J., Wang, H., O'Sullivan, B. P., Larkin, V., Allen, R. C., and Fink, M. P. (1994). Prior exposure to endotoxin exacerbates LPS-induced hypoxemia and alveolitis in anesthetized swine. *Shock* **2**, 362–369.

Yamamoto, S., Nakadate, T., Aizu, E., and Kato, R. (1990). Anti-tumor promoting action of pthalic acid mono-n-butyl ester cupric salt. *Carcinogenesis* **11**, 749–754.

PART **III**

Antioxidant Enzyme Induction and Pathophysiology

Anthony C. Allison

Dawa Corporation
Belmont, California 94002

Antioxidant Drug Targeting

I. Introduction

Antioxidants have been claimed to have efficacy in the prevention and treatment of many disorders. Sometimes the claims are broad, such as the beneficial effects of antioxidant mixtures in cosmetics and the supposed contribution of antioxidants in red wine to protection against coronary heart disease. Other reports are more specific and well established, such as the protection by aspirin and other nonsteroidal anti-inflammatory drugs (NSAIDs) against colon cancer (Giovannucci et al., 1994) and the cardioprotective effects of dietary α-tocopherol (Stamper et al., 1993; Rimm et al., 1993). Usually the mechanisms by which antioxidants exert these effects are uncertain. For many years antioxidants have been known to prevent lipid peroxidation, the depletion of reduced glutathione and other intracellular protectants against oxidation, and lipoprotein oxidation. During the past few years it has become apparent that antioxidants have other important

Advances in Pharmacology, Volume 38

effects, including the regulation of signal transduction, kinase and phosphatase activation, gene expression, and enzyme activation. Isoforms of functionally important oxidant enzymes such as prostaglandin endoperoxide (PGH) synthase have been identified.

A valuable method for establishing these effects has been the genetic targeting of individual oxidant enzymes and kinases, either overproducing or eliminating them in cell lines or in transgenic mice. Such genetic experiments point the way to the development of similarly targeted antioxidants. They also illuminate mechanisms by which well-established findings are exerted.

Developing antioxidant drugs requires their targeting first to organs, second to desired cell types, and third to particular intracellular metabolic pathways or signal transduction systems. This is a large subject, and only a few examples of each category are discussed here. Organ and tissue targeting is illustrated by the delivery of antioxidants to the skin and to the large intestine, using pro-drugs to overcome technical problems that have limited the usefulness of drugs so far developed. Targeting to particular cell types is illustrated by selective delivery to macrophages and to epithelial cells of hair follicles. The first intracellular targets discussed are enzymes, including phospholipase A_2 (PLA_2) and PGH synthase. Evidence is then reviewed that antioxidants can affect signal transduction systems, including the activation of kinases and of transcription factors. This makes possible coordinate inhibition, using small molecule antioxidants, of the expression of the genes for the pro-inflammatory cytokines tumor necrosis factor (TNF)-α and interleukin (IL)-1 in monocytes and macrophages. Antioxidants can also be used to block endothelial cell activation in response to TNF-α and other stimulators. Inhibition of the production of pro-inflammatory cytokines and endothelial cell activation provides a novel strategy for treating inflammatory conditions, which could avoid the side effects of glucocorticoids. Selective inhibition of the type 2 isoform of PGH synthase reduces the gastrointestinal side effects of NSAIDs, and clinical trials now in progress should establish whether these novel anti-inflammatory drugs are efficacious while having fewer side effects than those of currently used therapies.

II. Delivering Antioxidants to Organs or Tissues _____

A. Overcoming Irritant Effects of Antioxidants on the Skin

To protect the skin against ultraviolet (UV) radiation (which generates reactive oxygen intermediates) (see Black, 1987) and for anti-inflammatory activity it is desirable to administer an antioxidant in a cream or lotion formulation. An example is the use of *meso*-nordihydroguiauretic acid (NDGA) for the treatment of actinic keratoses (Olsen *et al.*, 1991). Actinic keratoses are produced by exposure to solar radiation over a long period. The epidermis shows parakeratosis, in which nuclei are retained in the superficial layers, and the dermis shows inflammatory infiltrates, with many basophils, as well as overproduction of elastin in the deeper layers. Redness and itching inconvenience the patient, and squamous carcinomas are much more likely to develop in actinic keratotic lesions than elsewhere in the skin. A double-blind, vehicle-controlled, multicenter trial showed that topical application of NDGA was an effective treatment for actinic keratosis (Olsen *et al.*, 1991). However, irritation of the skin, manifested by erythema or flaking, occurred in most patients receiving NDGA, thus limiting patient compliance.

The reason for the irritant effects of NDGA and some other antioxidants on the skin is not fully understood. However, as in the well-known case of ascorbic acid, antioxidants can have oxidant effects in the presence of redox-active metals. The skin contains readily detectable iron (Bisset *et al.*, 1994), some of which may be available to participate in the formation of oxidants. These could exert irritant effects, e.g., activating phospholipase (Robison *et al.*, 1990) and releasing prostaglandins and leukotrienes. Evidence has been presented that iron is a factor in skin photodamage: an iron chelator (2-furildioxime), topically applied to the skins of experimental animals and humans, was found to provide protection against ultraviolet radiation-induced erythema, infiltration of inflammatory cells, epidermal thickening, and induction of ornithine decarboxylase (Bisset *et al.*, 1994).

Thus one way to prevent irritant effects of antioxidants in the skin is to chelate iron and copper effectively. An alternative strategy is to use an ester pro-drug, which is not itself antioxidant. Esterases in the skin liberate the active drug. It has been found that high concentrations of the tetraacetate of the dicatechol rooperol (Van der Merwe *et al.*, 1993) can be applied to the skins of experimental animals and humans without irritant effects. Active dicatechol is released in the skin, and metabolites are detectable in the circulation. Thus an important problem in the targeting of drugs to the skin has been solved.

B. Delivering Antioxidants to the Large Intestine

To treat ulcerative colitis it is desirable to deliver antioxidants that can suppress the formation of pro-inflammatory cytokines (TNF-α and IL-1β) as well as leukotrienes. As discussed later an example of such a compound is the dicatechol rooperol (Van der Merwe *et al.*, 1993). However, the active dicatechol is relatively unstable and absorbed in the small intestine. As recovered from the plant *Hypoxis rooperi*, rooperol is a diglucoside (Fig. 1). This compound is stable and suitable for pharmaceutical formulation. It is not absorbed or degraded in the small intestine. However, in the human large intestine, bacterial β-glucosidases release the active dicatechol almost

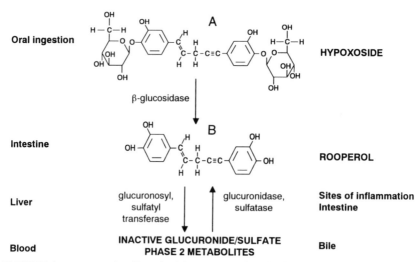

FIGURE I Hypoxoside efficiently delivers rooperol to the large intestine, where bacterial glucosidases liberate the active dicatechol. Rooperol tetraacetate, in which all four hydroxyl groups are acetylated, efficiently delivers rooperol to the skin, where esterases liberate the active dicatechol.

quantitatively. In the reducing environment of the colon, the dicatechol is stable and is delivered efficiently to the colonic mucosa. When absorbed into the bloodstream, rooperol is conjugated on the first pass through the liver to glucuronide and sulfate phase 2 metabolites with low biological activity. Thus the diglucoside pro-drug is an efficient system for the oral delivery of rooperol to the large intestine, with the objective of exerting anti-inflammatory activity in the mucosa of the large intestine while also reducing the risk of systemic side effects.

III. Targeting Antioxidants to Particular Cell Types

A. Delivering Antioxidants to Macrophages

As discussed later, some antioxidants inhibit the production of the pro-inflammatory cytokines TNF-α and IL-1β by macrophages. These cytokines play an important role in the pathogenesis of septic shock (Allison and Eugui, 1995). To prevent shock, it would be useful to deliver an antioxidant to macrophages residing in positions close to circulating blood. A convenient delivery system is provided by liposomes. Unless liposomes are coated with polyethylene glycol or another hydrophilic molecule, when injected into the bloodstream they are rapidly taken up by macrophages, especially in the liver and spleen (Weldon *et al.*, 1983). This property of liposomes has been used to target drugs active against *Leishmania* parasites, such as Amphotericin and antimony compounds, to macrophages (Weldon *et al.*, 1983). Agents activating macrophges have also been delivered to those cells in liposomes (Nacy *et al.*, 1989). In a similar way, liposomes could be used to achieve the opposite effect, targeting antioxidants to macrophages so as to prevent their activation, thereby suppressing the release of cytokines and lipid mediators in septic shock.

B. Delivering Antioxidants to Hair Follicles

Cancer chemotherapy exerts effects on both malignant cells and normal cells that have high mitotic activity, including hair follicles. Alopecia is severe in patients treated with the oxidant anthracycline antibiotic doxorubicin. This effect of doxorubicin can be reproduced in rats and is prevented by monoclonal antibodies against the drug entrapped in liposomes (Balsari *et al.*, 1994). Liposomes of appropriate composition provide a convenient way of targeting drugs to hair follicles (Lieb *et al.*, 1992). Because antibodies are relatively unstable and inconvenient to formulate, an alternative strategy is to formulate in liposomes an antioxidant that can prevent the effects of doxorubicin on rapidly dividing skin epithelial cells in hair follicles.

IV. Effects of Antioxidants on Enzymes

A. Oxidant Activation of Phospholipase and Prevention by Antioxidants

Phospholipase A_2 activation is pharmacologically important because the release of arachidonic acid is the rate-limiting step in the formation of lipid mediators of inflammation. Peroxidation of cell membrane phospholipids is accompanied by increased phospholipase activity, principally due to PLA_2 (Robison *et al.*, 1990). Phospholipases are not actually activated by peroxidation, but oxidized lipids affect the cell membrane so that it becomes a better substrate for PLA_2 and other phospholipases. Thus phospholipase activity on membranes might, in principle, be decreased by lipophilic antioxidants that inhibit free radical-induced oxidation of membrane phospholipids. Nordihydroguaiaretic acid (4',4'-(2,3-dimethyl-1,4-butanediyl)bis [1,2-benzenediol]), a 5-lipoxygenase inhibitor, has been classified as an antioxidant and radical trapping agent (Van der Zee *et al.*, 1989). Reactions catalyzed by fatty acyl peroxidases resemble the autoxidation of unsaturated lipids, suggesting that in both processes the mechanism of inhibition by antioxidants may be due to a termination of free radical chain reactions. Robison *et al.* (1990) found that *t*-butyl hydroperoxide augmented the release of arachidonic acid from rat alveolar macrophages and that linoleic acid hydroperoxide augmented arachidonic acid release from Chinese hamster lung fibroblasts. Both effects were blocked by NDGA in the concentration range $1-10$ μM. Analogous effects of antioxidants, preventing oxidant-induced activation of phospholipases, could be useful therapeutically.

B. Isoforms of PGH Synthase

Prostaglandin endoperoxide (PGH) synthase catalyzes the committed step in prostaglandin synthesis. During the past few years it has been shown that there are two different isoforms of PGH synthase, PGHS-1 and PGHS-2, encoded by separate genes (Smith *et al.*, 1994). PGHS-1 is expressed constitutively in most tissues, including gastric mucosa, renal tubules, and smooth muscle, whereas PGHS-2 is not. However, expression of the PGHS-2 gene is markedly increased in cell types such as macrophages, endothelial cells, and fibroblasts by bacterial products such as lipopolysaccharide (LPS), pro-inflammatory cytokines such as TNF and IL-1, and mitogenic ligands. Thus PGHS-1 is regarded as the enzyme producing prostaglandins that regulate gastrointestinal functions and vascular homeostasis, whereas PGHS-2 regulates prostaglandins involved in inflammation and mitogenesis (Smith *et al.*, 1994). Glucocorticoids have little effect on the expression of the PGHS-1 gene but inhibit the expression of PGHS-2.

The amino acid sequences of PGH synthase isoenzymes in the human and mouse show approximately 50% identity; in addition, all the residues required for catalysis by PGHS-1 are conserved in PGHS-2. Because enzymic activities (V_{max}) and affinity for arachidonic acid (K_m) are indistinguishable, both enzymes catalyze reactions that are mechanistically identical. However, there are differences in the mode of action of inhibitors, as well as sensitivity to inhibitors, which illustrate the possibilities for drug targeting. Aspirin acetylates the homologous residue in the active site of the two enzymes (Shimokawa and Smith, 1992; Smith *et al.*, 1994). However, aspirin irreversibly inhibits PGHS-1, preventing this enzyme from forming any other oxygenated product; in contrast, aspirin treatment of PGHS-2 results in the formation of 15-hydroxy-5c,8c,11c,13t-eicosatetraenoic acid (15-HETE) instead of PGH_2. Several drugs that selectively inhibit PGHS-2, with little or no effect on PGHS-1, have been identified (Smith *et al.*, 1994). These compounds show activity in experimental animal models of inflammation and are currently entering into clinical trials. It is hoped that they will display clinically useful anti-inflammatory activity without side effects on the gastrointestinal tract.

Selectively targeting genes encoding the isoforms of PGH synthase by genetic manipulation has produced interesting, and sometimes surprising, information. Mice homozygous for a disrupted *Ptgs 1* gene survive well and have no gastric pathology, even though their gastric PGE_2 levels are about 1% of those in wild-type mice (Langerbach *et al.*, 1995). The homozygous mutant mice show less indomethacin-induced gastric ulceration than wild-type mice, have reduced platelet aggregation, and a decreased inflammatory response to arachidonic acid, but not to 12-O-tetradecanoylphorbol-13-acetate (TPA). Mice homozygous for a disrupted *Pghs 2* gene have normal gastric mucosas and normal inflammatory responses to treatments with arachidonic acid or TPA (Morham *et al.*, 1995). However, their kidney development is abnormal, suggesting that PGHS-2 is required for postnatal maturation of the subcapsular nephrogenic zone.

However, overexpression of PGHS-2 leads to phenotypic changes in intestinal epithelial cells that could increase their tumorigenic potential. Humans who take aspirin or other NSAIDs regularly have a 40–50% lower relative risk of colorectal cancer when compared with persons not taking those medications (Giovannucci *et al.*, 1994). NSAIDs are also potent inhibitors of tumor formation in rodent models of chemically induced colon cancer (Reddy *et al.*, 1993). The expression of PGHS-1 is unaffected in colorectal carcinogenesis, but PGHS-2 mRNA and protein levels are increased in a significant number of human colon cancers (Eberhart, 1994; Kargman *et al.*, 1995). Tsuji and Du Bois (1995) overexpressed PGHS-2 in rat intestinal epithelial cells and found increased adhesion of the cells to the extracellular matrix, increased BCL2 protein expression, and resistance to butyrate-induced apoptosis. These effects, which are characteristic of malignant cells, were reversed by sulindac sulfide, a PGHS inhibitor. Sulindac is reported to increase the rate of programmed cell death in patients with familial adenomatous polyposis (Pasricka *et al.*, 1995). If these findings can be reproduced with selective PGHS-2 inhibitors, protection against colorectal cancer may be achieved without augmenting the risk of gastric or small intestinal damage.

V. Reactive Oxygen Intermediates as Signal Transducers

The role of NO as a signal transducer in vascular smooth muscle, neuronal, and other cell types is generally accepted. For example, NO activates guanylate cyclase, and cyclic GMP is a second messenger of smooth muscle relaxation (Moncada and Higgs, 1993). Evidence is accumulating that reactive oxygen intermediates can also transduce signals in plants, bacteria, and mammalian cells. A burst of H_2O_2 is produced by plants in response to injury and can increase their resistance to viral pathogens by serving as a small diffusible molecule that activates the coordinate expression of genes involved in the protective response (Levine *et al.*, 1994). Salicyclic acid binds to and inactivates catalase in tobacco plants, leading to a rise in H_2O_2 and activation of the expression of protective genes (Chen *et al.*, 1993). Bacteria have a DNA-binding protein (oxyR) that, in response to H_2O_2, activates the transcription of several genes encoding proteins that protect bacterial cells from oxidative stress (Storz *et al.*, 1990). The possible role of reactive oxygen intermediates in mammalian cell signal transduction is intriguing but less well established.

A. Insulin-like Effects of H_2O_2

It has long been known that H_2O_2 and vanadate mimic several of the metabolic and growth-promoting effects of insulin and related growth factors. Incubation of intact adipocytes or hepatoma cells with H_2O_2 increases phosphorylation of the insulin receptor and activates the insulin receptor kinase (Heffetz *et al.*, 1990). The same authors found that H_2O_2 rapidly augments tyrosine phosphorylation of several proteins in Fao cells that respond to insulin, and this effect was potentiated by vanadate. The enhanced phosphorylation was accompanied by decreased phosphotyrosine phosphatase activity. As B3CH-1 cells differentiated from myoblasts to myocytes, proteins undergoing tyrosine phosphorylation in response to H_2O_2 and vanadate appeared, in parallel with responses to insulin.

B. H_2O_2 Induction by Platelet-Derived Growth Factor

It has been reported that H_2O_2 generation is required for the transduction of signals following the binding of platelet-derived growth factor (PDGF) to rat vascular smooth muscle cells (Sundaresan *et al.*, 1995). Stimulation of the cells with PDGF was found to increase the intracellular concentration of H_2O_2. This increase could be attenuated by increasing the intracellular concentration of catalase or by the addition of N-acetylcysteine. The response of vascular smooth muscle cells to PDGF includes tyrosine phosphorylation, stimulation of mitogen-activated protein kinase, DNA synthesis, and chemotaxis. All these effects were inhibited when the PDGF-stimulated rise in H_2O_2 concentration was blocked. Exogenously added H_2O_2, in a narrow concentration range, stimulated tyrosine phosphorylation in and proliferation of vascular smooth muscle cells. The authors suggest that one mechanism by which cardioprotective effects of antioxidants (Stamper *et al.*, 1993; Rimm *et al.*, 1993) may be mediated is blocking the putative role of PDGF in atherosclerosis. Another effect of PDGF binding to its receptor is activation of the serine-threonine kinase c-Raf-1 (Morrison *et al.*, 1998), which phosphorylates IκBα and accelerates its degradation (Li and Sedivy, 1993). This is one mechanism by which oxidants activate the transcription nuclear factor (NF)-κB.

C. Redox Effects in Responses of T Lymphocytes

Schieven *et al.* (1993) reported that in human lymphocytes reactive oxygen intermediates, in combination with vanadate, activate the p56[lck] and p59[lyn] tyrosine kinases, which are important components of pathways signaling responses to antigenic stimulation. Anderson *et al.* (1994) found oxidative stimuli triggered NF-κB activation by one T-cell line tested but not another. In both cell lines TNF-α induced NF-κB activation. In all situations analyzed by Schieven and colleagues (1993), activation of NF-κB in lymphocytes and lymphocytic cell lines was tyrosine kinase dependent.

These redox effects are obviously complex and vary in different cell types and signal transduction systems. However, one of the ways in which antioxidants could exert useful pharmacological effects is by modifying such signal-transduction systems.

VI. Effects of Antioxidants on Transcription Factors _____

A. Nuclear Factor-κB

Although many transcription factors are inducible and arise by *de novo* synthesis, others pre-exist in cells. The activity of such primary transcription factors is regulated by posttranscriptional events, including phosphorylation, dephosphorylation, ligand binding, and proteolytic cleavage. NF-κB can be activated by several mechanisms, including growth factors, phorbol myristate acetate (PMA), and oxidants (Baeuerle and Henkel, 1994). The question arises whether it is possible to manipulate pharmacologically the function of a widely expressed primary transcription factor that can be activated in so many ways. Selectivity can be achieved by antioxidants because reactive oxygen intermediates can be generated in different cellular compartments and exert selective effects on signal transduction systems.

Lipopolysaccharides of gram-negative bacteria are potent inducers of the production of the pro-inflammatory cytokines TNF-α and IL-1β by cells of the monocyte-macrophage lineage. When human monocytes or rodent macrophages are exposed to LPS, a respiratory burst is recorded (Johnston, 1994). Although the magnitude of this burst is much less than that in neutrophils, it appears to be mediated by the same NADPH oxidase that transfers an electron to dioxygen. The burst is not observed in monocytes from patients with the sex-linked form of chronic granulomatous disease in which the primary genetic defect is in cytochrome b_{-245}, a membrane-associated enzyme required for superoxide production (Orkin, 1989). In the presence or absence of superoxide dismutase, superoxide is dismuted to hydrogen peroxide, which activates NF-κB, at least in some cells such as sublines of Jurkat T cells and HeLa cells (see Anderson et al., 1994; Baeuerle and Henkel, 1994).

Another mechanism of oxidant production appears to be involved in the TNF-α-mediated activation of NF-κB and killing of target cells. In this case, mitochondrial production of reactive oxygen intermediates seems to play a major role.

Baeuerle and Henkel (1994) conclude that oxidants represent a common intracellular messenger used for NF-κB activation by many inducing conditions. A variety of antioxidative agents suppress the activation of NF-κB, including N-acetyl-L-cysteine, L-cysteine, 2-mercaptoethanol, dithiocarbamates, α-lipoic acid, butylated hydroxyanisole, vitamin E and derivatives, and chelators of iron and copper ions. The chemical diversity of these compounds suggests that it is their shared antioxidative activity, rather than a particular structure-related activity, that suppresses NF-κB activation. The antioxidant pyrrollidone dithiocarbamate was found to prevent the degradation of IκB in response to PMA, suggesting that IκBα proteolysis is under control of the redox status of cells. According to a convenient model, H_2O_2 or other oxidants can activate kinases required for the site-specific phosphorylation of IκBα. This does not dissociate IκB from the trimer complex, but targets IκB for degradation by the ubiquitin proteasome pathway (Palombella et al., 1994; Brown et al., 1995; Lin et al., 1995). The p50/p65 transcriptionally active heterodimer then translocates to the nucleus.

A cytosolic serine-threonine kinase, Raf-1 kinase, is one candidate for this role. Cellular Raf-1 is an essential transducer of mitogenic signals from receptors for growth factors such as PDGF; c-Raf-1 is a substrate of the activated PDGF receptor (Morrison et al., 1988). Evidence now shows that H_2O_2 plays a role in the transduction of signals from the PDGF receptor (Sundaresan et al., 1995) and that Raf-1 kinase phosphorylates IκB (Li and Sedivy, 1993), resulting in the dissociation of IκB and the release of active NF-κB. Okadaic acid, which inhibits phosphatases 1 and 2A, activates NF-κB (Thevénin et al., 1991).

Tyrosine phosphorylation may also facilitate IκB degradation. Hypoxia (which depletes cellular antioxidants) increases the phosphorylation of IκB on tyrosine residues (Koong et al., 1994). Pretreatment of the cells with tyrosine kinase inhibitors or a dominant negative allele of Raf inhibits IκB degradation. IL-1 has been reported to induce tyrosine phosphorylation (Munoz et al., 1992), and the tyrosine kinase inhibitor herbimycin A can block the activation of NF-κB by IL-1 (Iwasaki et al., 1992). Activation of tyrosine phosphorylation by H_2O_2 has been described (Heffetz et al., 1990).

Redox events can also affect the binding of the p50/p65 heterodimer to DNA. Oxidation of NF-κB subunits in vitro abolishes DNA-binding activity, whereas the addition of reducing agents restores it. Site-specific mutagenesis showed that cysteine 62 in p53 is required for this effect (Matthews et al., 1992). This cysteine residue is conserved in all members of the rel family and is thought to be involved in protein–DNA interactions. Whether this redox-regulated event is important in vivo remains to be established.

B. Activator Protein-1 (AP-1)

Insights into the regulation of cell proliferation and the molecular basis of cancer have been gained through genetic and biochemical analyses of protooncogenes, the progenitors of retroviral-transforming genes. These include growth factors, cell surface receptors, G-proteins, and protein kinases. A subset of protooncogenes, illustrated by c-*fos* and c-*jun*, are expressed at low basal levels, but they can be induced rapidly and transiently by several extracellular stimuli. The proteins encoded by this family, and members of the related ATF/CREB family, form heterodimeric complexes through a coiled-coil structure called the leucine zipper. These complexes bind to AP-1 and the related cyclic AMP-responsive (CRE) motifs. Fos–Jun heterodimers bind DNA with high affinity, Jun homodimers bind DNA with lower affinity, and Fos does not form homodimers and binds DNA by itself.

The DNA-binding activity of Fos and Jun is regulated *in vitro* by a post-translational mechanism involving oxidation–reduction. Reduction of a conserved cysteine residue located in the DNA-binding domain of Fos and Jun by chemical reducing agents, or enzymically, stimulates AP-1 DNA binding *in vitro*, whereas oxidation or chemical modification of this residue inhibits DNA binding. The enzyme mediating this reduction is a product of the ref-1 gene, which also posseses apurinic/apyrimidinic endonuclease DNA repair activity (Xanthoudakis *et al.*, 1992). The same enzyme stimulates DNA binding by several other transcription factors, including NF-κB, Myb, and members of the ATF/CREB family. Thus one enzyme that may be involved in the redox control of transcription factor function, by maintaining a reducing environment in the nucleus, is identified.

In contrast, oxidative events in the cytoplasm can augment AP-1 activity in the nucleus. Exposure of cells to UV radiation damages DNA by the formation of pyrimidine dimers and oxidant effects, notably the formation of 8-hydroxydeoxyguanosine and 8-oxodeoxyguanosine. On Hela cells, both UV radiation and H_2O_2 were found to induce rapidly the expression of c-*jun* and c-*fos*, resulting in a long-lasting increase in AP-1 nuclear binding activity (Devary *et al.*, 1991). In the human promonocytic cell line U937, H_2O_2 was found to prime the cells for expression of the IL-1β gene (Lee and Ilnicka, 1993). This effect required protein synthesis and was not inhibited by staurosporine A, unlike PMA priming. Mobility shift assays using nuclear extracts showed that AP-1 activity was essentially unchanged after 0.5 hr but was markedly increased after 4 hr. In nuclei of unstimulated human monocytes, AP-1 activity was demonstrable, but it was significantly increased by LPS stimulation; in the presence of antioxidants, such as tetrahydropapaveroline, AP-1 activity was markedly decreased (Eugui *et al.*, 1994).

Comparable effects are not observed with all antioxidants in all cells. Thioredoxin reduces thiols and functions as a protein oxidoreductase inside cells. Transient expression or the exogenous addition of thioredoxin was found to inhibit NF-κB activity in the mouse fibrosarcoma cell line L929 and in the human cervical carcinoma cell line HeLa, but to increase AP-1 activity (Schenk *et al.*, 1994). This effect involved the *de novo* transcription of c-*jun* and c-*fos*, independently of protein kinase C. Structurally unrelated antioxidants such as pyrrolidine dithiocarbamate and butylated hydroxyanisole had similar effects. Thus effects on AP-1 depend on the antioxidant and cell type being studied.

Kinases are also required for AP-1 activation by mechanisms that cannot be reviewed here. As in the case of NF-κB, there could be redox effects on these kinases.

VII. Effects of Antioxidants on Gene Expression ⸻

A. Inhibition by Antioxidants of Endothelial Cell Activation

In the healthy vasculature, endothelial cells form a barrier preventing the egress of leukocytes into perivascular connective tissues and maintain an anticoagulant microenvironment. However, during inflammatory responses, endothelial cells become activated so that leukocytes are bound to their surfaces and can then migrate through the vessel wall. This process is accompanied by the expression of genes encoding adhesion molecules (E-selectin, VCAM-1, ICAM-1, and P-selectin). Genes for prothrombotic molecules (tissue factor and plasminogen activator inhibitor-1) also are expressed in activated endothelial cells, whereas the antithrombotic molecules thrombomodulin and heparan sulfate are downregulated; thus the probability of coagulation is increased.

Marui *et al.* (1993) and Ferran *et al.* (1995) have shown in endothelial cells that the antioxidant pyrrolidine dithiocarbamate (PDTC) inhibits the induction by TNF of adhesion molecules (E-selectin, VCAM-1, ICAM-1) and prothrombotic molecules (tissue factor and PAI-1). Molecular biological studies have shown that NF-κB is essential for E-selectin and ICAM-1 expression and that overexpression of IκB by genetic manipulation can prevent endothelial cell activation (Cheng *et al.*, 1994). PDTC inhibits TNF-induced NF-κB translocation to the nucleus in endothelial cells (Ferran *et al.*, 1995). Flavonoids also prevent endothelial cell activation, with the expression of adhesion molecules (Gerritsen *et al.*, 1995). Such inhibition could be therapeutically useful in inflammatory disorders, reperfusion injury, vasculitis, and allograft rejection.

B. Coordinate Inhibition by Antioxidants of the Production of Pro-inflammatory Cytokines

TNF-α and IL-1β are major cytokines produced by cells of the monocyte–macrophage lineage. It is generally accepted that TNF and IL-1 are pro-inflammatory and can exert effects independently as well as sequentially (Allison and Eugui, 1995); thus it is desirable to inhibit the production of both cytokines. Several laboratories have initiated research programs directed toward the identification of small molecules that inhibit the production and/or effects of TNF-α and IL-1β. The candidate compounds inhibit the production of both TNF-α and IL-1β at the levels of transcription (Eugui *et al.*, 1994) or translation (Lee *et al.*, 1994). Our strategy (Fig. 2) has been to identify antioxidants that inhibit, in a coordinate fashion, the activation of transcription factors required for the induced expression of the TNF-α and IL-1β genes in cells of the monocyte–macrophage lineage (Eugui *et al.*, 1994). The same compounds augment the expression of the gene for the IL-1 receptor antagonist (IL-1ra) in monocytes (Allison and Eugui, 1995).

Several reports have been published on the inhibitory effects of antioxidants on the production of individual cytokines by LPS-activated mouse peritoneal macrophages, human monocytes, THP-1 cells, and human whole blood. The phenomenon has been investigated systematically by comparing the effects of different types of antioxidants on the production of TNF-α, IL-1β, and IL-6 in cultured human peripheral blood mononuclear cells (PBMC) activated by LPS and in other ways (Eugui *et al.*, 1994). Some antioxidants, but not others, were found to be potent inhibitors of the production of all three cytokines. The molecular basis of antioxidant-mediated inhibition was analyzed, showing that antioxidants can inhibit the activation of transcription factors NF-κB and AP-1 (see earlier) and the transcription of cytokine genes. The *in vitro* observations were then confirmed by *in vivo* experiments in mice.

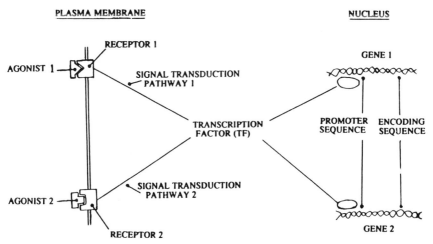

FIGURE 2   The strategic position of transcription factors at the point of convergence of different signal transduction pathways. By inhibiting the activation of transcription factors, it is possible to block the expression of several genes in a coordinate fashion.

Several moderately lipophilic antioxidants, including tetrahydropapaveroline (THP), butylated hydroxyanisole (BHA), and NDGA, were found to be potent inhibitors of IL-1β production (IC$_{50}$ 4 μM or lower, Table I). NDGA is an inhibitor of 5-lipoxygenase, as well as of lipid peroxidation. The structurally related dicatechol, rooperol, was subsequently shown to suppress the formation of IL-1β and TNF-α (Fig. 3). However, another redox 5-lipoxygenase inhibitor, zileuton, did not affect IL-1β formation, suggesting that 5-lipoxygenase products are not involved in the signal transduction system leading to cytokine production in LPS-activated monocytes.

The more hydrophilic antioxidants tested, ascorbic acid and trolox, had no effect on IL-1β production in concentrations up to 200 μM. Mannitol, a hydroxyl radical scavenger, was inactive at a 100 mM concentration. The same was true of the physiological lipophilic antioxidant α-tocopherol, as well as some classical antioxidants: butylated hydroxytoluene (BHT), quercetin, and N,N'-diphenyl-p-phenylenediamine. N-acetylcysteine had some inhibitory effect (IC$_{50}$ 42 mM), but was much lower than that of several lipophilic antioxidants (Table I). BHA, THP, 10,11-dihydroxyaporphine, and NDGA were further tested for effects on production of TNF-α and IL-6 by LPS-stimulated human PBMC. All three compounds were found to be approximately equipotent as inhibitors of the production of the three cytokines. However, the same compounds had no effect on IL-1-induced IL-6 production by fibroblasts.

Some of the compounds in Table I were selected for analysis of *in vivo* effects on cytokine production. TNF-α and IL-1β were measured in serum following a lethal challenge with LPS (200μg/mouse). Levels of TNF-α in circulating blood peak at about 1.5 hr, whereas those of IL-1β peak 4 hr after LPS injection. Subcutaneous administration of a single dose of 10,11-

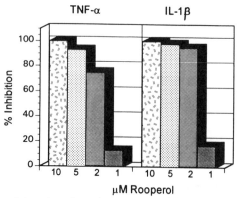

FIGURE 3   Rooperol, in a dose-dependent manner, inhibits the production of TNF-α and IL-1β by LPS-activated human peripheral blood mononuclear cells. Methods are described by Eugui *et al.* (1994).

TABLE I   Antioxidants Vary Widely in
Potency as Inhibitors of Cytokine Formation[a]

	$IC_{50}(\mu M)$
High activity	
Butylated hydroxyanisole	2.9
Tetrahydropapaveroline	1.0
Apomorphine	2.6
Norapomorphine	1.6
Nordihydroguaiauretic acid	1.3
Mepacrine	3.0
Low activity (insignificant inhibition in the range 50–200 μM)	
Ascorbic acid	
α-Tocopherol	
Mannitol	
Trolox	
Butylated hydroxytoluene	
Quercetin	
N,N'-Diphenyl-p-phenylene diamine	
Zileuton (5-lipoxygenase inhibitor)	

[a] Data from Eugui et al. (1994).

dihydroxyaporphine (100 mg/kg), given 30 min before challenge, inhibited TNF-α production by 95%. Before the IL-1β peak level, mice were given a second dose of dihydroxyaporphine (50 mg/kg), 2 hr after LPS and 2 hr before blood sampling. This treatment reduced the circulating levels of IL-1β by 88%. Thus in vivo cytokine production is strongly inhibited by THP and by dihydroxyaporphine.

To analyze the mechanism by which antioxidants suppress cytokine formation, levels of cytokine mRNAs were determined in LPS-stimulated PBMC (Eugui et al., 1994). LPS increases levels of IL-1β, TNF-α, IL-6, and IL-8 mRNAs in PBMC. The antioxidants tested decreased TNF-α, IL-1β, and IL-6 mRNA levels to baseline expression, but had less effect on IL-8 mRNA. To ascertain whether these effects are due to changes in transcription or changes in mRNA stability, nuclear transcription assays were performed. LPS markedly stimulated transcription of the IL-1β gene, and THP antagonized the stimulatory effect of LPS on transcription.

The experiments just described suggest that antioxidants suppress the LPS-stimulated transcription of some cytokine genes. To ascertain whether this is correlated with the inhibited activation of transcription factors, electrophoretic mobility shift assays were performed. NF-κB and AP-1 were analyzed first because of reports that they are subject to redox regulation.

Nuclear extracts of unstimulated PBMC show NF-κB activity, which increases following LPS stimulation. Specificity of binding was shown by competition with unlabeled oligonucleotides. When PBMC were treated with THP, in the presence or absence of LPS, nuclear NF-κB activity was markedly decreased or eliminated altogether. A higher concentration of THP did not affect the binding of NF-κB proteins to the DNA conserved sequence. Similar observations were made with AP-1 (Eugui et al., 1994). Thus THP inhibited the activation of NF-κB and AP-1 in intact cells but had no effect on the binding of these transcription complexes to cognate DNA recognition sequences. In cells treated with THP, no effect on several other transcription factors (SP-1, CRE, CTF/NF-1, OCT) could be demonstrated.

Inhibition by antioxidants of the activation of transcription factors, including NF-κB and AP-1, could explain their suppression of cytokine gene transcription. It is generally accepted that a c-fos serum-responsive element and NF-κB play major roles in promoting expression of the IL-6 gene (Hirano et al., 1990). Evidence implicating κB sequences in LPS-induced expression of the TNF-α gene in macrophages also has been presented (Shakhov et al., 1990). NF-κB has also been shown to be required for the induced expression of IL-1β genes in cells of the monocyte–macrophage lineage (Hiscott et al., 1993).

VIII. Discussion

The pathogenesis of inflammation is complex, involving proteins such as cytokines, complement components, and adhesion molecules; peptide mediators such as kinins; and lipid mediators such as prostaglandins, leukotrienes, and platelet-activating factor. To achieve therapeutically useful anti-inflammatory activity, it is often necessary to use drugs with a broad range of effects. The classical example is glucocorticoids, which augment IκB production, thereby preventing NF-κB activation; inhibit TNF-α and IL-1β formation; suppress the expression of PGHS-2; and prevent phospholipase A$_2$ activation, thereby suppressing the production of lipid mediators (see Allison and Eugui, 1995). Unfortunately, glucocorticoids also produce metabolic bone disease and other limiting side effects.

As reviewed in this chapter, some antioxidants also have broad effects: on signal transduction systems, including kinases and phosphatases; on the activation of transcription factors and gene expression; and on phospholipase A$_2$ activation. Indeed, it is possible to select antioxidants with a desired combination of these activities, as well as on prostaglandin synthases and/or 5-lipoxygenase. This has opened up a novel therapeutic strategy for inflammatory disorders, shock, reperfusion injury, and other applications based on the organ, tissue, cellular, and intracellular targeting of antioxidant drugs.

References

Allison, A. C., and Eugui, E. M (1995). Induction of cytokine formation by bacteria and their products. In "Virulence Mechanisms of Bacterial Pathogens" (J. A. Roth, ed.), 2nd Ed., pp. 303–332. American Society for Microbiology, Washington, DC.

Anderson, M. T., Staal, F. J. T., Gitler, et al. (1994). Separation of oxidant-initiated and redox-regulated steps in the NF-κB signal transduction pathway. Proc. Natl. Acad. Sci. USA 91, 11527–11531.

Baeuerle, P. A., and Henkel, T. (1994). Function and activities of NF-κB in the immune system. Annu. Rev. Immunol. 12, 141–179.

Balsari, A. L., Morelli, D., Ménard, S., Veronesi, U., and Colnaghi, M. I. (1994). Protection against doxorubicin-induced alopecia in rats by liposome-entrapped monoclonal antibodies. FASEB J. 8, 226–230.

Beg, A. A., Finco, T. S., Nanternet, P. V., and Baldwin, A. S., Jr. (1993). Tumor necrosis factor and interleukin-1 lead to phosphorylation and less of IκB: A mechanism for NF-κB activation. Mol. Cell. Biol. 13, 3301–3310.

Bissett, D. L., Oelrich, M., and Hannon, D. P. (1994). Evaluation of a topical iron chelator in animals and human beings: Short-term photoprotection by 2-furildioxime. J. Am. Acad. Dermatol. 31, 572–578.

Black, H. S. (1987). Potential involvement of free radical reactions in ultraviolet light-mediated cutaneous damage. Photochem. Photobiol. 46, 213–221.

Brown, K., Gerstberger, S., Carleon, L., et al. (1995). Control of IκB proteolysis by site-specific signal induced phosphorylation. Science 267, 1485–1488.

Chen, Z., Silva, H., and Klessig, D. F. (1993). Active oxygen species in the induction of plant systemic acquired resistance by salicylic acid. Science 262, 1883–1886.

Cheng, Q., Cant, C. A., Moll, T., et al. (1994). NF-κB subunit-specific regulation of the IκB promoter. J. Biol. Chem. 269, 13551–13557.

Devary, Y., Gottlieb, R. A., Lau, L. F., and Karin, M. (1991). Rapid and preferential activation of the c-jun gene during the mammalian UV response. Mol. Cell. Biol. 11, 2804–2811.

Eberhart, C. E., Coffey, R. J., Radhika, A., et al. (1994). Up-regulation of cyclooxygenase 2 gene expression in human colorectal adenomas and adenocarcinomas. Gastroenterology 107, 1183–1188.

Eugui, E. M., De Lustro, B., Rouhafza, S., et al. (1994). Some antioxidants inhibit, in a coordinate fashion, the production of tumor necrosis factor-α, IL-1β, and IL-6 by human peripheral blood mononuclear cells. Int. Immunol. 6, 409–422.

Ferran, C., Millan, M. T., Csizmadia, V., et al. (1995). Inhibition of NF-κB by pyrrolidine dithiocarbamate blocks endothelial cell activation. Biochem. Biophys. Res. Commun. 214, 212–223.

Gerritsen, M. E., Carley, W. W., Ranges, G. E., et al. (1995). Flavonoids inhibit cytokine-induced endothelial cell adhesion protein gene expression. Am. J. Pathol. 147, 278–292.

Giovannucci, E., Rimm, E. B., Stampfer, M. J., et al. (1994). Aspirin use and the risk for colorectal cancer and adenoma in male health professionals. Ann. Intern. Med. 121, 241–246.

Heffetz, D., Bushkin, H., and Zuk, Y. (1990). The insulinomimetic agents H_2O_2 and vanadate stimulate protein tyrosine phosphorylation in intact cells. J. Biol. Chem. 265, 2896–2902.

Hirano, T., Akira, S., Taga, T., and Kishimoto, T. (1990). Biological and clinical aspects of interleukin-6. Immunol. Today 11, 443–449.

Hiscott, J., Marois, J., Garaufolis, J., et al. (1993). Characterization of a functional NF-κB site in the human interleukin 1β gene. Mol. Cell Biol. 13, 6231–6240.

Iwasaki, T., Uehara, Y., Graves, L., et al. (1992). Herbimycin A blocks IL-1 induced NF-κB DNA-binding activity in lymphoid cell lines. FEBS Lett. 298, 240–244.

Johnston, R. B., Jr. (1994). Measurements of O_2^- secreted by monocytes and macrophages. Methods Enzymol. 105, 365–380.

Kargman, S., O'Neill, G., Vickers, P., *et al.* (1995). Expression of prostaglandin G/H synthase-1 and -2 protein in human colon cancer. *Cancer Res.* **55**, 2556–2559.

Koong, R. C., Chen, E. Y., and Giaccia, A. J. (1994). Hypoxia causes the activation of nuclear factor κB through the phosphorylation of IκB on tyrosine residues. *Cancer Res.* **54**, 1425–1430.

Langerbach, R., Morham, S. G., Tiano, H. F., *et al.* (1995). Prostaglandin synthase 1 gene disruption in mice reduces arachidonic acid-induced inflammation and indomethacin-induced gastric ulceration. *Cell* **83**, 483–492.

Lee, J. C., Badger, A. M., Griswold, D. E. *et al.* (1994). Bicyclic imidazoles as a novel class of cytokine synthesis inhibitors. *Ann. N.Y. Acad. Sci.* **696**, 149–170.

Lee, S. W., and Ilnicka, M. (1993). Hydrogen peroxide primes U937 cells to produce IL-1β. *Ann. N.Y. Acad. Sci.* **696**, 399–400.

Levine, A., Tenhaken, R., Dixon, R., and Lamb, C. (1994). H_2O_2 from the oxidative burst orchestrates the plant hypersensitive disease resistance response. *Cell* **79**, 583–593.

Li, S., and Sedivy, J. M. (1993). Raf-1 protein kinase activates the NF-κB transcription factor by dissociation of the cytoplasmic NF-κB–IκB complex. *Proc. Natl. Acad. Sci. U.S.A.* **90**, 9247–9251.

Lieb, L. M., Ramchendron, C., Egbaria, K., and Weiner, N. (1992). Topical delivery enhancement with multilamellar liposomes into pilosebacous units. 1. *In vitro* evaluation using fluorescent techniques with hamster ear model. *J. Invest Dermatol.* **99**, 108–113.

Lin, Y.-C., Brown, K., and Siebenlist, V. (1995). Activation of NF-κB requires proteolysis of the inhibitor IκBα; signal-induced phosphorylation of IκBα alone does not release active NF-κB. *Proc. Natl. Acad. Sci. U.S.A.* **92**, 552–556.

Marui, N., Offermann, M. K., Swerlick, R., *et al.* (1993). Vascular cell adhesion molecule-1 (VCAM-1) gene transcription and expression are regulated through an antioxidant-sensitive mechanism in human vascular endothelial cells. *J. Clin. Invest.* **92**, 1866–1874.

Matthews, J. R., Wakasugi, N., Virelizier, J.-L., *et al.* (1992). Thioredoxin regulates the DNA binding activity of NF-κB by reduction of a disulphide bond involving cysteine 62. *Nucleic Acids Res.* **20**, 3821–3830.

Moncada, S., and Higgs, A. (1993). The L-arginine-nitric oxide pathway. *N. Engl. J. Med.* **329**, 2002–2112.

Morham, S. G., Langerbach, R., Loftin, C. D., *et al.* (1995). Prostaglandin synthase 2 gene disruption causes severe renal pathology in the mouse. *Cell* **83**, 473–482.

Morrison, D. K., Kaplan, D. R., Rapp, U., and Roberts, T. M. (1988). Signal transduction from membrane to cytoplasm: Growth factors and membrane-bound oncogene products increase Raf-2 phosphorylation and associated protein kinase activity. *Proc. Natl. Acad. Sci. U.S.A.* **85**, 8855–8859.

Munoz, E., Zubiaga, A. M. Huang, C. K., and Huber, B. T. (1992). Interleukin-1 induces protein tyrosine phosphorylation. *Eur. J. Immunol.* **22**, 2101–2106.

Nacy, C. A., Gilbreath, M. J., Swartz, G. M., *et al.* (1989). Liposome composition and activation of macrophages for antimicrobial activities against leishmania parasites. *In* "Liposomes in the Therapy of Infectious Diseases and Cancer" (G. Lopez-Berestein and I. Fidler, eds.). pp. 167–176. A. R. Liss, New York.

Olsen, E. A., Abernathy, L., Kulp-Shorten, C., *et al.* (1991). A double-blind, vehicle-controlled study evaluating masoprocol cream in the treatment of actinic keratoses on the head and neck. *J. Am. Acad. Dermatol.* **24**, 738–743.

Orkin, S. H. (1989). Molecular genetics of chronic granulomatous disease. *Annu. Rev. Immunol.* **7**, 277–307.

Palombella, V. J., Rando, O. J., Goldberg, A. L., and Maniatis, T. (1994). The ubiquitin-proteasome pathway is required for processing the NF-κB1 precursor protein and the activation of NF-κB. *Cell* **76**, 773–785.

Pasricka, P. J., Redi, A., O'Connor, K., *et al.* (1995). The effects of sulindac on colorectal proliferation and apoptosis in familial adenomatous polyposis. *Gastroenterology* **109**, 994–998.

Reddy, B. S., Rao, C. V., Riensan, A., and Kelloff, G. (1993). Inhibitory effect of aspirin on azoxymethane-induced colon carcinogenesis in F344 rats. *Carcinogenesis* **14**, 1493–1497.

Rimm, E. B., Stampfer, M. J., Ascherio, A., *et al.* (1993). Vitamin E consumption and the risk of coronary heart disease in men. *N. Engl. J. Med.* **328**, 1450–1456.

Robison, T. W., Sevanian, A., and Forman, H. J. (1990). Inhibition of arachidonic acid release by nordihydroguiaretic acid and its antioxidant action in rat alveolar macrophages and Chinese hamster lung fibroblasts. *Toxicol. Appl. Pharmacol.* **105**, 113–122.

Schenk, H., Klein, M., Erdbrügger, W., *et al.* (1994). Distinct effects of thioredoxin and antioxidants on the activation of transcription factors NF-κB and AP-1. *Proc. Natl. Acad. Sci. U.S.A.* **91**, 1672–1676.

Schieven, G. L., Kirihara, J. M., Myers, J. A., *et al.* (1993). Reactive oxygen intermediates activate NF-κB in a tyrosine-kinase dependent mechanism, and in combination with vanadate activate the $p56^{kk}$ and $p59^{lyn}$ tyrosine kinases in human lymphocytes. *Blood* **82**, 1212–1220.

Schultze-Osthoff, K., Bakker, A. C., Vanhaesebroeck, B., *et al.* (1992). Cytotoxic activity of tumor necrosis factor is mediated by early damage to mitochondrial functions: Evidence for the involvement of mitochondrial radical generation. *J. Biol. Chem.* **267**, 5317–5322.

Schultze-Osthoff, K., Beyaert, R., Vandervoorde, V., *et al.* (1993). Depletion of mitochondrial electron transport abrogates the cytotoxic and gene induction effects of TNF. *EMBO J.* **12**, 3095–3104.

Shakhov, A. N., Collart, M. A., Vassali, P., *et al.* (1990). Kappa B-type enhancers are involved in lipopolysaccharide-mediated transcriptional activation of the tumor necrosis factor alpha gene in primary macrophages. *J. Exp. Med.* **171**, 35–47.

Shimokawa, T., and Smith, W. L. (1992). Prostaglandin endoperoxide synthase: The aspirin acetylation region. *J. Biol. Chem.* **267**, 12387–12392.

Smith, W. L., Meade, E. A., and De Witt, D. L. (1994). Interactions of PGH synthase isozymes-1 and -2 with NSAIDs. *Ann. N.Y. Acad. Sci.* **744**, 50–57.

Stamper, M. J., Hennekens, C. H., Manson, J. E., *et al.* (1993). Vitamin E consumption and the risk of coronary heart disease in women. *N. Engl. J. Med.* **328**, 1444–1449.

Storz, G., Tartaglia, L. A., and Ames, B. N. (1990). Transcriptional regulator of oxidative inducible genes: Direct activation by oxidation. *Science* **248**, 189–194.

Sundaresan, M., Yu, Z.-X., Ferrans, V. J., *et al.* (1995). Requirement for generation of H_2O_2 for platelet-derived growth factor signal transduction. *Science* **270**, 296–298.

Thévenin, C., Kim, S.-J., Riechmann, P., *et al.* (1991). Induction of nuclear factor κB and the human immunodeficiency virus long terminal repeat by okadaic acid, a specific inhibitor of phosphatases 1 and 2A. *New Biol.* **2**, 793–800.

Tsuji, M., and Du Bois, R. N. (1995). Alterations in cellular adhesion and apoptosis in epithelial cells over-expressing prostaglandin endoperoxide synthase 2. *Cell* **83**, 493–501.

Van der Merwe, M. J., Jenkins, K., Theron, E., and Van der Walt, B. J. (1993). Interaction of the di-cathechols rooperol and nordihycroguiauretic acid with oxidative systems in the human blood. *Biochem. Pharmacol.* **45**, 303–311.

Van der Zee, J., Eling, T. E., and Mason, R. P. (1989). Formation of free radical metabolites in the reaction between soybean lipoxygenase and its inhibitors: An ESR study. *Biochemistry* **28**, 8363–8367.

Waters, R. V., Webster, D., and Allison, A. C. (1993). Mycophenolic acid and some antioxidants induce differeentiation of monocytic lineage cells and augment production of the IL-1 receptor antagonist. *Ann. N.Y. Acad. Sci.* **696**, 185–196.

Weldon, J. S., Nunwell, J. F., Hansen, W. L., and Alving, C. R. (1983). Liposomal chemotherapy in visceral leishmaniasis: An ultrastructural study of an intracellular pathway. *Z. Parasitekd.* **69**, 415–424.

Wong, G. H. W., Elwell, J. H., Oberley, L. W., and Goeddel, D. V. (1989). Marginosis superoxide dismutase is essential for cellular resistance to cytotoxicity of tumor necrosis factor. *Cell* **58**, 923–931.

Xanthoudakis, S., Miao, G., Wong, F., *et al.* (1982). Redox activation of Fos-Jun DNA binding activity is mediated by a DNA repair enzyme. *EMBO J.* **11**, 3323–3335.

Thomas Primiano
Thomas R. Sutter
Thomas W. Kensler
Department of Environmental Health Sciences
The Johns Hopkins School of Hygiene and Public Health
Baltimore, Maryland 21205

Antioxidant-Inducible Genes

I. Introduction

Antioxidants inhibit the propagation of free radical reactions and therefore they have been utilized to prevent food spoilage (Sies, 1986; Halliwell and Gutteridge, 1989). Because of their widespread use as preservatives in processed foods, their biochemical effects have been investigated. Early feeding studies indicated that butylated hydroxytoluene (BHT) increased liver weight, induced proliferation of smooth endoplasmic reticulum, and elevated several hepatic microsomal mono-oxygenase activities typical of phase 1 metabolism (Gilbert and Goldberg, 1965; Botham *et al.*, 1970; Parke *et al.*, 1974a,b; Kahl and Netter, 1977; Benson *et al.*, 1978). In addition to phase 1, antioxidants also induce phase 2 xenobiotic-metabolizing enzymes. Phase 2 enzymes detoxify activated electrophilic metabolites of xenobiotics via conjugation of endogenous substrates such as glutathione (GSH). Induction of phase 2 xenobiotic-metabolizing enzymes was demonstrated to enhance the detoxication of potentially toxic electrophiles. For example, the levels of oxidized metabolites of benzo[*a*]pyrene were substantially reduced or virtually eliminated by the addition of either reduced GSH or rodent liver cytosol xenobiotic-metabolizing enzymes, such as glutathione *S*-transferases (GSTs), to a *Salmonella* mutagenicity test system (Benson *et al.*, 1978). Thus, antioxidants act not only to directly terminate the propagation of free radical reactions, but also to increase the activity of enzymes which readily metabolize and aid in the elimination of potential cytotoxic chemicals. Evidence that dietary administration of antioxidants induced phase 2 xenobiotic-metabolizing enzymes was first presented by Benson *et al.* (1978) and Cha *et al.* (1978). Butylated hydroxyanisole (BHA) caused large increases in the activity of hepatic microsomal epoxide hydrolase (mEH) and cytosolic GSTs in CD-1 mice and Sprague–Dawley rats. Activities of phase 2 xenobiotic-metabolizing enzymes were induced from 2- to 10-fold depending on the dose and specific antioxidant utilized (Cha *et al.*, 1979; Kensler *et al.*, 1985). Induction of phase 2 enzymes by antioxidants was found in many organs and tissues, such as liver, lung, kidney, small intestine, colon, and spleen, thereby affording protection at many anatomical sites (Benson *et al.*, 1979; Ansher *et al.*, 1986; Pearson *et al.*, 1988). This chapter presents evidence for several newly identified antioxidant-inducible genes, describes the proposed mechanisms of antioxidant signal transduction leading to enhanced expression of these enzymes, and complements the information presented in related reviews concerning the mechanisms and consequences of induction of phase 2 xenobiotic-metabolizing enzymes by antioxidants (Sies, 1986; Talalay, 1989; Kensler *et al.*, 1991, 1992a, 1993).

II. Genes Induced by Antioxidants

Antioxidant-inducible genes are expressed in most mammalian cells and tissues and have usually been found to be phase 1 and phase 2 xenobiotic-metabolizing enzymes. Expression of other proteins such as transcription factors, ferritin, and ribosomal proteins have been shown to be increased following antioxidant administration, and a comprehensive list of genes induced by antioxidants and references to the pertinent evidence for their inductions are presented in Table I.

A. Cytochrome P450s

Some antioxidants such as ethoxyquin and BHT induce certain cytochrome P450s. The cytochrome P450s are a superfamily of heme-containing enzymes that utilize molecular oxygen to hydroxylate a variety of organic substrates (for reviews, see Gonzalez, 1988; Okey, 1990; Nebert, 1991; Guengerich, 1992). The P450s can also transfer reducing equivalents from

NADPH via a specific P450 reductase to dealkylate a number of aliphatic compounds. The activity of P450 often results in the formation of electrophilic intermediary metabolites that may covalently modify DNA, proteins, and lipids. The amount of toxic metabolite available for interaction with its targets is represented, in part, by a balance between competing activation and detoxication reactions. Antioxidants typically provide protection by inducing the phase 2 conjugating systems so that xenobiotic detoxication overcompensates for the phase 1 activating systems. The methoxyresorufin. O-deethylation activity specific for the CYP1A family of enzymes was enhanced 2- to 4-fold by the dietary administration of ethoxyquin and oltipraz [4-methyl-5-(2-pyrazinyl)-1,2-dithiole-3-thione], but not BHA (Buetler *et al.*, 1995). An elevation in activity was associated with 2- to 3-fold increases of CYP1A2 and CYP3A2 mRNA levels. The mechanism of this induction of mRNA has not been identified to date. In comparison, ethoxyquin was an efficacious inducer of CYP2B-related benzyloxyresofurin de-ethylation (135-fold), whereas neither BHA nor oltipraz induced CYB2B activity (Putt *et al.*, 1991; Buetler *et al.*, 1995). The induction of cytochrome P450s is generally thought to metabolically activate procarcinogens, such as the formation of aflatoxin B_1-8,9-epoxide by CYP1A2 and CYP3A2 (Raney *et al.*, 1992). However, reductions in aflatoxin B_1–DNA and protein adducts, as well as the formation of preneoplastic foci, occur by pretreatment of the exposed animals with antioxidants (Kensler *et al.*, 1985; 1992b; Roebuck *et al.*, 1991; Egner *et al.*, 1995). These results suggest that the coinduction of both phase 1 and 2 enzymes, in the case of aflatoxin B_1, may favor a more rapid detoxication and consequent inhibition of tumorigenesis. Induction of these cytochrome P450s, therefore, may play an important role in activating xenobiotics so that they become much better substrates for the phase 2 conjugating systems, which are also induced. Alternatively, some antioxidants, such as oltipraz, have been reported to inhibit CYP1A2 activities *in vitro* (Putt *et al.*, 1991; Langouët *et al.*, 1995). This inhibition of the P450s could reduce the production of reactive metabolites, although it would appear to be transient in nature and dependent on the continued presence of the inhibitor (Primiano *et al.*, 1995a). Whether antioxidants inhibit or accelerate the production of reactive metabolites through modulation of the activities of cytochrome P450s needs to be clearly defined. Nevertheless, the integrated metabolic effects are generally to protect cells from toxicity and mutagenicity.

B. Glutathione S-Transferases

One of the most studied families of antioxidant-inducible genes are the GSTs. The GSTs are a family of dimeric enzymes that possess a multitude of functions, including the enzymatic conjugation of glutathione to electrophilic xenobiotics, the binding of endogenous ligands such as bilirubin, heme, and

TABLE I Induction of Genes by Antioxidants

Antioxidant-induced genes	Typical inducers	Species/tissue	Reference[b]
GST Ya1 and GST Ya2	BHA[a]	Rat L	Benson et al., 1978[1,2]; Kensler et al., 1985[1]; Meyer et al., 1993a[2]
	BHA	Mouse liver	Benson et al., 1978[1]; Pearson et al., 1983[3]; Friling et al., 1992a[5]; Prestera et al., 1993a[5]
	BHA	Rat and mouse L, Lg, K, SI	Benson et al., 1979[1]; Pearson et al., 1988
	BHT	Rat L	Kensler et al., 1985[1]
	EQ	Rat and mouse L	Benson et al., 1978[1,2]
	OLT	Rat L	Kensler et al., 1985, 1987[1]; Hayes et al., 1994[2]; Davidson et al., 1990[4]; Primiano et al., 1995a[3]
	OLT	Rat L, Lg, K, SI, St	Ansher et al., 1986[1]; Meyer et al., 1993a[2]
	OLT	Mouse L	Prestera et al., 1993[5]; Clapper et al., 1994[4]
	OLT	Human hepatocytes	Morel et al., 1993[3]
	D3T	Rat L	Kensler et al., 1987; Primiano et al., 1995a[4]
	D3T	Rat L, Lg, K, SI, St	Ansher et al., 1986[1]; Meyer et al., 1993a[3]
	DS	Mouse L, Lg, K, SI, St, Co	Benson and Barreto, 1985[1]
GST Yb1 and GST Yb2	BHA	Rat, L, Lg, K, SI, St	Benson et al., 1978[1,2]; Kensler et al., 1985[1]; Meyer et al., 1993a[2]
	BHA	Mouse L, Lg	Benson et al., 1979[1]
	BHT	K, SI	Kensler et al., 1985[1]
	EQ	Rat L	Benson et al., 1978[1,2]; Kensler et al., 1985[1]; Hayes et al., 1994[2]
	OLT	Rat L	Kensler et al., 1985, 1987[1]; Davidson et al., 1990[4]; Hayes et al., 1994[2]; Primiano et al., 1995a[3]
	OLT	Rat L, Lg, K, SI, St	Ansher et al., 1986[1]; Meyer et al., 1993a[2]
	OLT	Mouse L	Clapper et al., 1994[4]
	D3T	Rat L	Hayes et al., 1994; Kensler et al., 1985, 1987; Primiano et al., 1995a[4]
	D3T	Rat L, Lg, K, SI, St	Ansher et al., 1986[1]; Meyer et al., 1993a[3]
	DS	Mouse, L, Lg, K, SI, St	Benson and Barreto, 1985[1]
GST Yc1	OLT	Rat L, K	Meyer et al., 1993a[2]
		Rat L	Hayes et al., 1994[2]
	D3T	Rat L, Lg, K, SI, St	Meyer et al., 1993a[2]

GST Yc2	BHA	Rat L	Hayes et al., 1994[2]; Primiano et al., 1995a[3]
	EQ	Rat L	Hayes et al., 1994[2]
	OLT	Rat L, K	Hayes et al., 1994[2]
	OLT	Rat L	Primiano et al., 1995a[2]
	D3T	Rat L	Meyer et al., 1993a[2]; Primiano et al., 1995b[4]
GST Yp	OLT	Rat L	Meyer et al., 1993a[2]; Primiano et al., 1995a[2]
	OLT	Mouse L	Clapper et al., 1994[4]
	D3T	Rat L	Primiano et al., 1995b[4]
QR	BHA	Rat L, Lg, K, SI, St	DeLong et al., 1986[1,2]
	BHT	Rat L	Kensler et al., 1985[1]
	EQ	Rat L	Kensler et al., 1985[1]
	tBHQ	Rat L	Favreau and Pickett, 1991[5]
	tBHQ	Human L	Li and Jaiswal, 1992a[5]
	OLT	Rat L	Kensler et al., 1985[1], 1987[1]; Prestera et al., 1993[5]; Primiano et al., 1995a[4]
	OLT	Mouse and rat L, Lg, K, SI	Ansher et al., 1986[1]
	D3T	Rat L	Kensler et al., 1987[1]; Primiano et al., 1995b[4]
UGT1	BHA	Mouse L	Benson et al., 1986[1]
		Mouse L, Lg, K, SI, St, Co	Cha and Bueding, 1979[1]
	OLT	Rat liver	Kensler et al., 1985[1]; Kashfi et al., 1994[3]; Buetler et al., 1995[4]
	D3T	Rat liver	Kensler et al., 1987[1]
UGT2	BHA	Mouse L	Benson et al., 1986[1]
	BHT	Rat L	Kensler et al., 1985[1]; Kashfi et al., 1994[3]
	EQ	Rat L	Kensler et al., 1985[1]; Kashfi et al., 1994[3]
EH	BHA	Rat and mouse L	Cha et al., 1978[1]; Benson et al., 1979[1]
	BHA	Rat L	Kensler et al., 1985[1]
	BHT	Rat L	Kensler et al., 1985[1]
	EQ	Rat L	Kensler et al., 1985[1]
	OLT	Rat L	Kensler et al., 1985[1]
	D3T	Rat L	Kensler et al., 1985[1], 1987[1]; Primiano et al., 1995b[3]

(continues)

TABLE I—*Continued*

Antioxidant-induced genes	Typical inducers	Species/tissue	Reference[b]
AFAR	BHA	Rat L	Hayes et al., 1994[3]; Ellis et al., 1993[3]
	OLT	Rat L	Ellis et al., 1993[3]
	D3T	Rat L	Primiano et al., 1995b[4]
G6PD	BHA	Rat L	Cha and Bueding, 1979[1]; Kensler et al., 1985[1]
	BHT	Rat L	Kensler et al., 1985[1]
	EQ	Rat L	Kensler et al., 1985[1]
	OLT	Rat L	Kensler et al., 1985[1], 1987[1]
	OLT	Mouse and rat L, Lg, K, SI	Ansher et al., 1986[1]
	D3T	Rat L	Kensler et al., 1987[1]
CYP1A1	BHT	Rat L	Putt et al., 1991[1]
	EQ	Rat L	Putt et al., 1991[1]
CYP1A2	BHT	Rat L	Kensler et al., 1985[1]; Buetler et al., 1995[4]
	EQ	Rat L	Botham et al., 1970[1]; Putt et al., 1991[1,2]; Buetler et al., 1995[3]
	OLT	Rat L	Putt et al., 1991[1,2]; Buetler et al., 1995[3]
CYP2B1	EQ	Rat L	Parke, 1974a,b[1]; Buetler et al., 1995[3]
	OLT	Rat L	Putt et al., 1991[1,2]; Buetler et al., 1995[3]
CYP2E1	OLT	Rat L	Putt et al., 1991[1,2]
CYP3A2	BHA	Rat L	Putt et al., 1991[1,2]; Buetler et al., 1995[3]
	EQ	Rat L	Putt et al., 1991[1,2]; Buetler et al., 1995[3]
	OLT	Rat L	Putt et al., 1991[1,2]; Buetler et al., 1995[3]

c-*fos*	BHA	HepG2	Choi and Moore, 1993[5]
	BHT	HepG2	Choi and Moore, 1993[5]
fra-1	BHA	HeLa, HepG2, F9	Yoshioka et al., 1995[5]
fra-2	BHA	HeLa, HepG2, F9	Yoshioka et al., 1995[5]
c-*jun*	BHA, BHT	HepG2, Hepa1c1c7	Choi and Moore, 1993[5]; Bergelson et al., 1994
γGCS	BHA	Mouse L	Eaton and Hamel, 1994[1]; Borroz et al., 1994[4]
	BHA, BHT	Human L	Mulcahy and Gipp, 1995[5]
GSH reductase	BHA	Rat L	Kensler et al., 1985[1]
	BHT	Rat L	Kensler et al., 1985[1]
	EQ	Rat L	Kensler et al., 1985[1]
	OLT	Rat L	Kensler et al., 1985[1]
	OLT	Mouse and rat L, Lg, K, SI	Ansher et al., 1986[1]
	D3T	Rat L	Kensler et al., 1987[4]
Ferritin L	D3T	Rat L	Primiano et al., 1995b[2,4]
Ferritin H	D3T	Rat L	Primiano et al., 1995b[4]
Ribosomal S16	D3T	Rat L	Primiano et al., 1995b[4]
Ribosomal L18a	D3T	Rat L	Primiano et al., 1995b[4]

[a] The following abbreviations are used: GST, glutathione S-transferase; QR, NAD(P)H:quinone reductase; mEH, microsomal epoxide hydrolase; UGT, UDP-glucuronosyltransferase; AFAR, aflatoxin B₁ aldehyde reductase; G6PD, glucose-6-phosphate dehydrogenase; γGCS, γ-glutamylcysteine synthetase; BHA, butylated hydroxyanisole; BHT, butylated hydroxytoluene; EQ, ethoxyquin; DS, disulfiram; OLT, oltipraz; D3T, 1,2-dithiole-3-thione; L, liver; Lg, lung; K, kidney; SI, small intestine; St, stomach; Co, colon.

[b] The superscript numbers indicate the measurements for analyzing the mechanisms of induction utilized in each reference: (1) increased enzymatic or functional activity; (2) increased protein levels; (3) increased level of mRNA transcripts; (4) increased rate of transcription; and (5) antioxidant response element-mediated induction.

steroids, and selenium-independent glutathione peroxidase activity, which is significant in the detoxication of lipid hydroperoxides (including prostaglandins) and nucleotide hydroperoxides (for reviews, see Ketterer et al., 1988; Tsuchida and Sato, 1992). Four major classes of mammalian cytosolic isozymes, α, μ, π, and θ, have been identified from the primary structures deduced from cDNA sequencing. Except for θ, all of the classes react strongly to catalyze the conjugation of 1,2-chloro-3-dinitrobenzene to reduced glutathione. The activities of GSTs are elevated by BHA, BHT, ethoxyquin, and dithiolethiones between 1.5- and 4-fold in many rodent tissues (Table I). The induction of GSTs is generally greater in mouse than in rat tissues (Ansher et al., 1986, Pearson et al., 1988). These increased enzymatic activities of GSTs were the result of enhanced GST subunit expression. Hepatic GST class α and μ subunits, as determined by quantitative separation using ion-exchange chromatography, were also shown to be induced in mice and rats by BHA, ethoxyquin, and disulfiram (Benson et al., 1978). HPLC analysis of affinity-purified fractions showed that hepatic GST class α (Ya2 and Yc2), μ (Yb1), and π (Yp) subunits were induced from 2- to 20-fold by dietary or single-dose administration of oltipraz in rats (Meyer et al., 1993a; Primiano et al., 1995a). The highly inducible GST Yc2 subunit has been purified as a heterodimer with GST Yc1 (Hayes et al., 1991). This GST isozyme has the highest activity for the conjugation of glutathione to activated aflatoxin B_1 yet measured. Addition of ethoxyquin to the diet induced aflatoxin B_1 conjugation activities in rat hepatic cytosol approximately 6-fold as evidenced by HPLC analysis (Hayes et al., 1991). Corresponding levels of GST Yc2 isozyme were elevated 6-fold by ethoxyquin, as evidenced by immunoblot analysis (Hayes et al., 1994). Other GST isozymes also conjugate the highly mutagenic 8,9-epoxide of aflatoxin B_1 (Hayes et al., 1991; Raney et al., 1992), but at lower levels. Conjugation of activated aflatoxin B_1 inhibits formation of DNA adducts that may lead to mutations in critical cell regulatory genes such as p53 and Ha-ras (Aguilar et al., 1994; Cerutti et al., 1994).

Increased expression of GST subunits is the result of increased synthesis of their RNA transcripts. Cloning of a class μ GST cDNA from BHA-treated mouse liver and Northern blot analysis indicated that the elevation of protein levels was accompanied by approximately 20-fold increases in the levels of steady-state RNA (Pearson et al., 1983). The BHA metabolite, tert-butylhydroquinone(tBHQ), was also found to be a potent inducer of mouse GST Ya by increasing mRNA levels of this subunit (Friling et al., 1990). Although GST Yp levels were not elevated greatly by BHA or ethoxyquin (Hayes et al., 1994; Buetler et al., 1995), GST Yp levels were induced by 1,2-dithiole-3-thione approximately 20-fold over extremely low levels of constitutive expression (Primiano et al., 1995b). The GST Yc2 subunit mRNA levels are also expressed at extremely low levels constitutively and are induced dramatically by ethoxyquin, BHA, or oltipraz (Buetler et al., 1995). In summary, the GSTs therefore represent a highly inducible family of isozymes that afford protection from cytotoxic and mutagenic consequences of metabolically activated xenobiotics.

C. NAD(P)H:Quinone Reductase

NAD(P)H:quinone reductase (QR) catalyzes the obligatory two-electron reduction of quinones and their derivatives (for a review see Lind *et al.*, 1990) and prevents their participation in redox cycling and oxidative stress (Iyanagi and Yamazaki, 1970; Thor *et al.*, 1982). The resulting reduced quinols can then be glucuronidated by the action of UDP-glucuronoslytransferases (UGTs) and thereby eliminated (Lind *et al.*, 1982). A single gene encoding QR has been identified for both rat and humans (Bayney *et al.*, 1987; Robertson *et al.*, 1986), although several isoforms generated by post-translational modifications have been found (Segura-Aguilar *et al.*, 1992). The QR isoforms show broad substrate specificities for different quinones, such as menadione, benzo(*a*)pyrene-3,6-dione, and vitamin K_1. The ability of QR to detoxify a broad spectrum of quinones underscores the importance of this enzyme in protecting cells from catechols, quinone imines, and redox-cycling quinones generated from polycyclic aromatic hydrocarbons, phenolic steroids, acetaminophen, phenytoin, and others (for reviews see Talalay and Benson, 1982; Prochaska and Talalay, 1991). The protective effects of QR have been defined in a number of *in vitro* systems. Direct addition of QR enhances the efficiency of glucuronidation of benzo[*a*]pyrenediones and menadione-induced O_2 consumption by microsomal fractions (Lind *et al.*, 1978, 1982). Also, redox cycling of diethylstilbestrol and estrogen is inhibited by QR (Roy and Leihr, 1988). The toxicity of quinones is enhanced by the addition of dicoumarol, a potent inhibitor of QR (Thor *et al.*, 1982; Galaris *et al.*, 1985; Atallah *et al.*, 1988). Reduced toxicity to hydroquinone in resistant murine fibroblastoid cells strains as opposed to that in macrophages was correlated to the presence of higher QR activities (Thomas *et al.*, 1990).

Dietary administration of the antioxidants BHA, BHT, and ethoxyquin induced the QR-mediated reduction of menadione between 3- and 10-fold in rat and murine liver. In addition, QR activities were enhanced in kidney, lung, and small intestine (Benson *et al.*, 1979). The induction of QR by antioxidants often mirrors that of the GST Ya, except that the levels of induction are generally higher than for the GSTs (Kensler *et al.*, 1985; DeLong *et al.*, 1986). Therefore, QR, along with the GSTs, is an antioxidant-inducible enzyme that aids in reducing toxicity and tumorigenesis elicited by chemicals (Kahl, 1991; Prochaska and Talalay, 1991).

D. UDP-Glucuronosyltransferases

UDP-glucuronosyltransferases catalyze the conjugation of glucuronic acid with many endogenous substrates, such as steroid hormones, bilirubin, fat-soluble vitamins, and biogenic amines. In addition, xenobiotics, such as hydroxylated metabolites of benzo[a]pyrene, and drugs, such as acetaminophen, are conjugated by these enzymes, thus enhancing their rate of excretion (Burchell and Coughtire, 1990; Mackenzie and Rodbourn, 1990; Tephly and Burchell, 1990). Like the GSTs, UGTs comprise a multienzyme gene family, which has varying specificities for hydroxylated substrates (Burchell et al., 1991). UGTs play an important role in cytoprotection by preventing the accumulation of toxic metabolites or their precursors. Reactive intermediates generated by the bioactivation of benzo[a]pyrene are detoxified substantially by UGTs as evidenced by the identification of numerous benzo[a] pyrene glucuronide conjugates generated by hepatic microsomes or fibroblasts (Hu and Wells, 1992; Vienneau et al., 1995). Further evidence of this cytoprotective role for UGTs is that enhanced acetaminophen and benzo[a]-pyrene toxicities were found in rats deficient in bilirubin UGT activity (de Morais and Wells, 1988; de Morais et al., 1992).

UGTs are also induced by antioxidants. Increases of two- to three-fold in the formation of p-nitrophenol glucuronide conjugates were found after 2 weeks of administration of BHA, BHT, ethoxyquin, or oltipraz (Kensler et al., 1985). Multiple forms of UGTs are apparently induced as evidenced by the enhanced formation of conjugated products of p-nitrophenol by the isozyme UGT1*06, 3-hydroxybenzo[a]pyrene by UGT1*0, and androsterone UGT2B1 (Kashfi et al., 1994). In a separate study, BHA was found to induce acetaminophen glucuronide accumulation (Goon and Klaassen, 1992). In addition, treatment with BHA, ethoxyquin, or oltipraz was found to dramatically elevate UGT1*06 mRNA levels in rat liver (Buetler et al., 1995). It has been proposed by these and other authors that increased UGT mRNA levels following antioxidant exposure may reflect a common mechanism of induction with those of GSTs and QR.

E. Microsomal Epoxide Hydrolase

The microsomal epoxide hydrolase (mEH) enzyme inactivates both alkyl and aryl epoxides to corresponding dihydrodiols (Seidegård and DePierre, 1983). These hydrolyzed products can then be conjugated to corresponding sulfate or glucuronide metabolites, thereby enhancing their elimination. Evidence for multiple immunologically distinct isoforms of mEH with styrene oxide activity has been suggested (Bresnick et al., 1977; Guengrich et al., 1979a,b), yet only a single gene for mEH has been isolated and sequenced to date (Falany et al., 1987). Cytoprotection by mEH, as evidenced by the inhibition of mutagenicity of several oxides of benzo[a]pyrene and other aromatic compounds, has been demonstrated with the purified enzyme (Oesch, 1972; Oesch et al., 1976). In addition, expression of the mEH gene was elevated in drug-resistant cell lines (Carr and Laishes, 1981) and in preneoplastic nodules resistant to xenobiotics (Fairchild et al., 1978). Therefore, increased activity of mEH affords protection to cells by enhanced detoxication and elimination of cytotoxic epoxidized reactants.

Antioxidants dramatically induce the activity of mEH. Dietary administration of BHA, BHT, EQ, or dithiolethiones has been demonstrated to induce mEH hydrolysis of styrene oxide by as much as 20-fold following 2 weeks of exposure (Benson *et al.*, 1979; Kensler *et al.*, 1985). Elevations in the activity of mEH by dietary BHA were found in liver, lung, kidney, and small intestine, as seen with GSTs and QR (Benson *et al.*, 1979). Treatment with the antioxidant 1,2-dithiole-3-thione increased mRNA levels of mEH of ≈ 8-fold (Primiano *et al.*, 1995b). This high level of inducibility of mEH by various antioxidants suggests that it may play a major role in the protection of cells from the toxicity of xenobiotics.

F. Aflatoxin B_1-Aldehyde Reductase

In addition to conjugation to GSH by the GSTs, the ultimate carcinogen 8,9-aflatoxin B_1 epoxide is metabolized to *trans*-dihydrodiols by hepatic cytosolic fractions purified from ethoxyquin-treated rats (Judah *et al.*, 1993). The enzymatic activity was associated with an enzyme named aflatoxin B_1-aldehyde reductase (AFAR) (Hayes *et al.*, 1993). Complementary DNA clones were isolated, and the deduced AFAR amino acid sequence was shown to have between 20 and 40% homology with members of an aldo-keto reductase supergene family (Ellis *et al.*, 1993). AFAR was also capable of using NADPH to reduce 4-nitrobenzaldehyde. AFAR activity was induced greater than 15-fold by BHA, oltipraz, or ethoxyquin (Hayes *et al.*, 1993; Sharma *et al.*, 1993). Induction of AFAR activity by antioxidants was coincident with increased levels of AFAR protein and mRNA (Ellis *et al.*, 1993; Hayes *et al.*, 1993). Treatment of rats with the chemopreventive agent 1,2-dithiole-3-thione resulted in a greater than 50-fold transcriptional activation of the AFAR gene (Primiano *et al.*, 1995b). This tremendous induction of AFAR evoked by dithiolethiones may accelerate the detoxication of aflatoxin B_1 and hence inhibit aflatoxin B_1-mediated toxicity and tumorigenesis.

G. Dihydrodiol Dehydrogenases

Dihydrodiol dehydrogenases (DDs) catalyze the reduction of aldehydes and ketones to their corresponding alcohols by utilizing NADH or NADPH (Penning, 1993). These enzymes have been implicated in the detoxication of proximate (*trans*-dihydrodiol) and ultimate carcinogenic (diolepoxide) metabolites of polycyclic aromatic hydrocarbons. At least three different DDs has been purified, including 3α-hydroxysteroid dehydrogenase. The cytoprotective role of DDs has been demonstrated by their ability to reduce the mutagenic activity of benzo[*a*]pyrene (Glatt *et al.*, 1979). Substrate specificity studies show that non-K-region *trans*-dihydrodiols (such as 7,8-

dihydroxybenzo[a]pyrene) are the preferred substrates. The highly reactive ortho-quinones products are then rapidly conjugated to GSH (Smithgall et al., 1988; Shou et al., 1992). Overproduction of DDs in human colon cells resulted in increased resistance to ethacrynic acid, doxorubicin, and mitomycin C (Ciaccio et al., 1993). DD activity was induced 7- and 20-fold in human colon cells by 80 μM concentrations of hydroquinone and tert-butylhydroquinone, respectively (Ciaccio et al., 1993). DD belongs to a supergene family of aldo-keto reductases (Qin and Cheng, 1994) which includes 3α-hydroxysteroid dehydrogenase and prostaglandin F synthase (Pawloski et al., 1991). Because of the potent and significant regulatory effects of hydroxysteroids and prostaglandins on cell growth and function, the activities of this family of enzymes will become increasingly important as their roles in modulating cellular actions become known.

H. Aldehyde Dehydrogenases

Aldehyde dehydrogenases (ALDHs) catalyze the oxidation of aldehydes to carboxylic acids that usually gives them a detoxication function (for a review see Lindahl, 1992). There are several types of ALDHs in tissues that have been classified into groups based on their deduced primary structures. In humans, five liver ALDH isozymes have been purified and characterized. Of these, the cytosolic ALDH1 and mitochondrial ALDH 2, the two major isozymes, exhibit specificity for acetaldehyde. In contrast, ALDH3 is normally expressed in lung and stomach, and has specificity for benzaldehyde. ALDH3 is inducible in liver during carcinogenesis and by polycyclic aromatic hydrocarbons (Lindahl and Evces, 1984; Dunn et al., 1989). The cytoprotective role of ALDH is suggested by the correlation of its overexpression to cyclophosphamide resistance in breast adenocarcinoma cells (Sreerama and Sladek, 1994). Treatment of multidrug-resistant MCF-7/0 cells with 30 μM BHA, BHT, tert-butylhydroquinone, and ethoxyquin for 5 days dramatically elevated ALDH activity (15- to 50-fold) (Sreerama et al., 1995). Elevation of ALDH3 mRNA was coincident with the increased activity to metabolize benzaldehyde and mafosfamide. Coinduction of GST and QR activities was also found, suggesting that ALDH is part of the battery of enzymes induced by antioxidants (Sreerama et al., 1995).

I. Enzymes of Glutathione and Reduced Nicotinamide Metabolism

Elevation of intracellular glutathione levels augments cellular protection against reactive intermediates (for a review see Meister, 1994). Glutathione deficiency sensitizes cells to radiation (Dethmers and Meister, 1981) and to toxic effects of nitrogen mustards (Suzakake et al., 1982), cyclophosphamide (Ishikawa et al., 1989), and heavy metals (Naganuma et al., 1990). Therefore, maintaining or elevating GSH levels can be an important means of preventing cytotoxicity. Dithiolethiones and other antioxidants not only induce phase 2 xenobiotic-metabolizing enzymes, but enhance GSH levels (Kensler *et al.*, 1985, 1987; Ansher *et al.*, 1986). Increased levels of reduced GSH resulted from enhanced γ-glutamylcysteine synthetase (γGCS) activity (Eaton and Hamel, 1994). BHA administered at 0.75% in the diet induces γGCS activity approximately two-fold in mouse liver, which was correlated to elevations in mRNA levels (Borroz *et al.*, 1994; Buetler *et al.*, 1995). The γGCS holoenzyme exists as a dimer composed of a heavy (73 kDa) and light (28 kDa) subunit that can be dissociated under nondenaturing conditions (Seelig *et al.*, 1984). Transfection of cDNA expression vectors for the light and heavy subunits of γGCS results in the elevation of intracellular GSH levels and a concomitant resistance to melphalan (Mulcahy *et al.*, 1995). The heavy subunit possesses all of the catalytic activity and is the site for GSH feedback. Inhibition of this subunit by buthionine sulfoximine has been utilized as an experimental tool in elucidating the effects of GSH on cytoprotection (Meister, 1991).

Glutathione reductase is a dimeric cytosolic enzyme which utilizes NADPH as a cofactor to catalyze the reduction of oxidized glutathione (Meister, 1994). Glutathione reductase then maintains the intracellular levels of reduced glutathione following its utilization by glutathione peroxidase. The activity of glutathione reductase is increased 1.5- to 2-fold by treatment with BHA, BHT, ethoxyquin, or oltipraz (Kensler *et al.*, 1985). This increased activity likely contributes to increased levels of reduced GSH found after antioxidant administration. Because QR uses NADPH, maintenance of reduced NADPH is important. Glucose-6-phosphate dehydrogenase, an enzyme that leads into the glycolytic pathway, generates NADPH by enzymatic hydrolysis of ATP. The activity of this enzyme was shown to be elevated in rat tissues following the administration of antioxidants (Kensler *et al.*, 1985). Thus, antioxidants elevate and maintain reduced levels of GSH, which aid in the protection of the cell from various environmental stresses.

J. Other Proteins and Enzymes

Because most of the genes isolated to date are induced by transcriptional mechanisms resulting in elevated expression of mRNA, the powerful molecular biological technique of differential hybridization screening has been employed to identify additional antioxidant-inducible genes (Primiano *et al.*, 1995b). By screening a library prepared from the liver mRNA of rats treated with 1,2-dithiole-3-thione, more than a dozen cDNAs were isolated. Subsequent sequencing and database analysis revealed that many of these cDNAs represented genes that were not previously known to be induced by antioxidants. Clones for ferritin heavy and light chain subunits and two ribosomal proteins, S16 and L18a, were isolated. In addition, two cDNA clones for genes with unknown function, termed dithiolethione inducible genes (DIG-1 and DIG-2), were isolated. The induction of the mRNA of these genes by dithiolethiones was confirmed by Northern blot analysis (shown in Fig. 1). The levels of mRNA for these genes were induced between 2- and 20-fold. Results of nuclear run-on experiments indicated that, in general, the genes were also transcriptionally activated.

Induction of ferritin heavy and light chain subunits by antioxidants presents a very intriguing mechanism for cytoprotection. Within most cells

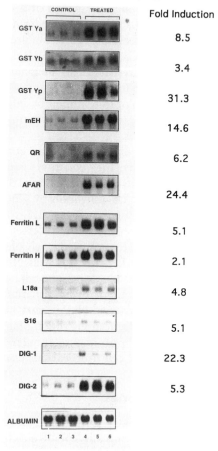

FIGURE I   Analysis of induction of RNA levels by RNA blot hybridization following 1,2-dithiole-3-thione treatment. Total RNA isolated from liver of vehicle-treated rats (lanes 1–3) and total RNA from 0.5 mmol/kg 1,2-dithiole-3-thione-treated rats (lanes 4–6) at 24 hr posttreatment were hybridized using GST Ya, GST Yb1, GST Yp, mEH, QR, AFAR, ferritin L, ferritin H, L18a, S16, DIG-1, DIG-2, and albumin standard cDNA probes, respectively. Corresponding ratios of treated to control values at 24 hr posttreatment are shown next to each blot. Quantitation was performed using a Fuji BAS1000 phosphorimaging system and was normalized for loading using signals from albumin-probed blots.

the major depot of nonmetabolic iron is ferritin, a multimeric (24 subunit) protein (heavy or H chain M_r = 21,000, light or L chain M_r = 20,700 in humans) with a total molecular weight of 450,000 that has a very high capacity for storing iron (4500 mol of iron/mol of ferritin). Thus, induction of the nascent ferritin L subunit protein, as observed following treatment with 1,2-dithiole-3-thione, could result in the substantial additional sequestration of iron. Also, as shown for the response to iron, the preferential production of L-rich forms of ferritin may provide an enhanced and prolonged capacity for iron storage (for review see Theil, 1990; Munro, 1993). Finally, knowledge that multimeric ferritin manifests ferroxidase activity (Levi *et al.*, 1988; Balla *et al.*, 1992) implies that ferritin-stored iron might resist cyclical reduction/oxidation reactions that tend to propagate and amplify oxidative damage.

Dithiolethiones induce ferritin L protein by approximately fourfold resulting from increased levels of ferritin mRNA through an enhanced rate of transcription (Primiano *et al.*, 1995b). This transcriptional activation of the ferritin L gene is in contrast to the known mechanism of translational activation of ferritins by a redox-sensitive iron-response element-binding protein. Reducing agents can modulate the levels of ferritin by coordinating with an iron–sulfur center on the iron-response element-binding protein (reviewed in Theil *et al.*, 1990). Therefore, intracellular antioxidant levels may modulate the level of iron at several regulatory steps, including the modulation of ferritin synthesis.

Additional proteins were found to be induced by 1,2-dithiole-3-thione. The ribosomal proteins S16 and L18a mRNAs were elevated approximately fivefold by a single dose of dithiolethione (Fig. 1). Corresponding increases in the ribosomal subunit proteins would imply that a general increase in translational activity may accompany the pleiotropic transcriptional responses (Primiano *et al.*, 1995b).

Noticeably lacking in induction are the antioxidant enzymes catalase, glutathione peroxidase, and superoxide dismutases. These enzymes are elevated in response to oxidative stress, such as that following tumor necrosis factor-α or hydrogen peroxide exposure (Wong and Goeddel, 1988; Mossman, 1991; Shull *et al.*, 1991). However, they are refractory to antioxidant administration in F344 rats (Kensler *et al.*, 1985). This separation of the oxidant and antioxidant response pathways implies that the generation of different signal transducers and modulation by specific transcription factors occurs so that cells respond differentially to oxidant or antioxidant stresses.

III. Mechanisms of Gene Induction by Antioxidants

Increased gene expression is often the result of an elevated rate of transcription (Darnell, 1982). For instance, nuclear run-on results showed that QR gene expression was transcriptionally activated between three- and eightfold in mice by *tert*-butylhydroquinone and oltipraz (Daniel *et al.*, 1986; Pinkus *et al.*, 1995). Similar studies in rats confirmed that oltipraz transcriptionally activated the GST Ya subunit gene within the first 48 hr (Davidson *et al.*, 1990). Activation of gene transcription was also found for GST Ya, Yb, Yp, QR, ferritin H and L subunits, ribosomal proteins S16 and L18a, AFAR, and the DIGs in response to 1,2-dithiole-3-thione (Primiano *et al.*, 1995b). Abundant evidence suggests that the genes encoding phase 2 xenobiotic-metabolizing enzymes and other antioxidant-inducible genes share a common mechanism of transcriptional regulation. A variety of totally unrelated chemical classes induce the same group of phase 2 xenobiotic-metabolizing enzymes. Prochaska *et al.* (1985) demonstrated that closely related structures that were susceptible to oxidation–reduction were better inducers than those that were resistant. The authors stated that this propensity for a compound to induce phase 2 enzymes was related to the efficiency of its olefinic bond to generate an electrophilic signal (Talalay *et al.*, 1988). The induction of the phase 2 enzymes by these compounds, including phenolic antioxidants, was independent of the Ah-receptor (Prochaska and Talalay, 1988). Because enhanced rates of transcription are often the result of specific binding of transcription factors to sites on the DNA, which increase the initiation of transcription (Mitchell and Tjian, 1989), a new *cis*-acting site on the DNA mediating antioxidant induction was proposed. Rushmore and Pickett (1990) identified a site on the GST Ya gene that mediated a fourfold increase in CAT activity in response to *tert*-butylhydroquinone, but not β-naphthoflavone. The site was termed the antioxidant-response element (ARE) (Rushmore *et al.*, 1991). These authors suggested that the common signal proposed by Prochaska *et al.* (1985) is mediated through the ARE.

As with all transcriptionally regulated genes, complex patterns of expression mediated by the ARE are possible and include (1) slight deviations from the consensus ARE sequence, either in primary structure or orientation of its component binding sites may exist for the various different inducible genes; (2) binding of different transcription factors in a tissue- or inducer-specific fashion may result in heterogeneous rates of initiation of transcription; and (3) signal transduction pathways may activate additional factors that are specific for subsets of inducible genes. Evidence for these possible mechanisms is described next.

A. The Antioxidant-Response Element

Transcriptional activation by antioxidants has been demonstrated to be the result of enhanced transcription factor binding to a *cis*-acting element known as the ARE, or EpRE, for the GST Ya and QR genes (for a review see Jaiswal, 1994). The sequences for the *cis*-acting ARE enhancer regions contain two or more copies of AP-1 or AP-1-like elements in a short stretch (40–45 nucleotides) of DNA (for a review see Jaiswal, 1994). The AP-1 element (5'-TGACTCA-3') is present in the upstream regions of several genes, including human metallothionein, SV40, and human collagenase genes, and is known to increase the transcription of these genes in response to 12-O-tetradecanoylphorbol-13-acetate (TPA) (Hai and Curran 1991; Hershman, 1991).

The number and orientation of AP-1-like elements in the enhancer region of antioxidant-inducible genes apparently affect their responsiveness to inducers. The AREs of rat and human orthologs of QR contain either one copy of the perfect consensus AP-1-binding site and two AP-1-like sequences or three AP-1-like sequences, respectively (Favreau and Pickett, 1991; Li and Jaiswal, 1992a). The rat and mouse GST Ya and the mouse GST Yp ARE sequences contain two AP-1-like sites arranged in direct repeat orientation at an interval of 8 bp (Rushmore et al., 1990; Friling et al., 1992a). At least two palindromic AP-1/AP-1-like elements are required for optimal expression and/or induction, with the 3' most element being obligatory (Xie et al., 1995). In addition to the ARE, the rat GST Yp gene also contains a silencer region (Okuda et al., 1989), which may account for its low constitutive level of expression. The minimum ARE sequence required for expression and induction of rat GST Ya and rat QR genes has been determined by mutational analysis to be 5'-TGACnnnGC-3' (Rushmore et al., 1991). The TGAC portion was required for basal expression. This finding is in contrast to the reports that mouse GST Ya subunit and human QR ARE require the presence of both of the AP-1-like elements for their proper function (Li and Jaiswal, 1992b; Prestera et al., 1993). The ARE has been shown to contribute to optimal expression and induction of GST Ya subunit genes. Two nucleotides, 'GC', immediately following the 3' AP-1-like element within the rat GST Ya gene and human QR ARE, are not required for basal expression but are essential for induction (Nguyen et al., 1994; Xie et al., 1995). These two nucleotides are part of the three nucleotides 'GCA', which are highly conserved in all AREs (Jaiswal, 1994).

The primary sequence of the ARE may vary between genes. For instance, a potential ARE sequence was found at position −305 to −279 of the rat ferritin L gene consisting of two 13-bp direct repeats, and this region may confer antioxidant inducibility (Leibold and Munro, 1987). The sequence is different from the consensus ARE in the 5-bases intervene between the TGAC and the GC motifs. The alternative sequence may bind transcription factors other than those of the rat GST Ya ARE. Additional DNA sequences resembling the ARE have been found for many genes, including p53 and homodimers or heterodimers of several leucine zipper transcription factors (Kataoka et al., 1994).

Neighboring cis elements may modify antioxidant inducibility. Prestera et al. (1993) reported the presence of an ETS protein-binding site just before the 5' AP-1-like element. Gel shift analysis using the ETS-binding site from the polyoma virus enhancer region ligated to a consensus AP-1 site, thus mimicking an ARE, was used as a probe (Bergelson and Daniel, 1994). This sequence was also ligated into the minimal promoter −187 upstream region of mouse GST Ya, which was fused to a chloramphenicol acetyltransferase (CAT) reporter gene. Inclusion of this sequence conferred increased binding of nuclear protein and increased transactivation in response to the inducer β-naphthoflavone. These results suggested that the ETS family of binding proteins (Elk-1/SAP-1/ERP) may be involved in optimal activation of the ARE, as is the case with the serum response element (Whitmarsh *et al.,* 1995). In contrast, the cooperative interaction between the ARE-binding proteins may not be universal in that the ETS-binding sites have not been found adjacent to either the AREs of human QR or rat GST Yp.

The ARE may be present in the regulatory region of most, if not all, of the genes encoding xenobiotic-metabolizing enzymes that are induced by antioxidants. For instance, an ARE has been reported in the human γGCS gene (Mulcahy and Gipp, 1995), which potentially mediates the induction elicited by BHA. Although induction of multiple forms of UGTs suggests that a common signal or transcription factor mediates an enhanced transcription, an ARE-like sequence has not yet been discovered. The gene for rat AFAR has yet to be cloned, but it would be interesting to determine whether a consensus ARE is present upstream of the transcription start site. The presence of consensus AREs in the genes for rat mEH (Falany *et al.,* 1987) or DD (Ciaccio *et al.,* 1993) have been suggested; however, investigations of ARE function are lacking.

In summary, the AREs in various genes usually contain two AP-1-like elements arranged in varying orientations, separated by either three or eight nucleotides. The orientations and spacing between the two AP-1-like elements could be important in determining the proper binding of specific transcription factors.

B. Proteins Binding to ARE

The transcription factors that bind to the ARE consensus sequence 5'-TGACnnnGC-3' have not been fully identified, but the Fos–Jun heterodimer or a putative leucine zipper AP-1-like binding protein have been suggested (Bergelson *et al.,* 1994; Kataoka *et al.,* 1994; Nguyen *et al.,* 1994). The assortment of proteins that can bind the AP-1 site continue to be discovered. Both the Fos family (c-Fos, Fos-B, Fra-1, Fra-2) and the Jun family (c-Jun, Jun-B, Jun-D) have been expanded (Hai and Curran, 1991; Nakabeppu and Nathans, 1991). All members of these families are capable of binding to AP-1 sites, although Fos subunits cannot form homodimers and can bind only in association with a Jun subunit. One common feature of all these factors is a leucine zipper domain, believed to be important in homo- or heterodimer formations. Gel mobility supershift assays were used to clearly establish that Jun-D, Jun-B, and c-Fos proteins bind to the AP-1 sequences contained within the ARE of human QR (Li and Jaiswal, 1992b). However, in a similar experiment, Jun and Fos proteins did not bind to the ARE of rat QR (Favreau and Pickett, 1993). Friling *et al.* (1992) also used gel mobility shift assays to demonstrate that *in vitro*-translated c-Jun and c-Fos bind to the ARE of the murine GST Ya gene. In addition, they have shown activation of CAT gene expression upon cotransfection of ARE-CAT with c-Jun and c-Fos into mouse embryonic F9 cells. In contrast to the work of Friling *et al.* (1992b), Nguyen and Pickett (1992) showed that *in vitro*-translated c-Jun and c-Fos do not bind to the ARE of the rat GST Ya gene and that the TPA-response element (TRE) from human collagenase gene did not compete for binding of the nuclear proteins to the ARE. These observations led them to conclude that *trans*-acting factors that bind to the ARE are not Jun–Fos heterodimers. The same authors used photochemical cross-linking techniques to demonstrate that a heterodimer with subunit molecular masses of approximately 28,000 and 45,000 Da binds to the ARE.

One recent study provides evidence that the major ARE-binding protein is not strictly the c-Fos/c-Jun heterodimer (Yoshioka *et al.*, 1995). Treatment of HeLa, HepG2, or F9 cells with tBHQ inhibited TPA induction of AP-1, but the appearance of Fra-1/Jun heterodimer DNA binding was increased. Therefore, the leucine zipper transcription factor Fra-1 can effectively titrate c-Jun and form complexes that inhibit transcriptional activation through the TRE. Indeed, Fra-1 and Fra-2 mRNA levels were dramatically increased in HeLa cells by tBHQ administration. Transactivation through the rat GST Ya ARE, but not the collagenase TRE, was increased by tBHQ. One interpretation of these results suggests that the Fra-1/Jun heterodimers may confer enhanced transcriptional activation to genes containing a consensus ARE versus those with a consensus TRE. However, a direct transcriptional activation of an ARE reporter construct by transfection of Fra-1 expression vector was not performed. Another leucine zipper transcription factor, the peroxisome proliferator-activated receptor α, can bind to c-Jun and inhibit constitutive expression of the GST Yp gene (Sakai *et al.*, 1995). Thus, these proteins and perhaps others can bind to the ARE and modulate its function. There is additional emerging evidence that several proteins may bind to the ARE. Favreau and Pickett (1993) have demonstrated by gel shift analysis that several bands are shifted when probed with radiolabeled ARE, most of which are not competed by the TRE. Wasserman and Fahl (1995) found that the 41-bp fragment containing the ARE of GST Ya could specifically bind at least seven different proteins. Furthermore, Wang and Williamson (1994) have purified an ARE-binding protein that has a gel mobility and chromatographic elution profile that is different from AP-1. The binding specificity of the transcription factor families Maf/Nrl and NF-E2 has been shown to be very similar to the consensus sequence for the ARE (Andrews *et al.*, 1993; Kerppola and Curran, 1994). The involvement of these families of transcription factors in mediating protection of cells from toxicity of xenobiotics through antioxidant-inducible mechanisms needs to be investigated.

In summary, several factors, which include AREs from different genes, species differences, orientations of the two AP-1 elements contained within the AREs, the presence and absence of ETS, and other unknown binding sites within ARE, may contribute to variability in ARE responsiveness. Further studies are needed to address several interesting questions regarding the complete identification of regulatory proteins in the ARE–nuclear protein complexes and how these proteins bind at AP-1 elements contained within the ARE.

C. Signal Transduction through the ARE

The means by which structurally diverse antioxidants increase the transcription of the xenobiotic-metabolizing enzyme genes and thereby inhibit cytotoxicity and mutagenicity of reactive electrophiles are of major importance for understanding both carcinogenic and anticarcinogenic processes. The exact mechanisms by which antioxidants and other compounds mediate transcriptional activation of the xenobiotic-metabolizing enzymes are not yet known. However, evidence does exist, however, to propose the role of a redox sensor. Such a sensor is influenced by the cellular oxidation state, translating it into a chemical signal that is transduced to the nucleus leading to transcriptional activation of the appropriate genes. A schematic diagram depicting the hypothetical signal transduction pathways and the transcription factors activated by these pathways is presented in Fig. 2. Using a series of redox-cycling quinones, Pinkus and colleagues (1995) have demonstrated that the degree of transactivation through the ARE is correlated to the level of production of reactive oxygen species. Further evidence in support of redox mechanisms mediating ARE activation is derived from the observation that sulfhydryl reagents modify the DNA-binding and transcriptional activation of the ARE of human QR by β-naphthoflavone (Li and Jaiswal, 1992a). Moreover, several compounds that induce the activation of the mouse GST Ya CAT reporter gene through ARE have the capacity to react with sulfhydryl groups (by oxido-reduction or alkylation), consistent with a mechanism involving protein thiol modifications (Prestera *et al.*, 1993). Other sulfhydryl-modifying compounds that enhance DNA binding and transactivation by AP-1 include the antioxidants pyrrolidine dithiocarbamate and N-acetylcysteine, as well as the antioxidant enzyme thioredoxin (Meyer *et al.*, 1993b) and the protein Ref-1 (Abate *et al.*, 1990). Ref-1 causes increased binding of Jun and Fos proteins to the AP-1 site by reducing the cysteine in their DNA-binding domain. *ref-1* transcription has been found to be elevated in human colon cells subjected to hypoxic stress in conjunction

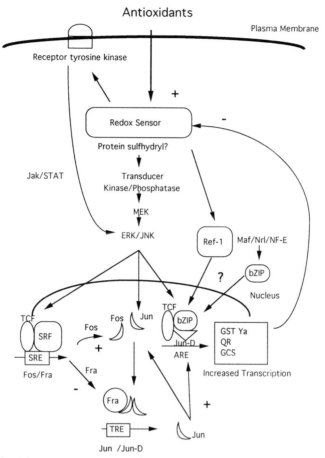

FIGURE 2   Schematic representation of the antioxidant signal transduction pathway. Antioxidants enter the cell and interact with a redox-sensing protein. Binding to this redox sensor activates redox-sensitive kinases (e.g., receptor tyrosine kinases), which then proceed to phosphorylate a cascade of kinases (MEK, ERK, JNK) to the transcription factor targets: Jun, Fos, serum response factor (SRF), ternary complex factor (TCF) families, or other basic leucine zipper (bZIP) proteins. DNA binding of these transcription factors is enhanced by phosphorylation and/or cysteine reduction such as thioredoxin or Ref-1. The transcription factors (Fra, bZIP) then regulate their own expression, as well as those of the xenobiotic-metabolizing enzymes though the ARE. Action of the induced enzymes ameliorates the antioxidant signal, returning the cell to its pretreated state.

with c-*jun* and *junD*, and the expression of the Ref-1 protein increased AP-1 binding within the ARE of the human QR gene (Yao *et al.*, 1994). The Ref-1 protein was induced prior to QR as assessed by a time-dependent nuclear run-on analysis. Further investigations are needed to understand the role of Ref-1 or a related protein(s) and/or protein kinases in signal transduction from antioxidants to Jun and Fos proteins and/or novel transcription factors that regulate QR and GST gene expression. It is possible that enhancement of transcription of these genes in response to antioxidants may involve the integration of two or more mechanisms that include the redox modification of cysteines in DNA-binding domains or protein kinases that result in the phosphorylation of DNA-binding proteins such as those of the Jun and Fos families and/or unknown proteins. For example, other transcription factors have been shown to be redox sensitive, such as Myb (Guehmann *et al.*, 1992), upstream stimulatory factor (Pognonec *et al.*, 1992), and papilloma type 1 E2 protein (McBride *et al.*, 1992). The kinase activity of the tyrosine kinase LTK was enhance by alkylating and thiol-oxidizing agents such as diamide, putatively because of an increased formation of disulfide-linked dimers.

The apparent paradox that both antioxidants and oxidants induce in the binding of Fos and Jun to the ARE has been addressed only recently. The protooncogenes Jun and Fos are activated transiently and rapidly by extracellular signals (for a review see Angel and Karin, 1991). This activation is believed to couple short-lived signals elicited by cell surface stimuli to long-term phenotypic changes by regulating the expression of specific target genes containing AP-1-binding sites. Cytokines, phorbol esters, metals, and other producers of oxidative stress such as hydrogen peroxide regulate Jun and Fos proteins at the AP-1-binding site by several mechanisms that include phosphorylation and cysteine modifications (Boyle *et al.*, 1991; Hunter and Karin, 1992; Xanthoudakis and Curran, 1992; Xanthoudakis *et al.*, 1992). The involvement of protein kinases in transcriptional activation of the QR and GST genes by antioxidants is speculated because promoters of these genes contain AP-1/AP-1-like binding sites. Further evidence that protein kinases are affected by antioxidants evolves from studies that TPA-inducible tyrosine kinase activity is inhibited by oltipraz (Sharma *et al.*, 1994). Additional evidence for redox activation of kinases acting in the antioxidant response signal transduction pathway comes from the transcriptional activation of c-fos by pyrrolidine dithiocarbamate. This activation was conferred by the serum response element, suggesting that the serum response factor and associated proteins function as primary antioxidant-response transcription factors. Transcriptional activation of the *c-fos* gene has been found to be exerted through two major *cis* elements: the serum response element that is recognized by a heterodimer of the serum response factor–ternary complex factor (Triesman, 1992). These results, coupled with those of Prestera *et al.* (1993) and Bergelson and Daniel (1994), provide additional evidence that the antioxidant signal transduction pathway may also involve ETS-binding proteins. The activity of the ternary complex factor (Elk-1), an ETS-binding protein, is rapidly increased in response to cell stimulation with various agents, such as growth factors, that lead to mitogen-activated protein (MAP) kinase activation (Whitmarsh *et al.*, 1995). Extracellular signal-regulated MAP kinases 1 (ERK1) and 2 (ERK2) appear to be responsible for the phosphorylation of ternary complex factors on sites that enhance its transcriptional activity (Marais *et al.*, 1993). It has been shown that ERK-2 is potently activated by hydrogen peroxide and BHT hydroperoxide (Guyton *et al.*, 1996). Activation of the Fos family of proteins may rely on ERK/ JNK regulation. JNKs activate c-*jun* (Devary *et al.*, 1992; Deng and Karin, 1994; Dérijard *et al.*, 1994), which is also necessary for the upregulation of the Fos gene family. Increased AP-1/ARE binding may then occur following heterodimer formation. Moreover, c-*cis* element-inducible factor activity is regulated by a different signaling pathway in which DNA binding and nuclear translocation are stimulated by phosphorylation caused by receptor tyrosine kinases and the Jak group of receptor-linked tyrosine kinases (Devary *et al.*, 1992; Hunter, 1993), suggesting that these may also function in the antioxidant signal transduction pathway. The combined activity of Jak-STAT kinases was found to potentiate the responses of ERK/JNK-phosphorylated MAPs (David *et al.*, 1995). The activation of a specific antioxidant signal transduction pathway may therefore be modulated at many steps by kinases and phosphatases.

Other pathways may also respond to antioxidant stimuli. In HeLa cells, the oxidative stress-responsive transcription factor NF-κB behaved in a manner strikingly opposite to AP-1 (Meyer *et al.*, 1993b). DNA binding and transactivation by NF-κB were strongly activated by H_2O_2, whereas the antioxidants were ineffective. In contrast, the induction of the human QR gene by oltipraz was correlated with increased bandshifts for NF-κB (Yao and O'Dwyer, 1995), suggesting that the redox- and kinase-associated pathways of NF-κB participate in antioxidant inductions. Conflicting data may be due to the different systems employed for investigation. Certain cell lines may lack or inappropriately express antioxidant redox sensors or phosphorylation pathways necessary for the activation of transcription factors resulting in antioxidant inducibility (Baeuerle, 1991).

Phenolic antioxidants may generate longer-lived phenoxyl radical metabolites of endogenous- or exogenous-reactive oxygen species (Kahl, 1991). The process of reduction–oxidation elicited by antioxidants or reactive oxygen species could directly or indirectly modulate proteins, such as Ref-1 or redox-sensitive kinases. Activation of Ref-1 binding or kinase phosphorylation of antioxidant-specific transcription factors, such as Fra-1, to the ARE may result in increased transcription of the gene. The future identification of redox-regulated kinases would bring new insights into mechanisms controlling gene expression in response to extracellular conditions.

IV. Consequences of Antioxidant Gene Induction _____

Most carcinogens need to be activated by endogenous enzyme systems to highly reactive species that bind to and alter the function of cellular macromolecules (Farber, 1980; Miller and Miller, 1981; Conney, 1982). The induction of phase 2 xenobiotic-metabolizing enzymes has been associated with an increased protection of affected tissues from cytotoxic and mutagenic reactivities of activated carcinogens (Talalay, 1989). Evidence in support of this hypothesis comes from studies that indicate that antioxidants are cancer chemopreventive agents. Chemoprevention of cancer is a means of cancer control in which the carcinogenic process is blocked, retarded, or reversed through the administration of one or several compounds (Wattenberg, 1985). Cancer chemoprevention by antioxidants was first described by Wattenberg. Detailed reviews highlighting the chemoprevention by antioxidants and the opportunities for clinical chemopreventive interventions have been published (Wattenberg, 1985; Kahl, 1991; Kensler and Helzlsouer, 1995).

Abundant evidence supports a major role of induction of phase 2 enzymes in chemoprevention. First, many chemopreventive agents are effective if given prior to and/or during carcinogen administration. Pretreatment of mice or rats with BHA, ethoxyquin, disulfiram, and oltipraz resulted in significant decreases in tumor number and incidence following dosing with a variety of carcinogens, such as aromatic polycyclic hydrocarbons, urethan, dimethylhydrazine, nitrosamines, azo dyes, and aminofluorenes (for a review see Wattenberg, 1985). These results suggest that elevations in phase 2 xenobiotic-metabolizing enzymes in the animals prior to exposure to carcinogen aided in the protection from tumorigenesis. Second, because chemoprevention is achieved against a wide variety of carcinogens, a mechanism with low specificity or broad activity must exist. The various families of xenobiotic-metabolizing enzymes display just such a broad specificity. Third, enzyme induction and chemoprevention are produced by the same compounds (Kahl, 1991). As summarized in Table I, most of the xenobiotic-metabolizing enzymes are induced by relatively similar doses of, BHA, BHT, ethoxyquin, or dithiolethiones. Additionally, inductions of similar magnitude are found for many tissues and by a variety of different antioxidants. Moreover, monitoring of enzyme induction has led to the recognition or isolation of a number of chemopreventive agents, including dithiolethiones (Wattenberg and Bueding, 1986).

More direct evidence that xenobiotic-metabolizing enzymes participate in cancer chemoprevention arises from knowledge of the genetic composition of GSTs in both cells and humans. Overexpression of recombinant GST α was found to protect cells from cytotoxicity produced by cyclophosphamide or melphalan (Manoharan *et al.*, 1987; for a review see Tew, 1994). Overexpression of recombinant GST Yc2 in mouse Hepa cells was found to reduce the number of DNA adducts formed after exposure to aflatoxin B_1 (Hayes *et al.*, 1994; Fields *et al.*, 1995). Thus, xenobiotic-metabolizing enzymes protect cells from toxic actions of activated metabolites, which also suggests that they prevent occurrence of mutagenic events and thereby tumorigenesis.

Additional evidence that xenobiotic-metabolizing enzymes provide protection from carcinogenesis has been inferred from studies of GST polymorphisms. Deficiencies in expression of GSTs in humans have been documented. Approximately half of the population lacks GST Mu expression (Harada *et al.*, 1987; Seidegård *et al.*, 1988; Hirvonen *et al.*, 1993), which is related to genetic polymorphism of the GST M1 or M3 gene. Homozygous loss of the GST M1 alleles (null genotype) was correlated to a reduction of GST Mu activity (null phenotype). The GST M null genotypes has been associated with predispositions for development of cancers of the breast (Zhong *et al.*, 1993), bladder (Bell *et al.*, 1993; Brockmoller *et al.*, 1994), and lung (Seidegård *et al.*, 1988; Nazar-Stewart *et al.*, 1993), although no association with GST M null genotype and susceptibility to colon cancer has been found (Lin *et al.*, 1995; Szarka *et al.*, 1995). The conflicting results may be explained in that persons with this genotype do not have altered expression of other classes of GSTs, including other Mu isozymes (Nakajima *et al.*, 1995). In addition, other polymorphisms (CYP1A1, CYP2D6) may combine to further heighten susceptibility to cancer.

V. Future Perspectives

The list of genes induced by antioxidants and other dietary factors is likely to grow in the near future. A cascade of genes encoding proteins that, once activated, prevent oxidative stress, mutagenicity, and carcinogenicity by subsequent exposure to drugs and carcinogens is envisioned. The regulation of antioxidant-inducible genes is of great importance to the understanding of the signals leading to cellular transformation and carcinogenesis. Antioxidants can be used as biological tools to decipher the encrypted signal transduction pathways that prevent cell transformation and carcinogenesis. The knowledge that antioxidants can induce many genes that are likely to contribute to cancer chemoprevention is of practical importance in that antioxidants are featured prominently in current strategies for chemoprevention (Kensler and Helzlsouer, 1995). Further identification and characterization into the types of naturally occurring and synthetic antioxidant compounds, which could act as potent inducers of the battery of xenobiotic-metabolizing enzyme genes, should aid in the development of effective strategies for cancer chemoprevention.

References

Abate C., Patel L., Rauscher, F. J., III, and Curran T. (1990). Redox regulation of Fos and Jun DNA-binding activity *in vitro*. *Science* **249**, 1157–1161.

Aguilar, F., Harris, C. C., Sun, T., Hollstein, M., and Cerutti, P. (1994). Geographic variation of p53 mutational profile in nonmalignant human liver. *Science* **264**, 1317–1319.

Andrews, N. C., Erdjument-Bromage, H., Davidson, M. B., Tempst, P., and Orkin, S. H. (1993). Erythroid transcription factor NF-E2 is a hematopoietic-specific basic-leucine zipper protein. *Nature* **362**, 722–728.

Angel, P., and Karin, M. (1991). The role of Jun, Fos and the AP-1 complex in cell-proliferation and transformation. *Biochim. Biophys. Acta* **1072**, 129–157.

Ansher, S. S., Dolan, P., and Bueding, E. (1986). Biochemical effects of dithiolethiones. *Food Chem. Toxicol.* **24**, 405–415.

Atallah, A. S., Landolph, J. R., Ernster, L., and Hochstein, P. (1988). DT-diaphorase activity and the cytotoxicity of quinones in C3H/10T1/2 mouse embryo cells. *Biochem. Pharmacol.* **37**, 2451–2459.

Baeuerle, P. A. (1991). The inducible transcription activator NF-κB: Regulation by distinct protein subunits. *Biochim. Biophys. Acta* **1072**, 63–80.

Balla, G., Jacob, H. S., Balla, J., Rosenberg, M., Nath, K., Apple, F., Eaton, J., and Vercellotti, G. (1992). Ferritin: A cytoprotective antioxidant stratagem for endothelium. *J. Biol. Chem.* **267**, 18148–18153.

Bauskin, A. R., Alkalay, I., Ben-Neriah, Y. (1991). Redox regulation of a protein tyrosine kinase in the endoplasmic reticulum. *Cell* **66**, 685–696.

Bayney, R. M., Rodkey, J. A., Bennet, C. D., Lu, A. Y. H., and Pickett, C. B. (1987). Rat liver NAD(P)H:quinone reductase nucleotide sequence analysis of a quinone reductase cDNA clone and prediction of the amino acid sequence of the corresponding protein. *J. Biol. Chem.* **262**, 572–575.

Bell, D. A., Taylor, J. A., Paulson, D. F., Robertson, C. N., Mohler, J. L., and Lucier, G. W. (1993). Genetic risk and carcinogen exposure: A common inherited defect of the carcinogen metabolism gene glutathione S-transferase M1 (GST M1) that increases susceptibility to bladder cancer. *J. Natl. Cancer Inst.* **85**, 1159–1164.

Bell, P., Falany, C. N., McQuiddy, P., and Kasper, C. B. (1990). Glucocorticoid repression and basal regulation of the epoxide hydrolase promoter. *Arch. Biochem. Biophys.* **279**, 363–369.

Benson, A. M., and Baretto, P. B. (1985). Effects of disulfiram, diethyldithiocarbamate, bisethyl-xanthogen, and benzylisothiocyanate on glutathione transferase activities in mouse organs. *Cancer Res.* **45**, 4219–4223.

Benson, A. M., Barreto, P. B., and Stanley, J. S. (1986). Induction of DT-diaphorase by anticarcinogenic sulfur compounds in mice. *J. Natl. Cancer Inst.* **76**, 467–473.

Benson, A. M., Batzinger, R. P., Ou, S-Y. L., Bueding, E., Cha, Y.-N., and Talalay, P. (1978). Elevation of hepatic glutathione S-transferase activities and protection against mutagenic metabolites of benzo[a]pyrene by dietary antioxidants. *Cancer Res.* **38**, 4486–4495.

Benson, A. M., Cha, Y.-N., Bueding, E., Heine, H. S., and Talalay, P. (1979). Elevation of extrahepatic glutathione S-transferase and epoxide hydratase activities by 2(3)-tert-butyl-4-hydroxyanisole. *Cancer Res.* **39**, 2971–2977.

Bergelson S., and Daniel, V. (1994). Cooperative interaction between ETS and AP-1 transcription factors regulates induction of glutathione S-transferase Ya gene expression. *Biochem. Biophys. Res. Commun.* **200**, 290–297.

Bergelson, S., Pinkus, R., and Daniel, V. (1994). Induction of AP-1 (Fos/Jun) by chemical agents mediates activation of glutathione S-transferase and quinone reductase gene expression. *Cancer Res.* **54**, 36–40.

Borroz, K. I., Buetler, T., and Eaton, D. L. (1994). Modulation of γglutamylcysteine synthetase large subunit mRNA expression by butylated hydroxyanisole. *Toxicol. Appl. Pharmacol.* **126**, 150–155.

Botham, C. M., Conning, D. M., Hayes, J., Litchfield, M. H., and McElligott, T. F. (1970). Effects of butylated hydroxytoluene on the enzymatic activity and ultrastructure of rat hepatocytes. *Food Cosmet. Toxicol.* **8**, 1–8.

Boyle, W. J., Smeal, T., Defize, L. H. K., Angel, P., Woodgett, J. R., Karin, M., and Hunter, T. (1991). Activation of protein kinase C decreases phosphorylation of c-Jun at sites that negatively regulate its DNA-binding activity. *Cell* **64**, 573–584.

Bresnick, E., Mukhtar, H., Stoming, T. A., Dansette, P. M., and Jerina, D. M. (1977). Effect of phenobarbital and 3-methylcholanthrene administration on epoxide hydrase levels in liver microsomes. *Biochem. Pharmacol.* **26**, 891–892.

Brockmoller, J., Kerb, R., Drakoulis, N., Staffeldt, B., and Roots, I. (1994). Glutathione S-transferase M1 and its variants A and B as host factors of bladder cancer susceptibility: A case control study. *Cancer Res.* **54**, 4103–4111.

Buetler, T. M., Gallagher, E. P., Wang, C., Stahl, D. L., Hayes, J. D., and Eaton, D. L. (1995). Induction of phase I and II drug-metabolizing enzyme mRNA, protein, and activity by BHA, ethoxyquin and oltipraz. *Toxicol. Appl. Pharmacol.* **135**, 45–57.

Burchell, B., and Coughtire, W. H. (1990). UDP-glucuronosyltransferases. *Pharmacol. Ther.* **43**, 261–289.

Burchell, B., Nebert, D. W., Nelson, D. R., Bock, K. W., Iyanagi, T. Jansen, P. L. M., Lancet, D., Mulder, G. J., Chowdhury, J. R., Siest, G., Tephly, T. R., and Mackenzie, P. I. (1991). The UDP glucuronosyl transferase gene superfamily: Suggested nomenclature based on evolutionary divergence. *DNA Cell Biol.* **10**, 487–494.

Carr, B. I., and Laishes, B. A. (1981). Carcinogen-induced drug resistance in rat hepatocytes. *Cancer Res.* **41**, 1715–1719.

Cerutti, P., Hussain, P., Pourzand, C., and Aguilar, F. (1994). Mutagenesis of the H-*ras* proto-oncogene and the p53 tumor suppressor gene. *Cancer Res.* **54**, 1934s–1938s.

Cha, Y.-N., and Bueding, E. (1979). Effect of 2(3)-tert-butyl-4-hydroxyanisole (BHA) administration on the activities of several hepatic microsomal and cytoplasmic enzymes in mice. *Biochem. Pharmacol.* **28**, 1917–1921.

Cha, Y.-N., Martz, F., and Bueding, E. (1978). Enhancement of liver microsome epoxide hydratase activity in rodents by treatment with 2(3)-tert-butyl-4-hydroxyanisole. *Cancer Res.* **38**, 4496–4498.

Choi, H.-S., and Moore, D. D. (1993). Induction of *c-fos* and *c-jun* gene expression by phenolic antioxidants. *Mol. Endocrinol.* **7**, 1596–1602.

Ciaccio, P. J., Stuart, J. E., and Tew, K. D. (1993). Overproduction of a 37.5-kDa cytosolic protein structurally related to prostaglandin F synthase in ethacrynic acid-resistant human colon cells. *Mol. Pharmacol.* **43**, 845–853.

Clapper, M. L., Everly, L. C., Strobel, L. A., Townsend, A. J., and Engstrom, P. F. (1994). Coordinate induction of glutathione S-transferase α, μ, and π expression in murine liver after a single administration of oltipraz. *Mol. Pharmacol.* **45**, 469–474.

Cobb, M. H., Boulton, T. G., and Robbins, D. J. (1991). Extracellular signal-regulated kinases: ERKs in progress. *Cell Regul.* **2**, 965–978.

Coles, B., and Ketterer, B. (1995). Metabolism of aflatoxin B1 by human hepatocytes in culture. *Proc. Amer. Assoc. Cancer Res.* **36**, 152.

Conney, A. H. (1982). Induction of microsomal enzymes by foreign chemicals and carcinogenesis by polycyclic aromatic hydrocarbons. *Cancer Res.* **42**, 4875–4917.

Daniel, V., Sharon, R., Tichauer, Y., and Sarid, S. (1986). Mouse glutathione S-transferase Ya subunit: Gene structure and sequence. *DNA* **6**, 317–324.

Darnell, J. E., Jr. (1982). Variety in the level of gene control in eukaryotic cells. *Nature* **297**, 365–371.

David, M., Petricoin, E., III, Benjamin, C., Pine, R., Weber, M. J., and Larner, A. C. (1995). Requirement for MAP kinase (ERK2) activity in interferon a- and interferon B-stimulated gene expression through STAT proteins. *Science* **269**, 1721–1723.

Davidson, N. E., Egner, P. A., and Kensler, T. W. (1990). Transcriptional control of glutathione S-transferase gene expression by the chemoprotective agent 5-(2-pyrazinyl)-4-methyl-1,2-dithiol-3-thione (oltipraz) in rat liver. *Cancer Res.* **50**, 2251–2257.

Davis, R. J. (1994). MAPKs: New JNK expands the group. *Trends Biochem. Sci.* **19**, 470–472.

Davis, R. J. (1993). The mitogen-activated protein kinase signal transduction pathway. *J. Biol. Chem.* **268**, 14553–14556.

DeLong, M. J., Dolan, P., Santamaria, A. B., and Bueding, E. (1986). 1,2-Dithiol-3-thione analogs: Effects on NAD(P)H:quinone reductase and glutathione levels in murine hepatoma cells. *Carcinogenesis* **7**, 977–980.

DeLong, M. J., Santamaria, A. B., and Talalay, P. (1987). Role of cytochrome P_{1-} 450 in the induction of NAD(P)H:quinone reductase in a murine hepatoma cell line and its mutants. *Carcinogenesis* **8**, 1549–1553.

de Morais, S. M. F., Chow, S. Y. M., and Wells, P. G. (1992). Biotransformation and toxicity of acetaminophen in congenic RHA rats with or without a hereditary deficiency in bilirubin UDP-glucuronosyltransferase. *Toxicol. Appl. Pharmacol.* **117**, 81–87.

de Morais, S. M. F., and Wells, P. G. (1988). Deficiency in bilirubin UDP-glucuronosyltransferase as a genetic determinant of acetaminophen toxicity. *J. Pharmacol. Exp. Ther.* **247**, 323–331.

Deng, T., and Karin, M. (1994). c-Fos transcriptional activity stimulated by H-Ras-activated protein kinase distinct from JNK and ERK. *Nature* **371**, 171–175.

Derijard, B., Hibi, M., Wu, I.-H., Barrett, T., Su, B., Deng, T., Karin, M., and Davis, R. J. (1994). JNK1: A protein kinase stimulated by UV light and Ha-Ras that binds and phosphorylates the c-Jun activation domain. *Cell* **76**, 1025–1037.

Dethmers, J. K., and Meister, A. (1981). Glutathione export by human lymphoid cells: Depletion of glutathione by inhibition of its synthesis decreases export and increases sensitivity to irradiation. *Proc. Natl. Acad. Sci. U.S.A.* **78**, 7492–7496.

Devary, Y., Gottlieb, R. A., Smeal, T., and Karin, M. (1992). The mammalian ultraviolet response is triggered by activation of Src tyrosine kinases. *Cell* **71**, 1081–1091.

Dunn, T. J., Koleske, A. J., Lindahl, R., and Pitot, H. C. (1989). Phenobarbital-inducible aldehyde dehydrogenase in the rat:cDNA sequence and regulation of the mRNA by phenobarbital in responsive rats. *J. Biol. Chem.* **264**, 13057–13063.

Eaton, D. L., and Hamel, D. M. (1994). Increase in γ-glutamylcysteine synthetase activity as a mechanism for butylated hydroxyanisole-mediated elevation of hepatic glutathione. *Toxicol. Appl. Pharmacol.* **126**, 145–149.

Egner, P. A., Gange, S. J., Dolan, P. M., Groopman, J. D., Muñoz, A., and Kensler, T. W. (1995). Levels of aflatoxin-albumin biomarkers in rat plasma are modulated by both long-term and transient interventions with oltipraz. *Carcinogenesis* **16**, 1769–1773.

Egner, P. A., Kensler, T. W., Prestera, T., Talalay, P., Libby, A. H., Joyner, H. H., and Curphey, T. J. (1994). Regulation of phase 2 enzyme induction by oltipraz and other dithiolethiones. *Carcinogenesis* **15**, 177–181.

Ellis, E. M., Judah, D. J., Neal, G. E., and Hayes, J. D. (1993). An ethoxyquin-inducible aldehyde reductase from rat liver that metabolizes aflatoxin B_1 defines a sub-family of aldo-keto reductases. *Proc. Natl. Acad. Sci. U.S.A.* **90**, 10350–10354.

Fairchild, C. R., Ivy, S. P., Rushmore, T., Lee, G., Koo, P., Goldsmith, M. E., Myers, C., Farber, E., and Cowan, K. H. (1978). Carcinogen-induced mdr over expression is associated with xenobiotic resistance in rat preneoplastic liver nodules and hepatocellular carcinomas. *Proc. Natl. Acad. Sci. U.S.A.* **84**, 7701–7705.

Falany, C. N., McQuiddy, P., and Kasper, C. B. (1987). Structure and organization of the microsomal xenobiotic epoxide hydrolase gene. *J. Biol. Chem.* **262**, 5924–5930.

Farber, E. (1980). The genesis of cancer with chemicals. *Arch. Pathol. Lab. Med.* **104**, 499–502.

Favreau, L., and Pickett, C. B. (1993). Transcriptional regulation of the rat NAD(P)H:quinone reductase gene: Characterization of a DNA–protein interaction at the antioxidant responsive element and induction by 12-O-tetradecanoylphorbol 13-acetate. *J. Biol. Chem.* **268**, 19875–19881.

Favreau, L. V., and Pickett, C. B. (1991). Transcriptional regulation of the rat NAD(P)H:quinone reductase gene: Identification of regulatory elements controlling basal level expression and inducible expression by planar aromatic compounds and phenolic antioxidants. *J. Biol. Chem.* **266**, 4556–4561.

Fields, W. R., Eaton, D., Doehmer, J., and Townsend, A. J. (1995). Transfected murine α-class glutathione S-transferase Yc2 protects against DNA damage by aflatoxin B1 in mammalian cells. *Proc. Amer. Assoc. Cancer Res.* **36**, 597.

Friling, R. S., Bensimon, A., Tichauer, Y., and Daniel, V. (1990). Xenobiotic-inducible expression of murine glutathione S-transferase Ya subunit gene is controlled by an electrophile-responsive element. *Proc. Natl. Acad. Sci. U.S.A.* **87**, 6258–6262.

Friling, R. S., Bergelson, S., and Daniel, V. (1992). Two adjacent AP-1-like binding sites form the electrophile-responsive element of the murine glutathione S-transferase Ya subunit gene. *Proc. Natl. Acad. Sci. U.S.A.* **89**, 668–672.

Galaris, D., Georgellis, A., and Rydstrom, J. (1985). Toxic effects of daunorubicin on isolated and cultured heart cells from neonatal rats. *Biochem. Pharmacol.* **34**, 989–995.

Gilbert, D., and Goldberg, L. (1965). Liver response tests. III. Liver enlargement and stimulation of microsomal processing enzyme activity. *Food Cosmet. Toxicol.* **3**, 417–432.

Glatt, H. R., Vogel, K., Bentley, P., and Oesch, F. (1979). Reduction of benzo(a)pyrene mutagenicity by dihydrodiol dehydrogenase. *Nature* **277**, 319–320.

Gonzalez, F. J. (1988). The molecular biology of cytochrome P-450s. *Pharmacol. Rev.* **40**, 243–288.

Goon, D., and Klaassen, C. D. (1992). Effects of microsomal enzyme inducers upon UDP-glucuronic acid concentration and UDP-glucuronosyltransferase activity in the rat intestine and liver. *Toxicol. Appl. Pharmacol.* **115**, 253–260.

Guehmann, S., Vorbrueggen, G., Kalkbrenner, F., and Moelling, K. (1992). Reduction of a conserved Cys is essential for Myb DNA-binding. *Nucleic Acids Res.* **20**, 2279–2286.

Guengerich, F. P. (1992). Metabolic activation of carcinogens. *Pharmacol. Ther.* **54**, 17–61.

Guengerich, F. P., Wang, P., Mason, P. S., and Mitchell, M. B. (1979a). Rat and human liver microsomal epoxide hydratase: Immunological characterization of various forms of the enzyme. *J. Biol. Chem.* **254**, 12255–12259.

Guengerich, F. P., Wang, P., Mitchell, M. B., and Mason, P. S. (1979b). Rat and human liver microsomal epoxide hydratase: Purification and evidence for the existence of multiple forms. *J. Biol. Chem.* **254**, 12248–12254.

Guyton, K. Z., Liu, Y.-L., Gorospe, M., Xu, Q., and Holbrook, N. J. (1996). Activation of mitogen-activated protein kinase by H_2O_2: Role in cell survival following oxidant injury. *J. Biol. Chem.* **271**, 4138–4142.

Hai T., and Curran T. (1991). Cross-family dimerization of transcription factors Fos/Jun and ATF/CREB alters DNA binding specificity. *Proc. Natl. Acad. Sci. U.S.A.* **88**, 3720–3724.

Hall, A. G., Matheson, E., Hickson, I. D., Foster, S. A., and Hogarth, L. (1994). Purification of an a-class glutathione S-transferase from melphalan resistant Chinese hamster ovary cells and demonstration of its ability to catalyze melphalan-glutathione adduct formation. *Cancer Res.* **54**, 3369–3372.

Halliwell, B., and Gutteridge, J. M. C. (1989). "Free Radicals in Biology and Medicine," 2nd Ed. Clarendon Press, Oxford.

Harada, S., Abai, M., Tanaka, N., Agrawal, D., and Goedde, H. W. (1987). Liver glutathione S-transferase polymorphism in Japanese and its pharmacogenetic importance. *Hum. Genet.* **75**, 322–325.

Hayes, J. D., Kerr, L. A., Peacock, S. D., Cronshaw, A. D., and McClellan, L. I. (1991). Hepatic glutathione S-transferases in mice fed on a diet containing the anticarcinogenic antioxidant butylated hydroxyanisole. *Biochem. J.* **277**, 501–512.

Hayes, J. D., Judah, D. J., and Neal, G. E. (1993). Resistance to aflatoxin B_1 is associated with the expression of a novel aldo-keto reductase which has catalytic activity towards a cytotoxicity aldehyde-containing metabolite of the toxin. *Cancer Res.* **53**, 3887–3894.

Hayes, J. D., Nguyen, T., Judah, D. J., Petersson, D. G., and Neal, G. E. (1994). Cloning of cDNAs from fetal rat liver encoding glutathione S-transferase Yc polypeptides. *J. Biol. Chem.* **269**, 20707–20717.

Hershman, H. R. (1991). Primary response genes induced by growth factors and tumor promoters. *Annu. Rev. Biochem.* **60**, 281–319.

Hibi, M., Lin, A., Smeal, T., Minden, A., and Karin, M. (1993). Identification of an oncoprotein- and UV-responsive protein kinase that binds and potentiates the c-Jun activation domain. *Genes Dev.* **7**, 2135–2148.

Hirvonen, A., Husgafvel-Pursiainen, K., Anttila, S., and Vainio, H. (1993). The GST M1 null genotypes as a potential risk modifier for squamous cell carcinoma of the lung. *Carcinogenesis* **14**, 1479–1481.

Hu, Z., and Wells, P. G. (1992). *In vitro* and *in vivo* biotransformation and covalent binding of benzo[*a*]pyrene in Gunn and RHA rats with genetic deficiency in bilirubin uridine diphosphate-glucuronosyltransferase. *J. Pharmacol. Exp. Ther.* **263**, 334–342.

Hunter, T. (1991). Activation of protein kinase C decreases phosphorylation of c-Jun at sites that negatively regulate its DNA-binding activity. *Cell* **64**, 573–584.

Hunter, T. (1993). Cytokine connections. *Nature* **366**, 114–116.

Hunter, T., and Karin M. (1992). The regulation of transcription by phosphorylation. *Cell* **70**, 375–387.

Ishikawa, M., Sasaki, K.-I., and Takayanagi, Y. (1989). Injurious effect of buthionine sulfoximine, an inhibitor of glutathione synthesis, on the lethality and urotoxicity of cyclophosphamide in mice. *J. Pharmacol. Jap.* **51**, 146–149.

Iyanagi, T., and Yamazaki, I. (1970). One-electron-transfer reactions in biochemical systems. V. Difference in the mechanism of quinone reduction by the NADH dehydrogenase and NAD(P)H dehydrogenase (DT-diaphorase). *Biochim. Biophys. Acta* **216**, 282–294.

Jaiswal, A. (1994). Antioxidant response element. *Biochem. Pharmacol.* **48**, 439–444.

Judah, D. J., Hayes, J. D., Yang, J.-C., Lian, L.-Y., Roberts, G. C. K., Farmer, P. B., Lamb, J. H., and Neal, G. E. (1993). A novel aldehyde reductase with activity towards a metabolite of aflatoxin B1 is expressed in rat liver during carcinogenesis and following administration of an antioxidant. *Biochem. J.* **292**, 13–18.

Kahl, R. (1991). Protective and hazardous properties of antioxidants. *In* "Oxidative Stress: Oxidants and Antioxidants" (H. Sies, ed.), pp. 245–273. Academic Press, London.

Kahl, R., and Netter, K. J. (1977). Ethoxyquin as an inducer and inhibitor of phenobarbital-type cytochrome P-450 in rat liver microsomes. *Toxicol. Appl. Pharmacol.* **40**, 473–483.

Kashfi, K., Yang, E. K., Chowdhury, J. R., Chowdhury, N. R., and Dannenberg, A. J. (1994). Regulation of uridine diphosphate glucuronosyltransferase expression by phenolic antioxidants. *Cancer Res.* **54**, 5856–5859.

Kataoka, K., Noda, M., and Nishizawa, M. (1994). Maf nuclear oncoprotein recognizes sequences related to AP-1 site and forms heterodimers with both Fos and Jun. *Mol. Cell. Biol.* **14**, 700–712.

Kaul, R., and Netter, K. J. (1977). Ethoxyquin as an inducer and inhibitor of phenobarbital type cytochrome P-450 in rat liver microsomes. *Toxicol. Appl. Pharmacol.* **40**, 473–483.

Kensler, T. W., Davidson, N. E., Groopman, J. D., Roebuck, B. D., Prochaska, H. J., and Talalay, P. (1993). Chemoprotection by inducers of electrophile detoxication enzymes. *In* "Antimutagenesis and Anticarcinogenesis Mechanisms" (G. Bronzetti *et al.*, eds.), pp. 127–136. Plenum Press, New York.

Kensler, T. W., Egner, P. A., Dolan, P., Groopman, J. D., and Roebuck, B. D. (1987). Mechanism of protection against aflatoxin tumorigenicity in rats fed 5-(2-pyrazinyl)-4-methyl-1,2-dithiol-3-thione (oltipraz) and related 1,2-dithiol-3-thiones and 1,2-dithiol-3-ones. *Cancer Res.* **47**, 4271–4277.

Kensler, T. W., Egner, P. A., Trush, M. A., Bueding, E., and Groopman, J. D. (1985). Modification of aflatoxin B1 binding to DNA in vivo in rats fed phenolic antioxidants, ethoxyquin and a dithiothione. *Carcinogenesis* **6**, 759–763.

Kensler, T. W., Groopman, J. D., Eaton, D. L., Curphey, T. J., and Roebuck, B. D. (1992a). Potent inhibition of aflatoxin-induced hepatic tumorigenesis by the monofunctional enzyme inducer 1,2-dithiol-3-thione. *Carcinogenesis* **13**, 95–100.

Kensler, T. W., Groopman, J. D., and Roebuck, B. D. (1991). Chemoprotection by oltipraz and other dithiolethiones. *In* "Cancer Chemoprevention" (L. Wattenberg, L. Lipkin, C. W. Boone, and G. J. Kelloff, eds.), pp. 205–226. CRC Press, Boca Raton, FL.

Kensler, T. W., and Helzlsouer, K. J. (1995). Oltipraz: Clinical opportunities for cancer chemoprevention. *J. Cell. Biochem.* **22S**, 101–107.

Kensler, T. W., Trush, M. A., and Guyton, K. Z. (1992b). Free radicals as targets for cancer protection: prospects and problems. *In* "Cellular and Molecular Targets for Chemoprevention" (V. E. Steele, G. D. Stoner, C. W. Boone, and G. J. Kelloff, eds.), pp. 173–192. CRC Press, Boca Raton, FL.

Kerppola, T. K., and Curran, T. (1994). A conserved region adjacent to the basic domain is required for recognition of an extended DNA binding site by Maf/Nrl family proteins. *Oncogene* **9**, 3149–3158.

Ketterer, B., Meyer, D. J., and Clark, A. G., (1988). Soluble glutathione transferase isozymes. *In* "Glutathione Conjugation" (B. Ketterer and H. Sies, eds.), pp. 74–135. Academic Press, London.

Kyriakis, J. M., Banerjee, P., Nikolakaki, E., Dai, T., Rubie, E. A., Ahmad, M. F., Avruch, J., and Woodgett, J.R. (1994). The stress-activated protein kinase subfamily of c-Jun kinases. *Nature* **369**, 156–160.

Langouët, Coles, B., Morel, F., Becquemont, L., Beaune, P., Guengrich, F. P., Ketterer, B., and Guillouzo, A. (1995). Inhibition of CYP1A2 and CYP3A4 by oltipraz results in reduction of aflatoxin B_1 metabolism in human hepatocytes in primary culture. *Cancer Res.* **55**, 5574–5579.

Leibold, E. A., and Munro, H. N. (1987). Characterization and evolution of the expressed rat ferritin light subunit gene and its pseudogene family: Conservation of sequences within noncoding regions of ferritin genes. *J. Biol. Chem.* **262**, 7335–7341.

Levi, S., Luzzago, A., Cesareni, G., Cozzi, A., Franceschinelli, F., Albertini, A., and Arosio, P. (1988). Mechanism of ferritin iron uptake: Activity of the H-chain and deletion mapping of the ferroxidase site. *J. Biol. Chem.* **263**, 18086–18092.

Li, Y., and Jaiswal, A. K. (1992a). Regulation of human NAD(P)H:quinone oxidoreductase gene: Role of AP-1 binding site contained within human antioxidant response element. *J. Biol. Chem.* **267**, 15097–15104.

Li, Y., and Jaiswal, A. K. (1992b). Identification of Jun-B as third member in human antioxidant response element-nuclear proteins complex. *Biochem. Biophys. Res. Commun.* **188**, 992–996.

Lin, H. J., Probst-Hensch, N. M., Ingles, S. A., Han C.-Y., Lin, B. K., Lee, D. B., Frankl, H. D., Lee, E. R., Longnecker, M. P., and Haile, R. W. (1995). Glutathione S-transferase (GST M1) null genotype, smoking, and prevalence of colorectal adenomas. *Cancer Res.* **55**, 1224–1226.

Lind, C., Cadenas, E., Hochstein, P., and Ernster, L. (1990). DT-diaphorase: Purification, properties and function. *Methods Enzymol.* **186**, 287–301.

Lind, C., Hochstein, P., and Ernster, L. (1982). DT-diaphorase as a quinone reductase: A cellular control device against semiquinone and superoxide radical formation. *Arch. Biochem. Biophys.* **216**, 178–185.

Lind, C., Vadi, H., and Ernster, L. (1978). Metabolism of benzo(*a*)pyrene-3,6-quinone and 3-hydroxybenzo(*a*)pyrene in liver microsomes from 3-methylcholanthrene-treated rats. *Arch. Biochem. Biophys.* **190**, 97–108.

Lindahl, R. (1992). Aldehyde dehydrogenases and their role in carcinogenesis. *Crit. Rev. Biochem. Mol. Biol.* **27**, 283–335.

Lindahl, R., and Evces, S. (1984). Rat aldehyde dehydrogenase. II. Isolation and characterization of four inducible enzymes. *J. Biol. Chem.* **259**, 11991–11996.

Mackenzie, P., and Rodbourn, L. (1990). Organization of the rat UDP-glucuronosyltransferase, UDPGTr-2, gene and characterization of its promoter. *J. Biol. Chem.* **265**, 11328–11332.

Manoharan, T. H., Puchalski, R. B., Burgess, J. A., Pickett, C. B., and Fahl, W. E. (1987). Promoter-glutathione S-transferase Ya cDNA hybrids: Expression and conferred resistance to an alkylating molecule in mammalian cells. *J. Biol. Chem.* **262**, 3739–3745.

Marais, R., Wynne, J., and Treisman R. (1993). The SRF accessory protein Elk-1 contains a growth factor-regulated transcriptional activation domain. *Cell* **73**, 381–393.

McBride, A. A., Klausner, R. D., and Howley, P. M. (1992). Conserved cysteine residue in the DNA-binding domain of the bovine papilloma type 1 E2 protein confers redox regulation of the DNA-binding activity *in vitro*. *Proc. Natl. Acad. Sci. U.S.A.* **89**, 7531–7535.

Meister, A. (1991). Glutathione deficiency produced by inhibition of its synthesis and its reversal; applications in research and therapy. *Pharmacol. Ther.* **51**, 155–194.

Meister, A. (1994). Glutathione, ascorbate, and cellular protection. *Cancer Res.* **54**, 1969s–1975s.

Meyer, D. J., Harris, J. M., Gilmore, K. S., Coles, B., Kensler, T. W., and Ketterer, B. (1993a). Quantitation of tissue- and sex-specific induction of rat GSH transferase subunits by dietary 1,2-dithiole-3-thiones. *Carcinogenesis* **14**, 567–572.

Meyer, M., Pahl, H. L., and Baeuerle, P. A. (1994). Regulation of the transcription factors NF-κB and AP-1. *Chem. Biol. Int.* **91**, 91–100.

Meyer M., Schreck R. and Baeuerle P.A. (1993b). H₂O₂ and antioxidants have opposite effects on activation of NF-kB and AP-1 in tact cells: AP-1 as secondary antioxidant-responsive factor. *EMBO J.* **12**, 2005–2015.

Miller, E. C., and Miller, J. A. (1981). Searches for ultimate chemical carcinogens and their reactions with cellular macromolecules. *Cancer Res.* **47**, 2327–2345.

Minden, A., Lin, A., Smeal, T., Derijard, B., Cobb, M., Davis, R., and Karin, M. (1994). c-Jun N-terminal phosphorylation correlates with activation of the JNK subgroup but not the ERK subgroup of mitogen activated protein kinases. *Mol. Cell. Biol.* **14**, 6683–6688.

Mitchell, P. J., and Tjian, R. (1989). Transcriptional regulation in mammalian cells by sequence-specific DNA binding proteins. *Science* **245**, 371–378.

Morel, F., Fardel, O., Meyer, D. J., Langouet, S., Gilmore, K. S., Meunier, B., Tu, C. P. D., Kensler, T. W., Ketterer, B., and Guillouzo, A. (1993). Preferential increase of glutathione S-transferase class α transcripts in cultured human hepatocytes by phenobarbital, 3-methylcholanthrene, and dithiolethiones. *Cancer Res.* **53**, 231–234.

Mossman, B. T. (1991). Differential regulation of antioxidant enzymes in response to oxidants. *J. Biol. Chem.* **266**, 24398–24403.

Mulcahy, R. T., Bailey, H. H., and Gipp, J. J. (1995). Transfection of complementary DNAs for the heavy and light subunits of human γ-glutamylcysteine synthetase results in an elevation of intracellular glutathione and resistance to melphalan. *Cancer Res.* **55**, 4771–4775.

Mulcahy, R. T., and Gipp, J. J. (1995). Identification of a putative antioxidant response element in the 5'-flanking region of the human γ-glutamylcyteine synthetase heavy subunit gene. *Biochem. Biophys. Res. Commun.* **209**, 227–233.

Munro, H. (1993). The ferritin genes: Their response to iron status. *Nutr. Rev.* **51**, 65–83.

Naganuma, A., Anderson, M. E., and Meister, A., (1990). Cellular glutathione is a determinant of sensitivity to mercuric chloride toxicity: Prevention of toxicity by giving glutathione monoester. *Biochem. Pharmacol.* **40**, 693–697.

Nakabeppu, Y., and Nathans, D. (1991). A naturally occurring truncated form of FosB that inhibits Fos/Jun transcriptional activity. *Cell* **64**, 751–759.

Nakajima, T., Elovarra, E., Anttila, S., Hirvonen, A., Camus, A.-M., Hayes, J. D., Ketterer, B., and Vainio, H. (1995). Expression and polymorphism of glutathione S-transferase in human lungs: Risk factors in smoking-related lung cancer. *Carcinogenesis* **16**, 707–711.

Nazar-Stewart, V., Motulsky, A. G., Eaton, D. L., White, E., Hornung, S. K., Leng, Z. T., Stapleton, P., and Weiss, N. S. (1993). The glutathione S-transferase mu polymorphism as a marker for susceptibility to lung carcinoma. *Cancer Res.* **53**, 2313–2318.

Nebert, D. W., Nelson, D. R., Coon, M. J., Estabrook, R. W., Feyerstein, R., Fuji-Kuriyama, Y., Gonzalez, F. J., Guengerich, F. P., Gunslaus, I. C., Johnson, E. F., Loper, J. C., Sato, R., Waterman, M. R., and Waxman, D. J. (1991). The P450 superfamily: Update on new sequences, gene mapping, and recommended nomenclature. *DNA Cell Biol.* **10**, 1–14.

Nguyen, T., and Pickett, C.B. (1992). Regulation of rat glutathione-S-transferase Ya subunit gene expression: DNA-protein interaction at the antioxidant response element. *J. Biol. Chem.* **267**, 13535–13539.

Nguyen, T., Rushmore, T. H., and Pickett, C. B. (1994). Transcriptional regulation of a rat liver glutathione *S*-transferase Ya subunit gene: Analysis of the antioxidant response element and its activation by phorbol ester 12-O-tetradecanoylphorbol-13-acetate. *J. Biol. Chem.* **269**, 13656–13662.

Oesch, F. (1972). Mammalian epoxide hydratases: Inducible enzymes catalyzing the inactivation of carcinogenic and cytotoxic metabolites derived from aromatic and olefinic compounds. *Xenobiotica* **3**, 305–340.

Oesch F., Bentley, P., and Glatt, H. R. (1976). Prevention of benzo(a)pyrene-induced mutagenicity by homogeneous epoxide hydratase. *Int. J. Cancer* **18**, 448–452.

Okey, A. B. (1990). Enzyme induction in the cytochrome P-450 system. *Pharmacol. Ther.* **45**, 241–298.

Okuda, A., Imagawa, M., Maeda, Y., Sakai, M., and Muramatsu, M. (1989). Structural and functional analysis of an enhancer GPEI having a phorbol 12-O-tetradecanoate 13-acetate responsive element-like sequence found in the rat glutathione transferase P gene. *J. Biol. Chem.* **264**, 16919–16926.

Okuda, A., Imagawa, M., Maeda, Y., Sakai, M., and Muramatsu, M. (1990). Functional cooperativity between two TPA responsive elements in undifferentiated F9 embryonic stem cells. *EMBO J.* **9**, 1131–1135.

Parke, D. V., Rahim, A., and Walker, R. (1974a). Inhibition of some rat hepatic microsomal enzymes by ethoxyquin. *Biochem. Pharmacol.* **23**, 3385–3394.

Parke, D. V., Rahim, A., and Walker, R. (1974b). Reversibility of hepatic changes caused by ethoxyquin. *Biochem. Pharmacol.* **23**, 1871–1876.

Paulson, K. E., Darnell, J. E., Jr., Rushmore, T., and Pickett, C. B. (1990). Analysis of the upstream elements of the xenobiotic compound-inducible and positionally regulated glutathione *S*-transferase Ya gene. *Mol. Cell Biol.* **101**, 1841–1852.

Pawloski, J. E., Huizinga, M., and Penning, T. M. (1991). Cloning and sequencing of the cDNA for rat liver 3a-hydoxysteroid/dihydrodiol dehydrogenase. *J. Biol. Chem.* **266**, 8820–8825.

Pearson, W. R., Reinhart, J., Sisk, S. C., Anderson, K. S., and Adler, P. N. (1988). Tissue-specific induction of murine glutathione S-transferase mRNAs by butylated hydroxyanisole. *J. Biol. Chem.* **26**, 13324–13332.

Pearson, W. R., Windle, J. F., Morrow, J. F., Benson, A. M., and Talalay, P. (1983). Increased synthesis of glutathione S-transferase in response to anticarcinogeneic antioxidants: Cloning and measurement of messenger RNA. *J. Biol. Chem.* **258**, 2052–2062.

Penning, T. M. (1993). Dihydrodiol dehydrogenase and its role in polycyclic aromatic hydrocarbon metabolism. *Chem. Biol. Int.* **89**, 1–34.

Pickett, C. B., Telakowski-Hopkins, C. A., Ding, G. J.-F., Argenbright, L., and Lu, A. Y.-H. (1984). Rat liver glutathione S-transferases: Complete nucleotide sequence of a glutathione S-transferase mRNA and the regulation of the Ya, Yb, and Yc mRNAs by 3-methylcholanthrene and phenobarbital. *J. Biol. Chem.* **259**, 5182–5188.

Pinkus, R., Bergelson, S., and Daniel, V. (1993). Phenobarbital induction of AP-1 binding activity mediates activation of glutathione-*S*-transferase and quinone reductase gene expression. *Biochem. J.* **290**, 637–640.

Pinkus, R., Weiner, L. M., and Daniel, V. (1995). Role of quinone-mediated generation of hydroxyl radicals in the induction of glutathione *S*-transferase gene expression. *Biochemistry* **34**, 81–88.

Pognonec, P., Kato, H., and Roeder, R. G. (1992). The helix-loop-helix/leucine repeat transcription factor USF can be functionally regulated in a redox-dependent manner. *J. Biol. Chem.* **267**, 24563–24567.

Prestera, T., Holtzclaw, W. D., Zhang, Y., and Talalay, P. (1993). Chemical and molecular regulation of enzymes that detoxify carcinogens. *Proc. Natl. Acad. Sci. U.S.A.* **90**, 2965–2969.

Primiano, T., Egner, P. A., Sutter, T. R., Kelloff, G. J., Roebuck, B. D., and Kensler, T. W. (1995a). Intermittent dosing with oltipraz: Relationship between chemoprevention of aflatoxin-induced tumorigenesis and induction of glutathione *S*-transferases. *Cancer Res.* **55**, 4319–4324.

Primiano, T., Gastel, J. A., Kensler, T. W., and Sutter, T. R. (1995b). Gene expression induced by chemoprotective agents. *Proc. Am. Assoc. Cancer Res.* **36**, 522.

Prochaska, H. J., De Long, M. J., and Talalay, P. (1985). On the mechanism of induction of cancer-protective enzymes: A unifying proposal. *Proc. Natl. Acad. Sci. U.S.A.* **82**, 8232–8236.

Prochaska, H., and Talalay, P. (1988). Regulatory mechanisms of monofunctional and bifunctional anticarcinogenic enzyme inducers in murine liver. *Cancer Res.* **48**, 4776–4782.

Prochaska, H., and Talalay, P. (1991). The role of NAD(P)H:quinone reductase in protection against the toxicity of quinones and related agents. *In* "Oxidative Stress: Oxidants and Antioxidants" (H. Sies, ed.), pp. 195–211. Academic Press, London.

Putt, D. A., Kensler, T. W., and Hollenberg, P. (1991). Effect of three chemoprotective antioxidants, ethoxyquin, oltipraz and 1,2-dithiole-3-thione on cytochrome P450 levels and aflatoxin B1 metabolism. *FASEB J.* **5**, A1517.

Qin, K.-N. and Cheng, K.-C. (1994). Structure and tissue specific expression of aldo-keto reductase superfamily. *Biochemistry* **33**, 3223–3228.

Raney, K. D., Meyer, J. D., Ketterer, B., Harris, T. M., and Guengrich, F. P. (1992). Glutathione conjugation of aflatoxin B1 *exo*- and *endo-epoxides* by rat and human glutathione S-transferases. *Chem. Res. Toxicol.* **5**, 470–478.

Robertson, J. A., Chen, H., and Nebert, D. W. (1986). NAD(P)H:menadione oxidoreductase: Novel purification of enzyme, cDNA and complete amino acid sequence and gene regulation. *J. Biol. Chem.* **261**, 15794–15799.

Roebuck, B. D., Liu, Y.-L., Rogers, A. R., Groopman, J. D., and Kensler, T. W. (1991). Protection against aflatoxin B[1]-induced hepatocarcinogenesis in F344 rats by 5-(2-pyrazinyl)-4-methyl-1,2-dithiole-3-thione(oltipraz): Predictive role for short-term molecular dosimetry. *Cancer Res.* **51**, 5501–5506.

Roy, D., and Liehr, J. G. (1988). Temporary decrease in renal quinone reductase activity induced by chronic administration of estradiol to male Syrian hamsters: Increased superoxide formation by redox cycling of estrogen. *J. Biol. Chem.* **263**, 3646–3651.

Rushmore, T. H., King, R. G., Paulson, K. E., and Pickett, C. B. (1990). Regulation of glutathione *S*-transferase Ya subunit gene expression: Identification of a unique xenobiotic-responsive element controlling inducible expression by planar aromatic compounds. *Proc. Natl. Acad. Sci. U.S.A.* **87**, 3826–3830.

Rushmore, T. H., Morton, M. R., and Pickett, C. B. (1991). The antioxidant responsive element: Activation by oxidative stress and identification of the DNA consensus sequence required for functional activity. *J. Biol. Chem.* **266**, 11632–11639.

Rushmore, T. H., and Pickett, C. B. (1990). Transcriptional regulation of the rat glutathione *S*-transferase Ya subunit gene: Characterization of a xenobiotic-responsive element controlling inducible expression by phenolic antioxidants. *J. Biol. Chem.* **265**, 14648–14653.

Sakai, M., Matsushima-Hibiya, Y., Nishizawa, M., and Nishi, S. (1995). Suppression of rat glutathione transferase P expression by peroxisome proliferators: Interaction between Jun and peroxisome proliferator-activated receptor α. *Cancer Res.* **55**, 5570–5376.

Sakai, M., Okuda, A., and Muramatsu, M. (1988). Multiple regulatory elements and phorbol 12-O-tetradecanoate 13-acetate responsiveness of the rat placental glutathione transferase gene. *Proc. Natl. Acad. Sci. U.S.A.* **85**, 9456–9460.

Schenk, H., Klein, M., Erdbrugger, W., Droge, W., and Schulze-Osthoff, K. (1994). Distinct effects of thioredoxin and antioxidants on the activation of transcription factors NF-κB and AP-1. *Proc. Natl. Acad. Sci. USA* **91**, 1672–1676.

Schreck, R., Albermann, K., and Baeuerle, P. A. (1992). Nuclear factor κB: An oxidative stress-responsive transcription factor of eukaryotic cells (a review). *Free Rad. Res. Commun.* **17**, 221–237.

Seelig, G. F., Simondsen, R. P., and Meister, A. (1984). Reversible dissociation of γglutamylcyteine synthetase into two subunits. *J. Biol. Chem.* **259**, 9345–9347.

Segura-Aguilar, J., Kaiser, R., and Lind, C. (1992). Separation and characterization of isoforms of DT-diaphorase from rat liver cytosol. *Biochim. Biophys. Acta* **1120**, 33–42.

Seidegård, J., and DePierre, J. W. (1983). Microsomal epoxide hydrolase, properties, regulation and function. *Biochim. Biophys. Acta* **695**, 307–320.

Seidegård, J., Vorachek, W. R., Pero, R. W., and Pearson, W. R. (1988). Hereditary differences in the expression of glutathione *S*-transferase active on trans-stilbene oxide are due to a gene deletion. *Proc. Natl. Acad. Sci. U.S.A.* **85**, 7293–7297.

Sharma, S., and Stutzman, J. (1993). A study of the candidate chemopreventive agent, oltipraz by gene regulation mapping. *Proc. Am. Assoc. Cancer Res.* **34**, 549.

Sharma, S., Stutzman, J. D., Keloff, G. J., and Steele, V. E. (1994). Screening of potential chemopreventive agents using biochemical markers of carcinogenesis. *Cancer Res.* **54**, 5848–5855.

Shou, M., Harvey, R. G., and Penning, T. M. (1992). Contribution of dihydrodioldehydrogenase to the metabolism of $(+/-)$-trans-7,8-dihydroxy-7,8-dihydrobenzo[a]pyrene in fortified rat liver subcellular fractions. *Carcinogenesis* **13**, 1814–1820.

Shull, S., Heintz, N. H., Periasamy, M., Manohar, M., Janssen, Y. M. W., Marsh, J. P., and Mossman, B. T. (1991). Differential regulation of antioxidant enzymes in response to oxidants. *J. Biol. Chem.* **266**, 24398–24403.

Sies, H. (1986). Biochemistry of oxidative stress. *Angew. Chem. Int. Ed. Eng.* **25**, 1058–1071.

Smithgall, T. E., Harvey, R. G., and Penning, T. M. (1988). Spectroscopic identification of *ortho*-quinones as the products of polycyclic aromatic trans-dihydrodiol oxidation catalyzed by dihydrodiol dehydrogenase: A potential route of proximate carcinogen metabolism. *J. Biol. Chem.* **263**, 1814–1820.

Sreerama, L., Rekha, G. K., and Sladek, N. E. (1995). Phenolic antioxidant-induced over-expression of class-3 aldehyde dehydrogenase and oxazaphosphorine-specific resistance. *Biochem. Pharmacol.* **49**, 669–675.

Sreerama, L., and Sladek, N. E. (1994). Identification of a methylcholanthrene-induced aldehyde dehydrogenase in a human breast adenocarcinoma cell line exhibiting oxazaphosphorine-specific acquired resistance. *Cancer Res.* **54**, 2176–2185.

Suzukake, K., Petro, B. J., and Vistica, D. T. (1982). Reduction in glutathione content of L-PAM resistant L1210 cells confers drug sensitivity. *Biochem. Pharmacol.* **31**, 121–124.

Szarka, C. E., Pfeiffer, G. R., Hum, S. T., Everley, L. C., Balshem, A. M., Moore, D. F., Litwin, S., Goosenberg, E. B., Frucht, H., Engstrom, P. F., and Clapper, M. L. (1995). Glutathione *S*-transferase activity and glutathione *S*-transferase mu expression in subjects with risk of colorectal cancer. *Cancer Res.* **55**, 2789–2793.

Talalay, P. (1989). Mechanisms of induction of enzymes that protect against chemical carcinogenesis. *Adv. Enzyme Regul.* **28**, 149–159.

Talalay, P., and Benson, A. M. (1982). Elevation of quinone reductase activity by anticarcinogenic antioxidants. *Adv. Enzyme Reg.* **20**, 287–300.

Talalay, P., De Long, M. J., and Prochaska, H. J. (1987). Molecular mechanisms in protection against carcinogenesis. *In* "Cancer Biology and Therapeutics" (J.G. Cory and A. Szentivanyi, eds.), pp. 197–216. Plenum Press, New York.

Talalay, P., De Long, M. J., and Prochaska, H. (1988). Identification of a common chemical signal regulating the induction of enzymes that protect against chemical carcinogenesis. *Proc. Natl. Acad. Sci. U.S.A.* **85**, 8261–8265.

Tephly, T. R., and Burchell, B. (1990). UDP-glucuronoslytransferases: A family of detoxifying enzymes. *Trends Pharmacol. Sci.* **11**, 276–279.

Tew, K. D. (1994). Glutathione-associated enzymes in anticancer drug resistance. *Cancer Res.* **54**, 4314–4320.

Theil, E. C. (1990). The ferritin family of iron storage proteins. *Adv. Enzymol.* **63**, 421–449.

Thomas, D. J., Sadler, A., Subrahmanyam, V. V., Siegel, D., Reasor, M. J., Wierda, D., and Ross, R. (1990). Bone marrow stromal cell bioactivation and detoxification of the benzene metabolite hydroquinone: Comparison of macrophages and fibroblastoid cells. *Mol. Pharmacol.* **37**, 255–262.

Thor, H., Smith, M. Y., Hartzell, P., Bellomo, G., Jewell, S. A., and Orrenius, S. (1982). The metabolism of menadione (2-methyl-1,4-naphthoquinone) by isolated hepatocytes: A study of the implications of oxidative stress in intact cells. *J. Biol. Chem.* **257**, 12419–12425.

Treisman, R. (1992). The serum response element. *Trends Biochem. Sci.* **17**, 423–426.

Tsuchida, S., and Sato, K. (1992). Glutathione transferases and cancer. *Crit. Rev. Biochem. Mol. Biol.* **27**, 337–384.

Vienneau, D. S., DeBoni, U., and Wells, P. G. (1995). Potential genoprotective role for UDP-glucuronosyltransferases in chemical carcinogenesis: Initiation of micronuclei by benzo(a)pyrene and benzo(e)pyrene in UDP-glucuronosyltransferase-deficient cultured rat skin fibroblasts. *Cancer Res.* **55**, 1045–1051.

Wang, B., and Williamson, G. (1994). Detection of a nuclear protein which binds specifically to the antioxidant responsive element (ARE) of the human NAD(P)H:quinone oxidoreductase gene. *Biochim. Biophys. Acta* **1219**, 645–652.

Wasserman, M., and Fahl, W. E. (1995). Factors interacting with the antioxidant response element of the murine glutathione S-transferase 1 (Ya) gene. *Proc. Am. Assoc. Cancer Res.* **36**, 598.

Wattenberg, L. W. (1985). Chemoprevention of cancer. *Cancer Res.* **45**, 1–8.

Wattenberg, L. W., and Bueding, E. (1986). Inhibitory effects of 5-(2-pyrazinyl)-4-methyl-1,2-dithiol-3-thione (oltipraz) on carcinogenesis induced by benzo(a)pyrene, diethylnitrosamine and uracil mustard. *Carcinogenesis* **7**, 1379–1388.

Whitmarsh, A. J., Shore, P., Sharrocks, A. D., and Davis, R. (1995). Integration of MAP kinase signal transduction pathways at the serum response element. *Science* **269**, 403–407.

Wong, G., and Goeddel, D. (1988). Induction of Mn-SOD by TNF: Possible protective mechanism. *Science* **242**, 941–944.

Xanthoudakis, S., and Curran, T. (1992). Identification and characterization of Ref-1, a nuclear protein that facilitates AP-1 DNA binding activity. *EMBO J.* **11**, 653–665.

Xanthoudakis, S., Miao, G., Wang, F., Pan, Y.-C. E., and Curran T. (1992). Redox activation of Fos-Jun DNA binding activity is mediated by a DNA repair enzyme. *EMBO J.* **11**, 3323–3335.

Xie, T., Belinsky, M., Xu, Y., and Jaiswal, A. K. (1995). ARE- and TRE-mediated regulation of gene expression. *J. Biol. Chem.* **270**, 6894–6900.

Yao, K.-S., and O'Dwyer, P. J. (1995). Involvement of NF-κB in the induction of NAD(P)H:oxidoquinone reductase (DT-diaphorase) by hypoxia, oltipraz and mitomycin C. *Biochem. Pharmacol.* **49**, 275–282.

Yao, K.-S., Xanthoudakis, S., Curran, T., and O'Dwyer, P. J. (1994). Activation of AP-1 and of a nuclear redox factor, Ref-1, in the response of HT29 colon cancer cells to hypoxia. *Mol. Cell. Biol.* **14**, 5997–6003.

Yoshioka, K., Deng, T., Cavigelli, M., and Karin, M. (1995). Antitumor promotion by phenolic antioxidants: Inhibition of AP-1 activity through induction of Fra expression. *Proc. Natl. Acad. Sci. U.S.A.* **92**, 4972–4976.

Zhong, S., Wyllie, A. H., Barnes, D., Wolf, C. R., and Spurr, N. K. (1993). Relationship between the GST M1 genetic polymorphism and susceptibility to bladder, breast, and colon cancer. *Carcinogenesis* **14**, 1821–1824.

Garth Powis
John R. Gasdaska
Amanda Baker

Arizona Cancer Center
University of Arizona
Tucson, Arizona 85724

Redox Signaling and the Control of Cell Growth and Death

I. Introduction

Cells maintain an intracellular environment that is reducing in the face of a highly oxidizing extracellular environment. Regulated alterations in the intracellular redox state (redox signaling) can modulate events such as DNA synthesis, enzyme activation, selective gene expression, and regulation of the cell cycle. The primary consequence of intracellular redox signaling is a change in the oxidation state of cysteine residues of key proteins. This form of posttranslational modification of protein is difficult to follow because it lacks a convenient marker and is readily reversed when the cell contents are exposed to extracellular oxidizing conditions. For this reason, knowledge of the role that redox signaling may play in cell function has lagged behind that of protein modification produced, for example, by the addition of a phosphate groups. This chapter discusses the components of the cellular machinery used for the redox regulation of protein activity and its consequences for cell growth and death. A summary is shown in Fig. 1.

II. Cellular Redox Systems

A. Glutathione/Glutathione Reductase

Reduced glutathione (GSH) is a tripeptide of γ-glutamate, cysteine, and glycine, which is found in all eukaryotic cells at a concentration between 1 and 10 mM (Kosower and Kosower, 1978). The intracellular environment is highly reducing and the ratio of GSH to oxidized glutathione (GSSG) under physiological conditions is 30 to 100:1 (Hwang *et al.*, 1992). Cellular GSH content is determined by the rates of import and synthesis vs the rates of efflux and conjugation. Most cells do not import GSH and synthesis occurs intracellularly in sequential, ATP-dependent reactions catalyzed by γ-glutamylcysteine synthetase and GSH synthetase (Meister, 1991). The rate of synthesis is influenced primarily by the availability of cysteine, but also by feedback inhibition by GSH. Conversion of GSH to its oxidized form, GSSG, occurs during the GSH peroxidase-catalyzed reduction of hydrogen peroxide and other peroxides and in spontaneous reactions with free radicals (Meister, 1994). GSSG is restored to a reduced form by the flavoprotein glutathione reductase (Lopez-Barea *et al.*, 1990). Irreversible loss of GSH occurs through conjugation to endogenous and exogenous electrophiles in reactions catalyzed by GSH transferases (Pickett and Lu, 1989; Morrow and Cowan, 1990). GSH turnover is also found in association with the import of γ-glutamyl amino acids and with the intercellular transport of cysteine (Meister, 1994).

Methods used to increase cellular GSH content include enhancing synthesis by the addition of precursors, including the cysteine analog L-2-oxothiazolidine-4-carboxylate (Williamson *et al.*, 1982) or GSH ester (Levy *et al.*, 1993). For cells able to import GSH, intracellular GSH content may be enhanced by the addition of GSH to the media (Lash *et al.*, 1986). GSH content can be lowered by irreversibly inhibiting γ-glutamylcysteine synthetase with L-buthionine-(SR)-sulfoximine (BSO) (Meister, 1988), increasing conjugation and efflux by diamide treatment (Reed, 1986), or by limiting the supply of cysteine. Levels of intracellular GSSG may be increased through inhibition of glutathione reductase with bis(2-chlorethyl)-nitrosourea (BCNU) (Babson *et al.*, 1981).

Redox reactions in which GSH plays a role include protein folding (Hwang *et al.*, 1992), conversion of ribonucleotides to deoxyribonucleotides (Holmgren, 1990), and maintenance of reduced pools of vitamins C and E (Winkler *et al.*, 1994). GSH can also undergo reversible thiol-disulfide exchange with proteins containing oxidized cysteine (i.e., cystine) residues (Ziegler, 1985).

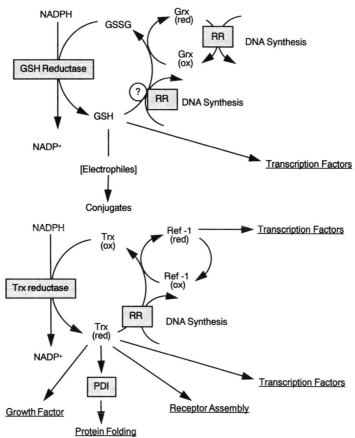

FIGURE I Redox systems affecting cell growth. NADPH reduces oxidized glutathione (GSSG) and thioredoxin [Trx(ox)] through glutathione reductase and thioredoxin reductase, respectively, giving reduced glutathione (GSH) and reduced thioredoxin Trx (red). GSH reduces oxidized glutaredoxin [Grx(ox)], and Trx reduces oxidized Ref-1. Trx and Grx provide reducing equivalents to ribonucleotide reductase (RR) for deoxyribonucleotide synthesis. A Grx-like or other protein thiol may transfer reducing equivalents to RR, shown here from GSH. GSH protects the cell against reactive electrophiles by forming conjugates. Trx reduces critical protein sulfhydryl residues to regulate protein folding, receptor assembly, transcription factor activity, and also acts as a growth factor. Protein folding is catalyzed by protein disulfide isomerase (PDI) which has Trx-like sequences. [Reprint by permission of the publisher from Powis *et al.*, (1995). Redox signalling and the control of cell growth and death. *Pharmacol. Ther.*, Vol. 68, pp. 149–173. Copyright 1995 by Elsevier Science Inc.]

B. Thioredoxin Reductase/Thioredoxin

The thioredoxin reductase (TR)/thioredoxin (Trx) couple is a ubiquitous redox system found in both prokaryotic and eukaryotic cells. TR is a FAD containing a flavoenzyme that uses NADPH to reduce Trx.

I. Thioredoxin Reductase

Escherichia coli is a 70-kDa homodimer (Thelander, 1967). The active site cysteine residues, Cys-135 and Cys-138, receive electrons from $FADH_2$ and transfer them to the active cystine bond of Trx (Prongay *et al.*, 1989). During reduction, TR undergoes a conformational change, which protects the reduced active site cysteines from the aqueous phase, preventing spontaneous oxidation. Upon binding of oxidized Trx to the active site, TR undergoes a conformational change to expose the active site cysteines, allowing reduction of the cystine bond of Trx (Waksman *et al.*, 1994). TrxB, the gene encoding for *E. coli* TR, has been cloned (Russel and Model, 1985), sequenced (Russel and Model, 1988), and its physical location identified (Delaney and Georgopoulos, 1992).

TR of higher-order organisms is a 116- to 120-kDa homodimer. Thioredoxin reductase has been cloned and sequenced from human placenta (Gasdaska *et al.*, 1995b). The predicted amino acid sequence for the monomer gives rise to a 495 residue 54-kDa protein. The two cysteine residues of the conserved catalytic site are located in the FAD-binding domain of the protein. Human thioredoxin reductase has a dimer interface domain that gives it strong similarity to human glutathione reductase and other NADPH-dependent reductases, but less similarity to prokaryotic and lower eukaryotic thioredoxin reductases. Studies of human TR have shown that it can reduce bacterial Trx, although at a decreased rate compared to human Trx (Oblong *et al.*, 1993). The converse has been found for bacterial TR using human Trx, where the bacterial homolog is a better substrate than mammalian Trx (Lozano *et al.*, 1994).

2. Thioredoxin

Trx is an 11- to 12-kDa protein with a highly conserved active site sequence of Trp-Cys-Gly-Pro-Cys-Lys. The two cysteine residues at the active site are readily oxidized and undergo NADPH-dependent reduction by TR. A single copy of the human Trx gene has been mapped to chromosome 3 at bands 3p11-p12 (Lafage-Pochitaloff-Huvale *et al.*, 1987). Analysis of genomic clones of Trx has found that the entire gene spans 13 kb and is composed of five exons (Tonissen and Wells, 1991; Kaghad *et al.*, 1994). In contrast, a murine Trx pseudogene has been identified on chromosome 1, in addition to a single copy of a Trx gene mapping to chromosome 4 (Taketo *et al.*, 1994). In yeast, two thioredoxin genes, TRX1 and TRX2, have been mapped to chromosome XII and VII, respectively (Muller, 1992). It has been suggested, on the basis of isoelectric focusing, that there are two separate forms of bovine Trx (Martinez-Galisteo *et al.*, 1993). A slightly larger mitochondrial form of Trx, compared to cytosolic Trx, has also been identified in pig heart based on electrophoretic mobility (Bodenstein-Lang *et al.*, 1989; Bodenstein and Follmann, 1991). However, small differences in electrophoretic mobility can be caused by different oxidation states of the Trx molecule (unpublished observations) or phosphorylation of Trx, which has been reported in *E. coli* (Conley and Pigiet, 1978).

Mutagenesis studies characterizing the conserved active site sequence of Trx have been reported. A residue that has drawn particular attention is a highly conserved lysine group adjacent to the C-terminal most active site cysteine. The presence of the positively charged amine group so close to the active site has led to the suggestion that it is critical in maintaining the thiolate anion of Cys-32 (Kallis and Holmgren, 1980). However, pH titration curves of human Trx analyzed by nuclear magnetic resonance spectroscopy have shown that the thiolate form of Cys-32 is stabilized by interaction of the S-γ of Cys-32 with the peptide backbone amide group of Cys-35 (Forman-Kay *et al.*, 1992). Mutagenesis of Lys-36 to Glu in both *E. coli* (Gleason *et al.*, 1990) and human Trx (Oblong *et al.*, 1995) did not affect the ability of the mutant Trx to be redox active. Thus, the highly conserved lysine group adjacent to the active site is not essential for the reduction of either prokaryotic or eukaryotic thioredoxin by TR, but is seemingly required for optimizing protein interactions with the flavoenzyme as evidenced by the increase in the K_m values for the mutant Lys-36 human Trx (Oblong *et al.*, 1995).

C. Protein Disulfide Isomerase

Protein disulfide isomerase (PDI) is a 57-kDa protein localized to the endoplasmic reticulum where it comprises 0.4% of the total protein. The principal function of PDI is to catalyze the correct folding of proteins during maturation and secretion from the endoplasmic reticulum (Edman *et al.*, 1985). In yeast, PDI is present as a 70-kDa protein (Mizunaga *et al.*, 1990) and is essential for cell viability (Farquhar *et al.*, 1991). The *E. coli* equivalent of PDI, dsbA, is only 23 kDa and catalyzes disulfide bond formation in the periplasmic space of the bacterium (Wunderlich and Glockshuber, 1993).

PDI contains two catalytic sites, with the sequence Cys-Gly-His-Cys-Lys, which varies from Trx only by substitution of His for Pro. PDI can substitute for Trx for reduction by TR (Lundström and Holmgren, 1990). Mutagenesis of the active site of Trx to the PDI sequence gives a mutant Trx, which is a better substrate for TR (Krause *et al.*, 1991) and has a 10-fold increase in PDI-like activity (Lundström *et al.*, 1992). Evidence suggests that the human gene for PDI can be divided into separate domains and that PDI is a modular protein (Freedman *et al.*, 1994).

PDI has been identified as the β-subunit of prolyl hydroxylase, a key enzyme involved in the maturation of collagen (Pihlajaniemi *et al.*, 1987). Prolylhydroxylase is a heterotetramer (a2b2) that catalyzes the conversion of proline to hydroxyproline in the consensus sequence of Pro-Pro-Gly found in nascent procollagen. The absence of PDI in prolylhydroxylase leads to an inactive form of the enzyme and blocks the secretion of collagen. Part of the function of PDI may be to help prevent the oxidation of collagen leading to a terminally blocked monomer species (Forster and Freedman, 1984). PDI has also been found as a component of the microsomal triacyl-glycerol transfer complex, which is required for the assembly and secretion of very low density lipoproteins and chylomicrons by the liver and intestine, respectively (Wettrau *et al.*, 1991). The presence of PDI in the microsomal triacylglycerol complex is essential for catalytic activity and for preventing the aggregation of the other 88-kDa subunit (Wettrau *et al.*, 1991). PDI, in conjunction with GSH and Grx, is able to reduce dehydroascorbate to ascorbic acid and may comprise the eukaryotic dehydroascorbate reductase activity (Wells *et al.*, 1990).

D. Glutaredoxin

Glutaredoxin (thioltransferase, Grx) is a 11- to 12-kDa redox protein (Wells *et al.*, 1993). Unlike Trx, Grx does not have a specific reductase but is reduced by interaction either with GSH and GSH reductase or with GSH and PDI. The physiological function of eukaryotic Grx remains unclear. A suggested function is the transfer of electrons to ribonucleotide reductase (RR), the first unique step of DNA synthesis (Xia *et al.*, 1992). However, GSH, Grx (Weckbecker and Cory, 1988), or Trx (Hopper and Iurlano, 1983) are not essential for the activation of RR, suggesting that an additional redox factor is involved.

E. Ref-I

Ref-1 is a 37-kDa protein originally isolated from HeLa cell nuclear extracts that catalyzes the reduction of the transcription protein complex AP-1 (Abate *et al.*, 1990; Xanthoudakis *et al.*, 1994). Two cysteine residues, Cys-65 and Cys-83, are critical for the redox activity of Ref-1 (Walker *et al.*, 1993). There is a direct cysteine-mediated interaction between Ref-1 and Jun, one of the components of AP-1 (Xanthoudakis *et al.*, 1994). Ref-1 also stimulates DNA binding by other transcription factors, including nuclear factor (NF)-κB (weakly), Myb, ATF, CREB (Xanthoudakis *et al.*, 1994), and EGR-1 (Huang and Adamson, 1993). Ref-1 can be regenerated by treatment with bacterial-reduced Trx, which itself does not reduce AP-1 (Abate *et al.*, 1990), so that the redox sequence in the cell may be NADPH to thioredoxin reductase, to Trx, to Ref-1, to AP-1. Ref-1 has also been found to have an apurine/apyrimidine endonuclease DNA repair activity and has a core domain that is highly conserved in a family of prokaryotic and eukaryotic DNA repair enzymes (Xanthoudakis *et al.*, 1992, 1994). However, the DNA repair activity of Ref-1 is separate from its redox activity. Sequences in the N-terminal domain of Ref-1 are required for redox activity whereas C-terminal sequences are necessary for DNA repair activity (Xanthoudakis *et al.*, 1994). Chemical alkylation or oxidation of Ref-1 inhibits its redox activity but does so without affecting its DNA repair activity (Xanthoudakis *et al.*, 1994). Site-directed mutagenesis has identified Cys-65 as the redox-active site of Ref-1 (Okuno *et al.*, 1993). Expression of antisense Ref-1 in Hela cells leads to decreased levels of Ref-1 protein and cells that have a normal morphology and grow normally, but that show increased cell killing under conditions of hypoxia or hyperoxia, and are hypersensitive to killing by a variety of DNA-damaging agents (Walker *et al.*, 1994).

F. Metallothionein

Metallothioneins (MTs) are cysteine-rich, metal-binding proteins found in most organisms studied (Ebadi and Iversen, 1994). There are three major classes (I, II, and III) of MT. Class I and II MTs are single-chain polypeptides whereas class III MTs are often oligomeric.

Mammalian MTs have 61 to 62 amino acid residues whereas invertebrates and fungi have smaller MTs, the smallest being only 25 amino acid residues found in *Neurospora crassa*. Mammalian MTs contain 20 cysteine residues, present as Cys-X-Cys or Cys-Cys sequences, which bind heavy metal and other metal ions through thiolate bonds. The amount of metal bound by MT is variable, but can be up to seven equivalents per polypeptide chain. MTs have no enzymatic activity and their physiological function appears to be to bind excess metal ions, thus limiting the free metal ion concentration. MT can also bind GSH, although how this might affect the biological properties of MT is not known (Brouwer *et al.*, 1993). MT acts as a free radical scavenger (Thornalley and Vasak, 1985), either through free radical reaction with cysteine residues (Thornalley and Vasak, 1985) or through the release of Zn from complexes with MT (Thomas *et al.*, 1986). MT may also sequester iron, thus preventing its participation in free radical reactions (Brookens *et al.*, 1995).

III. Redox Targets

A. Ribonucleotide Reductase

One of the earliest cellular functions ascribed to bacterial Trx was the redox regulation of RR. RR catalyzes the synthesis of deoxyribonucleotides from ribonucleotides, which is the first unique reaction of DNA synthesis and an essential step for cellular proliferation (Laurent *et al.*, 1964). Thus, Trx has been presumed to function as a key component in DNA synthesis through its ability to reduce RR. It was presumed that after each reaction,

RR had to be reduced by Trx to become active. This, in turn, led to the hypothesis that Trx was essential for cell viability. However, studies of a mutant of *E. coli* lacking Trx that was able to proliferate led to the identification of Grx (Holmgren, 1976). Double *E. coli* mutants lacking both Trx and Grx will not proliferate due to the accumulation of 3′-phosphoadenosine 5′-phosphosulfate (PAPS), an intermediate in the sulfate assimilation pathway (Russel and Holmgren, 1988, 1990). Inhibiting the formation of PAPS by the use of high concentrations of cystine or mutations in the genes required for PAPS synthesis overcomes the inhibition of growth (Russel and Holmgren, 1990). This indicates that there may be a third redox component in bacteria involved in the redox regulation of RR and DNA synthesis. Studies have suggested that this occurs in a GSH-dependent manner (Miranda-Vizuete *et al.*, 1994), and two proteins with Grx-like activity and typical active sites of Cys-Pro-Tyr-Cys, of 10 and 27 kDa, have been isolated from the *E. coli* mutant (Aslund *et al.*, 1994).

The importance of Trx for eukaryotic RR is less well understood. In mammalian cells, inhibition of TR by antitumor quinone drugs is associated with a decrease in the activity of RR (Mau and Powis, 1992). In yeast, it has been shown that Trx is not essential for RR activity because yeast mutants lacking both copies of the Trx gene are viable (Muller, 1991). However, the absence of Trx leads to an increase in the mitotic cycle and DNA replication rates. A homolog of Trx is found in Drosophila encoded in the deadhead locus (Salz *et al.*, 1994). Mutation of the locus has shown that the Trx homolog is not essential for cell viability but is essential for female meiosis.

B. Receptor Proteins

Trx reduces and activates the glucocorticoid receptor (Grippo *et al.*, 1983) and the interferon-γ receptor (Fountoulakis, 1992), whereas GSH can stabilize and activate the insulin–receptor complex (Cotgreave *et al.*, 1994). Pretreatment of the interleukin (IL)-2-dependent cytotoxic T-cell lines with high concentrations of GSH enhances IL-2 binding and internalization, and DNA synthesis (Liang *et al.*, 1989). Pretreatment of the cells with the GSH synthesis inhibitor BSO attenuated the increase in DNA synthesis in response to IL-2. Site-directed mutagenesis of the lymphokine, interleukin-2, has demonstrated the importance of a cysteine-containing region for biological activity that may be involved in the GSH response (Liang *et al.*, 1986). The β-subunits of both follitropin (FSH) and luteotropin contain a Trx-like domain (Boniface and Reichert, 1990). A 20 amino acid peptide derived from this domain binds to human follicle-stimulating hormone (FSH) receptors and propagates a signal characteristic of human FSH. Substitution of the active-site cysteines with serines had no effect on the ability of the peptides to bind to the receptors, but abolished the transduction of the signal across the membrane (Grasso *et al.*, 1993).

C. Transcription Factors

1. OxyR

The OxyR transcription factor found in *E. coli* and *S. typhimurium* functions specifically in response to H_2O_2. It remains tightly bound to its DNA promoter site while undergoing a conformational change, and negatively regulates its own transcription and activates the expression of genes involved in the bacterial response to H_2O_2-mediated oxidative stress (Storz *et al.*, 1990a,b). Of six cysteine residues in OxyR, only Cys-199 is involved in the conformational change, thus ruling out the possibility of disulfide bond formation, meaning that oxidation to a cysteine sulfenic acid may be involved (Storz *et al.*, 1990a).

2. NF-κB/Rel

NF-κB is a ubiquitous transcription factor that is important for the cellular response to oxidative stress and tumorigenesis (Higgins *et al.*, 1993; Meyer *et al.*, 1993). NF-κB is a heterodimer composed of p50 and p65 subunits. Activation of NF-κB occurs independently of protein synthesis and involves posttranslational mobilization from an inactive cytoplasmic complex with the inhibitory subunit, IκB. A protein closely related to p65 of NF-κB is the protooncogene c-Rel (Dyson *et al.*, 1990). Several different forms of IκB have been identified, including IκBα, β, γ, and Bcl-3, and all have multiple ankrin repeats which mediate their interaction with NF-κB. Upon activation, IκB is cleaved from the complex and the p50/p65 heterodimer is translocated to the nucleus where p50 binds to the decameric κB motif found in the enhancers of numerous cellular genes.

The binding of NF-κB to DNA, measured by gel-shift assays, is irreversibly inhibited by N-ethylmaleimide (NEM), an alkylator of sulfhydryl groups (Tolendano and Leonard, 1991), and is reversibly inhibited by diamide, which oxidizes sulfydryl groups (Tolendano and Leonard, 1991; Hayashi *et al.*, 1993). The thiol reagents dithiothreitol (DTT), 2-mercaptoethanol, but not reduced GSH increase the DNA binding of NF-κB (Hayashi *et al.*, 1993). The DNA-binding activity of the NF-κB p50 subunit is also increased by DTT and is abolished by NEM (Matthews *et al.*, 1992). The critical residue in p50 is Cys-62 and its deletion decreases the affinity and specificity of the DNA binding.

Studies in whole cells have shown an increase in NF-κB binding and transactivation under oxidizing conditions that can be blocked by antioxidants. Treatment of HeLa cells with H_2O_2 causes an increase in NF-κB DNA binding and transactivation as measured by a reporter construct (Meyer *et al.*, 1993). These effects of H_2O_2 are synergistic with similar effects produced by 12-*O*-tetradecanoylphorbol-13-acetate (TPA). Cells treated with TPA or TNF-α show a decrease in cellular GSH and an increase in a NF-κB reporter activity, whereas *N*-acetylcysteine (NAC), a cysteine derivative and a GSH precursor, protects against TPA and TNF-α-dependent NF-κB activation (Staal *et al.*, 1990; Shibanuma *et al.*, 1994). Antioxidants such as butylated hydroxyanisole (BHT), nordihydroquaretic acid, and α-tocopherol also block NF-κB transactivation in unstimulated, TPA, and TNF-α-stimulated cells (Israel *et al.*, 1992). TPA has been suggested to produce its effects on NF-κB by the protein kinase C (PKC)-dependent promotion of oxidant stress in the cells (Meyer *et al.*, 1993). In cells treated with the PKC inhibitor bryostatin to deplete PKC, TPA had no effect on NF-κB binding and transactivation (Schenk *et al.*, 1994).

However, Trx increases the DNA binding of NF-κB and is more potent than L-cysteine or GSH (Galter *et al.*, 1994) or nonphysiological reducing agents such as NAC, 2-mercaptoethanol, or DTT (Hayashi *et al.*, 1993; Sorachi *et al.*, 1992; Galter *et al.*, 1994). Oxidized Trx and GSSG inhibit the DNA binding of NF-κB (Galter *et al.*, 1994). Trx also increases the DNA binding of the NF-κB p50 subunit but not a redox-inactive form of p50 (Matthews *et al.*, 1992). Transient transfection of cells with Trx cDNA results in a large increase in transactivation by NF-κB measured by a reporter construct. However, in other studies, transient transfection of HeLa cells with Trx cDNA resulted in a dose-dependent decrease in TPA-stimulated, as well as nonstimulated, NF-κB DNA binding and transactivation (Schenk *et al.*, 1994). Exposure of cells to exogenous Trx also resulted in a dose-dependent inhibition of TPA-stimulated NF-κB activity (Schenk *et al.*, 1994).

3. AP-1

AP-1 is a transcription factor whose activation is a prerequisite for growth factor and TPA-stimulated cell growth. AP-1 is composed of the *jun* and *fos* gene products, which form homodimeric (Jun/Jun) or heterodimeric (Jun/Fos) complexes. DNA binding of the Fos–Jun homodimer (AP-1) is increased by the reduction of a single conserved cysteine in the DNA-binding domain of each of the proteins (Abate *et al.*, 1990; Walker *et al.*, 1993). Mutant Fos and Jun proteins where this cysteine residue is replaced by serine show constitutive DNA binding (Abate *et al.*, 1990). The naturally occurring v-*jun* oncogene has a point mutation that gives a serine instead of a Cys-154 residue (Bohmann *et al.*, 1987). Experimentally replacing Cys-154 with serine in Fos increases its transforming ability (Lafage-Pochitaloff-Huvale *et al.*, 1987). When expressed in chicken fibroblasts, this mutant Fos shows an increased AP-1 DNA-binding activity that is resistant to treatment with the oxidizing agent diamide (Okuno *et al.*, 1993). From these results it appears that the level of functional Fos–Jun complexes is redox regulated and that escape from redox control enhances the transforming activity.

Antioxidants such as NAC, DTT, and BHA increase unstimulated and TPA-stimulated AP-1 DNA binding and transactivation in cells (Meyer *et al.*, 1993; Schenk *et al.*, 1994) which is associated with an increase in c-*fos* and c-*jun* transcription (Meyer *et al.*, 1993; Schenk *et al.*, 1994). Oxidant stress caused by treatment of cells with BSO or diamide also stimulates the endogenous and inducible expression of c-*fos* and c-*jun*, and increases AP-1 binding (Bergelson *et al.*, 1994). AP-1 proteins can be reduced by a nuclear redox protein, Ref-1, but apparently not by Trx, or GSH and glutathione reductase. DNA binding of AP-1 is inhibited by oxidized Trx more efficiently than GSSG. In contrast, GSSG is more effective than oxidized Trx in inhibiting NF-κB DNA-binding activity (Galter *et al.*, 1994).

4. Other Transcription Factors

Diamide oxidation of the tumor suppressor protein p53 inhibits its DNA binding (Hainaut and Milner, 1993). Cysteine residues are necessary for the binding of zinc, which stabilizes the DNA-binding domain of p53 (Cho *et al.*, 1994), and for the binding of copper that also regulates DNA binding of p53 through a redox mechanism (Hainaut *et al.*, 1995). Site-directed mutagenesis to replace the zinc-binding Cys-173, Cys-235, or Cys-239 residues with Ser markedly decreases DNA binding and transcriptional activation of p53 and enhances the suppression of transformation by p53 (Rainwater *et al.*, 1995). Replacement of Cys-121, Cys-132, Cys-138, and Cys-272 of the DNA-binding domain of p53 partially blocks transactivation and the suppression of transformation. Thus, p53 exhibits more than one form of redox modulation of its conformation.

The transcription factor κU is a heterodimer composed of p70 and p86 subunits (Zhang and Yaneva, 1993). κU can be regulated through the activity of DNA-dependent protein kinase, and its DNA binding requires reduction of intrinsic cysteine residues mediated, perhaps, by a nuclear factor. v-Ets is a component of the avian acute leukemia virus 26-derived Gag-v-Myb-v-Ets fusion protein. The binding of v-Ets to DNA requires reducing conditions (Wasylyk and Wasylyk, 1993) and is inhibited by the oxidizing agents diamide and NEM. DNA binding of v-ETS probably involves the conserved Cys-394 and is more sensitive than c-Ets to inactivation by oxidizing agents because of the loss of the C-terminal sequence in v-Ets (Wasylyk and Wasylyk, 1993). The DNA-binding activity of the E2 polypeptides encoded by the bovine papillomavirus type-1 (BPV-1), which function as dimers in viral DNA replication and the regulation of viral gene expression, is redox regulated. Thiol-oxidizing agents lead to a loss of activity that can be reversed by DTT (McBride *et al.*, 1992). Although the DNA-binding domain of each E2 protein has three cysteine residues, the oxidation of only Cys-340 per dimer is sufficient to inhibit DNA binding (McBride *et al.*, 1992). Finally, Trx, in the presence of nuclear extract, restores the DNA binding of the oxidized Ah receptor (Ireland *et al.*, 1995).

D. Protein Folding and Degradation

As previously discussed, a major function of PDI is to catalyze correct protein folding during protein maturation and secretion from the endoplasmic reticulum (Edman et al., 1985). PDI has been suggested to have a chaperon function in the endoplasmic reticulum (Wang and Tsou, 1993). Trx also catalyzes protein folding by a mechanism similar to that of PDI (Lundström and Holmgren, 1990). Reduction by Trx can alter the protein structure so much that it renders the protein susceptible to proteolysis. Rat glandular kallikrein, a serine-type protease, is capable of cleaving Trx-treated prolactin in vitro to a form that is found in vivo (Hatala et al., 1991). Trx is active in reducing toxic venoms from such species as scorpion, wasp, and snakes (Lozano et al., 1994). This activity may lead to increased susceptibility of the toxins to proteolysis. Tetanus toxin has diminished toxicity when the interchain disulfide bond is reduced by Trx, but not by GSH or DTT (Kistner et al., 1993). Oxidized Trx reforms the cleaved interchain disulfide bond, thus restoring full potency to the tetanus toxin.

IV. Cellular Responses to Redox Changes ⸺⸺⸺⸺⸺

A. Cell Proliferation

The proliferation of some tumor cell lines is stimulated by relatively high concentrations of thiols such as cysteine and 2-mercaptoethanol (Broome and Jeng, 1973; Hewllet et al., 1977). This has been suggested to occur through an enhanced response to growth factors (Fanger et al., 1970).

Treatment of cells with platelet-derived growth factor (PDGF) results in a transient increase in intracellular H_2O_2 (Sundaresan et al., 1995). The antioxidants NAC and catalase prevent the increase in H_2O_2 and block the tyrosine phosphorylation and DNA synthesis caused by PDGF. This has led to the suggestion that H_2O_2 may be part of a signal cascade activated by PDGF (Sundaresan et al., 1995). H_2O_2 is produced endogenously in HeLa cells in response to PKC activation (Bhimani et al., 1993). Because exogenous H_2O_2 and superoxide can stimulate the growth of cells, it has been suggested that they may play a role in normal cell growth (Burdon and Gill, 1993).

A number of antioxidant enzymes show an increase in activity when cells exhibit contact inhibition of growth. They include PDI (Clive and Greene, 1994; Greene and Brophy, 1995), MnSOD but not CuZnSOD, catalase (Oberley et al., 1995), and DT-diaphorase (Schlager et al., 1993). It has been shown for DT-diaphorase that the increase in activity is directly related to the extent of contact inhibition of growth and decreases with increasing transformation and loss of contact inhibition of the cell lines (Schlager et al., 1993).

B. Thioredoxin and Cell Proliferation

A small molecular weight protein termed adult T-cell-derived leukemic growth factor (ADF) was isolated during the search for novel leukemic growth factors (Tagaya *et al.*, 1988, 1989). ADF was cloned and found to have nearly complete homology with the predicted amino acid sequence for Trx (Wollman *et al.*, 1988). Since then it has been determined by a number of groups that ADF and Trx are one and the same species, due to a discrepancy in the reported amino acid sequence of human Trx (Tagaya *et al.*, 1989; Deiss and Kimchi, 1991; Gasdaska *et al.*, 1994). Trx increases cellular DNA synthesis in synergy with a number of cytokines, including IL-1, IL-2 (Wakasugi *et al.*, 1990; Yodoi and Tursz, 1991), IL-4 (Darr and Fridovich, 1986), and IL-6 (Ifversen *et al.*, 1993).

Studies in our laboratory have focused on the ability of Trx to stimulate the proliferation of nonlymphoid cells in culture. We have found that exogenously added Trx stimulates mouse fibroblasts (Oblong *et al.*, 1994) and a number of human solid tumor cell lines (Wang and Semenza, 1995). In our studies, cell growth is rigidly defined as an increase in cell number, unlike many studies that characterize cell growth on the basis of an increase in DNA synthesis. Trx stimulates cell growth up to 90% as effectively as 10% fetal bovine serum stimulation. This is a characteristic exhibited by few other growth factors. One exception to this appears to be HepG2 cells whose proliferation is stimulated by Trx in serum-free medium but is inhibited in the presence of 0.5% serum.

Mechanistically, Trx does not appear to stimulate cell growth along classical lines by acting on a specific cell surface receptor. We could find no evidence for saturable binding of ^{125}I-labeled Trx to the surface of MCF-7 breast cancer cells, and there was minimal uptake of the ^{125}I-labeled Trx into the cells (Gasdaska *et al.*, 1995a). Instead, Trx appears to exert its cell growth-stimulating effect by sensitizing cells to growth factors produced by the cell itself. Replacing the medium each day with fresh medium and Trx completely abolishes the increase in cell proliferation. Such a process presumably removes the factors secreted by cells that are necessary for Trx-induced cell growth (Gasdaska *et al.*, 1995a). There was a significant correlation between the ability of a number of human cancer cell lines to proliferate in the absence of serum, which is presumably due to the autocrine production of growth factors by the cell, and the extent of stimulation by Trx. This helper, or voitocrine, mechanism appears to be unique to Trx and requires the redox-active form of the protein, in addition to other features present in human Trx but absent in *E. coli* Trx. Both the redox-inactive C32S/C35S Trx mutant and *E. coli* Trx fail to stimulate cell growth (Fig. 2). *E. coli* Trx has, however, been reported to stimulate the proliferation of virally transformed lymphoid cells and DNA synthesis in the presence of 2% serum (Wakasugi *et al.*, 1990; Biguet *et al.*, 1994). In our studies, *E.*

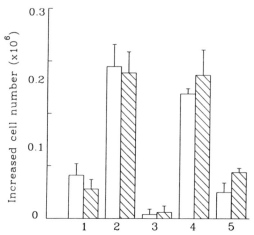

FIGURE 2 Stimulation of MCF-7 breast cancer cell proliferation by Trx. Cells were growth arrested in medium containing 0.5% bovine calf serum for 2 days before adding 1, nothing; 2, 1 μM human Trx; 3; 50 μM *E. coli* Trx; 4, 1 μM human Trx and 50 μM *E. coli* Trx; and 5, 1 μM redox-inactive C32S/C35S mutant human Trx. The increase in cell number was then measured over 2 days. Open boxes represent medium with 0.5% bovine calf serum, whereas hatched boxes represent medium without serum. Values are the mean of three determinations and bars are SE. [Reproduced with permission from Gasdaska *et al.* (1995a). Copyright American Association for Cancer Research, Inc.]

coli Trx had no effect on cell growth at concentrations 50-fold higher than human Trx, regardless of whether it was reduced enzymatically by *E. coli* TR or dithiothreitol. Other thiols such as GSH and DTT, at equally high concentrations, had no effect on cell growth under these conditions. It is not clear from our studies whether Trx is acting directly on endogenously produced growth factors or is increasing the sensitivity of growth factor receptors on the cell surface. We have ruled out the involvement of EGF and TGF-α, known autocrine growth factors secreted by MCF-7 breast cancer cells, as Trx growth partners. Trx does not appear to be involved in serum-stimulated growth. Trx produces no increase in cell proliferation when added with 10% fetal bovine serum. In addition, treatment of serum-stimulated cells with a polyclonal antibody to Trx, which blocks Trx-stimulated growth, has no effect on cell growth.

Trx loses its ability to stimulate cell growth relatively rapidly, but this activity is stable in the presence of DTT. We have found that freshly prepared Trx exists as a mixture of partly oxidized monomeric forms as determined by electrophoresis. Oxidation of human Trx leads to reversible homodimer formation (Ren *et al.*, 1993), a process that occurs during natural aging of the protein. Trx dimer formation has been directly related to decreased reduction by TR (Ren *et al.*, 1993).

The physiological significance of the ability of Trx to stimulate cell growth is unknown at the present time. The EC_{50} for Trx growth stimulation in MCF-7 breast cancer cells is 350 nM, which is considerably higher than the 4–18 nM concentration of Trx found in serum (Ren *et al.,* 1993). Higher Trx concentrations exist in tissues, 1 to 10 mM, which, if released extracellularly, might stimulate cell proliferation (Holmgren and Luthman, 1978). Trx can be secreted by cells (Ericson *et al.,* 1992; Rubartelli *et al.,* 1992, 1995) by a nonclassical leaderless pathway (Rubartelli *et al.,* 1992). The role Trx may play in tumorigenesis is also unknown. Trx protein levels are elevated in human cervical neoplastic squamous epithelial cells (Schall-reuter *et al.,* 1990) and hepatocellular carcinoma (Nakamura *et al.,* 1992). We have found that a significant number of human primary lung and colon cancers have increased Trx mRNA and protein levels compared to paired normal tissue (Gasdaska *et al.,* 1994; unpublished data). NIH/3T3 cells stably transfected with Trx cDNA show an increase in growth rate, whereas MCF-7 breast cancer cells transfected with cDNA for the redox-inactive C32S/C35S mutant Trx will no longer form colonies in soft agar (unpublished observations). These results suggest that increased Trx gene expression could contribute, in part, to the increased growth rate and transformed phenotype of some human tumors.

Trx has been identified as a component in the early pregnancy factor (EPF) system, a complex array of factors present in the sera of pregnant mammals (Clarke *et al.,* 1991). The binding of lymphocytes to red blood cells, i.e., rosette bud formation, by EPF occurs during the initial onset of pregnancy and several proteins of the EPF complex may act synergistically or in combination. A mutagenesis study of human Trx showed that the redox-active, catalytic site Cys-32 and Cys-35 residues were not essential for this function, but that Cys-74 was (Tonissen *et al.,* 1993).

C. Oxidant Signaling and Apoptosis

Programmed cell death is an important event in the normal processes of development and tissue remodeling (Wyllie, 1992; Cohen, 1993; Vaux *et al.,* 1994; Williams, 1994). Apoptosis is a form of programmed cell death characterized by membrane blebbing, chromatin margination, and breakdown of chromosomal DNA into nucleosome-sized fragments (Schwartz *et al.,* 1993). Apoptosis plays an important role in normal tissue development and remodeling and in protecting the organism against carcinogenesis. When a cell sustains substantial genetic damage that cannot be repaired by the normal DNA repair processes, this is recognized by sensory mechanisms in the cell and a sequence of events is initiated that leads to the death of the cell (Hickman *et al.,* 1994; Kerr *et al.,* 1994). Apoptosis results in the death of individual damaged cells and protects the organism from potentially harmful genetic changes that could lead to unregulated cell growth and cancer and is, therefore, a beneficial event. If apoptosis is impaired, there is an increased risk that cells will pass genetic damage on to their daughter cells, thus increasing the chances of neoplastic transformation. There are now well-documented instances where inhibition of apoptosis by the abnormal expression of the *bcl-2* oncogene (Korsmeyer, 1992a) or the loss of the *Rb* or *p53* tumor suppressor genes (Hollingsworth and Lee, 1991) are closely associated with malignancy. It also appears that as cells develop from a nontransformed state through a premalignant to a fully transformed state they progressively lose their ability to undergo apoptosis. For example, Bedi *et al.* (1995) have shown that the transformation of colorectal epithelium to carcinoma is associated with a decrease in the rate of spontaneous apoptosis. It may even be that deletion of apoptosis is a requisite event for the development of cancer (McDonnell *et al.*, 1995). Recent interest in apoptosis stems, in part, from a corollary to this view of apoptosis that if the surveillance mechanism for DNA damage or the signaling events leading to apoptosis can be enhanced, then it might be possible to decrease the risk of the development of cancer in high risk groups, or even in the normal population.

Considerable evidence now exists that an increase in reactive oxygen species constitutes an intracellular signal that can lead to apoptosis. Apoptosis can be induced in a number of cell systems by H_2O_2 (Lennon *et al.*, 1991), reactive oxygen species generated by the redox cycling of quinones (Jacobson and Raff, 1995), and radiation (Chang *et al.*, 1992). *c-myc*, which is essential for apoptosis in many systems, is induced by H_2O_2 and reactive oxygen species (Rao and Berk, 1992; Luna *et al.*, 1994). Hypoxia (Jacobson and Raft, 1995) and antioxidants (Chang *et al.*, 1992; Jacobson and Raff, 1995) inhibit apoptosis induced by these treatments. Trx protects U937 lymphoma cells against TNF-α-mediated cell killing (Matsuda *et al.*, 1991). The survival of embryonic mouse neurons is enhanced by Trx, as well as by 2-mercaptoethanol and NAC (Hori *et al.*, 1994). In these same studies, U251 astrocytoma cells were seen to produce increased levels of Trx in response to H_2O_2 treatment. Elevated Trx levels have also been observed in glial cells of the gerbil brain during reperfusion after ischemia (Tomimoto *et al.*, 1993). Thus, Trx secreted by glial cells may protect neurons, *in vivo*, from oxidative stress-induced cell death.

The *bcl-2* oncogene blocks apoptosis in diverse systems and protects cells against oxidative stress-induced damage (Hockenberry *et al.*, 1993) and killing (Kane *et al.*, 1993). *bcl-2* was first identified in association with the 14;18 translocation in B-cell lymphomas, which places the Bcl-2 protein-coding sequence under the control of an immunoglobulin promoter [for Bcl-2 reviews see McDonnell *et al.* (1993) and Korsmeyer (1992b)]. The oncogene encodes a 25-kDa integral membrane protein, whose precise function remains unclear. The studies of Hockenberry *et al.* (1993) demonstrate that expression of Bcl-2 in a T-cell hybridoma prevents apoptosis and lipid peroxidation that occur when unprotected cells are treated with glucocorticoids. Kane *et al.* (1993) have shown that Bcl-2 blocks GT1-7 neural cell death brought on by depletion of GSH. In this case, the mode of cell death is not apoptosis, but necrosis, suggesting that Bcl-2 counteracts a cellular process that may lead to either apoptosis or necrotic cell death. Bcl-2 expression in the GT1-7 cells resulted in an approximate doubling of the intracellular GSH content. Reoxygenation of HT-29 colon cancer cells after a period of hypoxia leads to a large increase in apoptosis that coincides with an increase in *ref*-1 mRNA, c-*myc* mRNA, and a late increase in *bcl*-2 mRNA (Yao *et al.*, 1994).

A model of apoptosis where oxidant signaling appears to play a major role is steroid hormone-induced apoptosis of the murine thymoma-derived WEHI7.2 cell (Briehl *et al.*, 1995; Baker *et al.*, 1995). Studies in our laboratory have shown that during dexamethasone-induced apoptosis there is a selective decrease in the transcript levels of a number of antioxidant defense enzymes, including both Mn and Cu/Zn superoxide dismutases, catalase, glutathione peroxidase, Trx, and DT-diaphorase (Baker *et al.*, 1996; Briehl *et al.*, 1995) (Fig. 3). The changes in antioxidant enzyme transcript levels are first seen 8 hr after dexamethasone treatment and precede apoptosis which is apparent by 24 hr and maximal by 48 hr. The superoxide dismutase, catalase, and DT-diaphorase activities of dexamethasone-treated WEHI7.2 cells show a fall by 24 hr and are maximally depressed by 48 hr (Baker *et al.*, 1996). In *bcl*-2-transfected WEHI7.2 cells that do not undergo dexamethasone-induced apoptosis, there are similar decreases in antioxidant enzyme transcript levels following dexamethasone treatment, but total superoxide dismutase and DT-diaphorase activities show a smaller decrease compared to wild-type WEHI7.2 cells, while the levels of catalase do not fall over 48 hr, despite a decrease in transcript levels. It thus appears that *bcl*-2 is able to prevent a decrease in antioxidant defense enzyme levels despite a decrease in gene transcription caused by dexamethasone treatment. This might be related to the inhibition by *bcl*-2 of proteases of the ICE family (Martin and Green, 1995), which might be able to degrade the antioxidant enzymes. Treatment of WEHI7.2 cells with the antioxidants Trolox (a water-soluble vitamin E derivative), catalase, and selenium, as well as hypoxic conditions, prevents dexamethasone-induced apoptosis (Baker *et al.*, 1996). This provides evidence that the decrease in antioxidant enzymes could lead to an increase in cellular reactive oxygen species responsible for signaling apoptosis. It is interesting that cysteine (Lee *et al.*, 1995) and catalase (Sandstrom and Buttke, 1993) have both been identified as factors spontaneously secreted by cells that inhibit apoptosis.

It should be emphasized that the increase in reactive oxygen species that follows a decrease in antioxidant enzymes is, most likely, a signaling event and not an effector mechanism for apoptosis. That is, oxygen radicals are not directly responsible for the DNA degradation and membrane damage seen during the final common pathway of apoptosis as has been proposed by some investigators (Jewell *et al.*, 1982; Buttke and Sandstrom, 1994).

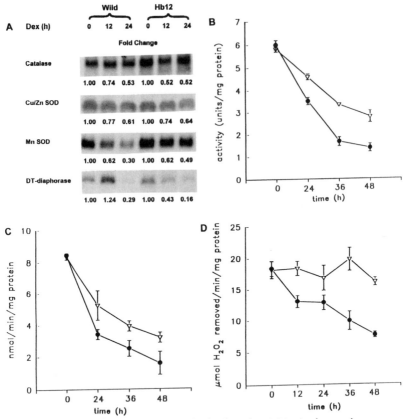

FIGURE 3 Antioxidant enzyme transcript levels and activities in dexamethasone-treated lymphoid cells. (A) Autoradiograms of Northern blots of total RNA before, at 12 hr, and at 24 hr after treatment of WEH17.2- and *bcl-2*-transfected W.Hb.12 cells with 1 μM dexamethasone. Numbers show the fold change in the message measured by densitometry normalized to β-actin mRNA. Enzyme activities in cell lysates from ($\bullet$) WEH17.2 and (∇) W.Hb.12 cells treated with 1 μM dexamethasone for up to 48 hr are shown: B, total SOD; C, DT-diaphorase; and D, catalase. Values are the mean of three determinations and bars are SE.

Furthermore, reactive oxygen species are probably only one of a number of signaling events that can initiate apoptosis. It is known, for example, that hypoxia does not inhibit apoptosis caused by staurosporine, a nonspecific PKC inhibitor, by the FAS receptor, by withdrawal of IL-3 from IL-3-dependent cells, or by the topoisomerase inhibitor camptothecin (Jacobson and Raff, 1995; Stefanelli *et al.*, 1995). The endogenous formation of reactive oxygen species could, however, be a constitutive factor that tends to drive cells to apoptosis even in the absence of exogenous stimuli. Such a model of apoptosis is consistent with the view, best expressed by Raff (1992), that the default state of cells is to die by programmed cell death unless kept alive by specific signals from other cells provided by growth factors and anti-apoptotic agents. It may be that cancerous cells deliver their own survival signals, thus becoming resistant to both intrinsic and induced apoptosis.

There are other disease states where weakened antioxidant defenses and oxidant stress are associated with inappropriate cell death. Genetic studies of individuals with amyotrophic lateral sclerosis have identified mutations in the gene coding for Cu,ZnSOD (Deng *et al.*, 1993). These mutations result in decreased enzyme activity, which may contribute to the observed pathology of motor neuron death. CD4$^+$ cells and lymph nodes from AIDS patients have decreased levels of GSH (Staal *et al.*, 1992) and Trx (Masutani *et al.*, 1992), respectively. Catalase, MnSOD, and glutathione peroxidase activities drop in T-cell lines grown *in vitro,* after infection with the HIV virus (Greenspan and Aruoma, 1994; Sandstrom *et al.*, 1994). β-Amyloid is a neurotoxic peptide that aggregates in the brain of Alzheimer's patients and has been found to generate free radical peptides (Hensley *et al.*, 1995). This has led to the hypothesis that oxidative stress, specifically membrane damage mediated by the β-amyloid-derived radicals, leads to the neurode-generation seen with Alzheimer's disease.

D. Oxidative Stress

Oxidant stress can be caused experimentally by exposing cells to agents such as diethylmaleate and H_2O_2. Treatment of cells with diethylmaleate results in transcriptional activation of c-*fos*, vimentin, cytochrome oxidase IV, ribosomal protein L4, and the guanine nucleotide regulatory protein nep1 (Ammendola *et al.*, 1995). Immediate early genes that are transcriptionally activated in cells treated with H_2O_2 include c-*fos*, c-*jun*, and egr-1 (Nose *et al.*, 1991; Amstad *et al.*, 1992; Datta *et al.*, 1993; Rao *et al.*, 1993; Ohba *et al.*, 1994). H_2O_2 causes tyrosine phosphorylation of multiple cellular proteins in lymphoma cells associated with the activation of p72^{Syk}, although there is no activation of Src family kinases (Schieven *et al.*, 1993). Levels of Trx in lymphocytes have been reported to be increased by H_2O_2 (Hayashi *et al.*, 1993).

MT genes are transcriptionally activated in cells during oxidative stress. MT levels are increased by X-irradiation (Koropatnick *et al.*, 1989), hyper-oxia (Veness-Meehan *et al.*, 1991), and drugs that increase free radical formation, such as paraquat (Bauman *et al.*, 1991). Overexpression of MT reduces the sensitivity of cells to oxidant-induced injury (Mello-Filho *et al.*, 1988; Sato and Bremner, 1993).

As already discussed, H_2O_2 treatment of cells results in the activation of NF-κB (Meyer *et al.*, 1993). Ras and mitogen-activated protein (MAP) kinase activities are activated by H_2O_2 and other oxidizing agents (Lander *et al.*, 1995). Inhibition of Ras with dominant negative mutant Ras or a Ras farnesyltransferase inhibitor blocks the activation of NF-κB by H_2O_2, suggesting that NF-κB is downstream of Ras activation (Lander *et al.*, 1995). In this study the guanine nucleotide exchange activity of Ras was found to be directly increased by H_2O_2.

In contrast to NF-κB, the DNA binding of AP-1 is only weakly induced by H_2O_2 (Amstad *et al.*, 1992; Meyer *et al.*, 1993; Rao *et al.*, 1993). The increase in the transcription of c-*jun* by H_2O_2 and reactive oxygen species (ROS) can be suppressed by inhibitors of PKC (Amstad *et al.*, 1992) and PLA$_2$ (Rao *et al.*, 1993) and requires poly(ADP)ribosylation of chromosomal proteins (Amstad *et al.*, 1992). Exposure of epithelial cells to pyrogallol, a generator of ROS, increases the DNA binding of NF-κB, which is followed by an increase in inducible nitric oxide synthase mRNA (Adcock *et al.*, 1994). Naphthoquinone and orthovanadate, both sources of ROS, stimulate protein tyrosine phosphorylation in cells, increase phosphatidylinositol-3-kinase activity, and increase DNA synthesis (Chen *et al.*, 1990; Chen and Chan, 1993).

E. Hypoxia

Exposure of a variety of cells to low oxygen causes the upregulation and DNA binding of a nuclear transcription factor, hypoxia inducible factor (HIF-1) (Beck *et al.*, 1993; Wang and Semenza, 1993a). Both *de novo* protein synthesis and protein phosphorylation appear to be necessary for HIF-1 upregulation (Wang and Semenza, 1993b). HIF-1 is a heterodimer of two basic helix–loop–helix subunits: 120-kDa HIF-1α and 92-kDa HIF-1β (Wang and Semenza, 1995). HIF-1 binds to the hypoxia enhancer regions of several genes encoding glycolytic enzymes (Semenza *et al.*, 1994) and to the erythropoietin hypoxia enhancer sequence (Beck *et al.*, 1993). Exposure of human tumor cells to hypoxia also causes increased transcription and protein levels of Ref-1 (Walker *et al.*, 1994; Yao *et al.*, 1994) and the selective induction of c-*jun* transcription leading to increased AP-1-binding activity (Yao and O'Dwyer, 1994). The levels of both c-*jun* and *junD* mRNA increase during hypoxia and fall on reoxygenation (Yao *et al.*, 1994). Increases in *fos* mRNA are less pronounced during hypoxia but are maintained on reoxygenation. During hypoxic exposure of HT-29 cells, increased binding to the AP-1 response element leads to transcriptional induction of the antioxidant enzyme DT-diaphorase (Yao *et al.*, 1994). There is also an increase in the activity of γ-glutamylcysteine synthetase, the rate-limiting event in GSH synthesis. Nucleosomal degradation, indicative of apoptosis, also increases after a few hours of hypoxia (Yao *et al.*, 1995). Exposure of Jurkat T cells to hypoxic conditions for several hours has been reported to result in increased NF-κB DNA binding and transactivation due to tyrosine phosphorylation and dissociation of the IκB inhibitory subunit (Koong *et al.*, 1994).

V. Conclusion

Compelling evidence from *in vitro* studies shows that alterations in the redox state of proteins involving key cysteine residues can lead to conformational changes that affect the biological function of the protein, either as an enzyme or as a DNA-binding transcription factor. The evidence that similar redox changes occur in intacts cell under physiological conditions is largely indirect and is based on the relatively nonselective effects of oxidizing agents and antioxidants. Recent studies have identified several redox systems in cells that appear to respond to external stimuli. They include GSH, Trx, Grx, Ref-1, and, almost certainly, other redox proteins yet to be discovered. There are a number of redox-sensitive effector systems in cells, including enzymes and transcription factors, that regulate cell growth, cell death, and transformation. Thus, the link between external stimuli and activation of processes involved in cell growth, death, and transformation, through redox modulation, is growing.

Acknowledgment

This work was supported by NIH Grants CA48725 and CA17094.

References

Abate, C., Patel, L., Rauscher, F. J., III, and Curran, T. (1990). Redox regulation of fos and jun DNA-binding activity *in vitro. Science* **249**, 1157–1161.

Adcock, I. M., Brown, C. R., Kwon, O., and Barnes, P. J. (1994). Oxidative stress induces NFκB DNA binding and inducible nos mRNA in human epithelial cells. *Biochem. Biophys. Res. Commun.* **199**, 1518–1524.

Ammendola, R., Fiore, F., Esposito, F., Caserta, G., Mesuraca, M., Russo, T., and Cimino, F. (1995). Differentially expressed mRNA as a consequence of oxidative stress in intact cells. *FEBS Lett.* **371**, 209–213.

Amstad, P. A., Krupitza, G., and Cerutti, P. A. (1992). Mechanism of c-*fos* induction by active oxygen. *Cancer Res.* **52**, 3952–3960.

Aslund, F., Ehn, B., Miranda-Vizuete, M., Pueyo, C., and Holmgren, A. (1994). Two additional glutaredoxins exist in *Escherichia coli:* Glutaredoxin 3 is a hydrogen donor for ribonucleotide reductase in a thioredoxin/glutaredoxin 1 double mutant. *Proc. Natl. Acad. Sci. U.S.A.* **91**, 9813–9817.

Babson, J. R., Abell, N. S., and Reed, D. J. (1981). Protective role of the glutathione redox cycle against adriamycin-mediated toxicity in isolated hepatocytes. *Biochem. Pharmacol.* **30**, 2299–2304.

Baker, A. F., Briehl, M. M., Dorr, R., and Powis, G. (1996). Decreased antioxidant defense and increased oxidant stress during dexamethasone-induced apoptosis: *bcl-2* selectively prevents the loss of catalase activity. *Cell Death Differ.* **3**, 207–213.

Bauman, J. W., Liu, L., Liu, Y. P., and Klaassen, C. D. (1991). Increase in metallothionein production by chemicals that induce oxidative stress. *Toxicol. Appl. Pharmacol.* **110**, 347–354.

Beck, I., Weinmann, R., and Caro, J. (1993). Characterisation of hypoxia-responsive enhancer in the human erythropoietin gene shows presence of hypoxia-inducible 120-kD nuclear DNA-binding protein in erythropoietin-producing and non producing cells. *Blood* **82,** 704–711.

Bedi, A., Pasricha, P. J., Akhtar, A. J., Barber, J. P., Bedi, G. C., Giardiello, F. M., Zehnbauer, B. A., Hamilton, S. R., and Jones, R. J. (1995). Inhibition of apoptosis during development of colorectal cancer. *Cancer Res.* **55,** 1811–1816.

Bergelson, S., Pinkus, R., and Daniel, V. (1994). Intracellular glutathione levels regulate fos/jun induction and activation of glutathione S-transferase gene expression. *Cancer Res.* **54,** 36–40.

Bhimani, R. S., Troll, W., Grunberger, D., and Frenkel, K. (1993). Inhibition of oxidative stress in HeLa cells by chemopreventive agents. *Cancer Res.* **53,** 4528–4533.

Biguet, C., Wakasugi, N., Mishal, Z., Holmgren, A., Chouaib, S., Tursz, T., and Wakasugi, H. (1994). Thioredoxin increases the proliferation of human B-cell lines through a protein kinase C-dependent mechanism. *J. Biol. Chem.* **269,** 28865–28870.

Bodenstein, J., and Follmann, H. (1991). Characterization of two thioredoxins in pig heart including a new mitochondrial protein. *Z. Naturforsch.* **46c,** 270–279.

Bodenstein-Lang, J., Buch, A., and Follmann, H. (1989). Animal and plant mitochondria contain specific thioredoxins. *FEBS Lett.* **258,** 22–26.

Bohmann, D., Bos, T. J., Admon, A., Nishimura, T., Vogt, P. K., and Tjian, R. (1987). Human proto-oncogene c-jun encodes a DNA binding protein with structural and functional properties of transcription factor AP-1. *Science* **238,** 1386–1392.

Boniface, J. J., and Reichert, L. E., Jr. (1990). Evidence for a novel thioredoxin-like catalytic property of gonadotropic hormones. *Science* **247,** 61–64.

Briehl, M. M., Cotgreave, I. A., and Powis, G. (1995). Downregulation of the antioxidant defense during glucocorticoid-mediated apoptosis. *Cell Death Differ.* **2,** 41–46.

Brookens, M. A., Lazo, J. S., Yalowich, J. C., Allen, W. P., and Whitmore, M. (1995). Metallothionein reduces the cytotoxicity and DNA damaging effects of nitric oxide. *Proc. Natl. Acad. Sci. U.S.A.,* in press.

Broome, J. D., and Jeng, M. W. (1973). Promotion of replication in lymphoid cells by specific thiols and disulfides in vitro. *J. Exp. Med.* **138,** 574–592.

Brouwer, M., Hoexum-Brouwer, T., and Cashon, R. E. (1993). A putative glutathione-binding site in CdZn-metallothionein identified by equilibrium binding and molecular-modeling studies. *Biochem. J.* **294,** 219–225.

Burdon, R. H., and Gill, V. (1993). Cellularly generated active oxygen species and HeLa cell proliferation. *Free Rad. Res. Commun.* **19,** 203–213.

Buttke, T. M., and Sandstrom, P. A. (1994). Oxidative stress as a mediator of apoptosis. *Immunol. Today* **15,** 7–10.

Chang, D. J., Ringold, G. M., and Heller, R. A. (1992). Cell killing and induction of manganous superoxide dismutase by tumor necrosis factor-α is mediated by lipoxygenase metabolites of arachidonic acid. *Biochem. Biophys. Res. Commun.* **188,** 538–546.

Chen, Y., and Chan, T. M. (1993). Orthovanadate and 2,3-dimethoxy-l,4-naphthoquinone augment growth factor-induced cell proliferation and c-fos gene expression in 3T3-L1 cells. *Arch. Biochem. Biophys.* **305,** 9–16.

Chen, Y., Yang, D.-C., Brown, A. B., Jeng, Y., Tatoyan, A., and Chan, T. M. (1990). Activation of a membrane-associated phosphatidylinositol kinase through tyrosine-protein phosphorylation by naphthoquinones and orthovanadate. *Arch. Biochem. Biophys.* **283,** 184–192.

Cho, Y., Gorina, S., Jeffrey, P. D., and Pavletich, N. P. (1994). Crystal structure of a p53 tumor suppressor-DNA complex: Understanding tumorigenic mutations. *Science* **265,** 346–354.

Clarke, F. M., Orozco, C., Perkins, A. V., Cock, I., Tonissen, K. F., Robins, A. J., and Wells, J. R. (1991). Identification of molecules involved in the "early pregnancy factor" phenomenon. *J. Reprod. Fertil.* **93,** 525–539.

Clive, D. R., and Greene, J. J. (1994). Association of protein disulfide isomerase activity and the induction of contact inhibition. *Exp. Cell Res.* **214**, 139–144.

Cohen, J. J. (1993). Apoptosis. *Immunol. Today* **14**, 126–130.

Conley, R. R., and Pigiet, V. (1978). *In vivo* distribution of phosphothioredoxin and thioredoxin in *Escherichia coli. J. Biol. Chem.* **253**, 5568–5572.

Cotgreave, I. A., Weis, M., Atzori, L., and Moldeus, P. (1994). Glutathione and protein function. *In* "Glutathione: Metabolism and Physiolgical Funtion" (J. Vina, ed.), pp. 155–175. CRC Press, Boca Raton, FL.

Darr, D., and Fridovich, I. (1986). Irreversible inactivation of catalase by 3-amino-1,2,4-triazole. *Biochem. Pharmacol.* **35**, 3642.

Datta, R., Taneja, N., Sukhatme, V. P., Qureshi, S. A., Weichselbaum, R., and Kufe, D. W. (1993). Reactive oxygen intermediates target CC(A/T)$_6$GG sequences to mediate activation of the early growth response 1 transcription factor gene by ionizing radiation. *Proc. Natl. Acad. Sci. U.S.A.* **90**, 2419–2422.

Deiss, L. P., and Kimchi, A. (1991). A genetic tool used to identify thioredoxin as a mediator of a growth inhibitory signal. *Science* **252**, 117–120.

Delaney, J. M., and Georgopoulos, C. (1992). Physical map locations of the trxB, htrD, cydC, and cydD genes of *Escherichia coli. J. Bacteriol.* **174**, 3824–3825.

Deng, H.-X., Hentati, A., Tainer, J. A., Iqbal, Z., Cayabyab, A., Hung, W.-Y., Getzoff, E. D., Hu, P., Herzfeldt, B., Roos, R. P., Warner, C., Deng, G., Soriano, E., Smyth, C., Parge, H. E., Ahmed, A., Roses, A. D., Hallewell, R. A., Pericak-Vance, M. A., and Siddique, T. (1993). Amyotrophic lateral sclerosis and structural defects in Cu,Zn superoxide dismutase. *Science* **261**, 1047–1051.

Dyson, H. J., Gippert, G. P., Case, D. A., Holmgren, A., and Wright, P. E. (1990). Three-dimensional solution structure of the reduced form of *Escherichia coli* thioredoxin determined by nuclear magnetic resonance spectroscopy. *Biochemistry* **29**, 4129–4136.

Ebadi, M., and Iversen, P. L. (1994). Metallothionein in carcinogenesis and cancer chemotherapy. *Gen. Pharmacol.* **25**, 1297–1310.

Edman, J. C., Ellis, L., Blacher, R. W., Roth, R. A., and Rutter, W. J. (1985). Sequence of protein disulphide isomerase and implications of its relationship to thioredoxin. *Nature* **317**, 267–270.

Ericson, M. L., Hörling, J., Wendel-Hansen, V., Holmgren, A., and Rosen, A. (1992). Secretion of thioredoxin after *in vitro* activation of human B cells. *Lymphokine Cytokine Res.* **11**, 201–207.

Fanger, M. W., Hart, D. A., Wells, J. V., and Nisonoff, A. (1970). Enhancement by reducing agents of the transformation of human and rabbit peripheral lymphocytes. *J. Immunol.* **105**, 1043–1045.

Farquhar, R., Homey, N., Murant, S. J., Bossier, P., Schultz, L., Montgomery, D., Ellis, R. W., Freedman, R. B., and Tuite, M. F. (1991). Protein disulfide isomerase is essential for viability in *Saccharomyces cerevisiae. Gene* **108**, 81–89.

Forman-Kay, J. D., Clore, G. M., and Gronenborn, A. M. (1992). Relationship between electrostatics and redox function in human thioredoxin: Characterization of pH titration shifts using two-dimensional homo- and heteronuclear NMR. *Biochemistry* **31**, 3442–3452.

Forster, S. J., and Freedman, R. B. (1984). Catalysis by protein disulphide-isomerase of the assembly of trimeric procollagen from procollagen polypoptide chains. *Biosci. Rep.* **4**, 223–229.

Fountoulakis, M. (1992). Unfolding intermediates of the extracellular domain of the interferon gamma receptor. *J. Biol. Chem.* **267**, 7095–7100.

Freedman, R. B., Hirst, T. R., and Tuite, M. F. (1994). Protein disulphide isomerase: Building bridges in protein folding. *Trends Biochem. Sci.* **19**, 331–336.

Galter, D., Mihm, S., and Dröge, W. (1994). Distinct effects of glutathione disulfide on the nuclear transcription factors κB and the activator protein-1. *Eur. J. Biochem.* **221**, 639–648.

Gasdaska, J. R., Berggren, M., and Powis, G. (1995a). Cell growth stimulation by the redox protein thioredoxin occurs by a novel helper mechanism. *Cell Growth Differ.* **6,** 1643–1650.

Gasdaska, P. Y., Gasdaska, J. R., Cochran, S., and Powis, G. (1995b). Cloning and sequencing of human thioredoxin reductase. *FEBS Lett.* **373,** 5–9.

Gasdaska, P. Y., Oblong, J. E., Cotgreave, I. A., and Powis, G. (1994). The predicted amino acid sequence of human thioredoxin is identical to that of the autocrine growth factor human adult T-cell derived factor (ADF): Thioredoxin mRNA is elevated in some human tumors. *Biochim. Biophys. Acta* **1218,** 292–296.

Gleason, F. K., Lim, C.-J., Gerami-Nejad, M., and Fuchs, J. A. (1990). Characterization of *Escherichia coli* thioredoxins with altered active site residues. *Biochemistry* **29,** 3701–3709.

Grasso, P., Crabb, J. W., and Reichert, L. E., Jr. (1993). An explanation for the disparate effects of synthetic peptides corresponding to human follicle-stimulating hormone beta-subunit receptor binding regions (33–53) and (81–95) and their serine analogs on steriodo-genesis in cultured rat Sertoli cells. *Biochem. Biophys. Res. Commun.* **190,** 56–62.

Greene, J. J., and Brophy, C. I. (1995). Induction of protein disulfide isomerase during proliferation arrest and differentiation of SH5Y neuroblastoma cells. *Cell. Mol. Biol.* **41,** 473–480.

Greenspan, H. C., and Aruoma, O. I. (1994). Oxidative stress and apoptosis in HIV infection: A role for plant-derived metabolites with synergistic antioxidant activity. *Immunol. Today* **15,** 209–213.

Grippo, J. F., Tienrungroj, W., Dahmer, M. K., Housley, P. R., and Pratt, W. B. (1983). Evidence that the endogenous heat-stable glucocorticoid receptor-activating factor is thioredoxin. *J. Biol. Chem.* **258,** 13658–13664.

Hainaut, P., and Milner, J. (1993). Redox modulation of p53 conformation and sequence-specific DNA binding *in vitro. Cancer Res.* **53,** 4469–4473.

Hainaut, P., Rolley, N., Davies, M., and Milner, J. (1995). Modulation by copper of p53 conformation and sequence-specific DNA binding: Role for Cu(II)/Cu(I) redox mechanism. *Oncogene* **10,** 27–32.

Hatala, M. A., DiPippo, V. A., and Powers, C. A. (1991). Biological thiols elicit prolactin proteolysis by glandular kallikrein and permit regulation by biochemical pathways linked to redox control. *Biochemistry* **30,** 7666–7672.

Hayashi, T., Ueno, Y., and Okamoto, T. (1993). Oxidoreductive regulation of nuclear factor κ B. *J. Biol. Chem.* **268,** 11380–11388.

Hensley, K., Carney, J. M., Mattson, M. P., Aksenova, M., Harris, M., Wu, J. F., Floyd, R. A., and Butterfield, D. A. (1995). A model for β-amyloid aggregation and neurotoxicity based on free radical generation by the peptide: Relevance to Alzheimer disease. *Proc. Natl. Acad. Sci. U.S.A.* **91,** 3270–3274.

Hewllet, G., Opitz, H. G., Schlumberger, H. D., and Lemke, H. (1977). Growth regulation of a murine lymphoma cell line by a 2-mercaptoethanol or macrophage-activated serum factor. *Eur. J. Immunol.* **7,** 781–785.

Hickman, J. A., Potten, C. S., Merritt, A. J., and Fisher, T. C. (1994). Apoptosis and cancer chemotherapy. *Philos. Trans. R. Soc. Lond. Series B: Biol. Sci.* **345,** 319–325.

Higgins, K. A., Perez, J. R., Coleman, T. A., Dorshkind, K., McComas, W. A., Sarmiento, U. M., Rosen, C. A., and Narayanan, R. (1993). Antisense inhibition of the p65 subunit of NF-κB blocks tumorigenicity and causes tumor regression. *Proc. Natl. Acad. Sci. U.S.A.* **90,** 9901–9905.

Hockenberry, D. M., Oltvai, Z. N., Yin, X.-M., Milliman, C. L., and Korsmeyer, S. J. (1993). Bcl-2 functions in an antioxidant pathway to prevent apoptosis. *Cell* **75,** 241–251.

Hollingsworth, R. E., and Lee, W.-H. (1991). Tumor suppressor genes: New prospects for cancer research. *J. Natl. Cancer Inst.* **83,** 91–96.

Holmgren, A. (1976). Hydrogen donor system for *Escherichia coli* ribonucleoside-diphosphate reductase dependent upon glutathione. *Proc. Natl. Acad. Sci. U.S.A.* **73,** 2275–2279.

Holmgren, A. (1990). Glutaredoxin: Structure and function. *In* "Glutathione: Metabolism and Physiological Function" (J. Vina, ed.), p. 145. CRC Press, Boca Raton, FL.

Holmgren, A., and Luthman, M. (1978). Tissue distribution and subcellular localization of bovine thioredoxin determined by radioimmunoassay. *Biochemistry* **17**, 4071–4077.

Hopper, S., and Iurlano, D. (1983). Properties of a thioredoxin purified from rabbit bone marrow which fails to serve as a hydrogen donor for the homologous ribonucleotide reductase. *J. Biol. Chem.* **258**, 13453–13457.

Hori, K., Katayama, M., Sato, N., Ishii, K., Waga, S., and Yodoi, J. (1994). Neuroprotection by glial cells through adult T cell leukemia-derived factor/human thioredoxin (ADF/TRX). *Brain Res.* **652**, 304–310.

Huang, R.-P., and Adamson, E. D. (1993). Characterization of the DNA-binding properties of the early growth response-1 (Egr-1) transcription factor: Evidence for modulation by a redox mechanism. *DNA Cell Biol.* **12**, 265–273.

Hwang, C., Sinskey, A. J., and Lodish, H. F. (1992). Oxidized redox state of glutathione in the endoplasmic reticulum. *Science* **257**, 1496–1502.

Ifversen, P., Zhang, X. M., Ohlin, M., Zeuthen, J., and Borrebaeck, C. A. (1993). Effect of cell-derived growth factors and cytokines on the clonal outgrowth of EBV-infected B cells and established lymphoblastoid cell lines. *Hum. Antibodies Hybridomas* **4**, 115–123.

Ireland, R. C., Li, S.-Y., and Dougherty, J. J. (1995). The DNA binding of purified Ah receptor heterodimer is regulated by redox conditions. *Arch. Biochem. Biophys.* **319**, 470–480.

Israel, N., Gougerot-Pocidalo, M. A., Aillet, F., and Virelizier, J. L. (1992). Redox status of cells influences constitutive or induced NF-κ B translocation and HIV long terminal repeat activity in human T and monocytic cell lines. *J. Immunol.* **149**, 3386–3393.

Jacobson, M. D., and Raff, M. C. (1995). Programmed cell death and Bcl-2 protection in very low oxygen. *Nature* **374**, 814–816.

Jewell, S. A., Bellomo, G., Thor, H., Orrenius, S., and Smith, M. (1982). Bleb formation in hepatocytes during drug metabolism is caused by disturbances in thiol and calcium ion homeostasis. *Science* **217**, 1257–1259.

Kaghad, M., Dessarps, F., Jacquemin-Sablon, H., Caput, D., Fradelizi, D., and Wollman, E. E. (1994). Genomic cloning of human thioredoxin-encoding gene: Mapping of the transcription start point and analysis of the promoter. *Gene* **140**, 273–278.

Kallis, G. B., and Holmgren, A. (1980). Differential reactivity of the functional sulfhydryl groups of cysteine-32 and cysteine-35 present in the reduced form of thioredoxin from *Escherichia coli. J. Biol. Chem.* **255**, 10261–10265.

Kane, D. J., Sarafian, T. A., Anton, R., Hahn, H., Gralla, E. B., Valentine, J. S., Örd, T., and Bredesen, D. E. (1993). Bcl-2 inhibition of neural death: Decreased generation of reactive oxygen species. *Science* **262**, 1274–1277.

Kerr, J. F. R., Winterford, C. M., and Harmon, B. V. (1994). Apoptosis: Its significance in cancer and cancer therapy. *Cancer* **73**, 2013–2026.

Kistner, A., Sanders, D., and Habermann, E. (1993). Disulfide formation in reduced tetanus toxin by thioredoxin: The pharmacological role of interchain covalent and noncovalent bonds. *Toxicon.* **31**, 1423–1434.

Koong, A. C., Chen, E. Y., and Giaccia, A. J. (1994). Hypoxia causes the activation of nuclear factor κB through the phosphorylation of IκBα on tyrosine residues. *Cancer Res.* **54**, 1425–1430.

Koropatnick, J., Leibbrandt, M., and Cherian, M. G. (1989). Organ-specific metallothionein induction by X-irradiation. *Radiat. Res.* **119**, 356–365.

Korsmeyer, S. J. (1992a). *Bcl-2:* An antidote to programmed cell death. *Cancer Surv.* **15**, 105–118.

Korsmeyer, S. J. (1992b). Bcl-2 initiates a new category of oncogenes: Regulators of cell death. *Blood* **80**, 879–886.

Kosower, N. S., and Kosower, E. M. (1978). The glutathione status of cells. *Int. Rev. Cytol.* **54**, 109–160.

Krause, G., Lundström, J., Barea, L., and Pueyo de la Cuesta, C. (1991). Mimicking the active site of protein disulfide-isomerase by substitution of proline 34 in *Escherichia coli* thioredoxin. *J. Biol. Chem.* **266,** 9494–9500.

Lafage-Pochitaloff-Huvale, M., Shaw, A., Dessarps, F., Mannoni, D., Fradelizi, D., and Wollman, E. E. (1987). The gene for human thioredoxin maps on the short arm of chromosome 3 at bands 3p11–p12. *FEBS Lett.* **255,** 89–91.

Lander, H. M., Ogiste, J. S., Teng, K. K., and Novogrodsky, A. (1995). p21^{ras} as a common signaling target of reactive free radicals and cellular redox stress. *J. Biol. Chem.* **270,** 21195–21198.

Lash, L. H., Hagen, T. M., and Jones, D. P. (1986). Exogenous glutathione protects intestinal epithelial cells from oxidative injury. *Proc. Natl. Acad. Sci. U.S.A.* **83,** 4641–4645.

Laurent, T. C., Moore, E. C., and Reichard, P. (1964). Enzymatic synthesis of deoxyribonucleotides. VI. Isolation and characterization of thioredoxin, the hydrogen donor from *Escherichia coli* B. *J. Biol. Chem.* **239,** 3436–3444.

Lee, S.-H., Fujita, N., Imai, K., and Tsuruo, T. (1995). Cysteine produced from lymph node stromal cells suppresses apoptosis of mouse malignant T-lymphoma cells. *Biochem. Biophys. Res. Commun.* **213,** 837–844.

Lennon, S. V., Martin, S. J., and Cotter, T. G. (1991). Dose-dependent induction of apoptosis in human tumor cell lines by widely diverging stimuli. *Cell. Prolif.* **24,** 203–214.

Levy, E. J., Anderson, M. E., and Meister, A. (1993). Transport of glutathione diethyl ester into human cells. *Proc. Natl. Acad. Sci. U.S.A.* **90,** 9171–9175.

Liang, C.-M., Lee, N., Cattell, D., and Liang, S.-M. (1989). Glutathione regulates interleukin-2 activity on cytotoxic T cells. *J. Biol. Chem.* **264,** 13519–13523.

Liang, S.-M., Thatcher, D. R., Liang, C.-M., and Bernard, A. (1986). Studies of structure–activity relationships of human interleukin-2. *J. Biol. Chem.* **261,** 334–337.

Lopez-Barea, J., Barcena, J. A., Bocanegra, J. A., Florindo, J., Garcia-Alfonso, C., Lopez-Ruiz, A., Martinez-Galisteo, E., and Peinado, J. (1990). Structure, mechanism, functions, and regulatory properties of glutathione reductase. *In* "Metabolism and Physiological Function" (J. Vina, ed.), p. 105. CRC Press, Boca Raton, FL.

Lozano, R. M., Yee, B. C., and Buchanan, B. B. (1994). Thioredoxin-linked reductive inactivation of venom neurotoxins. *Arch. Biochem. Biophys.* **309,** 356–362.

Luna, M. C., Wong, S., and Gomer, C. J. (1994). Photodynamic therapy mediated induction of early response genes. *Cancer Res.* **54,** 1374–1380.

Lundström, J., and Holmgren, A. (1990). Protein disulfide-isomerase is a substrate for thioredoxin reductase and has thioredoxin-like activity. *J. Biol. Chem.* **265,** 9114–9120.

Lundström, J., Krause, G., and Holmgren, A. (1992). A Pro to His mutation in active site of thioredoxin increases its disulfide-isomerase activity 10-fold: New refolding systems for reduced or randomly oxidized ribonuclease. *J. Biol. Chem.* **267,** 9047–9052.

Martin, S. J., and Green, D. R. (1995). Protease activation during apoptosis: Death by a thousand cuts? *Cell* **82,** 349–352.

Martinez-Galisteo, F., Padilla, C. A., Garcia-Alfonso, C., Lopez-Barea, J., and Barcena, J. A. (1993). Purification and properties of bovine thioredoxin system. *Biochimie* **75,** 803–809.

Masutani, H., Naito, M., Takahashi, K., Hattori, T., Koito, A., Takatsuki, K., Go, T., Nakamura, H., Fujii, S., Yoshida, Y., Okuma, M., and Yodoi, J. (1992). Dysregulation of adult T-cell leukemia-derived factor (ADF)/thioredoxin in HIV infection: Loss of ADF high-producer cells in lymphoid tissues of AIDS patients. *AIDS Res. Hum. Retrovir.* **8,** 1707–1715.

Matsuda, M., Masutani, H., Nakamura, H., Miyajima, S., Yamauchi, A., Yonehara, S., Uchida, A., Irimajiri, K., Horiuchi, A., and Yodoi, J. (1991). Protective activity of adult T cell leukemia-derived factor (ADF) against tumor necrosis factor-dependent cytotoxicity on U937 cells. *J. Immunol.* **147,** 3837–3841.

Matthews, J. R., Wakasugi, N., Virelizier, J.-L., Yodoi, J., and Hay, R. T. (1992). Thioredoxin regulates the DNA binding activity of NF-κB by reduction of a disulphide bond involving cysteine 62. *Nucleic Acids Res.* **20,** 3821–3830.

Mau, B.-L. and Powis, G. (1992). Mechanism-based inhibition of thioredoxin reductase by antitumor quinoid compounds. *Biochem. Pharmacol.* 43, 1613–1620.

McBride, A. A., Klausner, R. D., and Howley, P. M. (1992). Conserved cysteine residue in the DNA-binding domain of the bovine papillomavirus type 1 E2 protein confers redox regulation of the DNA-binding activity *in vitro*. *Proc. Natl. Acad. Sci. U.S.A.* 89, 7531–7535.

McDonnell, T. J., Marin, M. C., Hsu, B., Brisbay, S. M., McConnell, K., Tu, S.-M., Campbell, M. L., and Rodriguez-Villanueva, J. (1993). Symposium: Apoptosis/programmed cell death. *Radiat. Res.* 136, 307–312.

McDonnell, T. J., Meyn, R. E., and Robertson, L. E. (1995). Implications of apoptotic cell death regulation in cancer therapy. *Sem. Cancer Biol.* 6, 53–60.

Meister, A. (1988). Glutathione metabolism and its selective modification. *J. Biol. Chem.* 263, 17205–17208.

Meister, A. (1991). Glutathione deficiency produced by inhibition of its synthesis, and its reversal: Applications in research and therapy. *Pharmacol. Ther.* 51, 155–194.

Meister, A. (1994). Glutathione, ascorbate, and cellular protection. *Cancer Res.* 54, 1969s–1975s.

Mello-Filho, A. C., Chubatsu, L. S., and Meneghini, R. (1988). Chinese hamster cells rendered resistant to high cadmium concentrations also become resistant to oxidative stress. *Biochem. J.* 256, 475–479.

Meyer, M., Schreck, R., and Baeuerle, P. A. (1993). H_2O_2 and antioxidants have opposite effects on activation of NF-κ B and AP-1 in intact cells: AP-1 as secondary antioxidant-responsive factor. *EMBO J.* 12, 2005–2015.

Miranda-Vizuete, A., Martinez-Galisteo, E., Aslund, F., Lopez-Barea, J., Pueyo, C., and Holmgren, A. (1994). Null thioredoxin and glutaredoxin *Escherichia coli* K-12 mutants have no enhanced sensitivity to mutagens due to a new GSH-dependent hydrogen donor and high increases in ribonucleotide reductase activity. *J. Biol. Chem.* 269, 16631–16637.

Mizunaga, T., Katakura, Y., Miura, T., and Maruyama, Y. (1990). Purification and characterization of yeast protein disulfide isomerase. *J. Biochem.* 108, 846–851.

Morrow, C. S., and Cowan, K. H. (1990). Glutathione S-transferases and drug resistance. *Cancer Cells* 2, 15–22.

Muller, E. G. (1992). Thioredoxin genes in *Saccharomyces cerevisiae*: Map positions of TRX1 and TRX2. *Yeast* 8, 117–120.

Muller, E. G. D. (1991). Thioredoxin deficiency in yeast prolongs S phase and shortens the G1 interval of the cell cycle. *J. Biol. Chem.* 266, 9194–9202.

Nakamura, H., Masutani, H., Tagaya, Y., Yamauchi, A., Inamoto, T., Nanbu, Y., Fujii, S., Ozawa, K., and Yodoi, J. (1992). Expression and growth-promoting effect of adult T-cell leukemia-derived factor: A human thioredoxin homologue in hepatocellular carcinoma. *Cancer* 69, 2091–2097.

Nose, K., Shibanuma, M., Kikuchi, K., Kageyama, H., Sakiyama, S., and Kuroki, T. (1991). Transcriptional activation of early-response genes by hydrogen peroxide in a mouse osteoblastic cell line. *Eur. J. Biochem.* 201, 99–106.

Oberley, T. D., Schultz, J. L., Li, N., and Oberley, L. W. (1995). Antioxidant enzyme levels as a function of growth state in cell culture. *Free Rad. Biol. Med.* 19, 53–65.

Oblong, J. E., Gasdaska, P. Y., Sherrill, K., and Powis, G. (1993). Purification of human thioredoxin reductase: Properties and characterization by absorption and circular dichroism spectroscopy. *Biochemistry* 32, 7271–7277.

Oblong, J. E., Berggren, M., Gasdaska, P. Y., and Powis, G. (1994). Site-directed mutagenesis of active site cysteines in human thioredoxin produces competitive inhibitors o† human thioredoxin reductase and elimination of mitogenic properties of thioredoxin. *J. Biol. Chem.* 269, 11714–11720.

Oblong, J. E., Berggren, M., Gasdaska, P. Y., Hill S. R., and Powis, G. (1995). Site-directed mutagenesis of Lys36 in human thioredoxin: The highly conserved residue affects the efficiency of reduction and growth stimulation but is not essential for the redox protein's biochemical and biological properties. *Biochemistry* **34**, 3319–3324.

Ohba,M., Shibanuma, M., Kuroki, T., and Nose, K. (1994). Production of hydrogen peroxide by transforming growth factor-beta 1 and its involvement in induction of egr-1 in mouse osteoblastic cells. *J. Cell Biol.* **126**, 1079–1088.

Okuno, H., Akahori, A., Sato, H., Xanthoudakis, S., Curran, T., and Iba, H. (1993). Escape from redox regulation enhances the transforming activity of Fos. *Oncogene* **8**, 695–701.

Pickett, C. B., and Lu, A. Y. H. (1989). Glutathione S-transferases: Gene structure, regulation, and biological function. *Annu. Rev. Biochem.* **58**, 743–764.

Pihlajaniemi, T., Helaakoski, T., Tasanen, K., Myllyla, R., Huhtala, M. L., Koivu, J., and Kivirikko, K. I. (1987). Molecular cloning of the beta-subunit of human prolyl 4-hydroxylase. *EMBO J.* **6**, 643–649.

Powis, G., Briehl, M., and Oblong, J. (1995). Redox signalling and the control of cell growth and death. *Pharmacol. Ther.* **68**, 149–173.

Prongay, A. J., Engelke, D. R., and Williams, C. H., Jr. (1989). Characterization of two active site mutations of thioredoxin reductase from *Escherichia coli. J. Biol. Chem.* **264**, 2656–2664.

Raff, M. C. (1992). Social controls on cell survival and cell death. *Nature* **356**, 397–400.

Rainwater, R., Parks, D., Anderson, M. E., Tegtmeyer, P., and Mann, K. (1995). Role of cysteine residues in regulation of p53 function. *Mol. Cell. Biol.* **15**, 3892–3903.

Rao, G. N., and Berk, B. C. (1992). Active oxygen species stimulate vascular smooth muscle cell growth and proto-oncogene expression. *Circulat. Res.* **70**, 593–599.

Rao, G. N., Lassègue, B., Griendling, K. K., and Alexander, R. W. (1993). Hydrogen peroxide stimulates transcription of c-jun in vascular smooth muscle cells: Role of arachinoid acid. *Oncogene* **8**, 2759–2764.

Reed, D. J. (1986). Regulation of reductive processes by glutathione. *Biochem. Pharmacol.* **35**, 7–13.

Ren, X., Bjornstedt, M., Shen, B., Ericson, M. L., and Holmgren, A. (1993). Mutagenesis of structural half-cystine residues in human thioredoxin and effects on the regulation of activity by selenodiglutathione. *Biochem.* **32**, 9701–9705.

Rubartelli, A., Bajetto, A., Allavena, G., Wollman, E., and Sitia, R. (1992). Secretion of thioredoxin by normal and neoplastic cells through a leaderless secretory pathway. *J. Biol. Chem.* **267**, 24161–24164.

Rubartelli, A., Bonifaci, N., and Sitia, R. (1995). High rates of thioredoxin secretion correlate with growth arrest in hepatoma cells. *Cancer Res.* **55**, 675–680.

Russel, M., and Holmgren, A. (1988). Construction and characterization of glutaredoxin-negative mutants of *Escherichia coli. Proc. Natl. Acad. Sci. U.S.A.* **85**, 990–994.

Russel, M., and Holmgren, A. (1990). Thioredoxin or glutaredoxin in *Eshcerichia coli* is essential for sulfate reduction but not for deoxyribonucleotide synthesis. *J. Bacteriol.* **172**, 1923–1929.

Russel, M., and Model, P. (1985). Direct cloning of the trxB gene that encodes thioredoxin reductase. *J. Bacteriol.* **163**, 238–242.

Russel, M., and Model, P. (1988). Sequence of thioredoxin reductase from *Escherichia coli. J. Biol. Chem.* **263**, 9015–9019.

Salz, H. K., Flickinger, T. W., Mittendorf, E., Pellicena-palle, A., Petschek, J. P., and Albrecht, E. B. (1994). The drosophila maternal effect locus deadhead encodes a thioredoxin homolog required for female meiosis and early embryonic development. *Genetics* **136**, 1075–1086.

Sandstrom, P. A., and Buttke, T. M. (1993). Autocrine production of extracellular catalase prevents apoptosis of the human CEM T-cell line in serum-free medium. *Proc. Natl. Acad. Sci. U.S.A.* **90**, 4708–4712.

Sandstrom, P. A., Tebbey, P. W., Van Cleave, S., and Buttke, T. M. (1994). Lipid hydroperoxides induce apoptosis in T cells displaying a HIV-associated glutathione peroxidase deficiency. *J. Biol. Chem.* **269**, 798–801.

Sato, M., and Bremner, I. (1993). Oxygen free radicals and metallothionein. *Free Rad. Biol. Med.* **14**, 325–337.

Schallreuter, K. U., Gleason, F. K., and Wood, J. M. (1990). The mechanism of action of the nitrosoureas anti-tumor drugs on thioredoxin reductase, glutathione reductase and ribonucleotide reductase. *Biochim. Biophys. Acta* **1054**, 14–20.

Schenk, H., Klein, M., Erdbrügger, W., Dröge, W., and Schulze-Osthoff, K. (1994). Distinct effects of thioredoxin and antioxidants on the activation of transcription factors NF-κB and AP-1. *Proc. Natl. Acad. Sci. U.S.A.* **91**, 1672–1676.

Schieven, G. L., Kirihara, J. M., Burg, D. L., Geahlen, R. L., and Ledbetter, J. A. (1993). p72^{syk} tyrosine kinase is activated by oxidizing conditions that induce lymphocyte tyrosine phosphorylation and Ca^{2+} signals. *J. Biol. Chem.* **268**, 16688–16692.

Schlager, J. J., Hoerl, B. J., Riebow, J., Scott, D. P., Gasdaska, P., Scott, R. E., and Powis, G. (1993). Increased NADPH:(quinone-acceptor) oxidoreductase (DT-diaphorase) activity is associated with density dependent growth inhibition of normal but not transformed cells. *Cancer Res.* **53**, 1338–1342.

Schwartz, L. M., Smith, S. W., Jones, M. E. E., and Osborne, B. A. (1993). Do all programmed cell deaths occur via apoptosis? *Proc. Natl. Acad. Sci. U.S.A.* **90**, 980–984.

Semenza, G. L., Roth, P. H., Frang, H., and Wang, G. L. (1994). Transcriptional regulation of genes encoding glycoylytic enzymes by hypoxia-inducible factor 1. *J. Biol. Chem.* **269**, 23757–23763.

Shibanuma, M., Kuroki, T., and Nose, K. (1994). Inhibition of N-acetyl-L-cysteine of interleukin-6 mRNA induction and activation of NFκB by tumor necrosis factor α in a mouse fibroblastic cell line, Balb/3T3. *FEBS Lett.* **353**, 62–66.

Sorachi, K.-I., Sugie, K., Maekawa, N., Takami, M., Kawabe, T., Kumagai, S., Imura, H., and Yodoi, J. (1992). Induction and function of FcERII on YT cells; possible role of ADF/thioredoxin in FcERII expression. *Immunobiology* **185**, 193–206.

Staal, F. J., Roederer, M., and Herzenberg, L. A. (1990). Intracellular thiols regulate activation of nuclear factor κ-B and transcription of human immunodeficiency virus. *Proc. Natl. Acad. Sci. U.S.A.* **87**, 9943–9947.

Staal, F. J. T., Roederer, M., Israelski, D. M., Bubp, J., Mole, L. A., McShane, D., Deresinski, S. C., Ross, W., Sussman, H., Raju, P. A., Anderson, M. T., Moore, W., Ela, S. W., Herzenberg, L. A., and Herzenberg, L. A. (1992). Intracellular glutathione levels in T cell subsets decrease in HIV-infected individuals. *AIDS Res. Hum. Retrovir.* **8**, 305–311.

Stefanelli, C., Stanic, I., Bonavita, F., Muscari, C., Pignatti, C., Rossoni, C., and Caldarera, C. M. (1995). Oxygen tension influences DNA fragmentation and cell death in glucocorticoid-treated thymocytes. *Biochem. Biophys, Res. Commun.* **212**, 300–306.

Storz, B., Tarraglia, L. A., and Ames, B. N. (1990a). Transcriptional regulator of oxidative stress-inducible genes: direct activation by oxidation. *Science* **248**, 189–194.

Storz, G., Tartaglia, L. A., Farr, S. B., and Ames, B. N. (1990b). Bacterial defenses against oxidative stress. *Trends Genet.* **6**, 363–368.

Sundaresan, M., Yu, Z.-X., Ferrans, V. J., Irani, K., and Finkel, T. (1995). Requirement for generation of H_2O_2 for platelet-derived growth factor signal transduction. *Science* **270**, 296–299.

Tagaya, Y., Maeda, Y., Mitsui, A., Kondo, N., Matsui, H., Hamuro, J., Brown, N., Arai, K.-I., Yokota, T., Wakasugi, H., and Yodoi, J. (1989). ATL-derived factor (ADF), an IL-2 receptor/Tac inducer homologous to thioredoxin: Possible involvement of dithiol-reduction in the IL-2 receptor induction. *EMBO J.* **8**, 757–764.

Tagaya, Y., Okada, M., Sugie, K., Kasahara, T., Kondo, N., Hamuro, J., Matsushima, K., Dinarello, C. A., and Yodoi, J. (1988). IL-2 receptor(p55)/TAC-inducing factor purification and characterization of adult T cell leukemia-derived factor. *J. Immunol.* **140**, 2614–2620.

Taketo, M., Matsui, M., Rochelle, J. M., Yodoi, J., and Seldin, M. F. (1994). Mouse thioredoxin gene maps on chromosome 4, whereas its pseudogene maps on chromosome 1. *Genomics* **21**, 251–253.

Thelander, L. (1967). Thioredoxin reductase: Characterization of a homogenous preparation from *Escherichia coli* B. *J. Biol. Chem.* **242**, 852–859.

Thomas, J. P., Bachowski, G. J., and Girotti, A. W. (1986). Inhibition of cell membrane lipid peroxidation by cadmium- and zinc-metallothionein. *Biochim. Biophys. Acta* **884**, 448–461.

Thornalley, P. J., and Vasak, M. (1985). Possible role for metallothionein in protection against radiation-induced oxidative stress: Kinetics and mechanism of its reaction with superoxide and hydroxyl radicals. *Biochim. Biophys. Acta* **827**, 36–44.

Tolendano, M. B., and Leonard, W. J. (1991). Modulation of transcription factor NF-κB binding activity by oxidation-reduction *in vitro*. *Proc. Natl. Acad. Sci. U.S.A.* **88**, 4328–4332.

Tomimoto, H., Akiguchi, I., Wakita, H., Kimura, J., Hori, K., and Yodoi, J. (1993). Astroglial expression of ATL-derived factor, a human thioredoxin homologue, in the gerbil brain after transient global ischemia. *Brain Res.* **625**, 1–8.

Tonissen, K., Wells, J., Cock, I., Perkins, A., Orozco, C., and Clarke, F. (1993). Site-directed mutagenesis of human thioredoxin: Identification of cysteine 74 as critical to its function in the "early pregnancy factor" system. *J. Biol. Chem.* **268**, 22485–22489.

Tonissen, K. F., and Wells, J. R. E. (1991). Isolation and characterization of human thioredoxin-encoding genes. *Gene* **102**, 221–228.

Vaux, D. L., Haecker, G., and Strasser, A. (1994). An evolutionary perspective on apoptosis. *Cell* **76**, 777–779.

Veness-Meehan, K. A., Cheng, E. R. Y., Mercier, C. E., Blixt, S. L., Johnston, C. J., Watkins, R. H., and Horowitz, S. (1991). Cell-specific alterations in expression of hyperoxia-induced mRNAs of lung. *Am. J. Respir. Cell Mol. Biol.* **5**, 516–521.

Wakasugi, N., Tagaya, Y., Wakasugi, A., Mitsui, M., Maeda, M., Yodoi, J., and Tursz, T. (1990). Adult T-cell leukemia-derived factor/Thioredoxin produced by both human T-lymphotropic virus type 1 and Epstein-Barr virus-transformed lymphocytes acts as an autocrine growth factor and synergized with interleukin-1 and interleukin-2. *Proc. Natl. Acad. Sci. U.S.A.* **87**, 8282–8286.

Waksman, G., Krishna, T. S. R., Williams, C. H., Jr., and Kuriyan, J. (1994). Crystal structure of *Escherichia coli* thioredoxin reductase at 2 Å resolution. *J. Mol. Biol.* **236**, 800–816.

Walker, L. J., Craig, R. B., Harris, A. L., and Hickson, I. D. (1994). A role for the human DNA repair enzyme HAP1 in cellular protection against DNA damaging agents and hypoxic stress. *Nucleic Acids Res.* **22**, 4884–4889.

Walker, L. J., Robson, C. N., Black, E., Gillespie, D., and Hickson, I. D. (1993). Identification of residues in the human DNA repair enzyme HAP1 (Ref-1) that are essential for redox regulation of Jun DAN binding. *Mol. Cell. Biol.* **13**, 5370–5379.

Wang, C. C., and Tsou, C. L. (1993). Protein disulfide isomerase is both an enzyme and a chaperone. *FASEB J.* **7**, 1515–1517.

Wang, G. L., and Semenza, G. L. (1993a). General involvement of hypoxia-inducible factor-1 in transcriptional response to hypoxia. *Proc. Natl. Acad. Sci. U.S.A.* **90**, 4303–4308.

Wang, G. L., and Semenza, G. L. (1993b). Characterization of hypoxia-inducible factor-1 and regulation of DNA-binding activity by hypoxia. *J. Biol. Chem.* **268**, 21513–21518.

Wang, G. L., and Semenza, G. L. (1995). Purification and characterisation of hypoxia inducible factor-1. *J. Biol. Chem.* **270**, 1230–1237.

Wasylyk, C., and Wasylyk, B. (1993). Oncogenic conversion of Ets affects redox regulation *in-vivo* and *in-vitro*. Nucleic Acids Res. **21**, 523–529.

Weckbecker, G., and Cory, J. G. (1988). Ribonucleotide reductase activity and growth of glutathione-depleted mouse leukemia L1210 cells *in vitro*. *Cancer Lett.* **40**, 257–264.

Wells, W. W., Xu, D. P., Yang, Y. F., and Rocque, P. A. (1990). Mammalian thioltransferase (glutaredoxin) and protein disulfide isomerase have dehydroascorbate reductase activity. *J. Biol. Chem.* **265**, 15361–15364.

Wells, W. W., Yang, Y., Deits, T. L., and Gan, Z. R. (1993). Thioltransferases. *Adv. Enzym. Rel. Areas Mol. Biol.* **66**, 149–201.

Wettrau, J. R., Aggerbeck, L. P., Laplaud, P. M., and McLean, L. R. (1991). Structural properties of the microsomal triglyceride-transfer protein complex. Biochemistry 30, 4406–4412.

Williams, G. T. (1994). Apoptosis in the immune system. *J. Pathol.* 173, 1–4.

Williamson, J. M., Boettcher, B., and Meister, A. (1982). Intracellular cysteine delivery system that protects against toxicity by promoting glutathione synthesis. *Proc. Natl. Acad. Sci. U.S.A.* 79, 6246–6249.

Winkler, B. S., Orselli, S. M., and Rex, T. S. (1994). The redox couple between glutathione and ascorbic acid: A chemical and physiological perspective. *Free Rad. Biol. Med.* 17, 333–349.

Wollman, E. E., d'Auriol, L., Rimsky, L., Shaw, A., Jacquot, J.-P., Wingfield, P., Graber, P., Dessarps, F., Robin, P., Galibert, F., Bertoglio, J., and Fradelizi, D. (1988). Cloning and expression of cDNA for human thioredoxin. *J. Biol. Chem.* 263, 15506–15512.

Wunderlich, M., and Glockshuber, R. (1993). *In vivo* control of redox potential during protein folding catalyzed by bacterial protein disulfide-isomerase (DsbA). *J. Biol. Chem.* 268, 24547–24550.

Wyllie, A. H. (1992). Apoptosis and the regulation of cell numbers in normal and neoplastic tissues: An overview. *Cancer Metast. Rev.* 11, 95–103.

Xanthoudakis, S., Miao, G., Wang, F., Pan, Y.-C. E., and Curran, T. (1992). Redox activation of Fos-Jun DNA binding activity is mediated by a DNA repair enzyme. *EMBO J.* 11, 3323–3335.

Xanthoudakis, S., Miao, G. G., and Curran, T. (1994). The redox and DNA-repair activities of ref-1 are encoded by nonoverlapping domains. *Proc. Natl. Acad. Sci. U.S.A.* 91, 23–27.

Xia, T. H., Bushweller, J. H., Sodano, P., Billeter, M., Bjornberg, O., Holmgren, A., and Wuthrich, K. (1992). NMR structure of oxidized *Escherichia coli* glutaredoxin: Comparison with reduced *E. coli* glutaredoxin and functionally related proteins. *Protein Sci.* 1, 310–321.

Yao, K.-S., Clayton, M., and O'Dwyer, P. J. (1995). Apoptosis in human adenocarcinoma HT29 cells induced by exposure to hypoxia. *J. Natl. Cancer Inst.* 87, 117–122.

Yao, K. S., and O'Dwyer, P. J. (1994). Specific induction of Jun binding to the AP-1 response element of DT diaphorase in hypoxic HT29 colon adenocarcinoma cells. *Proc. Am. Assoc. Cancer Res.* 35, 3341.

Yao, K.-S., Xanthoudakis, S., Curran, T., and O'Dwyer, P. J. (1994). Activation of AP-1 and of a nuclear redox factor, ref-1, in the response of HT29 colon cancer cells to hypoxia. *Mol. Cell. Biol.* 14, 5997–6003.

Yodoi, J., and Tursz, T. (1991). ADF, a growth-promoting factor dervied from adult T cell leukemia and homologous to thioredoxin: Involvement in lymphocyte immortalization by HTLV-1 and EBV. *Adv. Cancer Res.* 57, 381–411.

Zhang, W.-W., and Yaneva, M. (1993). Reduced sulphydryl groups are required for DNA binding of KU protein. *Biochem. J.* 293, 769–774.

Ziegler, D. M. (1985). Role of reversible oxidation-reduction of enzyme thiols-disulfides in metabolic regulation. *Annu. Rev. Biochem.* 54, 305–329.

Robert A. Floyd

Free Radical Biology & Aging Research Program
Oklahoma Medical Research Foundation
Oklahoma City, Oklahoma 73104

Protective Action of Nitrone-Based Free Radical Traps against Oxidative Damage to the Central Nervous System

I. Introduction

The central nervous system (CNS) is sensitive to oxidative damage. Considerable evidence indicates that oxidative damage contributes to the etiology of several neurodegenerative diseases. α-Phenyl-*tert*-butyl nitrone (PBN), one of the most commonly used nitrone-based free radical traps (NFTs), has been shown to protect in several experimental models of neurodegenerative disease. Trapping of crucial directly damaging reactive free radicals and rendering them unreactive probably only partially explains the action of PBN. Recent observations demonstrate that PBN prevents the induction of inducible nitric oxide synthase (iNOS) by reactive oxygen species, thus preventing the formation of large tissue levels of nitric oxide (NO). This key observation provides a cogent rationale to explain its potent neuroprotective action, especially when it is considered that higher levels of NO are neurotoxic. Many neurodegenerative problems that develop with age appear to be related to increased or altered cytokine changes in brain, known to cause increased NO formation. This implicates that the suppression of iNOS gene expression by PBN and related compounds or the prevention of excess glutamate-mediated NO formation may prove therapeutically useful to protect in a wide range of neurodegenerative diseases. Mechanistic studies to address this concept are supportive of this notion.

II. CNS Is Sensitive to Oxidative Damage

Brain homogenate, contrary to most other tissue homogenate, rapidly peroxidizes when incubated at physiological temperature (Zaleska *et al.*, 1989). This led us to conclude that brain is abnormally sensitive to oxidative damage. This has proven to be true *in vivo* also. Willmore *et al.* (1978) and Triggs and Willmore (1984) demonstrated that the injection of Fe ions directly into brain caused the death of neurons at the site of injection, which resulted in formation of permanent epileptic foci.

It has been very difficult to assess reactive oxygen species (ROS) *in vivo*, but the use of salicylate trapping of hydroxyl-free radicals made it possible to show that increased salicylate hydroxylation occurs during a brain ischemia/reperfusion insult (IRI) (Cao *et al.*, 1988). Oxygen reentry into the ischemic brain is required for increased salicylate hydroxylation. Increased hydroxyl-free radical formation is also brain region specific and is in proportion to the degree of tissue injury brought on by the IRI (Carney *et al.*, 1992). General brain protein oxidation increases and there is a marked loss of glutamine synthase activity, an oxidative sensitive enzyme, after IRI (Oliver *et al.*, 1990). We have reviewed the field and concluded that oxidative damage and neurodegenerative diseases are closely linked (Floyd and Carney, 1992).

The reasons considered for the enhanced sensitivity of brain to oxidative damage include its high content of easily peroxidizable fatty acids, its modest to low levels of antioxidant protective enzymes, its high rate of oxygen utilization, its high content of Fe in certain regions, and the fact that it has high levels of ascorbate that will cause peroxidative damage when the tissue is disrupted (Floyd and Carney, 1992). Inherent oxidative stress imposed on aerobic biological systems is represented as the balance between the oxidative damage potential (P_0) imposed as counterbalanced by the antioxidant defense capacity (A_c) of the system (i.e., $P_0 \rightleftarrows A_c$). Thus, because oxygen is necessary and being constantly utilized presents an oxidative stress simply because some of the oxygen is converted to semireduced oxygen species ($O_2^{\cdot-}$, H_2O_2, $\dot{O}H$) and other ROS rather than being completely reduced to H_2O (Floyd, 1995). It appears that most of the ROS produced in aerobic metabolism occurs as by-products of mitochondrial respiration and microsomal metabolism (Chance *et al.*, 1979; Floyd, 1995). It is expected that there is a certain small amount of ROS present at all times simply because it is unlikely that all of the ROS formed could be scavenged.

An important fact is that older brain is more sensitive to oxidative damage than younger brain. This was demonstrated by the observation that 10 min of global brain ischemia resulted in death to approximately 85% of old (18 months) gerbils whereas 15 min of global brain ischemia caused death to only one-half of the younger (3 months) gerbils (Floyd, 1990; Floyd and Carney, 1995). It is not known why older brain is more sensitive to oxidative damage, but it has been concluded that age-dependent lesions in brain mitochondria are important factors (Floyd, 1996a). Two important facts support this statement: (1) real time nuclear magnetic resonance monitoring of the cortex of gerbils during a global brain ischemia and the subsequent reperfusion phase showed that recovery of both high energy phosphate content and normal pH values, both of which reflect mitochondrial function, were much faster in the younger animals than the older animals (Funahashi *et al.,* 1994) and (2) isolated brain synaptosomes subjected to an *in vitro* oxidative insult showed that those from older animals were more severely effected in energy metabolism parameters than those from younger animals (Floyd and Carney, 1991).

Thus far three important points have been put forward: (1) brain is sensitive to oxidative damage, (2) older brain is more sensitive to oxidative damage than younger brain, and (3) age-related lesions in brain mitochondria are related to the increased sensitivity of older brain to oxidative damage.

III. Spin-Trapping Reactions of Nitrone-Based Free Radical Traps

PBN and related NFTs have been used as analytical reagents to ascertain if free radical intermediates are involved, first in chemical reactions (Janzen, 1971) and then later in biochemical and biological systems. Janzen and Blackburn (1969) first coined the term spin-trap to denote the fact that the NFT reaction with a free radical produces a stable-free radical, termed the spin adduct. The reaction of PBN with a free radical is shown in the following equation:

PBN   Free radical   PBN–radical adduct
                             (reactive)   (stable)

Thus, in general, the nitroxyl-free radical product formed is much more stable than the more reactive free radical. In principle, the electron spin resonance spectrum of the trapped radical will allow one to deduce the nature of Ṙ that has been added to PBN. The free electron spin is preserved in the trapped species and hence the term spin-trapping.

The technique of spin-trapping has been very useful in ascertaining the nature of free radical intermediates under many circumstances, but there are conditions where it does not yield straightforward answers. The reasons seem to be related to the following facts: (1) not all free radicals are trapped with equal efficiency, (2) trapped free radicals can decay to yield products that are different than the original trapped species, and (3) trapped free radicals can be reduced by biological systems to nonparamagnetic species (Floyd, 1983).

IV. Antioxidant Activity of PBN

The antioxidant properties of PBN have been studied by Janzen *et al.* (1994). They used a rat liver microsomal system where lipid peroxidation was initiated using three different methods. A summary of the results is presented in Table I. It can be seen that as an antioxidant PBN acts as a "preventative" type rather than as a "chain-breaking" type (Janzen *et. al.*, 1994). Its antioxidant activity is relatively low compared to butylated hydroxytoluene (BHT) or Trolox.

TABLE I   Summary of Antioxidant Properties of PBN[a]

Type of antioxidant	*Examples*
Preventative	*PBN*
chain-breaking	*BHT, Trolox*
Comparisons in model systems	IC_{50}
A. Di-*tert*-butyl peroxyloxalate-initiated lipid peroxidation of rat liver microsomes	PBN $\sim$ 5 mM Trolox $\sim$ 40 μM BHT $\sim$ 6 μM
B. Azo-bis-2-amidino propane-initiated lipid peroxidation of rat liver microsomes	PBN $\sim$ 1.6 mM Trolox $\sim$ 17 μM BHT $\sim$ 3.4 μM
C. NADPH-driven reduction of CCl_4-mediated lipid peroxidation of rat liver microsomes	PBN $\sim$ 20 mM Trolox $\sim$ 1 mM BHT $\sim$ 50 μM

[a] Data from Janzen *et al.* (1994).

V. PBN Exhibits Protective Activity in Septic Shock Models

The first indication that NFTs exhibited pharmacological activity was demonstrated by Novelli in 1985 who showed that PBN protected rats from shock induced either by superior mesenteric artery occlusion or by lipopolysaccharide (LPS) injection as well as by shock trauma caused by drum rotation of the animals (Novelli *et al.*, 1985, 1986a,b). These observations were followed up and independently shown to be true by McKechnie *et al.* (1986), Hamburger and McCay (1989), and Progrebniak *et al.* (1992).

VI. PBN and Related Compounds Are Neuroprotective

We were the first to show that PBN had neuroprotective activity (Floyd, 1990). This discovery arose out of attempts to utilize PBN to trap putative-free radicals during brain IRI in the gerbil (Oliver *et. al.*, 1990). Although triplet electron spin resonance signals were obtained (Oliver *et al.*, 1990), we showed that these signals did not arise from di-*tert*-butyl nitroxide-free radical; but it was difficult to ascertain their origin, especially in light of reports of spurious nitroxides being present as contaminants (Buettner *et al.*, 1991). During the course of these experiments it was discovered that gerbils given PBN prior to a brain IRI were significantly protected from the lethality that normally occurs (Floyd, 1990). This serendipitous observation then lead us to ask if chronic PBN treatment would alter brain protein oxidation and behavioral parameters in older gerbils. These studies showed that chronic PBN dosing for 14 days lowered the otherwise age-related increased brain protein oxidation levels back to those observed in younger animals and that the protein oxidation levels gradually increased back to original higher levels within a week after ceasing PBN administration (Carney *et al.*, 1991). In addition to changes in protein oxidation, chronic PBN administration caused the older gerbils to regain their youthful competence in a short-term spatial memory test (Carney *et al.*, 1991). These two sets of observations (i.e., protection from brain IRI-mediated lethality and improvement in memory behavioral parameters) spurred a large amount of research effort from other laboratories, the end result of which has confirmed the neuroprotective activity of PBN.

Table II summarizes work demonstrating the neuroprotective activity of PBN and related compounds (Floyd *et al.*, 1996). Most of the research has been done in stroke models where NFTs (mostly PBN) have been shown to protect in the gerbil model if given either before or within 1 or 2 hr after

TABLE II Experimental Neurodegenerative Models where PBN and Related
Compounds Are Protective[a]

Model and Observations

Brain stroke
  Lethality prevented if PBN given before (A) or shortly after stroke (B–D)[b]
  Administration of PBN prevents CA_1 neuron loss if given before or after stroke (E)[b]
  PBN administration chronically to older gerbils prevents stroke lethality long after
    ceasing PBN dosing (F)[b]
  PBN, before or after stroke, protects from brain necrosis (G, H), edema and loss of
    behavioral scores (G)[c]
Brain aging
  PBN administered chronically decreased age-related brain protein oxidation and
    improved memory (I)[b]
Brain concussion
  PBN administered before prevented concussion-induced hydroxyl free radical
    formation (J)
Brain excitotoxicity, lowered energy production, and dopamine depletion models
  2-Sulfo-PBN suppressed neuronal death by NMDA, AMPA, malonate, or MPP$^+$ (K)

[a] Reproduced with permission of the publisher and Floyd *et al.* (1996). References: (A) Floyd,
1990; (B) Phillis and Clough-Helfman, 1990a; (C) Phillis and Clough-Helfman, 1990b;
(D) Yue *et al.*, 1992; (E) Clough-Helfman and Phillis, 1991; (F) Floyd and Carney, 1995;
(G) Cao and Phillis, 1994; (H) Zhao *et al.*, 1994; (I) Carney *et al.*, 1991; (J) Sen *et al.*, 1994;
and (K) Shulz *et al.*, 1995.
[b] Experiments done in Mongolian gerbils.
[c] Permanent (G) or transient (H) middle cerebral artery occlusion in rats.

IRI. This has been extended to the rat middle cerebral artery occlusion
(MCAO) stroke model, where PBN has been shown to protect in the permanent as well as in transient models if given as much as several hours after
the occlusion. In models of brain aging, brain concussion and experimental
models of Parkinson's disease, Huntington's disease, and excitotoxicity-
mediated or lowered energy production-mediated neurotoxicity, NFTs have
also been shown to protect, although fewer studies have been done in these
areas. Thus it is quite clear that PBN is neuroprotective. In a broader context,
PBN has been shown to prolong the life span by 50% in the senescent
accelerated mouse model (Edamatsu *et al.*, 1995).

VII. Pharmacology Studies of PBN and Related Compounds

There have been some, but not extensive, studies on the pharmacology
of NFTs (Table III). Chen *et al.* (1990) demonstrated that PBN after an ip
injection is rapidly taken up by all tissues, peaking in about 20 min. Relative
to the question as to the amount of NFT reaching the brain in relation to

TABLE III Summary of Studies on Pharmacology of PBN and Related Compounds

A. PBN is rapidly taken up by all tissues and it peaks 20 min after ip injection (Chen *et al.*, 1990)

B. PBN is almost equally distributed in all tissues, where its half-life is about 134 min in rats (Chen *et al.*, 1990)

C. PBN is metabolized in liver by mixed function oxidases and is excreted in urine as a metabolite (Chen *et al.*, 1990, 1991)

D. PBN appears in brain as authentic PBN (Chen *et al.*, 1990; Yue *et al.*, 1992; Cheng *et al.*, 1993)[b]

E. PBN brain levels reach about 500 μM after 150 mg/kg ip dose and decrease gradually (Cheng *et al.*, 1993)[b]

F. PBN and POBN[a] brain and blood levels reflect lipid solubility differences (Cheng *et al.*, 1993)

G. PBN and POBN brain levels exhibit significantly different kinetics (Cheng *et al.*, 1993)

[a] α-4-Pyridyl-N-oxide-N-*tert*-butyl nitrone.
[b] Studies were done in rats and gerbils.

the dose delivered and its proportion to the tissue weight, it is clear that PBN distributes rapidly and roughly equal to all tissues in proportion to the tissue/body weight ratio. Also, based on its demonstrated liver metabolism, it is lost from tissues rather rapidly (half-time about 134 min in rats). Extracellular PBN reaches nearly 500 μM in brain of a 60- to 80-g gerbil within 20 min after a 150-mg/kg ip injection (Yue *et al.* 1992) as compared to achieving nearly 550 μM in 20 min in the extracellular space of caudate nucleus of a 300-g rat after a 150-mg/kg ip injection (Cheng *et al.* 1993). After the 20-min peak of PBN in brain, it decreases by about 60% in 80 min paralleling, although slightly higher in concentration than, the blood content. Similar to PBN, extracellular brain POBN almost exactly replicates the blood levels on a time course basis, but exhibits overall different general kinetics. Thus, whereas brain PBN levels peaked in 20 min and then rapidly dropped, brain POBN only reached maximum levels at about 100 min.

VIII. PBN Prevents LPS-Mediated NO Formation in Liver

Miyajima and Kotake (1995) made an observation that has important implications in terms of understanding the neuroprotective action of PBN. They noted that PBN prevented the LPS-mediated formation of NO in liver of treated mice. The tissue NO was directly assessed using an NO-trapping agent (N-methyl-D-glucamine dithiocarbamate–Fe complex, MGD_2-Fe) combined with electron spin resonance methodology (Kotake *et al.*, 1995). A summary of the results is shown in Table IV. Data show that PBN given before LPS prevented the death of the animals and also prevented NO

TABLE IV   Effect of PBN and NOS Inhibitors on
LPS-Mediated NO Formation[a]

Treatment	Survival (%)	NO trapped (%)[b]
A. Saline (−0.5 hr) LPS (0 hr)	0	100 ± 26
B. PBN (0.5 hr) LPS (0 hr)	100	21 ± 2
C. Saline (−0.5 hr) LPS (0 hr)	20	115 ± 24
D. NMMA (−0.5 hr) LPS (0 hr)	—	65 ± 11
E. Saline (−0.5 hr) LPS (0 hr) NMMA (3 hr)	—	16 ± 3

[a] Data from Miyajima and Kotake (1995).
[b] NO trapped in liver by MGD-Fe as a percentage of treatment A at
6 hr after LPS; survivor data were taken at 24 hr.

formation in liver. The nitric oxide synthase (NOS) inhibitor NMMA (N^G-monomethyl L-arginine) if given before LPS depressed the amount of NO formed somewhat, but was more effective if given 3 hr prior to NO measurement, which was obtained 6 hr after LPS administration. The administration of PBN at 3 hr after LPS administration had perhaps some protective effect on survival, but had no effect on the amount of NO formed. Miyajima and Kotake (1995) showed that PBN did not directly act competitively on the NOS enzyme reaction per se to prevent NO formation.

IX. NO Chemistry and Its Potential Neurotoxic Action

Three different NOS enzymes mediate the formation of NO in brain. Two enzymes are constitutive [i.e., brain NOS (bNOS) and endothelial NOS (eNOS)] and the NO formed occurs in bursts because free Ca^{2+} is required. NO synthesized by eNOS is very important in blood pressure regulation and NO formed by bNOS acts in many cases as a retrograde neurotransmitter involved in long-term potentiation, an important event in memory encoding. Evidence shows, however, that glutamate neurotoxicity involves NO formation by bNOS (Dawson et al., 1993). In contrast, NO formed by the third NOS enzyme, iNOS, which normally exists at very low levels in the brain, can be synthesized at much higher levels by microglia, astrocytes, invading macrophages, and possibly a few neurons. Because iNOS-mediated NO formation is not Ca^{2+} dependent, it may remain active for hours or days to mediate the formation of high levels (up to 10 μM) of tissue NO.

NO should be written as NȮ, reflecting the odd number of electrons in the outer π^*2p orbital of the molecule. NO can lose and gain electrons as well as react with superoxide to form peroxynitrite in a diffusion-limited reaction rate. Protonation of peroxynitrite occurs with a pK_a of 6.8 to yield peroxynitrous acid, which decomposes spontaneously (half-time < 1 sec) to yield hydroxyl-free radical and ȮNO_2.

Many observations clearly demonstrate that NO at higher levels is toxic to neurons. These observations have been made in cell culture systems, but it is highly likely that NO at higher concentrations is also toxic to neurons in the living brain. It is not known which form of NO (i.e., NȮ, NO^+, O_2NO^-, etc.) species is toxic, but careful work in tissue culture suggests that peroxynitrite rather than NȮ is the toxic species. The mechanistic basis of NO-mediated neuronal killing is not known. But it is known that NO causes rather drastic alterations in tissue, including inhibition of mitochondrial respiration (Granger *et al.*, 1980), inactivation of aconitase (Drapier and Hibbs, 1986), and activation of PARS [poly (ADP) ribosylase] enzymes leading to the selective ribosylation of glyceraldehyde phosphate dehydrogenase (GAPDH), which is expected to shut down activity of this enzyme, thus drastically reducing glycolysis (Dawson *et al.*, 1993; Zhang *et al.*, 1994; Troncoso and Costello, 1995). It has also been shown that NO inhibits ribonucleotide reductase (Lepoivre *et al.*, 1992) and reacts with DNA to cause mutational events (Arroyo *et al.*, 1992). NO is also known to cause neuronal growth cone collapse by inhibiting fatty acylation (Hess *et al.*, 1993).

It appears highly likely that increased levels of NO act in the living brain to slow down mitochondrial activity in neurons. This rationale is based on the knowledge of biologically produced NO on target cells (i.e., on the extensive studies of how activated macrophages kill tumor cells). It was shown early that activated macrophages killed tumor cells by impairing mitochondrial respiration in the target cells (Granger *et al.*, 1980). In a classic study, Granger and Lehninger (1982) showed that the primary lesion site in the tumor cells was at complex I and complex II of the mitochondrial respiratory chain. Drapier and Hibbs (1986) also showed there was a loss of aconitase activity in the tumor cells. Aconitase as well as complex I and complex II of mitochondria have essential Fe-S sites. It was shown that macrophage-mediated killing was dependent on L-arginine (Drapier and Hibbs, 1988). Hibbs *et al.* (1988) then showed that authentic NO gas would mediate tumor killing and that NO was produced by the macrophages. Further studies using electron paramagnetic resonance showing that Fe–NO complexes were formed in macrophage-attacked tumor cells (Lancaster and Hibbs, 1990; Drapier *et al.*, 1991) helped to demonstrate the mechanistic basis in a more rigorous fashion. Adding to this concept is the fact that NO has been shown to inhibit mitochondrial respiration in hepatocytes (Stadler *et al.* 1992) and hepatoma cells where NO is produced by nearby Kupffer cells (Kurose *et al.* 1993), which are considered resident macrophages. Pertinent to the argument here is the recent demonstration that NO inhibits mitochondrial function in brain cells (Mitrovic *et al.*, 1994).

A recent observation is particularly pertinent. NO at a low concentration (<1 μM) reversibly inhibits mitochondrial respiration and oxidative phosphorylation in isolated mitochondria (Takehara *et al.*, 1995) if the oxygen tension is low (i.e., <50 μM). In contrast, NO at 1 μM had very little effect on isolated mitochondria respiration or oxidative phosphorylation at higher oxygen tensions. Because most tissues normally function at quite low oxygen tensions (<50 μM), NO levels of 1–10 μM become particularly important in terms of potentially reversibly inhibiting tissue mitochondrial function. The NO produced during the macrophage-mediated killing of tumor cells is due to the action of iNOS. It is reasonable to postulate that the mitochondrial function will be generally depressed in brain where iNOS is producing an increased flux of NO causing formation of 1 μM or greater amounts (i.e., much higher than the tissue would normally experience). It is also expected that the prevention of NO formation by inhibiting iNOS induction, thus lowering the tissue NO levels back to the normal extremely low levels, should allow restoration of proper functioning of the mitochondria.

X. Hypothesis to Explain the Neuroprotective Action of PBN

One can postulate that PBN exhibits neuroprotective action in brain because it prevents the induction of iNOS. iNOS induction is caused by events that result in increased levels of ROS and other mediators that act in a signal transduction-mediated gene induction pathway. Thus PBN may act by trapping the ROS and/or by preventing the rise of other mediators that trigger gene induction. The research of Baeuerle and colleagues (Meyer *et. al.*, 1993) clearly shows that gene induction has an antioxidant sensitive step that appears to involve ROS which mediates the breakdown of the inhibitory factor (IκB) that holds nuclear factor-κB (NF-κB) in check, thus preventing it from moving to the nucleus initiating gene expression. Certain proinflammatory cytokines cause gene induction, including iNOS induction, and thus PBN could also mediate protection by preventing the increase in cytokines. Cytokine synthesis is mediated by NF-κB-mediated processes where PBN might be expected to act.

There are several conditions where brain proinflammatory cytokine levels are known to increase. For instance, it is known that the upregulation of several brain cytokines occurs in the processes involved in the etiology of several neurological diseases (Merrill, 1992; Rothwell and Relton, 1993; Merrill and Jonakait, 1995). The cytokine interleukin-1 (IL-2) has been shown to increase 30-fold in the brains of Alzheimer's disease (AD) patients as well as in Down's syndrome (Griffin *et al.*, 1989). IL-6 has been shown to be increased in the plaques of AD patients (Ershler *et al.*, 1994). IL-6 is a proinflammatory cytokine that generally increases with age (Daynes *et al.*, 1993; Ershler, 1993; Ershler *et al.*, 1993; Sun *et al.*, 1993), even in individuals where there is no known general inflammatory condition present. The mechanisms of neurodegeneration and the protective action of PBN have been discussed (Floyd and Carney, 1995). Thus with age the brain has a tendency to take on a proinflammatory state which is a condition where induction of the iNOS gene is expected to increase. This condition then would tend to enhance the general tissue NO concentration which, under low oxygen tension, could inhibit the capacity of mitochondria to perform at maximum efficiency under conditions of large energy demands. Large energy demands are required during recovery from ischemia (Floyd and Carney, 1991). Thus, according to these arguments, the older brain, in contrast to the younger brain, would be less capable of responding to a massive demand in energy production because the mitochondria are functionally lesioned due to the low levels of NO present caused by the increased activity of iNOS. This could then help explain the reason why older animals are more susceptible to an IRI than younger animals (Floyd, 1990).

PBN has been shown to be protective if administered within a few hours after an IRI and to protect from an IRI if it has been administered in a chronic fashion for 2 weeks prior to the IRI. In the latter case its protective effect remains for several days after ceasing administration of PBN and at a time when no PBN remains in the brains (Floyd and Carney, 1995). Thus it is necessary to rationalize the neuroprotective action of PBN under these two extreme conditions. Table V presents the characteristics of these two extreme conditions in a framework incorporating the ideas developed earlier. As Table V illustrates, the old brain is considered to be in a smoldering proinflammatory state where cytokines are induced and extensive cellular remodeling has occurred. This is a state that has developed slowly over the lifetime of the individual. The increased level of cytokines act as triggering events that cause the induction of genes, iNOS and possibly others, that lead to the production of neurotoxic products, most likely at low levels, that act by generally decreasing the capability of the system to handle a large oxidative stress. The toxic products are considered to be the combined effect of increased oxidative damage and increased levels of NO. Following this scenario, chronic administration of PBN or related NFT is expected to shut down the induction of genes forming the toxic products, thus allowing the system to return to the normal "youthful" state which is more capable of handling a large oxidative stress. The sequence of events noted in the upper portion of Fig. 1 attempts to summarize these ideas where NO is the toxic product, but the lower part of Fig. 1 is an attempt to reflect the fact that other toxins are possibly contributing to the neurological damage. Figure 1 also shows ROS to be very important in the processes involved.

TABLE V Postulated Characteristics of Two Extreme Conditions Where Nitrone-Based Free Radical Traps Have Proven Neuroprotective

---

A. Smoldering proinflammatory state of brain
  1. Before NFT treatment
     a. Develops slowly over many months/years
     b. Increased cytokines triggering induction of gene(s) producing toxic products
     c. General low levels of toxic species produced
     d. General inefficiency of system to handle oxidative stress
  2. After chronic effective low-level NFT treatment
     a. Gradual shut-down of toxic gene(s) induction
     b. Lower levels of toxic species decrease and disappear
     c. General efficiency of system to handle oxidative stress returns
     d. Gradually (weeks) regains normal status
B. Rapid increase in oxidative damage
  1. Before NFT treatment
     a. Develops rapidly in minutes/hours
     b. Rapid increase in toxic species (ROS, lactate, etc.)
     c. Rapid induction of some genes
     d. Shut-down in energy production
  2. After NFT treatment
     a. Decrease in toxic species
     b. Shut-down of inducible gene(s)
     c. Restoration of energy production
     d. Prevention of rapid death

Table V also attempts to address the conditions that occur in an IRI where a rapid increase in oxidative damage occurs and where NFTs have proven protective. A question remains as to whether the iNOS gene induction is rapid enough to account for energy production shutdown and PBN-mediated reversal early (2 to 4 hr) after an IRI. Perusal of data collected by Siesjo's group (Folbergrova *et al.*, 1995) has shown that in the focal area of the transient MCAO rat brain stroke model, where PBN had significant

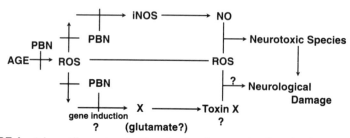

FIGURE I Scheme illustrating the ideas developed concerning the role of reactive oxygen species and nitric oxide in neurological damage that develops with age as well as during an ischemia/reperfusion insult to brain and the possible protective role of α-phenyl-*tert*-butyl nitrone.

protective activity even if given 3 hr after restoration of blood flow, lactate metabolism, an index of mitochondrial functionality (Funahashi *et al.*, 1994), is restored if PBN has been given whereas this is not the case in PBN-devoid animals (Folbergrova *et al.*, 1995). Thus IRI-mediated lesions in brain mitochondria that occur within a 2- to 6-hr period after an IRI seem to be too rapid to be accounted for by iNOS-mediated NO production. It is possible, however, that NO still may account for these rapid mitochondrial lesions. It has been demonstrated that glutamate neurotoxicity is mediated by NO produced by the action of bNOS (Dawson *et al.*, 1993). For this reason, glutamate has been put as a possible toxin in the lower portion of Fig. 1. PBN has been shown to prevent glutamate-induced neuronal killing in cell culture systems (Yue *et al.*, 1992). It should also be noted that Iadecola *et al.* (1996) demonstrated that the iNOS gene is turned on in the brain of rats given MCAO. The appearance of iNOS mRNA and enzyme activity began at 12 hr and peaked at 48 hr. They also noted that treatment with aminoguanidine, an iNOS competitive inhibitor, prevented neuronal death (Iadecola *et al.*, 1995). Processes triggering iNOS gene induction most likely began within minutes after the IRI. These stimuli, with perhaps the combined action of several other mediators, may have started irreversible gene induction events as late as 3 hr after the IRI. PBN may suppress the triggering stimuli (most likely ROS and secondary free radicals) and may prevent the formation of mediators (most likely certain cytokines), thus preventing the induction of iNOS and other genes involved in producing toxic products. Therefore if PBN is present in the first few hours when these key events are occurring, then it prevents the induction of iNOS that is involved in the production of NO that eventually causes neurological damage.

Figure 2 is presented in a schematic fashion to help clarify the concepts regarding the role of brain energy production capacity under normal condi-

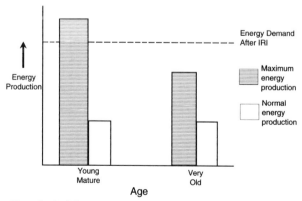

FIGURE 2   Hypothetical demonstration of energy production capacity of the brain of young mature versus very old individual under normal conditions as well as under maximum energy demand. The energy demand after a brain ischemia/reperfusion insult is noted.

tions as well as that which is maximally possible in relation to that which is needed to meet the challenge of a large IRI. Thus, in the young mature brain, its maximum energy production far exceeds that required to meet the needs following an IRI, which is also much higher than that which is normally produced in a nonchallenged state. In contrast to the young mature brain, the very old brain does not have the capacity to achieve the energy production level needed to meet the challenge imposed by an IRI, even though it has no problem in meeting the energy needs required for normal functioning. The lack of ability of the very old brain to achieve much higher levels of energy production may be due in part to an increased tissue level of NO acting to suppress the mitochondrial energy capacity. Total energy production also involves the rate of glycolysis; therefore suppression of GAPDH activity by NO-mediated activation of PARS leading to ribosylation of the enzyme may also contribute to a decrease in energy production capacity. If with time the toxins responsible for suppressing the maximum energy production are removed, then the ability to meet the challenge of an IRI should again be possible in the very old brain. If iNOS activity is responsible for the increased tissue NO and if PBN inhibits iNOS induction, then chronic treatment with PBN is expected to lower iNOS gene expression and thus lower the general tissue level of NO.

Acknowledgment

Research conducted leading to the concepts in this chapter was funded in part by NIH Grant AG09690.

References

Arroyo, P. L., Hatch-Pigott, V., Mower, H. F., and Cooney, R. V. (1992). Mutagenicity of nitric oxide and its inhibition by antioxidants. *Mutat. Res* **281**, 193–202.

Buettner, G. R., Scott, B. D., Kerber R. E., and Mugge, A. (1991). Free radicals from plastic syringes. *Free Rad. Biol. Med.* **11**, 69–70.

Cao, W., Carney, J. M., Duchon, A., Floyd, R. A., and Chevion, M. (1988). Oxygen free radical involvement in ischemia and reperfusion injury to brain. *Neurosci. Lett.* **88**, 233–238.

Cao, X., and Phillis, J. W. (1994). α-Phenyl-tert-butyl-nitrone reduces cortical infarct and edema in rats subjected to focal ischemia. *Brain Res.* **644**, 267–272.

Carney, J. M., Starke-Reed, P. E., Oliver, C. N., Landrum, R. W., Chen, M. S., Wu, J. F., and Floyd, R. A. (1991). Reversal of age-related increase in brain protein oxidation, decrease in enzyme activity, and loss in temporal and spacial memory by chronic administration of the spin-trapping compound N-tert-butyl-α-phenylnitrone. *Proc. Natl. Acad. Sci. U.S.A.* **88**, 3633–3636.

Carney, J. M., Tatsuno, T., and Floyd, R. A. (1992). The role of oxygen radicals in ischemic brain damage: Free radical production, protein oxidation and tissue dysfunction. *In* "Pharmacology of Cerebral Ischemia" (J. Krieglstein and H. Oberpichler-Schwenk, eds.), pp. 321–331. Wissenschaftliche Verlagsgesellshchaft, Stuttgart.

Chance, B., Sies, H., and Boveris, A. (1979). Hydroperoxide metabolism in mammalian organs. *Physiol. Rev.* 59, 527–605.

Chen, G., Bray, T. M., Janzen, E. G., and McCay, P. B. (1990). Excretion, metabolism and tissue distribution of a spin trapping agent, α-phenyl-N-*tert*-butyl-nitrone (PBN) in rats. *Free Rad. Res. Commun.* 9, 317–323.

Chen, G., Bray, T. M., Janzen, E. G., and McCay, P. B. (1991). The role of mixed function oxidase (MFO) in the metabolism of the spin trapping agent α-phenyl-N-tert-butyl-nitrone (PBN) in rats. *Free Rad. Res. Commun.* 14, 9–16.

Cheng, H.-Y., Liu, T., Feuerstein, G., and Barone, F. C. (1993). Distribution of spin-trapping compounds in rat blood and brain: In vivo microdialysis determination. *Free Rad. Biol. Med.* 14, 243–250.

Clough-Helfman, C., and Phillis, J. W. (1991). The free radical trapping agent N-*tert*-butyl-a-phenylnitrone (PBN) attenuates cerebral ischaemic injury in gerbils. *Free Rad. Res. Commun.* 15, 177–186.

Dawson, V. L., Dawson, T. M., Bartley, D. A., Uhl, G. R., and Snyder, S. H. (1993). Mechanisms of nitric oxide-mediated neurotoxicity in primary brain cultures. *J. Neurosci.* 13, 2651–2661.

Daynes, R. A., Araneo, B. A., Ershler, W. B., Maloney, C., Li, G.-Z., and Ryu, S.-Y. (1993). Altered regulation of IL-6 production with normal aging. *J. Immunol.* 150, 5219–5230.

Drapier, J.-C., and Hibbs, J. B., Jr. (1986). Murine cytotoxic activated macrophages inhibit aconitase in tumor cells. *J. Clin. Invest.* 78, 790–797.

Drapier, J.-C., and Hibbs, J. B., Jr. (1988). Differentiation of murine macrophages to express nonspecific cytotoxicity for tumor cells results in L-arginine-dependent inhibition of mitochondrial iron-sulfur enzymes in the macrophage effector cells. *J. Immunol.* 140, 2829–2838.

Drapier, J.-C., Pellat, C., and Henry, Y. (1991). Generation of EPR-detectable nitrosyl-iron complexes in tumor target cells cocultured with activated macrophages. *J. Biol. Chem.* 266, 10162–10167.

Edamatsu, R. Mori, A., and Packer, L. (1995). The spin-trap N-*tert*-α-phenyl-butylnitrone prolongs the life span of the senescence accelerated mouse. *Biochem. Biophys. Res. Commun.* 211, 847–849.

Ershler, W. B. (1993). Interleukin-6: A cytokine for gerontologists. *J. Am. Geriatr. Soc.* 41, 176–181.

Ershler, W. B., Sun, W. H., Binkley, N., Gravenstein, S., Volk, M. J., Kamoske, G., Klopp, R. G., Roecker, E. B., Daynes, R. A., and Weindruch, R. (1993). Interleukin-6 and aging: Blood levels and mononuclear cell production increase with advancing aging and *in vitro* production is modifiable by dietary restriction. *Lymphokine Cytokine Res.* 12, 225–230.

Ershler, W. B., Sun, W. H., and Binkley, N. (1994). The role of interleukin–6 in certain age-related diseases. *Drugs Aging* 5, 358–365.

Floyd, R. A. (1983). Hydroxyl free-radical spin-adduct in rat brain synaptosomes: Observations on the reduction of the nitroxide. *Biochim. Biophys. Acta* 756, 204–216.

Floyd, R. A. (1990). Role of oxygen free radicals in carcinogenesis and brain ischemia. *FASEB J.* 4, 2587–2597.

Floyd, R. A. (1995). Measurement of oxidative stress in vivo. In "The Oxygen Paradox" (K. J. A. Davies and F. Ursini, eds.), pp. 89–103. CLEUP University Press, Italy.

Floyd, R. A. (1996a). Mitochondrial damage in neurodegenerative disease. In "Free Radicals in Brain Physiology and Disorders" (L. Packer, M. Hiramatsu, and T. Yoshikawa, eds.). Academic Press, Orlando, in press.

Floyd, R. A., and Carney, J. M. (1991). Age influence on oxidative events during brain ischemia/reperfusion. *Arch. Gerontol. Geriatr.* 12, 155–177.

Floyd, R. A., and Carney, J. M. (1992). Free radical damage to protein and DNA: Mechanisms involved and relevant observations on brain undergoing oxidative stress. *Ann. Neurol.* 32, S22–S27.

Floyd, R. A., and Carney, J. M. (1995). Nitrone radicals traps (NRTs) protect in experimental neurodegenerative diseases. *In* "Neuroprotective Approaches to the Treatment of Parkinson's Disease and Other Neurodegenerative Disorders" (C. A. Chapman, C. W. Olanow, P. Jenner, and M. Youssim, eds.), pp. 69–90. Academic Press, London.

Floyd, R. A., Liu, G.-J., and Wong, P. K. (1996). Nitrone radical traps as protectors of oxidative damage in central nervous system. *In* "Handbook of Synthetic Antioxidants" (E. Cadenas and L. Packer, eds.). Dekker, New York, in press.

Folbergrova, J., Zhao, Q., Katsura, K.-I., and Siesjo, B. K. (1995). N-*tert*-butyl-α-phenylnitrone improves recovery of brain energy state in rats following transient focal ischemia. *Proc. Natl. Acad. Sci. U.S.A.* **92**, 5057–5061.

Funahashi, T., Floyd, R. A., and Carney, J. M. (1994). Age effect on brain pH during ischemia/ reperfusion and pH influence on peroxidation. *Neurobiol. Aging* **15**, 161–167.

Granger, D. L., and Lehninger, A. L. (1982). Sites of inhibition of mitochondrial electron transport in macrophage-injured neoplastic cells. *J. Cell Biol.* **95**, 527–535.

Granger, D. L., Taintor, R. R., Cook, J. L., and Hibbs, J. B., Jr. (1980). Injury of neoplastic cells by murine macrophages leads to inhibition of mitochondrial respiration. *J. Clin. Invest.* **65**, 357–370.

Griffin, W. S. T., Stanley, L. C., Ling, C., White, L., MacLeod, V., Perrot, L. J., White, C. L., III, and Araoz, C. (1989). Brain interleukin 1 and S-100 immunoreactivity are elevated in Down syndrome and Alzheimer disease. *Proc. Natl. Acad. Sci. U.S.A.* **86**, 7611–7615.

Hamburger, S. A., and McCay, P. B. (1989). Endotoxin-induced mortality in rats is reduced by nitrones. *Circ. Shock* **29**, 329–334.

Hess, D. T., Patterson, S. I., Smith, D. S., and Skene, J. H. P. (1993). Neuronal growth cone collapse and inhibition of protein fatty acylation by nitric oxide. *Nature* **366**, 562–565.

Hibbs, J. B., Jr., Taintor, R. R., Vavrin, Z., and Rachlin, E. M. (1988). Nitric oxide: A cytotoxic activated macrophage effector molecule. *Biochem. Biophys. Res. Commun.* **157**, 87–94.

Iadecola, C., Zhang, F., and Xu, X. (1995). Inhibition of inducible nitric oxide synthase ameliorates cerebral ischemic damage. *Am. J. Physiol.* **268**, R286–R292.

Iadecola, C., Zhang, F., Xu, S., Casey, R., and Ross, M. E. (1996). Inducible nitric oxide synthase gene expression in brain following cerebral ischemia. *J. Cereb. Blood Flow Metab.* **15**, 378–384.

Janzen, E. G. (1971). Spin trapping. *Acc. Chem. Res.* **4**, 31–40.

Janzen, E. G., and Blackburn, B. J. (1969). Detection and identification of short–lived free radicals by electron spin resonance trapping techniques (spin trapping): Photolysis of organolead, -tin, and -mercury compounds. *J. Am. Chem. Soc.* **91**, 4481–4490.

Janzen, E. G., West, M. S., and Poyer, J. L. (1994). Comparison of antioxidant activity of PBN with hindered phenols in initiated rat liver microsomal lipid peroxidation. *In* "Frontiers of Reactive Oxygen Species in Biology and Medicine" (K. Asada and T. Yoshikawa, eds.), pp. 431–434. Elsevier Science, Amsterdam.

Kotake, Y., Tanigawa, T., Tanigawa, M., and Ueno, I. (1995). Spin trapping isotopically-labelled nitric oxide produced from [^{15}N]L-arginine and [^{17}O]dioxygen by activated macrophages using a water soluble Fe^{++}-dithiocarbamate spin trap. *Free Rad. Res. Commun.* **23**, 287–295.

Kurose, I., Miura, S., Fukumura, D., Yonei, Y. I., Saito, H., Tada, S., Suematsu, M., and Tsuchiya, M. (1993). Nitric oxide mediates Kupffer cell-induced reduction of mitochondrial energization in hepatoma cells: A comparison with oxidative burst. *Cancer Res.* **53**, 2676–2682.

Lancaster, J. R., Jr., and Hibbs, J. B., Jr. (1990). EPR demonstration of iron-nitrosyl complex formation by cytotoxic activated macrophages. *Proc. Natl. Acad. Sci. U.S.A.* **87**, 1223–1227.

Lepoivre, M., Flaman, J.-M., and Henry, Y. (1992). Early loss of the tyrosyl radical in ribonucleotide reductase of adenocarcinoma cells producing nitric oxide. *J. Biol. Chem.* **267**, 22994–23000.

McKechnie, K., Furman, B. L., and Parratt, J. R. (1986). Modification by oxygen free radical scavengers of the metabolic and cardiovascular effects of endotoxin infusion in conscious rats. *Circ. Shock* **19**, 429–439.

Merrill, J. E. (1992). Tumor necrosis factor alpha, interleukin 1 and related cytokines in brain development: Normal and pathological. *Dev. Neurosci.* **14**, 1–10.

Merrill, J. E., and Jonakait, G. M. (1995). Interactions of the nervous and immune systems in development, normal brain homeostasis, and disease. *FASEB J.* **9**, 611–618.

Meyer, M., Schreck, R., and Baeuerle, P. A. (1993). H_2O_2 and antioxidants have opposite effects on activation of NF-KB and AP-1 in intact cells: AP-1 as secondary antioxidant-responsive factor. *EMBO J.* **12**, 2005–2015.

Mitrovic, B., Ignarro, L. J., Montestruque, S., Small, A., and Merrill, J. E. (1994). Nitric oxide as a potential pathological mechanism in demyelination: Its differential effects on primary glial cells in vitro. *Neuroscience* **61**, 575–585.

Miyajima, T., and Kotake, Y. (1995). Spin trapping agent, phenyl N-*tert*-butyl nitrone, inhibits induction of nitric oxide synthase in endotoxin-induced shock in mice. *Biochem. Biophys. Res. Commun.* **215**, 114–121.

Novelli, G. P., Angiolini, P., Consales, G., Lippi, R., and Tani, R. (1986a). Anti-shock action of phenyl-t-butyl-nitrone, a spin trapper. *In* "Oxygen Free Radicals in Shock" (G. P. Novelli and F. Ursini. eds.), pp. 119–124. Karger, Basel Florence.

Novelli, G. P., Anglolini, P., and Tani, R. (1986b). The spin trap phenyl butyl nitrone prevents lethal shock in the rat. *In* "Free Radicals in Liver Injury" (G. Poll, K. H. Cheeseman, M. U. Dianzani, and T. F. Slater, eds.), pp. 225–228. IRL Press Limited, Oxford.

Novelli, G. P., Angiolini, P., Tani, R., Consales, G., and Bordi, L. (1985). Phenyl-T-butyl-nitrone is active against traumatic shock in rats. *Free Rad. Res. Commun.* **1**, 321–327.

Oliver, C. N., Starke-Reed, P. E., Stadtman, E. R., Liu, G. J., Carney, J. M., and Floyd, R. A. (1990). Oxidative damage to brain proteins, loss of glutamine synthetase activity, and production of free radicals during ischemia/reperfusion-induced injury to gerbil brain. *Proc. Natl. Acad. Sci. U.S.A.* **87**, 5144–5147.

Phillis, J. W., and Clough-Helfman, C. (1990a). Protection from cerebral ischemic injury in gerbils with the spin trap agent N-tert-butyl-α-phenylnitrone (PBN). *Neurosci. Lett.* **116**, 315–319.

Phillis, J. W., and Clough-Helfman, C. (1990b). Free radicals and ischaemic brain injury: Protection by the spin trap agent PBN. *Med. Sci. Res.* **18**, 403–404.

Pogrebniak, H. W., Merino, M. J., Hahn, S. M., Mitchell, J. B., and Pass, H. I. (1992). Spin trap salvage from endotoxemia: The role of cytokine down-regulation. *Surgery* **112**, 130–139.

Rothwell, N. J., and Relton, J. K. (1993). Involvement of cytokines in acute neurodegeneration in the CNS. *Neurosci. Biobehav. Rev.* **17**, 217–227.

Schulz, J. B., Henshaw, D. R., Siwek, D., Jenkins, B. G., Ferrante, R. J., Cipolloni, P. B., Kowall, N. W., Rosen, B. R., and Beal, M. F. (1995). Involvement of free radicals in excitotoxicity in vivo. *J. Neurochem.* **64**, 2239–2247.

Sen, S., Goldman, H., Morehead, M., Murphy, S., Phillis, J. W. (1994). α-Phenyl-*tert*-butyl-nitrone inhibits free radical release in brain concussion. *Free Rad. Biol. Med.* **16**, 685–691.

Stadler, J., Billiar, T. R., Curran, R. D., Stuehr, D. J., Ochoa, J. B., and Simmons, L. (1992). Effect of exogenous and endogenous nitric oxide on mitochondrial respiration of rat hepatocytes. *Am. J. Physiol.* **260**, C910–C916.

Sun, W. H., Binkley, N., Bidwell, D. W., and Ershler, W. B. (1993). The influence of recombinant human interleukin-6 on blood and immune parameters in middle-aged and old Rhesus monkeys. *Lymphokine Cytokine Res.* **12**, 449–455.

Takehara, Y., Kanno, T., Yoshioka, T., Inoue, M., and Utsumi, K. (1995). Oxygen-dependent regulation of mitochondrial energy metabolism by nitric oxide. *Arch. Biochem. Biophys.* **323**, 27–32.

Triggs, W. J., and Willmore, L. J. (1984). *In vivo* lipid peroxidation in rat brain following intracortical Fe^{2+} injection. *J. Neurochem.* **42**, 976–980.

Troncoso, J. C., and Costello, A. C. (1995). Metal-catalyzed oxidation of bovine neurofilaments in vitro. *Free Rad. Biol. Med.* **18**, 891–899.

Willmore, L. J., Sypert, G. W., Munson, J. B., and Hurd, R. W. (1978). Chronic focal epileptiform discharges induced by injection of iron into rat and cat cortex. *Science* **200**, 1501–1503.

Yue, T.-L., Gu, J.-L., Lysko, P. G., Cheng, H.-Y. Barone, F. C., and Feuerstein, G. (1992). Neuroprotective effects of phenyl-*t*-butyl-nitrone in gerbil global brain ischemia and in cultured rat cerebellar neurons. *Brain Res.* **574**, 193–197.

Zaleska, M. M., Nagy, K., and Floyd, R. A. (1989). Iron-induced lipid peroxidation and inhibition of dopamine synthesis in striatum synaptosomes. *Neurochem. Res.* **14**, 597–605.

Zhang, J., Dawson, V. L., Dawson, T. M., and Snyder, S. H. (1994). Nitric oxide activation of poly(ADP-ribose) synthetase in neurotoxicity. *Science* **263**, 687–689.

Zhao, Q., Pahlmark, K., Smith, M.-I., and Siesjo, B. K. (1994). Delayed treatment with the spin trap α-phenyl-N-*tert*-butyl nitrone (PBN) reduces infarct size following transient middle cerebral artery occlusion in rats. *Acta Physiol. Scand.* **152**, 349–350.

Judy B. de Haan*
Ernst J. Wolvetang*
Francesca Cristiano*
Rocco Iannello*
Cecile Bladier†,1
Michael J. Kelner‡
Ismail Kola*

*Molecular Genetics and Development Group
Institute of Reproduction and Development
Monash University
Clayton Vic 3168, Australia

†Commissariat a l'Energie Atomique
Centre de Cadarache
Laboratoire de Phytotechnologie
13108 St-Paul-lez-Durance, France

‡Department of Pathology
University of California
San Diego, California 92103

Reactive Oxygen Species and Their Contribution to Pathology in Down Syndrome

I. Introduction

Down syndrome occurs at a rate of 1 in 700–1000 live births, and of all the cytogenetic abnormalities it is the one that most frequently comes to term and is responsible for the greatest number of individuals with mental retardation (Patterson, 1987). In addition to mental retardation, individuals with Down syndrome suffer from a wide range of abnormalities, including congenital heart defects (Rehder, 1981); *in utero* growth retardation, resulting in a reduced birth weight of approximately 10%; increased susceptibility to infection; and a 20- to 50-fold higher incidence of leukemia (Fong and Brodeur, 1987). These individuals show abnormalities of the viscerocranium

1 Present address: Molecular Genetics and Development Group, Institute of Reproduction and Development, Monash University, Clayton Vic 3168, Australia

that results in the characteristic facial features of Down syndrome (Caffey, 1978; Sumarsono *et al.*, 1996) and also display features of premature aging (Tam and Walford, 1980). Furthermore, all individuals develop Alzheimer-type neuronal pathology (Wisniewski *et al.*, 1985) by the third to fourth decade of life.

The involvement of reactive oxygen species (ROS) in some of the Down syndrome pathologies was initiated after assignment of the gene coding for Cu/Zn-superoxide dismutase (also known as Sodl) to chromosome 21 in humans (Tan *et al.*, 1973). All individuals with Down syndrome are either trisomic for the entire chromosome or part thereof (this includes transloca-tion and mosaic forms of Down syndrome). Although the Sodl gene is located just above the minimum critical Down syndrome region (Korenberg *et al.*, 1990), more than 99% of individuals with Down syndrome have an extra copy of the entire chromosome (95% are trisomies and the remainder are mosaics or have translocations of the entire chromosome 21) and are therefore trisomic for Sodl. Indeed, Sodl activity has been shown to be elevated by approximately 50% in red blood cells (Gilles *et al.*, 1976), platelets (Sinet *et al.*, 1975a), lymphocytes and polymorphonuclear granulo-cytes (Feaster *et al.*, 1977), and fibroblasts (Feaster *et al.*, 1977) of individuals with Down syndrome. It is this increased gene dosage for Sodl that has been proposed to contribute to the premature aging and/or mental retardation that occurs as part of the syndrome (Avraham *et al.*, 1991).

In gaining an understanding of how elevated Sodl levels may contribute to various pathologies in Down syndrome, it is important to first delineate the physiological role for Sodl within cells. The superoxide dismutase family of enzymes functions to remove superoxide radicals ($^-O_2\cdot$). These radicals are generated via a number of processes, the most important of which is oxidative metabolism. The process of $^-O_2\cdot$ dismutation generates hydrogen peroxide (H_2O_2), which in turn is removed via glutathione peroxidase (Gpx) and catalase. This two-step process leads to the elimination of H_2O_2 and noxious radicals that would otherwise interact with macromolecules such as DNA, protein, and lipids, thereby affecting their structure and function. However, any perturbation in the balance between the first and the second step of this antioxidant pathway results in a further reaction (known as the Fenton or Harber–Weiss reaction), where hydroxyl radicals ($\cdot OH$) are produced through the interaction of H_2O_2 with transition metals (Imlay *et al.*, 1988). These radicals are the most noxious of the radical species and quickly interact with macromolecules in their immediate vicinity (Badwey and Karnovsky, 1980). In this way, the hydroxyl radical is capable of initiating chains of peroxidation, resulting in damage to biologically impor-tant membranes and organelles.

Down syndrome is one such situation where the balance between the first and the second step of the antioxidant pathway is perturbed. Sodl is elevated as a consequence of gene dosage, whereas the genes coding for selenium-dependent (Gpxl) and independent glutathione peroxidases and catalase are not localized on chromosome 21 (Wieacker *et al.*, 1980; Chada *et al.*, 1990). Thus elevated Sodl activity in Down syndrome could result in the accumulation of H_2O_2, which through the Fenton reaction could lead to a loss of cellular function through damage to macromolecules.

II. Sodl and Gpxl Expression Levels in Down Syndrome _____

To date, limited information is available on the levels of both Sodl and Gpxl expression in Down syndrome tissues and organs. Most data have focused on erythrocytes, lymphoid cells, and fibroblasts derived from individuals with Down syndrome. One study has investigated Sodl and Gpxl activity in brains of Down syndrome-aborted conceptuses (Brooksbank and Balazs, 1984). In all cases, Sodl activity has been shown to be elevated in agreement with gene dosage. Gpxl activity, however, remains unaltered in Down syndrome brains (Brooksbank and Balazs, 1984). However, Gpxl activity is elevated in trisomy 21 erythrocytes (by approximately 40%) and lymphoid cells when compared with controls (Sinet *et al.*, 1975b; Frischer *et al.*, 1981; Neve *et al.*, 1983; Bjorksten *et al.*, 1984). Thus, from limited data, it would appear that the Gpxl gene is differentially regulated in Down syndrome tissues. Indeed, increased Gpxl levels in response to elevated Sodl activity have been termed "adaptive responses" (Ceballos *et al.*, 1988; Kelner and Bagnell, 1990a). Such an elevation in Gpxl levels in response to increased Sodl levels could restore the antioxidant balance, thereby reducing the potential damage to cellular structure and/or function. Thus it becomes important to establish the antioxidant profile in a range of Down syndrome organs and to compare these with age-matched controls.

Five organs (brain, heart, lung, liver, and thymus) were obtained from aborted control and Down syndrome fetuses, and total RNA was extracted. Northern blot analysis was performed on 20 μg of total RNA after hybridization with Sodl, Gpxl, and β-actin probes. Data for Sodl and Gpxl were corrected relative to β-actin, and Sodl and Gpxl expression was calculated for each sample. In most cases (the exception was the Down syndrome liver where a significant decrease in Sodl expression was observed), Sodl expression was elevated in rough agreement with gene dosage (data not shown). In most cases (again the exception was the Down syndrome liver where a significant decrease in Gpxl expression was observed) there was no significant difference in Gpxl expression between control and Down syndrome organs (data not shown). Importantly, all five organs investigated had elevated ratios of Sodl/Gpxl mRNA levels when the ratios were analyzed within individual organs (Fig. 1). Interestingly, even Down syndrome livers had elevated Sodl/Gpxl mRNA levels, as the median decrease in Gpxl mRNA was greater than the median decrease in Sodl mRNA. Thus the Sodl to Gpxl

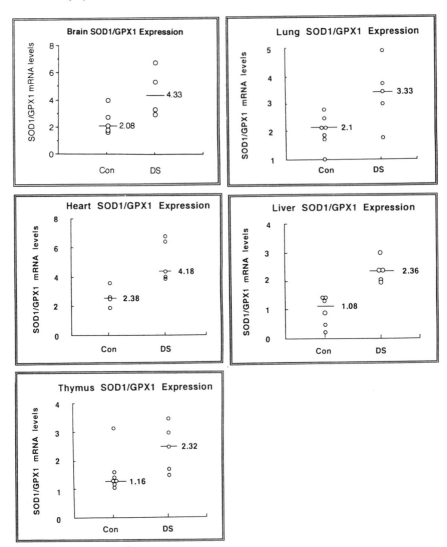

FIGURE I   Analysis of Sod1 and Gpx1 mRNA levels in control and Down syndrome organs. Northern blots (not shown) were densitometrically scanned and the results were corrected relative to β-actin. Median values (horizontal bar) are indicated for both control and Down syndrome groups. All five Down syndrome organs have a statistically significant elevation (approximately 2.0-fold) in Sod1/Gpx1 levels compared with controls ($p < 0.05$ for all five organs tested).

ratio, which is an important determinant of cellular damage, is elevated by approximately twofold in all organs investigated in this study.

The imbalance in the antioxidant enzymatic pathway places the developing Down syndrome fetus in an environment of increased oxidative stress.

Indeed, Brooksbank and Balazs (1984) have shown increased lipid peroxidation in Down syndrome fetal brains. Thus cellular damage may begin accumulating *in utero* in the Down syndrome fetus and continue during the life span of the individual with Down syndrome, resulting in premature aging changes such as the Alzheimer-type neuronal pathology.

III. Premature Aging of Down Syndrome Individuals _____

Premature aging changes in individuals with Down syndrome are evident visually [premature graying or loss of hair (Murdock and Evans, 1978)] as well as in detailed biochemical analysis. Of particular note is the granulovascular degeneration of neurons and the appearance of Alzheimer's disease pathology (senile plaques and neurofibrillary tangles), amyloidosis, hypogonadism, and degenerative vascular disease (Martin, 1977).

To invoke a perturbed antioxidant balance as a contributor to the premature aging changes in Down syndrome, it becomes important to first establish the profile of these enzymes during normal aging. de Haan *et al.* (1992) have shown that the ratio of Sodl to Gpxl (Gpxl or selenium-dependent glutathione peroxidase accounts for more than 90% of the removal of H_2O_2 in murine brains) is significantly elevated at both the mRNA (data not shown) and the protein level (Fig. 2) during aging of murine brains. Furthermore, catalase levels (the other enzyme capable of removing H_2O_2, albeit to a lesser extent) show no significant change with age in murine brains (data not shown) (de Haan *et al.*, 1992). The susceptibility to lipid peroxidation also increases with age in murine brains (Fig. 3). Thus an altered Sodl to Gpxl (and catalase) ratio exists in aging murine brains that is correlated with increased susceptibility to lipid damage.

The Sodl to Gpxl (and catalase) mRNA and enzymatic ratio was also investigated in a range of other murine tissues (Cristiano *et al.*, 1995; de Haan *et al.*, 1995) and their susceptibility to lipid damage with aging was assessed. Interestingly, most other organs show an adaptive rise in both Gpxl activity (Fig. 4) and catalase levels (data not shown) (de Haan *et al.*, 1995) in response to elevated Sodl activity. Indeed, most fail to show enhanced lipid peroxidation with advancing age (Fig. 5). These results imply that increased peroxidative damage is reduced or less in organs where there is an adaptive rise in the enzymes that remove H_2O_2. However, these data do not imply that "adapted" organs do not age. These organs still show constant levels of lipid peroxidation; however, lipid damage does not occur at an accelerated rate in these organs (as seen during aging in the brain).

The notion that an altered ratio of Sodl to Gpxl and catalase activity leads to aging changes is supported by Yarom *et al.* (1988). Their data show neuropathological changes (namely, withdrawal and destruction of some terminal axons and the development of multiple small terminals) in the

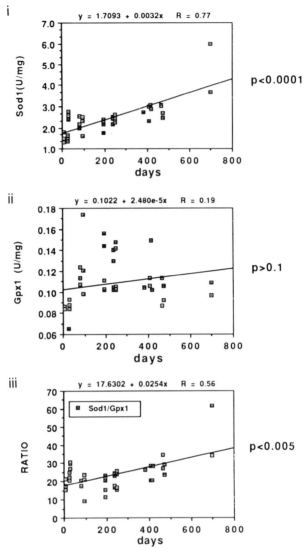

FIGURE 2 Analysis of Sod1 and Gpx1 protein levels in mice brains ranging from 129 to 315 days. (i) Scatter plot of Sod1 activity (U/mg protein) in murine brains versus age. Sod1 activity increases significantly with age ($p < 0.0001$). (ii) Scatter plot of Gpx1 activity in murine brains versus age. Gpx1 activity does not increase significantly with age ($p > 0.1$). (iii) The scatter plot of Sod1/Gpx1 brain activity versus age of mice shows a signficant increase with age ($p < 0.005$). (Reprinted from *Mol. Brain Res.*, **13**, J. B. de Haan *et al.* Cu/Zn superoxide dismutase mRNA and enzyme activity, and susceptibility to lipid peroxidation, increases with aging in murine brains, pp. 179–186. Copyright 1992 with kind permission of Elsevier Science–NL, Sara Burgerhartstraat 25, 1055 KV Amsterdam, The Netherlands.)

STIMULATED LIPID PEROXIDATION

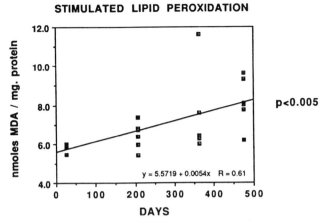

$y = 5.5719 + 0.0054x$ $R = 0.61$

p<0.005

FIGURE 3 Susceptibility to lipid peroxidation in aging murine brains. A statistically significant increase in lipid peroxidation (measured as nmol MDA/mg protein) occurred with age ($p < 0.005$). (Reprinted from *Mol. Brain Res.*, **13**, J. B. de Haan *et al.* Cu/Zn superoxide dismutase mRNA and enzyme activity, and susceptibility to lipid peroxidation, increases with aging in murine brains, pp. 179–186. Copyright 1992 with kind permission of Elsevier Science–NL, Sara Burgerhartstraat 25, 1055 KV Amsterdam, The Netherlands.)

neuromuscular junction of tongue muscles from Sodl transgenic mice, which are similar to those seen in tongue muscles of aging mice (Fahim and Robins, 1982) as well as in individuals with Down syndrome (Yarom *et al.*, 1986, 1987). Interestingly, these changes were observed in transgenic mice that overexpress Sodl by as little as 2-fold (the lowest fold increase obtained in their transgenic mice), thus emphasizing the physiological relevance of a 1.5-fold gene dosage increase in Sodl in Down syndrome. Furthermore, Ceballos *et al.* (1991) have shown enhanced lipid peroxidation in the brains of Sodl transgenic mice.

To directly test the consequences of a perturbation in the balance of antioxidant enzymes and how this contributes to the aging process, the effects of an altered antioxidant balance in murine NIH/3T3 cells were investigated (Kelner *et al.*, 1995; de Haan *et al.*, 1996). Cells were transfected with human Sodl cDNA and cell clones established that had elevated Sodl activity compared with parental controls. Two groups of Sodl transfectants were obtained. One group, representing the majority of clones (97%), had upregulated Gpxl activity in response to increased Sodl activity [the so-called "adaptive response" to elevated Sodl levels (Kelner and Bagnell, 1990a)]. These cells are referred to as "adapted" cells. The 3% that failed to upregulate Gpxl in response to increased Sodl are referred to as "non-adapted" cells. Furthermore, catalase levels are similar in adapted, non-adapted, and parental cells (data not shown). Interestingly, the nonadapted cell clones show many of the characteristic features of senescence in culture (Pendergrass *et al.*, 1989; Smith and Perierra-Smith, 1989; Kumazaki *et al.*,

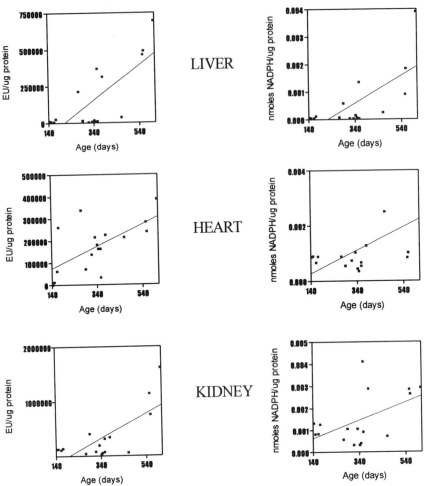

FIGURE 4   Sod1 and Gpx1 activity in a range of aging muring organs. (Left) Sod1 activity during aging in liver, heart, kidney, ovary, and bone. Significant rises were found in Sod1 activity in all of these organs ($p < 0.01$ for all five organs). (Right) Gpx1 activity measured in the same organs. Significant rises in Gpx1 were also found in all of these organs ($p < 0.05$ for all five organs). (Reprinted from *Mech. Ageing Dev.*, 80, Cristiano *et al.*, Changes in the levels of enzymes which modulate the antioxidant balance occur during aging and correlate with cellular damage, pp. 93–105, Copyright 1995, with kind permission from Elsevier Science Ireland Ltd., Bay 15K, Shannan Industrial Estate Co., Clare, Ireland.)

1991) whereas adapted cell clones are virtually identical to the parental cells from which they are derived. This is evident under light microscopy (Fig. 6) where nonadapted cells display altered morphology and size, namely enlarged cell and nucleus volume and increased adherent surface area, whereas adapted cells resemble parental controls. Furthermore, nonadapted cells show slower growth kinetics (Fig. 7) and produce more collagen and

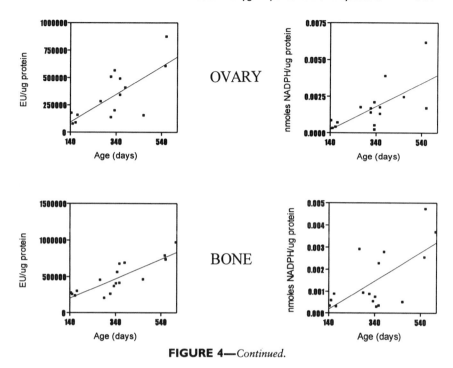

FIGURE 4—*Continued.*

fibronectin (data not shown) (Kelner and Bagnell, 1990b) than adapted cell clones or parental controls. Nonadapted cells are also less responsive to the stimulatory properties of growth factors (data not shown). Nonadapted cells also have higher levels of Cipl (a biochemical marker of senescence; Noda *et al.*, 1994) than adapted cells or parental controls (Fig. 8). These data therefore clearly show that an altered ratio of Sodl to Gpxl (and catalase) activity results in senescent features in cultured cells.

An altered Sodl to Gpxl (and catalase) activity ratio leads to the accumulation of H_2O_2, the intermediate product of the first and second step of the antioxidant pathway. Indeed, an analysis of the H_2O_2 produced and secreted by nonadapted cell clones demonstrates that these cells have higher intracellular levels and secrete higher levels of H_2O_2 than either adapted or parental cells (data not shown) (de Haan *et al.*, 1996). Increased levels of H_2O_2, via the Fenton reaction, can lead to the production of $\cdot OH$ and resultant damage to macromolecules as detailed previously. In this way senescent changes can be brought about through damage to important organelles, e.g., mitochondria and to DNA itself.

If an elevated level of H_2O_2, produced as a result of an altered ratio of Sodl to Gpxl (and catalase), is an important determinant of cellular aging, then administration of H_2O_2 per se to cells should induce many of the features of senescence seen in the NIH/3T3 nonadapted cell clones. Exposure of NIH/3T3 cells and primary diploid fibroblast cells (CCD45, data not

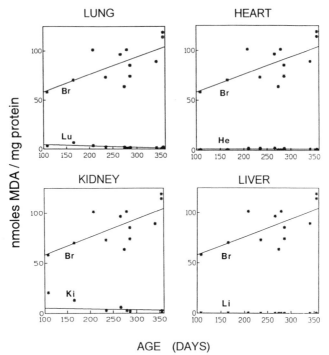

AGE   (DAYS)

FIGURE 5   Susceptibility to lipid peroxidation in a range of aging murine organs. None showed a statistically significant linear relationship of lipid peroxidation with age ($p > 0.1$ for all four organs). Lipid peroxidation in aging murine brains (Br) has been included on each graph as a comparison.

shown) (de Haan *et al.*, 1996) to increasing concentrations of H_2O_2 inhibited cell proliferation in a dose-dependent fashion (Fig. 9a) and induced changes in cell morphology (Fig. 9b) and stereology (data not shown) (de Haan *et al.*, 1996) in a manner analogous to the nonadapted Sodl transfectants. Furthermore, Cipl mRNA levels were elevated in NIH/3T3 cells after exposure to 150 μm H_2O_2 for 5 hr (Fig. 8, lanes 1 and 2) and 10 and 30μm H_2O_2 for 24 hr (Fig. 10). Interestingly, an analysis of cells derived from children with Down syndrome showed that these cells also display features of senescence compared with age-matched controls, namely (i) slower cell growth (Fig. 11a), (ii) altered cell morphology and stereology (data not shown) (de Haan *et al.*, 1996), and (iii) increased expression of Cipl (Fig. 11b). From these data it is clear that Down syndrome cells (where an alteration in Sodl to Gpxl exists as a consequence of gene dosage, see Fig. 1) behave similarly to both cells transfected with Sodl (and therefore having elevated Sodl to Gpxl activity) and cells treated with H_2O_2. Premature aging changes, as seen in Down syndrome, may therefore result from an altered

(i)

(ii)

(iii)

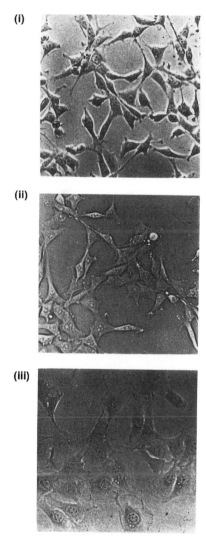

FIGURE 6 Cellular morphology of (i) control (C), (ii) adapted (AD), and (iii) nonadapted (NON-AD) Sod1-transfected cell lines (× 250). NON-AD cells have larger cell size, are more transluscent, and have larger nuclei volume. [Reprinted with permission of Oxford University Press from de Haan *et al.* (1996). Elevation in the ratio of Cu/Zn-superoxide dismutase to glutathione peroxidase activity induces features of cellular senescence and this effect is mediated by hydrogen peroxide. *Hum. Mol. Genet.* 5, 283–292.

ratio of Sodl to Gpxl and catalase activity through the production of increased H_2O_2 levels.

It is noteworthy that Cipl levels (a biochemical marker for senescence) are elevated in cells with an altered Sodl to Gpxl and catalase activity ratio

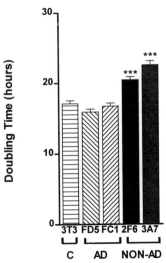

FIGURE 7   Cell proliferation of control (C), adapted (AD), and nonadapted (NON-AD) Sod1-transfected cell lines. Doubling times of AD, NON-AD, and parental cells lines were determined from nine replicates for each respective cell line. NON-AD cells have significantly elevated cell doubling times compared with AD and control cell lines. *** $p < 0.001$. [Reprinted with permission of Oxford University Press from de Haan *et al.* (1996). Elevation in the ratio of Cu/Zn-superoxide dismutase to glutathione peroxidase activity induces features of cellular senescence and this effect is mediated by hydrogen peroxide. *Hum. Mol. Genet.* 5, 283–292.

(e.g., nonadapted NIH/3T3 cells) and cells treated with H_2O_2 (NIH/3T3 cells). The Cipl protein (p21) is a cyclin-dependent kinase (Cdk) inhibitor (Harper *et al.*, 1993) that forms inhibitory complexes with many cyclins and their associated Cdks (Xiong *et al.*, 1993). In doing so, p21 inhibits the passage of cells through specific phases of the cell cycle, e.g., at G_1/S and G_2/M checkpoints. Our data may shed light on how H_2O_2 acts to induce some of the features of senescence. By upregulating Cipl in this way, cells are arrested at various points in the cell cycle; this could explain the slower growth kinetics observed. Furthermore, we investigated the levels of Cipl in adult murine brains and compared these with Cipl levels in neonatal and embryonic brains. As shown previously (de Haan *et al.*, 1992), a statistically significant linear increase in the ratio of Sodl to Gpx1 occurs in aging murine brains (see Fig. 2). Cipl mRNA levels are higher in the brains of adult mice compared with neonatal and/or embryonic mice, both by Northern and dot blot analysis (Figs. 12a and 12b). These data provide further evidence that Cipl mRNA levels are elevated in cells that have higher levels of Sodl relative to Gpx1 activity.

Our data therefore show that Cipl mRNA levels are elevated in cell systems where an alteration in the antioxidant balance exists (as a consequence of gene dosage in Down syndrome and during aging in murine brains), in nonadapted cells that have and secrete higher levels of H_2O_2, and

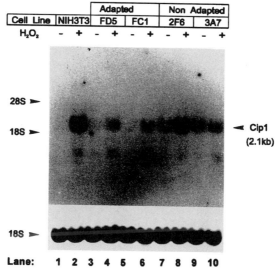

Cell Line	NIH3T3	Adapted		Non Adapted	
		FD5	FC1	2F6	3A7
H₂O₂	- +	- +	- +	- +	- +

28S ►

18S ►   ◄ Cip1 (2.1kb)

18S ►

Lane: 1 2 3 4 5 6 7 8 9 10

FIGURE 8   Northern blot analysis of Cip1 mRNA levels in parental, adapted, and non-adapted Sod1 transgenic cell lines. Nonadapted cells show higher levels of constitutive Cip1 expression (lanes 7 and 9) than control (lane 1) or adapted cells (lanes 3 and 5). Cip1 mRNA levels are induced in cell lines treated with 150 μM H₂O₂ for 5 hr compared with untreated counterparts (NIH/3T3 cells, lane 2; FD5, lane 4; FC1, lane 6; and 2F6, lane 8), except in 3A7. [Reprinted with permission of Oxford University Press from de Haan *et al.* (1996). Elevation in the ratio of Cu/Zn-superoxide dismutase to glutathione peroxidase activity induces features of cellular sensescence and this effect is mediated by hydrogen peroxide. *Hum. Mol. Genet.* 5, 283–292.

in cells treated with H₂O₂. Our data strongly suggest that this effect is being mediated through H₂O₂. However, the exact mechanism(s) whereby elevated H₂O₂ levels induce Cipl mRNA is unknown and remains to be elucidated. H₂O₂ is known to affect the expression of certain genes, e.g., AP-1 and fos (Abate *et al.*, 1990), and has been proposed to act as a messenger molecule in the activation of the NF-κB transcription factor (de Haan *et al.*, 1995; Schmidt *et al.*, 1995). H₂O₂ may be acting in a similar manner to upregulate Cipl (and/or other genes, e.g., fibronectin and collagen whose expression is altered during senescence). Understanding how H₂O₂ leads to altered gene expression may prove useful in developing strategies that prevent H₂O₂ production and/or scavenge H₂O₂. In this way it may be possible to slow or even reverse the aging process.

IV. Apoptosis and Down Syndrome

Apoptosis is a highly regulated form of cell death characterized by cell shrinkage, nuclear condensation, fragmentation of nuclear DNA between the internucleosomal breaks, and disintegration of the cell into apoptotic

b

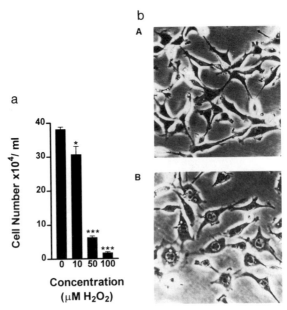

a

FIGURE 9   Proliferation and morphological changes after treatment with H_2O_2. (a) Proliferation of cells after treatment with increasing concentrations of H_2O_2. The histogram shows the number of viable cells (mean $\pm$ SEM) after a 24-hr culture in the presence of increasing concentrations of H_2O_2 ($n = 3$). (b) Morphological changes in cells after treatment with H_2O_2. Photomicrographs show the effects of 25 μM H_2O_2 on the morphology of NIH/3T3 cells after 15 hr (X250). (A) Untreated NIH/3T3 cells, and (B) H_2O_2-treated cells. H_2O_2-treated cells are larger, have bigger nuclei, and have a different shape from control cells. [Reprinted with permission of Oxford University Press from de Haan *et al.* (1996). Elevation in the ratio of Cu/Zn-superoxide dismutase to glutathione peroxidase activity induces features of cellular sensescence and this effect is mediated by hydrogen peroxide. *Hum. Mol. Genet.* 5, 283–292.

bodies (Kerr *et al.*, 1972; Wyllie, 1980). Both vertebrates and invertebrates use apoptosis for removing potential rogue or superfluous cells (Ellis and Horvitz, 1986). Apoptosis therefore plays a key role in negative selection in the thymus and in morphogenesis during development (Oppenheim, 1991; Raff *et al.*, 1993).

Evidence that oxidative stress plays a role in apoptosis comes from Allsopp *et al.* (1993). These authors found that the expression of Bcl-2 protects neurons from oxidative stress and makes them less growth factor dependent. Bcl-2 is the prototype of a family of proteins localized to intracellular membranes which homo- and heterodimerize to regulate apoptosis (Oltvai *et al.*, 1993). Indeed, the protooncogene Bcl-2 is able to prevent or inhibit apoptosis in most experimental systems and is commonly deregulated or mutated in many malignancies (Hockenbery *et al.*, 1991; Hockenbery, 1994). Although the exact mechanism of action of Bcl-2 is not known, it is clear that Bcl-2 inhibits apoptosis by modulating the flux of intracellular ROS. To this end, it has been suggested that Bcl-2 has peroxidase-like

H$_2$O$_2$ (μM)   0 10 30

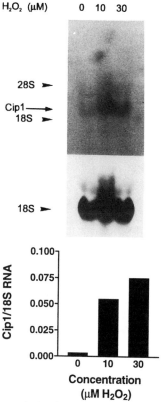

28S ➤

Cip1 ➤
18S ➤

18S ➤

FIGURE 10 Northern blot analysis of the effect of increasing concentrations of H$_2$O$_2$ on Cip1 expression in NIH/3T3 cells. The amount of Cip1 (relative to 18S rRNA) in control and H$_2$O$_2$-treated cells is shown at the bottom of the figure. Cip1 mRNA levels increase in a dose-dependent fashion after exposure of cells to 10 and 30 μM H$_2$O$_2$ for 24 hr. [Reprinted with permission of Oxford University Press from de Haan *et al.* (1996). Elevation in the ratio of Cu/Zn-superoxide dismutase to glutathione peroxidase activity induces features of cellular senescence and this effect is mediated by hydrogen peroxide. *Hum. Mol. Genet.* 5, 283–292.

activity and functions in the antioxidant pathway to prevent apoptosis (Hockenbery *et al.*, 1993; Kane *et al.*, 1993). Further evidence for this notion comes from Bcl-2 knockout mice that display polycystic kidney disease, thymic hypoplasia, and hair hypopigmentation, consistent with oxidative stress-mediated cell death (Veis *et al.*, 1993). Furthermore, oxidative stress-induced apoptosis has also been found to occur in the brain of patients with a form of amyotropic lateral sclerosis caused by point mutations in Cu/Zn-SOD (Deng *et al.*, 1993; Rosen *et al.*, 1993). Similarly, in Alzheimer's disease, oxidative stress leads to the aberrant expression of the β-amyloid protein and plaque formation in the brain which triggers apoptosis in neurons both *in vitro* (Loo *et al.*, 1993) and *in vivo* (Takashima *et al.*, 1993). In Parkinson's

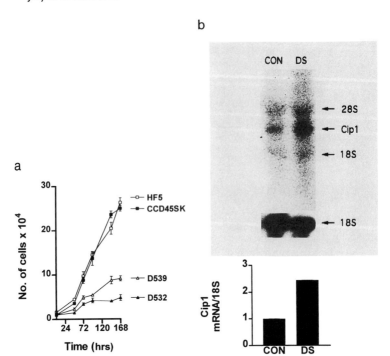

FIGURE 11   (a) Growth curves of Down syndrome and control cell lines. Each point represents the mean ± SEM for three replicates. Down syndrome cell lines (△, D539; ▲, D532) grow slower than control cell lines (□, HF5; ■, CCD45) over a 168-hr time period in culture. (b) Northern blot analysis of Cip1 mRNA levels in fibroblasts of a Down syndrome cell line compared with a control line. The amount of Cip1 (relative to 18S rRNA) in Down syndrome and control cells is shown at the bottom of the figure (data have been normalized relative to the controls). Down syndrome fibroblasts show increased levels of Cipl mRNA compared with control cells. [Reprinted with permission of Oxford University Press from de Haan *et al.* (1996). Elevation in the ratio of Cu/Zn-superoxide dismutase to glutathione peroxidase activity induces features of cellular senescence and this effect is mediated by hydrogen peroxide. *Hum. Mol. Genet.* 5, 283–292.

disease the destruction of the substantia nigra is caused by oxidative stress-induced apoptosis. Indeed, MPTP, a drug which induces Parkinson's disease, triggers apoptosis of the substantia nigra (Dipasquale *et al.*, 1991; Mitchell *et al.*, 1994). Antioxidants such as *N*-acetylcysteine effectively inhibit or prevent apoptotic cell death in the previously mentioned cases (Mayer and Noble, 1994). However, strong evidence for a role of ROS in apoptosis comes from Peled-Kamar *et al.* (1995) who investigated apoptosis in mice overexpressing Sodl. They found that the thymus of Sodl transgenic mice was smaller due to increased apoptosis and that T cells displayed increased hydrogen and lipid peroxide production and were more susceptible to apoptosis when challenged with lipopolysaccharides (LPS).

In order to test this, the previously described mouse NIH/3T3 fibroblast cells were treated with increasing concentrations of hydrogen peroxide (up

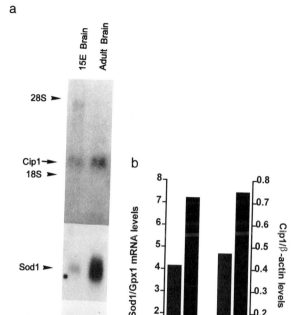

FIGURE 12   Analysis of Cip1 and Sod1 mRNA levels in brains derived from 15-day embryonic (15E), 1-day neonatal (neo), and adult (Ad) mice. (a) Northern blots of total RNA extracted from embryonic and adult brains showing elevated Sod1 and Cip1 mRNA levels in adult compared with embryonic brains. (b) Histogram depicting the Sod1/Gpx1 mRNA ratio and the Cip1/β-actin mRNA ratio in neonatal and adult mouse brains, as determined by dot blot analysis. Both the Sod1/Gpx1 and the Cip1/β-actin mRNA levels are elevated in adult murine brains compared with neonatal brains.

to 400 μM) and were assayed for apoptosis by nuclear morphology and FACS analysis after 48 hr. As shown in Fig. 13a, approximately 60% of the nonadapted 3A7 cell nuclei showed the condensed fragmented appearance of apoptosis after treatment with 400 μM H_2O_2. At this concentration, both the adapted cells and the control cells hardly displayed any apoptosis. Furthermore, Fig. 13b shows that the addition of redox cyclers dimethoxybenzoquinone (DMBQ) and menadione (data not shown), which favor mitochondrial superoxide formation and thus hydrogen peroxide formation, only induced apoptosis in the nonadapted 3A7 cell line. Based on these results and the pathology of the Sod1 transgenic mice (Peled-Kamar *et al.*, 1995), some of the neurodegenerative symptoms and thymic destruction in Down syndrome may be due to the oxidative stress caused by the imbalance in the Sod1/Gpx1 ratio, which in turn triggers apoptosis directly or confers an increased susceptibility of cells to apoptotic stimuli. Indeed, this notion is

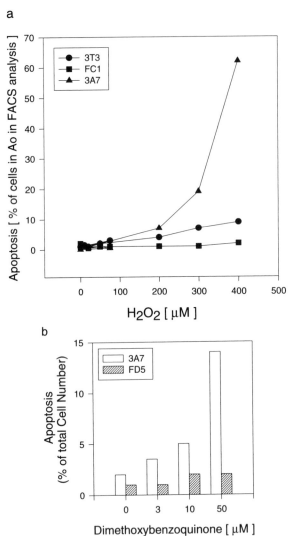

a

b

FIGURE 13 Approximately 1×10^5 cells were incubated in the presence of the indicated amounts of (a) hydrogen peroxide and (b) dimethoxybenzoquinone. After 48 hr, supernatants of the cell cultures were centrifuged to collect the detached apoptotic bodies. (a) Cell pellets were resuspended, pooled with the harvested cells, and fixed in EtOH for 24 hr at $-20°C$. After centrifugation, cells were stained with 20 μg/ml propidium iodide in the presence of 0.4 mg RNAse for 1 hr. Routine FACS analysis for cell cycle distribution shows the condensed apoptotic bodies in the Ao region just before the G_1 peak. The results are expressed as the amount of cells in this region compared with the total number of cells analyzed. (b) An aliquot of the cell suspension was cytospun onto polylysine-coated slides, fixed for 10 min in ethanol/acetic acid, washed once in phosphate-buffered saline, and stained with the DNA-binding dye DAPI. The amount of condensed fragmented apoptotic nuclei was scored under fluorescence microscopy and expressed as the percentage of the total number of nuclei on the slide.

strengthened by data of Busciglio and Yanker (1995), who show that cortical neurons from Down syndrome fetal brains have higher levels of H_2O_2, degenerate, and undergo apoptosis in culture when compared with age-matched control brains. Furthermore, the degeneration of Down syndrome neurons was prevented by treatment with free radical scavengers or catalase, suggesting that apoptosis was mediated by elevated H_2O_2 in these Down syndrome neurons.

Although the extra gene load of Sodl has been shown to be an important determinant in some of the pathologies associated with Down syndrome (Groner *et al.*, 1990), it is not the only gene on chromosome 21 that contributes to the syndrome. It has been shown that mice which overexpress the Ets-2 transcription factor display many of the typical bone and thymic abnormalities observed in Down syndrome (Sumarsono *et al.*, 1996). Preliminary evidence suggests that this is also due to an increased apoptotic index in the affected organs (results not shown). It is tempting to speculate that Ets-2 overexpression downregulates Bcl-2 expression through three negative ets-binding regulatory elements present in the Bcl-2 promotor (Chen and Boxer, 1995), thereby promoting apoptosis.

V. Conclusion and Discussion

Our data support the notion that ROS are involved in the premature aging of individuals with Down syndrome. The levels of Sodl and Gpx1 (the major enzyme involved in the removal of H_2O_2) have been investigated in Down syndrome fetal tissues, and a disproportionate increase in the ratio of Sodl/Gpx1 mRNA has been found in all five organs examined. It has been shown that a disproportionate increase in Sodl/Gpxl activity exists in murine brains as they age and that this is correlated with an increase in lipid damage with age. However, the strongest evidence that an altered antioxidant balance may contribute to changes associated with aging comes from data in cell lines, where elevated Sodl activity is achieved through transfection with human Sodl cDNA. Cells with an altered ratio of Sodl to Gpxl and catalase activity display many features of senescent cells in culture. Interestingly, those cells where Gpxl activity was upregulated in an adaptive response to elevated Sodl levels were morphologically and biochemically similar to the parental cells. These cells seemed to "naturally rescue" the aging phenotype.

Data from our study and that of others (Amstad *et al.*, 1991) support the notion that it is the balance in antioxidant enzymes that is an important determinant of cellular senescence and/or apoptosis. Cells overexpressing Sodl as a consequence of gene dosage (e.g., Down syndrome) or transfection with extra copies of Sodl (our nonadapted NIH/3T3 cells), as well as cells treated with H_2O_2, all display many features of senescence and/or apoptosis.

The fact that H_2O_2 per se is capable of inducing many of the features of senescence and apoptosis indicates that this molecule may be acting as an initiator of the senescent and/or apoptotic phenotype. It is already established that H_2O_2 acts as a messenger molecule in the activation of the NF-κB transcription factor (Schreck et al., 1991; de Haan et al., 1995). By affecting the binding and/or availability of transcription factors, H_2O_2 may affect the expression of senescent genes. Indeed this study has shown that a number of "senescent" genes are affected by H_2O_2 (e.g., Cipl, fibronection, and collagen).

Data from this study also indicate that a delicate antioxidant enzyme balance exists in cells. Too much Sod relative to either Gpx or catalase results in the accumulation of H_2O_2, which in turn, through the Fenton reaction, leads to the production of ·OH and resultant cellular damage. However, too little Sod enzyme is also not favorable because $-O_2$· radicals in themselves are toxic to cells (Tamura et al., 1992). Therefore fine-tuning of the antioxidant enzymes [together with the nonenzymatic antioxidants, e.g., vitamins and other small molecules that quench free radicals (Aruoma and Halliwell, 1987)] becomes imperative if the cell is to function successfully in an oxygen-rich environment. Down syndrome is one situation where the antioxidant balance is affected due to gene dosage. The developing fetus therefore has to cope with a highly pro-oxidant environment. Damage to biologically important macromolecules results due to the inability to prevent oxidative interactions. This accumulated macromolecular damage may be partly responsible for some of the abnormalities that are seen as part of the syndrome. Furthermore, the activation of transcription factors such as NF-κB, AP-1, and fos (Abate et al., 1990; Schreck et al., 1991) by H_2O_2 allows speculation that the expression of genes not residing on chromosome 21 may be altered and, in this way, may contribute to some of the pathologies in Down syndrome.

Acknowledgments

This work was funded in part by SmithKline Beecham Pharmaceuticals, the National Health and Medical Research Council of Australia (NH and MRC), and an Apex Foundation grant to JBdH. CB was supported by a grant from the French Minister of National Education, Higher Education, Research, and Employment. MJK was supported by grants from the National Institute of Environmental Health Sciences (ESO4989), National Cancer Institute (CA52310), and National Institute of Aging (AG10746).

References

Abate, C., Patel, L., Rauscher, F. J., III, and Curren, T. (1990). Redox regulation of Fos and Jun DNA-binding activity in vitro. Science **249**, 1157–1161.

Allsopp, T. E., Wyatt, S., Paterson, H. F., and Davies, A. M. (1993). The proto-oncogene bcl-2 can selectively rescue neurotrophic factor-dependent neurons from apoptosis. *Cell* **73**, 295–307.

Amstad, P., Peskin, A., Shah, G., Mirault, M. E., Moret, R., Zbinden I., and Cerutti, P. (1991). The balance between Cu,Zn-superoxide dismutase and catalase affects the sensitivity of mouse epidermal cells to oxidative stress. *Biochemistry* **30**, 9305–9313.

Aruoma, O. I., and Halliwell, B. (1987). Superoxide-dependent and ascorbate-dependent formation of hydroxyl radicals from hydrogen peroxide in the presence of iron: Are lactoferrin and transferrin promoters of hydroxyl-radical generation? *Biochem. J.* **241**, 273–278.

Avraham, K. B., Sugerman, H., Rotshenker, S., and Groner, Y. (1991). Down's syndrome: Morphological remodelling and increased complexity in the neuromuscular junction of transgenic CuZn-superoxide dismutase mice. *J. Neurocytol.* **20**, 208–215.

Badwey, J. A., and Karnovsky, M. L. (1980). Active oxygen species and the functions of phagocytic leukocytes. *Annu. Rev. Biochem.* **49**, 695–726.

Bjorksten, B., Marklund, S., and Hagglof, B. (1984). Enzymes of leucocyte oxidative metabolism in Down's syndrome. *Acta Paediatr. Scand.* **73**, 97–101.

Brooksbank, B. W. L., and Balazs, R. (1984). Superoxide dismutase, glutathione peroxidase and lipoperoxidation in Down's syndrome fetal brain. *Brain Res.* **318**, 37–44.

Busciglio, J., and Yanker, B. A. (1995). Apoptosis and increased generation of reactive oxygen species in Down's syndrome neurons *in vitro*. *Nature* **378**, 776–779.

Caffey, J. (1978). In "Paediatric X-Ray Diagnosis," Vol. 1, 7th Ed., pp. 155–157. Yearbook Medical Publishers Inc., Chicago.

Ceballos, I., Delabar, J. M., Nicole, A., Lynch, R. E., Hallewell, R. A., Kamoun, P., and Sinet, P. M. (1988). Expression of transfected human CuZn superoxide dismutase gene in mouse L cells and NS20Y neuroblastoma cells induces enhancement of glutathione peroxidase activity. *Biochim. Biophys. Acta* **949**, 58–64.

Ceballos, I., Nicole, A., Briand, P., Grimber, G., Delacourte, A., Flament, S., Blouin, J. L., Thevenin, M., Kamoun, P., and Sinet, P. M. (1991). Expression of human Cu-Zn superoxide dismutase gene in transgenic mice: Model for gene dosage effect in Down syndrome. *Free Rad. Res. Commun.* **12–13**, 581–589.

Chada, S., Le Beau, M. M., Casey, L., and Newburger, P. E. (1990). Isolation and chromosomal localization of the human glutathione peroxidase gene. *Genomics* **6**, 268–271.

Chen, H.-M., and Boxer, L. M. (1995). Il1-binding sites are negative regulators of bcl-2 expression in pre-B cells. *Mol. Cell. Biol.* **15**, 3840–3847.

Cristiano, F., de Haan, J. B., Iannello, R., and Kola, I. (1995). Changes in the levels of enzymes which modulate the antioxidant balance occur during aging and correlate with cellular damage. *Mech. Ageing Dev.* **80**, 93–105.

de Haan, J. B., Cristiano, F., Iannello R. C., Bladier, C., Kelner, M. J., and Kola, I. (1996). Elevation in the ratio of Cu/Zn-superoxide dismutase to glutathione peroxidase activity induces features of cellular senescence and this effect is mediated by hydrogen peroxide. *Hum. Mol. Genet.* **5**, 283–292.

de Haan, J. B., Cristiano, F., Iannelio, R. C., and Kola, I. (1995). Cu/Zn-superoxide dismutase and glutathione peroxidase during aging. *Biochem. Mol. Biol. Int.* **35**, 1281–1297.

de Haan, J. B., Newman, J. D., and Kola I. (1992). Cu/Zn superoxide dismutase mRNA and enzyme activity, and susceptibility to lipid peroxidation, increases with aging in murine brains. *Mol. Brain Res.* **13**, 179–186.

Deng, H.-X., Hentati, A., Tainer, J. A., Iqbal, Z., Cayabyab, A., Hung, W. Y., Getzoff, E. D., Hu, P., Herzfeldt, B., and Roos, R. P. (1993). Amyotrophic lateral sclerosis and structural defects in Cu,Zn superoxide dismutase. *Science* **261**, 1047.

Dipasquale, B., Marini, A. M., and Youle, R. J. (1991). Apoptosis and DNA degradation induced by 1-methyl-4-phenylpyridinium in neurons. *Biochem. Biophys. Res. Commun.* **181**, 1442–1448.

Ellis, H. M., and Horvitz, H. R. (1986). Genetic control of programmed cell death in the nematode C. elegans. Cell **44**, 817–829.

Fahim, M. A., and Robins, N. (1982). Ultrastructural studies of young and old mouse-neuromuscular junctions. *J. Neurocytol.* **11**, 641–656.

Feaster, W. W., Kwok, L. W., and Epstein, C. (1977). Dosage effects for superoxide dismutase-1 in nucleated cells aneuploid for chromosome 21. *Am. J. Hum. Genet.* **29**, 563–570.

Fong, C., and Brodeur, G. M., (1987). Down's syndrome and leukemia: Epidemiology, genetics, cytogenetics and mechanisms of leukomogenesis. *Cancer Genet. Cytogenet.* **28**, 55–76.

Frischer, H., Chu, L. K., Ahmad, T., Justice, P., and Smith, G. F. (1981). *In* "The Red Cell: Fifth Ann Arbor Conference" (G. J. Brewer, ed.), pp. 269–283. Alan R. Liss, New York.

Gilles, L., Ferradini, C., Foos, J., Pucheault, J., Allard, D., Sinet, P. M., and Jerome, H. (1976). The estimation of red blood cell superoxide dismutase activity by pulse radiolysis in normal and trisomic cells. *Hum. Genet.* **31**, 197–202.

Groner, Y., Elroy-Stein, O., Avraham, K. B., Yarom, R., Schickler, M., Knobler, H., and Rotman, G. (1990). Down syndrome clinical symptoms are manifested in transfected cells and transgenic mice overexpressing the human Cu/Zn-superoxide dismutase gene. *J. Physiol.–Paris* **84**, 53–77.

Harper, J. W., Adami, G. R., Wei, N., Keyomarsi, K., and Elledge, S. J. (1993). The p21 Cdk-interacting protein is a potent inhibitor of G1 cyclin-dependent kinases. *Cell* **75**, 805–816.

Hockenbery, D. M. (1994). bcl-2 in cancer, development and apoptosis. *J. Cell Sci.* **18**, 51–55.

Hockenbery, D. M., Oltvai, Z. N., Yin, X-M., Milliman, C. L., and Korsmeyer, S. J. (1993). Bcl-2 functions in an anti-oxidant pathway to prevent apoptosis. *Cell* **75**, 241–251

Hockenbery, D. M., Zutter, M., Hickey, W., Nahm, M., and Korsmeyer, S. J. (1991). Bcl-2 protein is topographically restricted in tissues characterised by apoptotic cell death. *Proc. Natl. Acad. Sci. U.S.A.* **88**, 6961–6965.

Imlay, J. A., Chin, S. M., and Linn, S. (1988). Toxic DNA damage by hydrogen peroxide through the Fenton reaction in vivo and in vitro. *Science* **240**, 640–642.

Kane, D. J., Sarafian, T. A., Anton, R., Hahn, H., Gralla-Butler, E., Selverstone-Valentine, J., Ord, T., and Bredesen, D. E. (1993). Bcl-2 inhibition of neural death: Decreased generation of reactive oxygen species. *Science* **262**, 1274–1277.

Kelner, M. J., and Bagnell, R. (1990a). Alteration of endogenous glutathione peroxidase, manganese superoxide dismutase, and glutathione transferase activity in cells transfected with a copper-zinc superoxide dismutase expression vector: Explanation for variations in paraquat resistance. *J. Biol. Chem.* **265**, 10872–10875.

Kelner, M. J., and Bagnell, R. (1990b). Alteration of growth rate and fibronectin by imbalances in superoxide dismutase and glutathione peroxidase activity. *Biol. React. Intermed.* **IV**, 305–309.

Kelner, M. J., Bagnell, R., Montoya, M., Estes, L., Uglik, S. F., and Cerutti, P. (1995). Transfection with human copper-zinc superoxide dismutase induces bidirectional alterations in other antioxidant enzymes, proteins, growth factor response, and paraquat resistence. *Free Rad. Biol. Med.* **18**, 497–506.

Kerr, J. F. R., Wyllie, A. H., and Currie, A. R. (1972). Apoptosis: A basic biological phenomenon with wide-ranging implications in tissue kinetics. *Br. J. Cancer* **26**, 239–257.

Korenberg, J. R., Kawashima, H., Pulst, S. M., Ikeuchi, T., Ogasawara, N., Yamamoto, K., Schonberg, S. A., West, R, Allen, L., Magenis, E., Ikawa K., Taniguchi, N., and Epstein, C. J. (1990). Molecular definition of a region of chromosome 21 that causes features of Down syndrome phenotype. *Am. J. Hum. Genet.* **47**, 236–246.

Kumazaki, T., Robetorye, R. S., Robetorye, S. C., and Smith, J. R. (1991). Fibronectin expression increases during in vitro cellular senescence: Correlation with increased cell area. *Exp. Cell Res.* **195**, 13–19.

Loo, D. T., Copani, A., Pike, C. J., Whitemore, E. R., Walencewicz, A. J., and Cotman, C. W. (1993). Apoptosis is induced by beta-amyloid in cultured central nervous system neurons. *Proc. Natl. Acad. Sci. U.S.A.* **90**, 7951–7955.

Martin, G. M. (1977). Genetic syndromes in man with potential relevence to the pathology of aging. *In* "Genetic Effects of Aging" (E. L. Schneider, ed.). National Foundation, March of Dimes.

Mayer, M., and Noble, M (1994). N-Acetyl cysteine is a pluripotent protector against cell death and enhancer of trophic factor-mediated cell survival *in vitro. Proc. Natl. Acad. Sci. U.S.A.* **91**, 7496–7500.

Mitchell, I. J., Lawson, S., Moser, B., Laidlaw, S. M., Cooper, A. J., Walkinshaw, G., and Waters, C. M. (1994). Glutamate-induced apoptosis results in a loss of striated neurons in the parkisonian rat. *Neuroscience* **63**, 1–5.

Murdock, J. C., and Evans, J. H. (1978). An objective *in vitro* study of ageing in the skin of patients with Down's syndrome. *J. Ment. Defic. Res.* **22**, 131–135.

Neve, J., Sinet, P. M., Molle, L., and Nicole, A. (1983). Selenium, zinc and copper levels in Down's syndrome (trisomy 21): Blood levels and relations with glutathione peroxidase and superoxide dismutase. *Clin. Chim. Acta* **133**, 209–214.

Noda, A., Ning, Y., Venable, S. F., Pereira-Smith, O. M., and Smith, J. R. (1994). Cloning of senescent cell-derived inhibitors of DNA synthesis using an expression screen. *Exp. Cell Biol.* **211**, 90–98.

Oltvai, Z. N., Milliman, C. L., and Korsmeyer, S. J. (1993). Bcl-2 heterodimerizes *in vivo* with a conserved homolog, Bax, that accelerates programmed cell death. *Cell* **74**, 609–619.

Oppenheim, R. W. (1991). Cell death during development of the nervous system *Annu. Rev. Neurosci.* **14**, 453–501.

Patterson, D. H. (1987). The causes of Down syndrome. *Sci. Am.* **257**, 42–49.

Peled-Kamar, M., Lotem, J., Okon, E., Sachs, L., and Groner, Y. (1995). Thymic abnormalities and enhanced apoptosis of thymocytes and bone marrow cells in transgenic mice overexpressing Cu/Zn-superoxide dismutase: Implications for Down syndrome. *EMBO J.* **14**, 4985–4993.

Pendergrass, W., Angello, J., and Norwood, T. H. (1989). The relationship between cell size, the activity of DNA polymerase alpha and proliferative activity in human diploid fibroblast-like cell cultures. *Exp. Gerontol.* **24**, 383–393.

Raff, M. C., Barres, B. A., Burne, J. F., Coles, H. S., Ishizake, Y., and Jacobson, M. D. (1993). Programmed cell death and the control of cell survival: Lessons from the nervous system. *Science* **262**, 695–700.

Rehder, H. (1981). Pathology of trisomy 21, with particular reference to persistent common atrioventricular canal of the heart. *In* "Trisomy 21: An International Symposium" (G. R. Burgio, M. Fraccaro, L. Tiepolo, and U. Wolf, eds.), pp. 57–73. Springer-Verlag, Berlin.

Rosen, D. R., Siddique, T., Patterson, D., Figlewicz, D. A., Sapp, P., Hentati, A., Donaldson, D., Goto, J., O'Regan, J. P., Deng, D. X., Rahmani, Z., Krizus, A., McKenna-Yasek, D., Cayabyab, A., Gaston, S. M., Berger, R., Tanzi, R. E., Halperrin, J. J., Hertzfeldt, B., Van den Bergh, R., Hung, W. Y., Bird, T., Deng, G., Mulder, D. W., Smyth, C., Laing, N. G., Soriano, E., Pericak-Vance, M. A., Haines, J., Rouleau, G. A., Gusella, J. S., Horvitz, H. R., and Brown, R. H., Jr. (1993). Mutations in Cu/Zn superoxide dismutase gene are associated with familial amyotrophic lateral sclerosis. *Nature* **362**, 59–62.

Schmidt, K. N., Amstad P., Cerutti, P., and Baeuerle, P. A. (1995). The roles of hydrogen peroxide and superoxide as messengers in the activation of transcription factor NF-κB. *Chem. Biol.* **2**, 13–22.

Schreck, R., Rieber, P., and Baeuerle, P. A. (1991). Reactive oxygen intermediates as apparently widely used messengers in the activation of the NF-kappa B transcription factor and HIV-1. *EMBO J.* **10**, 2247–2258.

Sinet, P. M., Michelson, A. M., Bazin, A., Lejeune, J., and Jerome, H. (1975a). Increase in glutathione peroxidase activity in erythrocytes from trisomy 21 subjects. *Biochem. Biophys. Res. Commun.* **67**, 910–915.

Sinet, P. M., Michelson, A. M., Bazin, A., Lejeune, J., and Jerome, H. (1975b). Superoxide dismutase activities of blood platelets in trisomy 21. *Biochem. Biophys. Res. Commun.* **67**, 904–909.

Smith, J. R., and Perierra-Smith, O. M. (1989). Altered gene expression during cellular aging. *Genome* **31**, 386–389.

Sumarsono, S. H., Wilson, T. J., Tymms, M. J., Venter, D., Corrick, C. M., Kola, R., Lahoud, M., Papas, T. S., Seth, A., and Kola, I (1996). Ets transgenic mice develop skeletal abnormalities analogous to those found in Down syndrome. *Nature* **379**, 534–537.

Takashima, A., Noguchi, K., Sato, K., Hoshino, T., and Imahori, K. (1993). Tau protein kinase I is essential for amyloid beta-protein-induced neurotoxicity. *Proc. Natl. Acad. Sci. U.S.A.* **90**, 7789–7793.

Tam, C. F., and Walford, R. L. (1980). Alteration in cyclic nucleotides and cyclase-specific activites in T lymphocytes of aging in normal humans and patients with Down's syndrome. *J. Immunol.* **125**, 1665–1670.

Tamura, K., Manabe, T., Andoh, K., Kyogoku, T., Oshio, G., and Tobe, T. (1992). Effect of intraarterial active oxygen species on the rat pancreas. *Hepatogastroenterology* **39**, 152–157.

Tan, Y. H., Tischfield, J., and Ruddle, F. H. (1973). The linkage of genes for the human interferon-induced antiviral protein and indophenol oxidase B traits to chromosome 21. *J. Exp. Med.* **137**, 317–330.

Veis, D. J., Sorenson, C. M., Shutter, J. R., and Korsmeyer, S. J. (1993). Bcl-2-deficient mice demonstrate fulminant lymphoid apoptosis, polycystic kidneys, and hypopigmented hair. *Cell* **75**, 229–240.

Wieacker, P., Mueller, D. R., Mayerova, A., Grzeschik, K. H., and Ropers, H. H. (1980). Assignment of the gene coding for human catalase to the short arm of chromosome 11. *Ann. Genet.* **23**, 73–77.

Wisniewski, K. E., Wisniewski, H. M., and Wen, G. Y. (1985). Occurrence of neuropathological changes and dementia of Alzheimer's disease in Down's syndrome. *Ann. Neurol.* **17**, 278–282.

Wyllie, A. H. (1980). Glucocorticoid-induced thymocyte apoptosis is associated with endogenous endonuclease activation. *Nature* **284**, 555–556.

Xiong, Y., Hannon, G. J., Zhang, H., Casso, D., Kobayashi, R., and Beach, D. (1993). p21 is a universal inhibitor of cyclin kinases. *Nature* **366**, 701–704.

Yarom, R., Sagher, U., Havivi, Y., Peled, I. J., and Wexler, M. R. (1986). Myofibres in tongues of Down's syndrome. *J. Neurol. Sci.* **73**, 279–287.

Yarom, R., Sherman, Y., Sagher, U., Peled, I. J., and Wexler, M. R. (1987). Elevated concentrations of elements and abnormalities of neuromuscular junctions in tongue muscles of Down's syndrome. *J. Neurol. Sci.* **88**, 41–53.

Yarom, R., Sapoznikov, D., Havivi, Y., Avraham, K. B., Schickler, M., and Groner, Y. (1988). Premature aging changes in neuromuscular junctions of transgenic mice with extra human CuZnSOD gene: A model for tongue pathology in Down's syndrome. *J. Neurol. Sci.* **88**, 41–53.

Paul G. Winyard
David R. Blake

Inflammation Research Group
St. Bartholomew's and the Royal London School of Medicine and Dentistry
London E1 1AD, United Kingdom

Antioxidants, Redox-Regulated Transcription Factors, and Inflammation

This chapter discusses evidence that many antioxidant compounds with anti-inflammatory properties are more likely to be acting through their effects on inflammatory cellular messenger systems than by direct prevention of free radical-mediated damage to biomolecules such as proteins. To exemplify this theme, this chapter concentrates on the role of reactive oxygen intermediates (ROI) and reactive nitrogen intermediates (RNI) in mediating the activation of the transcription factor, nuclear factor-κB (NF-κB), as this protein appears to play a central role in the inflammatory cellular response and has been an area of intensive research activity.

I. Background

A revolution in our understanding of the role of ROI/RNI in human inflammatory diseases is underway. During the 1980s, interest in the role

403

of free radicals in diseases with an inflammatory component, such as rheumatoid arthritis, inflammatory bowel disease, pulmonary emphysema, atherosclerosis, and neurodegenerative disorders, grew rapidly. Most research into the role of free radicals in inflammation was directed toward the putative role of ROI as damaging agents. Many studies centered on the oxidative modification of biomolecules such as α_1-antitrypsin, cartilage proteoglycans and collagen, low-density lipoprotein, DNA, lipids, immunoglobulin G, and so on (reviewed by Halliwell and Gutteridge, 1985; Merry *et al.*, 1989). The cytotoxicity of free radicals and their reaction products was also investigated, particularly in relation to endothelial cell function (reviewed by Kus *et al.*, 1995).

Thus, free radicals were studied within the context of a direct, destructive role in biology. This may be true for high concentrations of certain ROI/RNI, particularly in relation to damage to biomolecules that exist as a single or as very few copies, a prime example being DNA (Williams, 1985; Wiseman and Halliwell, 1996). However, recent evidence suggests that some ROI/RNI play physiologically important, cellular messenger roles at relatively low concentrations (Saran and Bors, 1989; Snyder, 1995).

In the course of the present decade, researchers have increasingly recognized this cellular control aspect of free radical action. Endothelium-derived relaxing factor (EDRF) has been identified as nitric oxide (NO·) (reviewed by Snyder, 1995), and ROI can activate apoptosis, a *programmed* form of cell death, at least in some circumstances (Jacobson, 1996). ROI appear to activate certain protein kinases involved in cellular signal transduction, such as protein kinase C (see, e.g., Brawn *et al.*, 1995; Whisler *et al.*, 1995) and mitogen-activated protein (MAP) kinase (Fialkow *et al.*, 1994). ROI also activate, or suppress, various protein phosphatases (Keyse and Emslie, 1992; Whisler *et al.*, 1995).

Furthermore, ROI have been implicated in the regulation of mammalian transcription factors such as nuclear factor (NF)-κB and activator protein-1 (AP-1) (Schreck *et al.*, 1992; Amstad *et al.*, 1992; Nose *et al.*, 1991) and certain "heat shock," or stress protein, transcription factors (HSTF; Morimoto, 1993). In these cases, ROI activity plays a critical role in regulating the expression of genes that encode either proteins with potential proinflammatory actions [such as tumor necrosis factor (TNF)], or proteins that may act in a protective fashion such as hsp (heat shock protein) 32, also known as heme oxygenase (Keyse and Tyrrell, 1990; Willis *et al.*, 1996).

In addition, the expression of genes that modulate the activity of transcription factors may, in turn, be induced by oxidative stress. An example is gadd153 (a gene induced by growth arrest and DNA damage; Fornace *et al.*, 1989), the murine product of which has be shown to bind to NF-IL6 (also known as C/EBPβ) and form a heterodimer that cannot bind to DNA (Ron and Habener, 1992). NF-IL6 upregulates the expression of many genes, including those encoding interleukins 6 and 8. Other examples of oxidative stress-inducible genes are gadd45, which regulates the effects of the transcription factor, p53, in growth arrest (Bae *et al.*, 1996), adapt15 (Crawford *et al.*, 1996), and early growth response-1 (egr-1) gene (Nose *et al.*, 1991). Fibroblast proliferation and collagen gene expression, associated with the late wound-healing phase of inflammation, have also been linked to intracellular ROI activity (Murrell *et al.*, 1990; Houglum *et al.*, 1994).

Perhaps the key to future antioxidant drug development for the treatment of human inflammatory diseases lies in our ability to modulate the previously mentioned ROI-mediated control pathways when they become perturbed as part of the inflammatory process.

The products of free radical reactions may also regulate gene expression, e.g., minimally modified LDL induces the activation of NF-κB (Liao *et al.*, 1993), which has been detected in human atherosclerotic plaques (Brand *et al.*, 1995). Atherosclerosis is a disease in which there is reasonably good evidence that lipoprotein oxidation plays a role. However, the effect of minimally oxidized LDL on NF-κB activation suggests that, ultimately, it is the effects of oxidation products on cell responses that is, perhaps, just as important in disease progression (Kus *et al.*, 1995).

The *inappropriate* activation of the responses referred to earlier, such as overexpression of particular genes, excess NO· production, and apoptosis, may be involved in disease progression. Nevertheless, it should be remembered that these processes, when properly controlled, are physiological. Therefore, ROI/RNI can no longer be regarded solely as damaging species whose complete elimination by antioxidants is bound to have beneficial effects on human health. Indeed, it can be argued that the converse may be the case.

In the case of treatment of human inflammatory diseases, it may be necessary to define the stage of the disease at which a particular antioxidant therapy may most effectively be used, e.g., during acute inflammatory episodes, or in chronic inflammation, or prophylaxis. Another factor that must be taken into account is the individual variation in sensitivity within the human population (Bashir *et al.*, 1992; Harris *et al.*, 1994), as well as the different susceptibilities to oxidative stress of different organs (Fraga *et al.*, 1990), tissues (Kristal *et al.*, 1994), and subcellular locations (Richter *et al.*, 1988).

The capacity of endogenous intracellular antioxidants or of repair mechanisms for oxidative DNA damage might determine the susceptibility of different cell populations to the induction of specific gene expression, killing, or even mutation. An increased susceptibility to the actions of ROI on target lymphocytes of patients with autoimmune disorders could thus be an important mechanism in the pathogenesis of diseases such as rheumatoid arthritis by inducing excessive inflammatory responses and increased clonal proliferation of autoreactive lymphocytes via the activation of transcription factors like NF-κB. Heliovaara *et al.* (1994) have suggested that lowered serum concentrations of vitamin E, β-carotene, and selenium (required for the antioxidant enzyme glutathione peroxidase) together may be a risk factor within the healthy population for the subsequent development of rheumatoid arthritis in humans. Thus, a lowered antioxidant status might interact with a genetically determined susceptibility to oxidative stress in the pathogenesis of this disease. It has also been found that the concentration of vitamin E was lowered in synovial fluid from inflamed joints when compared to matched plasma samples (Fairburn et al., 1992), even if corrected for cholesterol concentration. Despite this, there is still no convincing evidence from the clinical studies conducted thus far that supplementation with antioxidant nutrients can influence the processes of ongoing joint inflammation (Kus et al., 1995).

These considerations pose a challenge in the search for appropriate antioxidant therapies in human inflammatory diseases. We suggest that the successful therapeutic manipulation of cellular responses by antioxidant drugs will necessitate the maintenance of a critical balance between free radical activity and antioxidant status within the relevant target cells. It may be necessary to deliver antioxidants selectively to specific cell types, or even defined subcellular locations, and within a narrow "concentration window" that blocks "inappropriate" cellular responses, but leaves the physiological levels of free radical activity necessary for normal cell function. The consumption of vast amounts of antioxidant nutrients by the healthy population, in an effort to prevent chronic diseases may be detrimental: adverse effects of antioxidant vitamin supplementation have been reported in humans (α-Tocopherol, β-Carotene Cancer Prevention Study Group, 1994; Rowe, 1996; see Halliwell, 1994).

II. Two Redox-Controlled Transcription Factors: AP-1 and NF-κB

A key feature of inflammatory diseases such as rheumatoid arthritis is the increased expression of certain genes that encode "inflammatory" proteins (such as cytokines) and proteinases involved in tissue destruction, such as collagenases, gelatinases, and stromelysins. In turn, an important characteristic of gene expression is the control of gene transcription by specific proteins, transcription factors, which bind to short DNA sequence elements located adjacent to the promotor or in enhancer regions of genes. Once bound to DNA, transcription factors interact with each other and with the proteins of the transcriptional apparatus itself (e.g., RNA polymerase) to regulate gene expression. Considerable interest has been generated in the idea that transcription factors may be useful targets for novel therapeutic strategies in the treatment of human diseases, including inflammatory diseases (McKay et al., 1994).

The two transcription factors, AP-1 and NF-κB, are influenced by cellular redox status (reviews: Baeuerle and Henkel, 1994; Blake *et al.*, 1994) and have been implicated in the transcriptional regulation of a wide range of genes involved in cellular inflammatory responses and tissue destruction.

AP-1 is a protein dimer composed of the protooncogene products Fos and Jun. mRNA levels for c-*jun* and c-*fos* are strongly induced in response to hydrogen peroxide, ultraviolet light, and ionizing radiation in both fibroblasts and T cells (Amstad *et al.*, 1992). In contrast, AP-1 binding activity is only weakly induced by hydrogen peroxide (Meyer *et al.*, 1993).

Treatment with antioxidants such as pyrrolidine dithiocarbamate activates AP-1, and it has been suggested that the AP-1 DNA-binding site is an antioxidant response element (Meyer *et al.*, 1993). The AP-1 site, also referred to as the tetradeconyl phorbol acetate responsive element (TRE), is found in various genes, including those encoding human collagenase, stromelysin, transforming growth factors (TGFs) α and β, and IL-2 (Karin, 1991). The matrix metalloproteinases collagenase and stromelysin are thought to play a key part in cartilage destruction and in the inactivation of proteinase inhibitors in inflammatory joint disease (Gravallese *et al.*, 1991; Winyard *et al.*, 1991) and, possibly, in atherosclerosis (Henney *et al.*, 1991). TGFβ induces the expression of collagen type I and III genes, downregulates collagenase production, and induces the tissue inhibitor of metalloproteinases-1 (TIMP-1), a physiological inhibitor of stromelysin and collagenase (Edwards *et al.*, 1987).

The target genes for NF-κB comprise a growing list of genes intrinsically linked to a coordinated inflammatory response. These include genes encoding TNF-α, interleukin-1, interleukin-6, interleukin-8, the interleukin-2 receptor α chain, inducible nitric oxide synthase (iNOS), MHC class I antigens, E-selectin, vascular cell adhesion molecule-1, serum amyloid A precursor, and c-Myc (Baeuerle and Henkel, 1994). A variety of apparently unrelated anti-rheumatic and anti-inflammatory drugs have been proposed to exert their effects through the inhibition of NF-κB activation. This aspect is discussed later.

The DNA-binding, nuclear form of NF-κB is a protein heterodimer made up of one Rel-A (p65) subunit and one p50 subunit. In nonstimulated cells, NF-κB exists in an inactive, cytosolic form bound to its inhibitor, IκB. Activators of NF-κB (such as TNF-α, IL-1, phorbol esters, viruses, lipopolysaccharide, calcium ionophores, cycloheximide, and ionizing radiation) induce the dissociation of IκB from the NF-κB–IκB complex, and positively charged nuclear location sequences (NLS) in Rel-A and p50 are unmasked. NF-κB is then translocated to the nucleus, where it controls gene expression.

The importance of ROI in the expression of the genes coding for these proteins was highlighted in the seminal work of Baeuerle and colleagues (Schreck *et al.*, 1991). They showed that hydrogen peroxide could induce

NF-κB activity in a human T-cell line. This effect was blocked by the antioxidant N-acetylcysteine. Other reported activators include oxidized LDL (Liao et al., 1993) and nitric oxide (endothelial-derived relaxing factor; Lander et al., 1993), although the latter effect is disputed (see Schreck et al., 1992).

In addition to N-acetylcysteine, other antioxidants such as pyrrolidine dithiocarbamate, diethyl dithiocarbamate, 2-mercaptoethanol, ortho-phenanthroline, and desferrioxamine are able to inhibit the activation of NF-κB by the recognized stimuli of this transcription factor mentioned earlier (Schreck et al., 1992). In addition, α-tocopherol has been reported to suppress NF-κB activation (see Schreck et al., 1992). However, some of the effects of α-tocopherol on cellular signaling may not be due solely, or at all, to an antioxidant effect, as α-tocopherol has been shown to inhibit protein kinase C activity via a mechanism that does not involve the antioxidant action of α-tocopherol (Tasinato, et al., 1995). Many clinically useful anti-rheumatic compounds contain thiol groups (e.g., D-penicillamine and gold compounds such as aurothioglucose and aurothiomalate; Drury et al., 1984) and some of these have been shown to inhibit NF-κB activation, as is discussed later.

IκB dissociation from NF-κB involves its phosphorylation-controlled proteolytic degradation (Traenckner et al., 1994; Brown et al., 1995; Di-Donato et al., 1995), whereas ROI appear to control this IκB phosphorylation. This explains why NF-κB activation is blocked by a host of antioxidants. The precise mechanism by which ROI control the phosphorylation of IκB is unclear (Traenckner et al., 1994). ROI might activate an IκB kinase or suppress a phosphatase activity. Certainly, ROI have been shown to activate protein kinase C (PKC), as mentioned earlier. In addition, PKC was capable of catalyzing the phosphorylation of IκB within a preparation of the NF-κB–IκB complex that had been partially purified from cytosolic extracts (Ghosh and Baltimore, 1990). However, ROI-mediated NF-κB activation within living cells does not involve the direct phosphorylation of IκB by PKC: The reagent phorbol 12-myristate 13-acetate (PMA) activates NF-κB (via PKC activation) in many cell types, and this effect is blocked by antioxidants, including pyrrolidine dithiocarbamate. However, the latter compound dose not affect PKC activation by PMA, suggesting that antioxidants act on a further step between PKC activation and IκB phosphorylation (Baeuerle and Henkel, 1994). Thus, the IκB kinase remains to be identified.

Experiments using specific inhibitors of different proteinase classes indicate that the intracellular proteinase activity responsible for the degradation of the phosphorylated form of IκB is a chymotrypsin-like proteinase activity associated with the so-called multicatalytic proteinase complex, or proteasome, of the cell cytosol (Baeuerle and Henkel, 1994; Traenckner et al., 1994; DiDonato et al., 1995). Calpains do not appear to be involved.

Another important control step in NF-κB activation appears to involve the redox-sensitive molecule thioredoxin (Yodoi and Uchiyama, 1992). Human thioredoxin, or adult T-cell leukemia-derived factor (ADF), is a 13-KDa protein with a strong reducing activity through its dithiol group. In a human T-cell line (Jurkat), this protein is induced by oxidative stress (e.g., hydrogen peroxide) and, *in vitro*, human thioredoxin redox regulates the DNA-binding activity of NF-κB, apparently *via* the chemical reduction of a cysteine 62 residue, which is critical for DNA binding (Ghosh *et al.*, 1995; Muller *et al.*, 1995), within the p50 subunit of NF-κB (Matthews *et al.*, 1992). Regulation of DNA-binding activity by thioredoxin has also been reported in the case of AP-1 (Abate *et al.*, 1990). Interestingly, thioredoxin has been implicated in activating the glucocorticoid receptor to a steroid-binding state (Grippo *et al.*, 1985).

In order to explain the observations that activation of NF-κB can occur in response to both ROI and thioredoxin, Hayashi *et al.*, (1993) pointed out that ROI induce the expression of thioredoxin, which may, in turn, activate the binding of NF-κB. Perhaps a *transient* flux of ROI is required for IκB phosphorylation (the step that is inhibited by antioxidants), whereas thioredoxin is required (at a later time point after the stimulus) for the efficient binding of NF-κB to nuclear DNA.

III. ROI/RNI and NF-κB in Human Diseases

The activation of NF-κB has been implicated in a wide range of human diseases in which there is an inflammatory component and in which ROI/ RNI have been implicated. These include AIDS, where the expression of HIV is NF-κB dependent and inhibited by N-acetylcysteine (Staal *et al.*, 1990; Toledano and Leonard, 1991); atheroclerosis, where minimally modified LDL may be an activator of NF-κB (Liao *et al.*, 1993); rheumatoid arthritis (Handel *et al.*, 1995; Marok *et al.*, 1996); osteoporosis (Stein and Yang, 1995); asthma (Barnes, 1995); renal disease (Satriano and Schlondorff, 1994); Alzheimer's disease, where β-amyloid may be an activator of NF-κB (Behl *et al.*, 1994); ischemic–reperfusion injury in the brain (Kaltschmidt *et al.*, 1993); and the animal model of multiple sclerosis, experimental allergic encephalomyelitis (Kaltschmidt *et al.*, 1994).

Because of the range of inflammatory genes that are induced by activated NF-κB, it has been suggested that the activation of this transcription factor by ROI, produced as an early event, could play a critical role in inflammatory reactions (Kaltschmidt *et al.*, 1994). As indicated earlier, some of the cytokines whose genes are switched on by activated NF-κB, such as TNF-α and IL-1, are themselves activators of NF-κB, giving the potential for a positive feedback cycle within the inflammatory response. Even if the initial cellular stimulus for NF-κB activation is not ROI/RNI, the apparent merging of all the pathways of signal transduction involving NF-κB on a ROI-dependent step suggests a possible therapeutic target.

IV. Ischemic–Reperfusion Injury, Rheumatoid Arthritis, and Redox-Regulated Transcription Factors _____

Although a range of agents can stimulate the DNA-binding activity of these transcription factors, AP-1 and NF-κB have been suggested to respond to cellular redox status in a "yin–yang" fashion (Meyer *et al.*, 1993). For example, during hypoxia in a cultured tumor cell line (HeLa), AP-1 was strongly induced in a cycloheximide-sensitive response, i.e., new protein synthesis was required. The response was biphasic, with different members of the Jun/Fos family of transcription factors being found at each of the two maxima of AP-1 activation. In contrast, the preexisting, non-DNA-binding, NF-κB protein was not activated to a DNA-binding form under hypoxic conditions. However, it was rapidly and strongly activated when cells were reoxygenated (Rupec and Baeuerle, 1995). It was suggested that the proliferative response to hypoxia of these tumor cells may be largely mediated by AP-1, whereas the induction of inflammatory genes in hypoxic reperfusion may be primarily mediated by NF-κB.

It seems feasible that the AP-1–NF-κB "yin–yang" relationship that occurs in tissue hypoxia/reperfusion may be relevant to the dual proliferative/inflammatory response characteristic of rheumatoid synovitis, which has sometimes been referred to as "tumor-like proliferation" (Fassbender, 1984). The chronic state of the rheumatoid joint is both hypoxic and acidotic (Merry *et al.*, 1989). In cells cultured from the synovial membrane, the intracellular redox potential is lowered relative to normal cells. Thus in rheumatoid synovitis, 88% of the sulfur-containing amino acids are reduced, in contrast to 63% in control synoviocytes (Butcher *et al.*, 1973).

However, evidence has shown that hypoxic reperfusion injury occurs in the inflamed human joint (reviewed by Merry *et al.*, 1989). Joint movement in patients with rheumatoid arthritis produces intraarticular pressures in excess of the synovial capillary perfusion pressure. This phenomenon does not occur in normal joints, where the pressure remains subatmospheric throughout a movement cycle. During exercise of the inflamed joint, the intraarticular pressure is transmitted directly to the synovial membrane vasculature, producing occlusion of the superficial synovial capillary bed and ischemia. Reperfusion of the synovial membrane occurs when exercise is stopped and, as a consequence, ROI are generated by a mechanism involving the enzyme, xanthine oxidase (see Section VI).

The chronic reducing environment of the rheumatoid joint may contribute to the activation of AP-1 and genes under its control. At the same time, a superimposed, intermittent generation of H_2O_2 by the dismutation of $O_2^{\cdot-}$, produced in the inflamed joint by hypoxia/reperfusion cycling, might ensure the persistent expression of the NF-κB-dependent genes.

An immunohistochemical and *in situ* hybridization study of synovial tissue from rheumatoid arthritis patients showed numerous cells expressing *jun*-B and c-*fos* (the genes encoding AP-1 subunits; Kinne *et al.*, 1995). These positive cells were within the lining layer and diffuse infiltrates of the synovial membrane and were identified as fibroblast like, using cell-specific markers. Expression of *jun*-B/c-*fos* was not detected in lymphocytes or macrophage-like cells, whereas the number of cells that were positive was considerably smaller in osteoarthritis and healthy control synovia. This cellular distribution of Jun/Fos within the rheumatoid synovium contrasts with that seen for the p50 and p65 subunits of NF-κB in a immunohistochemical study by Handel *et al.* (1995). These authors found that p50/p65 was present in the nuclei of macrophage-like cells of the synovial lining and sublining areas, as well as in endothelial cells. There was little staining in normal control synovium samples.

Antibodies specific for the DNA-binding, nuclear form of NF-κB have been used to perform immunohistochemical studies in the rheumatoid arthritis and osteoarthritis synovium (Marok *et al.*, 1996). The basis of such "activity specific" antibodies is that they were raised against the NLS of Rel-A. Binding of IκB to the p50–Rel-A heterodimer sterically masks the NLS (Kaltschmidt *et al.*, 1995; see also earlier). In agreement with Handel *et al.* (1995), it was found that both vascular endothelial cells and macrophage-like cells in the synovium of rheumatoid arthritis patients show the presence of the activated form of NF-κB. Again, staining was nuclear, as would be expected, and synovial tissue from controls exhibited either no staining or only weak staining. The relative amounts of staining in these two cell types was also dependent on the chronicity of the disease.

In our study, staining for activated NF-κB was not associated with synovial lymphocyte aggregates. This result is consistent with previous observations of an apparent lack of recent activation of T lymphocytes within the rheumatoid synovium. Thus, whereas synovial T cells have been implicated in the pathology of rheumatoid arthritis, T-cell-derived cytokines and other markers of recent T-cell activation, such as the IL-2 receptor, have only been demonstrated at low levels in the rheumatoid joint (Salmon and Gaston, 1995).

V. NF-κB and Anti-inflammatory Drugs

The anti-inflammatory action of a wide range of well-known drugs, including sodium salicylate and aspirin (acetylsalicylic acid; Kopp and Ghosh, 1994) and gold(I) thiolate compounds such as aurothioglucose (Yang *et al.*, 1995), has now been attributed to the inhibition of NF-κB activation. The DNA-binding activity of NF-κB is Zn^{2+} dependent. Yang *et al.* (1995) suggested that cysteine residues may be involved in the binding of Zn^{2+}. Thus, in the case of aurothioglucose, Yang *et al.* (1995) proposed that Au(I) oxidizes the Zn^{2+}-associated thiolate anions to disulfides, thereby preventing the binding of NF-κB to DNA.

It has been suggested that glucocorticoids exert their immunosuppressive activity by inducing the synthesis of IκB (Auphan *et al.*, 1995; Scheinman *et al.*, 1995), which blocks the NF-κB-mediated expression of the genes described earlier. Another mechanism that may contribute to immunosuppresion involves the direct interaction of the glucocorticoid–receptor complex with NF-κB, thereby preventing its association with DNA [see references in Scheinman *et al*, (1995) and Auphan *et al*, (1995)]. A similar situation has been shown for the estrogen receptor (Stein and Yang, 1995). This receptor physically interacts with NF-κB and another transcription factor, C/EBPβ (NF-IL6). Both NF-κB and C/EBPβ regulate IL-6 gene expression in human osteoblasts. Their interaction with the estrogen receptor results in inhibition of the IL-6 promotor. This estrogen effect may be involved in the bone resorption associated with osteoporosis (the most common form is postmenopausal in women) and, possibly, in the marked preponderance of rheumatoid arthritis in females.

As discussed earlier, another important NF-κB-regulated gene is that encoding the cytokine, TNF-α. It also was mentioned that ROI are involved in TNF-induced activation of NF-κB, whereas ROI have also been shown to play a role in TNF-induced apoptosis (Larrick and Wright, 1990). TNF appears to play a central role in several inflammatory conditions, including rheumatoid arthritis (Chernajovsky *et al.*, 1995) and septic shock (Suitters *et al.*, 1994). Indeed, a marked feature of the pathology of the TNF transgenic mouse is synovitis (Keffer *et al.*, 1991). Clinical studies have shown that intravenous administration of antibodies to TNF produce dramatic anti-inflammatory effects in rheumatoid arthritis patients (Chernajovsky *et al.*, 1995).

VI. What Is the Cellular Source of ROI Involved in NF-κB Activation?

Many potential sources of ROI/RNI may be involved in inflammatory cell signaling. These include NADPH oxidase, nitric oxide synthases, cyclooxygenases and lipoxygenases, the mitochondrial respiratory chain, and xanthine oxidase. Because these systems have been extensively discussed elsewhere in the context of their radical-generating properties (see some of the references given in the following paragraphs), they are only briefly described here in relation to their possible roles in cell signaling.

The plasma membrane-bound NADPH oxidase of polymorphonuclear leukocytes (PMNs) contains cytochrome b.$_{245}$ and catalyzes the univalent reduction of molecular oxygen to generate the superoxide anion radical ($O_2 \cdot^-$; Forman and Thomas, 1986; Segal, 1991). In addition to PMNs, other inflammatory cells, e.g., lymphocytes and macrophages, possess a membrane NADPH oxidase (Hancock *et al.*, 1989), as do some noninflammatory cell types such as mesangial cells (see Satriano and Schlondorff, 1994). Indeed, in glomerular mesangial cells, the NADPH oxidase inhibitor 4'-hydroxy-3'-methoxyacetophenone (apocynin) inhibited the activation of NF-κB in response to TNF-α or aggregated IgG (Satriano and Schlondorff, 1994).

The free radical nitric oxide (NO·) has been shown to be an important factor in bioregulation, initially by the demonstration of its role as an endothelium-derived relaxing factor (EDRF) and subsequently by the observation of constitutive NO·-synthase enzymes in numerous other cell types and tissues (Snyder, 1995). In addition to the generation of NO· by the constitutive enzymes, NO· is generated from arginine by macrophages, neutrophils, and lymphocytes via the action of a calcium-independent, cytokine-inducible NO·-synthase, the expression of which results in the release of NO· in amounts greatly exceeding those from the constitutive pathway. The transcription of the iNOS gene is itself induced by NF-κB (see Baeuerle and Henkel, 1994; Barnes, 1995).

The arachidonic acid cascade is a central process in generating the important inflammatory mediators, prostaglandins and leukotrienes (Lewis and Keft, 1995). Arachidonic acid is released from the plasma membrane of the cell by the action of phospholipase A$_2$. The released lipid is then metabolized in a process that can give rise to intracellular ROI (Cadenas, 1989). Thus, arachidonic acid can be acted on either by cyclooxygenases (also known as prostaglandin H synthases), which catalyze conversion to prostaglandins, or by lipoxygenases, which catalyze the formation of lenkotrienes and hydroxyeicosatetraenoic acids. Both a constitutive form and an inducible form of cyclooxygenase have been identified (see Vane, 1994; Lewis and Keft, 1995), known as COX-1 (prostaglandin H$_2$ synthase-1; PGHS-1) and COX-2 (PGHS-2), respectively. The mode of action of nonsteroidal anti-inflammatory drugs involves the inhibition of cyclooxygenase activity, and an area of current interest is in the development of new drugs with fewer side effects, which selectively inhibit COX-2 but not COX-1. A number of new anti-inflammatory drugs have been targeted toward the selective inhibition of 5-lipoxygenase, an enzyme that catalyzes the formation of the proinflammatory molecule leukotriene B4 (LTB4; Lewis and Keft, 1995). However, 15-lipoxygenase has been implicated in the oxidative modification of LDL in atherogenesis (Jessup *et al.*, 1991; Yla-Herttuala *et al.*, 1991). As mentioned earlier, aspirin, a potent cyclooxygenase inhibitor, has been shown to block the activation of NF-κB in the human Jurkat T-cell line (Kopp and Ghosh, 1994). However, the related compound, salicylic acid, also inhibits NF-κB activation. Salicylic acid has anti-inflammatory activity, but does not inhibit cyclooxygenase. This is consistent with the idea that the anti-inflammatory activity of aspirin may not be wholly attributable to its ability to inhibit cyclooxygenase and that inhibition of NF-κB activation may play a role at high concentrations of aspirin.

Most of the oxygen consumed by mammalian cells is converted to water via the mitochondrial electron transport system, in which electrons flow from NADH to sequentially reduce flavoproteins, ubiquinone, mitochondrial cytochromes, and finally molecular oxygen. The last reaction of this respiratory chain, catalyzed by cytochrome oxidase, is the donation of four electrons to each O_2 molecule to form water. However, up to 5% of the electrons entering the mitochondrial electron transport chain can become uncoupled from it and singly leak out onto O_2 to form $O_2{}^{\cdot-}$ (Fridovich, 1979). This source of free radicals is thought to be involved in TNF-mediated cytotoxicity (Schulze-Osthoff *et al.*, 1992).

The enzyme xanthine dehydrogenase is present within the cell cytosol, but may also be bound on the endothelial cell surface. It catalyzes the oxidation of hypoxanthine and xanthine to uric acid. Xanthine dehydrogenase is thought to be located predominantly in the liver, small intestine, and capillary endothelium in humans (Jarasch *et al.*, 1986). However, the distribution is different in other species. In healthy tissue, most of the enzyme is present as the "D form," which transfers electrons to NAD⁺:

$$\text{xanthine} + H_2O + NAD^+ \rightarrow \text{uric acid} + NADH + H^+$$

However, approximately 10% of the enzyme is present as an oxidase ("type O") form, which transfers electrons to molecular oxygen to form $O_2{}^{\cdot-}$:

$$\text{xanthine} + H_2O + 2O_2 \rightarrow \text{uric acid} + 2O_2{}^{\cdot-} + 2H^+$$

Both reactions are inhibited by oxypurinol, the principal metabolite of allopurinol. The original mechanism proposed by McCord (1985) for the production of $O_2{}^{\cdot-}$ in ischemic tissues involves changes in purine metabolism within ischemic cells. During temporary ischemia, low oxygen concentrations cause a decline in mitochondrial oxidative phosphorylation by decreasing the capacity of the respiratory chain for coupled electron transport. This increases the dependence of the cell on ATP production via anaerobic glycolysis. Anaerobic glycolysis is an inefficient means of ATP production from glucose and leads to raised concentrations of adenosine and of its breakdown products, including hypoxanthine and xanthine, which are substrates for the xanthine dehydrogenase enzyme system. Cellular levels of ATP fall. Cells are no longer able to maintain proper ion gradients across their membranes, precipitating a redistribution of Ca^{2+} ions. The elevated cytosolic Ca^{2+} concentration activates a protease, possibly a calpain, which catalyzes the conversion of xanthine dehydrogenase to xanthine oxidase. Reperfusion of the temporarily ischemic organ restores a supply of the remaining substrate required for xanthine oxidase activity (i.e., O_2) and $O_2{}^{\cdot-}$ is generated.

The conversion of xanthine dehydrogenase to its oxidase form may also be induced by exposure of cells to TNF (see Larrick and Wright, 1990). Thus, some studies have demonstrated a protective effect of xanthine oxidase inhibitors such as allopurinol toward TNF-mediated cytotoxicity (Adamson and Billings, 1994; Olah et al., 1994). In contrast, many other studies have shown that the cytotoxicity of TNF was mainly related to uncoupling of the mitochondrial respiratory chain (Schulze-Osthoff et al., 1992; see later). Therefore, it appears that the contribution of different free radical sources to TNF cytotoxicity may be cell-type specific.

In the same way, it appears that the relative importance of different sources of ROI, and therefore the potential source of ROI involved in activating NF-κB, may be dependent on the type of cell, as will be discussed. It also seems plausible that the environment of the cell, e.g., oxygen tension or prior exposure to cytokines, as well as the nature of the NF-κB-activating stimulus, may play a role. This suggests that ROI-mediated NF-κB activation may not be important in every cell type, in all environments (Brennan and O'Neill, 1995).

Some studies have begun to address these issues. Los et al. (1995) studied the activation of the CD28-responsive complex, which has been shown to consist of protein subunits of the NF-κB family, as a consequence of the triggering of the CD28 surface receptor of isolated human peripheral blood T lymphocytes. It was shown that the intracellular formation of ROI was a required step in this process. To identify the source of the ROI, the effects of inhibitors of the ROI-generating enzymes discussed earlier were tested. No significant effects were seen with the NADPH oxidase inhibitor, diphenylene iodonium, or the respiratory chain inhibitor, rotenone, or the xanthine oxidase inhibitor, allopurinol. However, specific inhibitors of either 5-lipoxygenase (such as ICI 230487) or phospholipase A_2 (such as p-bromophenacyl bromide) prevented the activation of the NF-κB/CD28-responsive complex.

These results are of particular interest because the experiments were carried out in primary human cells. Many experiments that have been carried out in relation to the role of ROI/RNI as cellular messengers that induce gene expression have been done in cell lines. In such cells, some signaling pathways may be redundant.

However, experiments in the TNF-α-sensitive, L929 fibrosarcoma cell line (Schulze-Osthoff et al., 1993) showed that the activation of NF-κB by TNF-α was blocked by rotenone, a specific inhibitor of the electron flow within the mitochondrial electron transport chain, which inhibits ROI generation. L929 subclones that lacked a functional respiratory chain were resistant to NF-κB activation by TNF-α. These subclones were likewise resistant to the cytotoxicity of TNF-α, which also appears to involve ROI production. In addition, whereas rotenone reduced TNF cytotoxicity, inhibitors of two of the other potential ROI sources (xanthine oxidase and NADPH oxidase) had no effect (Schulze-Osthoff *et al.,* 1992). Likewise, mitochondrial respiratory chain uncoupling in isolated rat hepatocytes was associated with NF-κB activation (Garciaruiz *et al.,* 1995).

VII. Conclusion

It is clear that ROI and RNI are involved as cellular messengers that play an important role in the inflammatory response. These proinflammatory pathways may be blocked using existing antioxidant agents, but perhaps not very selectively. The challenge is to develop clinically useful compounds which will inhibit selected ROI- and/or RNI-dependent steps, while leaving physiological free radical activity intact. Perhaps specific inhibitors of free radical-generating enzymes may be the most promising class of compounds for research up to, and into, the new millenium.

Acknowledgments

The authors are grateful to the Arthritis and Rheumatism Council and the British Technology Group for financial assistance.

References

Abate, C., Patel, L., Rauscher, F. J., and Curran, T. (1990). Redox regulation of Fos and Jun DNA-binding activity in vitro. *Science* **249,** 1157–1161.

Adamson, G. M., and Billings, R. E. (1994). The role of xanthine oxidase in oxidative damage caused by cytokines in cultured mouse hepatocytes. *Life Sci.* **55,** 1701–1709.

Alpha-Tocopherol, β-Carotene Cancer Prevention Study Group. (1994). The effect of vitamin E and β-carotene on the incidence of lung cancer and other cancers in male smokers. *N. Engl. J. Med.* **330,** 1029–1035.

Amstad, P. A., Krupitza, G., and Cerutti, P. A. (1992). Mechanism of c-fos induction by active oxygen. *Cancer Res.* **52,** 3952–3960.

Auphan, N., DiDonato, J. A., Rosette, C., Helmberg, A., and Karin, M. (1995). Immunosuppression by glucocorticoids: Inhibition of NF-κB activity through induction of IκB synthesis. *Science* **270,** 286–290.

Bae, I., Smith, M. L., Sheikh, M. S., Zhan, Q. M., Scudiero, D. A., Friend, S. H., O'Connor, P. M., and Fornace, A. J. (1996). An abnormality in the p53 pathway following gamma-irradiation in many wild-type p53 human melanoma lines. *Cancer Res.* **56,** 840–847.

Barnes, P. J. (1995). Anti-inflammatory mechanisms of glucocorticoids. *Biochem. Soc. Trans.* **23,** 940–945.

Baeuerle, P. A., and Henkel, T. (1994). Function and activation of NF-κB in the immune system. *Annu. Rev. Immunol.* **12,** 141–179.

Bashir, S., Harris, G., Denman, A. M., Blake, D. R., and Winyard, P. G. (1992). Oxidative DNA damage and cellular sensitivity to oxidative stress in human autoimmune diseases. *Ann. Rheum. Dis.* **52,** 659–666.

Behl, C., Davis, J. B., Lesley, R., and Schubert, D. (1994). Hydrogen peroxide mediates amyloid β protein toxicity. *Cell* 77, 817–827.

Blake, D. R., Winyard, P. G., and Marok, R. (1994). The contribution of hypoxia-reperfusion injury to inflammatory synovitis: The influence of reactive oxygen intermediates on the transcriptional control of inflammation. *N.Y. Acad. Sci.* 723, 308–317.

Brand, K., Rogler, G., Bartsch, A., Knuechel, R., Page, M., Kaltschmidt, C., Baeuerle, P. A., and Neumeier, D. (1995). Nuclear factor-kappa B/Rel is activated in the atherosclerotic lesion. *Circulation* 92, 759. [Abstract]

Brawn, M. K., Chiou, W. J., and Leach, K. L. (1995). Oxidant-induced activation of protein-kinase-C in UC11MG cells. *Free Rad. Res.* 22, 23–37.

Brennan, P., and O'Neill, L. A. J. (1995). Effects of oxidants and antioxidants on nuclear factor κB activation in three different cell lines: Evidence against a universal hypothesis involving oxygen radicals. *Biochim. Biophys. Acta* 1260, 167–175.

Brown, K., Gerstberger, S., Carlson, L., Franzoso, G., and Siebenlist, U. (1995). Control of IκB-α proteolysis by site-specific, signal-induced phosphorylation. *Science* 267, 1485–1488.

Butcher, R. G., Bitensky, L., Cashman, B., and Chayen, J. (1973). Differences in the redox balance in human rheumatoid and non-rheumatoid synovial lining cells. *Beitr. Path. Bd.* 148, 265–274.

Cadenas, E. (1989). Biochemistry of oxygen toxicity. *Annu. Rev. Biochem.* 58, 79–110.

Chernajovsky, Y., Feldmann, M., and Maini, R. N. (1995). Gene therapy of rheumatoid arthritis via cytokine regulation: Future perspectives. *Br. Med. Bull.* 51, 503–516.

Crawford, D. R., Schools, G. P., Salmon, S. L., and Davies, K. J. A. (1996). Hydrogen peroxide induces the expression of adapt15, a novel RNA associated with polysomes in hamster HA-1 cells. *Arch. Biochem. Biophys.* 325, 256–264.

DiDonato, J. A., Mercurio, F., and Karin, M. (1995). Phosphorylation of IκBα precedes but is not sufficient for its dissociation from NF-κB. *Mol. Cell. Biol.* 15, 1302–1311.

Drury, P. L., Rudge, S. R., and Perrett, D. (1984). Structural requirements for activity of certain "specific" antirheumatic drugs: More than a simple thiol group? *Br. J. Rheumatol.* 23, 100–106.

Edwards, D. R., Murphy, G., Reynolds, J. J., Whitlam, S. E., Docherty, A. J. P., Angel, P., and Heath, J. K. (1987). Transforming growth factor beta modulates the expression of collagenase and metalloproteinase inhibitor. *EMBO J.* 6, 1899–1904.

Fairburn, K., Grootveld, M., Ward, R. J., Abiuka, C., Kus, M., Williams, R. B., Winyard, P. G., and Blake, D. R. (1992). α-Tocopherol, lipids and lipoproteins in knee-joint synovial fluid and serum from patients with inflammatory joint disease. *Clin. Sci.* 83, 657–664.

Fassbender, H. G. (1984). Current understanding of rheumatoid arthritis. *Inflammation* 8 (Suppl.), S27–S42.

Fialkow, L., Chan, C. K., Rotin, D., Grinstein, S., and Downey, G. P. (1994). Activation of the mitogen-activated protein-kinase signaling pathway in neutrophils: Role of oxidants. *J. Biol. Chem.* 269, 31234–31242.

Forman, H. J., and Thomas, M. J. (1986). Oxidant production and bactericidal activity of phagocytes. *Annu. Rev. Physiol.* 48, 669–680.

Fornace, A. J., Nebert, D. W., Hollander, M. C., Luethy, J. D., Papathanasiou, M., Fargnoli, J., and Holbrook, N. J. (1989). Mammalian genes coordinately regulated by growth arrest signals and DNA-damaging agents. *Mol. Cell. Biol.* 9, 4196–4203.

Fraga, C. G., Shigenaga, M. K., Park, J.-W., Degan, P., and Ames, B. N. (1990). Oxidative damage to DNA during aging: 8-hydroxy-2'-deoxyguanosine in rat organ DNA and urine. *Proc. Natl. Acad. Sci. U.S.A.* 87, 4533–4537.

Fridovich, I. (1979). Hypoxia and oxygen toxicity. *Adv. Neurol.* 26, 255–259.

Garciaruiz, C., Colell, A., Morales, A., and Kaplowitz, N. (1995). Role of oxidative stress generated from the mitochondrial electron-transport chain and mitochondrial glutathione status in loss of mitochondrial-function and activation of transcription factor nuclear factor-kappa-B: Studies with isolated mitochondria and rat hepatocytes. *Mol. Pharmacol.* **48**, 825–834.

Ghosh, S., and Baltimore, D. (1990). Activation in vitro of NF-κB by phosphorylation of its inhibitor IκB. *Nature* **344**, 678–682.

Ghosh, G., van Duyne, G., Ghosh, S., and Sigler, P. B. (1995). Structure of NF-κB p50 homodimer bound to a κB site. *Nature* **373**, 303–310.

Gravallese, E. M., Darling, J. M., Ladd, A. L., Katz, J. N., and Glimcher, L. H. (1991). In situ hybridization studies of stromelysin and collagenase messenger RNA expression in rheumatoid synovium. *Arthritis Rheum.* **34**, 1076–1084.

Grippo, J. F., Holmgren, A., and Pratt, W. B. (1985). Proof that the endogenous, heat-stable glucocorticoid receptor-activating factor is thioredoxin. *J. Biol. Chem.* **260**, 93–97.

Halliwell, B. (1994). Vitamin C: The key to health or a slow-acting carcinogen? *Redox Rep.* **1**, 5–9.

Halliwell, B., and Gutteridge, J. M. C. (1985). The importance of free radicals and catalytic metal ions in human disease. *Mol. Asp. Med.* **8**, 89–193.

Hancock, J. T., Maly, F. E., and Jones, O. T. G. (1989). Properties of the superoxide-generating oxidase of B-lymphocyte cell lines: Determination of Michaelis parameters. *Biochem. J.* **262**, 373–375.

Handel, M. L., McMorrow, L. B., and Gravallese, E. M. (1995). Nuclear factor-κB in rheumatoid synovium: Localisation of p50 and p65. *Arthritis Rheum.* **38**, 1762–1770.

Harris, G., Bashir, S., and Winyard, P. G. (1994). 7,8-Dihydro-8-oxo-2′-deoxyguanosine present in DNA is not simply an artefact of isolation. *Carcinogenesis* **15**, 411–413.

Hayashi, T., Ueno, Y., and Okamato, T. (1993). Oxidoreductive regulation of nuclear factor κB: Involvement of a cellular reducing catalyst thioredoxin. *J. Biol. Chem.* **268**, 11380–11388.

Heliovaara, M., Knekt, P., Aaran, R. K., Alfthan, G., and Aromaa, A. (1994). Serum antioxidants and risk of rheumatoid arthritis. *Ann. Rheum. Dis.* **53**, 51–53.

Henney, A. M., Wakeley, P. R., Davies, M. J., Foster, K., Hembry, R., Murphy, G., and Humphries, S. (1991). Localization of stromelysin gene expression in atherosclerotic plaques by *in situ* hybridization. *Proc. Natl. Acad. Sci. U.S.A.* **88**, 8154–8158.

Houglum, K., Bedossa, P., and Chojkier, M. (1994). TGF-beta and collagen-alpha 1 (I) gene expression are increased in hepatic acinar zone 1 of rats with iron overload. *Am. J. Physiol.* **267**, G908–G913.

Jacobson, M. D. (1996). Reactive oxygen species and programmed cell death. *TIBS* **21**, 83–86.

Jarasch, E.-D., Bruder, G., and Heid, H. W. (1986). Significance of xanthine oxidase in capillary endothelial cells. *Acta Physiol. Scand.* (Suppl.) **548**, 39–46.

Jessup, W., Darley-Usmar, V., O'Leary, V., and Bedwell, S. (1991). 5-Lipoxygenase is not essential in macrophage-mediated oxidation of low-density lipoprotein. *Biochem. J.* **278**, 163–169.

Kaltschmidt, B., Baeuerle, P. A., and Kaltschmidt, C. (1993). Potential involvement of the transcription factor NF-κB in neurological disorders. *Mol. Aspects Med.* **14**, 171–190.

Kaltschmidt, C., Kaltschmidt, B., Lannes-Vieira, J., Kreutzberg, G. W., Wekerle, H., Baeuerle, P. A., and Gehrmann, J. (1994). Transcription factor NF-κB is activated in microglia during experimental autoimmune encephalomyelitis. *J. Neuroimmunol.* **55**, 99–106.

Kaltschmidt, C., Kaltschmidt, B., Henkel, T., Stockinger, H., and Baeuerle, P. A. (1995). Selective recognition of the activated form of transcription factor NF-κB by a monoclonal antibody. *Biol. Chem. Hoppe-Seyler* **376**, 9–16.

Karin, M. (1991). The AP-1 complex and its role in transcriptional control by protein kinase C. *In* "Molecular Aspects of Cellular Regulation" (P. Cohen and J. G. Foulkes, eds.), Vol. 6, pp. 235–253. Elsevier, Amsterdam.

Keffer, J., Probert, L., Cazlaris, H., Georgopoulos, S., Kaslaris, E., Kioussis, D., and Kollias, G. (1991). Transgenic mice expressing human tumour necrosis factor: A predictive genetic model of arthritis. *EMBO J.* **10**, 4025–4031.

Keyse, S. M., and Emslie, E. A. (1992). Oxidative stress and heat shock induce a human gene encoding a protein-tyrosine phosphatase. *Nature* 359, 644–647.

Keyse, S. M., and Tyrrell, R. M. (1990). Induction of the heme oxygenase gene in human skin fibroblasts by hydrogen peroxide and UVA (365 nm) radiation: Evidence for the involvement of the hydroxyl radical. *Carcinogenesis* 11, 787–791.

Kinne, R. W., Boehm, S., Iftner, T., Aigner, T., Vomehm, S., Weseloh, G., Bravo, R., Emmrich, F., and Kroczek, R. A. (1995). Synovial fibroblast-like cells strongly express *jun*-B and c-*fos* proto-oncogenes in rheumatoid- and osteoarthritis. *Scand. J. Rheumatol.* 24 (Suppl. 101), 121–125.

Kopp, E., and Ghosh, S. (1994). Inhibition of NF-κB by sodium salicylate and aspirin. *Science* 265, 956–959.

Kristal, B. S., Kim, J. D., and Yu, B. P. (1994). Tissue-specific susceptibility to peroxyl radical-mediated inhibition of mitochondrial transcription. *Redox Rep.* 1, 51–55.

Kus, M. L., Fairburn, K., Blake, D. R., and Winyard, P. G. (1995). A vascular basis for free radical involvement in inflammatory joint disease. *In* "Immunopharmacology of Free Radical Species" (D. Blake and P. G. Winyard, eds.), pp. 97–112. Academic Press, London.

Lander, H. M., Sehajpal, P., Levine, D. M., and Novogrodsky, A. (1993). Activation of human peripheral blood mononuclear cells by nitric oxide-generating compounds. *J. Immunol.* 150, 1509–1516.

Larrick, J. W., and Wright, S. C. (1990). Cytotoxic mechanism of tumor necrosis factor-α. *FASEB J.* 4, 3215–3223.

Lewis, A. J., and Keft, A. F. (1995). A review on the strategies for the development and application of new anti-arthritic agents. *Immunopharmacol. Immunotoxicol.* 17, 607–663.

Liao, F., Andalibi, A., deBeer, F. C., Fogelman, A. M., and Lusis, A. J. (1993). Genetic control of inflammatory gene induction and NF-kB-like transcription factor activation in response to an atherogenic diet in mice. *J. Clin. Invest.* 91, 2572–2579.

Los, M., Schenk, H., Hexel, K., Baeuerle, P. A., Droge, W., and Schulze-Osthoff, K. (1995). IL-2 gene expression and NF-κB activation through CD28 requires reactive oxygen production by 5-lipoxygenase. *EMBO J.* 14, 3731–3740.

Marok, R., Winyard, P. G., Coumbe, A., Kus, M. L., Gaffney, K., Mapp, P. I., Blades, S., Morris, C. J., Blake, D. R., Kaltschmidt, C., and Baeuerle, P. A. (1996). Activation of the transcription factor NF-κB in the inflamed human synovium. *Arthritis Rheum.* 39, 583–591.

Matthews, J. R., Wakasugi, N., Virelizier, J.-L., Yodoi, J., and Hay, R. T. (1992). Thioredoxin regulates the DNA binding activity of NF-κB by reduction of a disulphide bond involving cysteine 62. *Nucleic Acids Res.* 20, 3821–3830.

McCord, J. M. (1985). Oxygen-derived free radicals in postischemic tissue injury. *N. Engl. J. Med.* 312, 159–163.

McKay, I. A., Winyard, P. G., Leigh, I. M., and Bustin, S. A. (1994). Nuclear transcription factors: Potential targets for new modes of intervention in skin disease. *Br. J. Dermatol.* 131, 591–597.

Merry, P., Winyard, P. G., Morris, C. J., Grootveld, M., and Blake, D. R. (1989). Oxygen free radicals, inflammation, and synovitis: The current status. *Ann. Rheum. Dis.* 48, 864–870.

Meyer, M. R., Schreck, R., and Baeuerle, P. A. (1993). Hydrogen peroxide and antioxidants have opposite effects on activation of NFκB and AP-1 in intact cells: AP-1 as secondary antioxidant-responsive factor. *EMBO J.* 12, 2005–2015.

Morimoto, R. I. (1993). Cells in stress: Transcriptional activation of heat shock genes. *Science* 259, 1409–1410.

Muller, C. W., Rey, F. A., Sodeoka, M., Verdine, G. L., and Harrison, S. C. (1995). Structure of the NF-κB p50 homodimer bound to DNA. *Nature* 373, 311–317.

Murrell, G., Francis, N., and Bromley, L. (1990). Modulation of fibroblast proliferation by oxygen free radicals. *Biochem. J.* 265, 659–665.

Nose, K., Shibanuma, M., Kikuchi, K., Kageyama, H., Sakiyama, S., and Kuroki, T. (1991). Transcriptional activation of early-response genes by hydrogen peroxide in a mouse osteoblastic cell line. *Eur. J. Biochem.* **201**, 99–106.

Olah, T., Regely, K., and Mandi, Y. (1994). The inhibitory effects of allopurinol on the production and cytotoxicity of tumor necrosis factor. *N.-S. Arch. Pharmacol.* **350**, 96–99.

Richter, C., Park, J.-W., and Ames, B. N. (1988). Normal oxidative damage to mitochondrial and nuclear DNA is extensive. *Proc. Natl. Acad. Sci. U.S.A.* **85**, 6465–6467.

Ron, D., and Habener, J. F. (1992). CHOP, a novel developmentally regulated nuclear protein that dimerizes with transcription factors C/EBP and LAP and functions as a dominant-negative inhibitor of gene transcription. *Genes Dev.* **6**, 439–453.

Rowe, P. M. (1996). Beta-carotene takes a collective beating. *Lancet* **347**, 249.

Rupec, R. A., and Baeuerle, P. A. (1995). The genomic response of tumor cells to hypoxia and reoxygenation: Differential activation of transcription factors AP-1 and NF-κB. *Eur. J. Biochem.* **234**, 632–640.

Salmon, M., and Gaston, J. S. H. (1995). The role of T-lymphocytes in rheumatoid arthritis. *Br. Med. Bull.* **51**, 332–345.

Saran, M., and Bors, W. (1989). Oxygen radicals acting as chemical messengers: A hypothesis. *Free Rad. Res. Commun.* **7**, 213–220.

Satriano, J., and Schlondorff, D. (1994). Activation and attenuation of transcription factor NF-κB in mouse glomerular measangial cells in response to tumor-necrosis-factor-alpha, immunoglobulin G, and adenosine 3′/5′-cyclic monophosphate: Evidence for involvement of reactive oxygen species. *J. Clin. Invest.* **94**, 1629–1636.

Scheinman, R. I., Cogswell, P. C., Lofquist, A. K., and Baldwin, A. S. (1995). Role of transcriptional activation of IκBα in mediation of immunosuppression by glucocorticoids. *Science* **270**, 283–286.

Schreck, R., Albermann, K., and Baeuerle, P. A. (1992). Nuclear factor κB: An oxidative stress-responsive transcription factor of eukaryotic cells (a review). *Free Rad. Res. Commun.* **17**, 221–237.

Schreck, R., Rieber, P., and Baeuerle, P. A. (1991). Reactive oxygen intermediates as apparently widely used messengers in the activation of the NF-KB transcription factor and HIV-1. *EMBO J.* **10**, 2247–2258.

Schulze-Osthoff, K., Bakker, A. C., Vanhaesebroeck, B., Beyaert, R., Jacob, W. A., and Fiers, W. (1992). Cytotoxic activity of tumor necrosis factor is mediated by early damage of mitochondrial functions: Evidence for the involvement of mitochondrial radical generation. *J. Biol. Chem.* **267**, 5317–5323.

Schulze-Osthoff, K., Beyaert, R., Vandevoorde, V., Haegeman, G., and Fiers, W. (1993). Depletion of the mitochondrial electron transport abrogates the cytotoxic and gene-inductive effects of TNF. *EMBO J.* **12**, 3095–3104.

Segal, A. W. (1991). Components of the microbicidal oxidase of phagocytes. *Biochem. Soc. Trans.* **49**, 49–50.

Snyder, S. H. (1995). No endothelial NO. *Nature* **377**, 196–197.

Staal, F. J. T., Roederer, M., and Herzenberg, L. A. (1990). Intracellular thiols regulate activation of nuclear factor κB and transcription of human immunodeficiency virus. *Proc. Natl. Acad. Sci. U.S.A.* **87**, 9943–9947.

Stein, B., and Yang, M. X. (1995). Repression of the interleukin-6 promotor by estrogen receptor is mediated by NF-κB and C/EBPβ. *Mol. Cell. Biol.* **15**, 4971–4979.

Suitters, A. J., Foulkes, R., Opal, S. M., Palardy, J. E., Emtage, J. S., Rolfe, M., Stephens, S., Morgan, A., Holt, A. R., Chaplin, L. C., Shaw, N. E., Nesbitt, A. M., and Bodmer, M. W. (1994). Differential effect of isotype on efficacy of anti-tumor necrosis factor α chimeric antibodies in experimental septic shock. *J. Exp. Med.* **179**, 849–856.

Tasinato, A., Boscoboinik, D., Bartoli, G.-M., Maroni, P., and Azzi, A. (1995). d-α-Tocopherol inhibition of vascular smooth muscle cell proliferation occurs at physiological concentrations, correlates with protein kinase C inhibition, and is independent of its antioxidant properties. *Proc. Natl. Acad. Sci. U.S.A.* **92**, 12190–12194.

Toledano, M. B., and Leonard, W. J. (1991). Modulation of transcription factor NF-κB binding activity by oxidation-reduction *in vitro. Proc. Natl. Acad. Sci. U.S.A.* **88**, 4328–4332.

Traenckner, E. B.-M., Wilk, S., and Baeuerle, P. A. (1994). A proteasome inhibitor prevents activation of NF-κB and stabilises a newly phosphorylated form of IκB-α that is still bound to NF-κB. *EMBO J.* **13**, 5433–5441.

Vane, J. (1994). Towards a better aspirin. *Nature* **367**, 215–216.

Whisler, R. L., Goyette, M. A., Grants, I. S., and Newhouse, Y. G. (1995). Sublethal levels of oxidant stress stimulate multiple serine threonine kinases and suppress protein phosphatases in Jurkat T-cells. *Arch. Biochem. Biophys.* **319**, 23–35.

Williams, R. J. P. (1985). The necessary and the desirable production of radicals in biology. *Phil. Trans. Roy. Soc. Lond.* B **311**, 593–603.

Willis, D., Moore, A. R., Frederick, R., and Willoughby, D. A. (1996). Heme oxygenase: A novel target for the modulation of the inflammatory response. *Nature Med.* **2**, 87–90.

Winyard, P. G., Zhang, Z., Chidwick, K., Blake, D. R., Carrell, R. W., and Murphy, G. (1991). Proteolytic inactivation of human $\alpha_1$antitrypsin by human stromelysin. *FEBS Lett.* **279**, 91–94.

Wiseman, H., and Halliwell, B. (1996). Damage to DNA by reactive oxygen and nitrogen species: Role in inflammatory disease and progression to cancer. *Biochem. J.* **313**, 17–29.

Yang, J. P., Merin, J. P., Nakano, T., Kato, T., Kitade, Y., and Okamoto, T. (1995). Inhibition of the DNA-binding activity of NF-κB by gold compounds in vitro. *FEBS Lett.* **361**, 89–96.

Yasumoto, K., Okamoto, S., Mukaida, N., Murakami, S., Mai, M., and Matsushima, K. (1992). Tumor necrosis factor α and interferon gamma synergistically induce interleukin 8 production in a human gastric cancer cell line through acting concurrently on AP-1 and NF-κB-like binding sites of the interleukin 8 gene. *J. Biol. Chem.* **267**, 22506–22511.

Yla-Herttuala, S., Rosenfeld, M. E., Parthasarathy, S., Sigal, E., Sarkioja, T., Witztum, J. L., and Steinberg, D. (1991). Gene expression in macrophage-rich human atherosclerotic lesions. 15. Lipoxygenase and acetyl low density lipoprotein receptor messenger RNA colocalize with oxidation specific lipid-protein adducts. *J. Clin. Invest.* **87**, 1146–1152.

Yodoi, J., and Uchiyama, T. (1992) Diseases associated with HTLV-I: Virus, IL-2 receptor dysregulation and redox regulation. *Immunol. Today* **13**, 405–410.

PART IV

Disease Processes

Hermann Esterbauer*
Reinhold Schmidt†
Marianne Hayn*

*Institute of Biochemistry
University of Graz
A-8010 Graz, Austria

†University Clinic of Neurology
University of Graz
A-8036 Graz, Austria

Relationships among Oxidation of Low-Density Lipoprotein, Antioxidant Protection, and Atherosclerosis

I. Introduction

Cardiovascular diseases (CVD), with an incidence of approximately 50%, are the main cause of death in most developed countries. To evaluate epidemiological studies and statistical data, one needs to keep in mind that CVD is a global term for a wide range of diseases. According to the World Health Organization (WHO) classification system ICD-9 and ICD-10, they are divided into 69 subgroups (No. 390 to 459), including ischemic heart disease (IHD), cerebrovascular disease, and other related diseases, e.g., acute myocardial infarction (AMI) (WHO, 1991). In 1990, the age-standardized cardiovascular disease mortality per 100,000 ranged from 752 (Romania) to 215 (France). Ischemic heart disease was the main cause of cardiovascular disease mortality (40–60%), but there are also exceptions such as Japan where the frequency of ischemic heart disease was only 16.7% (Table I).

The underlying primary cause of most cardiovascular diseases is believed to be atherosclerosis, a progressive multifactorial disease of the artery wall

425

TABLE I   Death Rate of Cardiovascular Disease (CVD) in Several Selected Countries According to WHO Statistics (1991)[a]

	Total CVD[b]	Percentage[c]		
		IHD	ZVD	Other
Romania	751.7	26.8	26.7	46.5
Scotland	435.1	59.3	25.9	14.8
Finland	418.5	58.1	23.5	18.4
Austria	382.4	39.5	26.3	35.2
Germany (west)	372.2	40.3	23.4	36.3
Sweden	368.1	56.8	18.7	24.5
United States	375.4	52.6	15.0	32.4
England/Wales	355.4	59.1	24.2	16.7
Italy	324.5	30.0	31.5	38.5
Switzerland	288.8	37.9	19.8	42.3
Japan	226.5	16.7	39.8	43.5
France	215.2	29.8	27.5	42.7

[a] From World Health Organization. (1991).
[b] Age standardized death rate/100,000.
[c] Percentage of total CVD (=100%) due to ischemic heart disease (IHD), zerebrovascular disease (ZVD), and other CVD, including acute myocardial infarction.

(Badimon *et al.*, 1993; Ross 1993; O'Brien and Chait, 1994; Navab *et al.*, 1995). The lipid hypothesis suggests that lipid deposits are formed in the subendothelial space due to increased plasma low-density lipoprotein (LDL), endothelial dysfunction and permeability, and an imbalance of cholesterol influx and efflux in the early stages of atherosclerosis. The development of early lipid deposits, fatty streaks, and advanced plaques is a slow process involving a complex interplay of many factors that are only partly understood at the cellular and molecular level. Many risk factors have been identified that influence atherosclerosis in several ways and at various points (Ross, 1993; O'Brien and Chait, 1994). The most important ones are dyslipidemia, hypertension, diabetes, and smoking.

The past decade saw a series of remarkable studies that suggested that oxidative stress, particularly the oxidation of low-density lipoprotein, is a risk factor and plays a role at several steps of atherosclerosis. Reviews on this subject have been published by Schwartz and Valente (1994), Witztum (1994), Panasenko and Sergienko (1994), Parthasarathy (1994), Esterbauer and Ramos (1995), Rice-Evans and Bruckendorfer (1995), Berliner *et al.* (1995), Alexander (1995), and Halliwell (1995).

A decrease of oxidative stress and antioxidant protection of LDL might therefore be a strategy with great promise for the prevention of atherosclerosis-associated cardiovascular disease. Compelling evidence shows that an improvement of the antioxidant status by simply eating antioxidant-rich food/vegetables and fruits more frequently could significantly reduce the risk of cancer (Block *et al.*, 1992). The evidence for a beneficial effect of antioxidants in cardiovascular disease is not yet as strong and stems primarily from a wide range of biochemical investigations (for reviews see Keaney and Frei, 1994; van Poppel *et al.*, 1994; Esterbauer and Ramos, 1995; Raal *et al.*, 1995), experimental animal studies (Steinberg *et al.*, 1992; Lynch and Frei, 1994), and epidemiological investigations (for reviews see Street *et al.*, 1994; Gaziano *et al.*, 1994; Duell, 1995; Hoffman and Garewal, 1995). Large-scale clinical intervention trials with the objective to test natural antioxidants in atherosclerosis, however, are not completed (Steinberg, 1995). Because such trials will ultimately have the power to assess the putative benefits of antioxidants in cardiovascular disease prevention, premature and perhaps unjustified conclusions should not be made until the results of placebo-controlled, double-blind intervention trials are available (Oliver, 1995; Steinberg, 1995).

This chapter describes the LDL oxidation hypothesis, the possible influence of oxidized LDL (oxLDL) and antioxidants on atherogenesis, and the assessment and relevance of *in vitro*-measured LDL oxidation indices. The last part of this contribution is a summary of epidemiological studies on relationships of antioxidant consumption and blood antioxidant concentration with cardiovascular disease.

II. The Low-Density Lipoprotein Oxidation Hypothesis _____

The arterial wall consists of three layers: intima, media, and adventitia. The innermost luminal part of the intima is a monolayer of endothelial cells lining the whole wall. The intact endothelial layer serves as a selective barrier for plasma lipids and also has antithrombotic and vasotonic properties (Badimon *et al.*, 1993; Ross, 1993). Early atherosclerosis is characterized by the massive accumulation of monocytes and cholesterol-containing lipid-laden foam cells in the subendothelial space, i.e., the layer between the endothelial cells and media. Most of the foam cells are derived from monocyte–macrophages, and much of the interest in oxidized LDL stems from the discovery that it exhibits *in vitro* properties, which could explain why monocytes immigrate from the circulation into the subendothelial space and differentiate there to macrophages and ultimately are converted to foam cells. A hallmark for the oxidation hypothesis was the work of Steinberg's group (for reviews see Steinberg *et al.*, 1989; Esterbauer *et al.*, 1992), who observed already in 1981 (Henriksen *et al.*, 1981) that LDL incubated for about 24 to 48 hr with endothelial cells is recognized by and avidly endocytozed via the scavenger receptor pathway of macrophages. In 1984, the same group (Steinbrecher *et al.*, 1984) discovered that the modification of LDL by endothelial cells involves lipid peroxidation and degradation of

LDL phospholipids. At the same time, Chisolm's group presented data showing that smooth muscle cells and neutrophils oxidize LDL by a free radical mechanism to a form with high cytotoxicity (Morel *et al.*, 1983, 1984; review by Chisolm, 1992). These discoveries offered for the first time a plausible explanation of how early lipid deposits are formed in the intima and how they could provoke a number of secondary deleterious effects. Release of cytotoxic lipid peroxidation products from oxidized LDL in the intima can be considered a constant irritant for the endothelial cell layer, leading to a cascade of events such as endothelial cell death, platelets aggregation, release of growth factors that promote smooth muscle cell proliferation, accumulation of inflammatory cells, and disturbances of eicosanoid homeostasis.

A number of subsequent studies revealed that oxidatively modified LDL processes, in addition to uncontrolled macrophage uptake and cytotoxicity, a number of other pro-atherogenic biological properties. Some of these properties are characteristic for fully oxidized LDL whereas others (e.g., endothelial cell expression of adhesion molecules) appear to be restricted to minimally modified LDL (MM-LDL). Investigations of pro-atherogenic effects of oxLDL and MM-LDL are summarized in Table II. The list is not exhaustive and the chronology may not always be precise. The purpose of Table II is to provide references useful in identifying findings that have contributed in a major way to our knowledge on pro-atherogenic properties of oxidatively modified LDL.

LDL oxidation is a lipid peroxidation chain reaction driven by free radical intermediates. As such, LDL oxidation, particularly if conducted *in vitro*, possesses the general characteristics of lipid peroxidation reactions and free radical reactions. What makes the process and its dynamics so complex in its chemistry is the fact that all components of LDL, i.e., apolipoprotein B, phospholipids, cholesteryl ester, free cholesterol, triglycerides, and antioxidants, participate at a certain step of the oxidation, leading to multiple secondary and tertiary reactions and reaction products.

Regarding the composition of oxidized LDL in comparison to native LDL, we refer to a review article by Esterbauer and Ramos (1995). It seems worthwhile to emphasize that compared to the extensive knowledge on biological properties of oxidized LDL, much less is known about the chemistry of the oxidation and the chemical structure of the oxidation products. Only in a few cases has an attempt been made to clearly identify the compounds ultimately responsible for a certain biological effect. For example, lysophosphatidyl-choline in oxLDL is assumed to mediate the chemotactic activity for monocytes and T lymphocytes and perhaps the expression of monocyte adhesion molecules by endothelial cells. Some aldehydes in oxLDL might be responsible for the interleukin-1β release from macrophages. Oxysterols and aldehydes can be highly cytotoxic and are likely candidates for the cytotoxic properties of oxLDL.

TABLE II   Biological Effects of Extensively Oxidized LDL and Minimally Oxidized LDL (MM-LDL)

---

1. Modification of LDL by endothelial cells (ECs) involves lipid peroxidation, oxLDL is recognized by the macrophage scavenger receptor (Steinbrecher *et al.*, 1984).
2. SMCs and ECs produce oxLDL by a free radical mechanism, oxLDL is cytotoxic (Morel *et al.*, 1983, 1994; reviewed by Chisolm, 1992).
3. Reduces motility of macrophages and may thus inhibit egression of macrophages and foam cells from the arterial lesion (Quinn *et al.*, 1987).
4. Chemotactic for monocytes (Quinn *et al.*, 1987) and smooth muscle cells (SMCs) (Autio *et al.*, 1990).
5. Inhibits production of platelet-derived growth factor (PDGF) by ECs (Fox *et al.*, 1987) and monocyte–macrophages (Malden *et al.*, 1991).
6. Is immunogenic and induces formation of antibodies (Palinski *et al.*, 1989).
7. MM-LDL stimulates release of monocyte-chemotactic protein-1 (MCP-1) from ECs (Cushing *et al.*, 1990).
8. MM-LDL increases adhesion of monocytes of ECs (Berliner *et al.*, 1990; Kim *et al.*, 1994).
9. MM-LDL stimulates expression of colony-stimulating factors (CSFs) for monocytes and granulocytes by ECs (Rajavashisth *et al.*, 1990).
10. Inhibits tumor necrosis factor expression by monocyte–macrophages (Hamilton *et al.*, 1990).
11. Stimulates (low concentration) or inhibits prostacyclin production by SMCs (Zhang *et al.*, 1990; Daret *et al.*, 1993).
12. Inhibits NO activation of guanylate cyclase (Schmidt *et al.*, 1990) and inhibits endothelial cell-dependent arterial relaxation (Ohgushi *et al.*, 1993).
13. Systemic administration into hamster initiates immediate leukocyte adhesion to capillary endothelium (Lehr *et al.*, 1991).
14. MM-LDL injected into mice increases serum and tissue levels of MCP-1 and CSF (Liao *et al.*, 1991).
15. Increases tissue factor expression by cultured ECs and suppresses protein C (Drake *et al.*, 1991; Weis *et al.*, 1991).
16. Increases PDGF expression by SMCs (Stiko-Rahm *et al.*, 1992; Zwijsen *et al.*, 1992).
17. Induces expression of interleukin-8 in monocytic cells and activates T lymphocytes (Frostegard *et al.*, 1992; Terkeltaub *et al.*, 1994).
18. Increases EC expression of plasminogen activator inhibitor-1 (Kugiyama *et al.*, 1993).
19. Induces increased expression of stress proteins by macrophages (Yamaguchi *et al.*, 1993).
20. Stimulates release of interleukin-1β from monocyte–macrophages (Thomas *et al.*, 1994).

Separate investigations of the protein and lipid part of oxLDL clearly demonstrated that both contribute to some extent to the pro-atherogenic properties of oxLDL. The interaction of oxLDL with the macrophage scavenger receptor is attributed to the oxidatively modified apolipoprotein B. The reaction of native LDL with aldehydes (e.g., malonaldehyde, 4-hydroxynonenal, or mixtures of aldehydes) produces an aldehyde-modified LDL showing a similar interaction with macrophages as oxLDL. This, together with the observation that aldehyde treatment as well as oxidation causes a loss of ε-amino groups of lysine residues in apolipoprotein B, has led to the assumption that aldehydes generated *in situ* within the oxidizing LDL particle interact with and become covalently linked to side chains of lysine and other nucleophilic amino acid residues (e.g., histidine) of apolipoprotein B. The consequence of such a modification is an increase of the net negative surface charge of LDL, which in turn results in an increased electrophoretic mobility of LDL. This is fully consistent with the fact that LDL oxidized *in vitro* or LDL extracted from the arterial wall exhibits a higher relative electrophoretic mobility (REM of about 2 to 4) compared to native LDL (reviewed by Esterbauer *et al.*, 1992). The modification of apolipoprotein B is likely indirectly responsible for the diminished interaction of oxLDL with the classical LDL receptor (B/E-receptor) which recognizes certain positively charged clusters of lysins in apolipoprotein B. The structure of the ligand(s) in oxLDL interacting with the scavenger receptor is unknown. The strong competition of acetylated LDL suggests that the ligands are preformed in apolipoprotein B and become exposed during oxidation rather than being generated *de novo* by the reaction of lipid oxidation products with amino acid side chains. The strongest evidence that oxidation of LDL does occur in the artery wall comes from immunological studies, which clearly demonstrated malonaldehyde and 4-hydroxynonenal modified apolipoprotein B in atherosclerotic lesions of rabbits and humans (Haberland *et al.*, 1988; Rosenfeld *et al.*, 1990; Juergens *et al.*, 1993).

LDL oxidation can be initiated *in vitro* by the incubation with macrophages, endothelial cells, smooth muscle cells, or neutrophils (for reviews see Gebicki *et al.*, 1991; Chait and Heinecke, 1994) or in a cell-free medium utilizing a variety of prooxidants such as lipoxygenase, myeloperoxidase, cholesterol oxidase, horseradish peroxidase/H_2O_2, defined oxygen radicals, UV light, γ-irradiation, heme/H_2O_2, copper ions, ceruloplasmin, thiols plus Fe^{2+}, peroxynitrite, and hypochlorite (reviewed by Gebicki *et al.*, 1991; Esterbauer and Ramos, 1995). Traces of transition metal ions in free or redox-active complexes are generally agreed to be essential for producing oxLDL showing high macrophage uptake or cytotoxicity.

The mechanism of initiation of LDL oxidation *in vivo* is largely unknown. It is believed to occur not in the circulation but in the arterial wall itself, where the LDL is sequestered by proteoglycans and other extracellular matrix components (Haberland and Steinbrecher, 1992). A putative candidate for oxidation of LDL in the artery wall is 15-lipoxygenase. It has been shown that this enzyme is present in human atherosclerotic lesions and is colocalized with deposits of oxLDL (Ylä-Herttuala *et al.*, 1990, 1991). Lipids extracted from pieces of thoracic aortas of subjects who died of acute heart failure (Kühn *et al.*, 1992) and of cholesterol-fed rabbits (Kühn *et al.*, 1994) contain considerable amounts of keto- and hydroxyoctadecadienoic acid esterified with cholesterol. Analysis of the positional and stereoisomers of the hydroxyoctadecadienoic acids suggests that 15-lipoxygenase may play a role during the early stages of plaque development whereas nonenzymatic peroxidation processes are relevant in later stages. These studies suggest that cellular 15-lipoxygenase of endothelial cells or monocyte–macrophages is involved in the initiation of the nonenzymatic oxidation process by providing seed hydroperoxides in LDL. That relatively small amounts (about 1 mol/ mol LDL) of 15-lipoxygenase-generated cholesteryl ester hydroperoxides render LDL highly susceptible to subsequent nonenzymatic oxidation by metal ions has been shown (Lass *et al.*, 1996). A striking enzymatic property of mammalian 15-lipoxygenase is that the cholesteryl-linoleate molecules of LDL become oxidized during the early stages and that this oxidation is not inhibited by α-tocopherol (Lass *et al.*, 1996). This observation may also have important consequences for the strategy of preventing LDL oxidation *in vivo*, as it suggests that 15-lipoxygenase inhibitors could be a promising approach in reducing the risk of oxidation and oxLDL-associated diseases such as cardiovascular disease.

III. Assessment of Oxidation Resistance of Low-Density Lipoprotein

Considerable interest has been focused on methods used to determine the oxidative modification of LDL and on ways to measure the inhibitory effects of antioxidants. The evaluation of lipoprotein oxidation *in vivo* or *ex vivo* is difficult, and suitable routine methods have not yet been developed. Attempts to measure the presence of oxidized LDL in the circulation have focused on the determination of so-called minimally oxidized LDL (LDL$^-$) in plasma (Cazzolato *et al.*, 1991). LDL$^-$ is assumed to be a certain type of minimally oxidized LDL. Employing protein chromatography (FPLC), the proportion of LDL$^-$ in total LDL has been estimated to be in the range of 0.5 to 9.8%. Indirect methods which could give some clues regarding the oxidation of LDL *in vivo* are the measurement of autoantibodies against oxLDL and oxLDL immune complexes (for a review see Tatzber and Esterbauer, 1995). Most of the investigations on oxidation resistance of LDL are *in vitro* studies where isolated LDL is oxidized by cells or in cell-free medium using one of the prooxidants mentioned previously. Up until now, no general and broad consensus exists as to how oxidation resistance and/or putative protective effects of antioxidants should be measured. In examining the recent literature (reviewed by Esterbauer *et al.*, 1995), it appears that most laboratories consider oxidation resistance as a kinetic parameter described by the progress of the LDL oxidation process. Because the oxidation of LDL *in vitro* is accompanied by characteristic changes of chemical, physicochemical, and biological properties, a variety of methods may be used for determining the rate and extent of oxidation (reviewed by Puhl *et al.*, 1994). They include measurement of the increase of thiobarbituric acid-reactive substances (TBARS), total lipid hydroperoxides, defined lipid hydroperoxides, hydroxy- and hydroperoxy fatty acids, conjugated dienes, oxysterols, lysophospholipids, aldehydes, and fluorescent chromophores as well as the measurement of disappearance of endogenous antioxidants and polyunsaturated fatty acids and oxygen uptake. The apolipoprotein B becomes progressively altered during oxidation; its loss of reactive amino groups and fragmentation to smaller peptides can be determined and also used as an index of oxidative modification. The net increase of the negative charge of the LDL particle due to oxidation can be analyzed as relative electrophoretic mobility by electrophoresis. Nuclear magnetic resonance, electron spin resonance, circular dichroism, and fluorescence polarization have also been applied to study certain aspects of the LDL oxidation process. The biological assays most frequently used for the assessment of the extent of oxidation are uptake of LDL by cultured macrophages and cytotoxicity toward cultured cells.

One model used to measure kinetic parameters of LDL oxidation that has become very popular is copper-mediated oxidation (about 0.1 μM LDL in phosphate-buffered saline plus 1 to 5 μM Cu^{2+}) in conjunction with a continuous measurement of the conjugated diene (CD) absorption at 234 nm (Esterbauer et al., 1992; Puhl et al., 1994). This CD method is highly reproducible, easy to perform, and useful for routine measurement. The oxidation indices that can readily be obtained from such diene measurements are length of lag time where propagation is inhibited, maximum rate of oxidation during propagation, and maximum amount of conjugated dienes formed. During the lag and propagation phase the time profiles for conjugated dienes, TBARS, and fluorescence at 430 nm are very similar. If performed under identical conditions, all three methods will give equivalent results for the susceptibility of LDL to oxidation as judged by the lag time.

Regnström et al. (1992) were the first who employed the CD method in a clinical situation and examined the oxidation resistance of LDL of patients who had an acute myocardial infarction. They found that the oxidation resistance of LDL, as measured by the duration of the length of the lag time, was inversely related to the severity of the infarction. Several small-scale case control studies using the CD method were later performed by various groups to evaluate whether cardiovascular disease patients possess an LDL with low oxidation resistance as judged by the lag time (Table III). The majority of the studies revealed a lower oxidation resistance for patients. Exceptions included patients suffering from intermittent claudication and severe carotid atherosclerosis, who showed no difference to age-matched controls. Table III also includes the measured lag time values for patients and controls in order to show that the difference was not strong but only in the range of about 10 to 30%. Whether such a relatively small decrease of the LDL oxidation resistance to Cu^{2+} ions indeed reflects an increased risk of oxidation *in vivo* in the arterial wall is uncertain. Studies (Abbey et al., 1993; Waeg et al., 1994) where the oxidation resistance of individual

TABLE III Case Control Studies of Relationship between Oxidation Resistance of LDL and Disease[a]

Major findings	Reference
In 35 acute myocardial infarction (AMI) patients, OR was inversely related to severity of infarction, lag varied between 40 and 90 min	Regnström et al. (1992)
Reduced OR in 12 bypass patients with progression compared to 16 controls (81 vs 94 min)	De Rijke et al. (1992)
Thirteen patients with intermittent claudication compared with 14 controls, no difference in OR (124 vs 124 min)	Zieden et al. (1992)
Reduced OR in 11 coronary artery disease (CAD) patients taking no β blockers compared to patients taking β blockers (114 vs 148 min), but the 20 CAD patients note different to 25 controls (132 vs 140 min)	Croft et al. (1992)
Reduced OR in 73 CAD patients compared to 71 controls (67 vs 116 min); lag from fluorescence	Cominacini et al. (1993)
Reduced OR in 36 hypertensive patients compared to 26 controls (124 vs 157 min)	Maggi et al. (1993)
Increased TBARS in Cu^{2+}-oxidized LDL of 48 CAD patients compared to 92 controls (285 vs 231 nmol)	Chiu et al. (1994)
Reduced OR in 25 hypertensive patients compared to 25 controls (TBARS, dienes)	Keidar et al. (1994)
Reduced OR in 20 patients with familial defective apoB compared to 20 controls (90 vs 108 min)	Stalenhoef et al. (1994)
Reduced OR in 37 uremic patients compared to 70 controls. (119 vs 142 min)	Maggi et al. (1994a)
Ninety-four patients with severe carotid atherosclerosis compared with 42 controls, no difference in OR (136 ± 27 vs 142 ± 23 min)	Maggi et al. (1994b)
Reduced OR in 25 hypercholesteremic patients compared with 15 controls (65 vs 93 min)	Cominacini et al. (1994)
Reduced OR in 27 patients with cystic fibrosis compared to 22 controls (48 vs 62 min)	Winklhofer-Roob et al. (1995)

[a] If not otherwise stated, oxidation resistance (OR) is the lag time (in minutes) measured principally as described by Puhl et al. (1994).

subjects was repeatedly measured over longer periods of time (up to 1 year) revealed that the lag time remains relatively constant, with variations of about 5 to 10%, suggesting it represents a subject specific feature of LDL. The intersubject variability in healthy controls is relatively high. In our studies, lag time values vary from 33 to 188 min (Esterbauer et al., 1993, 1995). Available data do not, however, allow firm conclusions whether persons with a short lag time are at a higher risk for developing atherosclerosis than persons with a long lag time. Consequently, the results of studies on OR measured by the CD method should not be overinterpreted until the clinical relevance of low LDL oxidation resistance, as judged by the CD method, becomes clear. An approach in this direction are studies (Tribble et al., 1992, 1994) showing that dense LDL is more susceptible to Cu^{2+} oxidation than buoyant LDL. This observation could perhaps explain in part why subjects with a predominance of dense LDL have a higher risk for atherosclerosis (Kraus, 1991).

The oxidative stress posed on LDL in assays employing 1 μM Cu^{2+} or more is relatively strong and causes complete oxidation within several hours. One inherent problem with the extrapolation of results from such in vitro measurements to the in vivo situation is that the strength of oxidative stress in the artery wall is unclear. It has been shown (Ziouzenkova et al., 1996) that the kinetic of Cu^{2+}-mediated LDL oxidation completely changes if the rate of initiation of oxidation is reduced by exposing LDL to low Cu^{2+} concentrations. With concentrations of about 0.5 μM Cu^{2+} or less, oxidation commences without a significant lag phase and LDL became heavily oxidized in a process apparently not inhibitable by vitamin E and the other antioxidants contained in LDL. This clearly indicates that oxidation indices which can be derived from in vitro assays are strongly influenced by the assay conditions and that much further work is needed to clarify the usefulness of such measurements for predicting the risk of atherosclerosis.

IV. Vitamin E Supplementation Increases Oxidation Resistance of Low-Density Lipoprotein as Judged by the Conjugated Diene Method

The oxidation resistance of LDL measured as lag time by the CD method shows a significant interindividual variability for clinically healthy, not vitamin E-supplemented subjects. This variation is caused partly by variations in the α-tocopherol content of LDL but also depends on not yet clearly defined other variables such as the content of polyunsaturated fatty acids, preformed peroxides, cholesterol–protein ratio, and the distribution of LDL subfractions. The group of Stocker repeatedly reported (Stocker et al., 1991; Bowry and Stocker, 1993) that ubiquinol-10 is more efficient in inhibiting LDL oxidation than α-tocopherol. The concentration of ubiquinol-10 in LDL is about one-tenth of the α-tocopherol concentration (Esterbauer and Ramos, 1995) and it is uncertain if such low concentrations indeed contribute to a significant extent to the lag phase in Cu^{2+} oxidation. Dietary supplementation with coenzyme Q_{10} increases the oxidation resistance of LDL to low fluxes of initiating radicals (Mohr et al., 1992). We found in our investigation of more than 200 LDL samples (reviewed in Esterbauer et al., 1995) that vitamin E (α- and γ-tocopherol) contribute in the statistical average 30% to the lag time in the CD method whereas 70% is due to vitamin E-independent variables. A similar relationship was found by Frei and Gaziano (1993). It should be noted, however, that there are very strong individual variations, and the majority of studies revealed that the vitamin E content of a given LDL per se is not predictive for the oxidative resistance measured by the CD method (reviewed in Esterbauer and Juergens, 1993). However, all of the vitamin E supplementation studies revealed that oral intake of α-tocopherol leads to a significant increase of the α-tocopherol content of LDL and that the increase correlates with the increase of the lag time measured *in vitro* by the CD method or slightly modified assays on LDL samples isolated from plasma of the supplemented subjects (Table IV). In our own study (reviewed in Esterbauer *et al.*, 1993), we found that daily doses of 150, 225, 800, or 1200 IU *RRR-α*-tocopherol over 3 weeks increased LDL α-tocopherol contents to 138 $\pm$ 12, 158 $\pm$ 32, 144 $\pm$ 12, and 215 $\pm$ 47% compared to the baseline value (= 100%) and that in parallel the oxidation resistance (= lag time) increased to 118 $\pm$ 17, 156 $\pm$ 22, 135 $\pm$ 23, and 175 $\pm$ 21% compared to baseline. This observation has been confirmed in several studies using different doses of vitamin E ranging from 60 to 1600 IU per day (Table IV). All of these studies found that the oral intake of vitamin E increases the oxidation resistance as judged by the CD method. Jialal *et al.* (1995) reported that a minimum dose of 400 IU vitamin E per day is needed to significantly increase the LDL oxidation resistance. Such a dose is about 10 times the recommended daily amount of adults (Machlin, 1991) and could hardly be reached without supplements.

Only one experimental animal study (Sasahara *et al.*, 1994) addressed the important question if whether the oxidation resistance of LDL is related to the extent of atherosclerosis. Pigtail monkeys were fed over 7.5 months with a cholesterol diet in combination with the synthetic antioxidant probucol at a daily dose of 60 mg/kg body weight. Consistent with previous probucol studies, it was found that the drug reduces intimal lesion size. Plasma LDL was isolated from probucol-treated and control animals, and lag times were measured with the CD assay (300 μg LDL cholesterol/ml plus 1.66 μM CuSO$_4$). The mean lag time of the control animals was 256 $\pm$ 53 min, the eight probucol-treated animals showed a significant prolongation with a mean value of 632 $\pm$ 286 min ($p = 0.0016$), and some animals had LDL with a lag time longer than 960 min. A plot of intimal lesion size versus lag time of the 16 animals showed that lesion size was inversely related to oxidation resistance of LDL as judged by the lag time; the relation was particularly evident for thoracic intimal lesions ($r^2 = 0.74$, $p = 0.0003$). In examining the individual data of this study, it is evident that this strong inverse relationship was mainly due to 3 animals that had lag times longer than 960 min, which is about four times the control value. Such a long lag time and such a strong increase cannot be obtained with vitamin E supplementation. Megadoses of 1200–2400 IU vitamin E/day gave roughly a twofold increase of the lag time (Table IV).

TABLE IV Studies Showing that Vitamin E Supplementation Increases Oxidation Resistance of LDL[a]

Doses and major findings	Reference
150, 225, 800, and 1200 IU αT per day over 3 weeks increase plasma α-tocopherol (αT) 1.46-, 1.65-, 1.83-, and 2.48-fold, LDL αT 1.38-, 1.58-, 1.44-, and 2.15-fold and lag time 1.18-, 1.56-, 1.35-, and 1.75-fold. Lag time 57 min in placebo control, 175 min with 1200 IU (for review see Esterbauer et al., 1993).	Dieber-Rotheneder et al. (1991)
1000 IU αT/day over 7 days increased plasma αT 3-fold and LDL αT 2.4-fold, lag time increased 1.4-fold (152 vs 108 min)	Princen et al. (1992)
A combination of daily 1000 mg vitC, 300 IU αT, and 18 mg β-carotene over 6 months, plasma vitC, αT, and β-carotene increased 1.3-, 1.6-, and 5-fold, lag time increase 1.35-fold from 51.8 to 70 min.	Abbey et al. (1993)
A combination of daily 1000 mg vitC, 800 IU αT, and 30 mg β-carotene over 3 months, plasma αT increased 5-fold, β-carotene 16.3-fold, vitC 2.6-fold, lag time increased 2.2-fold (114 vs 52 min), a group receiving 500 IU αT only had the same increase of lag time.	Jialal et al. (1993)
Daily doses of 800 IU α-tocopherol acetate over 10 days, LDL αT increased 1.95-fold, lag time judged by TBARS was 120 min for controls, supplemented subjects showed slower increase of TBARS with no clear lag time.	Rifici et al. (1993)
Daily doses of 1600 mg αT (= 2400 IU) plus 60 mg β-carotene over 3 months, plasma αT and β-carotene increased 2.4- and 9-fold, LDL αT and β-carotene increased 1.3- and 22-fold, significant reduced TBARS in Cu^{2+} and endothelial cell-oxidized LDL.	Reavan et al. (1993)
Subjects received 1200 mg αT (= 2400 IU) in combination with a standard diet, an oleate-rich diet, or a linoleate-rich diet. Dense LDL from the linoleate group showed lower oxidation resistance as judged by lag time and levels of conjugated dienes, oleate group showed higher oxidation resistance, the dense and boyant LDL αT values were 6.2, 4.8, and 5.2 μg/mg protein and 13.1, 11.0, and 12.3 μg/mg protein for the standard diet, oleate diet, and linoleate diet group.	Reavan et al. (1994)
Daily doses of 60, 200, 400, 800, and 1200 IU αT over 8 weeks compared to baseline plasma αT increased by 61, 80.97, 153, and 286%, LDL αT increased by 80, 116, 96, 151, and 158%, lag times increased by 9, 23, 25, 60, and 71%. Statistical analysis revealed that at least 400 IU are required to reduce oxidation resistance significantly.	Jialal et al. (1995)

[a] With the exception of the study by Reavan et al. (1993), oxidation resistance was assessed by lag time measurements of Cu^{2+}-mediated LDL oxidation.

V. Plasma Low-Density Lipoprotein Antioxidant Levels ———

Much has been written about vitamin E, and several books (Machlin, 1980; Diplock *et al.*, 1989; Ong and Packer, 1992; Packer and Fuchs, 1993) and many reviews have been published dealing with the various aspects of vitamin E biochemistry and clinical applications. It is striking, however, that comparative data on the vitamin E concentration of human plasma of larger populations in different countries are hardly available. Many of the previous studies refer to the review by Farrell (1980) who summarized earlier data from Harris *et al.* (1961) and Bieri *et al.* (1964) on serum levels of total vitamin E, i.e., α-plus γ-tocopherol. A large collection of data regarding plasma α-tocopherol levels in samples (each about 100 subjects) of European populations should be available from the MONICA study published by Gey (1992, 1993), Gey and Puska (1989), and Gey *et al.* (1991, 1993). These data were, however, not presented in table form but as graphs showing that ischemic heart disease mortality is in the study populations inversely related to the median of the absolute α-tocopherol values ($r^2 = 0.63$) and to the median of lipid standardized α-tocopherol values ($r^2 = 0.73$). The median α-tocopherol concentrations, estimated from one of these graphs (Gey, 1993), ranged from 20.1 (Scotland–Edinburgh) to 30.6 μM (Germany–Schleiz).

Table V attempts to summarize published data on α-tocopherol. Although it may not be complete, it should serve as a useful reference for studies where plasma α-tocopherol was examined in a relatively large number of subjects. The highest values (36.5 μM) are clearly those reported by Stähelin *et al.* (1984) for the Basel population, whereas the lowest values (15.9 and 17.0 μM) are those reported by Chen *et al.* (1992) for a population in China. The later study also reported the total plasma cholesterol levels, which were relatively low (127 mg/dl) for both male and female. The α-tocopherol : cholesterol ratio ($\mu M/mM$) of this Chinese population is therefore acceptable (0.48 to 0.52) and well within the range found, for example, in Austria (Table VI). Serious problems in accurate analyses can arise from inadequate sample storage as significant loss of α-tocopherol may occur if plasma or serum is stored over many years without EDTA at temperatures higher than $-70°C$ (Gey, 1995).

Similar to α-tocopherol, comprehensive and comparative data on the other plasma antioxidants are difficult to find. Within the Austrian Stroke Prevention Study (ASPS), which has in part been completed (Schmidt *et al.*, 1996a,b), the plasma concentrations of α-tocopherol, γ-tocopherol, retinol, β-carotene, α-carotene, lycopene, canthaxanthin, cryptoxanthin, and zeaxanthin/lutein have been examined for 1778 male and female subjects living in the area of the city of Graz. Plasma ascorbate and urate were also determined in a subgroup of 1458 subjects. All subjects were also examined regarding the lipid status, i.e., total cholesterol, triglycerides, HDL

TABLE V Summary of Studies of Plasma Vitamin E Concentration[a]

Study group	Mean ± SD	Comments	Reference
329, male + female, age 17–64 years, United States	24.4 ± 12.8	Total vitE, evaluation of data from Harris et al. (1961) and Bieri et al. (1964)	Farrell (1980)
701, male + female, age 16–71 years, Basel	36.5 ± 10.7	Plasma analyzed freshly, values of controls	Stähelin et al. (1984)
260, male, age 52–75 years, Hawaii	31.7 ± 12.2	Plasma stored at −75°C, 10 years, values of controls	Nomura et al. (1985)
55, male, age 35–64 years, London	24.8 ± 11.3	Plasma stored at −40°C, 6 years, values of controls	Wald et al. (1987)
143, male + female, age 36–79 years, Maryland	29.5 ± 12.4	Plasma stored at −75°C, 9 years	Schober et al. (1987)
108, male + female, age 15–99 years, Finland	22.8 ± 9.6	Plasma stored at −20°C, 14 years, values of controls	Knekt et al. (1988)
3250, female, age 35–64 years, China	17.0 ± 3.0	Total plasma cholesterol was 127 mg/dl for female and male (0.328 mM)	Chen et al. (1992)
3250, male, age 35–64 years, China	15.9 ± 3.0		
3480, male + female, age 4–93 years, United States	25.7	Serum stored at −70°C for up to 1 year	Sowell et al. (1994)
178, male, age 22–60 years, Maryland	26.4 ± 8.5	Serum stored at −70°C, 16 years	Street et al. (1994)
57, male, age 25–59 years, Spain	32.0 ± 10.0	Plasma stored at −24°C	Olmedilla et al. (1994)
54, female, age 25–59 years, Spain	32.6 ± 7.4		
14564 finish men (smoker), mean age 57.2 participants in the ATBC study	26.7	Median; 20% 21.5 μM, 80% 32.9 μM. Serum stored at −70°C	ATBC et al. (1994)
1951, male + female, age 18–65 years, Germany	30.6 ± 9.3	SD calculated from the reported SEM, median 29.1 μM, plasma stored at 8°C for 30 hr and then at −84°C.	Heseker et al. (1994)
355, male + female, age 45–74 years, Graz, Austria	30.3 ± 9.4	EDTA plasma stored at −70°C, 0.1 to 2 years, γ-tocopherol was 2.43 ± 1.22 μM	Schmidt et al. (1996a)

[a] The concentrations are given in μM and refer, if not otherwise stated, to α-tocopherol only. In case control studies, we computed the mean ±SD for all controls.

TABLE VI Plasma Antioxidants, Cholesterol, and Triglycerides of the Study Population of the Austrian Stroke Prevention Study[a]

	Mean ± SD	Median
α-Tocopherol (μM)	30.55 ± 10.41	29.03
γ-Tocopherol (μM)	2.45 ± 1.27	2.23
Retinol (μM)	1.93 ± 0.74	1.84
β-Carotene (μM)	0.49 ± 0.45	0.38
α-Carotene (μM)	0.09 ± 0.11	0.06
Lycopene (μM)	0.20 ± 0.18	0.16
Canthaxanthin (μM)	0.12 ± 0.08	0.10
Cryptoxanthin (μM)	0.25 ± 0.23	0.19
Zeaxanthin/lutein (μM)	0.56 ± 0.24	0.52
Ascorbate (μM)	56.83 ± 20.38	57.00
Urate (μM)	280.04 ± 75.21	271.7
Cholesterol (mg/dl)	230.07 ± 40.65	228
Cholesterol (mM)	5.95 ± 1.05	5.89
Triglycerides (mg/dl)	146.57 ± 102.95	124
α-Tocopherol/cholesterol (μM/mM)	5.16 ± 1.41	4.94

[a] Vitamin E, retinol, carotenoids, total cholesterol, and triglycerides were determined for 1778 participants, ascorbate and urate were determined for 1458 participants. The study included male (42%) and female (58%), ages 41 to 86 years, mean age ±SD: 62 ± 6.2 and 61 ± 6.3 years for males and females respectively. Values for a subgroup of 355 participants published by Schmidt *et al.* (1996a,b).

cholesterol, LDL cholesterol, lipoprotein(a), apolipoprotein A_1, and apolipoprotein B. A summary of the antioxidants, cholesterol, and triglycerides values found in the ASPS population is contained in Table VI. Figure 1 shows the frequency distribution of α-tocopherol, retinol, β-carotene, and ascorbate.

Heseker *et al.* (1994) published data representative for the German population comprising nearly 2000 male and female subjects in the age of 18 to >65 years. The mean ± SD values obtained in this study for plasma antioxidants are α-tocopherol, 30.6 ± 9.3 μM; retinol, 1.95 ± 0.53 μM; β-carotene, 0.67 ± 0.57 μM; and ascorbate, 76 ± 29.9 μM. Note that Heseker *et al.* (1994) reported the mean ± SEM and *n;* from these values we calculated the standard deviations (SDs) in order to facilitate a comparison with data from the ASPS and other studies (Tables V and VI). With the exception of ascorbate, the mean ± SD values are very close to the values obtained in the ASPS study. The reason why higher ascorbate values were found in the German population could be due to the fact that a photometric analysis was used in the Heseker study, whereas HPLC analysis was used in the ASPS study.

According to Gey (1995), the threshold levels of plasma antioxidants that can be considered as optimal levels regarding risk of cardiovascular disease and cancer are vitamin C, ≥ 50 μM; α-tocopherol, ≥ 30 μM (lipid

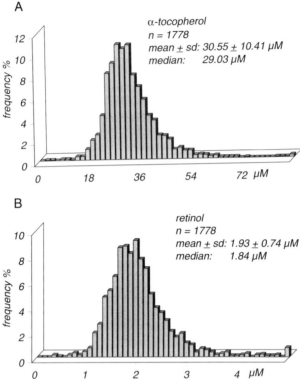

FIGURE 1 Frequency distribution of plasma concentrations of α-tocopherol (A) retinol (B), β-carotene (C), and ascorbate (D) in the study population of the Austrian Stroke Prevention Study. The study included male (42%) and female (58%) with a mean age of 61 ± 6.3 years; *n* gives the number of measured plasma samples.

standardized); β-carotene, >0.4 μM; vitamin A, >2.2 μM (lipid standardized); lycopene, >0.4 to 0.5 μM; total carotenoids, >3.2 μM; and α-tocopherol/cholesterol ratio (μM/mM), >5.1 to 5.2.

VI. Epidemiological Studies

A comprehensive review on epidemiological studies and intervention trials in relation to cardiovascular disease has been presented by Gaziano *et al.* (1994). Other reports in this context have been presented by Steinberg *et al.* (1992), van Poppel *et al.* (1994), Duell (1995), Steinberg (1995), Oliver (1995), and Hoffman and Garewal (1995). Several studies have examined the relationships between dietary intake of the antioxidants vitamin C, vitamin E, and carotenoids and cardiovascular disease (Table VII). Two studies (Vollset and Bjelke, 1983; Enstrom *et al.*, 1992) found an inverse relationship between vitamin C intake as judged by food questionnaires and

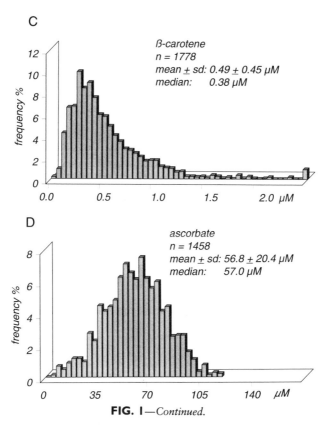

C

β-carotene
n = 1778
mean ± sd: 0.49 ± 0.45 μM
median: 0.38 μM

D

ascorbate
n = 1458
mean ± sd: 56.8 ± 20.4 μM
median: 57.0 μM

FIG. 1—*Continued.*

risk of cardiovascular disease. Other prospective studies found either no association between vitamin C intake and risk (Lapidus *et al.*, 1986; Rimm *et al.*, 1993) or only a weak and insignificant trend (Stampfer *et al.*, 1993; Knekt *et al.*, 1994). The two largest prospective dietary intake studies (Rimm *et al.*, 1993; Stampfer *et al.*, 1993) found a significant inverse relationship between the intake of vitamin E and the incidence of cardiovascular disease. The largest of these studies is the Nurses Health Study (NHS) which is based on 87,245 female nurses (United States) aged between 35 and 69 years and a detailed questionnaire comprising 116 items including intake of vitamin C, vitamin E, and β-carotene from food and supplements. The 8-year follow-up revealed that nurses in the highest quintile of vitamin E intake (100 IU vitamin E/day for more than 2 years) had a relative risk (RR) of 0.66 compared to the quintile of the lowest intake. A significant inverse relationship was also evident (RR = 0.76) for stroke (Stampfer *et al.*, 1993).

The Health Professional Follow-Up Study (HPFS) had a similar design as the NHS and involved 39,910 male dentists, pharmacists, and veterinarians between 40 and 75 years of age. The 4-year follow-up revealed a significant inverse relationship between vitamin E intake and risk of coronary heart

TABLE VII Epidemiological Studies on Relationship between Dietary Intake of Vitamin E, Vitamin C, and β-Carotene and Cardiovascular Disease

Study design	Major findings	Reference
16,713 men/women (Finland), periodic questionnaire on vitC intake from vegetables/fruits, 11-year follow-up, 438 cases of fatal cardiovascular disease (CVD)	Inverse relationship between vitC intake and risk	Vollset and Bjelke (1983)
1462 women (Sweden), questionnaire (24-hr recall) on vitC intake, 12-year follow-up, 28 cases of myocardial infarction, 13 cases of stroke	No association between vitC intake and risk	Lapidus *et al.* (1986)
Massachusetts Elderly Cohort Study: 1299 men/women questionnaire on β-carotene intake from vegetables/fruits, average of about 5-year follow-up, 151 cases of fatal CVD	Inverse relationship between β-carotene intake and risk for fatal myocardial infarction. RR = 0.32 in highest quartile	Gaziano *et al.* (1992)
National Health and Nutrition Examination Survey (NHANES-1): 11,349 middle-aged men/women (U.S.), food questionnaire of vitC intake, 10-year follow-up, 929 cases of fatal CVD	Inverse relationship between vitC intake and risk. RR = 0.66 with intake of 50 mg or more	Enstrom *et al.* (1992)
Health Professionals Follow-Up Study (HPFS): 39,910 health professionals (U.S.), questionnaire regarding vitC, vitE, β-carotene intake from food or supplements and other items, 4-year follow-up, 667 major cases of coronary heart disease (CHD), 106 fatal myocardial infarctions, 201 nonfatal myocardial infarctions, 360 revascularizations	Inverse relationship between vitE and β-carotene intake and risk. RR = 0.68 and 0.75 for highest quintile of vitE and β-carotene. No relationship with vitC intake	Rimm *et al.* (1993)
Nurses Health Study (NHS): 87,245 female nurses (U.S.), questionnaire as in the HPFS, 8-year follow-up, 552 major cases of CHD, including 115 fatal and 437 nonfatal myocardial infarctions, in addition 183 cases of ischemic stroke	Inverse relationship between vitE and β-carotene intake and risk for CHD. RR = 0.66 and 0.78 for highest quintile of vitE and β-carotene. RR = 0.80 for vitC but not significant. RR for stroke = 0.76 and 0.61 for highest quintile of vitE and β-carotene.	Stampfer *et al.* (1993)
5133 men and women (Finland), food questionnaire regarding intake of vitC, vitE, and β-carotene, 14-year follow-up, 244 fatal cases of CVD, including 186 men and 58 women	For both sexes inverse relationship between vitE intake and risk, for women similar trend for vitC. β-Carotene appears to potentiate vitE effect	Knekt *et al.* (1994)

disease. Subjects in the highest quintile of intake (100 IU/day over more than 2 years) compared with the lowest quintile had a relative risk of 0.68 (Rimm *et al.*, 1993). Consistent with the NHS and HPFS is the outcome of a Finnish study (Knekt *et al.*, 1994), which also revealed an inverse relationship between vitamin E intake and risk of cardiovascular disease during a 14-year follow-up period. The NHS and HPFS also evaluated β-carotene intake. The relative risk of subjects in the highest quintile compared to the subjects in the lowest quintile was 0.75 (HPFS) and 0.78 (NHS).

Gaziano *et al.* (1992) examined the effect of dietary β-carotene on cardiovascular disease (the Massachusetts Elderly Cohort Study) and found that the relative risks for fatal myocardial infarction were 1.00, 0.77, 0.59, and 0.32 from the lowest to the highest quintile of intake.

Questionnaires on dietary intake can give only a rough estimate on the true antioxidant status of subjects. More precise are studies in which the antioxidant status is measured (Table VIII). The MONICA study (Gey and Puska, 1989; Gey, 1993; Gey *et al.*, 1991) examined the plasma concentrations of vitamin E, vitamin C, vitamin A, and β-carotene in 16 European regions with each of about 100 healthy males ages 40–49 years. This study revealed that the incidence of coronary heart disease is inversely related to the concentration of vitamin E and, to a lesser extent, to vitamin A and vitamin C. The coronary heart disease mortality is about six times higher in some northern European countries (Finland, Scotland, Northern Ireland) than in the Mediterranean countries, and 60% of this difference can be explained by the difference in plasma vitamin E concentrations. Up to 90% of the difference can statistically be explained by the combined antioxidants, i.e., vitamin E, vitamin A, and vitamin C.

The Basel Prospective Study (Gey *et al.*, 1993) involved 2000 Swiss male with a mean age of 62 years. Plasma levels of α-tocopherol, vitamin C, and β-carotene were measured at baseline. The 12-year follow-up revealed no relationship between the incidence of coronary heart disease or stroke and the vitamin E concentration, probably because the Basel population had exceptionally high (see also Table V) plasma α-tocopherol levels with a median of 34.6 μM. The Basel study, however, showed that subjects with low β-carotene and vitamin C were at a higher risk for ischemic heart disease and stroke. An inverse relationship between plasma β-carotene and risk of myocardial infarction was also obtained in the prospective study presented by Morris *et al.* (1994). Several nested case control studies examined the relationship between antioxidants and coronary vascular disease and reported that patients have lower plasma β-carotene concentrations (Riemersma *et al.*, 1991; Street *et al.*, 1994). and lower vitamin C and vitamin E concentrations (Riemersma *et al.*, 1991). The EURAMIC study (Kardinaal *et al.*, 1993) measured vitamin E and β-carotene in adipose tissue of patients who had an acute myocardial infarction and in controls. This case control study revealed significantly lower β-carotene levels.

TABLE VIII Epidemiological Studies on Relationship between Antioxidants in Plasma or Adipose Tissue and Cardiovascular Disease

Study design	Major findings	Reference
WHO/MONICA study: Cross-cultural comparison of coronary heart disease mortality in 16 European populations regarding major plasma antioxidants. Study sample in each population consists of about 100 healthy males 40–49 years of age. The 16 populations show a six-fold difference in coronary heart disease mortality.	Significant inverse relationship between vitE and coronary heart disease mortality ($r^2 = 0.63$). About 60% of the difference in coronary heart disease can be explained by differences in vitE alone, up to 90% by vitE + vitA + vitC	Gey (1989, 1993) Gey et al. (1991)
Case control study with males, 35–54 years of age (Edinburgh, Scotland) regarding relationship between angina pectoris and plasma antioxidants (vitE, A, C, β-carotene); 110 cases, 394 controls	Patients had lower values of vitC, β-carotene, and vitE. RR = 2.68 in lowest quintile of vitE ($< 19\ \mu M$). RR = 1.41 and 1.63 for lowest quintile of β-carotene and vitC, respectively	Riemersma et al. (1991)
EURAMIC study: Case control study including 12 European countries regarding relationship between acute myocardial infarction and vitE and β-carotene in adipose tissue, 683 cases of acute myocardial infarction, 727 controls	Significantly lower levels of β-carotene in patients, RR = 1.98 for lowest percentile, no relationship to vitE	Kardinaal et al. (1993)
Basel Prospective Study: 2000 Swiss males with mean age of 62 years. Major plasma antioxidants measured shortly after blood sampling, 12-year follow-up for ischemic heart disease (87 fatal cases) and stroke (31 fatal cases)	Subjects with low β-carotene and vitC have an increased risk for ischemia heart disease (RR = 1.96) and stroke (RR = 4.17). No relationship with vitE, likely because vitE Basel is high, median 34.6 μM.	Gey et al. (1993)
Lipid Research Clinics Coronary Primary Prevention Trial (LRC-LPPT): 1899 subjects (40 to 59 years). Plasma β-carotene measured at baseline 13-year follow-up regarding coronary heart disease	Inverse relationship between β-carotene and risk, RR = 0.64 in highest quartile of β-carotene	Morris et al. (1994)
Case control study, 123 cases with myocardial infarction (23 to 58 years) and 246 controls, serum collected 7 to 14 years before diagnosis. α-Tocopherol, β-carotene, lycopene, lutein, and zeaxanthin were analyzed.	Increasing risk with decreasing β-carotene (p value for trend 0.02), a suggestive trend for lutein ($p = 0.09$)	Street et al. (1994)
Austrian Stroke Prevention Study (ASPS): A subgroup of 355 males and females (Graz, Austria), aged 45–74 years, was investigated by magnetic resonance imaging regarding occurrence of white matter hyperintensities in the brain. Also measured were plasma levels of vitE, vitC carotenoids, and lipids (see also Table VI)	Significant inverse relationship between α-tocopherol and occurrence of WWMH in brain. Similar trend for lycopene. RR = 7.11 and 5.20 for lowest quintile of α-tocopherol and lycopene, respectively. No relationship to vitC and β-carotene.	Schmidt et al. (1996a,b)

In the Austrian Stroke Prevention Study (ASPS) (Schmidt *et al.*, 1994, 1996a,b), 355 male and female subjects between 45 and 74 years of age were examined for relationships between the occurrence of white matter hyperintensities in the brain and plasma concentrations of vitamin E, vitamin C, β-carotene, and other carotenoids. White matter hyperintensities detectable by magnetic resonance imaging occur in approximately 50% of elderly subjects, and it is assumed that they represent a subclinical form of ischemic brain damage due to an early form of subcortical atherosclerotic encephalopathy. The ASPS revealed that the plasma levels of α-tocopherol and lycopene are significantly lower in subjects with early confluent and confluent white matter hyperintensities. The relative risk of persons in the lowest quintile of the α-tocopherol and lycopene distribution compared to those in the highest quintile was 7.11 and 5.20, respectively.

Large-scale, controlled chemoprevention trials used to evaluate specifically the effect of vitamin E, vitamin C, or β-carotene on the incidence of cardiovascular disease have not yet been published (Table IX). The Finnish ATBC Study (ATBC Study Group, 1994) examined supplements of dl-α-tocopheryl acetate and synthetic β-carotene in about 29,000 male smokers who were 50 to 69 years old. The primary objective was to evaluate the effect of the supplements on lung cancer incidence. Over a period of 5 to 8 years, the four groups participating in this study received 50 mg tocopheryl acetate/day, 50 mg tocopheryl acetate plus 20 mg β-carotene/day, 20 mg β-carotene/day, or a placebo. Although statistically not significant, the tocopherol group had somewhat fewer cases of fatal ischemic heart disease, but the cases of hemorrhagic stroke were slightly increased. In the β-carotene group, slightly more death occurred from ischemic heart disease than in the control group without β-carotene (mortality rate 77.1 versus 68.9 per 10.000 person years). The tocopherol group had a lower incidence of prostata cancer (rate 11.7 versus 17.8). A striking result of the ATBC was that more death occurred due to lung cancer, ischemic heart disease, and hemorrhagic stroke in the group receiving β-carotene.

A very large antioxidant supplementation trial with 29,000 males and females was performed in Linxian (China). The population of Linxian has a very high incidence of stomach and esophagus cancer, but a relatively low incidence of cardiovascular disease. The objective of the study was to evaluate the effects of vitamins and selenium on cancer incidence. The best effects were seen in the group receiving a combination of β-carotene (15 mg/day), α-tocopherol (30 IU/day), and selenium (50 μg/day) over a 5-year period. This combination reduced cancer mortality by 13% and slightly reduced the risk of stroke (RR = 0.90), but this trend did not reach statistical significance.

In a subgroup of the U.S. Physician Health Study, the therapeutic effect of β-carotene supplements (50 mg/day) on 330 cardiovascular disease patients was examined (Gaziano *et al.*, 1990). The 5-year follow-up revealed a significant reduction of cases of myocardial infarction and stroke. Another

TABLE IX Randomized Primary Intervention Trials on Prevention of Cardiovascular Disease by Antioxidants[a]

Study design	Major findings	Reference
20 angina pectoris patients received 3200 IU vitE/day over 9 weeks	No pain reduction	Anderson and Reid (1974)
52 angina pectoris patients received 1600 IU vitE/day over 6 months	No significant improvement	Gillian et al. (1977)
Physicians Health Study: Subgroup of 333 doctors with angina pectoris or bypass surgery or balloon angiography were treated with 50 mg β-carotene on alternate days for 5 years.	β-Carotene led to a significant reduction of fatal and nonfatal myocardial infarctions and fatal and nonfatal stroke. RR = 0.49 and 0.46 for infarction and stroke, respectively	Gaziano et al. (1990)
100 patients received following angioplasty 50 mg β-carotene on alternate days over 5 years	No significant reduction of restenosis	DeMaio et al. (1992)
29,000 clinically healthy men and women, age 40 to 69 years (Linxian, China), received a combination of 30 IU vitE, 15 mg β-carotene, and 50 μg selenium daily for 5 years	Trend, although not significant to a reduction of stroke; total mortality decreased by 9%, cancer death decreased by 13%, no data for heart disease.	Blot et al. (1993)
29,133 male smoker (Finnland) received 20 mg β-carotene and 50 mg vitE–acetate daily for 5–8 years	A trend to reduction (5%) of coronary heart disease; a 34% reduction of prostata carcinoma, but total cancer death rate increased.	ATBC study (1994)

[a] Note that the primary objective of the study by Blot et al. (1993) and the ATBC study was cancer prevention.

study (DeMaio *et al.*, 1992) found no effect of β-carotene on the incidence of restenosis in angioplasty patients. Two small studies with angina pectoris patients (Anderson and Reid, 1974; Gillian *et al.*, 1977) revealed that relatively large doses of vitamin E (1000 to 3200 IU/day) have no therapeutic effects.

VII. Conclusion

The LDL oxidation hypothesis has brought a great impact for free radical research in general and lipid peroxidation in particular. Many laboratories previously only marginally interested have moved into this interdisciplinary field of research. Consequently, significant progress has been made in the elucidation of basic principles of oxidative stress and antioxidant action.

A long list of putative pro-atherogenic properties of minimally and fully oxidized LDL has been identified by *in vitro* studies. The relative importance of these properties for the development of early lipid deposits, fatty streaks, advanced plaques, and occlusion is, however, to a large extent uncertain. That oxidation of LDL can take place *in vivo* in the artery wall has been clearly proven by immunohistochemical studies and investigations of arterial wall LDL. These observations, however, do not prove that LDL oxidation plays in any way or at any point a causal role in the development of atherosclerosis. With the current knowledge it cannot be excluded that LDL oxidation is a disease-associated epiphenomenon and rather the consequence than the cause of atherosclerosis. Several experimental animal studies not discussed in this chapter have revealed that relatively large doses of natural or synthetic antioxidants can reduce the rate of progression of atherosclerotic lesions. Epidemiological studies on the relationship between dietary antioxidant intake and plasma concentrations of antioxidants and cardiovascular disease also support that low levels of antioxidants, particularly vitamin E, are associated with an increased incidence of atherosclerosis. The putative positive effect of dietary vitamin E, vitamin C, and β-carotene may, however, not be related to antioxidant effects but to other anti-atherogenic properties. Moreover, in dietary intake studies, other food constituents (e.g., flavonoids) could have an important confounding beneficial effect. Chemoprevention trials are now needed to evaluate the effects of vitamin E, vitamin C, and β-carotene specifically in cardiovascular disease. The available epidemiological data strongly suggest that a clear benefit of these antioxidants cannot be expected for all but rather for such persons who have suboptimal levels. This needs to be considered in designing the strategy of intervention trials.

Acknowledgments ———————————————————————————————

The ASPS study has been supported by the Jubiläumsfonds der Österreichischen National-bank (Project 4484). Support by the Austrian Science Foundation (Project F709) is also acknowledged. We thank Karin Gollner for technical assistance in antioxidant analysis for ASPS.

References ———————————————————————————————

Abbey, M., Nestel, P. J., and Baghurst, P. A. (1993). Antioxidant vitamins and low-density-lipoprotein oxidation. *Am. J. Clin. Nutr.* **58**, 525–532.

Alexander, R. W. (1995). Hypertension and the pathogenesis of atherosclerosis. Oxidative stress and the mediation of arterial inflammatory response: A new perspective. *Hypertension* **25**, 155–161.

Anderson, T. W., and Reid, W. (1974). A double-blind trial of vitamin E in the treatment of angina pectoris. *Am. Heart J.* **93**, 444–449.

ATBC Cancer Prevention Study Group. (1994). The alpha-tocopherol, beta-carotene lung cancer prevention study: Initial results from a controlled trial. *N. Engl. J. Med.* **330**, 1029–1035.

Autio, I., Jaakkola, O., Solakivi, T., and Nikkari, T. (1990). Oxidized low-density lipoprotein is chemotactic for arterial smooth muscle cells in culture. *FEBS Lett.* **277**, 274–249.

Badimon, J. J., Fuster, V., Chesobro, J. H., and Badimon, L. (1993). Coronary atherosclerosis: A multifactorial disease. *Circulation* **87** (Suppl. II), 3–16.

Berliner, J. A., Navab, M., Fogelman, A. M., Frank, J. S., Demer, L. L., Edwards, P. A., Watson, A. D., and Lusis, A. (1995). Atherosclerosis: Basic mechanism, oxidation, inflammation, and genetics. *Circulation* **91**, 2488–2496.

Berliner, J. A., Territo, M. C., Sevanian, A., Ramin, S., Kim, J. A., Ramshad, B., Sterson, M., and Fogelman, A. M. (1990). Minimally modified low density lipoprotein stimulates monocyte endothelial interactions. *J. Clin. Invest.* **85**, 1260–1266.

Bieri, J. G., Teets, L., Beavady, B., and Andrews, E. L. (1964). Serum vitamin E levels in a normal adult population in the Washington, D.C., area. *Proc. Soc. Exp. Biol.* **117**, 131–133.

Block, G., Patterson, B., and Subar, A. (1992). Fruit, vegetables, and cancer prevention: A review of the epidemiological evidence. *Nutr. Cancer* **18**, 1–29.

Blot, W. J., Li, J., Taylor, P. R., Guo, W., Dawsey, S., Wang, G., Yang, C. S., Zheng, S., Gail, M., Li, G., Yu, Y., Liu, B., Tangrea, J., Sun, Y., Liu, F., Fraumeni, J. F., Zhang, Y., and Li, B. (1993). Nutritional intervention trials in Linxian, China: Supplementation with specific vitamin/mineral combinations, cancer incidence, and disease specific mortality in the general population. *J. Natl. Cancer Inst.* **85**, 1483–1492.

Bowry, V. W., and Stocker, R. (1993). Tocopherol-mediated peroxidation: The prooxidant effect of vitamin E on the radical initiated oxidation of human low-density lipoprotein. *J. Am. Chem. Soc.* **115**, 6029–6044.

Cazzolato, G., Avogaro, P., And Bittolo-Bon, G. (1991). Characterization of a more electronegatively charged LDL subfraction by ion exchange HPLC. *Free Rad. Bio. Med.* **11**, 247–253.

Chait, A., and Heinecke, J. W. (1994). Lipoprotein modification: Cellular mechanisms. *Curr. Opin. Lipid.* **5**, 365–370.

Chen, J., Geissler, C., Parpia, B., Li, J., and Campbell, T. C. (1992). Antioxidant status and cancer mortality in China. *Int. J. Epidemiol.* **21**, 625–635.

Chisolm, G. M. (1992). The oxidation of lipoproteins: Implications for atherosclerosis. *In* "Biological Consequences of Oxidative Stress: Implications for Cardiovascular Disease and Carcinogenesis" (L. Spatz and A. D. Bloom, eds.), pp. 78–106. Oxford Univ. Press, New York.

Chiu, H. C., Jeng, J. R., and Shieh, S. M. (1994). Increased oxidizability of plasma low density lipoprotein from patients with coronary artery disease. *Biochim. Biophys. Acta* **1225**, 200–208.

Cominacini, L., Garbin, U., Pastorini, A. M., Davoli, A., Campagnola, M., De Santis, A., Pasini, C., Faccini, G. B., Trevisan M. T., Bertozzi, L., Pasini, F., and Lo Cascio, V. (1993). Predisposition to LDL oxidation in patients with and without angiographically established coronary artery disease. *Atherosclerosis* **99**, 63–70.

Cominacini, L., Pastorino, A. M., Garbin, U., Campagnola, M., Desantis, A., Davoli, A., Faccini, G., Bertozzo, L., Pasini, F., Pasini, A. F., and Locascio, V. (1994). The susceptibility of low-density lipoprotein to in vitro oxidation is increased in hypercholesterolemic patients. *Nutrition* **10**, 527–531.

Croft, K. D., Dimmitt, S. B., Moulton, C., and Lawrence, J. B. (1992). Low density lipoprotein composition and oxidizability in coronary disease-apparent favourable effect of beta blockers. *Atherosclerosis* **97**, 123–130.

Cushing, S. D., Berliner, J. A., Valente, A. J., Territo, M. C., Navab, M., Perhami, R., Gerrity, R., Schartz, C. J., and Fogelman, A. M. (1990). Minimally modified low density lipoprotein induces monocyte chemotactic protein 1 in human endothelial cells and smooth muscle cells. *Proc. Natl. Acad. Sci. U.S.A.* **87**, 5134–5138.

Daret, D., Blin, P., Dorian, B., Rigaud, M., and Larrue, J. (1993). Synthesis of monohydroxylated fatty acids from linoleic acid by rat aortic smooth muscle cells and tisues: Influence on prostacyclin production. *J. Lipid. Res.* **34**, 1473–1482.

DeMaio, S. J., King, S. B., III, Lembo, N. J., Roubin, G. S., Hearn, J. A., Bhagavan, H. N., and Sgoutas, D. S. (1992). Vitamin E supplementation, plasma lipids and incidence of restenosis after percutaneous transluminal coronary angioplasty (PTCA). *J. Am. Coll. Nutr.* **11**, 131–138.

De Rijke, Y. B., Vogelezang, C. J. M., Van Berkel, T. J., Princen, H. M., Verwey, H. F., Van der Laarse, A., and Bruschke, A. V. G. (1992). Susceptibility of low-density lipoproteins to oxidation in coronary bypass patients. *Lancet* **340**, 858–859.

Dieber-Rotheneder, M., Puhl, H., Waeg, G., Striegl, G., and Esterbauer, H. (1991). Effects of oral supplementation with d-alpha-tocopherol on the vitamin E content of human low density lipoproteins and its oxidation resistance. *J. Lipid Res.* **8**, 1325–1332.

Diplock, A. T., Machlin, L. J., Packer, L., and Pryor, W. A., eds. (1989). Vitamin E: Biochemistry and health implications. *Ann. N.Y. Acad. Sci.* **570**.

Drake, T. A., Hannani, K., Fei, H., Lavi, S., and Berliner, J. A. (1991). Minimally oxidized low-density lipoprotein induces tissue factor expression in cultured human endothelial cells. *Am. J. Pathol.* **138**, 601–607.

Duell, P. B. (1995). The role of dietary antioxidants in prevention of atherosclerosis. *Endocrinologist* **5**, 347–356.

Enstrom, J. E., Kanim, L. E., and Klein, M. A. (1992). Vitamin C intake and mortality among a sample of the United States population. *Epidemiology* **3**, 194–202.

Esterbauer, H., Gebicki, J., Puhl, H., and Juergens, G. (1992). The role of lipid peroxidation and antioxidants in oxidative modification of LDL. *Free Rad. Bio. Med.* **13**, 341–390.

Esterbauer, H., Gieseg, S. P., Giessauf, A., Ziouzenkova, O., and Ramos, P. (1995). Role of natural antioxidants in inhibiting Cu^{++} mediated oxidation of LDL. *In* "Free Radicals, Lipoprotein Oxidation and Atherosclerosis" (G. Bellomo, G. Finardi, E. Maggi and C. Rice-Evans, eds.), pp. 11–26. Richelieu Press, London.

Esterbauer, H., and Juergens, G. (1993). Mechanistic and genetic aspects of susceptibility of LDL to oxidation. *Curr. Opin. Lipidol.* **4**, 114–124.

Esterbauer, H., Puhl, H., Waeg, G., Krebs, A., and Dieber-Rotheneder, M. (1993). The role of vitamin E in lipoprotein oxidation. *In* "Vitamin E in Health and Disease" (L. Packer and J. Fuchs, eds.), pp. 649–671. Dekker, New York.

Esterbauer, H., and Ramos, P. (1995). Chemistry and pathophysiology of oxidaton of LDL. *Rev. Physiol. Biochem. Pharmacol.* **127**, 31–64.

Farrell, P. M. (1980). Deficiency states, pharmacological effects, and nutrient requirements. *In* "Vitamin E: A Comprehensive Treatise" (L. J. Machlin, ed.), pp. 520–620. Dekker, New York.

Fox, P. L., Chisolm, G. M., and DiCorleto, P. E. (1987). Lipoprotein-mediated inhibition of endothelial cell production of platelet-derived growth factor-like protein depends on free radical lipid peroxidation. *J. Biol. Chem.* **262**, 6046–6054.

Frei, B., and Gaziano, J. M. (1993). Content of antioxidants, preformed lipid hydroperoxides, and cholesterol as predictors of the susceptibility of human LDL to metal ion-dependent and -independent oxidation. *J. Lipid Res.* **34**, 2135–2145.

Frostegard, J., Wu, R., Giscombe, R., Holm, G., Lefvert, A. K., and Nilsson, J. (1992). Induction of T-cell activation by oxidized low density lipoproteins. *Arterioscler. Thromb.* **12**, 461–467.

Gaziano, J. M., Manson, J. E., Branch, L. G., LaMott, F., Colditz, G. A., Buring, J. E., and Hennekens, C. H. (1992). Dietary beta carotene and decreased cardiovascular mortality in an elderly cohort. *J. Am. Coll. Cardiol.* **19**, 377.

Gaziano, J. M., Manson, J. E., and Hennekens, C. H. (1994). Natural antioxidants and cardiovascular disease: Observational epidemiologic studies and randomized trials. *In* "Natural Antioxidants in Human Health and Disease" (B. Frei, ed.), pp. 387–409. Academic Press, San Diego.

Gaziano, J. M., Manson, J. E., Ridker, P. M., Buring, J. E., and Hennekens, C. H. (1990). Beta carotene therapy for chronic stable angina. *Circulation* **82** (Suppl. III), 202.

Gebicki, J., Juergens, G., and Esterbauer, H. (1991). Oxidation of low density lipoprotein *in vitro*. *In* "Oxidative stress" (H. Sies, ed.), pp. 371–397. Academic Press, London.

Gey, K. F. (1992). Epidemiological correlations between poor plasma levels of essential antioxidants and the risk of coronary heart disease and cancer. *In* "Lipid-Soluble Antioxidants: Biochemistry and Clinical Applications" (A. S. H. Ong and L. Packer, eds.), pp. 442–456. Birkhäuser Verlag, Basel.

Gey, K. F. (1993). Vitamin E and other essential antioxidants regarding coronary heart disease: Risk assessment studies. *In* "Vitamin E in Health and Disease" (L. Packer and J. Fuchs, eds.), pp. 589–633. Dekker, New York.

Gey, K. F. (1995). Prevention of early stages of cardiovascular disease and cancer may require concurrent optimization of all major antioxidants and other nutrients: An update and reevaluation of observational data and intervention trials. *In* "Free Radicals, Lipoprotein Oxidation and Atherosclerosis" (G. Bellome, E. Maggi, and C. Rice-Evans, eds.). Richelieu Press, London.

Gey, K. F., and Puska, P. (1989). Plasma vitamin E and A inversely correlated to mortality from ischemic heart disease in cross-cultural epidemiology. *Ann. N.Y. Acad. Sci.* **570**, 254–282.

Gey, K. F., Puska, P., Jordan, P., and Moser, U. K. (1991). Inverse correlating between plasma vitamin E and mortality from ischemic heart disease in cross-cultural epidemiology. *Am. J. Clin. Nutr.* **53**, 326S–334S.

Gey, K. F., Stahelin, H. B., and Eichholzer, M. (1993). Poor plasma status of carotene and vitamin C is associated with higher mortality from ischemic heart disease and stroke: Prospective Basel Study. *Clin. Invest.* **71**, 3–6.

Gillilan, R. E., Mandell, B., and Warbasse, J. R. (1977). Quantitative evaluation of vitamin E in the treatment of angina pectoris. *Am. Heart J.* **93**, 444–449.

Haberland, M. E., Fong, D., and Cheng, L. (1988). Malondialdehyde-altered protein occurs in atheroma of Watanabe heritable hyperlipidemic rabbits. *Science* **241**, 215–218.

Haberland, M. E., and Steinbrecher, U. P. (1992). Modified low-density lipoproteins: Diversity and biological relevance in atherogenesis. *In* "Molecular Genetics of Coronary Artery disease: Candidate Genes and Processes in Atherosclerosis" (A. J. Lusis, J. I. Rotter, and R. S. Sparks, eds.) pp. 35–61. Monogr. Hum. Genet. 14.

Halliwell, B. (1995). Oxidation of low-density lipoproteins: Questions of initiation, propagation, and the effect of antioxidants. *Am. J. Clin. Nutr.* **61**, 670S–677S.

Hamilton, T. A., Ma, G. P., and Chisolm, G. M. (1990). Oxidized low density lipoprotein suppresses the expression of tumor necrosis factor-alpha mRNA in stimulated murine peritoneal macrophages. *J. Immunol.* **144**, 2343–2350.

Harris, P. L., Hardenbrock, E. G., Dean, F. P., Cusack, E. R., and Jensen, J. L. (1961). Blood tocopherol values in normal human adults and incidence of vitamin E deficiency. *Proc. Soc. Exp. Biol. Med.* **107**, 381–383.

Henriksen, T., Mahoney, E. M., and Steinberg, D. (1981). Enhanced macrophage degradation of low density lipoprotein previously incubated with cultured endothelial cells: Recognition by receptors for acetylated low density lipoproteins. *Proc. Natl. Acad. Sci. U.S.A.* **78**, 6499–6503.

Heseker, H., Schneider, R., Moch, K. J., Kohlmeier, M., and Kübler, W. (1994). Vitaminversorgung Erwachsener in der Bundesrepublik Deutschland. *In* "VERA-Schriftenreihe Band IV." Wissenschaftlicher Fachverlag Dr. Fleck.

Hoffmann, R. M., and Garewal, H. S. (1995). Antioxidants and the prevention of coronary heart disease. *Arch. Intern. Med.* **155**, 241–246.

Jialal, I., Fuller, C. J., and Huet, B. A. (1995). The effect of α-tocopherol supplementation on LDL oxidation: A dose-response study. *Arterioscler. Thromb. Vasc. Biol.* **15**, 190–198.

Jialal, I., and Grundy, S. M. (1993). Effect of combined supplementation with α-tocopherol, ascorbate, and beta carotene on low-density lipoprotein oxidation. *Circulation* **88**, 2780–2786.

Juergens, G., Chen, Q., Esterbauer, H., Mair, S., Ledinski, G., and Dinges, H. P. (1993). Immunostaining of human autopsy aortas with antibodies to modified apolipoprotein B and apoprotein (a). *Arterioscl. Thromb.* **13**, 1689–1699.

Kardinaal, A. F. M., Kok, F. J., Ringstad, J., Gomez-Aracena, J., Mazaev, V. P., Kohlmeier, L., Martin, B. C., Aro, A., Kark, J. D., Delgado-Rodriguez, M., Riemersma, R. A., Huttunen, J. K., and Martin-Moreno, J. M. (1993). Antioxidants in adipose tissue and risk of myocardial infarction: The EURAMIC study. *Lancet* **342**, 1379–1384.

Keaney, J. F., Jr., and Frei, B. (1994). Antioxidant protection of low-density lipoprotein and its role in the prevention of atherosclerotic vascular disease. *In* "Natural Antioxidants in Human Health and Disease" (B. Frei, ed.), pp. 303–357. Academic Press, San Diego.

Keidar, S., Kaplan, M., Shapira, C., Brook, J. G., and Aviram, M. (1994). Low density lipoprotein isolated from patients with essential hypertension exhibits increased propensity for oxidation and enhanced uptake by macrophages: A possible role for angiotensin II. *Atherosclerosis* **107**, 71–84.

Kim, J. A., Territo, M. C., Wayner, E., Carlos, T. M., Parhami, F., Smith, C. W., Haberland, M. W., Fogelman, A. M., and Berliner, J. A. (1994). Partial characterization of leukocyte binding molecules on endothelial cells induced by minimally oxidized LDL. *Arterioscler. Thromb.* **14**, 427–433.

Knekt, P., Aromaa, A., and Maatela, J., *et al.* (1988). Serum vitamin E, serum selenium and the risk of gastrointestinal cancer. *Int. J. Cancer*, **42**, 846–850.

Knekt, P., Reunanen, A., Järvinen, R., Seppänen, R., Heliövaara, M., and Aromaa, A. (1994). Antioxidant vitamin intake and coronary mortality in a longitudinal population study. *Am. J. Epidemiol.* **139**, 1180–1189.

Krauss, R. M. (1991). Low-density lipoprotein subclasses and risk of coronary artery disease. *Curr. Opin. Lipidol.* **2**, 248–252.

Kühn, H., Belkner, J., Wiesner, R., Schewe, T., Lakin, V. Z., and Tikhaze, A. K. (1992). Structure elucidation of oxygenated lipids in human atherosclerotic lesions. *Eicosanoids* **5**, 17–22

Kühn, H., Belkner, J., Zaiss, S., Fährenklemper, T., and Wohlfeil, S. (1994). Involvement of 15-lipoxygenase in early stages of atherogenesis. *J. Exp. Med.* **179**, 1903–1911.

Kugiyama, K., Sakamoto, T., Misumi, I., Sugiyama, S., Ohgushi, M., Ogawa, H., Horiguchi, M., and Yasue, H. (1993). Transferable lipids in oxidized low-density lipoprotein stimulate plasminogen activator inhibitor-1 and inhibit tissue-type plasminogen activator release from endothelial cells. *Circ. Res.* **73,** 335–343.

Lapidus, L., Anderson, H., Bengtson, C., and Bosceus, I. (1986). Dietary habits in relation to incidence of cardiovascular disease and death in women, a 12 year follow-up of participants in the study of women in Gothenberg, Sweden. *Am. J. Clin. Nutr.* **44,** 444–448.

Lass, A., Belkner, J., Esterbauer, H., and Kühn, H. (1996). Lipoxygenase treatment renders low-density lipoprotein susceptible to Cu^{2+}-catalysed oxidation. *Biochem. J.* **314,** 577–585.

Lehr, H. A., Hübner, C., Finckh, G., Angermüller, S., Nolte, D., Beisiegel, U., Kohschütter, A., and Messmer, K. (1991). Role of leukotrienes in leukocyte adhesion following systemic administration of oxidatively modified human low density lipoprotein in hamsters. *J. Clin. Invest.* **88,** 9–14.

Liao, F., Berliner, J. A., Mehrabian, M., Navab, M., Demer, L. L., Lusis, A. J., and Fogelman, A. M. (1991). Minimally modified low density lipoprotein is biologically active in vivo in mice. *J. Clin. Invest.* **87,** 2253–2257.

Lynch, S. M., and Frei, B. (1994). Antioxidants as antiatherogens: Animal studies. *In* "Natural Antioxidants in Human Health and Disease" (B. Frei, ed.), pp. 353–385. Academic Press, San Diego.

Machlin, L. J., ed. (1980). "Vitamin E: A Comprehensive Treatise." Dekker, New York.

Machlin, L. J. (1991). Vitamin E. *In* "Handbook of Vitamins" (L. J. Machlin, ed.), pp. 99–144. Dekker, New York.

Maggi, E., Belazzi, R., Falaschi, F., Frattoni, A., Perani, G., Finardi, G., Gazo, A., Nai, M., Romanini, D., and Bellomo, G. (1994a). Enhanced LDL oxidation in uremic patients—an additional mechanism for accelerated atherosclerosis. *Kidney Int.* **45,** 876–883.

Maggi, E., Chiesa, R., Melissani, G., Castellani, R., Astore, D., Grossi, A., Finardi, G., and Bellomo, G. (1994b). LDL oxidation in patients with severe carotid atherosclerosis: A study of in vitro and in vivo oxidation markers. *Arterioscler. Thromb.* **14,** 1892–1899.

Maggi, E., Marchesi, E., Ravetta, V., Falaschi, F., Finardi, G., and Bellomo, G. (1993). Low-density lipoprotein oxidation in essential hypertension. *J. Hyperten.* **11,** 1103–1111.

Malden, L. T., Chait, A., Raines, E. W., and Ross, R. (1991). The influence of oxidatively modified low density lipoproteins on expression of platelet-derived growth factor by human monocyte-derived macrophages. *J. Biol. Chem.* **266,** 13901–13907.

Mohr, D., Bowry, V. W., and Stocker, R. (1992). Dietary supplementation with coenzyme Q_{10} results in increased levels of ubiquinol-10 within circulating lipoproteins and increased resistance of human low-density lipoprotein to the initiation of lipid peroxidation. *Biochim. Biophys. Acta* **1126,** 247–254.

Morel, D. W., DiCorlet, P. E., and Chisolm, G. M. (1984). Endothelial and smooth muscle cells alter low density lipoprotein in vitro by free radical oxidation. *Atherosclerosis* **4,** 357–364.

Morel, D. W., Hessler, J. R., and Chisolm, G. M. (1983). Low density of lipoprotein cytotoxicity induced by free radical peroxidation of lipid. *J. Lipid Res.* **24,** 1070–1076.

Morris, D. L., Kritchevsky, S. B., and Davis, C. E. (1994). Serum carotenoids and coronary heart disease; The lipid research clinics coronary primary prevention trial and follow-up study. *J. Am. Med. Assoc.* **272,** 1439–1441.

Nomura, A. M., Stemmermann, G. N., Heilbrun, L. K., *et al.* (1985). Serum vitamin levels and the risk of cancer of specific sites in men of Japanese ancestry in Hawaii. *Cancer Res.* **45,** 2369–2372.

Navab, M., Fogelman, A. M., Berliner, J. A., Territo, M. C., Demer, L. L., Frank, J. S., Watson, A. D., Edwards, P. A., and Lusis, A. J. (1995). Pathogenesis of atherosclerosis. *Am. J. Cardiol.* **76,** 18C–23C.

O'Brien, K. D., and Chair, A. (1994). The biology of the artery wall in atherogenesis. *Am. J. Med. Clin. Nutr.* **78,** 41–67.

Ohgushi, M., Kugiyama, K., Fukunaga, K., Murohara, T., Sugiyama, S., Miyamot, E., and Yasue, H. (1993). Protein kinase C inhibitors prevent impairment of endothelium-dependent relaxation by oxidatively modified LDL. *Arterioscler. Thromb.* 13, 1525–1532.

Oliver, M. F. (1995). Antioxidant nutrients, atherosclerosis, and coronary heart disease. *Br. Heart J.* 73, 299–301.

Olmedilla, B., Granado, F., Blanco, I., and Rojas-Hidalgo, E. (1994). Seasonal and sex-related variations in six serum carotenoids, retinol, and α-tocopherol. *Am. J. Clin. Nutr.* 60, 106–110.

Ong, A. S. H., and Packer, L., eds. (1992)."Lipid-Soluble Antioxidants: Biochemistry and Clinical Applications." Birkhäuser Verlag, Basel.

Packer, L., and Fuchs, J., eds. (1993)."Vitamin E in Health and Disease." Dekker, New York.

Palinski, W., Rosenfeld, M. E., Ylä-Herttuala, S., Gurtner, G. C., Socher, S. S., Butler, S. W., Parthasarathy, S., Carew, T. E., Steinberg, D., and Witztum, J. L. (1989). Low density lipoprotein undergoes oxidative modification in vivo. *Proc. Natl. Acad. Sci. U.S.A.* 86, 1372–1376.

Panasenko, O. M., and Sergienko, V. I. (1994). Free-radical modification of blood lipoproteins and atherosclerosis. *Biol. Med.* 7, 323–364.

Parthasarathy, S., ed. (1994). "Modified Lipoproteins in the Pathogenesis of Atherosclerosis." CRC Press, Boca Raton, FL.

Princen, H. M. G., Van Poppel, G., Vogelezang, C., Buytenhek, R., and Kok, F. J. (1992). Supplementation with vitamin E but no β-carotene in vivo protects low density lipoprotein from lipid peroxidation in vitro. *Arterioscler. Thromb.* 12, 554–562.

Puhl, H., Waeg, G., and Esterbauer, H. (1994). Methods to determine oxidation of low-density lipoproteins. *Methods Enzymol.* 233, 425–441.

Quinn, M. T., Parthasarathy, S., Fong, L. G., and Steinberg, D. (1987). Oxidatively modified low density lipoproteins: A potential role in recruitment and retention of monocyte/macrophages during atherogenesis. *Proc. Natl. Acad. Sci. U.S.A.* 84, 2995–2998.

Raal, F. J., Areias, A. J., and Joffe, B. I. (1995). Low density lipoproteins and atherosclerosis-quantity or quality? *Redox Rep.* 1, 171–176.

Rajavashisth, T. B., Andalibi, A., Territo, M. C., Berliner, J. A., Navab, M., Fogelman, A. M., and Lusis, A. J. (1990). Induction of endothelial cell expression of granulocyte and macropage colonystimulating factors by modified low-density lipoproteins. *Nature* 344, 254–257.

Reaven, P. D., Grasse, B. J., and Tribble, D. L. (1994). Effects of linoleate-enriched and oleate-enriched diets in combination with α-tocopherol on the subfractions to oxidative modification in humans. *Arterioscler. Thromb.* 14, 557–566.

Reaven, P. D., Khouw, A., Beltz, W. F., Parthasarathy, S., and Witztum, J. L. (1993). Effect of dietary antioxidant combinations in humans. *Arterioscler. Thromb.* 13, 590–600.

Regnström, J., Nilsson, J., Tornvall, P., Landou, C., and Hamsten, A. (1992). Susceptibility to low-density lipoprotein oxidation and coronary atherosclerosis in man. *Lancet* 339, 1183–1186.

Rice-Evans, C., and Bruckendorfer, K. R., eds. (1995). "Oxidative Stress, Lipoproteins and Cardiovascular Dysfunction." Portland Press.

Riemersma, R. A., Wood, D. A., Macintyre, C. H. H., Elton, R. A., Gey, K. F., and Oliver, M. F. (1991). Risk of angina pectoris and plasma concentrations of vitamins A, C, E and carotene. *Lancet* 337, 1–5.

Rifici, V. A., and Khachadurian, A. K. (1993). Dietary supplementation with vitamins C and E inhibits in vitro oxidation of lipoproteins. *J. Am. Col. Nutr.* 12, 631–637.

Rimm, E. B., Stampfer, M. J., Ascherio, A., Giovannucci, E., Colditz, G. A., and Willett, W. C. (1993). Dietary intake and risk of coronary heart disease among men. *N. Engl. J. Med.* 328, 1450–1456.

Rosenfeld, M. E., Palinski, W., Ylä-Herttuala, S., Butler, S., and Witztum, J. L. (1990). Distribution of oxidation specific lipid-protein adducts and apolipoprotein B in atherosclerotic lesions of varying servity from WHHL rabbits. *Atherosclerosis* 10, 336–349.

Ross, R. (1993). The pathogenesis of atherosclerosis: A perspective for the 1990s. *Nature* **362**, 801–809.

Sasahara, M., Raines, E. W., Chait, A., Carew, T. E., Steinberg, D., Wahl, P. W., and Ross, R. (1994). Inhibition of hypercholesterolemia-induced atherosclerosis in the nonhuman primate by probucol. *J. Clin. Invest.* **94**, 155–164.

Schectman, G., Byrd, J. C., and Gruchow, H. W. (1989). The influence of smoking on vitamin C status in adults. *Am. J. Public Health* **79**, 158–162.

Schmidt, K., Graier, W. F., Kostner, G. M., Mayer, B., and Kukovetz, W. R. (1990). Activation of soluble guanylate cyclase by nitrovasodilators is inhibited by oxidized low-density lipoprotein. *Biochem. Biophys. Res. Commun.* **172**, 614–619.

Schmidt, R., Fazekas, F., Hayn, M., Kapeller, P., Schmidt, H., Kostner, G. M., Lechner, H., and Esterbauer, H. (1996a). Prevalence and risk factors for silent ischemic brain damage (SIBD) on MRI: The Austrian Stroke Prevention Study. *Neurology* **46**, A288.

Schmidt, R., Hayn, M., Fazekas, F., Kapeller, P., Eber, B., Lechner, H., and Esterbauer, H. (1996b). Magnetic resonance imaging white matter hyperintensities in clinically normal elderly individuals: Correlations with plasma concentrations of naturally occurring antioxidants. *Stroke* (in press).

Schmidt, R., Lechner, H., Fazekas, F., Niederkorn, K., Reinhart, B., Grieshofer, P., Homer, S., Offenbacher, H., Koch, M., Eber, B., Schumacher, M., Kapeller, O., Freidl, W., and Dusek, T. (1994). Assessment of cerebrovascular risk profiles in healthy persons: Definition of research goals and the Austrian Stroke Prevention Study (ASPS). *Neuroepidemiology* **13**, 308–313.

Schober, S. E., Comstock, G. W., Helsing, K. J., *et al.* (1987). Serologic precursors of cancer. I. Prediagnostic serum nutrients and colon cancer risk. *Am. J. Epidemiol.* **126**, 1033–1041.

Schwartz, C. J., and Valente, A. J. (1994). The pathogenesis of atherosclerosis. *In* "Natural Antioxidants in Human Health and Disease" (B. Frei, ed.), pp. 287–302. Academic Press, San Diego.

Sowell, A. L., Huff, D. L., Yeager, P. R, Caudill, S. P., and Gunter, E. W. (1994). Retinol, α-tocopherol, lutein/zeaxanthin, β-cryptoxanthin, lycopene, α-carotene, trans-β-carotene, and four retinyl esters in serum determined simultaneously by reversed-phase HPLC with multiwavelength detection. *Clin. Chem.* **40**, 411–416.

Stahelin, H. B., Rösel, F., Buess, E., Brubacher, G., *et al.* (1984). Cancer, vitamins, and plasma lipids: Prospective Basel study. *J. Natl. Cancer Inst.* **73**, 1463–1468.

Stalenhoef, A. F. H., Defesche, J. C., Kleinveld, H. A., Demacker, P. N. M., and Kastelein, J. J. P. (1994). Decreased resistance against in vitro oxidation of LDL from patients with familial defective apolipoprotein B-100. *Arterioscler. Thromb.* **14**, 489–493.

Stampfer, M. J., Hennekens, C. H., Manson, J. E., Colditz, G. A., Rosner, B., and Willett, W. C. (1993). Vitamin E consumption and the risk of coronary disease in women. *N. Engl. J. Med.* **328**, 1444–1449.

Steinberg, D. (1995). Clinical trials of antioxidants in atherosclerosis: Are we doing the right thing? *Lancet* **346**, 36–38.

Steinberg, D., Parthasarathy, S., Carew, T. E., Khoo, J. C., and Witztum, J. S. (1989). Beyond cholesterol: Modifications of low density lipoprotein that increase its atherogenicity. *N. Engl. J. Med.* **320**, 915–924.

Steinberg, D., and workshop participants (1992). Antioxidants in the prevention of human atherosclerosis: Summary of the proceedings of a national heart, lung, and blood institute workshop: September 5–6, 1991, Bethesda, Maryland. *Circulation* **85**, 2338–2344.

Steinbrecher, U. P., Parthasarathy, S., Leake, D. S., Witztum, J. L., and Steinberg, D. (1984). Modification of low density lipoprotein by endothelial cells involves lipid peroxidation and degradation of low density lipoprotein phospholipids. *Proc. Natl. Acad. Sci. U.S.A.* **81**, 3883–3887.

Stiko-Rahm, A., Hultegardh-Nilsson, A., Regnström, J., Hamsten, A., and Nilsson, J. (1992). Native and oxidized LDL enhances production of PDGF AA and surface expression of

PDGF receptors in cultured human smooth muscle cells. *Arterioscler. Thromb.* **12**, 1099–1109.

Street, D. A., Comstock, G. W., Salkfeld, R. M., Schuep, W., and Klag, M. (1991). A population based case-control study of serum antioxidants and myocardial infarction. *Am. J. Epidemiol.* **134**, 719–720.

Street, D. A., Comstock, G. W., Salkeld, R. M., Schüep, W., and Klag, M. J. (1994). Serum antioxidants and myocardial infarction: Are low levels of carotenoids and α-tocopherol risk factors for myocardial infarction? *Circulation* **90**, 1154–1161.

Stocker, R., Bowry, V. W., and Frei, B. (1991). Ubiquinol-10 protects human low density lipoprotein more efficiently against lipid peroxidation than does α-tocopherol. *Proc. Natl. Acad. Sci. U.S.A.* **38**, 1646–1650.

Tatzber, G., and Esterbauer, H. (1995). Autoantibodies to oxidized low-density lipoprotein. *In* "Free Radicals, Lipoprotein Oxidation and Atherosclerosis" (G. Bellomo, G. Finardi, E. Maggi, and C. Rice-Evans, eds.), pp. 245–259. Richelieu Press, London.

Terkeltaub, R., Banka, C. L., Solan, J., Santoro, D., Brand, K., and Curtiss, L. K. (1994). Oxidized LDL induces monocytic cell expression of interleukin-8, a chemokine with T-lymphocyte chemotactic activity. *Arterioscler. Thromb.* **14**, 47–53.

Thomas, C. E., Jackson, R. L., Ohweiler, D. F., and Ku, G. (1994). Multiple lipid oxidation products in low-density lipoproteins induce interleukin-1 beta release from human blood mononuclear cells. *J. Lipid. Res.* **35**, 417–427.

Tribble, D. L., Holl, L. G., Wood, P. D., and Krauss, R. M. (1992). Variations in oxidative susceptibility among 6 low density lipoprotein subfractions of differing density and particle size. *Atherosclerosis* **93**, 189–199.

Tribble, D. L., Vandenberg, J. J. M., Motchnik, P. A., Ames, B. N., Lewis, D. M., Chait, A., and Krauss, R. M. (1994). Oxidative susceptibility of low-density lipoprotein subfractions is related to their ubiquinol-10 and alpha-tocopherol content. *Proc. Natl. Acad. Sci. U.S.A.* **91**, 1183–1187.

Van Poppel, G., Kardinaal, A., Princen, H., and Kok, F. J. (1994). Antioxidants and coronary heart disease. *Ann. Med.* **26**, 429–434.

Vollset, S. E., and Bjelke, E. (1983). Does consumption of fruit and vegetables protect against stroke? *Lancet* **2**, 742.

Waeg, G., Puhl, H., and Esterbauer, H. (1994). LDL oxidation: Results and relevance for atherogenesis and possible clinical consequences. *Fat. Sci. Technol.* **96**, 20–22.

Wald, N. J., Thompson, S. G., Densem, J. W., *et al.* (1987). Serum vitamin E and subsequent risk of cancer. *Br. J. Cancer* **56**, 69–72.

Weis, J. R., Pitas, R. E., Wilson, B. D., and Rodgers, G. M. (1991). Oxidized low-density lipoprotein increases cultured human endothelial cell tissue factor activity and reduces protein C activation. *FASEB J.* **5**, 2459–2465.

World Health Organization (WHO). (1991). World Health Statistics, Annual. Genova.

Winkelhofer-Roob, B. M., Puhl, H., Khoschsurur, G., Vant Hof, M. A., Esterbauer, H., and Shmerling, D. H. (1995). Enhanced resistance to oxidation of low density lipoproteins and decreased lipid peroxide formation during β-carotene supplementation in cystic fibrosis. *Free Rad. Bio. Med.* **18**, 849–859.

Witztum, J. L. (1994). The oxidation hypothesis of atherosclerosis. *Lancet* **344**, 793–795.

Yamaguchi, M., Sato, H., and Bannai, S. (1993). Induction of stress proteins in mouse peritoneal macrophages by oxidized low density lipoprotein. *Biochem. Biophys. Res. Commun.* **193**, 118–1201.

Ylä-Herttuala, S., Rosenfeld, M. E., Parthasarathy, S., Glass, C. K., Sigal, E., Särkioja, T., Witztum, J. L., and Steinberg, D. (1991). Gene expression in macrophage-rich human atherosclerotic lesions: 15-Lypoxygenase and acetyl low density lipoprotein receptor messenger RNA colocalize with oxidation specific lipid-protein adducts. *J. Clin. Invest.* **87**, 1146–1152.

Ylä-Herttuala, S., Rosenfeld, M. E., Parthasarathy, S., Glass, C. K., Sigal, E., Witztum, J. L., and Steinberg, D. (1990). Colocalization of 15-lipoxygenase mRNA and protein with isotopes of oxidized low density lipoprotein in macrophage-rich areas of atherosclerotic lesions. *Proc. Natl. Acad. Sci. U.S.A.* **87**, 6959–6963.

Zhang, H., Basra, H. J. K., and Steinbrecher, U. P. (1990). Effects of oxidatively modified LDL on cholesterol esterification in cultured macrophages. *J. Lipid Res.* **31**, 1361–1369.

Zieden, B., Molgaard, J., and Olsson, A. (1992). Low-density lipoprotein oxidation and coronary atherosclerosis. *Lancet* **340**, 727–728.

Ziouzenkova, O., Gieseg, S. P., Ramos, P., and Esterbauer, H. (1996). Factors affecting resistance of LDL to oxidation. *Lipids* **31**, 571–576.

Zwijsen, R. M. L., Japenga, S. C., Heijen, A. M. P., Van den Bos, R. C., and Koeman, J. H. (1992). Induction of platelet-derived growth factor chain a gene expression in human smooth muscle cell by oxidized low density lipoproteins. *Biochem. Biophys. Res. Commun.* **186**, 1410–1416.

Samuel Louie*
Barry Halliwell*,†
Carroll Edward Cross*

*Division of Pulmonary-Critical Care Medicine
Department of Internal Medicine
University of California, Davis
Sacramento, California 95817

†Pharmacology Group
University of London King's College
London SW3 6LX, United Kingdom

Adult Respiratory Distress Syndrome: A Radical Perspective

I. Introduction

Adult respiratory distress syndrome (ARDS) describes a life-threatening acute lung injury with an incidence of between 50,000 and 150,000 cases annually in the United States (NHLBI, 1972), although more recent figures argue for substantially lower incidence, possibly by an order of magnitude (Thomsen and Morris, 1995). In the original description, 12 adult patients developed clinical features very similar to infantile respiratory distress syndrome within 96 hr after the associated injury or illness, e.g., sepsis, aspiration of gastric contents or major trauma, heralded by the onset of severe dyspnea, tachypnea, hypoxemia refractory to oxygen supplementation, loss of lung compliance, and diffuse alveolar infiltrates on chest X-ray. Despite the use of mechanical ventilation with positive end expiratory pressure (PEEP), only 5 survived. Autopsy findings in the 7 who died revealed diffuse alveolar damage with atelectasis, interstitial and intraalveolar hemorrhage,

Advances in Pharmacology, Volume 38

hyaline membrane formation, and numerous alveolar macrophages (Asbaugh *et al.*, 1967).

The original description of ARDS identified clinical features that constitute the most commonly used criteria to diagnose ARDS (Table I). The National Heart, Lung, and Blood Institute uses slightly different criteria. An expanded definition of ARDS has been proposed to stratify patients by the extent of their acute lung injury by evaluating radiographic appearance, severity of the hypoxemia, PEEP requirements, and lung compliance (Murray *et al.*, 1988; Bernard *et al.*, 1994).

Collectively, these definitions identify a heterogenous group of critically ill patients with hypoxemic respiratory failure who require mechanical ventilation, PEEP, and a high partial pressure of inspired O_2. Prospective clinical studies have established sepsis, aspiration of gastric contents, and major remote organ trauma as the most common disorders associated with high risk for ARDS. Mortality associated with ARDS has remained between 50 and 70% and reaches 90% when associated with sepsis syndrome or multiorgan failure (Maunder and Hudson, 1991), although there is some evidence that this mortality may be decreasing (Milberg *et al.*, 1995).

The treatment of ARDS has remained largely supportive for almost a quarter of a century, using O_2, volume-controlled mechanical ventilation,

TABLE I   Criteria Used for the Diagnosis of ARDS

Clinical criteria for ARDS diagnosis
  1. Appropriate risk factor(s)
  2. Respiratory distress with labored breathing
  3. Severe hypoxemia
  4. Chest X-ray evidence of pulmonary edema
  5. Loss of lung compliance
  6. Absence of primary heart failure

NHLB Institute criteria for ARDS diagnosis
  1. Widespread, bilateral infiltrates on chest X-ray less than 7 days of duration
  2. Hypoxemia with $PaO_2/FiO_2 < 150$ off positive end expiratory pressure (PEEP) or < 200 on PEEP
  3. Pulmonary artery occlusion pressure < 18 mm Hg

Expanded definition of ARDS (Murray *et al.*, 1988)
  1. Chest X-ray score between 0 and 4
  2. Hypoxemia score between 0 and 4
  3. PEEP score between 0 and 4
  4. Lung compliance score between 0 and 4
The lung injury score is obtained by dividing the aggregate sum by the number of components used:
  No lung injury is a score of 0
  Mild to moderate lung injury, between 0.1 and 2.5
  Severe lung injury (ARDS) greater than 2.5

PEEP, antibiotics for infections and careful management of intravenous fluids, and nutrition. Therapeutic interventions in ARDS have been hindered by our inability to clearly identify the basic underlying mechanism(s) of lung injury in ARDS. Despite the widespread belief that inflammatory and immune processes are fundamental to its pathobiology, recent clinical trials with a number of agents designed to modify selected aspects of inflammatory and immune mediator cascades have been ineffective in changing clinical outcome in ARDS. This is probably because of the complex temporal and interactive nature of these processes and the failure to comprehend the exact contribution of each process to the development of ARDS (Fig. 1).

Pathologically, ARDS is a common response of the lung to a variety of different and frequently unrelated insults (Bachofen and Weibel, 1982). One important consequence is the alteration in permeability of alveolar epithelial and microvascular endothelial membrane barriers, and the upregulation of phagocytic, endothelial, and epithelial cell adhesion molecules. The net physiologic result is severe hypoxemic acute respiratory failure due to lung ventilation-perfusion mismatching and extensive intrapulmonary shunts.

The acute phase of ARDS (0–5 days) is characterized by a sudden wave of inflammatory-immune system activation. Activated alveolar macrophages (AMs) and neutrophils (PMNs) recruited to the lung from the peripheral circulation by inflammatory chemokines, cytokines, and other humoral mediators appear to combine to cause varying degrees of injury to both lung capillaries and alveolar epithelium. Injury to these cells results in interstitial and alveolar edema and severe derangements in lung gas exchange. Reactive

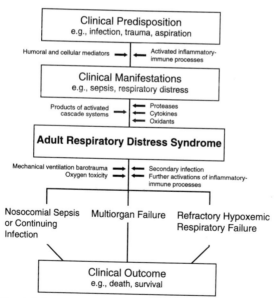

FIGURE I   Simplified view of events initiating adult respiratory distress syndrome.

O_2 species (ROS) and possibly reactive nitrogen species (RNS)[1] are likely responsible for at least part of the phagocyte-induced lung cell injury. Type I epithelial cells are damaged and do not appear to regenerate. The subacute phase of ARDS (5–10 days) is characterized by persistent interstitial edema with fewer PMNs and more AMs. A network of cytokines and growth factors mediates a complex subacute or early reparative stage of lung injury characterized by proliferation of alveolar type II cells, which repopulate the denuded alveolar epithelium. Evidence of fibroblast proliferation and collagen deposition are also present. Patients who survive from ARDS at this stage can either recover near normal lung function (Fowler *et al.*, 1985) or enter a chronic phase (1–2 weeks) characterized by fibrosing alveolitis, pulmonary vascular microthrombosis, and loss of normal lung architecture (Meduri *et al.*, 1995).

Although in the past the foundation of knowledge for all disease lay in understanding its descriptive pathology, scientists and clinicians today resort to molecular and cellular biology as their main experimental tools to gain knowledge of the basic mechanisms of disease. Recent advances in understanding of inflammation–immune system biology combined with an appreciation of potential toxicities of ROS and RNS have prompted the investigation of new therapeutic approaches to ARDS. This chapter provides a perspective of the mechanisms underlying ARDS based on recent basic science and clinical investigations.

II. Pathobiologic Considerations

A. Injury and Inflammation

The extent to which acute inflammation causes ARDS or is a consequence of acute lung injury remains controversial, as temporal relationships among the cellular, humoral, and other mechanisms of injury and inflammation are complex and not clearly understood. It is evident from the initial description of ARDS that a brief latent period of 24 to 72 hr exists before the development of frank hypoxemic respiratory failure and, conceivably, it is during this period that the key elements of acute injury and the coexisting inflammatory and immune processes may be identified and studied.

One logical sequence of events proposed to lead to ARDS has been the activation of complement, soon accompanied by the upregulation of adhesion molecules on phagocytes and endothelial cells and the recruitment of

---

[1] ROS is a general term that includes oxygen radicals, e.g., superoxide radical ($O_2\cdot^-$) and hydroxyl radical ($\cdot OH$) and nonradicals, e.g., hydrogen peroxide (H_2O_2) and hypochlorous acid ($HOCl/OCl^-$). RNS include molecules such as nitric oxide ($\cdot NO$), nitrogen dioxide ($\cdot NO_2$), and peroxynitrite ($ONOO^-$).

activated PMNs to the lung from the circulation in response to direct pulmonary injury (e.g., aspiration or toxic gas exposure), gram-negative bacterial sepsis, or major trauma. Activated PMNs can then initiate and potentiate lung damage by releasing their arsenal of potentially toxic agents, which includes arachidonic acid metabolites, proteases, and ROS (and, at least in some species, RNS).

Although gastric aspiration, sepsis, and major trauma most commonly cause ARDS, the majority of patients with these predispositions do not develop ARDS (Hudson *et al.*, 1995). In part, this may be due to a genetic or nutrition-related predisposition to such lung injury. Recent investigations have linked a prior history of chronic alcoholism with a significantly increased risk of developing ARDS in critically ill patients. (Moss *et al.*, 1996). In hospitals, mortality was significantly higher in patients with a prior history of alcohol abuse (65%) than in patients who developed ARDS without a history of alcohol abuse (36%). PMNs isolated from the serum of intoxicated adults demonstrated upregulation of adhesion molecules and were already primed for the production of oxidative products (Balla *et al.*, 1993). Furthermore, as neutropenic patients can develop ARDS (Ognibene *et al.*, 1986), non-PMN-dependent pathways (including those involving activated AMs) must exist and be capable of the initiation and propagation of ARDS. The hypothesis that one orderly sequence of events exists during acute lung injury, which leads to ARDS, seems very unlikely. Table II lists some important pathobiologic considerations believed to be operative in ARDS.

TABLE II Pathobiologic Considerations in ARDS

Humoral aspects	Cellular aspects	Oxidant aspects	Other aspects
Complement	Neutrophils	Oxygen toxicity	Sepsis syndrome
C5a, C3a	Monocytes and	Reactive oxygen species	Multiorgan failure
Arachidonic acid	macrophages	Superoxide radical	
metabolites	Eosinophils	Hydrogen peroxide	
Thromboxane A$_2$	Platelets	Hydroxyl radical	
Leukotrienes	Endothelial cells	Singlet oxygen	
Prostaglandins	Epithelial cells	Hypochlorous acid	
Platelet-activating	Fibroblasts	Reactive nitrogen	
factor		species	
Endotoxin		Nitric oxide	
Cytokines		Peroxynitrite	
TNFα, IL-1β, IL-6,		Transition metal ion	
IL-8		catalysts	
		Hypoxia–reoxygenation	
		phenomena	
		Xanthine oxidase	
		Antioxidant depletion	

B. Humoral Aspects

Systemic complement activation, specifically C5a, C3a, and other related peptides, can produce acute pulmonary edema in mice that is mediated by activated PMNs and their arsenal of proteases and ROS/RNS-generating abilities (Till *et al.*, 1982). Antibodies to C5a have prevented pulmonary edema and improved survival in primates with *Escherichia coli*-induced sepsis (Stevens *et al.*, 1986). However, high levels of activated complement in 40 patients with sepsis syndrome did not predict the development of ARDS (Weinberg *et al.*, 1984). A prospective study of activated complement in 100 patients at high risk for developing ARDS found normal plasma C5a levels, whether or not ARDS developed (Maunder and Hudson, 1991). C3a levels were high in most patients whether or not ARDS developed. Consequently, the role of activated complement in causing ARDS is not clear. It may promote PMN sequestration and activation, but there are several other mediators that can do the same, perhaps with a more sustained effect.

Eicosanoids are potent humoral mediators of inflammation that are derived from arachidonic acid through reactions catalyzed by cyclooxygenase and lipoxygenase enzymes. Phospholipase A_2 (PLA_2) can play an important role in inflammation by releasing arachidonic acid from membrane phospholipids (Holtzman, 1991). Increased PLA_2 activities have been measured in human bronchoalveolar lavage fluid (BALF) from ARDS patients when compared to normal controls and correlated positively with lung injury scores (Kim *et al.*, 1995). Some eicosanoids are capable of causing PMN chemotaxis and aggregation. For example, thromboxane A_2 is a major product of the cyclooxygenase pathway that may be involved in acute lung injury by activated complement (Gee *et al.*, 1985). When produced by activated phagocytes, thromboxane A_2 can promote PMN adherence to endothelial surfaces, an important aspect of phagocyte-related endothelial damage.

Leukotrienes are important products of the lipoxygenase pathway that can cause PMN chemotaxis and aggregation in the lung (e.g., LTB4) as well as cause increased microvascular permeability (e.g., LTC4, LTD4, and LTE4) (Samuelsson *et al.*, 1987; Perlman *et al.*, 1989). Elevated levels of LTC4 and LTD4 in BALF from patients with ARDS and those at risk have been reported. The increased concentrations of leukotrienes correlated with lavage protein content, an indicator of altered epithelial permeability, perhaps suggesting a causal relationship (Stephenson *et al.*, 1988).

Another mediator derived from arachidonic acid is platelet-activating factor (PAF). PAF is synthesized and released by PMNs, AMs, and pulmonary microvascular endothelium (Pinkard, 1988). It activates not only platelets but also phagocytes; for example, PAF can cause PMNs to produce thromboxane A_2, leukotrienes, and $O_2^{\cdot-}$ (Braquet, 1987). Other PAF actions include increasing pulmonary microvascular permeability and causing systemic hypotension.

Cytokines are extracellular cell-signaling molecules that cause local and systemic inflammation (Billiau and Vandekerckhove, 1991). Derived mainly from cells of monocyte–macrophage lineage, cytokines can transmit information to target cells through receptor interactions that in turn modulate the regulation of many physiologic and inflammatory-immune processes. Numerous actions of cytokines are mediated indirectly, for example, the activation, priming[2] and chemotaxis of circulating PMNs [e.g., by interleukin (IL)-1α and tumor necrosis factor (TNF)-α]. Many cytokines have concentration-dependent effects on cell physiology. For example IL-2 infusions to treat cancer patients have been shown to induce a dose-dependent lung injury that resembles ARDS (Rabinovici et al., 1994). The proinflammatory cytokines IL-1β and TNF-α are recognized as early mediators during the acute phase of ARDS (Le and Vilcek, 1987). Endotoxin may mediate its effects in sepsis syndrome primarily through the synthesis and release of IL-1β (Movat, 1987). IL-1β can increase the expression of endothelial and PMN adhesion molecules, facilitating PMN sequestration and activation. IL-1β can produce sepsis syndrome and increase pulmonary microvascular permeability when infused into rabbits (Goldblum, 1987). AMs isolated from ARDS patients have been shown to release significant amounts of IL-1β, which could intensify and prolong damaging inflammatory immune processes (Jacobs et al., 1989).

TNF-α has been shown by many to be a pivotal immunomodulatory cytokine in gram-negative sepsis induced by endotoxin. Experimentally, TNF-α stimulates macrophages and vascular endothelium to produce IL-1β. Like IL-1β, TNF-α can induce endothelial and phagocytic cell expression of adhesion molecules. It can cause increased pulmonary microvascular permeability and edema. TNF-α is detectable in plasma from patients in septic shock at high risk for ARDS (Marks et al., 1990). Significantly more patients with septic shock and measurable TNF-α developed ARDS compared to patients with septic shock only. The hypothesis that TNF-α is involved in the pathogenesis of ARDS is further supported by several studies demonstrating its presence in BALF of ARDS patients (Millar et al., 1989; Hyers et al., 1991; Suter et al., 1992). Indeed, the local release and accumulation of TNF-α and IL-1β in the lung by macrophages or other cells is associated with the development of ARDS (Rothstein and Schreiber, 1988). In the same group of ARDS patients, an increased level of interferon and elastase in bronchoalveolar lavage fluid was evidence of intrapulmonary activation and degranulation of phagocytes (Suter et al., 1992).

However, the adverse effects of TNF-α may not be related to plasma levels or the duration of the plasma increases (Marks et al., 1990), suggesting

---

[2] Priming is defined as the exposure of phagocytes to a stimulus, usually at very low doses, so that there is a markedly increased response to a second stimulus. PMNs can be "primed" for enhanced releases of both oxidative metabolites and nonoxidative products in response to a second stimulus.

that concentrations in a local tissue microenvironment are probably more relevant than systemic levels. Direct systemic administration of TNF-α does not cause all the features of sepsis syndrome and shock (Rothstein and Schreiber, 1988). Levels of TNF-α comparable to those seen in the plasma of patients with septic shock when applied to suspension cell cultures cause small or no effects (Michie *et al.*, 1988). This has led investigators to hypothesize that these inflammatory mediators were part of a larger interactive and probably synergizing tissue cytokine network (Cannon *et al.*, 1990; Billiau and Vandekerckhove, 1991).

Serum levels of IL-6 often peak later than IL-1β and TNF-α. As the principal hepatocyte-stimulating factor, IL-6 may cause the induction and release of acute-phase proteins, e.g., protease inhibitors, into the circulation to counteract tissue damage activated by inflammatory-immune processes. However, high serum levels of IL-6 appear to correlate significantly with fatal outcomes in patients with septic shock from endotoxemia (Yoshimoto *et al.*, 1992). Further investigations are needed to ascertain whether IL-6 is an important mediator of inflammation or a consequence of the inflammatory events triggered by IL-1β and TNF-α.

The potent PMN chemoattractant and activator IL-8 has also been implicated in the pathogenesis of ARDS (Groeneveld *et al.*, 1995) since levels in blood or BALF are elevated in patients at risk for ARDS and have been linked to mortality in some studies (Miller *et al.*, 1992).

Cytokines may also exert protective effects by inducing increases in intracellular levels of superoxide dismutase (SOD), ferritin, and other protective agents (Larry and Wright, 1990). Both IL-1β and TNF-α have been shown to cause a dramatic induction of Mn-SOD mRNA levels in rat pulmonary epithelial cells (Visner *et al.*, 1990). Pretreatment with a combination of TNF-α and IL-Iβ increased the activities of the pulmonary antioxidant enzymes catalase, Mn-SOD, glutathione reductase, and glutathione peroxidase and improved survival in rats continuously exposed to hyperoxia for 72 hr (White *et al.*, 1987; 1989). Finally, a new group of "counterinflammatory" cytokines [e.g., IL-10, IL-4, and transforming growth factor β (TGF-β)] can be expected to play a role but has been little studied in ARDS. Because of its ability to modulate inflammatory-immune homeostasis and to stimulate the production of extracellular–matrix proteins, TGF-β can be expected to play a key role in ARDS-related reparative and fibrogenic responses (Kulkarni *et al.*, 1993).

C. Cellular Aspects

Considerable evidence links PMNs and AMs to the acute lung injury evident in many forms of ARDS (Donnelly *et al.*, 1994). Increased numbers of PMNs can be recovered from the BALFs of ARDS patients (Weiland *et al.*, 1986). The expression of multiple intercellular adhesion molecules on both phagocytes and endothelial surfaces represents the initial and requisite steps in phagocyte recruitment into the lung. Endotoxins and/or cytokines are often presumed to be involved in the upregulation of adhesion molecules. Once activated, PMNs in the lung microenvironment can synthesize and release eicosanoids, proteases, and ROS and possibly RNS, which can directly or indirectly cause increased pulmonary microvascular permeability and edema (Table III).

One detrimental action of proteases released by adherent, PMNs may be the degradation of extracellular SOD (McCord *et al.*, 1994). The production of RNS by rodent PMNs is well established but the extent of their production by human PMNs is still a subject of debate (Miles *et al.*, 1995). BALF from ARDS patients contains increased elastolytic activity that could mirror a deficiency in antiprotease activity (Lee *et al.*, 1981). PMNs isolated from the blood of ARDS patients are often activated and demonstrate enhanced O_2.$^-$ production (Weiland *et al.*, 1986). PMNs may also release hypochlorous HOCl/OCl$^-$, which can oxidize antiproteases and essential sulfhydryl (-SH) groups as well as chlorinate tyrosine residues, perhaps interfering with cell signal transduction (Weiss, 1989; Smith, 1994). Activated PMNs can decrease the surface tension properties of surfactant *in vitro,* probably through effects of both direct oxidative and proteolytic mechanisms (Ryan *et al.*, 1991).

However, the PMN must not necessarily be regarded as the *bête noire* of ARDS. Activated PMNs release the protein lactoferrin, which binds iron

TABLE III   Selected Neutrophil and Alveolar Macrophage Products

---

Neutrophil products
  Cyclooxygenase metabolites, e.g., thromboxane A_2
  Lipoxygenase metabolites, e.g., LTB4
  Platelet-activating factor
  Reactive oxygen species, e.g., superoxide radical, hydrogen peroxide
  Myeloperoxidase
  Proteases, e.g., elastases
  Lactoferrin
  Cytokines
  Nitric oxide?*

Alveolar macrophage products
  Cyclooxygenase metabolites, e.g., thromboxane A_2
  Lipoxygenase metabolites, e.g., LTB4, C4, D4
  Platelet-activating factor
  Cytokines, e.g., IL-1β, TNFα, IL-8
  Reactive oxygen species, e.g., superoxide radical, hydrogen peroxide
  Proteases
  Transferrin
  Nitric oxide?*

---

* Amount of ˙NO produced depends on species.

and prevents iron-dependent free radical reactions (Halliwell and Gutteridge, 1989). PMNs also have beneficial functions in combating pulmonary infections. Bactericidal activity, ROS production, and chemotaxis are depressed in PMNs recovered in BAL from ARDS patients (Martin *et al.*, 1991). While the explanation for this observation remains unclear, the defect in alveolar PMNs (which may represent a population of "spent" PMNs) may increase susceptibility to pulmonary infection and cripple the ability of the lung to initiate reparative processes that may be mediated by PMNs (Repine and Beehler, 1991). Thus, it is uncertain whether suppression of PMN function would be beneficial or deleterious in ARDS.

AMs, the predominant resident inflammatory cells in the normal lung, play a central role in regulating lung function in health and disease. They appear to be a key effector cell in events leading to acute pulmonary injury. AMs are capable of being primed and activated with the subsequent release of cytokines, proteases, ROS, RNS, and other potentially toxic products (Table III) (Sibille and Reynolds, 1990), although the conditions under which human AMs can produce RNS are yet to be clarified (Miles *et al.*, 1995). AMs recovered from ARDS patients are activated and capable of releasing cytokines, some of which can induce PMN degranulation and prime PMN cytotoxicity (Billiau and Vandekerckhove, 1991). Circulating monocytes in the lung may also participate in lung inflammation (Dehring and Wismar, 1989). Platelets also sequester within the lung microvasculature in ARDS (Heffner *et al.*, 1987). The degree to which they actively participate in lung inflammation is not clear.

Type II alveolar epithelial cells are important because they produce surfactant and are necessary for repopulating the alveolar epithelium after acute lung injury (Mason, 1985). They may be important in the removal of alveolar edema fluid and severe injury to them may delay the resolution of ARDS. Ineffective surfactant has been recovered from BALF of ARDS patients (Mason, 1985). Surfactant abnormalities resulting from PMN-mediated damage (Enhorning, 1989) can promote widespread congestive atelectasis with ventilation–perfusion mismatching and intrapulmonary shunting leading to hypoxemia, thus requiring the administration of higher levels of inspired O_2.

Finally, diffuse damage to microvascular permeability barriers is often considered the cardinal event in ARDS. Endothelial cells are active participants in lung injury and inflammation (Curzen, 1994). For example, they produce the adhesion glycoproteins CD-11b and CD-18 necessary for PMNs to attach to lung endothelial cells (Donnelly *et al.*, 1994; Laurent *et al.*, 1994), elaborate arachidonic acid metabolites and PAF (Pinkard *et al.*, 1988), and generate ROS and RNS (Darley-Usmar *et al.*, 1995).

D. Reactive Oxygen and Nitrogen Species

Numerous investigations have suggested that ROS, and to a lesser degree RNS, may be important mediators of pulmonary damage in ARDS (McCord *et al.,* 1994). Molecular targets of ROS and RNS include lipids, proteins, and nucleic acids. $O_2\cdot^-$ is generated from the one electron reduction of O_2. Activated PMNs and macrophages generate $O_2\cdot^-$ using the NADPH oxidase system in membranes. $O_2\cdot^-$ is quickly converted to H_2O_2 intracellularly by SOD and extracellularly by nonenzymatic dismutation of $O_2\cdot^-$. Neither $O_2\cdot^-$ nor H_2O_2 is very reactive. However, much of the damage done by $O_2\cdot^-$ or H_2O_2 *in vivo* may be due to their conversion into highly reactive hydroxyl radicals ($\cdot$OH). Formation of $\cdot$OH radicals from $O_2\cdot^-$ and H_2O_2 can be achieved in the presence of trace amounts of catalytic transition metal ions such as iron and copper ions. Lung damage mediated by $\cdot$OH can occur by the Fenton reaction in which iron ions from injured endothelial or epithelial cells or erythrocytes react with H_2O_2 released by activated phagocytes to form $\cdot$OH:

$$Fe^{2+} + H_2O_2 \rightarrow Fe^{3+} + \cdot OH + OH^- \tag{1}$$

In the Fenton reaction, $\cdot$OH is generated from H_2O_2 and Fe^{2+} is oxidized to Fe^{3+} (in the presence of an iron species that can undergo redox cycling). Several reducing agents, such as $O_2\cdot^-$, thiols, and ascorbate, could assist this reaction by recycling Fe^{3+}:

$$Fe^{3+} + O_2\cdot^- \rightarrow Fe^{2+} + O_2 \tag{2}$$

$O_2\cdot^-$ might additionally mobilize trace amounts of iron from ferritin (present in lung lining fluids) to catalyze these reactions. Activated PMNs contain and release the enzyme myeloperoxidase, which can catalyze the formation of HOCl/OCl$-$, a major PMN-derived oxidizing and chlorinating agent that can react with many biological molecules:

$$H_2O_2 + Cl^- + H^+ \rightarrow HOCl + H_2O \tag{3}$$

HOCl can react with $O_2\cdot^-$ to generate $\cdot$OH (Folkes *et al.,* 1995):

$$HCOl + O_2\cdot^- \rightarrow Cl^- + \cdot OH + O_2 \tag{4}$$

HOCl also depletes antioxidants, e.g., by oxidizing $-$SH groups and ascorbate (Hu *et al.,* 1993), and it chlorinates tyrosine residues in proteins.

Stimulation by various cytokines or lipopolysaccharide causes AMs and endothelial and respiratory tract epithelial cells to produce significant amounts of $\cdot$NO and $O_2\cdot^-$ for prolonged periods of time in close proximity to the blood–gas interface (Ischiropoulos *et al.,* 1992). Indeed, increased levels of $\cdot$NO have been measured in exhaled gas in an animal model of sepsis and appears to be an early marker of acute lung inflammation (Stewart *et al.,* 1995). Further, $\cdot$NO can react with $O_2\cdot^-$ to produce the very potent RNS peroxynitrite ($ONOO^-$):

$$O_2\cdot^- + \cdot NO \rightarrow ONOO^- \tag{5}$$

This reaction occurs at almost a diffusion controlled rate (Huie and Padmaja, 1993). $ONOO^-$ is directly cytotoxic and at physiologic pH protonates and decomposes to a range of reactive species, one of which has a reactivity similar to that of ·OH (Beckman *et al.*, 1990). ONOO⁻ and/or its derivatives are capable of oxidizing lipids (Radi et al., 1991a), −SH groups (Radi *et al.*, 1991b), and antioxidants such as ascorbate (Bartlett *et al.*, 1995) and urate (van der Vliet *et al.*, 1994). Furthermore, ONOO⁻ can damage alveolar type II epithelial cells (Hu *et al.*, 1994) and specific surfactant apoproteins (Haddad *et al.*, 1994) to a considerable extent. It also inactivates α-1 proteinase inhibitor (Moreno and Pryor, 1992).

The experimental evidence suggesting a role for ROS and, more recently, RNS in ARDS consists of identifying the presence of such species or the products of their reactions (Haddad *et al.*, 1994) and/or the depletion of lung antioxidants (Pacht *et al.*, 1991). ROS and RNS may be directly responsible for causing molecular damage to important intracellular and extracellular molecules in ARDS. Detecting primary damage by ROS and RNS is often complicated because after tissue injury of any kind, it is likely that the injured cells will undergo oxidative damage more readily than normal cells (Halliwell *et al.*, 1992). Consequently, even if products of oxidative damage are identified, it is difficult to ascertain whether they cause cell injury or are a consequence of it. Nonetheless, one can hypothesize that ARDS is mediated to a significant degree by ROS and/or RNS and that activated phagocytes amplify the initial lung injury by releasing ROS or RNS.

Evidence for a role of ROS in ARDS includes the recovery of myeloperoxidase, oxidized antiprotease, cholesterol hydroperoxides, and *meta*-tyrosine, *ortho*-tyrosine, and kynurenine (an oxidation product of tryptophan) from bronchoalveolar surfaces (Cochrane *et al.*, 1983; Weiland *et al.*, 1986; Fantone *et al.*, 1987; Cross *et al.*, 1990; Quinlan *et al.*, 1995) and the presence of H_2O_2 in the expired breath of ARDS patients (Baldwin *et al.*, 1986). Although plasma lipid hydroperoxides have been reported as elevated in some animal models (Ward *et al.*, 1985; Takeda *et al.*, 1986; Demling *et al.*, 1988) and in some critically ill patients (Takeda *et al.*, 1984; Richard and Lemonnier, 1987), this has not been verified in all ARDS patients (Cross *et al.*, 1990). Of course, oxidative damage can occur without lipid peroxidation (Halliwell *et al.*, 1992). Hence, the lack of detectable lipid hydroperoxides may mean that lipid peroxidation is not a major mechanism of cell injury or else the products of lipid peroxidation are rapidly cleared from the systemic circulation.

Several lung cells, including airway epithelial cells and possibly macrophages, have the capacity to express inducible nitric oxide synthase after stimulation with proinflammatory cytokines. There is a growing body of both clinical and experimental data demonstrating that activation of the inflammatory-immune system, whether locally or systemically, is associated with an overproduction of ·NO. ·NO has a wide spectrum of effects ranging from physiological control and antioxidant defenses to cytotoxicity. Evidence for a role of RNS in ARDS includes demonstrations that nitric oxide synthase activity is present and increased amounts of ·NO and ONOO⁻ are produced in lungs of experimental animals in models of sepsis (Wizemann *et al.*, 1994; Stewart *et al.*, 1995), and the detection of 3-nitrotyrosine, a stable by-product of RNS reaction with tyrosine residues, in the lungs of patients with ARDS and animals with acute lung injury utilizing a polyclonal antibody to nitrotyrosine (Haddad *et al.*, 1994; Wizemann *et al.*, 1994). $ONOO^-$, but not $O_2\cdot^-$ or $\cdot NO$, is capable of nitrating phenolic rings and tyrosine residues of proteins and other biological molecules, including surfactant protein A, and decreases the function of surfactant protein A (Haddad *et al.*, 1996).

The precise role of $\cdot NO$ and $ONOO^-$ in ARDS is unknown. While some studies have emphasized $\cdot NO$-mediated damage during endotoxemia (e.g., severe hypotension), others have suggested that $\cdot NO$ may play a protective role in antioxidant defenses by being a potent pulmonary vasodilator, by its ability to decrease PMN and platelet sequestration, and by its ability to scavenge other free radicals (Wizemann *et al.*, 1994; Darley-Usmar *et al.*, 1995). The ratio of ROS to $\cdot NO$ may be of importance in that $\cdot NO$ may block some of the effects of $ONOO^-$, e.g., on lipid peroxidation (Rubbo *et al.*, 1994). Further studies utilizing nitric oxide synthase inhibitors are required to ascertain the role of RNS in ARDS. It should, however, be noted that rodents lacking inducible nitric oxide synthase are not resistant to endotoxin-induced death (Laubach *et al.*, 1995).

Indirect evidence of oxidative stress includes decreased plasma α-tocopherol and ascorbic acid levels in ARDS patients (Richard *et al.*, 1986; Cross *et al.*, 1990). Decreased levels of glutathione (GSH) and increased levels of oxidized GSSG in the epithelial lining fluid of patients with ARDS have also been reported (Bunnell and Pacht, 1993). The release of ROS by PMNs and AMs in the lung may injure endothelial cells (Varani *et al.*, 1985; Dobrina and Patriaraca, 1986; Lewis and Granger, 1986; Ryan and Vann, 1987). Multiorgan failure in ARDS could be related to ROS released from activated phagocytes upon endothelial cells of extrapulmonary organs (Mizer *et al.*, 1989).

Hypoxia–reoxygenation injury, perhaps involving xanthine oxidase-derived $O_2\cdot^-$ and H_2O_2 (McCord, 1985), may also occur in the lungs in ARDS (Koyama *et al.*, 1987; Lynch *et al.*, 1988). Circumstantial evidence includes the finding of pulmonary microthrombosis in ARDS patients (Vesconi *et al.*, 1988) that can be inhomogeneous in distribution (Maunder *et al.*, 1986). Moreover, elevated levels of plasma xanthine oxidase activity and its hypoxanthine substrate have been reported in ARDS patients (Grum *et al.*, 1987), the latter being consistent with systemic defects in O_2 transport and delivery in ARDS. This enzyme oxidizes hypoxanthine and xanthine to make both $O_2\cdot^-$ and H_2O_2.

O_2 toxicity often occurs in ARDS as an unavoidable complication of essential treatment. Hyperoxia can cause further lung damage by augmenting ROS production (Gerschman *et al.*, 1954; Freeman and Crapo, 1981; Jamieson, 1989). Lung previously injured by any of the inciting events discussed earlier, e.g., sepsis or gastric aspiration, may be at greater risk of developing O_2 toxicity. If acute inflammation has depleted antioxidant defenses in the lung, the toxic effects of increased O_2 administration may be increased. As is the case with humoral and cellular mediators, increased pulmonary microvascular permeability and edema are features of O_2 toxicity, making it difficult to ascertain its importance to the overall pathobiology of ARDS. The presence of increased protein levels in BALF and the elevated bronchoalveolar clearance of labeled tracer molecules indicate early injury to the lung parenchyma during O_2 toxicity (Davis *et al.*, 1983; Griffith *et al.*, 1986). PMNs may compound the injury associated with hyperoxia by sequestering in the lung, as illustrated by animal models of O_2 toxicity (Rinaldo *et al.*, 1988).

In summary, there is considerable evidence for oxidative stress in ARDS. A destructive synergy among inflammatory cells, humoral mediators, and ROS/RNS can occur, resulting in pulmonary microvascular endothelial and alveolar epithelial injury in ARDS (Fig. 2). Epithelial injury can be further exacerbated by pulmonary O_2 toxicity. Increased generation of $O_2{\cdot}^-$, H_2O_2, and the decompartmentalization of iron and copper can create the conditions

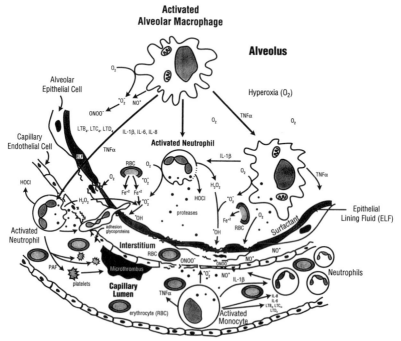

FIGURE 2   Selected interactions of humoral and cellular mediators, reactive oxygen species, and reactive nitrogen species in the acute injury of pulmonary epithelial and endothelial cell barriers in ARDS.

that favor ·OH and H_2O_2 formation and damage of molecular targets in the lung and extrapulmonary organs. The view that ARDS is primarily a lung disorder has more recently evolved into a broader concept that ARDS is a systemic disorder characterized by inflammation and oxidative stress, which is often associated with sepsis syndrome and multiorgan failure.

E. Sepsis Syndrome

ARDS is almost invariably associated with sepsis syndrome, either as the inciting event leading to ARDS or as a secondary complication (Maunder and Hudson, 1991). Sepsis syndrome defines a specific group of critically ill patients with documented serious infection who are febrile or hypothermic, tachycardic, and tachypneic who have evidence of inadequate organ perfusion, e.g., altered cerebral consciousness, hypoxemia, oliguria, or lactic acidosis (evidence of inadequate hepatic perfusion). As sepsis becomes more severe and/or more protracted, acute lung injury may occur, which is clinically recognized as ARDS. Unlike other forms of ARDS (e.g., fat embolism syndrome), mortality from acute respiratory failure in the setting of gram-negative sepsis can approach 90% (Rubin et al., 1990). Recent studies show an alarming increase in the incidence of gram-positive sepsis (Bone, 1994). Although not all patients with sepsis develop ARDS, the same mediators that are responsible for the humoral and cellular inflammatory response in sepsis are involved in the pathogenesis of ARDS, e.g., involvement of cytokines (Damas et al., 1989).

The consequences of untreated sepsis syndrome caused by gram-negative bacteria can be linked to the activation of inflammatory cascades by endotoxin. In the case of gram-positive bacteria, other cell wall constituents and exotoxins activate inflammatory cascades. Sepsis-related activation of macrophages to release TNF-α, IL-1, IL-2, IL-6, and IL-8 can recruit PMNs to lung microvascular surfaces and initiate inflammation by upregulation of adhesive receptors on phagocytes and endothelium (Laurent et al., 1994). These "sticky PMNs" can release toxic proteases, ROS, and RNS into the microvascular environment, causing injury that results in increased capillary permeability and pulmonary edema.

Early and late mortality in ARDS is related to or uncontrolled sepsis and multiorgan failure, not refractory hypoxemic respiratory failure, which accounts for only 5 to 15% of deaths (Doyle et al., 1995). Death can ensue despite adequate antibiotic coverage of infection (Seidenfeld et al., 1986). Patients with ARDS are at a greater risk for secondary infections by virtue of their underlying disease, need for mechanical ventilation, and intravascular lines. Continuing sepsis may also emanate from damage to intestinal villi, which has been observed in a model of TNF-α-induced acute lung injury (Remick et al., 1987). If this occurs in humans, it may serve as a portal for the translocation of gram-negative bacteria into the systemic circulation and lead to continued sepsis and progressive dysfunction of one or more vital organs (Matuschak and Rinaldo, 1988). Consequently, preventing sepsis could significantly reduce the incidence of ARDS and successful treatment of sepsis could reduce early and late mortality in ARDS.

The onset of acute renal failure, gastrointestinal bleeding, hepatobiliary dysfunction, coagulation abnormalities including disseminated intravascular coagulation, or central nervous system impairment in patients with ARDS defines multiorgan failure (Beal and Cerra, 1994). The incidence of a demonstrable degree of organ failure in ARDS patients ranges from 20% for the gastrointestinal tract and central nervous system to as high as 95% for the liver. Mortality rises progressively as more organs fail, with a 40% mortality with lung only, 56% with one extrapulmonary organ, 73% with two, 82% with three, and approximates 100% mortality with four (NHLBI, 1979). The onset of multiorgan failure in ARDS is frequently delayed for many days or weeks, suggesting that other factors may be involved in its pathobiology such as severe malnutrition (Moss et al., 1996), including deficiencies of micronutrient antioxidants. Supportive interventions such as mechanical ventilation, dialysis, and total parenteral nutrition appear to have little impact on overall survival once multiorgan failure develops and may, by virtue of their implementation, invite further medical complications. Although inflammatory cells may not have the most critical role in the pathogenesis of multiorgan failure, these cells, their products, and mediators probably contribute to the development of structural and functional changes in extrapulmonary organs. Survival from ARDS may be dependent on the return of normal physiologic function in extrapulmonary organs such as the liver and kidney which may provide important defenses and regulatory mechanisms to clear bacteria and their products, to counteract ROS/RNS, and modulate the humoral and cellular mediators of inflammation.

III. Therapeutic Strategies

Current therapeutic measures for patients with ARDS are complex and logically include: (i) treatment of the underlying condition(s) that precipitated ARDS; (ii) supportive physiologic measures during ARDS to improve gas exchange and O_2 delivery; (iii) strategies to ameliorate and reverse the pathobiological processes underlying ARDS; and (iv) strategies designed to intensify helpful reparative mechanisms. Most of the newer therapies focus on inhibiting the presumed pathobiological causes related to the activation of inflammatory-immune processes (Hudson, 1995) and few have addressed reparative processes (see Table IV).

TABLE IV Therapeutic Strategies in ARDS

Current approach	Experimental approaches
Prevention of ARDS risk factors	Anti-inflammatory agents
Treatment of underlying condition	Ibuprofen
Mechanical ventilation	Prostaglandin E_1
Positive end expiratory pressure	Pentoxifylline
Oxygen	Lipoxygenase inhibitors
Fluids	Antibody therapy
Antibiotics	Anti-C5a
Hemodynamic monitoring	Anti-J5 antigen
Measures to minimize risk of sepsis and/or	Anti-HA-1A
multiorgan failure	E5 antibody
	Anti-TNFα
	IL-1 receptor antagonist
	Anti-PMN adhesive protein
	Antioxidant therapy
	Superoxide dismutase
	N-Acetylcysteine
	Ascorbic acid
	Iron chelators
	Catalase
	α-Tocopherol
	Antiproteases
	Allopurinol
	Exogenous surfactant therapy
	New modes of mechanical ventilation
	Pressure control–inverse ratio ventilation
	Extracorporeal CO_2 removal
	Other
	Lung transplantation

The heterogeneity of ARDS patients is a potentially serious problem for therapeutic clinical trials. The extent to which the various pathobiological aspects of ARDS are involved may be different in direct forms of lung injury such as aspiration, in sepsis, and in major trauma. Therapeutic trials may selectively benefit one group and have an adverse or no effect on others, thus giving little overall evidence of clinical benefit. The identification and timing of studies of homogeneous populations with one substantial risk factor for ARDS may well be critical to the design and testing of specific protocols based on sound therapeutic strategies (Bernard *et al.*, 1994).

A. Recent Advances in Mechanical Ventilation

The mainstay of supportive treatment in ARDS remains O_2 therapy in conjunction with optimized ventilation managements. By recruiting lung volume and keeping alveoli open to improve intrapulmonary shunting, PEEP can improve pulmonary gas exchange and reduce inspired O_2 concentrations. Normal alveoli experience higher than normal distension during PEEP and may be damaged by peak inspiratory pressures as low as 30 cm H_2O (Tsuno *et al.*, 1990). Alternative methods of mechanical ventilation endeavor to reduce lung barotrauma and volutrauma (Slutsky, 1988). Besides reducing PEEP whenever possible, extracorporeal CO_2 removal (Morris *et al.*, 1994), high-frequency ventilation (Holzapfel *et al.*, 1987), and the use of exogenous surfactant to augment conventional mechanical ventilation have been evaluated with disappointing results (Anzueto *et al.*, 1994). Clinical studies have not demonstrated any advantage with pressure-controlled inverse ratio ventilation in ARDS. The improvement in gas exchange with this mode presumably results from prolonging the inspiratory time and pressure plateau to recruit alveoli, and decreased expiratory time to prevent their collapse during exhalation. The normal inspiratory/expiratory ratio (I/E ratio) of 1 : 2 during volume-controlled mechanical ventilation is reversed to 2 : 1 or 3 : 1 during pressure-controlled inverse ratio ventilation. However, no difference in arterial oxygenation or improvement in clinical outcome was found in ARDS patients receiving this ventilation mode when a constant total PEEP was maintained (Mercat *et al.*, 1993).

B. Surfactant

Surfactant is abnormal in ARDS as reflected in phospholipid composition as well as in surface tension properties (Pison *et al.*, 1989). Intrabronchial instillation of surfactant into each lobar bronchus using a fiberoptic bronchoscope or aerosolization of various surfactant preparations has generally been well tolerated and causes a modest improvement in gas exchange in ARDS patients. However, no significant changes in lung compliance or chest X-ray have been found and its use has generally not been shown to significantly alter the natural history of ARDS. Clinical trials of exogenous surfactant in ARDS in the United States and Europe have not shown efficacy, and most clinical studies of its use in adults have been terminated (Anzueto *et al.*, 1994). Any benefit of newer modes of mechanical ventilation or exogenous surfactant in the treatment of the gas exchange abnormalities in ARDS may be negligible since the majority of ARDS patients die from sepsis syndrome and/or multiorgan failure, and not refractory hypoxemic respiratory failure (Doyle *et al.*, 1995).

C. Nitric Oxide (·NO)

·NO can decrease the pulmonary artery pressure and increase arterial oxygenation by selectively dilating pulmonary vessels in ventilated areas of lung. ·NO causes smooth muscle relaxation by activating soluble guanylate cyclase and inhibits platelet aggregation and adhesion to endothelium by increasing cGMP. Because of these properties, ·NO has been proposed as a means of selectively reducing pulmonary hypertension and improving ventilation/perfusion mismatching in patients with ARDS. In a study of 10 ARDS patients, continuous inhalation of ·NO at 18 and 36 ppm for 3 to 53 days decreased pulmonary artery pressure and intrapulmonary shunting, and improved oxygenation without apparent lung injury from ·NO or systemic methemoglobinemia (Rossaint *et al.*, 1993; Krafft *et al.*, 1996). Inhaled ·NO is rapidly taken up in the pulmonary capillaries by hemoglobin, thus limiting any prolonged vasodilating effects to the pulmonary vasculature. Theoretically, the rate of uptake in the lung of concentrations below 100 ppm is several times faster than its rate of oxidation to more toxic species (e.g., ·NO_2), hence it is possible to inhale NO safely. Moreover, ·NO permitted a decrease in inspired O_2 by 15% in the study cited. Overall, 8 patients survived. Five patients received extracorporeal support. However, improved survival in ARDS treated with inhaled ·NO has not been demonstrated in subsequent uncontrolled studies (Rossaint *et al.*, 1995; Krafft *et al.*, 1996). A multicenter, randomized, double-blind clinical trial to determine whether inhaled ·NO will improve morbidity and mortality in ARDS is currently underway.

The potential toxicity of inhaled ·NO could limit its usefulness in ARDS. The optimal dose of ·NO and guidelines for when to begin treatment and when to stop treatment in ARDS have not been established. Unrecognized methemoglobinemia from ·NO inhalation could potentially decrease O_2 delivery in critically ill patients. In high concentrations, ·NO may inactivate critical cellular enzymes by forming iron-sulfur-nitrosyl derivatives and by inhibiting both DNA and protein synthesis.

Because ·NO is a free radical, it can react with $O_2^{·-}$ to produce $ONOO^-$, a potent oxidizing agent known to initiate lipid peroxidation in biological membranes, hydroxylation and nitration of aromatic amino acid residues, and sulfhydryl oxidation of proteins. Furthermore, $ONOO^-$ can decompose to yield ·NO_2-like and ·OH-like species, which are potent toxic nitrating and hydroxylating species (Darley-Usmar *et al.*, 1995).

Finally, data are accumulating to show that ·NO can play a protective role under oxidative stress resulting from peroxides, including the termination of lipid peroxidation chain propagation (e.g., Wink *et al.*, 1995). It is likely that ·NO-related cytotoxic and/or protective chemical reaction are dependent on the relative extracellular fluid fluxes and/or concentrations of ·NO and other ROS and that temporal and spatial considerations play a significant role (Rubbo *et al.*, 1994; Moncada and Higgs, 1995).

D. Anti-inflammatory Agents

Anti-inflammatory agents may act by (i) blocking the production of humoral mediators, e.g., ibuprofen; (ii) blocking cytokine receptors with specific receptor-binding ligands, e.g., TNF-α and receptor antagonists; (iii) blocking humoral mediator action with nonspecific drugs, e.g., corticosteroids and pentoxifylline; or (iv) blocking signal transduction and intracellular pathways through which humoral mediators act, e.g., the nuclear factor-κ B (NF-κB).

No therapeutic benefit was found in a prospective, placebo-controlled trial of high-dose corticosteroids in established ARDS (Bernard *et al.*, 1987). Similar treatment of sepsis syndrome did not prevent the development of ARDS (Luce *et al.*, 1988). Instead, the duration of ARDS was significantly prolonged and the mortality rate was increased in patients treated with corticosteroids, due, in part, to nosocomial infection.

Prostaglandin E_1 (PGE$_1$) can block platelet aggregation and modulate inflammation by suppressing PMN chemotaxis and release of lysosomal enzymes (Zurier *et al.*, 1974). However, PGE$_1$ in a large multicenter study involving 100 patients with established ARDS produced no survival advantage (Bone *et al.*, 1989).

Ibuprofen was evaluated because of its ability to inhibit cyclooxygenase. Ibuprofen significantly reduced pulmonary microvascular permeability and edema in sheep with septic peritonitis (Gnidec *et al.*, 1988). It prevented leakage of plasma proteins into the alveolar space and attenuated PMN migration in a porcine model of sepsis-induced lung injury studied by repetitive BAL (Jenkins *et al.*, 1991). However, a randomized, double-blind multicenter clinical trial of ibuprofen in patients with sepsis failed to show significant improvement in outcome (Haupt *et al.*, 1991).

Ketoconazole, an antifungal agent, is also an inhibitor of leukotriene production and thromboxane synthesis. Preliminary clinical studies have shown that ketoconazole may prevent ARDS in patients who are at high risk, typically those with sepsis or major trauma (Yu and Tomasa, 1993). Definitive recommendations on its use as a prophylactic or therapeutic agent in ARDS await further clinical trials.

Pentoxifylline is a phosphodiesterase inhibitor that can decrease PMN chemotaxis, phagocytic activity, and $O_2{\cdot}^-$ production (Sear *et al.*, 1990). It can inhibit the effects of endotoxin and endotoxin-induced cytokines on PMN function (Sullivan *et al.*, 1988) and inhibit the release of TNF-α by macrophages (Schade, 1990). Pentoxifylline also protects against TNF-α-induced increases in pulmonary microvascular permeability in an animal model of acute long injury (Lilly *et al.*, 1989). However, there are no published studies in humans yet to recommend pentoxifylline as a therapeutic agent in ARDS.

E. Immunologic Therapy

Since most of the detrimental changes seen in ARDS are associated with the activation of systemic and localized inflammatory-immune processes, it follows that antibodies against specific mediators of these processes would be expected to blunt the development of ARDS. In animal studies, antibodies against complement (Stevens *et al.*, 1986) or major PMN adhesive glycoproteins (Vedder *et al.*, 1988) decreased the adherence of activated PMNs to endothelium and prevented lung injury and inflammation. IL-1 receptor antagonist (IL-1ra) given to rabbits before *E. coli* infusion attenuated the clinical features of sepsis syndrome, decreased the infiltration of PMNs into the lung, and prevented death in treated animals (Wakabayashi *et al.*, 1991). Passive immunotherapy with antibody to TNF-α prevented septic shock in mice (Beutler *et al.*, 1985). Baboons given a lethal intravenous dose of live *E. coli* 2 hr after receiving TNF-α antibodies were protected from septic shock and the development of ARDS (Tracey *et al.*, 1987).

However, clinical trials with antiendotoxin and anticytokine (e.g., IL-1ra and anti-TNFα antibodies) in human patients with sepsis and/or ARDS have been either negative or inconclusive (Abraham *et al.*, 1995). Antibodies against various selectin and integrin adhesion molecules have diminished lung damage in a variety of animal models (Faull, 1995), but results of their use in controlled studies of ARDS patients are yet to be completed. The same can be said for the case of PMN NADPH oxidase inhibitors (Wang *et al.*, 1994).

Clinical trials using human polyclonal antiserum to heat-inactivated J5 mutant *E. coli* suggested that passive immunotherapy with anti-J5 antiserum reduced mortality in patients with gram-negative bacteremia (Ziegler *et al.*, 1982) and prevented septic shock and death in high-risk surgical patients (Baumgartner *et al.*, 1985). However, a convincing benefit from the use of monoclonal antibodies HA-1A and E5 to endotoxin in humans with sepsis syndrome has not been demonstrated despite randomized, placebo-controlled clinical trials (McCloskey *et al.*, 1994; Bone *et al.*, 1995). Neither antibody, as might be expected, improved survival in patients with sepsis not due to gram-negative bacteria. The large number of patients not responding to monoclonal antibodies suggests that they alone are probably unable to prevent death from gram-negative sepsis (Halliwell *et al.*, 1992; Cross, 1994).

A combination of therapies directed at different temporal stages of the septic process may be required (Halliwell *et al.*, 1992; Cross *et al.*, 1993). Another approach actively being explored involves therapeutic modulations of multifunctional signal transduction pathways such as activations of NF-κB, a nuclear factor that modifies promotor sites controlling the transcription of numerous critical inflammatory-immune system regulatory proteins including various adhesion molecules and cytokines. Finally, natural modulators of the proinflammatory cytokines (e.g., TNF-α, IL-1), such as IL-10 and TGF-β, may play a role in therapies designed to modify prooxidant cytokine networks (Armstrong *et al.*, 1996).

F. Antioxidant Therapy

It is possible that the initial event in the pathobiology of ARDS from a variety of causes involves ROS/RNS. Consequently, antioxidants that either scavenge ROS/RNS or prevent their function, for example, by chelating transition metal ions catalytic for free radical reactions, may be effective in modulating lung injury and inflammation (Halliwell, 1991).

Some simple therapeutic measures to limit oxidant damage in ARDS are routinely applied, e.g., reduction in the fraction of inspired O_2 concentration whenever possible. This is important not only because hyperoxia gives rise to ROS, but also because PMN ROS production appears increased at elevated O_2 tensions (Krieger et al., 1984). Although the rationale for the therapeutic use of antioxidants exists, few extensive controlled clinical trials have been done. It is possible that antioxidants modulating PMN activation could increase the incidence of infection, as has been implicated in use of corticosteroids in ARDS.

Antioxidants can be nonenzymatic or enzymatic in their actions and extracellular or intracellular in location. The possibility of increasing extracellular or intracellular concentrations of antioxidants maybe an important therapeutic approach in counteracting the adverse effects of ROS/RNS. Antioxidant therapy with liposome-encapsulated, antioxidant intracellular enzymes such as catalase and superoxide dismutase ameliorated experimental lung injury due to hyperoxia (Turrens et al., 1984). Endotoxin (Frank et al., 1980) and even small quantities of TNF-α and/or IL-1β (White et al., 1987) can confer protection against oxidative lung injury in certain model systems, possibly by inducing the production of increased levels of intracellular antioxidant enzymes. This can present a therapeutic enigma in that nonlethal doses of these cytokines can be expected to induce a relative tolerance to subsequent higher lethal levels of not only the inducing cytokine but also to different kinds of stresses (e.g., a broad range of oxidative stress). Thus, it is not necessarily obvious that antibodies against proinflammatory cytokines (e.g., TNF-α, IL-1) would be beneficial to administrate at all stages of ARDS.

Nonenzymatic antioxidant intervention with N-acetylcysteine (NAC) (Bernard et al., 1984), α-tocopherol (McMillan and Boyd, 1982), ascorbic acid (McMillan and Boyd, 1982), lazaroids (Louie et al., 1991), and allopurinol (Jenkinson et al., 1991) have demonstrated protective effects in some animal models of acute lung injury. NAC is a good scavenger of ·OH, $HOCl/OCl^-$, and probably $ONOO^-$ and also helps to maintain cellular GSH levels (Moldeus et al., 1986). In addition to its direct antioxidant effects, NAC may blunt PMN activation and sequestration in the lung (Lucht et al., 1987). In animal studies, NAC protected against O_2 toxicity in the lung (Wagner et al., 1989). However, in randomized, double-blind clinical trials in ARDS patients, NAC failed to improve respiratory physiology or survival when compared to placebo-treated controls (Jepsen et al., 1992; Suter et al., 1994).

α-Tocopherol is the major chain-breaking antioxidant residing in the lipid phase of cellular membranes. A partially but not rigorously controlled trial in which α-tocopherol was administered by gastric tube to patients with threatening or manifest ARDS claimed a beneficial effect of α-tocopherol in ARDS, as measured by survival rate [39% for those patients receiving high enteric α-tocopherol supplements, $n = 87$, vs 17% for controls, $n = 89$ (Wolf and Seeger, 1982)]. Augmenting respiratory tract epithelial cells with α-tocopherol by incorporating the vitamin into liposomes and administrating the preparation directly into the lungs (Freeman and Panus, 1988) may facilitate further clinical testing of α-tocopherol supplementation in ARDS. A problem with both enteral and parenteral α-tocopherol supplements is that α-tocopherol has to be administered for several days to raise tissue levels.

Ascorbic acid has a range of antioxidant properties, e.g., it can scavenge $O_2 \cdot^-$, $ONOO^-$, $\cdot OH$, aqueous peroxyl radicals ($LOO\cdot$), and myeloperoxidase-derived $HOCl/OCl^-$ and regenerate α-tocopherol (Halliwell et al., 1992), but it can also be pro-oxidant in the presence of transition metal ions by promoting the redox cycling of "free" iron and/or copper (Halliwell, 1991). In vitro studies with activated PMNs and ascorbic acid suggested that very low plasma ascorbate levels found in ARDS patients could reflect in vivo PMN activation (Cross et al., 1990). Consequently, supplemental ascorbic acid may be effective in reducing oxidative stress that in turn may ameliorate injury and inflammation in ARDS and improve survival (Sawyer et al., 1989). However, the potential pro-oxidant interaction of ascorbate with iron ions and/or copper may curtail any beneficial effects of ascorbate in ARDS. Patients with sepsis rapidly consume IV-infused ascorbate (Galley et al., 1996), presumably either as a result of increased radical scavenging or via the production of redox cycling of iron and/or copper. Important in this regard is the finding of increased bleomycin-detectable "free" iron in the plasma of patients with sepsis (Gutteridge et al., 1994; Galley et al., 1996).

Iron ion chelators, of which the prototype is deferoxamine, are effective extracellular antioxidants because they restrict the availability of iron ions to participate in the Fenton reaction. Deferoxamine is not only a powerful chelator of iron ions, but also rapidly scavenges $\cdot OH$ radicals and slowly scavenges peroxyl radicals in aqueous solutions (Halliwell, 1991). Deferoxamine also has been reported to decrease acute lung injury after complement activation (Ward et al., 1983), thermal injury (Till et al., 1985), O_3 exposure (Louie et al., 1993), reperfusion lung injury (Kennedy et al., 1989), and smoke inhalation (LaLonde et al., 1994),suggesting the involvement of iron ions in these conditions. Other chelators and inhibitors of iron-catalyzed free radical reactions are under investigation, e.g., the lazaroids (Braughler et al., 1989), but specific recommendations on the use of iron ion chelators in ARDS cannot be made at present.

Therapeutic strategies involving antioxidant therapy must identify the molecular targets of ROS, e.g., proteins, DNA, or lipids, and deal with the problem of delivering enough antioxidants to the microenvironments where ROS/RNS are formed (e.g., between PMN and endothelial cells). Recent advances in gene technology have allowed utilization of plasmids to increase transiently the intracellular production of antioxidants, e.g., SOD and catalase. Similar technology may allow transient inhibition of intracellular metabolic processes that generate and release ROS from inflammatory cells. Antioxidant therapy in combination with other therapeutic interventions may provide the protection necessary for endogenous reparative processes to occur in patients with ARDS and avoid the development of multiorgan failure. For example, in a preliminary report, ARDS patients received an antioxidant regimen of N-acetylcysteine (po), α-tocopherol (po), selenium (iv), and ascorbic acid (iv) within 24 hr of diagnosis for 3 days and experienced a significant reduction in mortality ($n = 25$; 20% mortality) compared to a control group($n = 20$; 65% mortality) (Millili and Marino, 1993). Further rigorously controlled studies are required before definitive recommendations regarding the routine use of antioxidants in ARDS can be made.

IV. Conclusion

The view that ARDS is primarily a lung disorder has evolved into the present concept that ARDS represents a systemic disorder characterized by the intense activation of inflammatory-immune processes that manifest most severely in the lung. The pathobiology is very complex but almost certainly includes injurious mechanisms involving ROS and probably RNS. Although there is strong evidence that ˙NO participates in many inflammatory-immune processes including ARDS, to date there is only circumstantial evidence that ˙NO is a pathogenic mediator participating in mediating lung tissue injury in ARDS. As explained in this chapter, it is evident that

- No single pathobiological pathway leads directly to the acute lung injury.
- Complex multifactorial humoral and cellular mediators acting at different time sequences are responsible for the microvascular and alveolar cell damage leading to the exudative pulmonary edema and respiratory failure that causes ARDS.
- Evidence links ROS (and possibly RNS) to ARDS, but the exact mechanisms and pathobiological significance of their involvement remain unclear.
- Useful therapeutic approaches focusing on antioxidant supplements remain to be validated in ARDS.

Although further elucidations of the exact mechanisms responsible for acute lung injury continue unabated in experimental models of ARDS, the ultimate model remains humans. Clinical investigations of therapeutic strategies must allow for the heterogeneity of ARDS patients and the multifactorial nature of the inciting pathobiology. Only then can the design and testing of specific protocols based on sound therapeutic strategies be successful. Single agent or combination therapy must specifically target the untoward humoral, cellular, and oxidant events without compromising normal or helpful adaptive cellular functions. Prospective randomized studies, with comparison to conventional therapies, are essential in evaluating any new intervention in ARDS but in most instances are probably unlikely to be undertaken. It is doubtful that any single therapeutic approach will be strikingly successful after the onset of ARDS. Combinations of approaches directed at different stages of ARDS and including antioxidant strategies are predictively what will be required (Halliwell *et al.*, 1992; Cross, 1994).

References

Abraham, E., Wunderink, R., Silverman, H., *et al.* (1995). Efficacy and safety of monoclonal antibody to human tumor necrosis factor α in patients with sepsis syndrome. *JAMA* **273**, 934–941.

Anzueto, A., Baughman, R., Guntupalli, K., *et al.* (1994). An international randomized placebo-controlled trial evaluating the safety and efficacy of aerosolized surfactant in patients with sepsis-induced ARDS. *Am. J. Respir. Crit. Care Med.* **149A**, A567.

Armstrong, L., Jordan, N., and Millar, A. (1996). Interleukin 10 (IL-10) regulation of tumour necrosis factor α (TNF-α) from human alveolar macrophages and peripheral blood monocytes. *Thorax* **51**, 143–149.

Asbaugh, D. G., Bigelow, D. B., Petty, T. L., and Levine, B. E. (1967). Acute respiratory distress in adults. *Lancet* **2**, 319–323.

Bachofen, M., and Weibel, E. R. (1982). Structural alterations of lung parenchyma in the adult respiratory distress syndrome. *Clin Chest Med.* **3**, 35–56.

Baldwin, S. R., Gun, C. M., Boxer, L. A., Simon, R. H., Ketai, L. H., and Devall, L. J. (1986). Oxidant activity in expired breath of patients with adult respiratory distress syndrome. *Lancet* **1**, 11–14.

Balla, A. K., Doi, E. M., Wunder, P. R., Ogle, J. D., and DeBault, L. E. (1993). Human polymorphonuclear leukocyte (PMN) priming and activation by acute ethanol intoxication. *Adv. Exp. Med. Biol.* **335**, 165–168.

Barlett, D., Church, D. F., Bounds, P. L., and Koppenol, W. H. (1995). The kinetics of the oxidation of L-ascorbic acid by peroxynitrite. *Free Rad. Biol. Med.* **18**, 85–92.

Baumgartner, J. D., Glauser, M. P., McCutchan, J. A., *et al.* (1985). Prevention of gram-negative shock and death in surgical patients by antibody to endotoxin core glycolipid. *Lancet* **2**, 59–63.

Beal, A. L., and Cerra, F. B. (1994). Multiple organ failure syndrome in the 1990s: Systemic inflammatory response and organ dysfunction. *JAMA* **271**, 226–233.

Beckman, J. S., Beckman, T. W., Chen, J., *et al.* (1990). Apparent hydroxyl radical production by peroxynitrite: Implications for endothelial injury from nitric oxide and superoxide. *Proc. Natl. Acad. Sci. U.S.A.* **87**, 1620–1624.

Bell, R. C., Coalson, J. J., Smith, J. D., and Johanson, W. G. (1983). Multiple organ system failure and infection in adult respiratory distress syndrome. *Ann. Intern. Med.* **99**, 293–298.

Bernard, G. R. (1990). Potential of N-acetylcysteine as treatment for the adult respiratory distress syndrome. *Eur. Respir. J.* **3**(11), 496s–498s.

Bernard, G. R., Artigas, A., Brigham, K. L., *et al.* (1994). The American-European consensus conference on ARDS definitions, mechanisms, relevant outcomes, and clinical trial coordination. *Am. J. Respir. Crit. Care Med.* **149**, 818–824.

Bernard, G. R., Luce, J. M., Sprung, C. L., *et al.* (1987). High dose corticosteroids in patients with adult respiratory distress syndrome. *N. Engl. J. Med.* **317**, 1565–1570.

Bernard, G. R., Lucht, W. D., Niedermeyer, M. E., *et al.* (1984). Effect of N-acetylcysteine on the pulmonary response to endotoxin in the awake sheep and upon *in vitro* granulocyte function. *J. Clin. Invest.* **73**, 1772–1784.

Beutler, B., Milsark, I. W., and Cerarni, A. (1985). Passive immunization against cachectin/tumor necrosis factor protects mice from the lethal effect of endotoxin. *Science* **229**, 869–871.

Billiau, A., and Vandekerckhove, F. (1991). Cytokines and their interactions with other inflammatory mediators in the pathogenesis of sepsis and septic shock. *Eur. J. Clin. Invest.* **21**, 559–573.

Bone, R. C. (1994). Gram positive organisms and sepsis. *Arch. Intern. Med.* **154**, 26–34.

Bone, R. C., Balk, R. A., Fein, A. M., *et al.* (1995). A second large controlled clinical study of E5, a monoclonal antibody to endotoxin: Results of a prospective, multicenter, randomized, controlled trial. *Crit. Care Med.* **23**, 994–1005.

Bone, R. C., Balk, R. A., Slotman, G., *et al.* (1992). Adult respiratory distress syndrome: Sequence and importance of development of multiple organ failure. *Chest* **101**, 320–326.

Bone, R. C., Slotman, G., Maunder, R., *et al.* (1989). Randomized double-blind multicenter study of prostaglandin E1 in patients with the adult respiratory distress syndrome. *Chest* **96**, 114–119.

Braquet, P., Touqui, L., Shen, T. Y., and Vargaftig, B. B. (1987). Perspectives in platelet-activating research. *Pharmacol. Rev.* **39**, 97–145.

Braughler, J. M., Hall, E. D., Jacobsen, E. J., McCall, J. M., and Means, E. D. (1989). The 21-aminosteroids: Potent inhibitors of lipid peroxidation for the treatment of central nervous system trauma and ischemia. *Drugs Future* **14**, 143–152.

Bunnell, E., and Pacht, E. R. (1993). Oxidized glutathione is increased in the alveolar fluid of patients with the adult respiratory distress syndrome. *Am. Rev. Respir. Dis.* **148**, 1174–1178.

Cannon, J. G., Tompkins, R. G., Gelfand, J. A., *et al.* (1990). Circulating interleukin-1 and tumor necrosis factor in septic shock and experimental endotoxin fever. *J. Infect. Dis.* **161**, 79–84.

Cochrane, C. G., Spragg, R. G., Revak, S. D., *et al.* (1983). The presence of neutrophil elastase and evidence of oxidation activity in bronchoalveolar lavage fluid of patients with adult respiratory distress syndrome. *Am. Rev. Respir. Dis.* **127**, 525–527.

Cross, A. S. (1994). Antiendotoxin antibodies: A dead end? *Ann. Intern. Med.* **121**, 58–60.

Cross, A. S., Opal, S. M., Palardy, J. E., *et al.* (1993). The efficacy of combination immunotherapy in experimental *Pseudomonas sepsis. J. Infect. Dis.* **176**, 112–118.

Cross, C. E., Forte, T., Stocker, R., *et al.* (1990). Oxidative stress and abnormal cholesterol metabolism in patients with adult respiratory distress syndrome. *J. Lab. Clin. Med.* **115**, 396–404.

Cross, C. E., van der Vliet, A., O'Neill, C. A., and Eiserich, J. P. (1984). Reactive oxygen species and the lung. *Lancet* **344**, 930–933.

Curzen, N. P., Griffiths, M. J. D., and Evans, T. W. (1994). Role of endothelium in modulating the vascular response to sepsis. *Clin. Sci.* **86**, 359–374.

Damas, P., Reuter, A., Gysen, P., Demonty, J., Lamy, M., and Franchimont, P. (1989). Tumor necrosis factor and interleukin-1 serum levels during severe sepsis in humans. *Crit. Care Med.* **17**, 975–978.

Darley-Usmar, V., Wiseman, H., and Halliwell, B . (1995). Nitric oxide and oxyen radicals: A question of balance. *FEBS Lett.* **369**, 131–135.

Davis, W. B., Rennard, S. I., Bitterman, P. B., and Crystal, R. G. (1983). Pulmonary oxygen toxicity: Early reversible changes in human alveolar structures induced by hyperoxia. *N. Engl. J. Med.* **309**, 878–883.

Dehring, D. J., and Wismar, B. L. (1989). Intravascular macrophages in pulmonary capillaries of humans. *Am. Rev. Respir. Dis.* **139**, 1027–1029.

Demling, R. H., Lalonde, C., Ryan, P., *et al.* (1988). Endotoxemia produces an increase in arterial but not venous lipid peroxides in sheep. *J. Appl. Physiol.* **64**, 592–598.

Dobrina, A., and Patriarca, P. (1986). Neutrophil–endothelial cell interaction. *J. Clin. Invest.* **78**, 462–471.

Donnelly, S. C. (1994). Role of selectins in development of adult respiratory distress syndrome. *Lancet* **344**, 215–219.

Dorinsky, P. M., and Gadek, J. E. (1989). Mechanisms of multiple nonpulmonary organ failure in ARDS. *Chest* **96**, 885–892.

Doyle, R. L., Szaflarski, N., Modin, G. W., *et al.* (1995). Identification of patients with acute lung injury: Predictors of mortality. *Am. J. Respir. Crit. Care Med.* **152**, 1818–1824.

Enhorning, G. (1989). Surfactant replacement in adult respiratory distress syndrome. *Am. Rev. Respir. Dis.* **140**, 281–283.

Faull, R. J. (1995). Adhesion molecules in health and disease. *Aust. N.Z. J. Med.* **25**, 720–730.

Folkes, L. K., Candeias, L. P., and Wardman, P. (1995). Kinetics and mechanism of hypochlorous acid reactions. *Arch. Biochem. Biophys.* **323**, 120–126.

Fowler, A. A., Hamman, R. F., Zerbe, G. O., Benson, K. N., and Hyers, T. M. (1985). Adult respiratory distress syndrome. Prognosis after onset. *Am. Rev. Respir. Dis.* **132**, 472–478.

Frank, L., Summerville, F. L., Massaro, D. (1980). Protection from oxygen toxicity with endotoxin: Role of the endogenous antioxidant enzymes of the lung. *J. Clin. Invest.* **65**, 1104–1110.

Freeman B. A., and Crapo, J. D. (1981). Hyperoxia increases oxygen radical production in rat lungs and lung mitochondria. *J. Biol. Chem.* **256**; 10986–10992.

Freeman, B. A., and Panus, P. C. (1988). Specific modulation of alveolar type II epithelial α-tocopherol content. *Am. Rev. Respir. Dis.* **137**, A84.

Galley, H. F., Davies, M. J., and Webster, N. R. (1996). Ascorbyl radical formation in patients with sepsis: Effect of ascorbate loading. *Free Rad. Biol. Med.* **20**, 139–143.

Gee, M. H., Perkowski, S. Z., Tahamont, M. V., Flynn, J. T. (1985). Arachidonate cyclooxygenase metabolites as mediators of complement-initiated lung injury. *Fed. Proc.* **44**, 46–52.

Gerschman, R., Gilbert, D. L., Nye, S. W., Dwyer, P., and Feen, W. O. (1954). Oxygen poisoning and x-irradiation: A mechanisms in common. *Science* **119**, 623–626.

Giroir, B. P. (1993). Mediators of septic shock: New approaches for interrupting the endogenous inflammatory cascade. *Crit. Care Med.* **21**, 780–789.

Ghofrani, H. A., Rosseau, S., Walmrath, D., Kadus, W., Kramer, A., Grimminger, F., Lohmeyer, J., and Seeger, W. (1996). Compartmentalized lung cytokine release in response to intravascular and alveolar endotoxin challenge. *Am. J. Physiol.* **270**, L62–L68.

Gnidec, A. G., Sibbald, W. J., Cheung, H., and Metz, C. A. (1988). Ibuprofen reduces the progression of permeability edema in an animal model of hyperdynamic sepsis. *J. Appl. Physiol.* **65**, 1024–1032.

Goldblum, S. E., Jay, M., Yoneda, K., Cohen, D. A., McClain, C. J., and Gillespie, M. N. (1987). Monokine-induced acute lung injury in rabbits. *J. Appl. Physiol.* **63**, 2093–2100.

Griffith, D. E., Holden, W. E., Morris, J. F., Min, L. K., and Krishnamurthy, G. T. (1986). Effect of common therapeutic concentrations of oxygen on lung clearance of Tc-DPTA and bronchoalveolar lavage albumin concentration. *Am. Rev. Respir. Dis.* **134**, 233–237.

Groeneveld, A. B. J., Raijmakers, P. G. H. M., and Hack, C. E. (1995). Interleukin 8-related neutrophil elastase and the severity of the adult respiratory tract syndrome. *Cytokine* **7**, 746–752.

Grum, C. M., Ragsdale, R. A., Ketai, L. H., *et al.* (1987). Plasma xanthine oxidase activity in patients with adult respiratory distress syndrome. *J. Crit. Care* **2**, 22–26.

Gutteridge, J. M. C., Quinlan, G. J., and Evans, T. W. (1994). Transient iron overload with bleomycin detectable iron in the plasma of patients with ARDS. *Thorax* **49**, 707–710.

Haddad, I. Y., Pataki, G., Hu, P., Galliani, C., Beckman, J. S., and Matalon, S. (1994). Quantitation of nitrotyrosine levels in lung sections of patients and animals with acute lung injury. *J. Clin. Invest.* **94**, 2407–2413.

Haddad, I. Y., Zhu, S., Ischiropoulos, H., and Matalon, S. (1996). Nitration of surfactant protein A results in decreased ability to aggregate lipids. *Am. J. Physiol.* **270**, L281–L288.

Halliwell, B. (1991). Drug antioxidant effects: A basis for drug selection? *Drugs* **42**, 569–605.

Halliwell, B., and Gutteridge, J. M. C. (1989). "Free Radicals in Biology and Medicine," 2nd Ed. Clarendon Press, Oxford.

Halliwell, B., Gutteridge, J. M. C., and Cross, C. E. (1992). Free radicals, antioxidants, and human disease: Where are we now? *J. Lab. Clin. Med.* **119**, 598–620.

Haupt, M. T., Jastremski, M. S., Clemmer, T. P., Metz, C. A., and Goris, G. B. (1991). The ibuprofen study group: Effect of ibuprofen in patients with severe sepsis. A randomized, double-blind, multicenter study. *Crit. Care Med.* **19**, 1339–1347.

Heffner, J. E., Sahn, S. A., Repine, J. E. (1987). The role of platelets in the adult respiratory distress syndrome. *Am. Rev. Respir. Dis.* **136**, 482–492.

Holtzman, M. J. (1991). State of the art: Arachidonic acid metabolism. *Am. Rev. Respir. Dis.* **143**, 188–203.

Holzapfel, L., Robert, D., Perrin, F., Gaussorgues, P., and Giudicelli, D. P. (1987). Comparison of high-frequency jet ventilation to conventional ventilation in adults with respiratory distress syndrome. *Intensive Care Med.* **13**, 100–105.

Hu, M. L., Louie, S., Cross, C. E., Motchnik, P., and Halliwell B. (1993). Antioxidant protection against hypochlorous acid in human plasma. *J. Lab. Clin. Med.* **121**, 257.

Hu, P., Ischiropoulos, H., Beckman, J. S., and Matalon, S. (1994). Peroxynitrite inhibition of oxygen consumption and ion transport in alveolar type II cells. *Am. J. Physiol.* **266**, L628–L634.

Hudson, L. D. (1995). New therapies for ARDS. *Chest* **108**(Suppl.), 74S–91S.

Hudson, L. D., Milberg, J., Anardi, D., *et al.* (1995). Clinical risks for development of the acute respiratory distress syndrome. *Am. J. Respir. Crit. Care Med.* **151**, 293–301.

Huie, R. E., and Padmaja, S. (1993). The reaction of nitric oxide with superoxide. *Free Rad. Res. Commun.* **18**, 195–199.

Hyers, T. M., Tricomi, S. M., Dettenmeier, P. A., and Fowler, A. A. (1991). Tumor necrosis factor levels in serum and bronchoalveolar lavage fluid of patients with the adult respiratory distress syndrome. *Am. Rev. Respir. Dis.* **144**, 268–271.

Ischiropoulos, H., Zhu, L., and Beckman, J. S. (1992). Peroxynitrite formation from macrophage-derived nitric oxide. *Arch. Biochem. Biophys.* **298**, 446–451.

Jacobs, R. F., Tabor, D. R., Burks, A. W., and Campbell, G. D. (1989). Elevated interleukin-1 release by human alveolar macrophages during the adult respiratory distress syndrome. *Am. Rev. Respir. Dis.* **140**, 1686–1692.

Jamieson, D. (1989). Oxygen toxicity and reactive oxygen metabolites in mammals. *Free Rad. Biol. Med.* **7**, 87–108.

Jenkins, J. K., Carey, P. D., Bryne, K., Sugerman, H. J., and Fowler, A. A. (1991). Sepsis-induced lung injury and the effects of ibuprofen pretreatment. *Am. Rev. Respir. Dis.* **143**, 155–161.

Jenkinson, S. G., Roberts, R. J., DeLemos, R. A., *et al.* (1991). Allopurinol-induced effects in premature baboons with respiratory distress syndrome. *J. Appl. Physiol.* **70**, 1160–1167.

Jepsen, S., Herlevsen, P., Knudsen, P., Bud, M. I., and Klausen, N. O. (1992). Antioxidant treatment with N-acetylcysteine during adult respiratory distress syndrome: A prospective randomized, placebo-controlled study. *Crit. Care Med.* **20**, 918–923.

Krafft, P., Fridrich, P., Fitzgerald, R. D., Koc, D., and Steltzer, H. (1996). Effectiveness of nitric oxide inhalation in septic ARDS. *Chest* **109**, 486–493.

Kennedy, T. P., Rao, N. V., Hopkins, C., Pennington, L., Tolley, E., and Hoidal, J. R. (1989). Role of reactive oxygen species in reperfusion injury of the rabbit lung. *J. Clin. Invest.* **83**, 1326–1335.

Kim, D. K., Fukuda, T., Thompson, T., *et al.* (1995). Bronchoalveolar lavage fluid phospholipid A2 activities are increased in human adult respiratory distress syndrome. *Am. J. Physiol.* **169**, L109–L118.

Koyama, I., Toung, T. J. K., Rogers, M. C., *et al.* (1987). Oxygen radicals mediate reperfusion lung injury in ischemic oxygen-ventilated canine pulmonary lobe. *J. Appl. Physiol.* **63**, 111–115.

Krafft, P., Fridrich, P., Fitzgerald, R. D. *et al.* (1996). Effectiveness of nitric oxide inhalation in septic ARDS. *Chest* **109**, 486–493.

Krieger, B. P., Loomis, W. H., and Spragg, R. G.(1984). Granulocytes and hyperoxia act synergistically in causing acute lung injury. *Exp. Lung Res.* **7**, 77–83.

Kulkarni, A. B., Huh, C.-G., Becker, D., *et al.* (1993). Transforming growth factor β_1 null mutation in mice causes excessive inflammatory response and early death. *Proc. Natl. Acad. Sci. U.S.A.* **90**, 770–774.

Kunkel, S. L., Standiford, T., Kasahara, K., and Strieter, R. M. (1991). Interleukin-8 (IL-8): The major neutrophil chemotactic factor in the lung. *Exp. Lung Res.* **17**, 17–23.

LaLonde, C., Ikegami, K., and Demling, R. (1994). Aerosolized deferoxamine prevents lung and systemic injury caused by smoke inhalation. *J. Appl. Physiol.* **77**, 2057–2064.

Larrick, J. W., and Wright, S. C. (1990). Cytotoxic mechanism of tumor necrosis factor-α. *FASEB J.* **4**, 3215–3223.

Laubach, V. E., Shesley, E. G., Smithies, O., and Sherman, P. A. (1995). Mice lacking inducible nitric oxide synthase are not resistant to lipopolysaccharide-induced death. *Proc. Natl. Acad. Sci. U.S.A.* **92**, 10688–10692.

Laurent, T., Markert, M., von Fliedner, V., *et al.* (1994). CD11b/CD18 expression adherence, and chemotaxis of granulocytes in adult respiratory distress syndrome. *Am. J. Respir. Crit. Care Med.* **149**, 1534–1538.

Le, J., and Vilcek, J. (1987). Tumor necrosis factor and interleukin-1: Cytokines with multiple overlapping biological activities. *Lab. Invest.* **56**, 234–248.

Lee, C. T., Fein, A. M., Lippman, M., *et al.* (1981). Elastolytic activity in pulmonary lavage fluid from patients with adult respiratory distress syndrome. *N. Engl. J. Med.* **340**, 109–196.

Lewis, R. E., and Granger, H. J. (1986). Neutrophil-dependent mediation of microvascular permeability. *Fed. Proc.* **45**, 109–113.

Lilly, G. M., Sandu, J. S., Ishizaka, A., *et al.* (1989). Pentoxifylline prevents tumor necrosis factor-induced lung injury. *Am. Rev. Respir. Dis,* **139**, 1361–1368.

Louie, S., Russell, L. A., Teague, S. V., and Cross, C. E. (1991). Nasal inhalation of aerosolized lazaroid U78517F ameliorates acute ozone lung injury in rat lungs. *Clin. Res.* **39**, 108A.

Louie, S., Arata, M. A., Offerdahl, S. D., and Halliwell, B. (1993). Effects of tracheal insufflation of deferoxamine upon acute ozone toxicity in rats. *J. Lab. Clin. Med.* **121**, 502–509.

Luce, J. M., Montgomery, A. B., Marks, J. D., *et al.* (1988). Ineffectiveness of high dose methylprednisolone in preventing parenchymal lung injury and improving mortality in patients with septic shock. *Am. Rev. Respir. Dis.* **138**, 62–68.

Lucht, B. D., English, D. K., Bernard, G. R., *et al.* (1987). Prevention of release or granulocyte aggregants into sheep lung lymph following endotoxemia by N-acetylcysteine. *Am. J. Med. Sci.* **249**, 161–167.

Lynch, M. J., Grum, C. M., Gallagher, K. P., *et al.* (1988). Toxic oxygen metabolites from oxidase systems cause rapid serum complement activation, RBC fragility and acute edematous oxidative lung injury in rats subjected to skin burn. *Clin. Res.* **36**, 194A.

Marks, J. D., Marks, C. B., Luce, J. M., *et al.* (1990). Plasma tumor necrosis factor in patients with septic shock. *Am. Rev. Respir. Dis.* **141**, 94–97.

Martin, T. R., Pistorese, B. P., Hudson, L. D., and Maunder, R. J. (1991). The function of lung and blood neutrophils in patients with the adult respiratory distress syndrome: Implications for the pathogenesis of lung infections. *Am. Rev. Respir. Dis.* **144**, 254–262.

Mason, R. J. (1985). Pulmonary alveolar type II epithelial cells and adult respiratory distress syndrome. *West. J. Med.* **143**, 611–615.

Matuschak, G. M., and Rinaldo, J. E. (1988). Organ interactions in the adult respiratory distress syndrome during sepsis: Role of the liver in host defense. *Chest* **94**, 400–406.

Maunder, R. J., and Hudson, L. D. (1991). Clinical risks associated with the adult respiratory distress syndrome. *In* "Adult Respiratory Distress Syndrome" (W. M. Zapol and F. Lemaire, eds.), pp. 1–18. Dekker, New York.

Maunder, R. J., Shuman, W. P., McHugh, J. W., *et al.* (1986). Preservation of normal lung regions in the adult respiratory distress syndrome: Analysis by computed tomography. *JAMA* **255**, 2463–2465.

McCloskey, M. D., Straube, R. C., Sanders, C., *et al.* (1994). Treatment of septic shock with human monoclonal antibody HA-1A: A randomized double-blind placebo-controlled trial. *Ann. Intern. Med.* **121**, 1–5.

McCord, J. M. (1985). Oxygen-derived free radicals in postischemic tissue injury. *N. Engl. J. Med.* **312**, 159–161.

McCord, J. M., Gao, B., Leff, J., and Flores, S. (1994). Neutrophil-generated free radicals: Possible mechanisms of injury in adult respiratory distress syndrome. *Environ. Health Perspect.* **102** (Suppl. 10), 57–60.

McMillan, D. D., and Boyd, G. N. (1982). The role of antioxidants and diet in the prevention or treatment of oxygen-induced lung microvascular injury. *Ann. N.Y. Acad. Sci.* **384**, 535–543.

Meduri, G. U., Eltorky, M., and Winer-Muram, H. T. (1995). The fibroproliferative phase of late adult respiratory distress syndrome. *Semin. Respir. Infect.* **10**, 154–175.

Mercat, A., Graini, L., Teboul, J. L., Lenique, F., and Richard, C. (1993). Cardiorespiratory effects of pressure controlled ventilation with and without inverse ratio in the adult respiratory distress syndrome. *Chest* **104**, 871–875.

Michie, H. R., Manogue, K. R., Spriggs, D. R., *et al.* (1988). Detection of circulating tumor necrosis factor after endotoxin administration. *N. Engl. J. Med.* **318**, 1481–1486.

Milberg, J. A., Davis, D. R., Steinberg, K. P., and Hudson, L. D. (1995). Improved survival of patients with acute respiratory distress syndrome (ARDS): 1983–1993. *JAMA* **273**, 306–309.

Miles, A. M., Owens, M. W., Milligan, S., Johnson, G. G., Fields, J. Z., Ing, T. S., Kottapalli, V., Keshavarzian, A., and Grisham, M. B. (1995). Nitric oxide synthase in circulating vs. extravasated polymorphonuclear leukocytes. *J. Leukocyte Biol.* **58**, 616–622

Millar, A. B., Foley, N. M., Singer, M., Johnson, N McI, Meager, A., and Rook, G. A. W. (1989). Tumour necrosis factor in bronchopulmonary secretions of patients with adult respiratory distress syndrome. *Lancet* **2**, 712–714.

Miller, E. J., Cohen, A. B., Nagao, S., *et al.* (1992). Elevated levels of NAP-1/interleukin-8 are present in the airspaces of patients with the adult respiratory distress syndrome and are associated with increased mortality. *Am. Rev. Respir. Dis.* **146**, 427–432.

Millili, J. J., and Marino, P. L. (1993). Improved clinical outcome in adult respiratory distress syndrome (ARDS) treated with antioxidants. *Free Rad. Bio. Med.* **15**, 512.

Mizer, L. A., Weisbrode, S. E., and Dorinsky, P. M. (1989). Neutrophil accumulation and structural changes in nonpulmonary organs after lung injury induced by phorbol myristate acetate. *Am. Rev. Respir. Dis.* **139**, 1017–1026.

Moldeus, P., Cotgreave, I. A., and Berggren, M. (1986). Lung protection by a thiol-containing antioxidant: N-acetylcysteine. *Respiration* **50**, 31–43.

Moncada, S., and Higgs, E. A. (1995). Molecular mechanisms and therapeutic strategies related to nitric oxide. *FASEB J.* **9**, 1319–1330.

Moreno, J. J. and Prior, W. A. (1992). Inactivation of α-1 proteinase inhibitor by peroxynitrite. *Chem. Res. Toxicol.* **5**, 425–431.

Morris, A. H., Wallace, C. J., Menlove, R. L., *et al.* (1994). Randomized clinical trial of pressure-controlled inverse ratio ventilation and extracorporeal CO_2 removal for adult respiratory distress syndrome. *Am. J. Respir. Crit. Care Med.* **149**, 295–305.

Moss, M., Bucher, B., Moore, F. A., Moore, E. E., and Parsons, P. E. (1996). The role of chronic alcohol abuse in the development of acute respiratory distress syndrome in adults. *JAMA* 275, 50–54.

Movat, H. Z. (1987). Tumor necrosis factor and interleukin-1: Role in acute inflammation and microvascular injury. *J. Lab. Clin. Med.* 110, 668–681.

Murray, J. F., Matthay, M. A., Luce, J. M., and Flick, M. R. (1988). An expanded definition of the adult respiratory distress syndrome. *Am. Rev. Respir. Dis.* 138, 720–723.

National Heart and Lung Institute. (1972). Respiratory diseases: Task force report on problems, research approaches, needs. DHEW Publication No (NIH) 73–432:167–80, U.S. Government Printing Office, Washington, D.C.

National Heart, Lung, and Blood Institute. (1979). Extracorporeal support for respiratory insufficiency: Collaborative study. United States Department of Health, Education and Welfare. Public Health Service.

Ognibene, F. P., Martin, S. E., Parker, M. M., *et al.* (1986). Adult respiratory distress syndrome in patients with severe neutropenia. *N. Engl. J. Med.* 315, 547–551.

Pacht, E. R., Timerman, A. P., Lykens, M. G., and Merola, A. J. (1991). Deficiency of alveolar fluid glutathione in patients with sepsis and the adult respiratory distress syndrome. *Chest* 100, 1397–1403.

Perlman, M. B., Johnson, A., Jubiz, W., *et al.* (1989). Lipoxygenase products induce neutrophil activation and increase endothelial permeability after thrombin-induced pulmonary microembolism. *Circ. Res.* 64, 62–73.

Pinkard, R. N., Ludwig, J. C., and McManus, L. M. (1988). Platelet activating factors. *In* "Inflammation: Basic Principles and Clinical Correlates" (J. J. Gallin, I. M. Goldstein, and R. Snyderman, eds.), pp. 139–167. Raven Press, New York.

Pison, U., Seeger, W., Buchhorn, R., *et al.* (1989). Surfactant abnormalities in patients with respiratory failure after multiple trauma. *Am. Rev. Respir. Dis.* 140, 1033–1039.

Quinlan, G. J., Evans, T. W., and Gutteridge, J. M. C. (1995). Measurement of metatyrosine, ortho-tyrosine and kynurene in BAL proteins of patients with ARDS: Evidence for the formation of hydroxyl radicals. *Thorax* 50 (Suppl. 2), A69.

Rabinovici, R., Sofronski, P., Borborglu, P., *et al.* (1994). Interleukin 2-induced lung injury: The role of complement. *Cir. Res.* 74, 329–335.

Radi, R., Beckman, J. S., Bush, K. M., and Freeman, B. A. (1991a). Peroxynitrite-induced membrane lipid peroxidation: The cytotoxic potential of superoxide and nitric oxide. *Arch. Biochem. Biophys.* 288, 481–487.

Radi, R., Beckman, J. S., Bush, K. M., and Freeman, B. A. (1991b). Peroxynitrite oxidation of sulfhydryls: The cytotoxic potential of superoxide and nitric oxide. *J. Biol. Chem.* 266, 4244–4250.

Remick, D. G., Kunkel, R. G., Larrick, J. W., and Kunkel, S. L. (1987). Acute *in vivo* effects of human recombinant tumor necrosis factor. *Lab. Invest.* 56, 583–590.

Repine, J. E., and Beehler, C. J. (1991). Neutrophils and adult respiratory distress syndrome: Two interlocking perspectives in 1991. *Am. Rev. Respir. Dis.* 144, 251–252.

Richard, C., Lemonnier, F., Conturier, M., Rion, B., and Auzepy, P. (1986). Vitamin E and selenium deficiency during acute respiratory distress syndrome. *Am. Rev. Resp. Dis.* 133, A203.

Richard, C., and Lemonnier, F. (1987). Lipoperoxidation and vitamin E consumption during adult respiratory distress syndrome. *Am. Rev. Respir. Dis.* 135, 425A.

Rinaldo, J. E., English, D., Levine, J., Stiller, R., and Henson, J. (1988). Increased intrapulmonary retention of radiolabeled neutrophils in early oxygen toxicity. *Am. Rev. Respir. Dis.* 137, 345–352.

Rossaint, R., Falke, K. J., Lopez, F., *et al.* (1993). Inhaled nitric oxide for the adult respiratory distress syndrome. *N. Engl. J. Med.* 328, 339–405.

Rossaint, R., Schmidt-Ruhnke, H., Steudel, W., *et al.* (1995). Retrospective analysis of survival in patients with severe acute respiratory distress syndrome treated with and without nitric oxide inhalation. *Crit. Care Med.* 23, A112.

Rothstein, J. L., and Schreiber, H. (1988). Synergy between tumor necrosis factor and bacterial products causes hemorrhagic necrosis and lethal shock in normal mice. *Proc. Natl. Acad. Sci. U.S.A.* **85**, 607–611.

Rubbo, H., Radi, R., Trujillo, M., Telleri, R., Kalyanaraman, B., Barnes, S., Kirk, M., and Freeman, B. A. (1994). Nitric oxide regulation of superoxide and peroxynitrite dependent lipid peroxidation: Formation of novel nitrogen containing oxidized lipid derivatives. *J. Biol. Chem.* **269**, 26066–26075.

Rubin, D. B., Wiener-Kronish, J. P., Murrary, J. F., *et al.* (1990). Elevated von Willebrand factor antigen is an early plasma predictor of acute lung injury in nonpulmonary sepsis syndrome. *J. Clin. Invest.* **86**, 474–480.

Ryan, S. F., Ghassibi, Y., and Liau, D. F. (1991). Effects of activated polymorphonuclear leukocytes upon pulmonary surfactant *in vitro. Am. J. Respir. Cell. Mol. Biol.* **4**, 33–41.

Ryan, U. S., and Vann, J. M. (1987). Endothelial cells: A source and target of oxidant damage. *In* "Oxygen Radicals in Biology and Medicine" (M. G., Simic, K. A. Taylor, and A. F. Ward, eds.), pp. 963–968. Plenum, New York.

Samuelsson, B., Dahlen, S. E., Lindgren, J. A., Rouzer, C. A., and Serhan, C. N. (1987). Leukotrienes and lipoxins: Structures biosynthesis, and biological effects. *Science* **237**, 1171–1176.

Sawyer, M. A. J., Mike, J. J., Chavin, K., and Marino, P. L. (1989). Antioxidant therapy and survival in ARDS. *Crit. Care Med.* **17**, S153.

Schade, U. F. (1990). Pentoxifylline increases survival in murine endotoxin shock and decreases formation of tumor necrosis factor. *Circ. Shock* **31**, 171–181.

Seear, M. D., Hannam, V. L., Kaapa, P., Rau, J. U., and O'Brodovich, H. M. (1990). Effect of pentoxifylline on hemodynamics, alveolar fluid reabsorption and pulmonary edema in a model of acute lung injury. *Am. Rev. Respir. Dis.* **142**, 1083–1087.

Seidenfeld, J. J., Pohl, D. F., Bell, R. C., *et al.* (1986). Incidence, site and outcome of infection in patients with adult respiratory distress syndrome. *Am. Rev. Respir. Dis.* **134**, 12–16.

Sibille, Y., and Reynolds, H. Y. (1990). State of the art: Macrophages and polymorphonuclear neutrophils in lung defense and injury. *Am. Rev. Respir. Dis.* **141**, 471–501.

Slutsky, A. S. (1988). Nonconventional modes of ventilation. *Am. Rev. Respir. Dis.* **138**, 175–183.

Smith, J. A. (1994). Neutrophils, host defense, and inflammation: A double-edged sword. *J. Leukocyte Biol.* **56**, 672–686.

Stephenson, A. H., Lonigro, A. J., Hyers, T. M., Webster, R. O., and Fowler, A. A. (1988). Increased concentration of leukotrienes in bronchoalveolar lavage fluid of patients with ARDS or at risk for ARDS. *Am. Rev. Respir. Dis.* **138**, 714–719.

Stevens, J. H., O'Hanley, P., Shapiro, J. M., *et al.* (1986). Effects of anti-C5a antibodies of the adult respiratory distress syndrome in septic primates. *J. Clin. Invest.* **77**, 1812–1816.

Stewart, T. E., Valenza, F., Ribeiro, S. P., *et al.* (1995). Increased nitric oxide in exhaled gas as an early marker of lung inflammation in a model of sepsis. *Am. J. Respir. Crit. Care Med.* **151**, 713–718.

Sullivan, G. W., Carper, H. T., Novick, W. J., and Mandell, G. L. (1988). Pentoxifylline inhibits the inflammatory effects of endotoxin and endotoxin-induced cytokines on neutrophil function. *In* "Pentoxifylline and Leukocyte Function" (G. L. Mandell and Novick, eds.), pp. 37–48. Haber and Flora, Weston, CT.

Suter, P. M., Domenighetti, G., Schaller, M. D., Laverriere, M. C., Ritz, R., and Perret, C. (1994). *N*-Acetylcysteine enhances recovery from acute lung injury in man: A randomized, double-blind, placebo-controlled clinical study. *Chest* **105**, 190–194.

Suter, P. M., Suter, S., Girardin, E., Roux-Lombard, P., Grau, G. E., and Dayer, J. M. (1992). High bronchoalveolar levels of tumor necrosis factor and its inhibitors, interleukin-1, interferon, and elastase, in patients with adult respiratory distress syndrome after trauma, shock, or sepsis. *Am. Rev. Respir. Dis.* **145**, 1016–1022.

Takeda, K. Y., and Shimada, M. (1984). Plasma lipid peroxides and α-tocopherol in critically ill patients. *Crit. Care Med.* **12**, 957–959.

Takeda, K. Y., Shimada, T., Okada, M., *et al.* (1986). Lipid peroxidation in experimental septic rats. *Crit. Care Med.* **14**, 719–723.

Thomsen, G. E., and Morris, A. H. (1995). Incidence of the adult respiratory distress syndrome in the state of Utah. *Am. J. Respir. Crit. Care Med.* **152**, 965–971.

Till, G. O., Hatherill, J. R., Tourtellotte, W. W., *et al.* (1985). Lipid peroxidation and acute lung injury following thermal trauma to skin: Evidence for role of hydroxyl radical. *Am. J. Pathol.* **119**, 376–383.

Till, G. O., Johnson, K. J., Kunkel, R., *et al.* (1982). Intravascular activation of complement and acute lung injury. *J. Clin. Invest.* **69**, 1126–1135.

Tracey, K. J., Fong, Y., Hesse, D. G., *et al.* (1987). Anti-cachectin/TNFα monoclonal antibodies prevent septic shock during lethal bacteremia. *Nature* **330**, 662–664.

Tsuno, K., Prato, P., and Kolobow, T. (1990). Acute lung injury from mechanical ventilation at moderately high airway pressures. *J. Appl. Physiol.* **69**, 956–961.

Turrens, J. F., Crapo, J. D., and Freeman, B. A. (1984). Protection against oxygen toxicity by intravenous injection of liposome-entrapped catalase and superoxide dismutase. *J. Clin. Invest.* **73**, 87–95.

Van der Vliet, A., Smith, D., O'Neill, C. A., Kaur, H., Darley-Usmar, V., Cross, C. E., and Halliwell, B. (1994). Interactions of peroxynitrite with human plasma and its constituents: Oxidative damage and antioxidant depletion. *Biochem. J.* **303**, 295–301.

Varani, J., Fligiel, S. E. G., Till, G. O., *et al.* (1985). Pulmonary endothelial cell killing by human neutrophils. *Lab. Invest,* **53**, 656–663.

Vedder, N. B., Winn, R. K., Chi, E. Y., *et al.* (1988). A monoclonal antibody to the adherence-promoting leukocyte glycoprotein, CD18, reduces organ injury and improves survival from hemorrhagic shock and resuscitation in rabbits. *J. Clin. Invest.* **81**, 939–944.

Vesconi, S., Rossi, G. P., Pesenti, A., *et al.* (1988). Pulmonary microthrombosis in severe adult respiratory distress syndrome. *Crit. Care Med.* **16**, 111–113.

Villar, J., and Slutsky, A. S. (1989). The incidence of ARDS. *Am. Rev. Respir. Dis.* **140**, 814–816.

Visner, G. A., Dougall, W. C., Wilson, J. M., Burr, I. A., and Nick, H. S. (1990). Regulation of manganese superoxide dismutase by lipopolysaccharide, interleukin-1, and tumor necrosis factor. *J. Biol. Chem.* **265**, 2856–2864.

Wagner, P. D., Mathieu-Costello, O., Bebout, D. E., *et al.* (1989). Protection against pulmonary oxygen toxicity by N-acetylcysteine. *Eur. Respir. J.* **2**, 116–126.

Wakabayashi, G., Gelfand, J. A., Burke, J. F., *et al.* (1991). A specific receptor antagonist for interleukin-1 prevents *Escherichia coli*-induced shock in rabbits. *FASEB J.* **5**, 338–343.

Wang, W., Suzuki, Y., Tanigaki, T., Rank, D. R., and Raffin, T. A. (1994). Effect of the NADPH oxidase inhibitor apocynin on septic lung injury in guinea pigs. *Am. J. Respir. Crit. Care Med.* **150**, 1449–1452.

Ward, P. A., Till, G. O., and Beauchamp, C. (1983). Evidence for role of hydroxyl radical in complement and neutrophil-dependent tissue injury. *J. Clin. Invest.* **72**, 789–801.

Ward, P. A., Till, G. O., Hatherill, J. R., *et al.* (1985). Systemic complement activation, lung injury, and products of lipid peroxidation. *J. Clin. Invest.* **76**, 517–527.

Warren, H. S., Danner, R. L., and Munford, R. S. (1992). Antiendotoxin monoclonal antibodies. *N. Engl. J. Med.* **326**, 1153–1156.

Weiland, J. E., Davis, W. B., Holter, J. F., *et al.* (1986). Lung neutrophils in the adult respiratory distress syndrome: Clinical and pathophysiological significance. *Am. Rev. Respir. Dis.* **133**, 218–225.

Weinberg, P. F., Matthay, M. A., Webster, R. O., *et al.* (1984). Biologically active products of complement and acute lung injury in patients with the sepsis syndrome. *Am. Rev. Respir. Dis.* **130**, 791–796.

Weiss, S. J. (1989). Tissue destruction by neutrophils. *N. Engl. J. Med.* **320**, 365–376.

White, C. W., Ghezzi, P., Dinarello, C. A., *et al.* (1987). Recombinant TNF/cachectin and IL-1 pretreatment decreases lung oxidized glutathione accumulation, lung injury, and mortality in rats exposed to hyperoxia. *J. Clin. Invest.* **79**, 1868–1873.

White, C. W., Ghezzi, P., McMahon, S., Dinarello, C. A., and Repine, J. E. (1989). Cytokines increase rat lung antioxidant enzymes during exposure to hyperoxia. *J. Appl. Physiol.* **66**, 1003–1007.

Wink, D. A., Cook, J. A., Pacelli, R., *et al.* (1995). Nitric oxide (NO) protects against cellular damage by reactive oxygen species. *Toxicol. Lett.* **82/83**, 221–226.

Wizemann, T. M., Gardner, C. R., Laskin, J. D., *et al.* (1994). Production of nitric oxide and peroxynitrite in the lung during acute endotoxemia. *J. Leukocyte Biol.* **56**, 759–768.

Wolf, H. R. D., and Seeger, H. W. (1982). Experimental and clinical results in shock lung treatment with vitamin E. *Ann. N.Y. Acad. Sci.* **393**, 392–409.

Yoshimoto, T., Kakanishi, K., Hirose, S., *et al.* (1992). High serum IL-6 level reflects susceptible status of the host to endotoxin and IL-1/tumor necrosis factor. *J. Immunol.* **148**, 3596–3603.

Yu, M., and Tomasa, G. (1993). A double-blind, prospective, randomized trial of ketoconazole, a thromboxane synthetase inhibitor, in the prophylaxis of the adult respiratory distress syndrome. *Crit. Care Med.* **21**, 1635–1642.

Ziegler, E. J., McCutchan, J. A., Fierer, J., *et al.* (1982). Treatment of gram-negative bacteremia and shock with human antiserum to a mutant *Escherichia coli*. *N. Engl. J. Med.* **307**, 1225–1230.

Ziegler, E. J., Teng, N. N. H., Douglas, H., *et al.* (1987). Treatment of Pseudomonas bacteria in neutropenic rabbits with human monoclonal IgM antibody against *E. coli* lipid A. *Clin. Res.* **35**, 619A.

Zurier, R. B., Weissmann, G., Hoffstein, S., *et al.* (1974). Mechanisms of lysosomal enzyme release from human leukocytes. *J. Clin. Invest.* **53**, 297–309.

Albert van der Vliet*
Jason P. Eiserich*
Gregory P. Marelich*,†
Barry Halliwell*,‡
Carroll Edward Cross*,†

*Division of Pulmonary and Critical Care Medicine and
†Adult Cystic Fibrosis Program
University of California Davis Medical Center
Sacramento, California 95817

‡Neurodegenerative Disease Research Centre
Pharmacology Group
University of London, Kings College
London SW3 6LX, United Kingdom

Oxidative Stress in Cystic Fibrosis: Does It Occur and Does It Matter?

I. Introduction

Cystic fibrosis (CF), one of the most common of the lethal autosomal recessive diseases, occurs in all ethnic groups and geographic locations. Its highest incidence is in individuals of Northern European heritage, where approximately 1 in every 2000 persons is afflicted (Boat *et al.*, 1989). CF is a generalized multiorgan system disease arising from a single biochemical abnormality in a protein encoded by a gene located on chromosome 7 (band q31–32) (Tsui and Buchwald, 1991). This protein (the cystic fibrosis transmembrane conductance regulator, CFTR) appears to couple ATP hydrolysis with transmembrane chloride (Cl^-) conductance across apical epithelial surfaces under the control of cAMP-dependent protein kinase A regulation (Gadsby and Nairn, 1994), but likely has other physiological functions that remain to be clarified, such as conductance of other anions [e.g., ATP (Casavola *et al.*, 1995; Schwiebert *et al.*, 1995)] and regulation of other distinct ion channels (Tsui, 1995).

Advances in Pharmacology, Volume 38
Copyright © 1997 by Academic Press, Inc. All rights of reproduction in any form reserved.

The defect in CF impairs the normal movement of water and electrolytes across various epithelial surfaces, most notably in the respiratory tract (RT) (Boucher, 1994a,b), but also in the pancreatohepatobililary system, the gastrointestinal tract, and the sweat excretion system. It remains to be clarified whether CFTR plays a role in the abnormalities in mucin (glycoconjugate) secretion (Mergey *et al.*, 1995) and in the hypersusceptibility to infections (Imundo *et al.*, 1995; Pier *et al.*, 1996) that is seen in CF patients.

RT abnormalities dominate the clinical manifestations of CF. There is decreased Cl^- secretion and an increased sodium absorption. There is inadequate hydration of the respiratory tract secretions (Boucher, 1994a,b; Widdicombe, 1994), making them more tenacious and difficult to clear. This predisposes the CF patient to impaired mucociliary clearance, chronic lung infection, and bronchiectasis (Tsui and Buchwald, 1991). The lung infection and the ensuing inflammatory-immune responses of the lung lead to the progressive lung destruction responsible for the pulmonary morbidity and mortality of CF (Marelich and Cross, 1996).

Recent scientific efforts in CF have included corrections of the basic defect in CFTR expression in human cell lines by CFTR gene transfer (Drumm *et al.*, 1990), the development of mouse models for CF (Dorin, 1995; Zeiher *et al.*, 1995) and gene transfer to transgenic mouse models (Drumm *et al.*, 1990), and *in vivo* gene transfer to the human nasal (Zabner *et al.*, 1993) and tracheobronchial (Crystal *et al.*, 1994) epithelia. However, there are still many obstacles to overcome before gene transfer therapy becomes a routine treatment in CF (Curiel *et al.*, 1996; Rosenfeld and Collins, 1996). Furthermore, although genetic treatments could be expected to restore RT water and electrolyte homeostasis in early CF and thus prevent the sequence of airway infection and injury, gene therapy is less likely to reverse lung injury and disease progression in patients who already have extensive lung damage.

The infectious inflammatory-immune processes in the lungs of CF patients cause severe oxidative stress (Salh *et al.*, 1989; Brown and Kelly, 1994a,b; Winklhofer-Roob, 1994; Portal *et al.*, 1995) and related effects on the antiprotease/protease balance (Bruce *et al.*, 1985; O'Connor *et al.*, 1993; Birrer *et al.*, 1994; Konstan *et al.*, 1994; Stone *et al.*, 1995). Thus, both antioxidant (Roum *et al.*, 1993) and antiprotease (McElvaney *et al.*, 1991; Birrer, 1995) approaches have been proposed for therapy in CF patients. This chapter discusses the evidence that oxidative stress occurs in CF, both systemically and in the lung. We attempt to assess whether oxidative stress contributes to the disease pathology and speculate on whether antioxidants might have useful therapeutic effects.

II. Role of Oxygen-Derived Species

The biomedical literature contains a multitude of claims that oxygen free radicals (such as $O_2\cdot^-$ and $\cdot OH$) and other oxygen-derived species (such as H_2O_2 and $HOCl/OCl^-$) are involved in a wide spectrum of human diseases (Halliwell and Gutteridge, 1989; Halliwell *et al.*, 1992). Oxidative damage has been implicated in many pathophysiological conditions ranging from cataracts to arthritis and from AIDS to age-related diseases such as stroke and cancer. This wide range of implicated involvement strongly suggests that oxidative stress is a common element in most, if not all, human diseases.

Figure 1 shows some of the reasons for this. Tissue injury increases free radical formation by such mechanisms as injury of mitochondria (so that they "leak" more electrons to O_2 to form $O_2 \cdot ^-$), release of "catalytic" iron ions and heme proteins, and increases in the activity of radical generating systems (e.g., releases of xanthine oxidase). This is accompanied by the activation of complex interacting components of inflammatory-immune processes. Phagocytes recruited and activated to the site of tissue injury represent an additional major source of radical and oxidative generating systems (e.g., activated O_2 species such as $O_2 \cdot ^-$ and H_2O_2 and myeloperoxidase). The problem is not so much the occurrence of oxidative stress, but whether it causes or contributes significantly to the tissue injury. In some diseases, oxidative stress may be a major mechanism of injury, amenable to therapy. In other diseases, it may make no significant contribution to disease pathology. Of course, exactly the same is true of other putative mediators of injury, such as prostaglandins, leukotrienes, nitric oxide ($\cdot$NO), and interleukins (Halliwell *et al.*, 1992).

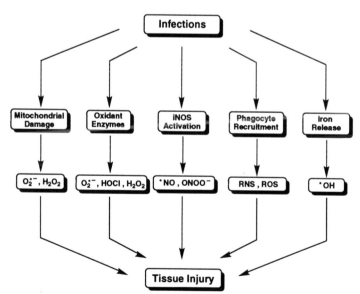

FIGURE I   Reactive oxygen species can be involved in the pathobiology of cell injury and tissue damage at various stages. Tissue injury (induced by infections in the context of CF respiratory tract disease) leads to the production of reactive oxygen and nitrogen species (ROS and RNS) by various mechanisms, which in turn result in further tissue injury.

TABLE I   Lung Conditions in which Oxidative Stress Could Be Involved

---

Cystic fibrosis
Acute respiratory distress syndrome
Oxygen toxicity
Bronchopulmonary dysplasia
Cigarette smoke effects
Lung cancer
Emphysema
Hypoxia/ischemia
Air pollutant exposure (O_3, NO_2, SO_2, N_2O_5, auto exhaust)
Mineral dust pneumoconiosis, including asbestos
Bleomycin/paraquat toxicity
Infant respiratory distress syndrome/bronchopulmonary dysplasia
Radiation toxicity
Reimplantation response of lung allografts

III. Oxidative Stress and Lung Disease

The lungs are particularly prone to oxidative insult in that the respiratory tract is exposed to the highest O_2 tension of any body tissue. It is also continuously exposed to oxidants in inhaled air and to blood-borne toxins. In addition, oxidant products are generated by activated phagocytic cells normally resident in the lung (mainly pulmonary alveolar macrophages) as well as those phagocytic cells recruited following lung injury (Cross *et al.*, 1994a), e.g., infiltrated neutrophils (PMNs) in cigarette smokers (Hunninghake and Crystal, 1983; Hogg, 1987). Table I shows some of the lung disorders in which oxidative stress is believed to play a role.

IV. Evidence for Oxidative Stress in Cystic Fibrosis

What evidence is available to suggest that oxidative stress occurs in CF? Table II presents a list of processes that could augment oxidative stress in

TABLE II   Potential Contributors to Systemic Oxidative Stress in Cystic Fibrosis

---

Increased metabolic rate (particularly in later stages of the disease)
Increased P450 oxidative reactions
Leukocytosis and phagocyte "priming" by such factors as cytokines
Chronic infection, inflammation, and increased immune responses
Increased amounts of cytokines TNFα and IL-1, which can induce oxidative stress
Fat malabsorption leading to malabsorption of fat-soluble vitamins (especially vitamin E)
Decreased absorption and perhaps increased turnover of carotenoids
Disordered iron metabolism, especially in the respiratory tract

---

CF patients (Salh *et al.*, 1989; Brown and Kelly, 1994a,b; Winklhofer-Roob, 1994; Portal *et al.*, 1995). The reported evidence for oxidative stress in CF is listed in Table III. It includes alleged increases in lipid peroxidation (Portal *et al.*, 1995) (see footnotes to Table III), increased susceptibility of plasma lipoproteins to peroxidation (Winklhofer-Roob *et al.*, 1995), and increased susceptibility of red blood cells to peroxide-induced hemolysis *in vitro* (James *et al.*, 1991). The latter two effects are partially reversible by dietary supplementation with vitamin E. Interestingly, red blood cell superoxide dismutase and catalase levels are reportedly increased in CF (Matkovics *et al.*, 1982), a finding that needs further documentation since almost any parameter measured in red blood cells is affected by mean red cell age.

Other forms of evidence for a role of oxidative stress in CF include elevated breath pentane levels (Kneepkens *et al.*, 1992; Bilton *et al.*, 1991), an arguable "marker" of increased lipid peroxidation (Kohlmuller and Kochen,

TABLE III   Reported Indices of Increased Oxidative Stress in CF

Elevated plasma TBARS levels (Portal *et al.*, 1995)[a]
Elevated levels of plasma organic hydroperoxides (Portal *et al.*, 1995)
Increased susceptibility of lipoproteins to peroxidation (Winklhofer-Roob *et al.*, 1995)
Increased susceptibility of RBC to peroxide-induced hemolysis (James *et al.*, 1991)
Increased plasma and tissue 9,11-dienoeic acid levels (Salh *et al.*, 1989)[b]
Elevated breath pentane levels (Kneepkens *et al.*, 1992; Bilton *et al.*, 1991)[c]
Increases in myeloperoxidase in RT sputum (Mohammed *et al.*, 1988; Niggemann *et al.*, 1995)
Increased PMN-derived long-lived oxidants in sputum (Witko-Sarsat *et al.*, 1995)
Increased "free" iron, catalytic for free radical reactions, in RT sputum (Britigan and Edeker, 1991; Haas *et al.*, 1991; Wolz *et al.*, 1994)
Evidences of protein oxidation in sputum (Birrer, 1995; Meyer *et al.*, 1995)
Decreased GSH levels in bronchoalveolar lavage (Roum *et al.*, 1993)
Increased amounts of prooxidant cytokines (e.g., TNF-α, IL-1) (Elborn *et al.*, 1993; Bonfield *et al.*, 1995)
Increased neutrophil and monocyte numbers and oxidase activities (Regelmann *et al.*, 1991; Danel *et al.*, 1996)
Decreases in plasma selenium (and plasma and RBC GP) levels (Castillo *et al.*, 1981; Portal *et al.*, 1993, 1995)
Decreases in plasma ascorbic acid and α-tocopherol levels in some patients (Castillo *et al.*, 1981; Winklhofer-Roob, 1994; Winklhofer-Roob *et al.*, 1995)
Decreased plasma β-carotene levels in many patients (Homnick *et al.*, 1993; Winklhofer-Roob, 1994; Portal *et al.*, 1995)[d]
Increased oxidative damage to DNA (Brown *et al.*, 1995)

---

[a] Often measured by nonspecific assays (Halliwell and Chirico, 1993; Gutteridge, 1995).
[b] Putative product of FR attack on lipids, but can often result from bacterial metabolism and/ or diet (Britton *et al.*, 1992).
[c] It has been suggested that most assays of pentane exhalation in human breath have been confounded by CO-elution of isoprene (Kohlmuller and Kochen, 1993).
[d] It must not be assumed that any beneficial effects of carotenoids in CF are necessarily related to antioxidant effects (Krinsky, 1993).

1993), and decreased GSH levels in bronchoalveolar lavage fluid (Roum *et al.*, 1993), but not in red blood cells (Mangione *et al.*, 1994). Observed increased levels of the enzyme myeloperoxidase (MPO) in RT secretions (Mohammed *et al.*, 1988; Niggemann *et al.*, 1995) are presumably secondary to the large numbers of neutrophils present in the RT of CF patients (Danel *et al.*, 1996). MPO utilizes H_2O_2 to oxidize Cl^- and Br^- to generate cytotoxic hypohalous acids that can inactivate α_1-antiprotease (O'Connor *et al.*, 1993; Birrer *et al.*, 1994). Hypochlorous acid ($HOCl/OCl^-$), a powerful oxidizing and chlorinating agent, depletes ascorbate and -SH-containing compounds (Hu *et al.*, 1992; Folkes *et al.*, 1995; Vissers and Winterbourn, 1995), presumably including those on cell surfaces, causing cellular damage. The increased activity of MPO in CF sputum also coincides with increased levels of chloramines [formed by reaction of $HOCl/OCl^-$ with amino compounds such as taurine (Witko-Sarsat *et al.*, 1995)].

Inactivation of α_1-antiprotease by MPO-generated hypohalous acids would be expected to lead to increased protease activity. Among its damaging effects, the augmented protease activity could degrade various iron-containing proteins present in lung secretions, including the endogenous transferrin and secreted lactoferrin (Britigan and Edeker, 1991; Wolz *et al.*, 1994).

Bacterial siderophores also contribute to the iron chelators present in the RT secretions of CF patients (Haas *et al.*, 1991). Most important are probably the siderophores of *Pseudomonas aeruginosa*, the most common of the CF pathogens in adults. These siderophores are part of their iron acquisition system and may also directly relate to their virulence (Litwin and Calderwood, 1993). At sites of microbial accumulation, both bacterial products and PMN proteases and oxidant species could be expected to break down both endogenous and exogenous iron chelators. This would lead to an increase in the levels of catalytically active iron in the RT, capable of reducing H_2O_2 to highly reactive ·OH (Britigan and Edeker, 1991; Britigan *et al.*, 1994).

In addition, CF patients have been shown to have increases in pro-inflammatory and pro-oxidant cytokines such as tumor necrosis factor (TNF)-α and interleukin (IL)-1 in RT secretions and, in smaller levels, in blood (Elborn *et al.*, 1993; Bonfield *et al.*, 1995). These cytokines could stimulate phagocyte ROS production (e.g., by "priming" mechanisms responsible for ROS generation) (Pfeffer *et al.*, 1993). A provocative finding is the report that monocyte-derived macrophages from CF patients appear to have an increased rate of TNF-α gene transcription (Pfeffer *et al.*, 1993). Many CF patients have also been shown to have decreases in selenium and in selenium-containing glutathione peroxidase activity in plasma and in red blood cells (Castillo *et al.*, 1981; Portal *et al.*, 1995; Winklhofer *et al.*, 1995). In select patients with CF, decreases in ascorbic acid and vitamin E have been reported (Castillo *et al.*, 1981; Winklhofer-Roob, 1994;

Winklhofer-Roob *et al.*, 1995). The decreases in α-tocopherol are of special interest because many investigators have shown that decreased levels of vitamin E potentiate lung injury in several models of oxidant-induced lung damage, including that caused by O_3 (Pryor, 1991).

Finally, markedly decreased amounts of β-carotene are present in almost all CF patients with active disease and who are not taking supplements (Winklhofer-Roob, 1994; Portal *et al.*, 1995). As shown by Homnick *et al.* (1993) (Table IV), this deficiency appears to extend to several different plasma carotenoids and appears more severe than the deficiencies in vitamins A and E.

V. The Special Case of β-Carotene Deficiency

The extremely low β-carotene levels in CF patients are particularly intriguing. β-Carotene can exert antioxidant effects in lipid systems *in vitro* under certain well-defined circumstances (Britton, 1995; Krinsky, 1993; Krinsky and Sies, 1995). The ability of β-carotene to do this *in vivo* is not certain (e.g., increased dietary β-carotene does not render plasma LDL less sensitive to oxidation) and, even when demonstrated, is considerably less than that seen for α-tocopherol (Tsuchihashi *et al.*, 1995). β-Carotene is best known as an efficient quencher of singlet O_2 (Britton, 1995). Because there is no known major source of singlet O_2 in CF patients, the overall role of β-carotene as an antioxidant in CF is somewhat uncertain.

One of the predominant biological functions of β-carotene in humans is as a precursor of retinol, vitamin A. There is little evidence that vitamin A acts as an antioxidant *in vivo*, but it is a key micronutrient in cell development, growth, and differentiation (Ross, 1993), as has been particularly well demonstrated in the case of RT cells (Zachman, 1995). Retinol is

TABLE IV Levels of Selected Lipophilic Antioxidants in CF Patients and Controls[a]

	CF patients (μM)	Controls (μM)
β-Carotene	0.04	0.20
Lutein	0.08	0.38
Lycopene	0.08	0.80
α-Carotene	0.01	0.03
Retinol (vitamin A)	0.97	1.42
α-Tocopherol (vitamin E)	15	18

Note. Values listed are means and are not corrected for LDL, cholesterol, or lipid levels.
[a] From Homnick *et al.* (1993).

also an essential micronutrient for normal immune function (van Poppel *et al.*, 1993).

The decrease in β-carotene levels seen in CF cannot be easily explained by decreased intake or by disordered fat absorption (Winklhofer-Roob, 1994), but may directly relate to the degree of disease severity, e.g., the more severe the disease, the lower the β-carotene level. It seems possible that this could relate to an increased turnover of carotenoids, possibly due to their reaction with O_2-derived species, analogous to the situation that has been proposed for cigarette smokers (van Antwerpen *et al.*, 1995). However, very few measurements of products of oxidized carotenoids have been reported and, to our knowledge, detailed studies of absorption, packaging, and distribution of β-carotene in CF patients have only just begun. To our knowledge, no studies have attempted to detail the effect of β-carotene supplementation on the clinical status of CF patients, and the levels of carotenoids in RT secretions and/or RT epithelial cells (RTECs) have not been reported. In this regard, a recent report that supplementation with carotenoids lowers malondialdehyde levels (a "marker" of lipid peroxidation) in the plasma of children with CF is provocative (Lepage *et al.*, 1996).

VI. Oxidative Stress at the Airway Surface

A major part of the pathophysiology of CF, including oxidative stress, takes place at the airway surface. First, there is chronic infection and inflammation, with the presence of large numbers of bacteria (especially *Pseudomonas* species) and 1000-fold increases in the numbers of neutrophils, intermixed with the respiratory tract lining fluids (RTLFs) (Boat *et al.*, 1989; Danel *et al.*, 1996). Many of the neutrophils are activated, releasing proteases and a range of oxygen-derived species, including $O_2\cdot^-$, H_2O_2, $HOCl/OCl^-$, and perhaps $\cdot NO$.

If it is true that a significant part of the RT damage in CF is mediated by oxidative stress and related phagocyte damage, then it can be suggested that augmentation of RT antioxidants could be used as a treatment (Crystal, 1991; Roum *et al.*, 1993). Let us first consider the normal antioxidant composition of the RT epithelial surface and how it could influence the RT pathobiology of CF.

The antioxidants in lower RTLFs (Slade *et al.*, 1993; Cross *et al.*, 1994b) are in general similar to those present in other extracellular fluids (Halliwell and Gutteridge, 1990; Dabbagh and Frei, 1995) except that mucus is present (Table V). However, there are several other significant differences. For example, when one compares the levels of distal RTLFs [also called epithelial lining fluids (ELFs)] to plasma, considerably more GSH is present (Table VI). This comparison is somewhat difficult because back calculations of levels measured in airway or bronchoalveolar lining fluids to those actually

TABLE V   Antioxidant Species in Respiratory Tract Lining Fluids

Low molecular mass antioxidants (e.g., uric acid, GSH, ascorbic acid)
Metal-binding proteins (e.g., lactoferrin, ceruloplasmin, transferrin)
Small quantities of antioxidant enzymes (e.g., catalase, glutathione peroxidase, superoxide
dismutase)
"Sacrificial" reactive proteins and unsaturated lipids
Mucus including various species of secreted glycoconjugates (especially mucin)

present in the RTLFs *in vivo* make several assumptions, the most important
of which include the uncertainties of the various methodologies used to
calculate the resident RTLF volume (Slade *et al.*, 1993; Cross *et al.*, 1994b).
However, it is clear that most antioxidants, including protein antioxidants
(such as protein-SH), are present in much smaller amounts in distal RTLFs
than in plasma. There are few if any accurate estimates of antioxidant
concentrations in upper RTLFs.

A major difference between plasma and upper RTLFs is the presence
of mucus, which may be an important antioxidant in both the respiratory
and the gastrointestinal tracts (Cross *et al.*, 1984; Grisham *et al.*, 1987).
Several mucus constituents (e.g., carbohydrates) are powerful scavengers of
oxygen-derived species, such as ·OH. Mucus contains an abundance of
protein thiols and disulfide bonds (Gum, 1996), both of which would be
excellent scavengers of ·OH and HOCl/OCl⁻ [e.g., both react fast with
thiols and disulfides (Aruoma *et al.*, 1989)]. Other interesting features of
RTLFs include the presence of uric acid, an important antioxidant that may
be cosecreted with mucus in the upper airways (Peden *et al.*, 1993). The
presence of GSH and ascorbic acid in lower RTLFs, at levels exceeding
those of plasma (at least for GSH), suggests the possibility of active RT
antioxidant secretory or regeneratory mechanisms by lower RTECs. Also
important is the presence of high concentrations of the iron-binding protein
lactoferrin in the upper RTLFs, which, like uric acid, is actively secreted by

TABLE VI   Antioxidant Species in Plasma and in Lower RTLFs

Antioxidant	Plasma (μM)	Lower RTLFs (μM)
Ascorbic acid	40	100[a]
Glutathione	1.5	100
Uric acid	300	90
α-Tocopherol	25	2.5
Albumin-SH	500	70

[a] Few data exist for ascorbic acid and it is not altogether clear that
lower RTLFs (ELF) ascorbate levels are significantly greater than those
of plasma in the human.

some populations of upper RTECs cells (Masson *et al.*, 1977). Iron species bound to lactoferrin are incapable of catalyzing free radical reactions (Aruoma and Halliwell, 1987), but may be a source of catalytic iron in CF because of oxidant/proteolytic degradation (see earlier discussion). Ceruloplasmin and transferrin are present in RTLFs, but at much lower concentrations than in plasma (Davis and Pacht, 1991).

VII. Potential Role of Nitric Oxide and Related Species in Cystic Fibrosis

·NO represents an important potential mediator in the pathobiology of CF. ·NO is produced constitutively in the lung, as evidenced by its detection in expired air of healthy animals and humans (Gustafsson *et al.*, 1991; Schedin *et al.*, 1995), and plays a critical role in many aspects of normal pulmonary function (Zapol *et al.*, 1994; Adnot *et al.*, 1995). Histochemical activity and immunoreactivity of both inducible and constitutive forms of nitric oxide synthetase (NOS) in cultured human and rat RTECs have been reported (Kobzik *et al.*, 1993; Asano *et al.*, 1994). Normal human airway epithelium continuously synthesizes significant quantities of ·NO by the inducible NOS isoform, and the concept that RTECs represent a key active participant in airway inflammatory-immune processes has been put forth (Guo *et al.*, 1995). Several studies showing significantly increased ·NO production by RTECs exposed to various cytokines further suggests the role of RTECs in pulmonary host defense and in pathways which may also be involved in free radical-mediated lung injury processes (Robbins *et al.*, 1994; Freeman *et al.*, 1995, Gutierrez *et al.*, 1995; Warner *et al.*, 1995). Nasal passage production of ·NO is seven-fold greater than lower RT ·NO production during tital breathing (Kimberly *et al.*, 1996).

There is a growing body of both experimental and clinical data that demonstrates that many forms of acute and/or chronic inflammation are associated with enhanced RT production of ·NO. This ·NO, in turn, may contribute to the cell and tissue injury seen in inflammatory-immune processes (Cattell and Jansen, 1995). The contributing role of ·NO in inflammatory diseases of the lung is supported by the fact that tissue levels of NOS activity are elevated in a variety of inflammatory lung diseases (Belvisi *et al.*, 1995). In fact, increased levels of ·NO in the expired breath have been reported in patients suffering from asthma (Kharitonov *et al.*, 1994), bronchiectasis (Kharitonov *et al.*, 1995a), sepsis (Stewart *et al.*, 1995), and in normal subjects with upper RT infections (Kharitonov *et al.*, 1995b). Elevated levels of ·NO (as measured by NO_2^-/NO_3^- production) have been reported in RT secretions of patients with CF (Francoeur and Denis, 1995), implicating ·NO as a potentially important mediator and/or modulator of tissue injury in this disease.

Recent evidence that ·NO plays a role in airway glycoconjugate secretion (Nagaki *et al.*, 1995) and in PMN recruitment (Gaboury *et al.*, 1993; Kurose *et al.*, 1995) may be relevant in CF, a disease characterized by excessive mucin secretion and PMN accumulations (Boat *et al.*, 1989). Although ·NO has many helpful physiological properties (Zapol *et al.*, 1994; Adnot *et al.*, 1995) and may be protective toward RTLF lipids and RTEC membranes by acting as an inhibitor of radical chain propagation reactions (Rubbo *et al.*, 1995), excesses are generally thought to be harmful (Darley-Usmar *et al.*, 1995). Significantly, one model of immune complex-mediated lung injury was ameliorated by inhibitors of ·NO formation (Mulligan *et al.*, 1991). This is important because immune complex depositions are present in the airways of CF patients (Boat *et al.*, 1989).

Although direct reactions of ·NO with most biological molecules are slow, its reaction with other free radicals is very rapid. In fact, ·NO reacts with $O_2\cdot^-$ at near diffusion-limited rates ($k = 6.7 \times 10^9\ M^{-1}sec^{-1}$; Huie and Padmaja, 1993) to produce the cytotoxic species peroxynitrite ($ONOO^-$), a species likely to be formed during inflammatory processes in the lung. $ONOO^-$ is capable of nitrating phenolic compounds, including the amino acid tyrosine (Beckman *et al.*, 1992; Ischiropoulos *et al.*, 1992a; van der Vliet *et al.*, 1995). $ONOO^-$ is also capable of oxidizing lipids (Radi *et al.*, 1991a), sulfhydryl groups (Radi *et al.*, 1991b), sulfides such as methionine (Pryor *et al.*, 1994), and antioxidants such as ascorbate (Bartlett *et al.*, 1995) and urate (van der Vliet *et al.*, 1994). Because reactive nitrogen species are capable of efficiently depleting sulfhydryls, these mechanisms may contribute to the low levels of glutathione observed in many inflammatory diseases of the lung (Pacht *et al.*, 1991; White *et al.*, 1994), including CF (Roum *et al.*, 1993). Although the potential biological importance of tyrosine nitration is relatively unknown, it has been shown that $ONOO^-$ is capable of nitrating surfactant proteins *in vitro*, leading to decreased surfactant function (Haddad *et al.*, 1993, 1994b). This type of mechanism could be of significance in CF.

The nitration reactions of $ONOO^-$ have received increasing attention over the past several years. The importance of such reactions has been underscored by the detection of 3-nitrotyrosine in a variety of degenerative diseases, including atherosclerosis (Beckman *et al.*, 1994), rheumatoid arthritis (Kaur and Halliwell, 1994), and amyotrophic lateral sclerosis (Abe *et al.*, 1995). More relevant to CF, 3-nitrotyrosine has been detected in lung tissues of patients with acute lung injury (Haddad *et al.*, 1994a; Kooy *et al.*, 1995) and in animal models following acute endotoxemia (Wizemann *et al.*, 1994) and rat lung ischemia (Ischiropoulos and Al-Mehdi, 1995). Based on tyrosine nitration assays and the formation of $ONOO^-$-specific luminescence, activated macrophages (Ischiropoulos *et al.*, 1992b), dependent (Carreras *et al.*, 1994; Salman-Tabcheh *et al.*, 1995), and endothelial cells (Kooy and Royall, 1994) have been proposed to form significant quantities of $ONOO^-$ *in vitro*, and it is thought that these cells are responsible for 3-nitrotyrosine formation *in vivo* via the production of $ONOO^-$. How-

502 Albert van der Vliet et al.

ever, there is still some question as to the amounts of ·NO that can be produced by activated macrophages and extravasated PMNs in humans (Miles et al., 1995).

In addition to tyrosine nitration, the massive recruitment and activation of neutrophils in the CF lung would lead to the production of HOCl/OCl⁻. Figure 2 illustrates the potential phagocyte and RTEC-mediated mechanisms by which protein and peptide tyrosine residues may be modified in the RT. Recently, dityrosine has been detected in elevated amounts in bronchoalveolar lavage fluids and in plasma of patients with CF, compared to normal subjects (Meyer et al., 1995). Initial studies from our laboratory have indicated that 3-nitrotyrosine can also be detected in expectorated sputum from adult CF patients. In fact, activated neutrophils can readily chlorinate tyrosine residues (Domigan et al., 1995; Kettle, 1996). In addition, dityrosine can be formed by reaction with myeloperoxidase/H₂O₂ (Heinecke et al., 1993; Marquez and Dunford, 1995). All three of these modified tyrosine products could be used as "markers" or "dosimeters" for the production of reactive nitrogen and reactive O₂ species in CF, and could potentially be useful in determining the effectiveness of antioxidants and other therapeutic strategies in ameliorating the oxidative component of tissue injury in CF.

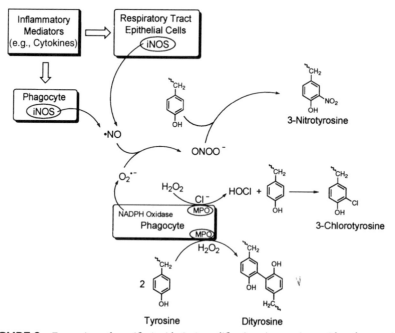

FIGURE 2 Formation of specific (oxidative) modifications in tyrosine residues by reactive oxygen and nitrogen species in the respiratory tract. It is arguable as to whether or not human phagocytes produce significant quantities of reactive nitrogen species (Miles et al., 1995).

VIII. Other Considerations ─────────────────────────────

A major consideration is that although RTLFs appear to contain higher concentrations of GSH than plasma, RTLF volumes are relatively small and antioxidant turnovers are unknown (Slade *et al.*, 1993; Cross *et al.*, 1994b). These facts make it difficult to judge their total capacity to protect RTECs against oxidative stress.

Second, the degree to which cell adhesion mechanisms may bring PMNs (and bacteria) in direct apposition to RTECs must be taken into account. These mechanisms would minimize the effect of RTLF antioxidants and raise in importance the contributing role of lipophilic membrane antioxidants, such as vitamin E.

Third, inflammation and its mediators activate RT neurohumoral mechanisms that result in the entry of increasing amounts of plasma constituents, including plasma antioxidant proteins, into RTLFs (Persson *et al.*, 1991). This congests airways and impedes gas exchange, but does considerably augment the antioxidants present within the airway lumen.

Fourth, and especially in the presence of airway RT inflammation, intracellular antioxidants released from shed RTECs or from inflammatory cells may well augment (to an unknown degree) the measured RTLF antioxidant capabilities. This may be particularly true for the antioxidant enzymes superoxide dismutase, catalase, and glutathione peroxidase.

IX. Potential for Antioxidant Therapy ─────────────────

Although antioxidant administration to CF patients will clearly not cure the disease, it may have benefits. Should CF patients be prescribed massive pharmacological doses of antioxidants? Should aerosolized antioxidants, including genes encoding for antioxidant defenses, be administered in conjunction with bronchodilators, aerosolized DNAase, and anti-inflammatory agents? Given the current limited efficiency of pharmacologic CFTR channel "optimizing" agents such as aerosolized uridine triphosphate and CFTR gene (Zabner *et al.*, 1993; Crystal *et al.*, 1994; Curiel *et al.*, 1996; Rosenfeld and Collins, 1996), antioxidants may have a favorable cost-benefit ratio.

It is probably prudent to suggest that vitamins E and C and carotenoids be administered orally in amounts that normalize or nearly normalize plasma levels in CF patients. The evidence may be particularly strong for vitamin E as experimental deficiencies of this antioxidant seem to predispose the lung to oxidant injury (Pryor, 1991). However, based on the paucity of scientific information available, we are not convinced that massive doses of such compounds should be administered, as effects are still unknown and, at least for vitamin A and possibly β-carotene, could be potentially harmful (Rowe, 1996). It is unknown whether levels of these antioxidants are increased substantially in RTLFs or RTECs following oral supplementation, and what effects they may have, if any, on pulmonary function.

For example, if antioxidants are to be aerosolized directly into the respiratory tract, how would the *Pseudomonas* be affected? Mucoid *Pseudomonas* phenotypes already secrete one antioxidant, their slimy alginate, which presumably protects them against neutrophil-derived antioxidants (Simpson *et al.*, 1993). Would aerosolized antioxidants protect the bacteria even more? Administration of augmented amounts of iron chelator substances presents a similar enigma. Would the augmented iron chelators, engineered so as to be resistant to proteolytic and oxidant destruction, be inhibitory toward bacterial growth or merely stimulate them to augment their own bacterial siderophore production and virulence (Litwin and Calderwood, 1993)? Because oxidants play an important role in providing host defenses against microorganisms, it is not intuitive that by augmenting RTLF antioxidants we would strengthen RT antimicrobial defense capabilities.

X. Dilemma of Administering Aerosolized Thiols or Ascorbic Acid

If iron does represent an important potential pro-oxidant constituent of CF RTLFs (see earlier discussion), thiols and ascorbic acid could potentiate its toxicity by keeping it in the Fe^{2+} state, thus facilitating pro-oxidant reactions, such as the Fenton reaction (Huston *et al.*, 1992; Korge and Campbell, 1993; Sparrow and Olszewski, 1993). Thiols produce a second potentially important risk, for in the presence of certain free radicals, thiyl radicals form (Munday, 1989); thiyl radicals and products of their reaction with O_2 may sometimes represent a more noxious species than those quenched by the thiol in the primary reaction (Munday, 1989). Finally, as disulfide bonds often play an important role in the conformational integrity, stability, and functional capabilities of receptor and plasma membrane proteins, it is possible that reductive stress may play a role as important as oxidative stress at the boundary between RTLFs and RTECs. Thus, some degree of caution should be exercised before recommending that large amounts of these reducing antioxidants be administered to patients with CF, particularly so if via the inhaled route and in the presence of airway sepsis.

Mention should be made of the fact that certain antibiotics administered to CF patients, including aminoglycosides (which are frequently administered to CF patients), are free radical scavengers (Cantin and Woods, 1993). It is not certain if these could play a role, even when administered by aerosol, in causing significant augmentations of RTLF antioxidants.

Increasing evidence shows that vitamin E and β-carotene may provide some protection against cancer and cardiovascular disease, although there is arguable evidence presented regarding β-carotene supplementation (Rowe,

1996). There is no reason, a priori, to suspect that they cannot be helpful to patients with CF. In fact, the nutritional guidelines available for caregivers of patients with CF are beginning to recognize the possible contributions that antioxidants might make to patient management (Ramsey *et al.*, 1992).

XI. Conclusion

This chapter focused on the possible roles that oxidative stress may contribute to the pathophysiology of CF. These oxidative processes, in the main, are related to the intense activation of RT inflammatory-immune processes that accompany the disease and are coupled with deficiencies in the lipophilic antioxidant micronutrients in CF. This chapter has focused on some of the emerging ideas of how the balance between reactive oxidative species and reactive nitrogen species may interact with biomolecular constituents at RT surfaces. An important area not addressed in any detail is how these events occurring at RT surfaces might effect the molecular events which control RTEC signaling and gene expression. Further understandings of these processes can be expected to play an important role in innovative future treatments of the devastating effects of CF on the RT of patients with this disorder.

Acknowledgments

This work was supported in part by grants from the Cystic Fibrosis Foundation and the Ministry of Agriculture, Fisheries, and Food (U.K.).

References

Abe, K., Pan, L.-H., Watanabe, M., Kato, T., and Itoyama, Y. (1995). Induction of nitrotyrosine-like immunoreactivity in the lower motor neuron of amyotrophic lateral sclerosis. *Neurosci. Lett.* **199**, 152–154.

Adnot, S., Raffestin, B., and Eddahibi, S. (1995). NO in the lung. *Resp. Physiol.* **101**, 109–120.

Aruoma, O. I., and Halliwell, B. (1987). Superoxide-dependent and ascorbate-dependent formation of hydroxyl radicals from hydrogen peroxide in the presence of iron: Are lactoferrin and transferrin promoters of hydroxyl-radical generation? *Biochem. J.* **241**, 273–278.

Aruoma, O. I., Halliwell, B., Hoey, B. M., and Butler, J. (1989). The antioxidant action of N-acetylcysteine: Its reaction with hydrogen peroxide, hydroxyl radical, superoxide and hypochlorous acid. *Free Rad. Biol. Med.* **6**, 593–597.

Asano, K., Chee, C. B. E., Gaston, B., Lilly, C. M., Gerard, C., Drazen, J. M., and Stamler, J. S. (1994). Constitutive and inducible nitric oxide synthase gene expression, regulation, and activity in human lung epithelial cells. *Proc. Natl. Acad. Sci. U.S.A.* **91**, 10089–10093.

Bartlett, D., Church, D. F., Bounds, P. L., and Koppenol, W. H. (1995). The kinetics of the oxidation of L-ascorbic acid by peroxynitrite. *Free Rad. Biol. Med.* **18**, 85–92.

Beckman, J. S., Ischiropoulos, H., Zhu, L., van der Woerd, M., Smith, C., Chen, J., Harrison, J., Martin, J. C., and Tsai, M. (1992). Kinetics of superoxide dismutase- and iron-catalyzed nitration of phenolics by peroxynitrite. *Arch. Biochem. Biophys.* **298**, 438–445.

Beckman, J. S., Ye, Y. Z., Anderson, P. G., Chen, J., Accavitti, M. A., Tarpey, M. M., and White C. R. (1994). Extensive nitration of protein tyrosines in human atherosclerosis detected by immunohistochemistry. *Biochem. Hoppe-Seyler* **375**, 81–88

Belvisi, M., Barnes, P. J., Larkin, S., Yacoub, M., Tadjkarimi, S., Williams, T. J., and Mitchell, J. A. (1995). Nitric oxide synthase activity is elevated in inflammatory lung disease in humans. *Eur. J. Pharmacol.* **283**, 255–258.

Bilton, D., Maddison, J., Webb, A. K., Seabra, L., Jones, M., and Braganza, J. M. (1991). Cystic fibrosis, breath pentane, and lipid peroxidation. *Lancet* **338**, 1420.

Bitter, P. (1995). Proteases and antiproteases in cystic fibrosis: Pathogenetic considerations and therapeutic strategies. *Resp.* **62**(Suppl. 1), 25–28.

Birrer, P., McElvaney, N. G., Rudeberg, A., Wirz Sommer, C., Liechti-Gallati, S., Kraemer, R., Hubbard, R., and Crystal, R. G. (1994). Protease-antiprotease imbalance in the lungs of children with cystic fibrosis. *Am. J. Respir. Crit. Care Med.* **150**, 207–213.

Boat, T. F., Welsh, M. J., and Beaudet, A. L. (1989). Cystic fibrosis. *In* "The Metabolic Basis of Inherited Diseases" (C. R. Scriver *et al.*, eds.), 6th Ed. pp. 2649–2680. McGraw-Hill, New York.

Bonfield, T. L., Panuska, J. R., Konstan, M. W., Hilliard, K. A., Hilliard, J. B., Ghnaim, H., and Berger, M. (1995). Inflammatory cytokines in cystic fibrosis lungs. *Am. J. Respir Crit. Care Med.* **152**, 2111–2118.

Boucher, R. C. (1994a). Human airway ion transport. Part one. *Am. J. Respir. Crit. Care Med.* **150**, 271–281.

Boucher, R. C. (1994b). Human airway ion transport. Part two. *Am. J. Respir. Crit. Care Med.* **150**, 581–593.

Britigan, B. E., and Edeker, B. L. (1991). Pseudomonas and neutrophil products modify transferrin and lactoferrin to create conditions that favor hydroxyl radical formation. *J. Clin. Invest.* **88**, 1092–1102.

Britigan, B. E., Rasmussen, G. T., and Cox, C. D. (1994). Pseudomonas siderophore pyochelin enhances neutrophil-mediated endothelial cell injury. *Am. J. Physiol.* **266**, L192–L198.

Britton, G. (1995). Structure and properties of carotenoids in relation to function. *FASEB J.* **9**, 1551–1558.

Britton, M., Fong, C., Wickens, D., and Yudkin, J. (1992). Diet as a source of phospholipid esterified 9,11-octadecadienoic acid in humans. *Clin. Sci.* **83**, 97–101.

Brown, R. K., and Kelly, F. J. (1994a). Evidence for increased oxidative damage in patients with cystic fibrosis. *Pediatr. Res.* **36**, 487–493.

Brown, R. K., and Kelly, F. J (1994b). Role of free radicals in the pathogenesis of cystic fibrosis. *Thorax* **49**, 738–742.

Brown, R. K., McBurney, A., Lunec, J., and Kelly, F. J. (1995). Oxidative damage to DNA in patients with cystic fibrosis. *Free Rad. Bio. Med.* **18**, 801–806.

Bruce, M. C., Poncz, L., Klinger, J. D., Stern, R. C., Tomashefski, J. F., Jr., and Dearborn D. G. (1985). Biochemical and pathologic evidence for proteolytic destruction of lung connective tissue in cystic fibrosis. *Am. Rev. Respir. Dis.* **132**, 529–535.

Cantin, A., and Woods, D. E. (1993). Protection by antibiotics against myeloperoxidase-dependent cytotoxicity to lung epithelial cells in vitro. *J. Clin. Invest.* **91**, 38–45.

Carreras, M. C., Pargament, G. A., Catz, S. D., Poderoso, J. J., and Boveris, A. (1994). Kinetics of nitric oxide and hydrogen peroxide production and formation of peroxynitrite during the respiratory burst of human neutrophils. *FEBS Lett.* **341**, 65–68.

Casavola, V., Turner, R. J., Guay-Broder, C., Jacobson, K. A., Eidelman, O., and Pollard, H. B. (1995). CPX, a selective A_1-adenosine-receptor antagonist, regulates intracellular pH in cystic fibrosis cells. *Am. J. Physiol.* **38**, C226–C233.

Castillo, R., Landon, C., Eckhardt, K., Morris, V., Levander, O., and Lewiston, N. (1981). Selenium and vitamin E status in cystic fibrosis. *J. Pediatr.* **99**, 583–585

Cattell, V., and Jansen, A. (1995). Inducible nitric oxide synthase in inflammation. *Histochem. J.* **27**, 777–784.

Cross, C. E., Halliwell, B., and Allen, A. (1984). Antioxidant protection: A function of tracheobronchial and gastrointestinal mucus. *Lancet* **1**, 1328–1330.

Cross, C. E., van der Vliet, A., O'Neill, C. A., and Eiserich, J. P. (1994a). Free radicals and antioxidants: Reactive oxygen species and the lung. *Lancet* **344**, 930–933.

Cross, C. E., van der Vliet, A., O'Neill, C. A., Louie, S., and Halliwell, B. (1994b). Oxidants, antioxidants, and respiratory tract lining fluids. *Environ. Health Perspect.* **102** (Suppl. 10), 185–192.

Crystal, R. G. (1991). Oxidants and respiratory tract epithelial injury: Pathogenesis and strategies for therapeutic intervention. *Am. J. Med.* **91** (Suppl. 3C), 39S–44S.

Crystal, R. G., McElvaney, N. G., Rosenfeld, M. A., Chu, C.-S, Mastrangeli, A., Hay, J. G., Brody, S. L., Jaffe, H. A., Eissa, N. T., and Danel, C. (1994). Administration of an adenovirus containing the human CFTR CDNA to the respiratory tract of individuals with cystic fibrosis. *Nature Genet* **8**, 42–51.

Curiel, D. T., Pilewski, J. M., and Albelda, S. M. (1996). Minireview: Gene approaches for inherited and acquired lung diseases. *Am. J. Respir. Cell Mol. Biol.* **14**, 1–8.

Dabbagh, A. J., and Frei, B. (1995). Human suction blister interstitial fluid prevents metal ion-dependent oxidation of low density lipoprotein by macrophages and in cell-free systems. *J. Clin. Invest.* **96**, 1958–1966.

Danel, C., Erzurum, S. C., McElvaney, N. G., and Crystal, R. G. (1996). Quantitative assessment of the epithelial and inflammatory cell populations in large airways of normals and individuals with cystic fibrosis. *Am. J. Respir. Crit. Care Med.* **153**, 362–368.

Darley-Usmar, V., Wiseman, H., and Halliwell, B. (1995). Nitric oxide and oxygen radicals: A question of balance. *FEBS Lett.* **369**, 131–135.

Davis, W. B., and Pacht, E. R. (1991). Extracellular antioxidant defenses. *In* "The Lungs: Scientific Foundations" (R. G. Crystal, J. B. West, P. J. Barnes, N. S. Cherrniack, and E. R. Weibel, eds.) pp: 1821–1826. Raven Press, New York.

Domigan, N. M., Charlton, T. S., Duncan, M. W., Winterbourne, C. C., and Kettle, A. J. (1995). Chlorination of tyrosyl residues in peptides by myeloperoxidase and human neutrophils. *J. Biol. Chem.* **270**, 16542–16548.

Dorin, J. R. (1995). Development of mouse models for cystic fibrosis. *J. Inherit. Metab. Dis.* **18**, 495–500.

Drumm, M. L., Pope, H. A., Cliff, W. H., Rommens, J. M., Marvin, S. A., Tsui, L. C., Collins, F. S., Frizzell, R. A., and Wilson, J. M. (1990). Correction of the cystic fibrosis defect *in vitro* by retrovirus-mediated gene transfer. *Cell* **62**, 1227–1233.

Elborn J. S., Cordon, S. M., Western, P. J., MacDonald, I. A., and Shale, D. J. (1993). Tumor necrosis factor-α, resting energy expenditure and cachexia in cystic fibrosis. *Clin. Sci.* **85**, 563–568.

Folkes, L. K., Candeias, L. P., and Wardman, P. (1995). Kinetics and mechanisms of hypochlorous acid reactions. *Arch. Biochem. Biophys.* **323**, 120–126.

Francoeur, C., and Denis, M. (1995). Nitric oxide and interleukin-8 as inflammatory components of cystic fibrosis. *Inflammation* **195**, 587–598.

Freeman, B. A., Gutierrez, H., and Rubbo, H. (1995). Nitric oxide: A central regulatory species in pulmonary oxidant reactions. *Am. J. Physiol.* **268**, L697–L698.

Gaboury, J., Woodman, R. C., Granger, D. N., Reinhardt, P., and Kubes, P. (1993). Nitric oxide prevents leukocyte adherence: Role of superoxide. *Am. J. Physiol.* **265**, H862–H867.

Gadsby, D. C., and Nairn, A. C. (1994). Regulation of CFTR channel gating. *Trends Biochem. Sci.* **19**, 513–518.

Grisham, M. B., Von Ritter, C., Smith, B. F., LaMont, J. T., and Granger, D. N. (1987). Interaction between oxygen radicals and gastric mucin. *Am. J. Physiol.* **253**, G93–G96.

Gum, J. R. (1996). Human mucin glycoproteins: Varied structures predict diverse properties and specific functions. *Biochem. Soc. Trans.* **23**, 795–799.

Guo, F. H., De Raeve, H. R., Rice, T. W., Stuehr D. J., Thunnissen, F. B. J. M., and Erzurum, S. C. (1995). Continuous nitric oxide synthesis by inducible nitric oxide synthase in normal human airway epithelium in vivo. *Proc. Natl. Acad. Sci. U.S.A.* **99**, 7809–7813.

Gustafsson, L. E., Leone, A. M., Persson, M. G., Wiklund, N. P., and Moncada, S. (1991). Endogenous nitric oxide is present in exhaled air of rabbits, guinea pigs and humans. *Biochem. Biophys. Res. Commun.* **181**, 852–857.

Gutierrez, H. H., Pitt, B. R., Schwarz, M., Watkins, S. C., Lowenstein, C., Caniggia, I., Chumley, P., and Freeman, B. A. (1995). Pulmonary alveolar epithelial inducible NO synthase gene expression: Regulation by inflammatory mediators. *Am. J. Physiol.* **268**, L501–L508.

Gutteridge, J. M. C. (1995). Lipid peroxidation and antioxidants as biomarkers of tissue damage. *Clin Chem* **41**, 1819–1828.

Haas, B., Kraut, J., Marks, J., Zanker, S. C., and Castignetti, D. (1991). Siderophore presence in sputa of cystic fibrosis patients. *Infect. Immun.* **59**, 3997–4000.

Haddad, I. Y., Crow, J. P., Hu, P., Ye, Y., Beckman, J., and Matalon, S. (1994a). Concurrent generation of nitric oxide and superoxide damages surfactant protein A. *Am. J. Physiol.* **267**, L242–L249.

Haddad, I. Y., Ischiropoulos, J., Holm, B. A., Beckman, J. S., Baker, J. R., and Matalon, S. (1993). Mechanisms of peroxynitrite-induced injury to pulmonary surfactants. *Am. J. Physiol.* **265**, L555–L564.

Haddad, I. Y., Pataki, G., Hu, P., Galliani, C., Beckman, J. S., and Matalon, S. (1994b). Quantitation of nitrotyrosine levels in lung sections of patients and animals with acute lung injury. *J. Clin. Invest.* **94**, 2407–2413.

Halliwell, B., and Chirico, S. (1993). Lipid peroxidation: Its mechanism, measurement and significance. *Am. J. Clin. Nutr.* **57**(Suppl.), 715S–725S.

Halliwell, B., Cross, C. E., and Gutteridge, J. M. C. (1992). Free radicals, antioxidants, and human disease: Where are we now? *J. Lab. Clin. Med.* **119**, 598–620.

Halliwell, B., and Gutteridge, J. M. C. (1989). "Free Radicals in Biology and Medicine," 2nd Ed. Clarendon Press, Oxford.

Halliwell, B., and Gutteridge, J. M. C. (1990). The antioxidants of human extracellular fluids. *Arch. Biochem. Biophys.* **280**, 1–8

Heinecke, J. W., Li, W., Francis, G. A., and Goldstein, J. A. (1993). Tyrosyl radical generated by myeloperoxidase catalyzes the oxidative cross-linking of proteins. *J. Clin. Invest.* **91**, 2866–2872.

Hogg, J. C. (1987). Neutrophil kinetics in lung injury. *Physiol. Rev.* **67**, 1249–1295.

Homnick, D. N., Cox, J. H., DeLoof, M. J. H., and Ringer, T. V. (1993). Carotenoid levels in normal children and in children with cystic fibrosis. *J. Pediatr.* **122**, 703–707.

Hu, M. L., Louie, S., Cross, C. E., Motchnik, P., and Halliwell, B. (1992). Antioxidant protection against hypochlorous acid in human plasma. *J. Lab. Clin. Med.* **121**, 257–262.

Huie, R. E., and Padmaja, S. (1993). The reaction of NO with superoxide. *Free Rad. Res. Commun.* **18**, 1995–199.

Hunninghake, G. W., and Crystal, R. G. (1983). Cigarette smoking and lung destruction: Accumulation of neutrophils in the lungs of cigarette smokers. *Am. Rev. Respir. Dis.* **128**, 833–838.

Huston, P., Espenson, J. H., and Bakac, A. (1992). Reactions of thiyl radicals with transition-metal complexes. *J. Am. Chem. Soc.* **114**, 9510–9516.

Imundo, L., Barasch, J., Prince, A., and Al-Awqati, Q. (1995). Cystic fibrosis epithelial cells have a receptor for pathogenic bacteria on their apical surface. *Proc. Natl. Acad. Sci. U.S.A.* **92**, 3019–3023.

Ischiropoulos, H., and Al-Mehdi, A. B. (1995). Reactive species in ischemic rat lung injury: Contribution of peroxynitrite. *Am. J. Physiol.* **269**, L158–L164.

Ischiropoulos, H., Zhu, L., and Beckman, J. S. (1992a). Peroxynitrite formation from macrophage-derived nitric oxide. *Arch. Biochem. Biophys.* **298**, 446,451.

Ischiropoulos, H., Zhu, L., Chen, J., Tsai, M., Martin, J. C., Smith, C. D., and Beckman, J. S. (1992b). Peroxynitrite-mediated tyrosine nitration catalyzed by superoxide dismutase. *Arch. Biochem. Biophys.* **298**, 431–437.

James, D. R., Alfaham, M., and Goodchild, M. C. (1991). Increased susceptibility to peroxide-induced haemolysis with normal vitamin E concentrations in cystic fibrosis. *Clin. Chim. Acta* **204**, 279–290.

Kaur, H., and Halliwell, B. (1994). Evidence for nitric oxide-mediative damage in chronic inflammation: Nitrotyrosine in serum and synovial fluid from rheumatoid patients. *FEBS Lett.* **350**, 9–12.

Kettle, A. J. (1996). Neutrophils convert tyrosyl residues in albumin to chlorotyrosine. *FEBS Lett.* **379**, 103–106.

Kharitonov, S. A., Wells, A. U., O'Connor, B. J., Cole, P. J., Hansell, D. M., Logan-Sinclair, R. B., and Barnes, P. J. (1995a). Elevated levels of exhaled nitric oxide in bronchiectasis. *Am. J. Respir. Crit. Care Med.* **151**, 1889–1893.

Kharitonov, S. A., Yates, D., and Barnes, P. J. (1995b). Increased nitric oxide in exhaled air of normal human subjects with upper respiratory tract infections. *Eur. Respir. J.* **8**, 295–297.

Kharitonov, S. A., Yates, D., Robbins, R. A., Logan-Sinclair, R., Shinebourne, E. A., and Barnes, P. J. (1994). Increased nitric oxide in exhaled air of asthmatic patients. *Lancet* **343**, 133–135.

Kimberly, B., Nejadnik, B., Giraud, G. D., and Holden, W. E. (1996). Nasal contribution to exhaled nitric oxide at rest and during breathholding in humans. *Am. J. Respir. Crit. Care Med.* **153**, 829–836.

Kneepkens, C. M. F., Ferreira, C., Lepage, G., and Roy, C. C. (1992). Hydrocarbon breath test in cystic fibrosis: Evidence for increased lipid peroxidation. *J. Pediatr. Gastroenterol. Nutr.* **15**, 344. [Abstract]

Kobzik, L., Bredt, D., Lowenstein, C., Snyder, S. H., Drazen, J. M., Sugarbaker, D., and Stamler, J. S. (1993). Nitric oxide synthase in human and rat lung. *Am. J. Respir. Cell Mol. Biol.* **9**, 371–377.

Kohlmuller, D., and Kochen, W. (1993). Is *n*-pentane really an index of lipid peroxidation in humans and animals? A methodological reevaluation. *Anal. Biochem.* **210**, 268–276.

Konstan, M. W., Hilliard, K. A., Norvell, T. M., and Berger, M. (1994). Bronchoalveolar lavage findings in cystic fibrosis patients with stable, clinically mild lung disease suggest ongoing infection and inflammation. *Am. J. Respir. Crit. Care Med.* **150**, 448–454.

Kooy, N. W., and Royall, J. A. (1994). Agonist-induced peroxynitrite production from endothelial cells. *Arch. Biochem. Biophys.* **310**, 352–359.

Kooy, N. W., Royall, J. A., Ye, Y. Z., Kelly, D. R., and Beckman, J. S. (1995). Evidence for in vivo peroxynitrite production in human acute lung injury. *Am. J. Respir. Crit. Care Med.* **151**, 1250–1254.

Korge, P., and Campbell, K. B. (1993). The effect of changes in iron redox state on the activity of enzymes sensitive to medication of SH groups. *Arch. Biochem. Biophys.* **304**, 420–428.

Krinsky, N. I. (1993). *Annu. Rev. Nutr.* **13**, 561–587 (and references cited therein).

Krinsky, N. I., and Sies, H. (1995). The present status of antioxidant vitamins and beta-carotene. *Am. J. Clin. Nutr.* **62**(Suppl.), 1299S–1300S.

Kurose, I., Wolf, R., Grisham, M. B., Ty, A., Specian, R. D., and Granger, D. N. (1995). Microvascular responses to inhibition of nitric oxide production: Role of active oxidants. *Circ. Res.* **76**, 30–39.

Lepage, G., Champagne, J., Ronco, N., Lamarre, A., Osberg, I., Sokol, R. J., and Roy, C. C. (1996). Supplementation with carotenoids corrects increased lipid peroxidation in children with cystic fibrosis. *Am. J. Clin. Nutr.* **64**, 87–93.

Litwin, C. M., and Calderwood, S. B. (1993). Role of iron in regulation of virulence genes. *Clin. Microbiol. Rev.* **6**, 137–149.

Mangione, S., Patel, D. D., Levin, B. R., and Fiel, S. B. (1994). Erythrocytic glutathione in cystic fibrosis. *Chest* **105**, 1470–1473.

Marelich, G. P., and Cross, C. E. (1996). Cystic fibrosis in adults: From researcher to practitioner. *West. J. Med.* **164**, 321–334.

Marquez, L. A., and Dunford, H. B. (1995). Kinetics of oxidation of tyrosine and dityrosine by myeloperoxidase compounds I and II. *J. Biol. Chem.* **270**, 30434–30440.

Masson, P. L., Heremans, J. F., Prignot, J. J., and Wauters, G. (1977). Immunohistochemical localization and bacteriostatic properties of an iron-binding protein from bronchial glands. *Thorax* **21**, 538–544.

Matkovics, B., Gyurkovits, K., László, A., and Szabó, L. (1982). Altered peroxide metabolism in erythrocytes from children with cystic fibrosis. *Clin. Chim. Acta* **125**, 59–62.

McElvaney, N. G., Hubbard, R. C., Birrer, P., *et al.* (1991). Aerosol α-1-antitrypsin treatment for cystic fibrosis. *Lancet* **337**, 392–394.

Mergey, M., Lemnaouar, M., Veissiere, D., Perricaudet, M., Gruenert, D. C., Picard, J., Capeau, J., Brahimi-Horn, M.-C., and Paul, A. (1995). CFTR gene transfer corrects defective glycoconjugate secretion in human CF epithelial tracheal cells. *Am. J. Physiol.* **269**, L855–L864.

Meyer, K., Brown, R., and Zimmerman, J. (1995). Myeloperoxidase and dityrosine in cystic fibrosis. *Am. J. Respir. Cell Mol. Biol.* **151**, A248.

Miles, A. M., Owens, M. W., Milligan, S., Johnson, G. G., Fields, J. Z., Ing, T. S., Kottapalli, V., Keshavarzian, A., and Grisham, M. B. (1995). Nitric oxide synthase in circulating vs. extravasated polymorphonuclear leukocytes. *J. Leukocyte Biol.* **58**, 616–622.

Mohammed, J. R., Mohammed, B. S., Pawluk, L. J., Bucci, D. M., Baker, N. R., and Davis, W. B. (1988). Purification and cytotoxic potential of myeloperoxidase in cystic fibrosis sputum. *J. Lab. Clin. Med.* **112**, 711–720.

Mulligan, M. S., Hevel, J. M., Marletta, M. A., and Ward, P. A. (1991). Tissue injury caused by deposition of immune complexes is L-arginine dependent. *Natl. Acad. Sci. U.S.A.* **88**, 6338–6342.

Munday, R. (1989). Toxicity of thiols and disulphides: Involvement of free-radical species. *Free Rad. Biol. Med.* **7**, 659–673.

Nagaki, M., Shimura, S., Irokawa, T., Saski, T., and Shirato, K. (1995). Nitric oxide regulation of glycoconjugate secretion from feline and human airways in vitro. *Respir. Physiol.* **102**, 79–88.

Niggemann, B., Stiller, T., Magdorf, K., and Wahn, U. (1995). Myeloperoxidase and eosinophil cationic protein in serum and sputum during antibiotic treatment in cystic fibrosis patients with *Pseudomonas aeruginosa* infection. *Mediat. Inflamm.* **4**, 282–288.

O'Connor, C. M., Gaffney, K., Keane, J., Southey, A., Byrne, N., O'Mahoney, S., and Fitzgerald, M. X. (1993). α_1-Proteinase inhibitor, elastase activity, and lung disease severity in cystic fibrosis. *Am. Rev. Respir. Dis.* **148**, 1665–1670.

Pacht, E. R., Timerman, A. P., Lykens, M. G., and Merola, A. J. (1991). Deficiency of alveolar fluid glutathione in patients with sepsis and the adult respiratory distress syndrome. *Chest* **100**, 1397–1403.

Peden, D. B., Swiersz, M., Ohkubo, K., and Hahn, B. (1993). Nasal secretion of the ozone scavenger uric acid. *Am. Rev. Respir. Dis.* **148**, 455–461.

Persson, C. G. A., Erjefält, I., Alkner, U., Baumgarten, C., Greiff, L., Gustafsson, B., Luts, A., Pipkorn U., Sundler, F., Svensson, C., and Wollmer, P. (1991). Plasma exudation as a first line respiratory mucosal defence. *Clin. Exp. Allergy* **21**, 17–24.

Pfeffer, K. D., Huecksteadt, T. P., and Hoidal, J. R. (1993). Expression and regulation of tumor necrosis factor in macrophages from cystic fibrosis patients. *Am. J. Respir. Cell Mol. Biol.* **9**, 511–519.

Pier, G. B., Grout, M., Zaidi, T. S., Olsen, J. C., Johnson, L. G., Yankaskas, J. R., and Goldberg, J. B. (1996). Role of mutant CFTR in hypersusceptibility of cystic fibrosis patients to lung infections. *Science* **271**, 64–67.

Portal, B., Richard, M. J., Ducros, V., Aguilaniu, B., Brunel, F., Faure, H., Gout, J. P., Bost, M., and Favier, A. (1993). Effect of double-blind crossover selenium supplementation on biological indices of selenium status in cystic fibrosis patients. *Clin. Chem.* **39**, 1023–1028.

Portal, B. C., Richard, M.-J., Faure, H. S., Hadjian, A. J., and Favier, A. E. (1995). Altered antioxidant status and increased lipid peroxidation in children with cystic fibrosis. *Am. J. Clin. Nutr.* **61**, 843–847.

Pryor, W. A. (1991). Can vitamin E protect us against the pathological effects of ozone in smog? *Am. J. Clin. Nutr.* **53**, 702–722.

Pryor, W. A., Jin, X., and Squadrito, G. L. (1994). One- and two-electron oxidations of methionine by peroxynitrite. *Proc. Natl. Acad. Sci. U.S.A.* **91**, 11173–11177.

Radi, R., Beckman, J. S., Bush, K. M., and Freeman, B. A. (1991a). Peroxynitrite oxidation of sulfhydryls: The cytotoxic potential of superoxide and nitric oxide. *J. Biol. Chem.* **266**, 4244–4250.

Radi, R., Beckman, J. S., Bush, K. M., and Freeman, B. A. (1991b). Peroxynitrite-induced membrane lipid peroxidation: The cytotoxic potential of superoxide and nitric oxide. *Arch. Biochem. Biophys.* **288**, 481–487.

Ramsey, B. W., Farrell, P. M., Pencharz, P., and the Consensus Committee. (1992). Nutritional assessment and management in cystic fibrosis: A consensus report. *Am. J. Clin. Nutr.* **55**, 108–116.

Regelmann, W. E., Skubitz, K. M., and Herron, J. M. (1991). Increased monocyte oxidase activity in cystic fibrosis heterozygotes and homozygotes. *Am. J. Respir. Cell Mol. Biol.* **5**, 27–33.

Robbins, R. A., Springall, D. R., Warren, J. B., Kwon, O. J., Buttery, L. D. K., Wilson, A. J., Adcock, I. M., Riveros-Moreno, V., Moncada, S., Polak, J., and Barnes, P. J. (1994). Inducible nitric oxide synthase is increased in murine lung epithelial cells by cytokine stimulation. *Biochem. Biophys. Res. Commun.* **198**, 835–843.

Rosenfeld, M. A., and Collins, F. S. (1996). Gene therapy for cystic fibrosis. *Chest* **109**, 241–252.

Ross, A. C. (1993). Cellular metabolism and activation of retinoids: Roles of cellular retinoid-binding proteins. *FASEB J.* **7**, 317–327.

Roum, J. H., Buhl, R., McElvaney, N. G., Borok, Z., and Crystal, R. G. (1993). Systemic deficiency of glutathione in cystic fibrosis. *J. Appl. Physiol.* **75**, 2419–2424.

Rowe, P. M. (1996). Beta-carotene takes a collective beating. *Lancet* **347**, 249.

Rubbo, H., Parthasarathy, S., Barnes, S., Kirk, M., Kalyanaraman, B., and Freeman, B. A. (1995). Nitric oxide inhibition of lipoxygenase-dependent liposome and low-density lipoprotein oxidation: Termination of radical chain propagation reactions and formation of nitrogen-containing oxidized lipid derivatives. *Arch. Biochem. Biophys.* **324**, 15–25.

Salh, B., Webb, K., Guyan, P. M., Day, J. P., Wickens, D., Griffin, J., Braganza, J. M., and Dormandy, T. L. (1989). Aberrant free radical activity in cystic fibrosis. *Clin. Chim. Acta* **181**, 65–74.

Salman-Tabcheh, S., Guerin, M.-C., and Torreilles, J. (1995). Nitration of tyrosyl-residues from extra- and intracellular proteins in human whole blood. *Free Rad. Biol. Med.* **19**, 695–698.

Schedin, U., Frostell, C., and Gustafsson, L. E. (1995). Nitric oxide occurs in high concentrations in monkey upper airways. *Acta Physiol. Scand.* **155**, 473–474.

Schwiebert, E. M., Egan, M. E., Hwang, T. H., Fulmer, S. B., Allen, S. S., Cutting, G. R., and Guggino, W. B. (1995). CFTR regulates outwardly rectifying chloride channels through an autocrine mechanism involving ATP. *Cell* **81**, 1063–1073.

Simpson, J. A., Smith, S. E., and Dean, R. T. (1993). Alginate may accumulate in cystic fibrosis lung because the enzymatic and free radical capacities of phagocytic cells are inadequate for its degradation. *Biochem. Mol. Biol. Inter.* **30**, 1021–1034.

Slade, R., Crissman, K., Norwood, J., and Hatch, G. (1993). Comparison of antioxidant substances in bronchoalveolar lavage cells and fluid from humans, guinea pigs, and rats. *Exp. Lung Res.* **19**, 469–484.

Sparrow, C. P., and Olszewski, J. (1993). Cellular oxidation of low density lipoprotein is caused by thiol production in media containing transition metal ions. *J. Lipid Res.* **34**, 1219–1228.

Stewart, T. E., Valenza, F., Ribeiro, S. P., Wener, A. D., Volgyesi, G., Mullen, J. B. M., and Slutsky, A. S. (1995). Increased nitric oxide in exhaled gas as an early marker of lung inflammation in a model of sepsis. *Am. J. Respir. Crit. Care Med.* **151**, 713–718.

Stone, P. J., Konstan, M. W., Berger, M., Dorkin, H. L., Franzblau, C., and Snider, G. L. (1995). Elastin and collagen degradation products in urine of patients with cystic fibrosis. *Am. J. Respir. Crit. Care Med.* **152**, 157–162.

Tsuchihashi, H., Kigoshi, M., Iwatsuki, M., and Niki, E. (1995). Action of β-carotene as an antioxidant against lipid peroxidation. *Arch. Biochem. Biophys.* **323**, 137–147.

Tsui, L-C. (1995). The cystic fibrosis transmembrane conductance regulator gene. *Am. J. Respir. Crit. Care Med.* **151**, S47–S53.

Tsui, L., and Buchwald, M. (1991). Biochemical and molecular genetics of cystic fibrosis. *In* "Advances in Human Genetics" (H. Harris and K. Hirschhorn, eds.), Vol. 20, pp. 153–266. Plenum, New York.

van Antwerpen, V. L., Theron, A. J., Richards, G. A., *et al.* (1995). Plasma levels of beta-carotene are inversely correlated with circulating neutrophil counts in young male cigarette smokers. *Inflammation* **19**, 405–414.

van der Vliet, A., Eiserich, J. P., O'Neill, C. A., Halliwell, B., and Cross, C. E. (1995). Tyrosine modification by reactive nitrogen species: A closer look. *Arch. Biochem. Biophys.* **319**, 341–349.

van der Vliet, A., Smith, D., O'Neill, C. A., Kaur, H., Darley-Usmar, V., Cross, C. E., and Halliwell, B. (1994). Interactions of peroxynitrite with human plasma and its constituents: Oxidative damage and antioxidant depletion. *Biochem. J.* **303**, 295–301.

van Poppel, G., Spanhaak, S., and Ockhuizen, T. (1993). Effect of β-carotene on immunological indexes in healthy male smokers. *Am. J. Clin. Nutr.* **57**, 402–407.

Vissers, M. C. M., and Winterbourn, C. C. (1995). Oxidation of intracellular glutathione after exposure of human red blood cells to hypochlorous acid. *Biochem. J.* **307**, 57–62.

Warner, R. L., Paine, R., Christensen, P. J., Marletta, M. A., Richards, M. K., Wilcoxen, S. E., and Ward, P. A. (1995). Lung sources and cytokine requirements for in vivo expression of inducible nitric oxide synthase. *Am. J. Respir. Cell Mol. Biol.* **12**, 649–661.

White, A. C., Thannickal, V. J., and Fanburg, B. L. (1994). Glutathione deficiency in human disease. *J. Nutr. Biochem.* **5**, 218226.

Widdicombe, J. H. (1994). Accumulation of airway mucus in cystic fibrosis. *Pulm. Pharmacol.* **7**, 225–233.

Winklhofer-Roob, B. M. (1994). Oxygen free radicals and antioxidants in cystic fibrosis: The concept of an oxidant-antioxidant imbalance. *Acta Paediatr. Suppl.* **395**, 49–57.

Winklhofer-Roob, B. M., Ziouzenkova, O., Puhl, H., Ellemunter, H., Greiner, P., Müller, G., van't Hof, M. A., Esterbauer, H., and Shmerling, D. H. (1995). Impaired resistance to oxidation of low density lipoprotein in cystic fibrosis: Improvement during vitamin E supplementation. *Free Rad. Biol. Med.* **19**, 725–733.

Witko-Sarsat, V., Delacourt, C., Rabier, D., Bardet, J., Nguyen, A. T., and Descamps-Latscha, B. (1995). Neutrophil-derived long-lived oxidants in cystic fibrosis sputum. *Am. J. Respir. Crit. Care Med.* **152**, 1910–1916.

Wizemann, T. M., Gardner, C. R., Laskin, J. D., Quinones, S., Durham, S. K., Goller, N. L., Ohnishi, S. T., and Laskin, D. L. (1994). Production of nitric oxide and peroxynitrite in the lung during acute endotoxemia. *J. Leukocyte Biol.* **56**, 759–768.

Wolz, C., Hohloch, K., Ocaktan, A., Poole, K., Evans, R. W., Rochel, N., Albrecht-Gary, A.-M., Abdallah, M. A., and Döring, G. (1994). Iron release from transferrin by pyoverdin and elastase from *Pseudomonas aeruginosa*. *Infect. Immun.* 4021–4027.

Zabner, J., Couture, L. A., Gregory, R. J., Graham, S. M., Smith, A. E., and Welsh, M. J. (1993). Adenovirus-mediated gene transfer transiently corrects the chloride transport defect in nasal epithelia of patients with cystic fibrosis. *Cell* **75**, 207–216.

Zachman, R. D. (1995). Role of vitamin A in lung development. *J. Nutr.* **125**, 1634S–1638S.

Zapol, W. M., Rimar, S., Gillis, N., Marletta, M., and Bosken, C. H. (1994). Nitric oxide and the lung. *Am. J. Respir. Crit. Care Med.* **149**, 1375–1380.

Zeiher, B. G., Eichwald, E., Zabner, J., Smith, J. J., Puga, A. P., *et al.* (1995). A mouse model for the ΔF508 allele of cystic fibrosis. *J. Clin. Invest.* **96**, 2051–2064.

Allen Taylor
Thomas Nowell
Jean Mayer USDA Human Nutrition Research Center
on Aging at Tufts University
Boston, Massachusetts 02111

Oxidative Stress and Antioxidant Function in Relation to Risk for Cataract

The lens is composed primarily of proteins and water. Half-lives of many of the proteins are measured in decades. The light and oxygen to which the lens is exposed are associated with extensive damage to the long-lived lens proteins and other constituents. With progressive damage, the altered proteins accumulate, aggregate, and precipitate in opacities, or cataracts. Cataracts, which impair vision, afflict ~50% of persons >75 years of age in the United States and account for the largest line item within the Medicaid budget. In the less-developed world, the age-adjusted risk for cataract is almost twice as high. The young lens has substantial reserves of antioxidants and antioxidant enzymes that may prevent damage. Proteases may selectively remove obsolete proteins and provide a second level of defense. Compromises of function of the lens upon aging are associated with and may be causally related to depleted or diminished primary antioxidant reserves, antioxidant enzyme capabilities, and diminished secondary defenses such as proteases. Environmental stresses such as smoking and excessive light exposure appear to provide an additional oxidative challenge associated with the depletion of antioxidants as well as with enhanced risk for cataract. Data from several recent epidemiological studies indicate that optimizing nutriture (particularly ascorbate, carotenoids, and tocopherol), as well as fruit and vegetable intake, may provide the least costly and most practicable means to delay cataract. Poor education and lower socioeconomic status are associated with poorer nutriture and are also significantly related to an increased risk for these debilities.

I. Introduction

The number of associations between nutriture and age-related eye diseases have burgeoned in the last decade, inspired in part by early studies regarding antioxidant properties of nutrients.

Studies regarding the etiology of cataract now include laboratory and epidemiological investigations. This chapter briefly reviews available data, as well as intervention trials, regarding associations between antioxidant nutrients and eye lens cataract in humans. Readers can refer to other reviews (Bunce *et al.*, 1990; Taylor, 1992; Taylor *et al.*, 1993; Jacques *et al.*, 1994) for more thorough treatments, particularly with respect to animal studies. Much of the rich body of pioneering work is, of necessity, given limited coverage here (Blondin *et al.*, 1986; Blondin and Taylor, 1987; Taylor, 1993).

The primary function of the eye lens is to collect and focus light on the retina. To do so it must remain clear throughout life (Figs. 1a and 1d). The lens is located posterior to the cornea and iris and receives nutriture from the aqueous humor. Although the lens appears to be free of structure, it is exquisitely designed. A single layer of epithelial cells is found directly under

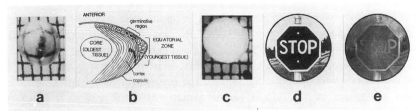

FIGURE I   Clear and cataractous lens. (a) Clear lens allows an unobstructed view of the wire grid placed behind it. (b) The structure of the lens. The anterior surface of the lens has a unicellular layer of epithelial cells (youngest tissue). Cells at the anterior equatorial region divide and migrate to the cortex as they are overlaid by less mature cells. These cells produce a majority of the crystallins. As development and maturation proceed, the cells denucleate and elongate. Tissue originally found in the embryonic lens is found in the core (oldest tissue) of the lens in a mature animal. (c) The cataractous lens prohibits viewing the wire grid behind it. (d) View through a clear uncolored young lens. The image is clear and crisp. (e) View through a lens with developing cataract. The image is partially obscured, and the field is darkened due to age-related browning of the lens.

the anterior surface of the collagenous membrane in which it is encapsulated (Fig. 1b). The epithelial cells at the germinative region divide, migrate posteriorly, and differentiate into lens fibers. The fibers elaborate as their primary gene products the predominant proteins of the lens called crystallins. They also lose their organelles. New cells are formed throughout life but older cells are usually not lost. They are compressed into the center or nucleus of the lens. There is a coincident dehydration of the proteins and the lens itself. Together with modifications of the protein (noted later) and other constituents, these changes result in a less flexible lens upon aging.

As the lens ages the proteins are photooxidatively damaged, aggregate, and accumulate in lens opacities (Figs. 1c and 1e). Dysfunction of the lens due to opacification is called cataract. The term "age-related cataract" is used to distinguish lens opacification associated with old age from opacification associated with other causes, such as congenital and metabolic disorders (Jacques and Taylor, 1991).

II. Public Health Issues Regarding Cataract _____

Cataract is one of the major causes of blindness throughout the world (Kupfer, 1984; Schwab, 1990; World Health Organization, 1991). In the United States, the prevalence of visually significant cataract increases from approximately 5% at age 65 to ~50% for persons older than 75 years (Leibowitz et al., 1980; Klein et al., 1992, 1993). In less developed countries, such as India (Chatterjee et al., 1982), China (Wang et al., 1990), and Kenya (Whitfield et al., 1990), cataracts are more common and develop earlier in life than in more developed countries. For example, by age 60, cataract with low vision or aphakia (i.e., absence of the lens, which is usually the result of cataract extraction) is approximately five times more common in India than in the United States (Leibowitz et al., 1980; Chatterjee et al., 1982). The impact of cataract on impaired vision is much greater in less-developed countries, where more than 90% of the cases of blindness and visual impairment are found (Dana et al., 1990; Chan and Billson, 1991; World Health Organization, 1991; Salive et al., 1992; Wormald et al., 1992) and where there is a dearth of ophthalmologists to perform lens extractions (Berger et al., 1989; Seddon et al., 1994). Such surgery is routinely successful in restoring sight.

Given both the extent of disability caused by age-related cataract and its costs, $5 billion/year[1] in the United States, it is urgent that we elucidate causes of cataract and identify strategies to slow the development of this disorder. It is estimated that a delay in cataract formation of about 10 years would reduce the prevalence of visually disabling cataract by about 45%

[1] Congressional Testimony of S. J. Ryan, May 5, 1993.

(Kupfer, 1984). Such a delay would enhance the quality of life for much of the world's older population and substantially reduce both the economic burden due to disability and surgery related to cataract.

III. Clinical Features of Cataract

There are several systems for evaluating and grading cataracts. Most of these employ an assessment of extent or density and location of the opacity (Chylack *et al.*, 1993). Coloration or brunescence is also quantified, as these diminish visual function (Chylack *et al.*, 1994; Wolfe *et al.*, 1993). Usually evaluated are opacities in the posterior subcapsular, nuclear, cortical, and multiple (mixed) locations (Fig. 1).

IV. Oxidation, Smoking, and Cataract Formation

The solid mass of the lens is approximately 98% protein. Because these proteins undergo minimal turnover as the lens ages, they are subject to the chronic stresses of exposure to light and oxygen. Consequently, it is not surprising that these proteins are extensively damaged in the aged lens. Lens opacities develop as the damaged proteins aggregate and precipitate (Taylor *et al.*, 1993). Fiber cell membrane lipid damage is also associated with lens opacities (Berman, 1991; Jacques and Chylack, 1991; Jacques *et al.*, 1994). Smoking and ultraviolet light, which appear to induce oxidative stress (Zigman, 1983; Schectman *et al.*, 1989), are also associated with cataractous changes to the lens *in vitro* (Shalini *et al.*, 1994; Rao *et al.*, 1995; Zigman *et al.*, 1995) with elevated cataract risk (Zigman *et al.*, 1979; Brilliant *et al.*, 1983; Taylor *et al.*, 1988; Christen *et al.*, 1992; Hankinson *et al.*, 1992b; Klein *et al.*, 1992; West, 1992; Wong *et al.*, 1993; Hirvela *et al.*, 1995; West and Valmadrid, 1995), as well as with depletion of plasma ascorbate and carotenoid levels (Russell-Briefel *et al.*, 1985; Robertson *et al.*, 1989; Schectman *et al.*, 1989; Leske *et al.*, 1991; Mares-Perlman *et al.*, 1995b).

In young lenses, damaged proteins are usually maintained at harmless levels by defense systems. Primary defenses that directly protect the lens against the initial oxidative insult include small molecule antioxidants (e.g., vitamins C and E, glutathione, and carotenoids) (Berger *et al.*, 1989; Pau *et al.*, 1990; Taylor *et al.*, 1991; Mune and Meydani, 1995; Yeum *et al.*, 1995) and antioxidant enzyme systems (e.g., superoxide dismutase, catalase, and the glutathione redox cycle) (Fridovich, 1984; Varma *et al.*, 1984; Zigler and Goosey, 1984; Rathbun *et al.*, 1990; Giblin *et al.*, 1992). The lens also has secondary defense systems, which include proteolytic enzymes that selectively identify and remove damaged or obsolete proteins (Taylor and Davies, 1987; Berger *et al.*, 1988; Jahngen-Hodge *et al.*, 1991; Jahngen-

Hodge *et al.*, 1992; Huang *et al.*, 1993; Taylor *et al.*, 1993; Obin *et al.*, 1994; Shang and Taylor, 1995). Accumulation of (photo)oxidized (and/or otherwise modified) proteins in older lenses indicates that protective systems are not keeping pace with the insults that damage lens proteins. This occurs in part because, like bulk proteins, enzymes that comprise some of the protective systems are damaged by photo-oxidation (Blondin and Taylor, 1987; Taylor and Davies, 1987; Taylor *et al.*, 1993). Interactions between the primary and secondary antioxidant defense systems and putative ramifications of these relationships on cataract risk are summarized in Fig. 2.

V. Associations between Antioxidants and Cataract _____

Many cell-free, *in vitro,* and animals studies addressed putative roles for antioxidants in maintenance of lens and retina function. These were

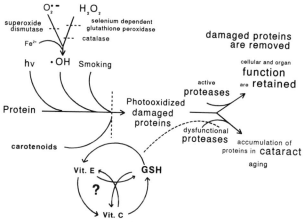

FIGURE 2   Proposed interactions among lens constituents, oxidants, light, smoking, antioxidants, antioxidant enzymes, and proteases. The constituent proteins in the lens are extremely long lived. In both lens and retina they are subject to damage by light and various forms of oxygen. Such damage is limited by antioxidant enzymes (superoxide dismutase, catalase, glutathione reductase/peroxidase) which convert active oxygen to less damaging species. Direct protection may also be offered by antioxidants [vitamin C, vitamin E, carotenoids, and glutathione (GSH)]. Levels of reduced forms of some, but not all, of these molecules maybe determined by interactions between them and with the environment (Wefers and Sies, 1988; Burton *et al.*, 1990; Meister, 1994). On aging, antioxidant levels in some eye tissues (lens) are diminished and the antioxidant enzymes may be at reduced catalytic competence, resulting in increased levels of damage. Proteins that are obsolete or damaged are reduced to their constituent amino acids if proteolytic activity is sufficient. If proteolytic activity is insufficient, damaged proteins may accumulate, aggregate, and precipitate in cataracts in the lens. Older tissues frequently show lower levels of proteolytic activities. H_2O_2, hydrogen peroxide; O_2^- superoxide radical; HO; hydroxyl radical; the dashed lines indicate sites where antioxidants might protect against damage to proteins and proteases.

reviewed recently and inspired the epidemiological work described later (Taylor *et al.*, 1993; Jacques *et al.*, 1994). In order to fully appreciate the data presented in this chapter, readers should be aware that the various studies used different lens classification schemes, different definitions of high and low levels of nutrients, and different age groups of subjects. Although cataracts frequently take years to develop, only two studies measured nutrient levels more than once (Hankinson *et al.*, 1992a; Taylor *et al.*, 1994).

More than 10 epidemiological studies examined the associations between cataract and antioxidant nutrients (Mohan *et al.*, 1989; Robertson *et al.*, 1989; The Italian–American Cataract Study Group, 1991; Jacques and Chylack, 1991; Leske *et al.*, 1991; Hankinson *et al.*, 1992a; Jacques *et al.*, 1992; Knekt *et al.*, 1992; Mares-Perlman *et al.*, 1994, 1995a,b; Seddon *et al.*, 1994; Vitale *et al.*, 1994). Seven of the studies were retrospective case control studies comparing the nutrient levels of cataract patients with that of similarly aged individuals with clear lenses (Mohan *et al.*, 1989; Robertson *et al.*, 1989; The Italian–American Cataract Study Group, 1991; Jacques and Chylack, 1991; Leske *et al.*, 1991; Mares-Perlman *et al.*, 1994; Vitale *et al.*, 1994). Our ability to interpret data from retrospective studies, such as these, is limited by the concurrent assessment of lens status and nutrient levels. Prior diagnosis of cataract might influence behavior of cases, including diet, and it might also bias reporting of usual diet. Three other studies assessed nutrient levels and/or supplement use, and then followed individuals with intact lenses for 8 (Hankinson *et al.*, 1992a; Knekt *et al.*, 1992) and 5 years (Seddon *et al.*, 1994), respectively. Prospective studies, such as these, are less prone to bias because the assessment of exposure is performed before the outcome is present. These latter studies did not directly assess lens status, but used cataract extraction or reported diagnosis of cataract as a measure of cataract risk. Extraction may not be a good measure of cataract incidence (development of new cataract) because it incorporates components of both incidence and progression in severity of existing cataract. However, extraction is the result of visually disabling cataract and is the end point that we wish to prevent. Although Hankinson *et al.* (1992a) measured nutrient intake over a 4-year period, Knekt *et al.* (1992) used only one measure of serum antioxidant status and Seddon *et al.* (1994) used only one measure of supplement use. One measure may not provide an accurate assessment of usual, long-term nutrient levels. Multiple measures may be the best nutritional correlate of cataract (Taylor *et al.*, 1994). Another study ($n = 367$), which monitored cataract *in vivo* and cataract extraction but did not find associations between nutriture and cataract, is not further described because the cataract classifications do not match those shown in Figs. 3–6 (Wong *et al.*, 1993).

A. Vitamin C

Vitamin C is probably the most effective, least toxic water-soluble antioxidant identified in mammalian systems (Levine, 1986; Frei *et al.*, 1988).

Lens concentrations of vitamin C ($\sim$mM) are many fold higher than in plasma (Taylor *et al.*, 1991). But vitamin C concentrations are compromised on aging and/or cataractogenesis (Berger *et al.*, 1988, 1989). Interest in the utility of vitamin C has been fueled by observations that (1) eye tissue levels of this vitamin are related to dietary intake in humans (Taylor *et al.*, 1991) and animals (Berger *et al.*, 1988, 1989) and (2) the concentration of vitamin C in the lens was increased with dietary supplements beyond levels achieved in persons who already consumed more than two times the RDA (60 mg/ day) for vitamin C (Taylor *et al.*, 1991). Although biochemically plausible, there are no data to demonstrate that vitamin C induces damage in the lens *in vitro* (Blondin and Taylor, 1987; Garland, 1991; Naraj and Monnier, 1992).

Vitamin C was considered in eight published studies (Mohan *et al.*, 1989; Robertson *et al.*, 1989; The Italian–American Cataract Study Group, 1991; Jacques and Chylack, 1991; Leske *et al.*, 1991; Hankinson *et al.*, 1992a; Mares-Perlman *et al.*, 1994, 1995a; Vitale *et al.*, 1994) and one preliminary report (Jacques *et al.*, 1992) and was observed to be inversely associated with at least one type of cataract in seven of these studies (Fig. 3). Jacques and Chylack (1991) observed that persons with high plasma vitamin C levels ($>$90 μM) had less than one-third the prevalence of early cataract as persons with low plasma vitamin C ($<$40 μM), although this difference was not statistically significant [risk ratio (RR): 0.29; 95% confidence interval (CI): 0.06–1.32] after adjustment for age, sex, race, and history of diabetes (Fig. 3a). They observed similar relationships between intake of vitamin C and cataract prevalence. Among persons with higher vitamin C intakes ($>$490 mg/day), the prevalence of cataract was 25% of the prevalence among persons with lower intakes ($<$125 mg/day) [RR: 0.25; CI: 0.06–1.09, Fig. 3a (Jacques and Chylack, 1991)].

This relationship is corroborated by data from other studies. Robertson and co-workers (1989) compared cases (with cataracts that impaired vision) to age- and sex-matched controls who were either free of cataract or had minimal opacities that did not impair vision. Results indicated that the prevalence of cataract in persons who consumed daily vitamin C supplements of $>$300 mg/day was approximately one-third the prevalence in persons who did not consume vitamin C supplements (RR: 0.30; CI: 0.24–0.77, Fig. 3b). Leske and co-workers (1991) observed that persons with vitamin C intake in the highest 20% of their population group had a 52% lower prevalence for nuclear cataract (RR: 0.48; CI:0.24–0.99) compared with persons who had intakes among the lowest 20% after controlling for age and sex (Fig. 3c). Weaker inverse associations were noted for other types of cataract (Figs. 3c–3f). After controlling for nine potential confounders, including age, diabetes, smoking, and energy intake, Hankinson and co-workers (1992a) did not observe an association between total vitamin C intake and rate of cataract surgery (RR: 0.98; CI: 0.72–1.32) in a large prospective study when they compared women with high intakes (median = 705 mg/day) to women with low intakes (median = 70 mg/day)

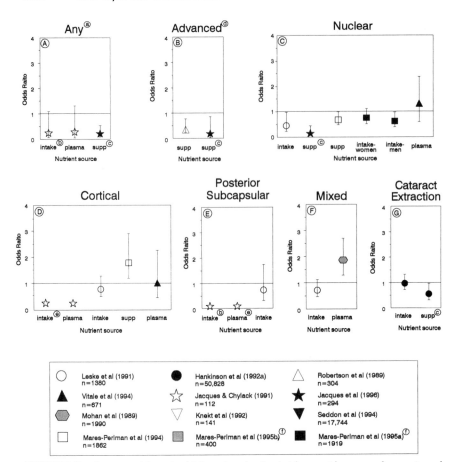

FIGURE 3 Cataract risk ratio, high vs low intake (with or without supplements), and plasma levels for vitamin C. ⓐ Any = cataract with LOCS II grade ≥ 1; ⓑ p ≤ 0.1; ⓒ supp use > 10 yrs; ⓓ Adv = cataract with LOCS II grade ≥ 2; ⓔ p ≤ 0.05; ⓕ When data was available for comparable groups of subjects, the most recent data from the largest sample was used for these displays.

(Fig. 3g). However, they did note that women who consumed vitamin C supplements for >10 years had a 45% reduction in rate of cataract surgery (RR: 0.55; Cl: 0.32–0.96). Age-adjusted analyses (Jacques *et al.*, 1996) based on 165 women with high vitamin C intake (mean = 294 mg/day) and 136 women with low vitamin C intake (mean = 77 mg/day) demon-strated that the women who took vitamin C supplements ≥10 years had >70% lower prevalence of early opacities (RR: 0.27; Cl 0.11–0.67, Fig. 3a) and >80% lower risk of advanced opacities (RR:0.19; Cl: 0.05–0.80,

Fig. 3b) at any site compared with women who did not use vitamin C supplements.

In comparison to these data, Mares-Perlman and co-workers (1994) report that the past use of supplements containing vitamin C was associated with a reduced prevalence of nuclear cataract (RR: 0.7; Cl: 0.5–1.0, Fig. 3c), but an increased prevalence of cortical cataract (adjusted RR: 1.8; Cl: 1.2–2.9) after controlling for age, sex, smoking, and history of heavy alcohol consumption (Fig. 3d). Mohan *et al.* (1989) also noted an 87% (RR: 1.87; Cl: 1.29–2.69) increased prevalence of mixed cataract with posterior subcapsular and nuclear involvement for each standard deviation increase in plasma vitamin C levels. Vitale and co-workers (1994) observed that persons with plasma levels greater than 80 μM and below 60 μM had similar prevalences of both nuclear (RR: 1.31; Cl: 0.61–2.39) and cortical (RR: 1.01; Cl: 0.45–2.26) cataract after controlling for age, sex, and diabetes. Similarly, no differences in cataract prevalence were observed between persons with high (>261 mg/day) and low (<115 mg/day) vitamin C intakes. One other study (The Italian–American Cataract Study Group, 1991) failed to observe any association between prevalence of cataract and vitamin C intake.

B. Vitamin E

Vitamin E, a natural lipid-soluble antioxidant, can inhibit lipid peroxidation (Machlin and Bendich, 1987) and appears to stabilize lens cell membranes (Libondi *et al.*, 1985). Vitamin E may be affected by ascorbate (see legend to Fig. 2) and also enhances glutathione recycling, perhaps helping to maintain reduced glutathione levels in the lens and aqueous humor (Costagliola *et al.*, 1986).

Three studies assessing plasma vitamin E levels also reported significant inverse associations with cataract (Fig. 4). Knekt and co-workers (1992) followed a cohort of 1419 Finns for 15 years and identified 47 patients admitted to ophthalmological wards for mature cataract. They selected two controls per patient matched for age, sex, and municipality. These investigators reported that persons with serum vitamin E concentrations above approximately 20 μM had about one-half the rate of subsequent cataract surgery (RR: 0.53; Cl: 0.24–1.1, Fig. 4g) compared with persons with vitamin E concentrations below this concentration. Vitale and co-workers (1994) observed the age-, sex-, and diabetes-adjusted prevalence of nuclear cataract to be about 50% less (RR: 0.52; Cl: 0.27–0.99, Fig. 4c) among persons with plasma vitamin E concentrations greater than 29.7 μM compared to persons with levels below 18.6 μM. A similar comparison showed that the prevalence of cortical cataract did not differ between those with high and low plasma vitamin E levels (RR: 0.96; Cl: 0.52–0.1.78, Fig. 4d). Jacques and Chylack (1991) also observed the prevalence of posterior

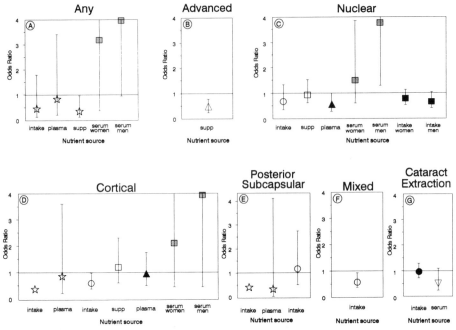

FIGURE 4   Same as Fig. 3, but for cataract risk ratio high vs low intake (with or without supplements) and levels for vitamin E, including γ-tocopherol and α-tocopherol (symbols as in Fig. 3).

subcapsular cataract to be 67% (RR: 0.33; Cl: 0.03–4.13, Fig. 4e) lower among persons with plasma vitamin E levels above 35 μM relative to persons with levels below 21 μM after adjustment for age, sex, race, and diabetes; however, the effect was not statistically significant. Prevalence of any early cataract (RR: 0.83; Cl: 0.20–3.40, Fig. 4a) or cortical cataract (RR: 0.84; Cl: 0.20–3.60, Fig. 4d) did not differ between those with high and low plasma levels. Plasma vitamin E was also inversely associated with the prevalence of cataract in a large Italian study after adjusting for age and sex, but the relationship was no longer statistically significant after adjusting for other factors such as education, sunlight exposure, and family history of cataract (The Italian–American Cataract Study Group, 1991). One other study failed to observe any association between cataract and plasma vitamin E levels (Mohan *et al.*, 1989). Mares-Perlman *et al.* (1995b) observed an inverse (nonsignificant) relationship (RR: 0.61; Cl: 0.32–1.19) between serum γ-tocopherol (which has lower biological vitamin E activity compared to α-tocopherol) and severity of nuclear sclerosis but a positive, significant relationship between elevated serum α-tocopherol levels and severity of nuclear cataract (RR: 2.13; Cl: 1.05–4.34, Fig. 4c). Serum γ-tocopherol was not significantly associated with cortical or any cataract in these studies, whereas serum α-tocopherol appeared to be associated with nonsignificant increases in risk for cortical or any cataract (Figs. 4d and 4a, respectively).

Vitamin E intake was inversely correlated with cataract risk in two studies (Fig. 4). Robertson and co-workers (1989) found among age- and sex-matched cases and controls that the prevalence of advanced cataract was 56% lower (RR: 0.44; CI: 0.24–0.77, Fig. 4b) in persons who consumed vitamin E supplements (>400 IU/day) than in persons not consuming supplements. Jacques and Chylack (unpublished observations) also observed a 67% (RR: 0.33: CI: 0.12–0.96) reduction in the prevalence of cataract for vitamin E supplement users after adjusting for age, sex, race, and diabetes. These investigators observed a nonsignificant inverse association when they related total vitamin E intake (combined dietary and supplemental intake) to cataract prevalence (Jacques and Chylack, 1991). Persons with vitamin E intake greater than 35.7 mg/day had a 55% lower prevalence of early cataract (RR: 0.45; CI: 0.12–1.79) than did persons with intakes less than 8.4 mg/day. Leske and colleagues (1991) also observed that vitamin E intake was inversely associated with the prevalence of cataract after controlling for age and sex. Persons with vitamin E intakes among the highest 20% had an approximately 40% lower prevalence of cortical (RR: 0.59; CI: 0.36–0.97, Fig. 4d) and mixed (RR: 0.58; CI: 0.37–0.93, Fig. 4d) cataract relative to persons with intakes among the lowest 20%.

In contrast to these studies, Mares-Perlman and co-workers (1994) observed only weak, nonsignificant associations between vitamin E supplement use and nuclear (RR: 0.9; CI: 0.6–1.5, Fig. 4c) and cortical (RR: 1.2; CI: 0.6–2.3, Fig. 4d) cataract. Hankinson *et al.* (1992a) found no association between vitamin E intake and cataract surgery. Women with high vitamin E intakes (median = 210 mg/day) had a similar rate of cataract surgery (RR: 0.96; CI: 0.72–1.29) as women with low intakes (median = 3.3 mg/day). In partial contrast with their positive correlations between serum α-tocopherol levels and cataract, Mares-Perlman *et al.* (1995a) found that dietary vitamin E was associated (nonsignificantly) with diminished risk for nuclear cataract in men, but not in women (Mares-Perlman *et al.*, 1995a).

C. Carotenoids

The carotenoids, like vitamin E, are natural lipid-soluble antioxidants (Machlin and Bendich, 1987). β-Carotene is the best known carotenoid because of its importance as a vitamin A precursor. It exhibits particularly strong antioxidant activity at low partial pressures of oxygen (15 torr) (Burton and Ingold, 1984). Partial pressure of oxygen in the core of the lens is approximately 20 torr (Kwan *et al.*, 1972). However, it is only 1 of ~400 naturally occurring carotenoids (Erdman, 1988) and other carotenoids may have similar or greater antioxidant potential (Krinsky and Deneke, 1982; Di Mascio *et al.*, 1991). In addition to β-carotene, α-carotene, lutein, and lycopene are important carotenoid components of the human diet (Micozzi *et al.*, 1990). Carotenoids have been identified in the lens in $\approx$10 ng/g wet weight concentrations (Daicker *et al.*, 1987; Yeum *et al.*, 1995), but there are no laboratory data relating carotenoids to cataract formation.

Jacques and Chylack (1991) observed that persons with high plasma total carotenoid concentrations (>3.3 μM) had less than one-fifth the prevalence of cataract compared to persons with low plasma carotenoid levels (<1.7 μM) (RR: 0.18; CI: 0.03–1.03) after adjustment for age, sex, race, and diabetes (Fig. 5a). However, they were unable to observe an association between carotene intake and cataract prevalence (Jacques and Chylack, 1991). Persons with carotene intakes above 18,700 IU/day had the same prevalence of cataract as those with intakes below 5677 IU/day (RR: 0.91; CI: 0.23–3.78). Knekt and co-workers (1992) reported that among age- and sex-matched cases and controls, persons with serum β-carotene concentrations above approximately 0.1 μM had a 40% reduction in the rate of cataract surgery compared with persons with concentrations below this level (RR: 0.59; CI: 0.26–1.25, Fig. 5e).

Hankinson and co-workers (1992a) reported that the multivariate-adjusted rate of cataract surgery was about 30% lower (RR: 0.73; CI: 0.55–0.97) for women with high carotene intakes (median = 14,558 IU/day) compared with women with low intakes of this nutrient (median = 2935 IU/day). However, although cataract surgery was inversely associated with total carotene intake, it was not strongly associated with consumption of carotene-rich foods, such as carrots. Rather, cataract surgery was associated with lower intakes of foods such as spinach that are rich in lutein and xanthin carotenoids rather than β-carotene. Unfortunately, cataract surgery was not an end point in other studies that considered xanthaphylls (Mares-Perlman *et al.*, 1994, 1995b). The most recent study that correlated serum carotenoids and severity of nuclear and cortical opacities (Mares-Perlman *et al.*, 1995b) indicates that higher levels of individual or total carotenoids in the serum were not associated with less severe nuclear or cortical cataract overall (Figs. 5f–5t). However, associations between risk for some forms of cataract and nutriture differed between men and women, for example, nuclear cataract and α-carotene intake (Fig. 5h). Other examples where cataract risk in women vs men showed opposing relationships to nutriture was in cortical or any cataract and serum β-carotene (Figs. 5i and 5k) and serum lycopene (Figs. 5l and 5n). Furthermore, a marginally significant trend for lower risk ratio for cortical opacity with increasing serum levels of β-carotene was observed in men, but not in women. Higher serum levels of α-carotene, β-cryptoxanthin, and lutein were significantly related to a lower risk for nuclear sclerosis only in men who smoked. In contrast, higher levels of some carotenoids were often directly associated with an elevated risk for nuclear sclerosis and cortical cataract (Figs. 5i and 5k), particularly in women. Vitale and colleagues (1994) also examined the relationships

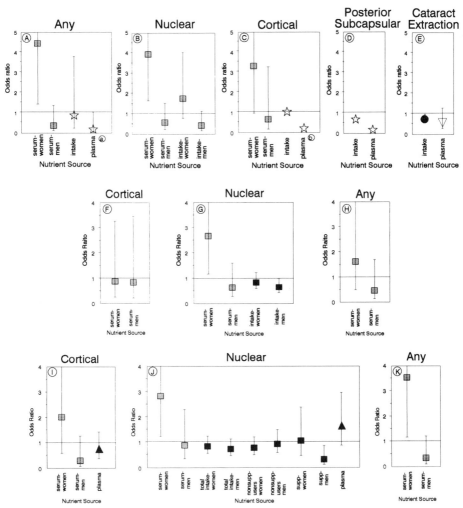

FIGURE 5   Same as Fig. 3, but in levels for carotenoids, including α-carotene, β-carotene, lycopene, β-cryptoxanthin, and lutein (symbols as in Fig. 3). ⓐ p < 0.1; ⓑ p < 0.05.

between plasma β-carotene levels and age-, sex-, and diabetes-adjusted prevalence of cortical and nuclear cataract (Fig. 5j). Although data suggested a weak inverse association between plasma β-carotene and cortical cataract and a weak positive association between this nutrient and nuclear cataract, neither association was statistically significant. Persons with plasma β-carotene concentrations above 0.88 μM had a 28% lower prevalence of cortical cataract (RR: 0.72: CI: 0.37–1.42) and a 57% (RR: 1.57; CI:

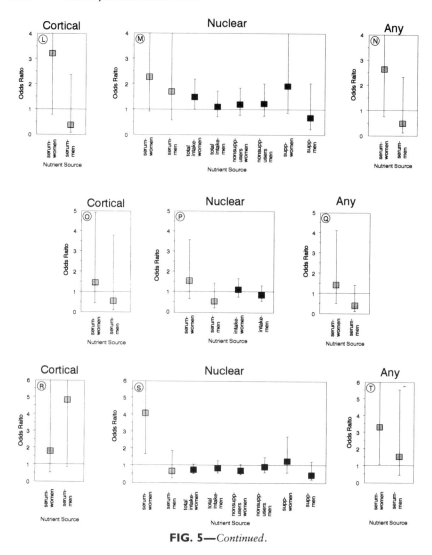

FIG. 5—*Continued*.

0.84–2.93) higher prevalence of nuclear cataract compared to persons with levels below 0.33 μM.

D. Smoking, Antioxidants, and Cataracts

Smoking is associated with diminished carotenoid (Russell-Briefel *et al.*, 1985) and ascorbate status (Schectman *et al.*, 1989). Smoking is also associated with an enhanced risk for cataract (West, 1992). Of interest are recent observations that for male smokers there appears to be an inverse relationship between serum levels of α-carotene, β-cryptoxanthin, lutein, and severity of nuclear sclerosis (Mares-Perlman *et al.*, 1995b) (the reverse may be true for women) and a diminished risk for cataract in smokers who use multivitamins (Seddon *et al.*, 1994).

E. Antioxidant Nutrient Combinations

Combinations of multiple antioxidant nutrients were also considered (Fig. 6) because of possible synergistic effects of the antioxidant nutrients

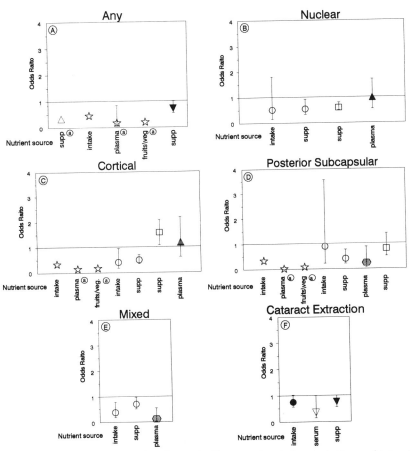

FIGURE 6 Same as Fig. 3, but for antioxidant nutrient index, including multivitamin supplements (symbols as in Fig. 3). ⓐ p ≤ 0.05.

on cataract risk (Fig. 2). The first, and perhaps most important, study in terms of revealing the utility of diet indicates a significant five-fold decrease in risk ratio for cataract between persons consuming ≥ 1.5 servings of fruits and/or vegetables (Jacques and Chylack, 1991). Jacques and Chylack (1991) also found that the adjusted prevalence of all types of cataract was 40% (RR: 0.62; CI: 0.12–1.77) and 80% (RR: 0.16; CI: 0.04–0.82, Fig. 6a) lower for persons with moderate and high antioxidant index scores (based on combined plasma vitamin C, vitamin E, and carotenoid levels) as compared with persons with low scores. Using a similar index based on combined antioxidant nutrient intakes (vitamin C, vitamin E, and carotene, as well as riboflavin), Leske and co-workers (1991) found that persons with high scores had 60% lower adjusted prevalence of cortical (RR: 0.42; CI: 0.18–0.97, Fig. 6c) and mixed (RR: 0.39; CI: 0.19–0.80, Fig. 6e) cataract compared to those who had low scores. However, Robertson and co-workers (1989) found no enhanced benefit to persons taking both vitamin E and vitamin C supplements compared with persons who only took either vitamin C or vitamin E. Mohan and co-workers (1989) constructed a somewhat more complex antioxidant scale that included red blood cell levels of glutathione peroxidase, glucose-6-phosphate dehydrogenase and plasma levels of vitamin C and vitamin E. Even though they failed to see any protective associations with any of these individual factors, and even reported a positive association between plasma vitamin C and prevalence of cataract, they found that persons with high antioxidant index scores had a substantially lower prevalence of cataracts involving the posterior subcapsular region (RR: 0.23, CI: 0.06–0.88, Fig. 6d) or mixed cataract with posterior subcapsular and nuclear components (RR: 0.12; CI: 0.03–0.56, Fig. 6e) after multivariate adjustment. Hankinson and co-workers (1992a) calculated an antioxidant score based on intakes of carotene, vitamin C, vitamin E, and riboflavin and observed a 24% reduction in the adjusted rate of cataract surgery among women with high-antioxidant scores relative to women with low scores (RR: 0.76; CI: 0.57–1.03, Fig. 6f). Knekt and co-workers (1992) observed that the rate of cataract surgery for persons with high levels of both serum vitamin E and β-carotene concentrations appeared lower than the rate for persons with either high vitamin E or high β-carotene levels. Persons with high serum levels of either nutrient had a rate of cataract surgery that was 40% less than persons with low levels of both nutrients (RR: 0.38; CI: 0.15–1.0, Fig. 6f). Vitale and co-workers (1994) also examined the relationship between antioxidant scores (based on plasma concentrations of vitamin C, vitamin E, and β-carotene) and prevalence of cataract, but did not see evidence of any association. The age-, sex-, and diabetes-adjusted risk ratios were close to one for both nuclear (RR: 0.96; CI: 0.54–1.70, Fig. 6b) and cortical (RR: 1.17; CI: 0.62–2.20, Fig. 6c) cataract. Relationships between multiple antioxidant nutrients and cataract risk are further supported by multivitamin use data. Leske and co-workers (1991) found that the use of multivitamin supplements was associated with a decreased prevalence for each type of cataract: 60, 48, 45, and 30%, respectively, for posterior subcapsular (RR: 0.40; CI: 0.21–0.77, Fig. 6d), cortical (RR: 0.52; CI: 0.36–0.72, Fig. 6c), nuclear (RR: 0.55; CI: 0.33–0.92, Fig. 6b), and mixed (RR: 0.70; CI: 0.51–0.97, Fig. 6e) cataracts. Seddon and co-workers (1994) also observed a reduced risk for incident cataract for users of multivitamins (RR: 0.73; CI: 0.54–0.99, Fig. 6f).

F. Intervention Trials

To date, only one intervention trial designed to assess the effect of vitamin supplements on cataract risk has been completed. Sperduto and co-workers (1993) took advantage of two ongoing, randomized, double-blinded vitamin and cancer trials to assess the impact of vitamin supplements on cataract prevalence. The trials were conducted among almost 4000 participants ages 45 to 74 years from rural communes in Linxian, China. Participants in one trial received either a multinutrient supplement or a placebo. In the second trial, a more complex factorial design was used to evaluate the effects of four different vitamin/mineral combinations: retinol (5000 IU) and zinc (22 mg); riboflavin (3 mg) and niacin (40 mg); vitamin C (120 mg) and molybdenum (30 μg); and vitamin E (30 mg), β-carotene (15 mg), and selenium (50 μg). At the end of the 5- to 6-year follow-up, the investigators conducted eye examinations to determine the prevalence of cataract.

In the first trial there was a significant 43% reduction in the prevalence of nuclear cataract for persons aged 65 to 74 years for persons receiving the multinutrient supplement (RR: 0.57; CI: 0.36–0.90). The second trial demonstrated a significantly reduced prevalence of nuclear cataract in persons receiving the riboflavin/niacin supplement relative to those persons not receiving this supplement (RR: 0.59; CI: 0.45–0.79). The effect was strongest in those aged 65 to 74 years (RR: 0.45; CI: 0.31–0.64). However, the riboflavin/niacin supplement appeared to increase the risk of posterior subcapsular cataract (RR: 2.64; CI: 1.31–5.35). The results further suggested a protective effect of the retinol/zinc supplement (RR: 0.77; CI: 0.58–1.02) and the vitamin C/molybdenum supplement (RR: 0.78; CI: 0.59–1.04) on prevalence of nuclear cataract.

VI. Conclusions

Age-related eye diseases can be devastating in terms of quality of life of our most frail and with respect to national public health and economics. The perils of oxidative and photo-oxidative insult, due in part to smoking, cannot be ignored. The impression one gets from examining the work done to date is that nutrient intake can be optimized to delay cataract. However, it is too early to declare that increased consumption or intake of specific levels of nutrients is associated with a diminished risk of cataract at any one location in the eye. Optimization of nutriture can be achieved through better diets and, perhaps, with the aid of supplements once appropriate levels of specifically beneficial nutrients are defined. This remains an important objective. In addition to quantifying optimal intake, it is essential to know for how long or when would intake of the nutrient be useful with respect to delaying cataract. The answer to these questions is almost at hand. Two studies indicated that persons who consumed supplements of ascorbate for more than 10 years have a decreased risk for cataract or cataract extraction. However, these results must be corroborated by other research, including quantitative information regarding the quantity and the term of intake of other nutrients as well. Epidemiological data also indicate that it is imperative to execute longitudinal studies and possibly intervention trials. Monitoring correlations between nutrient status and intermediate markers of cataract (when they are defined) may allow anticipation of relationships between nutrient status and cataract.

Acknowledgments

We acknowledge the assistance of Dr. Xin Gong in the preparation of figures and Ms. E. Epstein for preparation of the manuscript. This work was supported in part by NEI Grant EY08566, USDA Contract 53-3K06-0-1, Henkel Corporation, and Hoffmann-La Roche Inc.

References

Berger, J., Shepard, D., Morrow, F., Sadowski, J., Haire, T., and Taylor A. (1988). Reduced and total ascorbate in guinea pig eye tissues in response to dietary intake. *Curr. Eye Res.* **7**, 681–686.

Berger, J., Shepard, D., Morrow, F., and Taylor, A. (1989). Relationship between dietary intake and tissue levels of reduced and total vitamin C in the guinea pig. *J. Nutr.* **119**, 1–7.

Berman, E. R. (1991). *In* "Biochemistry of the Eye," pp. 210–308. Plenum Press, New York.

Blondin, J., Baragi, V. J., Schwartz, E., Sadowski, J., and Taylor, A. (1986). Delay of UV-induced eye lens protein damage in guinea pigs by dietary ascorbate. *Free Rad. Biol. Med.* **2**, 275–281.

Blondin, J., and Taylor, A. (1987). Measures of leucine aminopeptidase can be used to anticipate UV-induced age-related damage to lens proteins: Ascorbate can delay this damage. *Mech. Ageing Dev.* **41**, 39–46.

Brilliant, L. B., Grasset, N. C., Pokhrel, R. P., Kolstad, A., and Lepkowski, J. M. (1983). Associations among cataract prevalence, sunlight hours, and altitude in the Himalayas. *Am. J. Epidemiol.* **118**, 250–264.

Bunce, G. E., Kinoshita, J., and Horwitz, J. (1990). Nutritional factors in cataract. *Annu. Rev. Nutr.* **10**, 233–254.

Burton, W., and Ingold, K. U. (1984). Beta-carotene: An unusual type of lipid antioxidant. *Science* **224**, 569–573.

Burton, G. W., Wronska, U., Stone, L., Foster, D. O., and Ingold, K. U. (1990). Biokinetics of dietary RRR-α-tocopherol in the male guinea pig at three dietary levels of vitamin C and two levels of vitamin E: Evidence that vitamin C does not "spare" vitamin E in vivo. *Lipids* **25**, 199–210.

Chan, C. W., and Billson, F. A. (1991). Visual disability and major causes of blindness in NSW: A study of people aged 50 and over attending the Royal Blind Society 1984 to 1989. *Aust. N. Z. J. Ophthalmol.* **19**, 321–325

Chatterjee, A., Milton, R. C., and Thyle, S. (1982). Prevalence and etiology of cataract in Punjab. *Brit. J. Ophthalmol.* **66**, 35–42.

Christen, W. G., Manson, J. E., Seddon, J. M., Glynn, R. J., Buring, J. E., Rosner, B., and Hennekens, C. H. (1992). A prospective study of cigarette smoking and risk of cataract in men. *JAMA* **268**, 989–993.

Chylack, L. T., Jr., Wolfe, J. K., Friend, J., Singer, D. M., Wu, S. Y., and Leske, M. C. (1994). Nuclear cataract: Relative contributions to vision loss of opalescence and brunescence. *Invest. Ophthalmol. Vis. Sci.* **35**, 1824. [Abstract]

Chylack, L. T., Jr., Wolfe, J. K., Singer, D. M., Leske, M. C., Bullimore, M. A., Bailey, I. L., Friend, J., McCarthy, D., and Wu, S. (1993). The lens opacities classification system III. *Arch. Ophthalmol.* **111**, 831–836.

Costagliola, C., Iuliano, G., Menzione, M., Rinaldi, E., Vito, P., and Auricchio, G. (1986). Effect of vitamin E on glutathione content in red blood cells, aqueous humor and lens of humans and other species. *Exp. Eye Res.* **43**, 905–914.

Daicker, B., Schiedt, K., Adnet, J. J., and Bermond, P. (1987). Canthaxamin retinopathy: An investigation by light and electron microscopy and physiochemical analyses. *Graefe's Arch. Clin. Exp. Ophthalmol.* **225**, 189–197.

Dana, M. R., Tielsch, J. M., Enger, C., Joyce, E., Santoli, J. M., and Taylor, H. R. (1990). Visual impairment in a rural Appalachian community: Prevalence and causes. *JAMA* **264**, 2400–2405.

Di Mascio P., Murphy, M. E., and Sies, H. (1991). Antioxidant defense systems: The role of carotenoids, tocopherols and thiols. *Am. J. Clin. Nutr.* **53**, 194S–200S.

Erdman, J. (1988). The physiologic chemistry of carotenes in man. *Clin. Nutr.* **7**, 101–106.

Frei, B., Stocker, R., and Ames, B. N. (1988). Antioxidant defenses and lipid peroxidation in human blood plasma. *Proc. Natl. Acad. Sci. U.S.A.* **85**, 9748–9752.

Fridovich, I. (1984). Oxygen: Aspects of its toxicity and elements of defense. *Curr. Eye Res* **3**, 1–2.

Garland, D. D. (1991). Ascorbic acid and the eye. *Am. J. Clin. Nutr.* **54**, 1198S–1202S.

Giblin, F. J., McReady, J. P., and Reddy, V. N. (1992). The role of glutathione metabolism in detoxification of H_2O_2 in rabbit lens. *Invest. Ophthalmol. Vis. Sci.* **22**, 330–335.

Hankinson, S. E., Stampfer, M. J., Seddon, J. M., Colditz, G. A., Rosner, B., Speizer, F. E., and Willett, W. C. (1992a). Nutrient intake and cataract extraction in women: A prospective study. *Br. Med. J.* **305**, 335–339.

Hankinson, S. E., Willett, W. C., Colditz, G. A., Seddon, J. M., Rosner, B., Speizer, F. E., and Stamper, M. J. (1992b). A prospective study of cigarette smoking and risk of cataract surgery in women. *JAMA* **268**, 994–998.

Hirvela, H., Luukinen, H., and Laatikainen, L. (1995). Prevalence and risk factors of lens opacities in the elderly in Finland: A population-based study. *Ophthalmology* **102**, 108–117.

Huang, L. L., Jahngen-Hodge, J., and Taylor, A. (1993). Bovine lens epithelial cells have a ubiquitin-dependent proteolysis system. *Biochim. Biophys. Acta* **1175**, 181–187.

Jacques, P. F., and Chylack, L. T., Jr. (1991). Epidemiologic evidence of a role for the antioxidant vitamins and carotenoids in cataract prevention. *Am. J. Clin. Nutr.* **53**, 352S–355S.

Jacques, P. F., Chylack, L. T., Jr., and Taylor A. (1994). Relationships between natural antioxidants and cataract formation. *In* "Natural Antioxidants in Human Health and Disease" (B. Frei, ed.), pp. 513–533. Academic Press, Orlando.

Jacques, P. F., Lahav, M., Willett, W. C., and Taylor, A. (1992). Relationship between long-term vitamin C intake and prevalence of cataract and macular degeneration. *Exp. Eye Res.* (Suppl. 1) 55, S152. [Abstract]

Jacques, P. F., and Taylor, A. (1991). Micronutrients and age-related cataracts. *In* "Micronutrients in Health and in Disease Prevention" (A. Bendich and C. E. Butterworth, eds.), pp. 359–379. Dekker, New York.

Jacques, P., Taylor, A., Lahav, M., Lee, Y., Hankinson, S., and Willett, W. (1996). Associations between risk for cataract and long-term ascorbate supplementation. *Invest. Ophthal. Vis. Sci.* 37, S236. [Abstract]

Jahngen-Hodge, J., Cyr, D., Laxman, E., and Taylor, A. (1992). Ubiquitin and ubiquitin conjugates in human lens. *Exp. Eye. Res.* 55, 897–902.

Jahngen-Hodge, J., Laxman, E., Zuliani, A., and Taylor A. (1991). Evidence for ATP ubiquitin-dependent degradation of proteins in cultured bovine lens epithelial cells. *Exp. Eye Res.* 52, 341–347.

Klein, B. E. K., Klein, R., and Linton, K. L. P. (1992). Prevalence of age-related lens opacities in a population: The Beaver Dam Eye Study. *Ophthalmology* 99, 546–552.

Klein, R., Klein, B. E., Linton, K. L., and DeMets, D. L. (1993). The Beaver Dam eye study: The relation of age-related maculopathy to smoking. *Am. J. Epidemiol.* 137, 190–200.

Knekt, P., Heliovaara, M., Rissanen, A., Aromaa, A., and Aaran, R. (1992). Serum antioxidant vitamins and risk of cataract. *Br. Med. J.* 305, 1392–1394.

Krinsky, N. I., and Deneke, S. S. (1982). Interaction of oxygen and oxy-radicals with carotenoids. *J. Natl. Cancer. Inst.* 69, 205–210.

Kupfer, C. (1984). The conquest of cataract: A global challenge. *Trans. Ophthal. Soc. UK* 104, 1–10.

Kwan, M., Niinikoski, J., and Hunt, T. K. (1972). *In vivo* measurement of oxygen tension in the cornea, aqueous humor, and the anterior lens of the open eye. *Invest. Ophthalmol.* 11, 108–114.

Leibowitz, H., Krueger, D., Maunder, C., Milton, R. C., Mohandas, M. K., Kahn, H. A., Nickerson, R. J., Pool, J., Colton, T. L., Ganley, J. P., Loewenstein, J. I., and Dawber, T. R. (1980). The Framingham Eye Study Monograph. *Surv. Ophthalmol.* (Suppl), 24, 335–610.

Leske, M. C., Chylack, L. T., Jr., and Wu, S. (1991). The lens opacities case-control study risk factors for cataract. *Arch. Ophthalmol.* 109, 244–251.

Levine, M. (1986) New concepts in the biology and biochemistry of ascorbic acid. *N. Engl. J. Med.* 314, 892–902.

Libondi, T., Menzione, M., and Auricchio, G. (1985). *In vitro* effect of alpha-tocopherol on lysophosphatidylcholine-induced lens damage. *Exp. Eye Res.* 40, 661–666.

Machlin, L. J., and Bendich, A. (1987). Free radical tissue damage: Protective role of antioxidant nutrients. *FASEB J.* 1, 441–445.

Mares-Perlman, J. A., Klein, B. E. K., Klein, R., and Ritter, L. L. (1994). Relationship beween lens opacities and vitamin and mineral supplement use. *Ophthalmology* 101, 315–355.

Mares-Perlman, J. A., Brady, W. E., Klein, B. E. K., Klein, R., Haus, G. J., Palta, M., Ritter, L. L., and Shoff, S. M. (1995a). Diet and nuclear lens opacities. *Am. J. Epidemiol.* 141, 322–334.

Mares-Perlman, J. A., Brady, W. E., Klein, B. E. K., Klein, R., Palta, M., Bowen, P., and Stacewicz-Sapuntzakis, M. (1995b). Serum carotenoids and tocopherols and severity of nuclear and cortical opacities. *Invest. Ophthalmol. Vis. Sci.* 36, 276–288.

Meister, A. (1994) Glutathione-ascorbic acid antioxidant system in animals. *J. Biol. Chem.* 269, 9397–9400.

Micozzi, M. S., Beecher, G. R., Taylor, H. R., and Khachik, F. (1990). Carotenoid analyses of selected raw and cooked foods associated with a lower risk for cancer. *J. Natl. Cancer Inst.* 82, 282–285.

Mohan, M., Sperduto, R. D., Angra, S. K., Milton, R. C., Mathur, R. L., Underwood, B., Jafery, N., and Pandya, C. B. (1989). India–US case-control study of age-related cataracts. *Arch. Ophthalmol.* **107**, 670–676.

Mune, M., and Meydani, M. (1995). Effect of calorie restriction on liver and kidney glutathione in aging Emory mice. *Age* **18**, 43–49.

Naraj, R. M., and Monnier, V. M. (1992). Isolation and characterization of a blue fluorophore from human eye lens crystallins: *In vitro* formation from Maillard action with ascorbate and ribose. *Biochim. Biophys. Acta* **1116**, 34–42.

Obin, M. S., Nowell, T., and Taylor, A. (1994). The photoreceptor G-protein transducin (G_t) is a substrate for ubiquitin-dependent proteolysis. *Biochem. Biophys. Res. Comm.* **200**, 1169–1176.

Pau, H., Graf, P., and Sies, H. (1990). Glutathione levels in human lens; regional distribution in different forms of cataract. *Exp. Eye Res.* **50**, 17–20.

Rao, C. M., Qin, C., Robison, W. G., Jr., and Zigler, J. S., Jr. (1995). Effect of smoke condensate on the physiological integrity and morphology of organ cultured rat lenses. *Curr. Eye Res.* **14**, 295–301.

Rathbun, W. B., Holleschau, A. M., Murray, D. L., Buchanan, A., Sawaguchi, S., and Tao, R. V. (1990). Glutathione synthesis and glutathione redox pathways in naphthalene cataract in the rat. *Curr. Eye Res.* **9**, 45–53.

Robertson, J., McD., Donner, A. P., and Trevithick, J. R. (1989). Vitamin E intake and risk for cataracts in humans. *Ann. N.Y. Aced. Sci.* **570**, 372–382.

Russell-Briefel, R., Bates, M. W., and Kuller, L. H. (1985). The relationship of plasma carotenoids to health and biochemical factors in middle-aged men. *Am. J. Epidemiol.* **122**, 741–749.

Salive, M. E., Guralnik, J., Christian, W., Glynn, R. J., Colsher, P., and Ostfeld, A. M. (1992). Functional blindness and visual impairment in older adults from three communities. *Ophalmology* **99**, 1840–1847.

Schectman, G., Byrd, J. C., and Gruchow, H. W. (1989). The influence of smoking on vitamin C status in adults. *Am. J. Health* **79**, 158–162.

Schwab, L. (1990). Cataract blindness in developing nations. *Int. Ophthalmol. Clin.* **30**, 16–18.

Seddon, J. M., Christen, W. G., Manson, J. E., LaMotte, F. S., Glynn, R. J., Buring, J. E., and Hennekens, C. H. (1994). The use of vitamin supplements and the risk of cataract among US male physicians. *Am. J. Public Health* **84**, 788–792.

Shalini, V. K., Luthra, M., Srinivas, L., Rao, S. H., Basti, S., Reddy, M., and Balasubramanian, D. (1994). Oxidative damage to the eye lens caused by cigarette smoke and fuel smoke condensates. *Ind. J. Biochem. Biophys.* **31**, 261–266.

Shang, F., and Taylor, A. (1995). Oxidative stress and recovery from oxidative stress are associated with altered ubiquitin conjugating and proteolytic activities in bovine lens epithelial cells. *Biochem. J.* **307**, 297–303.

Sperduto, R. D., Hu, T.-S., Milton, R. C., Zhao, J., Everett, D. F., Cheng, Q., Blot, W. J., Bing, L., Taylor, P. R., Jun-Yao, L., Dawsey, S., and Guo, W. (1993). The Linxian Cataract Studies: Two nutrition intervention trials. *Arch. Ophthalmol.* **111**, 1246–1253.

Taylor, A. (1992). Vitamin C. *In* "Nutrition in the Elderly: The Boston Nutritional Status Survey" (S. C. Hartz, R. M. Russell, and I. H. Rosenberg, eds.), pp. 147–50. Smith Gordon Limited, London.

Taylor, A. (1993). Cataract: Relationships between nutrition and oxidation. *J. Am. Coll. Nutr.* **12**, 138–146.

Taylor, A., and Davies, K. J. A. (1987). Protein oxidation and loss of protease activity may lead to cataract formation in the aged lens. *Free Rad. Biol. Med.* **3**, 371–377.

Taylor, A., Jacques, P., Lahav, M., Hankinson, S., Lee, Y., and Willett, W. (1994). Relationship between long-term dietary and supplement ascorbate intake and risk of cataract. *Exp. Eye Res.* Suppl. 1, 59, S133. [Abstract]

Taylor, A., Jacques, P. F., and Dorey, C. K. (1993). Oxidation and aging: Impact on vision. *J. Toxicol. Indust. Health* **9**, 349–371.

Taylor, A., Jacques, P. F., Nadler, D., Morrow, F., Sulsky, S. I., and Shepard, D. (1991). Relationship in humans between ascorbic acid consumption and levels of total and reduced ascorbic acid in lens, aqueous humor, and plasma. *Curr. Eye Res.* **10**, 751–759.

Taylor, H. R., West, S. K., Rosenthal, F. S., Newland, H. S., Abbey, H., and Emmett, E. A. (1988). Effect of ultraviolet radiation on cataract formation. *N. Engl. J. Med.* **319**, 1429–1433.

The Italian-American Cataract Study Group. (1991). Risk factors for age-related cortical, nuclear, and posterior subcapsular cataracts. *Am. J. Epidemiol.* **133**, 541–553.

Varma, S. D., Chand, O., Sharma, Y. R., Kuck, J. F., and Richards, K. D. (1984). Oxidative stress on lens and cataract formation: Role of light and oxygen. *Curr. Eye Res.* **3**, 35–57.

Vitale, S., West, S., Hallfrisch, J., Alston, C., Wang, F., Moorman, C., Muller, D., Singh, V., and Taylor, H. R. (1994). Plasma antioxidants and risk of cortical and nuclear cataract. *Epidemiology,* **4**, 195–203.

Wang, G.-M., Spector, A., Luo, C.-Q., Tang, L. Q., Xu, L. H., Guo, W. Y., and Huang, Y. Q. (1990). Prevalence of age-related cataract in Ganzi and Shanghai: The Epidemiological Study Group. *Chinese Med. J.* **103**, 945–951.

Wefers, H., and Sies, H. (1988). The protection by ascorbate and glutathione against microsomal lipid peroxidation is dependent on vitamin E. *FEBS Lett.* **174**, 353–357.

West, S. (1992). Does smoke get in your eyes? *JAMA* **268**, 1025–1026.

West, S. K., Munoz, B., Emmett, E. A., and Taylor, H. R. (1992). Cigarette smoking and risk of nuclear cataracts. *Arch. Ophthalmol.* **107**, 1166–1169.

West, S. K., and Valmadrid, C. T. (1995). Epidemiology of risk factors for age-related cataract. *Surv. Ophthalmol.* **39**, 323–334.

Whitfield, R., Schwab, L., Ross-Degnan, D., Steinkuller, P., and Swartwood, J. (1990). Blindness and eye disease in Kenya: Ocular status survey results from the Kenya Rural Blindness Prevention Project. *Brit. J. Ophthalmol.* **74**, 333–340.

Wolfe, J. K., Chylack, L. T., Jr., Leske, M. C., Wu, S. Y., and LSC Group. (1993). Lens nuclear color and visual function. *Invest. Ophthalmol. Vis. Sci.* 34, 1223. [Abstract]

Wong, L., Ho, S. C., Coggon, D., Cruddas, A. M., Hwang, C. H., Ho, C. P., Robertshaw, A. M., and MacDonald, D. M. (1993). Sunlight exposure, antioxidant status, and cataract in Hong Kong fishermen. *J. Epidemiol. Community Health* 47, 46–49.

World Health Organization. (1991). Use of intraocular lenses in cataract surgery in developing countries. *Bull. World Health Organ.* **69**, 657–666.

Wormald, R. P. L., Wright, L. A., Courtney, P., Beaumont, B., and Haines, A. P. (1992). Visual problems in the elderly population and implications for services. *BMJ* **304**, 1226–1229.

Yeum, K.-J., Taylor, A., Tang, G., and Russell, R. M. (1995). Measurement of carotenoids, retinoids, and tocopherols in human lenses. *Invest. Ophthalmol. Vis. Sci.* **36**, 2756–2761.

Zigler, J. S., and Goosey, J. D. (1984). Singlet oxygen as a possible factor in human senile nuclear cataract development. *Curr. Eye Res.* **3**, 59–65.

Zigman, S. (1983). Effects of near ultraviolet radiation on the lens and retina. *Doc. Ophthalmol.* **55**, 375–391.

Zigman, S., Datiles, M., and Torczynski, E. (1979). Sunlight and human cataract. *Invest. Ophthalmol. Vis. Sci.* **18**, 462–467.

Zigman, S., McDaniel, T., Schultz, J. B., Reddan, J., and Meydani, M. (1995). Damage to cultured lens epithelial cells of squirrels and rabbits by UV-A (99.9%) plus UV-B (0.1%) radiation and alpha tocopherol protection. *Mol. Cell. Biochem.* **143**, 35–46.

John T. Landrum*
Richard A. Bone†
Mark D. Kilburn*

Departments of *Chemistry and †Physics
Florida International University
Miami, Florida 33199

The Macular Pigment: A Possible Role in Protection from Age-Related Macular Degeneration

I. Introduction

The macula is the anatomical region of the retina that is responsible for central vision. Centered on the fovea, where the visual axis meets the retina, it extends radially outward to a distance of about 2.75 mm (Davson, 1990). The macula is divided into the inner macula and the outer macula. The inner macula extends radially out to a distance of 1.5 mm whereas the outer macula is defined by the surrounding annular ring. The central 2 to 3 mm of the fovea is easily recognizable due to its yellow coloration which results from the presence of the "macular pigment," a feature found only in the primate retina. Despite its small size, this region of the retina projects onto a disproportionately large area of the visual cortex. Coupled with a very high density of cone photoreceptors, this endows the macula with the highest degree of visual acuity. It is therefore not surprising that considerable effort is devoted to understanding and, when possible, treating diseases

537

which disrupt the normal functioning of the macula. One such disease is age-related macular degeneration (AMD) which occurs in 19.7% of the population above the age of 65 (Kahn *et al.*, 1977; Hyman, 1992). It is the leading cause of visual impaiment in the United States and is an irreversible condition. Factors with an apparent positive correlation with AMD include light skin color, light iris color, high exposure to ambient light, low levels of dietary xanthophylis, and low levels of serum xanthophylls (Hyman, 1992; Cruickshanks *et al.*, 1993; Eye Diseases Case-Control Study Group, 1993; Seddon *et al.*, 1994).

This chapter briefly reviews the current state of knowledge of the macular pigment, AMD, and the possible link between them and presents data which indicate that lower than normal amounts of macular pigment are found in persons with AMD (Landrum *et al.*, 1995). This suggests that the macular pigment may play a protective role in the eye. This chapter also reports the preliminary results of a study designed to determine whether the amount of macular pigment can be enhanced by dietary supplementation with carotenoids.

II. The Macular Pigment

The macula is composed of several functionally and structurally distinct layers overlaying and supporting the cone and rod photoreceptors found in this small region. Microspectrophotometry of transverse sections of the macula of the monkey retina has revealed the location of the macular pigment (Snodderly *et al.*, 1984). It is found mainly in the photoreceptor axons—the Henle fiber layer—of the macula, with a lesser amount in the inner plexiform layer in many eyes. Evidence that the same situation exists in the human eye has been reported (Bone and Landrum, 1984) based on the dichroic properties of the xanthophylls (Bone and Landrum, 1983). Because both of the pigmented layers are anterior to the light-sensitive photoreceptor outer segments, the macular pigment represents a color filter through which light must pass before detection. The optical density at the 460-nm wavelength maximum of this filter varies in normal individuals from zero to over 1.0 (Bone and Sparrock, 1971; Werner *et al.*, 1987). This variation results in a 10-fold range in the possible exposure of the underlying tissues to blue light.

Wald (1945) found that pooled extracts of the macular pigment have a carotenoid-like visible absorbance spectrum which appeared to match that of the xanthophyll lutein. In 1963 Brown and Wald suggested that a mixture of lutein and its *cis* isomers better matched the visible spectrum. Our own interest in the macular pigment in the early 1980s led us to show, using high-performance liquid chromatography (HPLC), that it consisted of the xanthophyll isomers, lutein and zeaxanthin (Bone *et al.*, 1985). This result has since been corroborated by Handelman *et al.* (1988). We were later able to obtain unequivocal identification using mass spectrometry (Bone *et al.*, 1993). By concentrically sectioning the retina around the fovea and analyzing for the two xanthophylls by HPLC, we have been able to show how their distribution varies with eccentricity (Bone *et al.*, 1988; Bone and Landrum, 1992). The chromatograms in Fig. 1 show that zeaxanthin is the predominant pigment in the center of the macula, usually being about twice as abundant as lutein. The concentration of both isomers diminishes radially away from the center of the macula and lutein becomes increasingly more dominant. In the outer retina, at distances exceeding about 1 cm from the fovea, lutein is about twice as abundant as zeaxanthin. The results of Handelman *et al.* (1988) are consistent with these observations.

More recent work in our laboratory has shown that the zeaxanthin component found in the human retina is itself composed of all three of the possible stereoisomers shown, together with lutein, in Fig. 2 (Bone *et al.*, 1993). One of these, SSZ, is present only as a relatively small component (Bone *et al.*, 1993). RRZ is of dietary origin, being relatively abundant in many foods, especially corn, whereas RSZ (or meso-Z) and SSZ are not common in the diet (Maoka *et al.*, 1986) and have yet to be detected in human serum despite considerable effort. Mapping studies of these stereoisomers have shown that relative to RRZ, the concentration of RSZ decreases with eccentricity, having its maximum value in the center of the macula (Bone *et al.*, 1994). This decrease in RSZ closely corresponds to the increase in lutein, relative to RRZ, with eccentricity. This can be seen in the chromatograms of Figs. 1 and 3. It has been suggested that the presence of RSZ may be the result of isomerization of lutein to RSZ by an enzyme (Bone *et al.*, 1993). The functional significance of this isomer distribution is not known.

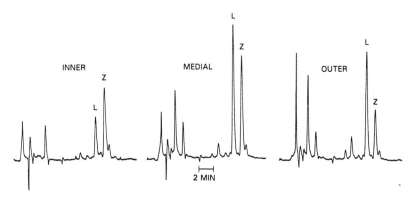

FIGURE 1   Reversed-phase HPLC chromatograms of extracts from the three concentric areas of the retina. Inner: a disk centered on the fovea obtained with a 3-mm trephine; medial: an annulus obtained with 3- and 11-mm trephines; and outer: an annulus obtained with 11- and 21-mm trephines. L, lutein: Z, zeaxanthin.

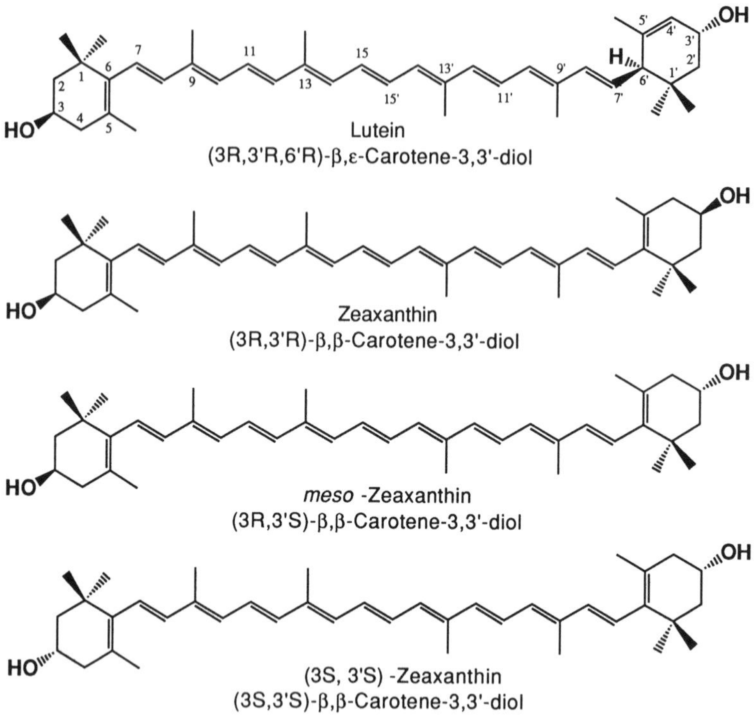

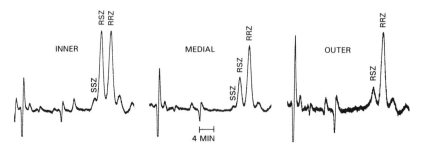

HO
OH
H
7
11
15
13'
9'
2
1
6
9
13
3
15'
11'
7'
6'
1'
2'
3'
4'
5'
4
5
Lutein
(3R,3'R,6'R)-β,ε-Carotene-3,3'-diol
Zeaxanthin
(3R,3'R)-β,β-Carotene-3,3'-diol
meso -Zeaxanthin
(3R,3'S)-β,β-Carotene-3,3'-diol
(3S, 3'S) -Zeaxanthin
(3S,3'S)-β,β-Carotene-3,3'-diol

FIGURE 2 Stereochemical structures of the macular pigment components. The 3′ hydroxy groups on lutein and *meso*-zeaxanthin have the same absolute configuration, making interconversion possible by a movement of the 4′–5′double bond (lutein) to the 5′–6′ position (*meso*-zeaxanthin).

INNER
SSZ
RSZ
RRZ
MEDIAL
SSZ
RSZ
RRZ
OUTER
RSZ
RRZ
4 MIN

FIGURE 3 Normal-phase HPLC chromatograms of dicarbamate derivatives of zeaxanthin collected during reversed-phase chromatography of extracts from the retina (see Fig. 1). The normal-phase column separates the diastereomeric zeaxanthin derivatives from each other: SSZ, (3S,3′S)-zeaxanthin; RSZ, (3R,3′S)-zeaxanthin; RRZ, (3R,3′R)-zeaxanthin.

Animals are incapable of *de novo* synthesis of carotenoids; carotenoids in animal tissues can be traced to dietary origins (Weedon, 1971; Goodwin, 1992). Malinow *et al.* (1980) showed that monkeys fed a xanthophyll-free diet for several years had undetectable levels of macular pigment. Through measurements on monozygotic twins, Hammond *et al.* (1995a) demonstrated that the amount of macular pigment in an individual is not completely determined by genetics. Hammond *et al.* (1995b) have also found that cigarette smoking is correlated with reduced amounts of macular pigment. These findings suggest the possibility of modifying the amount of macular pigment, either through dietary supplementation or through a change of life-style.

The function of the macular pigment has not been unequivocally determined. In the past it was proposed that one function might be to reduce the adverse effects of chromatic aberration in the ocular media, thereby increasing acuity (Walls, 1967; Reading and Weale, 1974). Currently, a more generally held view is that the pigment probably acts in a protective capacity against the damaging effects of blue light (Ditchburn, 1973; Kirschfeld, 1982; Bone *et al.*, 1984). This essential role of carotenoids to protect against photo-oxidation in photosynthetic plants by blue light has been long established (Krinsky, 1968, 1971, 1979). Ultraviolet light, also potentially damaging to the retina, is effectively attenuated through absorption by the ocular media. Blue light can induce the formation of reactive radicals, triplet excited states, superoxide, and singlet oxygen within the retina, specifically the choriocapillaris, Bruch's membrane, and the retinal pigment epithelium (RPE) (Gottsch *et al.*, 1990). The formation of such species may be greatly reduced in individuals having a high level of macular pigmentation. The xanthophylls which comprise the macular pigment are effective quenchers of excited triplet states and are reactive toward singlet oxygen and radicals (Foote *et al.*, 1970). They may serve actively to protect the nerve tissue of the macula in which they are incorporated from the degradative effects of these species. They may also serve passively by shielding those tissues posterior to the outer plexiform layer from excessive blue light. These include those tissues which are most adversely affected in AMD: the photoreceptors, Bruch's membrane, and the RPE.

III. Age-Related Macular Degeneration

AMD is a progessive disease involving the RPE–photoreceptor complex and is identified by a group of clinically observable characteristics (Bird, 1992). The disease is broadly classified into two forms: the dry, atrophic and the wet, neovascular (Bird, 1992; Kincaid, 1992). The disease develops gradually over a period of many years with loss of sight being the ultimate result. Vision may be little impaired in the early dry form because the degenerative process is restricted largely to the outer macular region and to the RPE with little effect on the photoreceptors (Sarks and Sarks, 1994; Klein et al., 1995) The disease may progress from the dry atrophic to the wet, neovascular form. Neovascularization from the choroid results in diminished visual acuity (Sarks, 1976).

The nature and severity of these anatomical changes in the macula are typically diagnosed through the use of a variety of fundus photographic techniques (Arnold et al., 1995). A uniform grading scale has been proposed for the diagnosis of AMD and includes identification of the extent, number, type, and size of observed drusen, the presence of choroidal neovascularization (CNV), and geographic atrophy (GA) (Bird et al., 1995). Drusen are observed as spots of higher than normal reflectivity in the fundus photograph and vary in size from 30 to 50 μm (the limit of detection) to over 125 μm. They occupy the region between the RPE and choriocapillaris, and range in thickness from about 10–20 μm for observable hard drusen to 85 μm for clusters and soft drusen. The thickening of the drusen results in thinning of the overlying RPE, RPE detachment from Bruch's membrane, degeneration of the RPE, and formation of a neovascular membrane (Sarks et al., 1994). Geographic atrophy of the RPE often evolves after the failure of the RPE (Bird, 1992). Drusen have an unusually high lipid content and it has been suggested that they form when lipofuscin builds up in the RPE and just below it (Feeney-Burns and Ellersieck, 1985; El Baba et al., 1986). Lipufuscin is the indigestable, fluorescent catabolic product of lipid oxidation resulting from the lifelong shedding of photoreceptor end segments (Weiter et al., 1986). It has been postulated that exposure of the retina to excessive blue light may increase the rate of lipofuscin formation (Feeney-Burns et al., 1990; Gottsch et al., 1990). A process that places oxidative stress on the neural retina would be expected to be a risk factor for early drusen formation and the subsequent damage to the RPE associated with AMD (Ham et al., 1984).

Our studies address two related but separate questions: (1) Do lower than normal levels of macular pigmentation represent a risk factor for the development of AMD? (2) Can dietary supplementation with lutein and/or zeaxanthin effectively change these pigment levels in the macula, thereby conferring additional protection against AMD, or slowing its development? This report includes data from two separate studies: a comparison of macular pigmentation in eyes of normal and AMD donors, and the effect of supplementation of lutein in the diet on macular pigmentation in human subjects.

IV. Methods

A. Analysis of Carotenoids from Eyes

I. Tissue Samples

Human donor eyes, both controls and those diagnosed with AMD, were obtained fom the National Disease Research Interchange (NDRI). Protocols required that they be enucleated within 12 hr of death and immediately fixed in formaldehyde. Each eye was dissected in 0.9% saline in order to remove the neural retina. The retina was supported on a 1-inch Lucite sphere and sectioned into a disk centered on the fovea and two concentric annuli using trephines of diameters 3, 11, and 21 mm.

2. Carotenoid Extraction and Analysis

Each tissue sample was ground in 2 ml of ethanol/deionized water (1 : 1) after the addition of 0.5 ml of an ethanol solution of monohexyl lutein ether (MHL) as an internal standard. This corresponded to an accurately known mass (ca. 12 ng) of the standard. MHL was prepared in a similar manner to that previously described for the monomethyl lutein ether (Jensen and Hertzberg, 1966). Our choice of MHL was based on a desire to use an internal standard with a suitable retention time and as structurally and chemically close to the analyte as possible. MHL was found to have the same stability as lutein on the chromatographic system.

The homogenate was transferred to a large culture tube after which the tissue grinder was rinsed with three 2-ml aliquots of the ethanol/water mixture, followed by two 5-ml aliquots of hexane; the rinses were added to the culture tube. The culture tube was placed in an ultrasonic bath for 1 min and then on a vortex mixer for 1 min. To facilitate separation of the aqueous phase from the xanthophyll-containing organic phase, about 1 ml of saturated sodium chloride solution was added to the culture tube followed by a further 10 sec on the vortex mixer. The mixture was then poured into centrifuge tubes and centrifuged for 3 min. Having transferred the upper (organic) layer to a pear-shaped flask with a syringe, the aqueous phase was returned to the culture tube and 10 ml of hexane, via the tissue grinder, was added. The mixture was again vortexed and centrifuged, and the organic layer was added to the pear-shaped flask. A third such procedure completed the extraction process. The hexane was evaporated to dryness under a stream of nitrogen and the solute was transferred to a 1-ml Wheaton vial using three 200-μl aliquots of ethanol to rinse the pear-shaped flask. The ethanol was evaporated to dryness with nitrogen prior to HPLC analysis.

HPLC was conducted on a reversed-phase system using a 250×2-mm column packed with Ultracarb 3 μm ODS (Phenomenex). The mobile phase was acetonitrile : methanol (85 : 15) with the addition of 0.1% triethylamine to minimize carotenoid degradation. The flow rate was 0.125 ml/min and detection was at 449 nm.

3. Stereoisomer Analysis

Zeaxanthin collected during HPLC was further analyzed, by modification of the methods of Rüttiman et al. (1983) and Schiedt et al. (1995), to determine its stereoisomer composition. The sample was dried under nitrogen in a siliconized microcentrifuge tube and transferred to a glove box containing a dry nitrogen atmosphere. Here it was dissolved in 50 μl of pyridine/benzene (1 : 1) to which was added 1 μl of (S)-$(+)$-1-[1-naphthyl]ethyl isocyanate. The reaction, which results in dicarbamate esters of zeaxanthin, was allowed to proceed at room temperature for 48 hr. To remove the excess isocyanate, 5 ml of hexane saturated with water was added 1 ml at a time to the sample tube which was shaken, placed in an ultrasonic bath, and the contents poured into a test tube. The isocyanate reacts with the water to produce a white precipitate. After adding an additional 5 ml of hexane and centrifuging, the hexane containing the zeaxanthin dicarbamate was removed with a syringe. The last step of adding hexane, centrifuging, and so on was repeated twice more. The hexane was then evaporated under a stream of nitrogen, and the solute was transferred and concentrated into the bottom of a clean microcentifuge tube.

HPLC of the derivatized samples was carried out on a normal-phase system using a 250 $\times$ 2-mm column packed with Prodigy 5 μm silica (Phenomenex). The mobile phase was isopropyl acetate/hexane (88 : 12) at a flow rate of 0.125 ml/min and detection was at 451 nm. Near baseline separation of the stereoisomers of zeaxanthin dicarbamates was obtained and their relative proportions were determined from peak areas.

B. Uptake of Lutein in Human Adults

I. Sampling

Two healthy adult males, ages 42 and 51, weighing 60 and 61 kg, respectively, ingested the equivalent of 30 mg of lutein per day in the form of lutein esters (source: marigold flowers) suspended in 2 ml of canola oil. The dose was taken daily in the early to midmorning after a normal light breakfast. The fasting serum lutein levels of both individuals were determined on the morning of the first dose as a measure of baseline. Blood samples were drawn at 2- to 3-hr intervals throughout the first day for both subjects and then daily for the next 3 days. Following the first week of supplementation, blood samples were drawn weekly. Blood was collected into a standard Vacutainer serum separator tube containing no anticoagulent. After allowing about 30 min for coagulation, the sample was centrifuged for 10 min and the serum was removed by pipette. Serum samples were stored at $-20°C$ prior to analysis.

2. Carotenoid Extraction

Carotenoids were extracted from the serum by a minor modification of widely used methods (Handelmann *et al.*, 1992; Guiliano *et al.*, 1993). Aliquots of serum (200 μl) were diluted with 2 ml of 50% ethanol/water to ensure the precipitation of protein components. Twenty microliters of an internal standard, monohexyl lutein ether, containing about 90 ng, was added to the solution at this point for quantification of the carotenoids by HPLC. This solution was extracted three times with 2-ml portions of hexane by vortexing the sample for 1 min followed by centrifuging for 5 min and pipetting off the hexane layer. The three portions of hexane were dried under a stream of nitrogen gas and stored under nitrogen at $-20°C$ until analysis was completed.

3. HPLC Analysis

Serum extracts were dissolved in 40 μl of ethanol prior to injection. Samples were vigorously agitated on a vortex mixer for 1 min to ensure dissolution of the sample. Two replicate analyses were carried out using 20-μl aliquots. Serum carotenoids were eluted with acetonitrile/methanol (85 : 15) at a flow rate of 1 ml/min through a 15 cm × 4.6 mm Adsorbosphere ODS 3-μm HS column (Alltech) coupled to a 25 cm × 4.6 mm Spherisorb ODS 5-μm column (Keystone Scientific) with detection of the carotenoids at 451 nm.

C. Macular Pigment Measurements

I. Heterochromatic Flicker Photometry

The optical density of the macular pigment was measured for each subject using the method of heterochomatic flicker photometry (Bone and Sparrock, 1971; Bone et al., 1992). The concentration of pigment in the macula is proportional to its optical density, and the actual amount of pigment was assumed to be proportional to concentration. Thus optical density was taken to be a measure of the total amount of pigment.

In the flicker method, a small visual stimulus is presented to the eye which alternates in wavelength between 460 nm, the peak absorbance wavelength of the macular pigment, and 540 nm where pigment absorbance is zero (Bone et al., 1992). Above a certain frequency, color fusion occurs but the stimulus continues to flicker. At a higher frequency, a critical condition can be reached where flicker can be eliminated only if the two wavelength components are matched in luminance. If the stimulus is viewed peripherally, so that the image falls outside the macula, neither wavelength is attenuated by the macular pigment. However, if the stimulus is viewed centrally, the intensity of the 460-nm light must be increased to compensate for the absorption by the macular pigment in order to achieve a luminance match. Thus it is possible to determine the optical density of a subject's macular pigment at the peak wavelength, or indeed any other wavelength.

The validity of this technique depends on the relative spectral reponse of the receptors being the same in the central and peripheral locations used. The flicker, which the subject seeks to eliminate, is one of luminance and, assuming photopic conditions, luminance is most likely due to an additive response from the long and middle wavelength sensitive cones (Guth et al., 1980). Evidence shows that these two cone types are present in equal ratios in the two locations used (Wooten and Wald, 1973). The short wavelength cones, whose relative abundances differ between the two locations, are generally not assumed to contribute to luminance (Guth *et al.* 1980), although others, using flicker techniques, have sought to eliminate their participation (Pease *et al.*, 1987; Werner *et al.*, 1987; Hammond *et al.*, 1995a). The ultimate justification for our procedure is found in the accurate reproduction of the macular pigment absorbance spectrum which it generates (Bone *et al.*, 1992).

2. Apparatus

The apparatus consisted of a two-channel Maxwellian view system based on a single light source, a 75-W xenon arc lamp. The wavelengths of the two channels were detemined by 460- and 540-nm interference filters, respectively, having half-widths of 7 and 9 nm. The channels were combined by a rotating semicircular mirror, and a circular aperture in a white screen provided a 1.5° diameter stimulus. Cross-hairs facilitated central fixation of the stimulus. The screen was 18° in diameter and was illuminated with white light from the same source. The illuminance of the screen was adjusted to provide the same retinal illuminance of 4 log Td as the stimulus. This was considered to be sufficiently high to minimize problems associated with rod intrusion which could otherwise differentially affect measurements in the macula and peripheral retina (Wysecki and Stiles, 1982). A small red LED was located 8° above the center of the stimulus to provide a fixation mark for peripheral viewing of the stimulus. The intensity of the 460-nm channel was adjustable by the subject through a neutral density, compensated wedge whose setting could be recorded by a push-button. The flicker frequency was also under the subject's control via a potentiometer. An adjustable dental impression bite ensured accurate and steady positioning of the subject's eye relative to the exit pupil.

3. Measurements

The ficker frequency was set to a predetermined value which, for central viewing by the subject, would allow flicker to be eliminated only over a very small range of wedge settings. This frequency was in the 25- to 30-Hz range. Having set the wedge to meet the no-flicker condition, the subject adapted to the viewing conditions by fixating with one eye on the stimulus cross-hairs for 2 min. The subject's other eye was occluded by an eye patch. At the end of this period, the subject proceeded to make a series of 10 to 15 wedge settings, attempting to obtain the center of the no-flicker range. The wedge was randomly offset after each setting. On occasions, the subject could not eliminate flicker entirely but instead sought a condition of minimum flicker. This was followed by another series of 10 to 15 settings while fixating on the LED, the frequency having been reduced to 12 to 16 Hz in order to reduce the range of no flicker. The whole procedure was then repeated for the subject's other eye. The optical density of the macular pigment of the subject was measured daily for a period of 1 week prior to the commencement of lutein supplementation, and daily thereafter.

V. Results

A. Control vs AMD Eyes

Data from 15 control and 22 AMD eyes have been obtained to date. The results are shown in Table I which, for each eye studied, depicts the total carotenoid present in each of the three sections of the retina in picomoles per square millimeter of tissue. Also included are the averages with associated standard deviations for the control and AMD sets. AMD eyes had on average approximately 70% of the total carotenoid found in controls, a figure that was very consistent across the retina. Seventeen (77%) of the 22 AMD eyes had total amounts of lutein and zeaxanthin in the central 3 mm of the retina which were below the mean (5.9 pmol/mm^2) for the control group. For the two annuli, having outer diameters of 11 and 21 mm, respectively, 15 (68%) of the AMD group were found to be lower in total carotenoid content than the mean of the corresponding regions in the control group. The distributions of lutein and zeaxanthin were similar in both groups. The proportion of RS-zeaxanthin decreased with increasing eccentricity as previously reported (Landrum *et al.*, 1995).

B. Lutein Uptake in Human Subjects

Figures 4a and 4b show the increase in serum lutein concentration in the two subjects during the time course of the supplementation experiment. The concentration of lutein in both subjects increased by a factor of about 10 times within the first week and remained high thereafter. Subjects JTL and RAB initially differed by a small but significant value with RAB having an approximately 25% lower concentration. Over the period of supplementation, a difference in levels appeared to remain but was no longer statistically significant.

C. Macular Pigmentation

Figures 5a and 5b show the absorbance of the macular pigment during the time course of the experiment in subjects JTL and RAB, respectively. An increase in the macular pigment level of subject JTL was first observable on about the 14th day of supplementation. This subject had macular pigment levels in both eyes that were experimentally determined by repeated measurements to be equal (±2%). The initial values of 0.570 ± 0.018 and 0.582 ± 0.028 for the right and left eyes, respectively, were determined

TABLE I   Distribution of Total Carotenoids in Individual Control and AMD Eyes[a]

	Donor no.	Total carotenoid/unit area (pmol/mm²)		
		Inner (7.1 mm²)	Medial (93 mm²)	Outer (343 mm²)
Control eyes	1	12.8	0.88	0.19
	2	10.5	0.51	0.10
	3	10.4	0.89	0.18
	4	9.3	1.35	0.36
	5	8.4	0.38	0.07
	6	5.8	0.19	0.06
	7	5.3	0.23	0.43
	8	5.1	0.48	0.21
	6	4.7	0.15	0.05
	9	4.6	0.21	0.06
	9	4.3	0.18	0.05
	10	2.5	0.07	0.03
	10	2.2	0.08	0.03
	11	2.0	0.26	0.20
	12	1.0	0.09	0.02
Control average ± SD		5.9 ± 3.4	0.40 ± 0.36	0.14 ± 0.12
AMD eyes	13	9.5	0.47	0.19
	13	9.4	0.78	0.23
	14	8.4	0.30	0.12
	14	7.7	0.56	0.13
	15	6.7	0.17	0.09
	16	5.7	0.24	0.08
	17	4.8	0.47	0.15
	18	4.5	0.34	0.07
	16	4.5	0.20	0.06
	19	4.5	0.14	0.05
	17	4.0	0.24	0.11
	19	3.8	0.05	0.09
	20	3.4	0.45	0.16
	21	3.4	0.20	0.09
	20	2.4	0.46	0.13
	22	2.3	0.13	0.05
	22	1.9	0.11	0.06
	23	1.2	0.03	0.02
	1	0.71	0.47	0.19
	23	0.46	0.03	0.02
	24	0.43	0.10	0.03
	24	0.32	0.07	0.02
AMD average ± SD		4.1 ± 2.8	0.27 ± 0.20	0.097 ± 0.058

[a] Averages and standard deviations for the two groups are also given.

a

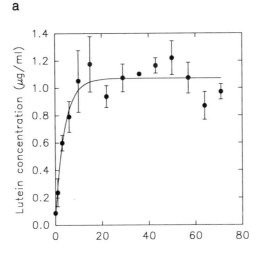

b

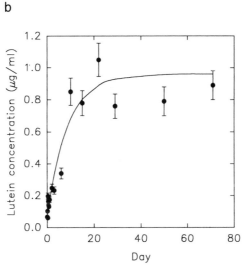

FIGURE 4 The time-dependent increase in the serum lutein level of subject JTL (a) and RAB (b). Error bars represent the standard deviations in the measurements. Day 0 represents the beginning of lutein supplementation.

by averaging 15 measurements obtained over a 17-day time period. This period began 9 days prior to the start of supplementation and included the first 8 days after supplementation. Comparison of these values with the averages of 15 measurements obtained over the 18-day period at the end of the experiment (0.67 ± 0.026 and 0.701 ± 0.034 for the right and left eyes, respectively) shows that the increase in optical density of the macular

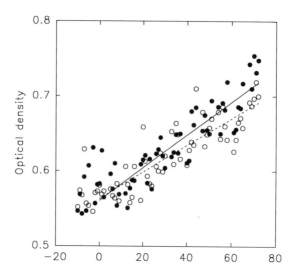

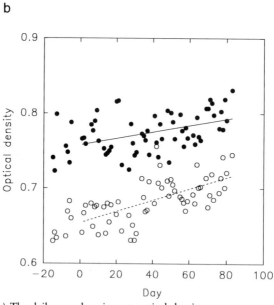

FIGURE 5 (a) The daily macular pigment optical density measurements for subject JTL from 7 days prior to the start (day 0) of the lutein supplementation through day 72. Left eye, solid circles; right eye, open circles. The solid line is the linear least-squares fit to the left eye data and has a slope of 15.3 × 10⁻³ absorbance units per week. The dashed line is a fit to the right eye data and has a slope of 12.5 × 10⁻³ absorbance units per week. (b) The daily macular pigment optical density measurements for subject RAB from 7 days prior to the start (day 0) of the lutein supplementation through day 83. Left eye, solid circles; right eye, open circles. The solid line is the linear least-squares fit to the left eye data and has a slope of 3.1 × 10⁻³ absorbance units per week. The dashed line is a fit to the right eye data and has a slope of 5.3 × 10⁻³ absorbance units per week.

pigment is highly significant ($p < 0.0005$ for both the right and left eyes based on a one-sided t test). The observed rate of increase in the optical density of the macular pigment in subject JTL was found to be nearly equal in both eyes with an average value of 13.9×10^{-3} absorbance units per week.

For subject RAB, right and left eyes were found to have significantly different macular pigment optical densities. The initial mean value for the right eye was 0.662 ± 0.016 whereas that of the left eye was 0.764 ± 0.016 as determined from nine measurements over the same time period as that used for subject JTL. This corresponded to a difference of 15% between the subject's left and right eyes. The eye having the lower pigmentation (right) showed an increase in optical density with a rate of 5.3×10^{-3} absorbance units per week over the course of the experiment. A positive slope of 3.1×10^{-3} absorbance units per week was obtained for the left eye. The increase in macular pigment, determined by comparing the initial averages for each eye and the final averages (0.704 ± 0.019 right; 0.792 ± 0.0195 left), computed as for subject JTL, was found to be highly significant ($p < 0.0005$ for the right eye and $p < 0.001$ for the left eye).

VI. Discussion and Conclusions

Few studies exist in which serum carotenoid concentrations and those in other tissues have been determined and correlated (Parker, 1989; Parker, 1993; Schmitz et al., 1993). Hammond et al. (1995a) have demonstrated that macular pigment levels in monozygotic twins are not entirely the result of genetic factors and that dietary xanthophyll levels may be one factor contributing to the observed level of pigmentation. This chapter has shown that serum levels increase significantly in subjects supplemented with lutein at a dosage level of 30 mg/day. These data are consistent with those of McLoone et al. (1995). Based on the concentration of lutein ($0.19 \ \mu g/ml$) in the serum 24 hr after administration of the first supplement dose, and assuming a serum volume of 2700 ml (45 ml/kg body weight), an approximate efficiency of uptake of <1.7% was estimated for the 24-hr period. The lutein concentration reached a plateau on about day 10 (JTL) and day 20 (RAB) of supplementation that was, within the experimental variability, the same for both subjects. The time dependence of the lutein concentration was reasonably modeled by an exponential of the form $y = A(1-Be^{-Ct})$ (Zeng et al, 1993).

Macular pigmentation increase appears to be a slow process. For subject JTL, the rate of 13.9×10^{-3} absorbance units/week was determined from a linear fit to the measured optical density data in Fig. 5a. Nevertheless, this amounted to a 15% increase in the pigment level after 72 days of lutein supplementation. The slow rate of change may be partly due to the need for lutein to diffuse into the avascular macular region of the retina. The increase is not always bilaterally symmetric as seen for subject RAB. Further, the rate of increase in macular pigmentation was very different for the two subjects in this study. The low rate observed for both eyes of subject RAB (5.3 and 3.1 × 10^{-3} absorbance units per week for the fight and left eyes, respectively) may have been due to an initially high level of pigmentation compared to that of subject JTL. Furthermore, the lower rate was observed in the eye having the higher macular pigment level. The possibility that the level of macular pigment will reach a plateau exists and is currently under investigation.

A variety of data are suggestive of a relationship between the pathology of AMD and macular pigmentation (Schalch, 1992). AMD initially affects the RPE, Bruch's membrane, and the choriocapillaris layers of the retina whereas the macular pigment lies in the overlaying nerve tissue. As such it would be expected that AMD at the outset would not necessarily affect the macular pigment. Our data comparing the amount of macular pigment present in control and AMD eyes show an average level of pigmentation throughout the retinas in the AMD group that is about 70% of that in the control group. This difference is not restricted to the macular region as would be expected if atrophy of the macula was the cause. The lowered level of pigmentation in AMD eyes as compared to controls is remarkably consistent across the retina, extending radially far beyond the macula where little or no degeneration occurs. The differences observed between control and AMD eyes in the inner, medial, and outer annuli were found to be statistically significant, based on a one-sided t test, with $p < 0.05, 0.1$, and 0.1, respectively. This observation is consistent with the hypothesis that lowered levels of macular pigmentation are a causitive factor in the development of macular degeneration and not necessarily a result of the degeneration process. In this regard, one donor (Donor 1, Table I) who had AMD in one eye but not in the other is of interest. The diseased eye had a very low macular pigment level in the inner 3-mm macular region, but both eyes were comparable in the surrounding retina. Donor 1 was diagnosed as having the wet neovascular form of the disease. In this case, the degenerative process was apparently restricted to the central region of the macular where loss of pigmentation occurred. For this individual, macular pigment loss would appear to be due to degenerative processes. This example serves to point out that multiple factors contribute to the development of the disease and that low pigmentation, while possibly an important risk factor, is not the sole determinant.

Our data and those of others (Weiter, 1988; Haegerstrom-Portnoy, 1988) suggest that macular pigmentation does function to protect the retina. An increased rate of photo-oxidation that might accompany lower macular pigment levels in some individuals could contribute to a more rapid buildup of the drusen characteristic of the atrophic regions of the macula and associated with AMD. A relationship has been established between increased serum levels of lutein and corresponding increases in the concentration of lutein in the macula of the human eye. Long-term lutein supplementation of individuals having low levels of macular pigmentation could result in a significant increase in the the level of pigmentation within the macula.

Acknowledgments

The authors thank Dr. Alan N. Howard, F.R.C.S., and The Howard Foundation for a generous gift of lutein used in the supplementation study and for valuable discussions; Christina Gomez and Hilda Joa for technical assistance; and The National Disease Research Interchange for supplying the human eyes. This work was (AMD/Control study) supported in part by a grant from the NIH (#GM 08205).

References

Arnold, J. J., Sarks, S. H., Killingsworth, M. C., and Sarks, J. P. (1995). Reticular psuedodrusen: A risk factor in age-related maculopathy. *Retina* 15, 183–191.

Bird, A. C. (1992). Pathophysiology of AMD. *In* "Age-Related Macular Degeneration: Principles and Practice" (G. R. Hampton and P. T. Nelson, eds.), pp. 63–84. Raven Press, New York.

Bird, A. C., Bressler, N. M., Bressler, S. B., Chisholm, I. H., Coscas, G., Davis, M. D., de Jong, P. T. V. M., Klaver, C. C. W., Klein, B. E. K., Klein, R., Mitchell, P., Sarks, J. P., Sarks, S. H., Sourbrane, G., Taylor, H. R., and Vingerling, J. R. (1995). An international classification and grading system for age-related maculopathy and age-related macular degeneration. *Surv. Ophthalmol.* 39, 367–374.

Bone, R. A., and Landrum, J. T. (1983). Dichroism of lutein: A possible basis for Haidinger's brushes. *Appl. Optics* 22, 775–776.

Bone, R. A., and Landrum, J. T. (1984). Macular pigment in Henle fiber membranes: A model for Haidinger's brushes. *Vis. Res.* 24, 103–108.

Bone, R. A., and Landrum, J. T. (1992). Distribution of the macular pigment components, zeaxanthin and lutein in human retina. *In* "Methods in Enzymology" (L. Packer, ed.), Vol. 213, pp. 360–366. Academic Press, San Diego.

Bone, R. A., Landrum, J. T., and Cains, A. (1992). Optical density of the macular pigment *in vivo* and *in vitro*. *Vis. Res.* 32, 105–110.

Bone, R. A., Landrum, J. T., Fernandez, L., and Tarsis, S. L. (1988). Analysis of the macular pigment by HPLC: Retinal distribution and age study. *Invest. Ophthalmol. Vis. Sci.* 29, 843–849.

Bone, R. A., Landrum, J. T., Hime, G. W., Cains, A. and Zamor, J. (1993). Stereochemistry of the human macular carotenoids. *Invest. Ophthalmol. Vis. Sci.* 34, 2033–2040.

Bone, R. A., Landrum, J. T., and Tarsis, S. L. (1985). Preliminary identification of the human macular pigment. *Vis. Res.* 25, 1531–1535.

Bone, R. A., Landrum, J. T., Vidal, I., and Menendez, E. (1994). Distribution of macular pigment stereomers in individual eyes, including those with age-related macular degeneration. *Invest. Ophthalmol. Vis. Sci.* 35 (Suppl.), 1502.

Bone, R. A., and Sparrock, J. M. B. (1971). Comparison of macular pigment densities in human eyes. *Vis. Res.* 11, 1057–1064.

Brown, P. K., and Wald, G. (1963). Visual pigments in human and monkey retinas. *Nature (London)* 200, 37–43.

Cruickshanks, K. J., Klein, R., and Klein, B. E. K. (1993). Sunlight and age-related macular degeneration. *Arch. Ophthalmol.* **111**, 514–518.

Davson, H. (1990). "Physiology of the Eye." Pergamon Press, New York.

Ditchburn, R. W. (1973). "Eye-Movements and Visual Perception." Clarendon Press, Oxford.

El Baba, F., Green, W. R., Fleischmann, J., Finkelstein, D., and de la Cruz, Z. C. (1986). Clinicopathologic correlation of lipidization and detachment of the retinal pigment epithelium. *Am. J. Ophthalmol.* **101**, 576–583.

Eye Disease Case-Control Study Group. (1993). Antioxidant status and neovascular age-related macular degeneration. *Arch. Ophthalmol.* **111**, 104–109.

Feeney-Burns, L., Burns, R. P., and Gao, C.-L (1990). Age-related macular changes in humans over 90 years old. *Am. J. Ophthalmol.* **109**, 265–278.

Feeney-Burns, L., and Ellersieck, M. R. (1985). Age-related changes in the ultrastructure of Bruch's membrane. *Am. J. Ophthalmol.* **100**, 686–697.

Foote, C. S., Chang, Y. C., and Denny, R. W. (1970). Chemistry of singlet oxygen. X. Carotenoid quenching parallels biological protection. *J. Am. Chem. Soc.* **92**, 5216–5218.

Goodwin, T. W. (1992). Distribution of carotenoids. *In* "Methods in Enzymology" (L. Packer, ed.), Vol. 213, pp. 167–172. Academic Press, San Diego.

Gottsch, J. D., Pou, S., Bynoa, L. A., and Rosen, G. M. (1990). Hematogenous photosensitization. A mechanism for development of age-related macular degeneration. *Invest. Ophthalmol. Vis. Sci.* **31**, 1674–1692.

Guiliano, A. R., Matzner, M. B., and Canfield, L. M. (1993). Assessing variability in quantitation of carotenoids in human plasma: Variance component model. *In* "Methods in Enzymology" (L. Packer, ed.), Vol. 214, pp. 94–101. Academic Press, San Diego.

Guth, S. L., Massof, R. W., and Benzschawel, T. (1980). Vector model for normal and dichromatic color vision. *J. Opt. Soc. Am.* **70**, 197–212.

Haegerstrom-Pornoy, G. (1988). Short-wavelength-sensitive-cone sensitivity loss with aging: Protective role for macular pigment? *J. Opt. Soc. Am. [A.]* **5**, 2140–2144.

Ham, W. T., Jr., Mueler, H. A., Ruffolo, J. J., Millen, J. E., Cleary, S. F., Guerry, R. K., and Guerry, D. (1984). Basic mechanisms underlying the production of photochemical lesions in the mammalian retina *Curr. Eye Res.* **3**, 165–174.

Hammond, B. R., Jr., Fuld, K., and Curran-Celentano, J. (1995a). Macular pigment density in monozygotic twins. *Invest. Ophthalmol. Vis. Sci.* **36**, 2531–2541.

Hammond, B. R., Snodderly, D. M., and Wooten, B. R. (1995b). The relationship between cigarette smoking and peak macular pigment density. *Invest. Ophthalmol. Vis. Sci.* **29**, 850–855.

Handelmann, G. J., Dratz, E. A., Reay, C. C., and van Kuijk, F. J. G. M. (1988). Carotenoids in the human macula and whole retina. *Invest. Ophthalmol. Vis. Sci.* **29**, 850–855.

Handelmann, G. J., Shen, B., and Krinsky, N. I. (1992). High resolution analysis of carotenoids in human plasma by high-performance liquid chromatography. In "Methods in Enzymology" (L. Packer, ed.), Vol. 213, pp. 336–346. Academic Press, San Diego.

Hyman, L. (1992). Epidemiology of AMD. *In* "Age-Related Macular Degeneration: Principles and Practice" (G. R. Hampton and D. T. Nelson, eds.), pp. 1–35. Raven Press, New York.

Jensen, S. L., and Hertzberg, S. (1966). Selective preparation of lutein monomethyl ethers. *Acta Chem. Scand.* **20**, 1703–1709.

Kahn, H. A., Leibowitz, H. M., Ganley, J. P., *et al.* (1977). The Framingham eye study. I. Outline and major prevalence findngs. *Am. J. Epidemiol.* **106**, 17–32.

Kincaid, M. C. (1992). Pathology of AMD. In "Age-Related Macular Degeneration: Principles and Practice" (G. R. Hampton and P. T. Nelson, eds.), pp. 37–61. Raven Press, New York.

Kirshfeld, K. (1982). Carotenoid pigments: Their possible role in protecting against photooxidation in eyes and photoreceptor cells. *Proc. R. Soc. Lond. [Biol.]* **216**, 71–85.

Klein, R., Wang, Q., Klein, B. E. K., Moss, S. E., and Meuer, S. M. (1995). The relationship of age-related maculopathy, cataract, and glaucoma to visual acuity. *Invest. Ophthalmol. Vis. Sci.* **36**, 182–191.

Krinsky, N. I. (1968). The protective function of carotenoid pigments. *Photophysiology* 3, 123–195.

Krinsky, N. I. (1971). Function. *In* "Carotenoids" (O. Isler, ed.), pp. 669–716. Birkhauser, Basel.

Krinsky, N. I. (1979). Carotenoid protection against oxidation. *Pure Appl. Chem.* 51, 649–660.

Landrum, J. T., Bone, R. A., Vidal, I., Menendez, E., and Kilburn, M. (1995). Macular pigment stereomers in individual eyes: A comparison between normals and those with age-related macular degeneration. *Invest. Ophthalmol. Vis. Sci.* 36, (Suppl.), 892.

Malinow, M. R., Feehey-Burns, L., Peterson, L. H., Klein, M. L., and Neuringer, M. (1980). Diet-related macular anomalies in monkeys. *Invest. Ophthalmol. Vis. Sci.* 19, 857–863.

Maoka, T., Arai, A., Shimizu, M., and Matsuno, T. (1986). The first isolation of enantiomeric and meso-zeaxanthin in nature. *Comp. Biochem. Physiol. B* 83, 121–123.

McLoone, U., Chopra, M., Williams, N. R., and Thurnham, D. I. (1995). The effect of *in vitro* and *in vivo* supplementation with lutein on LDL oxidation *in vitro*. *Proc. Nutr. Soc.* 54, 168A.

Parker, R. S. (1989). Carotenoids in human blood and tissues. *J. Nutr.* 119, 101–104.

Parker, R. S. (1993). Carotenoids in human blood and tissues. *In* "Methods in Enzymology" (L. Packer, ed.), Vol. 214, pp. 86–93. Academic Press, San Diego.

Pease, P. L., Adams, A. J., and Nuccio, E. (1987). Optical density of human macular pigment. *Vis. Res.* 27, 705–710.

Reading, V. M., and Weale, R. A. (1974). Macular pigment and chromatic aberration. *J. Opt. Soc. Am. [A.]* 64, 231–234.

Rüttimann, A., Schiedt, K., and Vecci, M. (1983). Separation of (3R,3'R)-, (3R,3'S; meso)- (3S,3'S)-zeaxanthin, (3R, 3'R, 6'R)-, (3R,3'S,6'S)-, and (3S,3'S,6'S)-lutein via the dicarbamates of (S)-(+)-l-[1-naphthyl]ethylisocyanate. *J. High. Res. Chrom. Commun.* 6, 612–616.

Sarks, S. H. (1976). Ageing and degeneration in the macular region: A clinico-pathological study. *Br. I. Ophthalmol.* 60, 324–341.

Sarks, S. H., and Sarks, J. P. (1994). Age-related macular degeneration: Atrophic form. *In* "Retina" (S. J. Ryan, A. P. Schachat, and R. M. Murphy, eds.), Vol. 2, pp. 1071–1102. Mosby, St. Louis.

Sarks, J. P., Sarks, S. H., and Killingsworth, M. C. (1994). Evolution of soft drusen in age-related macular degeneration. *Eye* 8, 269–283.

Schalch, W. (1992). Carotenoids in the retina: A review of their possible role in preventing or limiting damage caused by light and oxygen. *In* "Free Radicals and Aging" (I. Emerit and B. Chance, eds.), pp. 280–298. Birkhauser Verlag, Basel.

Scheidt, K., Bischof, S., and Glinz, E. (1995). Example 5: Fish—Isolation of astaxanthin and its metabolites from skin of Atlantic Salmon (Salmo salor). *In* "Carotenoids" (G. Britton, S. Liaaen-Jensen, and H. Pfander, eds.), pp. 243–252. Birkhauser Verlag, Basel.

Schmitz, H. H., Poor, C. L., Gugger, E. T., and Erdman, J. W., Jr. (1993). Analysis of carotenoids in human and animal tissues. *In* "Methods in Enzymology" (L. Packer, ed.), Vol. 214, pp. 102–116. Academic Press, San Diego.

Seddon, J. M., Ajani, U. A., Sperduto, R. D., Hiller, R., Blair, N., Burton, T. C., Farber, M. D., Gragoudas, E. S., Haller, J., Miller, D. T. Yannuzzi, L. A., and Willett, W. (1994). Dietary carotenoids, vitamins A, C, and E and advanced age-related macular degeneration. *JAMA* 272, 1413–1420.

Snodderly, D. M., Auran, J. D., and Delori, F. C. (1984). The macular pigment. II. Spatial distribution in primate retinas. *Invest. Ophthalmol. Vis. Sci.* 25, 674–685.

Wald, G. (1945). Human vision and the spectrum. *Nature (London)* 101, 653–658.

Walls, G. L. (1967). "The Vertebrate Eye and Its Adaptive Radiation." Hafner, New York.

Weedon, B. C. L. (1971). Occurence. *In* "Carotenoids" (O. Isler, ed.), pp. 267–324. Birkhauser, Basel.

556 John T. Landrum *et al.*

Weiter, J. J., Delori, F. C., and Dorey, C. K. (1988). Central sparing in annular macular degeneration. *Am. J. Ophthalmol.* **106**, 286–292.

Weiter, J. J., Delori, F. C., Wing, G. L., and Fitch, K. A. (1986). Retinal pigment epithelial lipofuscin and melanin and choroidal melanin in human eyes. *Invest. Ophthalmol. Vis. Sci.* **27**, 145–152.

Werner, J. S., Donnelly, S. K., and Kliegl, R. (1987). Aging and human macular pigment density: Appended with translations from the work of Max Schutze and Ewald Hering. *Vis. Res.* **27**, 257–268.

Wooten, B. R., and Wald, G. (1973). Color-vision mechanism in the peripheral retinas of normal and dichromatic observers. *J. Gen. Physiol.* **61**, 125–145.

Wysecki, G., and Stiles, W. S. (1982). "Color Science: Concepts and Methods, Quantitative Data and Formulae." Wiley, New York.

Zeng, S., Furr, H. C., and Olson, J. A. (1993). Human metabolism of carotenoid analogs and apocarotenoids. *In* "Methods in Enzymology" (L. Packer, ed.), Vol. 214, pp. 137–147. Academic Press, San Diego.

David P. R. Muller

Division of Biochemistry and Genetics
Institute of Child Health
London WC1N 1EH, United Kingdom

Neurological Disease

I. Introduction

From theoretical considerations, the nervous system (which in this chapter includes the retina) is likely to be particularly vulnerable to oxidative stress. Thus the brain contains high concentrations of polyunsaturated fatty acids that are susceptible to lipid peroxidation, receives a disproportionately large percentage of oxygen, is relatively deficient in antioxidant systems, and contains specific regions that have high concentrations of iron (see Olanow, 1990). The retina is also served with a plentiful supply of oxygen, has an abundant supply of mitochondria, and has an unusually high rate of oxidative metabolism (Handelman and Dratz, 1986). The retinal rod outer segments are particularly vulnerable to lipid peroxidation as more than 65% of the membrane fatty acids are polyunsaturated, which is the highest proportion found in any vertebrate tissue examined to date. The retina is also frequently exposed to intense light, which can be phototoxic.

There are essentially two situations where oxidative stress may be involved in neural and retinal dysfunction: where there is a deficiency of an antioxidant and where antioxidant protection is compromised by increased oxidative stress. In both of these situations treatment with added antioxidants may be beneficial. These two aspects are considered in turn.

II. Antioxidant Deficiency

Much of the work involving the role of antioxidants in neurological function and disease has been concerned with vitamin E. The most active form of this fat-soluble vitamin is α-tocopherol which also appears to be the major lipid-soluble chain-breaking antioxidant *in vivo* (Burton *et al.*, 1983). There are at least three situations where vitamin E status is reduced compared to that of the normal adult: (i) the newborn and particularly the premature infant, (ii) patients with fat malabsorptive conditions, and (iii) patients with familial isolated vitamin E deficiency.

A. Newborn and Premature Infants

Many studies have documented decreased serum concentrations of vitamin E in the newborn (see Muller, 1987). In addition to this decrease in antioxidant defenses, the newborn and particularly the premature infant are also at risk from oxidative stress. Thus at birth an infant abruptly enters a relatively hyperoxic extrauterine environment with the alveoli of the lung being exposed to an oxygen tension approximately five times greater than that during intrauterine development. The situation is further exacerbated in many small premature infants, where as part of the management of the respiratory distress syndrome, they are exposed to increased concentrations (up to 100%) of oxygen.

The role of antioxidant, particularly vitamin E therapy, in the premature infant remains uncertain and controversial. It has been suggested for the prevention of a number of conditions where oxidative stress has been implicated, including the neurological conditions of intraventricular hemorrhage and retinopathy of prematurity.

I. Intraventricular Hemorrhage (IVH)

Since the first suggestion by Chiswick *et al.* (1983) that supplements of vitamin E might decrease the incidence of IVH in preterm babies, there have been a number of studies. Law *et al.* (1990) reviewed the results of four trials of vitamin E supplementation on intracranial hemorrhage and calculated that the vitamin decreased the incidence of IVH by approximately 50%. They pointed out that the major neurological disability associated with IVH alone affects approximately 3% of all very low birth weight babies. If vitamin E reduces its incidence by a mean of 50% with an upper 95% confidence limit of 75%, then at best vitamin E supplementation of all low birth weight babies will reduce the prevalence of neurological disability by approximately 2.5% of all treated infants. This would be acceptable if all potential hazards could be ruled out (see later).

2. Retinopathy of Prematurity (ROP)

ROP or retrolental fibroplasia is a potentially blinding disorder seen in premature infants, which was first described by Terry (1942). The first nonrandomized trial of vitamin E in ROP was carried out by Owens and Owens (1949), and this suggested that the vitamin might have a beneficial effect. Since then, numerous trials of vitamin E supplementation have been carried out with conflicting results (see Muller, 1992). Law et al. (1990) have analyzed and combined seven randomized trials of vitamin E supplementation in ROP. They calculated that there was no statistically significant reduction in the risk of ROP following vitamin E treatment for either all infants with ROP or those with the severe form of the condition. From studies carried out by Phelps (1982) and Ng et al. (1988), it has been calculated that if prophylactic vitamin E was given to all premature infants, a very large percentage (approximately 90%) would be given the vitamin who would not develop the severe form of the condition. This is acceptable if such supplementation was completely safe or if the benefits strongly outweigh the risks. Reports have, however, appeared that suggest that necrotizing enterocolitis (a severe gastrointestinal condition affecting predominantly premature infants) and sepsis may be associated with the prophylactic administration of vitamin E in the premature infant (see Muller, 1992).

The efficacy of prophylactic supplements of vitamin E in the premature infant remains controversial and there are a number of possible hazards associated with supplementation. The author, therefore, still agrees with the recommendation of the American Committee on Fetus and Newborn (1985), which concluded that the prophylactic use of pharmacological vitamin E was experimental and that high doses of vitamin E should not be given routinely to infants weighing less than 1500 g, even if such use is limited to infants who require supplemental oxygen.

The possible effects on ROP of modulating water-soluble antioxidant systems such as vitamin C or selenium concentrations have been discussed (Kretzer and Hittner, 1988) but trials do not appear to have been undertaken.

B. Chronic Fat Malabsorptive States

I. Abetalipoproteinemia

All untreated patients with abetalipoproteinemia, an inborn error of lipoprotein metabolism involving a defect in the microsomal triglyceride transfer protein (Wetterau *et al.*, 1992; Sharp *et al.*, 1993), have undetectable serum concentrations of vitamin E from birth (Muller *et al.*, 1974). This condition, therefore, provides an ideal model for the study of the role of vitamin E in human nutrition. Among the clinical features of abetalipoproteinemia is the development of a characteristic neurological syndrome during the second decade of life (Muller *et al.*, 1983). This is progressive, leading eventually to crippling and blindness, and has been described as "devastating" (Herbert *et al.*, 1978). No cases of spontaneous improvement have been reported.

A study was undertaken to treat children with abetalipoproteinemia with very large oral doses of vitamin E (approximately 100 mg/kg/day all-*rac-α*-tocopheryl acetate; Ephynal, supplied by Hoffmann-La Roche and Co. Ltd.). The original cohort of eight patients has now been receiving vitamin E for 24–29 years. The clinical results have shown definite benefit and can be summarized as follows. If started sufficiently early, vitamin E supplementation totally prevents the development of all neurological and retinal features. If supplementation is commenced after the onset of signs and symptoms, progression is invariably halted and, in some cases, reversed (Muller *et al.*, 1977, 1983). Other groups have reported similar results (Azizi *et al.*, 1978; Miller *et al.*, 1980; Hegele and Angel, 1985; Kane and Havel, 1989).

2. Other Fat Malabsorptive States

Although vitamin E deficiency can occur in any chronic disorder of fat absorption, it is likely to be particularly severe in cholestatic liver disease. This is because of impaired bile flow that results in a decreased concentration of bile salts in the intestinal lumen and, therefore, impaired solubilization and absorption of the vitamin (Harries and Muller, 1971). Neurological features very similar to those found in abetalipoproteinemia have been described in a number of studies of patients with cholestatic liver disease (Elias *et al.*, 1981; Rosenblum *et al.*, 1981; Guggenheim *et al.*, 1982; Alvarez *et al.*, 1983; Sokol *et al.*, 1985). The typical neurological syndrome associated with severe and chronic vitamin E deficiency has also been described in patients with extensive intestinal resection (Harding *et al.*, 1982; Howard *et al.*, 1982) and cystic fibrosis (Elias *et al.*, 1981; Willison *et al.*, 1985). Improvement in neurological function following treatment with appropriate supplements (dose and type) of vitamin E has also been reported in patients from all these groups (Elias *et al.*, 1981; Harding *et al.*, 1982; Guggenheim *et al.*, 1982; Howard *et al.*, 1982; Sokol *et al.*, 1985). In cholestatic liver disease it is generally necessary to overcome the problems of solubilization by either giving intramuscular injections of the vitamin (Harries and Muller, 1971; Guggenheim *et al.*, 1982; Sokol *et al.*, 1985) or providing oral vitamin E in the form of α-tocopheryl polyethylene glycol-1000 succinate which is water soluble and can, therefore, be absorbed (Sokol *et al.*, 1987, 1993). Evidence for a causal relationship between vitamin E deficiency and the neurological sequelae was provided by the study of Sokol *et al.* (1985) who followed the clinical results of vitamin E supplementation in children with cholestatic liver disease.

3. Retinal Involvement in Fat Malabsorptive States

At one time it was thought that retinal involvement was a feature of abetalipoproteinemia rather than of vitamin E deficiency, but it has now been reported in patients with cholestatic liver disease (Alvarez *et al.*, 1983) and ileal resection (Howard *et al.*, 1982). There is much discussion in the literature as to whether these retinal abnormalities result from a deficiency of both vitamins A and E or whether they can be prevented by administration of either vitamin alone (Wolff *et al.*, 1964; Gouras *et al.*, 1971; Bohlmann *et al.*, 1972; Sperling *et al.*, 1972; Azizi *et al.*, 1978; Bishara *et al.*, 1982; Howard *et al.*, 1982; Alvarez *et al.*, 1983). Vitamin E does, however, appear to be involved, as progression of the retinal changes has been reported in vitamin E-deficient patients who have been receiving sufficient vitamin A to maintain normal serum vitamin A concentrations (Wolff *et al.*, 1964; Bohlmann *et al.*,1972; Howard *et al.*, 1982; Alvarez *et al.*, 1983). In one patient there was objective improvement of vision (visual fields and electroretinogram) after 2 years of "aggressive" treatment with vitamin E (Howard *et al.*, 1982).

C. Familial Isolated Vitamin E Deficiency

A number of patients have now been reported with a familial isolated deficiency of vitamin E without generalized fat malabsorption (see Sokol *et al.*, 1988). They have similar neurological features to the patients described earlier with various fat malabsorptive disorders and have responded to treatment in a similar way. These patients lack a functional hepatic-binding protein for α-tocopherol (Ouahchi *et al.*, 1995) that is necessary for its transfer to very low-density lipoproteins (VLDL). This results in impaired secretion of vitamin E back into the circulation and its sequestration in the liver. There is no evidence of a deficiency of any other nutrient in these patients and they, therefore, provide further evidence for a causal relationship between a deficiency of vitamin E and the neurological findings.

D. Neuropathology Associated with Vitamin E Deficiency

Another line of evidence suggesting that a deficiency of vitamin E can result in neurological sequelae is that the neuropathological features found in vitamin E-deficient rats, monkeys, and humans are very similar (Einarson, 1952; Machlin *et al.*, 1977; Nelson *et al.*, 1978, 1981; Sung *et al.*, 1980; Rosenblum *et al.*, 1981; Towfighi, 1981; Southam *et al.*, 1991). In general the central nervous system is more severely affected than the peripheral, with sensory axons more involved than the motor. The neuropathology of both the central and the peripheral nervous system is suggestive of a "dying back" process (Nelson et al., 1981) which is caused by a primary damage to the axon of the neuron followed by secondary demyelination (Thomas et al., 1984; Wichman et al., 1985).

E. Animal Models

To answer fundamental questions arising from the clinical and pathological observations described earlier, functional and biochemical studies in a rat model of vitamin E deficiency have been carried out.

I. Functional Studies

Detailed electrophysiological studies have been undertaken in vitamin E-deficient humans, which have reported abnormalities of sensory (particularly central conduction) and visual-evoked potentials but no effect on brain stem auditory-evoked potentials or peripheral sensory motor responses (see Muller and Goss-Sampson, 1990). Similar results were obtained in the vitamin E-deficient rat (Goss-Sampson et al., 1988, 1990, 1991), which proves the validity of the model. The neuropathological findings, which include an accumulation of organelles at the distal ends of nerves (Towfighi, 1981; Southam et al., 1991), strongly suggest an abnormality of "turnaround." This is the mechanism whereby materials that have descended the axon in the anterograde transport system are packaged into lysosomes for their return to the cell body in the retrograde transport system. That "turnaround" may be affected was further supported by finding reduced velocities of both fast anterograde and retrograde transport of endogenous acetylcholinesterase in the vitamin E-deficient rat (Southam et al., 1991).

2. Biochemical Studies

Detailed longitudinal studies have been carried out in the rat (Goss-Sampson et al., 1988) to examine the loss of α-tocopherol from neural and other tissues during the course of vitamin E deficiency. The decrease in α-tocopherol concentrations was found to be less rapid in neural (brain, cord, and nerve) than in nonneural tissues (serum, liver, and adipose tissue). Similar results have been reported in the guinea pig (Vatassery and Younoszai, 1978) and mouse (Vatassery et al., 1984). They are also consistent with the estimated half-lives of natural (RRR) α-tocopherol in various tissues (Ingold et al., 1987). All of these studies suggest that neural tissues preferentially conserve vitamin E, which may reflect a reduced rate of turnover compared with other tissues.

3. Possible Mechanism of Action of Vitamin E

As a result of its antioxidant properties, it is assumed that vitamin E is able to terminate lipid peroxidation in membranes of neural tissues and thereby prevent the characteristic neurological abnormalities associated with vitamin E deficiency. Some evidence for this hypothesis was provided by Nelson (1987) who showed that the characteristic neuropathology of vitamin E deficiency in the rat could be prevented by the addition of antioxidants such as ethoxyquin and promethazine. Further evidence was provided by the study of Southam et al. (1991), who reported that the addition of excess peroxidizable substrate in the form of polyunsaturated fat markedly accelerated the rate of development of the neurological syndrome in vitamin E-deficient rats.

It has been postulated that mitochondria may be particularly suceptible to peroxidation as their membrane lipids are highly polyunsaturated (Molenaar et al., 1972) and that mitochondria produce increased concentrations of oxygen-derived free radicals as a by-product of oxidative phosphorylation. Functional impairment of axonal mitochondria could lead to abnormalities in fast anterograde and retrograde transport, which are energy-dependent processes, and thus to defective "turnaround." The resultant accumulation of organelles may then "plug off" the terminal axons so that they become isolated from the cell body and ultimately degenerate. This process could then spread centripetally in a "dying back" manner. Support for this hypothesis comes from two studies. First, it has been shown from fractionation studies of myelinated nerves that the organelles of the axon, including the mitochondria, are particularly susceptible to oxidative stress during severe and chronic vitamin E deficiency (MacEvilly and Muller, 1996). Second, muscle mitochondria from vitamin E-deficient rats showed significant decreases in the activities of complexes I and IV of the respiratory chain, a reduction in the respiratory control ratio (indicative of membrane damage), and increased membrane fluidity (Thomas et al., 1993).

III. Increased Concentrations of Reactive Oxygen Species

Oxygen-derived free radicals have been implicated in the pathogenesis of a number of neurological conditions and, as a result, antioxidant supplementation has been suggested for the prevention and treatment of many of them. Some of these conditions are now discussed.

A. Parkinson's Disease

1. Etiology

The etiology of Parkinson's disease (PD) is unknown but there are theoretical reasons to suggest that oxidative stress may be involved. First, 1-methyl-4-phenyl-1,2,3,6-tetrahydropyridine (MPTP), a contaminant of a designer drug, selectively destroys the dopamine-producing cells of the substantia nigra (SN) and produces a parkinsonian syndrome in humans and experimental animals (Davis *et al.*, 1979; Langston *et al.*, 1983). The toxicity of MPTP results from its oxidation to the active metabolite 1-methyl-4-phenyl pyridine (MPP^+) which is taken up by the mitochondria of dopaminergic neurons where it inhibits complex 1 of the respiratory chain (Schapira *et al.*, 1992) and generates an oxidative stress (Cleeter *et al.*, 1992). There is also evidence that an oxidation–reduction reaction may occur between MPP^+ and another metabolite $MPDP^+$ (1-methyl-4-phenyl-2,3-dihydropyridine) to produce superoxide (Rossetti *et al.*, 1988). Second, oxidative stress may arise from the metabolism of dopamine. Dopamine is normally stored in vesicles where it is inert, but when it is released into the cytosol, it is metabolized either enzymatically (by monoamine oxidase) or by autoxidation; both processes result in the formation of hydrogen peroxide. Evidence shows that the turnover of dopamine is increased in PD (Mogi *et al.*, 1988) and thus the cytosolic concentrations of dopamine and hydrogen peroxide may be higher than normal. Hydrogen peroxide is normally cleared by catalase or reduced glutathione and glutathione peroxidase, but any excess hydrogen peroxide can be converted in the presence of iron and superoxide to the highly toxic hydroxyl radical, which may then cause or contribute to the damage of dopaminergic neurons (Olanow, 1990).

2. Evidence for Increased Oxidative Stress

Several lines of evidence suggest that there may be increased oxidative stress in the SN of patients with PD. The following observations refer to specific changes seen in the SN but not in other brain regions of patients with PD nor are they seen in the SN of controls with or without neurodegeneration.

1. An increase in lipid peroxidation (Dexter *et al.*, 1989a).

2. A decrease in the concentration of reduced glutathione (Riederer *et al.*, 1989; Jenner *et al.*, 1992). To investigate this observation further, Sian *et al.* (1994) investigated the activities of enzymes involved in glutathione metabolism. They found that the activity of γ-glutamyltranspeptidase, which is involved in the breakdown and translocation of glutathione, was selectively increased which could result in an increased efflux and, therefore, depletion of glutathione.

3. An increase in iron in the brain of patients with PD has been reported by Earle (1968), Riederer *et al.* (1989), and Dexter *et al.* (1989b). Iron is normally bound to ferritin and is not available to catalyze oxidation reactions, but Dexter *et al.* (1990) reported a decrease in ferritin throughout the brain in PD. The presence of increased iron and decreased ferritin concentrations in the SN could result in a proportion of the iron being in a free form and thus able to catalyze the formation of reactive oxygen species (ROS).

4. A decrease in complex I activity has been reported (Schapira *et al.*, 1990) which could lead to an increased production of ROS and in turn cause an irreversible and selective inhibition of complex 1 (Cleeter *et al.*, 1992).

5. Neuromelanin is believed to be produced as a result of the autoxidation of catecholamines, a process which also results in the production of ROS (Graham, 1979). This is relevant to the pathogenesis of PD as there is a positive relationship between the percentage loss of dopaminergic neurons in different cell groups and their melanin content (Mann and Yates, 1983; Hirsch, 1992). It is, therefore, of interest that the accumulation of iron in PD brain appears to be primarily localized to the neuromelanin granules within the neurons of the SN (Olanow, 1993). Aluminium was also shown to accumulate in these neuromelanin granules (Olanow, 1993) and this metal is known to substantially enhance iron-induced lipid peroxidation (Gutteridge *et al.*, 1985). These observations again suggest that the neurons which degenerate in PD are those in which high levels of ROS are produced.

To gain an understanding of the sequence of events involved in the oxidative stress and pathogenesis of PD, Jenner *et al.* (1992) investigated autopsy material from individuals with incidental Lewy body disease which is thought to represent the early presymptomatic stages of PD. Such individuals were found to have a normal dopamine, iron, ferritin, and zinc status, whereas glutathione concentrations in the SN were reduced to a similar extent to that seen in advanced PD. The activity of complex 1 was intermediate between that seen in PD and controls. These findings suggest that the earliest defect may be defective protection against oxidative stress and that the mitochondrial defect may be important in the chain of events leading to dopaminergic cell death in the SN of parkinsonian patients.

Despite increasing evidence suggesting that oxidative stress is involved in the pathogenesis of PD, it is not possible to distinguish whether this is part of a primary degenerative process or a secondary consequence of cellular degeneration. Even if it is secondary, it is still important because once oxidative stress starts it could potentiate cell damage. Thus therapeutic interventions designed to reduce oxidative stress have been suggested for the treatment of PD.

3. Therapeutic Interventions

The results of a large, multicenter controlled clinical trial involving 800 patients receiving vitamin E either with placebo or with deprenyl (a monoamine oxidase type B inhibitor) have been reported (Parkinson Study Group, 1993). No beneficial effects of vitamin E were found and there did not appear to be any interaction between vitamin E and deprenyl. The failure of vitamin E to influence the progression of PD in this study does not necessarily mean that antioxidants may not be effective. Vitamin E is a secondary antioxidant that halts the chain reaction of lipid peroxidation and may be less effective than primary antioxidants that prevent the formation of oxygen-derived free radicals and the initiation of lipid peroxidation. It is also not known whether sufficient vitamin E accumulated in the appropriate regions of the central nervous system of these patients. Two prospective double-blind controlled trials have investigated the possibility that L-deprenyl could diminish the oxidative metabolism of dopamine and thus influence the rate of progression of the condition (Parkinson Study Group, 1989, 1993; Tetrud and Langston, 1989). Both studies concluded that deprenyl significantly delayed the development of disability and prolonged the time until symptomatic treatment with L-DOPA was required. This is consistent with the inhibition of dopamine metabolism resulting in reduced free radical production, although a symptomatic effect of deprenyl cannot be ruled out.

Other therapeutic strategies involving a reduction of oxidative stress have been suggested, including the removal of excess iron by chelators (Halliwell, 1989; Hall, 1992). Thus if oxidative stress does play a role in the pathogenesis of PD, there may be strategies available to alter the natural history of the condition.

B. Tardive Dyskinesia

Tardive Dyskinesia (TD) is a major complication of the long-term use of neuroleptic drugs for the control of acute psychotic behavior and ROS have been implicated in its etiology (Cadet et al., 1986). The use of neuroleptics results in an increased turnover of catecholamines and particularly dopamine in the brain (Korpi and Wyatt, 1984) and, as discussed earlier, the oxidation of dopamine produces ROS which could lead to lipid peroxidation and the destabilization of neuronal membranes. This hypothesis is supported by a number of lines of evidence. First, changes in membrane fluidity have been reported both *in vitro* (Zubenko and Cohen, 1985) and *in vivo* (Cohen and Zubenko, 1985) when cells are exposed to neuroleptics. Second, a significant increase in free radical activity and lipid peroxidation has been reported in the cerebrospinal fluid of patients with TD compared to controls (Lohr et al., 1990). Third, a number of studies have shown that vitamin E can be of benefit in this condition (Lohr et al., 1987; Elkashef et al., 1990; Egan et al., 1992; Adler et al., 1993; Dabiri et al., 1994).

C. Down Syndrome

The pathological mechanisms involved in Down syndrome (DS) are unknown, but oxidative stress as a result of alterations in the activity of cellular antioxidant systems has been implicated. DS generally results from a trisomy of chromosome 21. The gene for Cu,Zn SOD (SOD1) is encoded on this chromosome and, as a result of gene dosage, its activity is increased by 50% in all DS tissues, including the brain. SOD catalyzes the dismutation of the superoxide-free radical to oxygen and hydrogen peroxide and, as discussed earlier for PD, excess hydrogen peroxide can result in an increased production of the highly reactive hydroxyl radical.

A number of lines of evidence support a link between increased activity of SOD1 and increased concentrations of ROS in the pathogenesis of DS. In erythrocytes and fibroblasts from patients with DS, the increased concentration of hydrogen peroxide resulting from the increased SOD1 activity is compensated by an increase in glutathione peroxidase activity (Sinet et al., 1975, 1979). This is not a gene dosage effect as the gene for glutathione peroxidase is on chromosome 3 (Wijner et al., 1978). This compensatory increase in glutathione peroxidase activity does not, however, occur in the fetal DS brain (Brooksbank and Balazs, 1984) and this may, therefore, render it unusually susceptible to oxidative stress. In addition Brooksbank and Balazs (1984) have shown that fetal brain from DS patients was more susceptible to in vitro peroxidative stress than control brain tissue, suggesting that enhanced lipid peroxidation could occur in vivo.

The link between gene dosage, an increase in SOD1 activity, and increased free radical production has been confirmed by studies where the human SOD1 gene has been transfected into cell lines in vitro and mice in vivo. The addition of an extra SOD1 gene into human and mouse cells in vitro resulted in an increase in glutathione peroxidase activity, presumably as a result of increased SOD1 activity, a consequent increase in hydrogen peroxide concentrations (Ceballos-Picot et al., 1988; Kellner and Bagnell, 1990), and increased lipid peroxidation (Elroy-Stein et al., 1986). Studies with transgenic mice with an extra copy of the human SOD1 gene showed (i) a twofold increase in SOD1 activity in the brain, (ii) similar neuronal specificity of the human SOD1 activity as observed in humans, (iii) no increase in the activity of brain glutathione peroxidase, (iv) increased lipid peroxidation in the brain (Ceballos-Picot et al., 1991), and (v) ultrastructural changes in the neuromuscular junction of the tongue muscle which are similar to those seen in the tongue of DS patients (Avraham et al., 1988). These results suggest that gene dosage of the SOD1 gene could have deleterious effects on the nervous system. The effect of antioxidant supplementation in these transgenic mice does not appear to have been studied.

Increased in vitro lipid peroxidation and decreased vitamin E concentrations in red cells, and decreased plasma concentrations of vitamins E and A have been reported in patients with DS (Shah and Johnson, 1989), which is consistent with the possible involvement of oxidative stress. Intervention studies using antioxidants such as vitamin E have been suggested but it does not appear that any trials have been carried out.

D. Alzheimer's Disease

Patients with Alzheimer's Disease (AD) have a characteristic neuropathology consisting of a loss of neurons, accumulation of neurofibrillary tangles, deposition of amorphous aggregates of protein (amyloid), and scattered foci of cellular debris and amyloid known as neuritic plaques, which is particularly evident in the cerebral cortex and hippocampus. The neuropathology is indistinguishable from that found in adults with DS (Wisniewski et al., 1985) and for this reason, together with other evidence (see later), oxidative stress has been implicated in the pathogenesis of AD (Volicer and Crino, 1990).

Evidence for increased oxidative stress in AD includes increased lipid peroxidation in the frontal cortex (Subbarao et al., 1990) and increased oxidative damage to glutamine synthetase, an enzyme particularly sensitive to oxidation (Smith et al., 1991). Increased concentrations of aluminium are found in the brains of patients with AD and, as discussed previously, this could enhance iron-induced lipid peroxidation. In common with DS, lowered plasma concentrations of vitamins A and E have been reported in AD (Jeandel et al., 1989; Zaman et al., 1992). Despite the changes in circulating concentrations of antioxidants, no differences were observed in vitamin E concentrations in the cortex of patients with AD compared to appropriate controls (Metcalfe et al., 1989). Intervention studies using antioxidants have been suggested in AD but again it does not appear that any trials have been carried out.

E. Motor Neuron Disease

Motor neuron disease (MND), also known as amyotrophic lateral sclerosis or Lou Gehrig's disease, is a condition in which there is a degeneration of the motor neurons of the spinal cord, brain stem, and cerebral cortex. Approximately 10% of cases are familial and inherited as an autosomal dominant trait with the remainder being sporadic. The genetic defect in familial MND has now been linked to SOD1 (Deng et al., 1993; Rosen et al., 1993), and to date more than 20 different missense mutations in the SOD1 gene have been identified. Despite this important observation, the pathogenesis of familial MND is not understood. It is not clear, for example, whether the condition arises simply from a loss of SOD1 activity or from the gain of a novel function. Current research into these and related questions has been reviewed by Brown (1995) and is summarized here.

Mutations of SOD1 in familial MND have been shown to result in a loss of enzyme activity in a number of tissues including red blood cells, lymphoblastoid cell lines, and brain, which appears to result principally from diminished stability of the mutant protein. Inhibition of SOD1 activity in cultured slices of rat spinal cord resulted in apoptotic degeneration of spinal neurons, including motor neurons (Rothstein et al., 1994). This toxicity could be prevented by the antioxidant N-acetylcysteine and potentiated by the inhibition of glutamate transport. These observations, therefore, suggest that the loss of motor neurons in MND could result from decreased SOD1 activity, possibly potentiated by impaired glutamate transport.

Evidence for a gain of function comes from studies in transgenic mice that express an extra copy of either the wild type or the mutant human gene for SOD1. Those with the wild-type gene showed no signs of MND, whereas those given the mutant gene developed a syndrome similar to that seen in MND patients (Gurney *et al.*, 1994). The mechanism for this apparent cytotoxicity of the mutant SOD1 is not understood. One hypothesis implicates peroxynitrite (Beckman, 1993), which is formed by the reaction of superoxide with nitric oxide. Beckman (1993) has suggested that the mutant SOD1 molecule may not only have a reduced ability to scavenge superoxide, but also an increased affinity for peroxynitrite. Reduced dismutation of superoxide could produce increased concentrations of peroxynitrite, which in turn could result in the nitration of critical cellular targets such as the neurofilaments of the motor neurons and tyrosine kinase receptors.

Because the familial and sporadic forms of MND are indistinguishable, both clinically and pathologically, it has been suggested that the mechanisms of cell injury are likely to be similar. No changes in SOD1 activity have been reported in patients with the sporadic form of the condition but evidence of oxidative stress in postmortem tissues, such as increased concentrations of carbonyl proteins, have been reported (Bowling *et al.*, 1993).

Because the genetic defect in familial MND affects an antioxidant enzyme, it is hoped that detailed studies in this condition will result in an improved understanding of the mechanisms involved not only in MND but also in other neurodegenerative conditions where oxidative stress has been implicated.

F. Ischemia/Reperfusion and Trauma

The overall outcome following a cardiac arrest or stroke is dependent on reducing damage to the brain. Although significant brain damage occurs during the period of ischemia, further damage appears to occur following reperfusion. Trauma to the central nervous system results in primary mechanical damage and also to secondary consequences which appear to be very similar to those seen in ischemia/reperfusion (I/R) (see Braughler and Hall, 1989). The biochemical mechanisms seen in I/R and trauma are, therefore, considered together.

I. Evidence for Increased Oxidative Stress

The mechanisms involved in the neuronal cell death and brain damage associated with I/R and trauma are likely to be multiple, but among them there is evidence that ROS are involved:

1. Increased indices of peroxidation have been reported in I/R and trauma (see, e.g., Hall and Braughler, 1989; Carney and Floyd, 1991; Sakamoto *et al.*, 1991).

2. A decrease in antioxidant concentrations (vitamins C and E, reduced glutathione, ubiquinols) in the central nervous system following trauma and I/R has been reported by a number of groups suggesting that these antioxidants are being utilized to protect against oxidative stress (Yoshida *et al.*, 1982; Pietronigro *et al.*, 1983; Saunders *et al.*, 1987; Lemke *et al.*, 1990; Mizui *et al.*, 1992). Depletion of reduced glutathione with buthionine sulfoximine, a specific inhibitor of glutathione synthesis, exacerbated the cortical infarction following ischemia (Mizui *et al.*, 1992).

3. Inhibition of lipid peroxidation by compounds with known antioxidant activity has been shown to attenuate the effects of traumatic injury and I/R. Thus the administration of α-tocopherol (in some studies together with selenium) before trauma or I/R resulted in a reduction in lipid peroxidation (Yamamoto *et al.*, 1983; Saunders *et al.*, 1987). The converse has also been shown, i.e., that vitamin E-deficient rats exposed to I/R showed evidence of increased lipid peroxidation (Yoshida *et al.*, 1984). Increasing the activity of SOD in the central nervous system can also be protective. Thus SOD conjugated to polyethylene glycol and entrapped within lysosomes reduced cortical infarction following focal cerebral ischemia (Imaizumi *et al.*, 1990; Traystman *et al.*, 1991). Further evidence comes from the use of transgenic mice, which over-express SOD1 activity (Kinouchi *et al.*, 1991). Following focal cerebral ischemia the transgenic animals had significantly decreased infarction and brain edema, with increased concentrations of reduced glutathione and ascorbate in the surrounding cortex and striatum compared with control mice. The addition of other agents known to have antioxidant activity such as methylprednisolone (Hall and Braughler, 1989), the 21-aminosteroids (Hall and Braughler, 1989), and spin traps (Carney and Floyd, 1991) have been shown to protect against the effects of trauma and I/R.

4. The deliberate induction of peroxidative damage, e.g., by the microinjection of ferrous chloride into the spinal cord or the infusion of ADP/iron/hypoxanthine/xanthine oxidase into brain has been shown to cause similar pathological changes to those seen in trauma and I/R (Anderson and Mearns, 1983; Chan *et al.*, 1984).

2. Possible Mechanisms for Increased Oxidative Stress

The mechanism(s) for the increased production of ROS remains speculative but a number of possibilities exist that are not mutually exclusive. These are summarized here as they have been reviewed in detail elsewhere (e.g., Braughler and Hall, 1989; Hall and Braughler, 1989; Traystman *et al.*, 1991). Traumatic injury or ischemia has been shown to lead to an accumulation of free fatty acids, particularly arachidonic acid. This is thought to result from a failure of ATP-dependent ionic pumps, resulting in an influx of calcium ions, subsequent activation of phospholipase C, and the liberation of free fatty acids from the membrane. Upon reperfusion the released arachidonic acid is metabolized via the lipoxygenase and cyclo-oxygenase pathways to prostaglandins and thromboxanes, and in the process superoxide is produced.

Another postulated mechanism involves the metabolism of adenine nucleotides which build up during ischemia. This mechanism is dependent on the fact that xanthine dehydrogenase is converted to xanthine oxidase by calcium-activated proteases during ischemia. The dehydrogenase transfers its electrons to NAD, whereas the oxidase uses oxygen and produces superoxide and hydrogen peroxide. The irreversible proteolytic conversion of xanthine dehydrogenase to the oxidase was first suggested by Granger *et al.* (1981) to explain the production of superoxide during I/R of the intestine of the cat. This conversion has since been described in many tissues (e.g., Engerson *et al.*, 1987). The situation in the brain is not, however, clear and Mink *et al.* (1990) were unable to find any evidence for this conversion during canine cerebral ischemia.

Other sources of increased concentrations of ROS in ischemic brain are (i) an increased leak of superoxide from the electron transport chain when supplies of oxygen are reduced, (ii) an increase in the respiratory burst as a result of neutrophil activation in response to tissue injury, (iii) autoxidation of catecholamines and an increase in monoamine oxidase activity as described previously in PD, and (iv) direct exposure of the central nervous system to hemoglobin, which in addition to providing a rich source of iron can itself undergo slow autoxidation resulting in the production of superoxide (see Braughler and Hall, 1989). This mechanism may be particularly relevant to trauma. ROS can react with most cellular macromolecules and there are, therefore, numerous mechanisms by which they can cause damage.

Another mechanism that may contribute to cerebral injury involves the excitatory amino acid neurotransmitters such as glutamate and aspartate. Concentrations of these amino acids have been shown to increase following ischemia (Beneveniste *et al.*, 1984) and this could lead to a stimulation of nitric oxide synthesis from arginine. In normal brain, nitric oxide is nontoxic and an important neuronal messenger (Dawson *et al.*, 1992), but following reperfusion it is able to react with superoxide to form peroxynitrite, which in turn decomposes to form the highly reactive hydroxyl radical or a radical with similar reactivity (see Beckman, 1991). These observations, therefore, provide a link between two apparently different mechanisms of brain injury, i.e., the generation of excitatory amino acids and ROS (Bondy, 1995).

G. Neuronal Ceroid Lipofuscinoses

Neuronal ceroid lipofuscinoses (NCL) or Batten's disease is a group of neurodegenerative conditions that can be divided into the infantile, late infantile, juvenile, and adult forms. In all these conditions there is a characteristic accumulation of storage material in lysosomes of both neuronal and nonneuronal cells that has the histochemical properties of ceroid or lipofuscin (Lake, 1992). It has been shown that a major component of this lipofuscin material in all forms of NCL, except infantile, is a proteolipid with the same amino acid sequence as subunit C of mitochondrial ATP synthase (Hall *et al.*, 1991). The accumulation of lipofuscin, which is thought to be a product of peroxidation, is likely to be a secondary phenomenon. In Finland, antioxidant therapy has been used in patients with juvenile NCL (Santavuori *et al.*, 1988), and although it did not prevent or correct the condition, the authors considered that some patients derived some benefit from this treatment. We have investigated a mouse model of NCL and found that vitamin E concentrations were significantly reduced (Martin *et al.*, 1994) and lipid peroxidation increased (unpublished observations) in moderately affected animals. The addition of antioxidant supplements normalized the biochemical parameters but did not appear to influence the clinicopathological process in the mutant mice (Martin *et al.*, 1995).

IV. Conclusions

Oxidative stress, resulting from either decreased protection from antioxidant systems or increased concentrations of ROS, has been implicated in a whole range of neurological conditions. In the majority of these conditions it is not known whether oxidative stress is part of the primary degenerative process or secondary to cellular degeneration following another underlying mechanism. In either situation, therapeutic interventions designed to reduce oxidative stress could be of benefit. The treatment of severe vitamin E deficiency with appropriate supplements of the vitamin can either prevent or at least halt the progression of the characteristic neurological features but in the majority of clinical neurological conditions the therapeutic benefit of antioxidant supplementation still requires to be proved.

Acknowledgments

I thank the Friedreich's Ataxia Group, F. Hoffmann-La Roche Ltd., the Leverhulme Trust, and the Szeben Peto Foundation for their generous financial support over the years.

References

Adler, L. A., Peselow, E., Rotroson, J., Duncan, E., Lee, M., Rosenthal, M., and Angrist, B. (1993). Vitamin E treatment of tardive dyskinesia. *Am. J. Psych.* 150, 1405–1407.
Alvarez, F., Landrieu, P., Laget, P., Lemonnier, F., Odievre, M., and Alagille, D. (1983). Nervous and ocular disorders in children with cholestasis and vitamin A and E deficiences. *Hepatology* 3, 410–414.

Anderson, D. K., and Mearns, E. D. (1983). Lipid peroxidation in spinal cord: FeCl2 induction and protection with antioxidants. *Neurochem. Pathol.* **1**, 249–264.

Avraham, K. B., Schickler, M., Sapoznikov, D., Yarom, R., and Groner, Y. (1988). Down's syndrome: Abnormal neuromuscular function in tongue of transgenic mice with elevated levels of human Cu/Zn superoxide dismutase. *Cell* **54**, 823–829.

Azizi, E., Zaidman, J. L., Eschar, J., and Szeinberg, A. (1978). Abetalipoproteinaemia treated with parenteral and oral vitamins A and E and with medium chain triglycerides. *Acta Pediatr. Scand.* **67**, 797–801.

Beckman, J. S. (1991). The double-edged role of nitric oxide in brain function and superoxide-mediated injury. *J. Dev. Physiol.* **15**, 53–59.

Beckman, J. S., Carson, M., and Smith, C. D. (1993). ALS, SOD and peroxinitrite. *Nature* **364**, 584.

Beneveniste, H., Drejer, J., Schousboe, A., and Diemer, N. H. (1984). Elevation of the extracellular concentrations of glutamate and aspartate in rat hippocampus during transient cerebral ischemia monitored by intracerebral microdialysis. *J. Neurochem.* **43**, 1369–1374.

Bishara, S., Merin, S., Cooper, M., Azizi, E., Delpre, G., and Deckelbaum, R. J. (1982). Combined vitamin A and E therapy prevents retinal electrophysiological deterioration in abetalipoproteinaemia. *Br. J. Ophthalmol.* **66**, 767–770.

Bohlmann, H. E., Thiede, H., Rosenstiel. K., Herdermenten, S., Panitz, D., and Tackmann, W. (1972). Abetalipoproteinamie bei drei Geschwiston. *Dtsch Med. Wochenschr.* **97**, 892–896.

Bondy, S. C. (1995). The relation of oxidative stress and hyperexcitation to neurological disease. *Proc. Soc. Exp. Biol. Med.* **208**, 337–345.

Bowling, A. C., Schulz, J. B., Brown, R. H. J., and Beal, M. F. (1993). Superoxide dismutase activity, oxidative damage and mitochondrial energy metabolism in familial and sporadic amyotrophic lateral sclerosis. *J. Neurochem.* **61**, 2322–2325.

Braughler, J. M., and Hall, E. D. (1989). Central nervous system trauma and stroke. I. Biochemical considerations for oxygen radical formation and lipid peroxidation. *Free Rad. Biol. Med.* **6**, 289–301.

Brooksbank, B. W. L., and Balazs, R. (1995). Superoxide dismutase, glutathione peroxidase and lipoperoxidation in Down's syndrome fetal brain. *Dev. Brain Res.* **16**, 37–44.

Brown, R. H. J. (1995). Amyotrophic lateral sclerosis: Recent insights from genetics and transgenic mice. *Cell* **80**, 687–692.

Burton, G. W., Joyce, A., and Ingold, K. U. (1983). Is vitamin E the only lipid soluble chain-breaking antioxidant in human blood plasma and erythrocyte membranes? *Arch. Biochem. Biophys.* **221**, 281–290.

Cadet, J. L., Lohr, J. B., and Jeste, D. V. (1986). Free radicals and tardive dyskinesia. *Trends Neurosci.* **9**, 107–108.

Carney, J. M., and Floyd, R. A. (1991). Protection against oxidative damage to CNS by a-phenyl-tert-butyl nitrone (PBN) and other spin-trapping agents: A novel series of nonlipid free radical scavengers. *J. Mol. Neurosci.* **3**, 47–57.

Ceballos-Picot, I., Delabas, J. M., Nicole, A., Lynch, R. E., Hallewell, R. A., Kamoun, P., and Sinet, P. M. (1988). Expression of transfected Cu/Zn superoxide dismutase gene in mouse L cells and NS 20Y neuroblastoma cells induces enhancement of glutathione peroxidase activity. *Biochim. Biophys. Acta* **949**, 58–64.

Ceballos-Picot, I., Nicole, A., Briand, P., Grimber, G., Delacourte, A., Defossez, A., Javoy-Agid, F., Lafon, M., Blouin, J. L., and Sinet, P. M. (1991). Neuronal specific expression of human copper-zinc superoxide dismutase in transgenic mice: animal model of gene dosage in Down's syndrome. *Brain Res.* **552**, 198–214.

Chan, P. H., Fishman, R. A., and Longar, S. M. (1984). Brain injury edema, and vascular permeability changes induced by oxygen-derived free radicals. *Neurology* **34**, 315–320.

Chiswick, M. L., Johnson, M., Woodall, C., Gowland, M., Davies, J., Toner, N., and Sims, D. G. (1983). Protective effect of vitamin E (DL-alpha-tocopherol) against intraventricular haemorrhage in premature babies. *BMJ* **287**, 81–84.

Cleeter, M. W. J., Cooper, J. M., and Schapira, A. H. V. (1992). Irreversible inhibition of mitochondrial complex 1 by 1-methyl-4-phenylpyridinium: evidence for free radical involvement. *J. Neurochem.* **58**, 786–789.

Cohen, B. M., and Zubenko, G. S. (1985). In vivo effects of psychotropic agents on the physical properties of cell membranes in the rat brain. *Psychopharmacology* **86**, 365–368.

Committee on Fetus and Newborn. (1985). Vitamin E and the prevention of retinopathy of prematurity. *Pediatrics* **76**, 315–316.

Dabiri, L. M., Pasta, D., Darby, J. K., and Mosbacher, D. (1994). Effectiveness of vitamin E for treatment of long-term tardive dyskinesia. *Am. J. Psych.* **151**, 925–926.

Davis, G. C., Williams, A. C., Markey, S. P., Ebert, M. H., Caine, E. D., Reichert, C. M., and Kopin, L. J. (1979). Chronic parkinsonism secondary to intravenous injection of merperidine analogues. *Psych. Res.* **1**, 249–254.

Dawson, T. M., Dawson, V. L., and Snyder, S. H. (1992). A novel neuronal messenger molecule in brain: The free radical, nitric oxide. *Ann. Neurol.* **32**, 297–311.

Deng, H., Hentati, A., Tainer, J. A., Iqbal, Z., Cayabyab, A., Hung, W., Getzoff, E. D., Hu, P., Herzfeldt, B., Roos, R. P., Warner, C., Deng, G., Soriano, E., Smyth, C., Parge, H. E., Ahmed, A., Roses, A. D., Hallewell, R. A., Pericak-Vance, M. A., and Siddique, T. (1993). Amyotrophic lateral sclerosis and structural defects in Cu,Zn superoxide dismutase. *Science* **261**, 1047–1051.

Dexter, D. T., Carayon, A., Vidailhet, M., Ruberg, M., Agid, F., Agid, Y., Lees, A. J., Wells, F. R., Jenner P., and Marsden, C. D. (1990). Decreased ferritin levels in brain in Parkinson's disease. *J. Neurochem.* **55**, 16–20.

Dexter, D. T., Carter, C. J., Wells. F. R., Javoy-Agid, F., Agid, Y., Lees, A., Jenner, P., and Marsden, C. D. (1989a). Basal lipid peroxidation in substantia nigra is increased in Parkinson's disease. *J. Neurochem.* **52**, 381–389.

Dexter, D. T., Wells, F. R., Lees, A. J., Agid, F., Agid, Y., Jenner, P., and Marsden, C. D. (1989b). Increased nigral iron content and alterations in other metal ions occurring in brain in Parkinson's disease. *J. Neurochem.* **52**, 1830–1836.

Earle, K. M. (1968). Studies on Parkinson's disease including X-ray, fluorescent spectroscopy of formalin-fixed brain tissue. *J. Neuropathol. Exp. Neurol.* **27**, 1–14.

Egan, M. F., Hyde, T. M, Albers, G. W., Elkashef, A., Alexander, R. C., Reeve, A., Blum, A., Saeng, R. E., and Wyatt, R. J. (1992). Treatment of tardive dyskinesia with vitamin E. *Am. J. Psych.* **149**, 773–777.

Einarson, L. (1952). Criticizing review of the concepts of the neuromuscular lesions in experimental vitamin E deficiency, preferably in adult rats. *Acta Psych. Scand.* **78**, 9–76.

Elias, E., Muller, D. P. R., and Scott, J. (1981). Association of spinocerebellar disorders with cystic fibrosis or chronic childhood cholestasis and very low serum vitamin E. *Lancet* **2**, 1319–1321.

Elkashef, A. M., Ruskin, P. E., Bacher, N., and Barrett, D. (1990). Vitamin E in the treatment of tardive dyskinesia. *Am. J. Psych.* **147**, 505–506.

Elroy-Stein, O., Bernstein, Y., and Groner, Y. (1986). Overproduction of human Cu-Zn superoxide dismutase in transfected cells: Extenuation of paraquat-mediated cytotoxicity and enhancement of lipid peroxidation. *EMBO J.* **5**, 615–622.

Engerson, T. D., McKelvey, T. G., Rhyme, D. B., Boggio, E. B., Snyder, S. J., and Jones, H. P. (1987). Conversion of xanthine dehydrogenase to oxidase in ischemic rat tissues. *J. Clin. Invest.* **79**, 1564–1570.

Goss-Sampson, M. A., Kriss, A., and Muller, D. P. R. (1990). A longitudinal study of somatosensory, brainstem auditory and peripheral sensory: Motor conduction during vitamin E deficiency in the rat. *J. Neurol. Sci.* **100**, 79–84.

Goss-Sampson, M. A., MacEvilly, C. J., and Muller, D. P. R. (1988). Longitudinal studies of the neurobiology of vitamin E and other antioxidant systems, and neurological function in the vitamin E deficient rat. *J. Neurol. Sci.* **87**, 25–35.

Goss-Sampson, M. A., Muller, D. P. R., and Kriss, A. (1991). Abnormalities of the electroretinogram and visual evoked potential in vitamin E deficient rats. *Exp. Eye Res.* 53, 623–627.

Gouras, P., Carr, R. E., and Gunkel, R. D. (1971). Retinitis pigmentosa in abetalipoproteinaemia: Effects of vitamin A. *Invest. Ophthalmol.* 10, 784–793.

Graham, D. G. (1979). On the origin and significance of neuromelanin. *Arch. Pathol. Lab. Med.* 103, 359–362.

Granger, D. N., Rutili, G., and McCord, J. M. (1981). Superoxide radicals in feline intestinal ischemia. *Gastroenterology* 81, 22–29.

Guggenheim, M. A., Ringel, S. P., Silverman, A., and Grabert, B. E. (1982). Progressive neuromuscular disease in children with chronic cholestasis and vitamin E deficiency: Diagnosis and treatment with alpha tocopherol. *J. Pediatr.* 100, 51–58.

Gurney, M. E., Pu, H., Chiu, A. Y., Canto, M. C. D., Polchow, C. Y., Alexander, D. D., Caliendo, J., Hentati, A., Kwon, Y. W., Deng, H., Chen, W., ZZhai, P., Sufit, R. L., and Siddique, T. (1994). Motor neuron degeneration in mice that express a human Cu,Zn superoxide dismutase mutation. *Science* 264, 1772–1775.

Gutteridge, J. M., Quinlan, G. J., Clark, I., and Halliwell, B. (1985). Aluminium salts accelerate peroxidation of membrane lipids stimulated by iron salts. *Biochim. Biophys. Acta* 835, 441–447.

Hall, E. D. (1992). Novel inhibitors of iron-dependent lipid peroxidation for neurodegenerative disorders. *Ann. Neurol.* 32, S137–S142.

Hall, E. D., and Braughler, J. M. (1989). Central nervous system trauma and stroke. I. Physiological and pharmacological evidence for involvement of oxygen radicals and lipid peroxidation. *Free Rad. Biol. Med.* 6, 303–313.

Hall, N. A., Lake, B. D., Dewji, N. N., and Patrick, A. D. (1991). Lysosomal storage of subunit C of mitochondrial ATP synthase in Batten's disease (ceroid-lipofuscinosis). *Biochem. J.* 275, 269–272.

Halliwell, B. (1989). Oxidants and the central nervous system: some fundamental questions. *Acta Neurol. Scand.* 126, 23–33.

Handelman, G. J., and Dratz, E. A. (1986). The role of antioxidants in the retina and retinal pigment epithelium and the nature of prooxdant-induced damage. *Adv. Free Rad. Biol. Med.* 2, 1–89.

Harding, A. E., Muller, D. P. R., Thomas, P. K., and Willison, H. J. (1982). Spinocerebellar degeneration secondary to chronic intestinal malabsorption: A vitamin E deficiency syndrome. *Ann. Neurol.* 12, 419–424.

Harries, J. T., and Muller, D. P. R. (1971). Absorption of vitamin E in children with biliary obstruction. *Gut* 12, 579–584.

Hegele, A., and Angel, A. (1985). Arrest of neuropathy and myopathy in abetalipoproteinemia with high-dose vitamin E therapy. *Can. Med. Assoc. J.* 132, 41–44.

Herbert, P. N., Gotto, A. M., and Fredrickson, D. S. (1978). Familial lipoprotein deficiency. *In* "The Metabolic Basis of Inherited Disease" (J. B. Stanbury, D. S. Wyngaarden, and D. S. Fredrickson, eds.), pp. 544–588. McGraw Hill, New York.

Hirsch, E. C. (1992). Why are nigral catecholaminergic neurons more vulnerable than other cells in Parkinson's disease? *Ann. Neurol.* 32, S88–S93.

Howard, L., Oveson, L., Satya-Murti, S., and Chu, R. (1982). Reversible neurological symptoms caused by vitamin E deficiency in a patient with short bowel syndrome. *Am. J. Clin. Nutr.* 36, 1243–1249.

Imaizumi, S., Woolworth, V, Fishman, R. A., and Chan, P. H. (1990). Liposome-entrapped superoxide dismutase reduces cerebral infarction in cerebral ischemia in rats. *Stroke* 21, 1312–1317.

Ingold, K. U., Burton, G. W., Foster, D. O., Hughes, L., Lindsay, D. A., and Webb, A. (1987). Biokinetics of and discrimination between dietary RRR- and SRR-α-tocopherols in the male rat. *Lipids* 22, 163–172.

Jeandel, C., Nicolas, M. B., Dubois, F., Nabet-Belleville, F., Penin, F., and Cury, G. (1989). Lipid peroxidation and free radical scavengers in Alzheimer's disease. *Gerentology* **35**, 275–282.

Jenner, P., Sian, J. D., D. T., Schapira, A. H. V., and Marsden, C. D. (1992). Oxidative stress as a cause of nigral cell death in Parkinson's disease and incidental Lewy body disease. *Ann. Neurol.* **32**, S82–S87.

Kane, J. P., and Havel, R. J. (1989). Disorders of the biogenesis and secretion of lipoproteins containing the B apolipoproteins. In "The Metabolic Basis of Inherited Disease" (C. R. Scriver, A. L. Beaudet, W. S. Sly, and D. Valle, eds.), pp. 1139–1164. McGraw Hill, New York.

Kellner, M. J., and Bagnell, R. (1990). Alteration of endogenous glutathione peroxidase, manganese superoxide dismutase and glutathione transferase activity in cells transfected with a copper-zinc superoxide dismutase expression vector. *J. Biol. Chem.* **265**, 10872–10875.

Kinouchi, H., Epstein, C. J., Mizui, T., Carlson, E., Chen, S. F., and Chan, P. H. (1991). Attenuation of focal cerebral ischemic injury in transgenic mice overexpressing Cu Zn superoxide dismutase. *Proc. Natl. Acad. Sci. U.S.A.* **88**, 11158–11162.

Korpi, E. R., and Wyatt, R. J. (1984). Reduced haloperidol: Effects on striatal dopamine metabolism and conversion to haloperidol in the rat. *Psychopharmacology* **83**, 34–37.

Kretzer, F. L., and Hittner, H. M. (1988). Retinopathy of prematurity: Clinical implications of retinal development. *Arch. Dis. Child.* **63**, 1151–1167.

Lake, B. D. (1992). Lysosomal and peroxisomal disorders. In "Greenfield's Neuropathology" (J. H. Adams and L. W. Duchen, eds.), pp. 709–810. Edward Arnold, London.

Langston, J. W., Ballard, P., Tetrud, J. W., and Irwin, I. (1983). Chronic parkinsonism in humans due to a product of merperidine-analog synthesis. *Science* **219**, 979–980.

Law, M. R., Wijewardene, K., and Wald, N. J. (1990). Is routine vitamin E administration justified in very low birth weight infants? *Dev. Med. Child. Neurol.* **32**, 442–450.

Lemke, M., Ames, B. N., B., and Faden, A. I. (1990). Decrease in tissue levels of ubiquinol-9 and -10, ascorbate and α-tocopherol following spinal cord impact trauma in rats. *Neurosci. Lett.* **108**, 201–206.

Lohr, J. B., Jeste, D. V. C., M. A, and Wyatt, R. J. (1987). Alpha-tocopherol in tardive dyskinesia. *Lancet* **1**, 913–914.

Lohr, J. B., Kuczenski, R., Bracha, H. S., Moir, M., and Jeste, D. V. (1990). Increased indices of free radical activity in the cerebrospinal fluid of patients with tardive dyskinesia. *Biol. Psych.* **28**, 535–539.

MacEvilly, C. J., and Muller, D. P. R. (1996). Lipid peroxidation in neural tissues and fractions from vitamin E deficient rats. *Free Rad. Biol. Med.* **20**, 639–648.

Machlin, L. J., Filipski, R., Nelson, J. S., Horn, L. R., and Brin M. (1977). Effects of a prolonged vitamin E deficiency. *J. Nutr.* **107**, 1200–1208.

Mann, D. M. A., and Yates, P. O. (1983). Possible role of neuromelanin in the pathogenesis of Parkinson's disease. *Mech. Ageing Dev.* **21**, 193–203.

Martin, J. E., Hindmarsh, A., Merryweather, I., and Muller, D. P. R. (1994). Antioxidant deficit in the Mnd mouse. *Brain Pathol.* **4**, 458.

Martin, J. E., Jowatt, V., Merryweather, I., and Muller, D. P. R. (1995). The effect of antioxidant supplementation in the Mnd mouse. *Nreuropathol. Appl. Neurobiol.* **21**, 448.

Metcalfe, T., Bowen, D. M., and Muller, D. P. R. (1989). Vitamin E concentrations in human brain of patients with Alzheimer's disease, fetuses with Down's syndrome, centenarians and controls. *Neurochem. Res.* **14**, 1209–1212.

Miller, R. G., Davis, C. J. F., Illingworth, D. R., and Bradley, W. (1980). The neuropathy of abetalipoproteinemia. *Neuropathy* **30**, 1286–1289.

Mink, R. B., Dutka, A. J., Kumaroo, K. K., and Hallenbeck, J. M. (1990). No conversion of xanthine dehydrogenase to oxidase in canine cerebral ischemia. *Am. J. Physiol.* **259**, H1655–H1659.

Mizui, T., Kinouchi, H., and Chan, P. H. (1992). Depletion of brain glutathione by buthionine sulfoximine enhances cerebral ischemic injury in rats. *Am. J. Physiol.* **262**, H313–H317.

Mogi, M., Harada, M., Kuichi, K., Kondo, T., Narabayashi, H., Rausch, D., Riederer, P., Jellinger, K., and Nagatsu, T. (1988). Homospecific activity (activity per enzyme protein) of tyrosine hydroxylase increases in Parkinson's brain. *J. Neural Transm.* **72**, 77–81.

Molenaar, I, Vos, J., and Hommes, F. A. (1972). Effect of vitamin E deficiency on cellular membranes. *Vits. Horms.* **30**, 45–82.

Muller, D. P. R. (1987). Free radical problems of the newborn. *Proc. Nutr. Soc.* **46**, 69–75.

Muller, D. P. R. (1992). Vitamin E therapy in retinopathy of prematurity. *Eye* **6**, 221–225.

Muller, D. P. R., and Goss-Sampson, M. A. (1990). Neurochemical, neurophysiological and neuropathological studies in vitamin E deficiency. *Crit. Rev. Neurobiol.* **5**, 239–263.

Muller, D. P. R., Harries, J. T., and Lloyd, J. K. (1974). The relative importance of the factors involved in the absorption of vitamin E in children. *Gut* **15**, 966–971.

Muller, D. P. R., Lloyd, J. K., and Bird, A. C. (1977). Long term management of abetalipoproteinaemia: Possible role for vitamin E. *Arch. Dis. Child.* **52**, 209–214.

Muller, D. P. R., Lloyd, J. K., and Wolff, O. H. (1983). Vitamin E and neurological function. *Lancet* **1**, 225–228.

Nelson, J. S. (1987). Effects of free radical scavengers on the neuropathology of mammalian vitamin E deficiency. *In* "Clinical and Nutritional Aspects of Vitamin E" (O. Hayaishi and M. Mino, eds.), pp. 157–159. Elsevier, Amsterdam.

Nelson, J. S., Fitch, C. D., Fischer, V. W., Broun, G. O., and Chou, A. C. (1981). Progressive neuropathologic lesions in vitamin E deficient Rhesus monkeys. *J. Neuropathol. Exp. Neurol.* **40**, 166–186.

Nelson, J. S., Fitch, C. D., Fischer, V. W., Rosenblum, J., Keating, J., Prensky, A., Celisia, G., Broun, G., Chou, A., Machlin, L., and Woolsey, R. (1978). Progressive neuropathologic lesions with vitamin E deficiency in mammals including man. *J. Neuropathol. Exp. Neurol.* **37**, 666.

Ng, Y. K., Fielder, A. R., Shaw, D. E., and Levene, M. I. (1988). Epidemiology of retinopathy of prematurity. *Lancet* **II**, 1235–1238.

Olanow, C. W. (1990). Oxidation reactions in Parkinson's disease. *Neurology* **40**, 32–37.

Olanow, C. W. (1993). A radical hypothesis for neurodegeneration. *Trends Neurosci.* **16**, 439–944.

Ouahchi, K., Arita, M., Kayden, H., Hentati, F., Hamida, M. B., Sokol, R., Arai, H., Inoue, K., Mandel, J., and Koenig, M. (1995). Ataxia with isolated vitamin E deficiency is caused by mutations in the α-tocopherol transfer protein. *Nature Genet.* **9**, 141–145.

Owens, W. C., and Owens, E. U. (1949). Retrolental fibroplasia in premature infants. II. Studies on the prophylaxis of the disease: the use of alpha-tocopheryl acetate. *Am. J. Ophthalmol.* **32**, 1631–1637.

Parkinson Study Group. (1989). Effect of deprenyl on the progression of disability in early Parkinson's disease. *N. Engl. J. Med.* **321**, 1364–1371.

Parkinson Study Group. (1993). Effects of tocopherol and deprenyl on the progression of disability in early Parkinson's disease. *N. Engl. J. Med.* **328**, 176–183.

Phelps, D. L. (1982). Vitamin E and retrolental fibroplasia in 1982. *Pediatrics* **70**, 420–425.

Pietronigro, D. D., Housepian, M., Demopoulos, H. B., and Flamm, E. S. (1983). Loss of ascorbic acid from injured feline spinal cord. *J. Neurochem.* **41**, 1072–1076.

Riederer, P., Sofic, E., Rausch, W., Schmidt, B., Reynolds, G. P., Jellinger, K., and Yaudim, M. B. H. (1989). Transition metals, ferritin, glutathione, and ascorbic acid in parkinsonian brains. *J. Neurochem.* **52**, 515–520.

Rosen, D. R., Siddique, T., Patterson, D., Figlewicz, D. A., Sapp, P. H. A, Donaldsdon, D., Goto, J., O'Regan, J. P., Deng, H., Rohmani, Z., Krizus, A., McKenna-Yasek, D., Cayabyab, A., Gaston, S. M., Berger, R., Tanzi, R. E., Halperin, J. J., Hertzfeldt, B., Van den Berg, R., Hung, W., Bird, T., Deng, G., Mulder, D. W., Smyth, C., Laing, N. G., Soriano, E.,

Pericak-Vance, M. A., Haines, J., Rouleau, G. A., Gusella, J. S., Horvitz, H. R., and Brown, R. H., Jr. (1993). Mutations in Cu/Zn superoxide dismutase gene are associated with familial amyotrophic lateral sclerosis. *Nature* **362**, 59–62.

Rosenblum, J. L., Keating, J. P., Prensky, A. L., and Nelson, J. S. (1981). A progressive neurologic syndrome in children with chronic liver disease. *N. Engl. J. Med.* **304**, 503–508.

Rossetti, Z. L., Sotgiu Sharp, D. E., A., Hadjiconstantinou, M., and Neff, N. H. (1988). 1-Methyl-4-phenyl-1,2,3,6-tetrahydropyridine (MPTP) and free radicals in vitro. *Biochem. Pharmacol.* **37**, 4573–4574.

Rothstein, J. D., Bristol, L. A., Hostler, B., Brown, R. H. J., and Kuncl, R. W. (1994). Chronic inhibition of superoxide dismutase produces apoptotic death of spinal neurons. *Proc. Natl. Acad. Sci. USA* **91**, 4155–4159.

Sakamoto, A., Ohniski, S. T., Ohnishi, T., and Ogawa, R. (1991). Relationship between free radical production and lipid peroxidation during ischemia-reperfusion injury in the rat brain. *Brain Res.* **554**, 186–192.

Santavuori, P., Heiskala, H., Westermarck, T., Sainio, K., and Moren, R. (1988). Experience over 17 years with antioxidant treatment in Spielmeyer-Sjogren disease. *Am. J. Med. Gen. Suppl.* **5**, 265–274.

Saunders, R. D., Dugan, L. L., Demeduik, P., Means, E. D., Horrocks, L. A., and Anderson, D. K. (1987). Effects of methylprednisolone and the combination of α-tocopherol and selenium on arachidonic acid metabolism and lipid peroxidation in traumatized spinal cord tissue. *J. Neurochem.* **49**, 24–31.

Schapira, A. H. V., Mann, V. M., Cooper, J. M., Krige, D., Jenner, P. J., and Marsden, C. D. (1992). Mitochondrial function in Parkinson's disease. *Ann. Neural.* **32**, S116–S124.

Schapira, A. H. V., Mann, V. M., Dexter, D., Cooper, J. M., Daniel, S. E., Jenner, P., Clark, J. B., and Marsden, C. D. (1990). Anatomic and disease specificity of NADH CoQ reductase (complex 1) deficiency in Parkinson's disease. *J. Neurochem.* **55**, 2142–2145.

Shah, S. N., and Johnson, R. C. (1989). Antioxidant vitamin (A and E) status of Down's syndrome subjects. *Nutr. Res.* **9**, 709–715.

Sharp, D., Blinderman, L., Combs, K. A., Kienzle, B., Ricci, B., Wager-Smith, K., Gil, C. M., Turck, C. W., Bouma, M., Rader, D. J., Aggerbeck, L. P., Gregg, R. E., Gordon, D. A., and Wetterau, J. R. (1993). Cloning and Gene defects in microsomal triglyceride transfer protein associated with abetalipoproteinaemia. *Nature* **365**, 65–69.

Sian, J., Dexter, D. T., Lees, A. L., Daniel, S., Jenner, P., and Marsden, C. D. (1994). Glutathione-related enzymes in brain in Parkinson's disease. *Ann. Neurol.* **36**, 356–361.

Sinet, P. M., Lejeune, J., and Jerome, H. (1979). Trisomy 21 (Down's syndrome), glutathione peroxidase, hexose monophosphate shunt and I.Q. *Life Sci.* **24**, 29–34.

Sinet, P. M., Michelson, A. M., Bazin, A., Lejeune, J., and Jerome, H. (1975). Increase in glutathione peroxidase activity in erythrocytes from trisomy 21 subjects. *Biochem. Biophys. Res. Commun.* **67**, 910–915.

Smith, C. D., Carney, J. M., Starke-Reed, P. E., Oliver, C. N., Stadtman, E. R., Floyd, R. A., and Markesberry, W. R. (1991). Excess brain protein oxidation and enzyme dysfunction in normal aging and Alzheimer's disease. *Proc. Natl. Acad. Sci. U.S.A.* **88**, 10540–10543.

Sokol, R. J., Butler-Simon, N., Conner, C., Huebi, J. E., Sinatra, F. R., Suchy, F. J., Heyman, M. B., Perrault, J., Rothbaum, R. J., Levy, J., Iannaccone, S. T., Shneider, B. L., Koch, T. K., and Narkewicz, M. R. (1993). Multicenter trial of d-α-tocopheryl polyethylene glycol 1000 succinate for treatment of vitamin E deficiency in children with chronic cholestasis. *Gastroenterology* **104**, 1727–1735.

Sokol, R. J., Guggenheim, M. A., Iannaccone, S. T., Barkhaus, P. E., Miller, C., Silverman, A., Ballistreri, W. F., and Heubi, J. E. (1985). Improved neurologic function after long term correction of vitamin E deficiency in children with chronic cholestasis. *N. Engl. J. Med.* **313**, 1580–1586.

Sokol, R. J., Heubi, J. E., Butler-Simon, N., McClung, H. J., Lilly, J., and Silverman, A. (1987). Treatment of vitamin E deficiency during chronic childhood cholestasis with oral d-α-tocopheryl polyethylene glycol-1000 succinate. *Gastroenterology* **93**, 975–985.

Sokol, R. J., Kayden, H. J., Bettis, D. B., Traber, M. G., Neville, H., Ringel, S., Wilson, W. B., and Stumpf, D. A. (1988). Isolated vitamin E deficiency in the absence of fat malabsorption-familiar and sporadic cases: Characterisation and investigation of causes. *J. Lab. Clin. Med.* 111, 548–559.

Southam, E., Thomas, P. K., King, R. H. M., Goss-Sampson, M. A., and Muller, D. P. R. (1991). Experimental vitamin E deficiency in rats: Morphological and functional evidence of abnormal axonal transport secondary to free radical damage. *Brain* 114, 915–936.

Sperling, M. A., Hiles, D. A., and Kennerdell, J. S. (1972). Electroretinographic responses following vitamin A therapy in a-beta-lipoproteinaemia. *Am. J. Opthalmol.* 73, 342–351.

Subbarao, K. V., Richardson, J. S., and Ang, L. C. (1990). Autopsy samples of Alzheimer's cortex show increased lipid peroxidation in vitro. *J. Neurochem.* 55, 342–345.

Sung, J. H., Park, S. H., Mastri, A. R., and Warwick, W. J. (1980). Axonal dystrophy in the gracile nucleus in congenital biliary atresia and cystic fibrosis (mucoviscidosis): Beneficial effect of vitamin E therapy. *J. Neuropathol. Exp. Neurol.* 39, 584–597.

Terry, T. L. (1942). Extreme prematurity and fibroblastic overgrowth of persistent vascular sheath behind each crystalline lens. I. Preliminary report. *Am. J. Ophthalmol.* 25, 203–204.

Tetrud, J. W., and Langston, J. W. (1989). The effect of deprenyl (Selegiline) on the natural history of Parkinson's disease. *Science* 245, 519–522.

Thomas, P. K., Cooper, J. M., King, R. H. M., Workman, J. M., Schapira, A. H. V., Goss-Sampson, M. A., and Muller, D. P. R. (1993). Myopathy in vitamin E deficient rats: Muscle fibre necrosis associated with disturbances of mitochondrial function. *J. Anat.* 183, 451–461.

Thomas, P. K., Landon, D. N., and King, R. H. M. (1984). Diseases of the peripheral nerve. *In* "Greenfield's Neuropathology" (J. Hume Adams, J. A. N. Corsellis, and L. W. Duchen, eds.). Edward Arnold, London.

Towfighi, J. (1981). Effects of chronic vitamin E deficiency on the nervous system of the rat. *Acta Neuropathol.* 54, 261–267.

Traystman, R. J., Kirsch, J. R., and Koehler, R. C. (1991). Oxygen radical mechanisms of brain injury following ischemia and reperfusion. *J. Appl. Physiol.* 71, 1185–1195.

Vatassery, G. T., Angerhofer, C. K., and Peterson, F. J. (1984). Vitamin E concentrations in the brains and some selected peripheral tissues of selenium-deficient and vitamin E-deficient mice. *J. Neurochem.* 42, 554–558.

Vatassery, G. T., and Younoszai, R. (1978). Alpha-tocopherol levels in various regions of the central nervous system of the rat and guinea pig. *Lipids* 13, 828–831.

Volicer, L., and Crino, P. B. (1990). Involvement of free radicals in dementia of the Alzheimer type: A hypothesis. *Neurobiol. Aging* 11, 567–571.

Wetterau, J. R., Aggerbeck, L. P., Bouma, M., Eisenberg, C., Munck, A., Hermier, M., Schmitz, J., Gay, G., Rader, D. J., and Gregg, R. E. (1992). Absence of microsomal triglyceride transfer protein in individuals with abetalipoproteinaemia. *Science* 258, 999–1001.

Wichman, A., Buchthal, F., Pezeshkpour, G. H., and Gregg, R. E. (1985). Peripheral neuropathy in abetalipoproteinemia. *Neurology* 35, 1279–1289.

Wijner, L. M. M., Menteba-Van Heusel, M., Pearson, P. L., and Khan, P. M. (1978). Assignment of a gene for glutathione peroxidase (GPX1) to human chromosome 3. *Cytogenet. Cell. Genet.* 22, 232–235.

Willison, H. J., Muller, D. P. R., Matthews, S., Jones, S., Kriss, A., Stead, R. J., Hodson, M. E., and Harding, A. E. (1985). A study of the relationship between neurological function and serum vitamin E concentrations in patients with cystic fibrosis. *J. Neurol. Neurosurg. Psych.* 48, 1097–1102.

Wisniewski, K. E., Wisniewski, H. M., and Wen, G. Y. (1985). Occurrence of neuropathological changes and dementia of Alzheimer's disease in Down's syndrome. *Ann. Neurol.* 17, 278–282.

Wolff, O. H., Lloyd, J. K., and Tonks, E. L. (1964). A-beta-lipoproteinaemia with special reference to the visual defect. *Exp. Eye Res.* 3, 439–442.

Yamamoto, M., Shima, T., Uozumi T, Sogabe, T., Yamada, K., and Kawasaki, K. (1983). A possible role of lipid peroxidation in cellular damages caused by cerebral ischemia and the protective effect of alpha-tocopherol administration. *Stroke* **14**, 977–982.

Yoshida, S., Abe, K., Busto, R., Watson, B. D., Kogure, K., and Ginsberg, M. D. (1982). Influence of transient ischemia on lipid-soluble antioxidants, free fatty acids and energy metabolites in rat brain. *Brain Res.* **245**, 307–316.

Yoshida, S., Busto, R., Santiso, M., and Ginsberg, M. D. (1984). Brain lipid peroxidation induced by postischemic reoxygenation in vitro: Effect of vitamin E. *J. Cereb. Blood Flow Metab.* **4**, 466–469.

Zaman, Z., Roche, S., Fielder, P., Frost, P. G., Niriella, D. C., and Cayley, A. C. D. (1992). Plasma concentrations of vitamins A and E and carotenoids in Alzheimer's disease. *Age Ageing* **21**, 91–94.

Zubenko, G. S., and Cohen, B. M. (1985). Effects of psychotropic agents on the physical properties of platelet membranes in vitro. *Psychopharmacology* **86**, 369–373.

Wulf Dröge
Andrea Gross
Volker Hack
Ralf Kinscherf
Michael Schykowski
Michael Bockstette
Sabine Mihm
Dagmar Galter

Division of Immunochemistry
Deutsches Krebsforschungszentrum
D-69120 Heidelberg, Germany

Role of Cysteine and Glutathione in HIV Infection and Cancer Cachexia: Therapeutic Intervention with N-Acetylcysteine

I. Introduction

Massive loss of skeletal muscle mass (cachexia) is one of the hallmarks of HIV infection (Kotler *et al.*, 1985; Grunfeld, 1991), sepsis, and trauma (Long *et al.*, 1976; Brennan, 1977) and is the single, most common cause of death among cancer patients (Warren, 1932; Harnett, 1952; Lawson *et al.*, 1982; Friedman, 1987; Pisters and Pearlstone, 1993). The pathogenetic mechanism of the skeletal muscle catabolism in HIV infection, cancer, and sepsis is largely unknown, and there is no satisfactory treatment for these catabolic conditions. Although the work of many laboratories on the pathogenetic mechanisms has provided a large body of phenomenological findings and inspired useful working hypotheses, the demonstration of cause-and-effect relationships *in vivo* has been difficult. Nevertheless, the analysis of the cachectic mechanisms is a tremendous challenge in view of the enormous medical, socioeconomical, and psychological implications of this debilitating condition.

A significant impairment of immunological functions is often associated with the catabolic process. In HIV infection, the immunological dysfunction develops progressively into a severe condition called acquired immunodeficiency syndrome (AIDS) (Pantaleo *et al.*, 1993). The relatively milder form of immunological dysfunction in cancer patients has been demonstrated among others by mitogenic stimulation *in vitro* (reviewed in Dröge *et al.*, 1988a).

Increasing evidence suggests that an abnormal cysteine and glutathione metabolism plays a decisive role in the development of catabolic conditions and associated immunological dysfunctions. This chapter gives a conceptual overview and a guideline for further experimentation. In this context, this chapter also discusses the chances for a therapeutic intervention with cysteine derivatives such as *N*-acetylcysteine (NAC). At this moment, there is no other satisfactory treatment available for these disease processes.

B. Methodological Aspects

Studies on biochemical and immunological parameters in patients yield mainly phenomenological findings. Certain diseases and conditions such as sepsis, chronic fatigue syndrome (CFS), and amyotrophic lateral sclerosis (ALS) were found to share some of the typical biochemical and immunopathological changes associated with HIV infection and cancer and can thus provide useful comparative information. A correlation between unrelated parameters cannot prove a cause-and-effect relationship but is often useful in encouraging or discouraging further experimentation. The formal proof of a cause-and-effect relationship, however, can rarely be obtained in studies on patients. Investigations on patients have, therefore, been complemented by studies in animal models, including tumor-bearing mice. The simultaneous analysis of liver and skeletal muscle tissue samples, for example, has been possible in this animal model. Despite some species-related differences, it was satisfying to see that the basic pathological changes in the various systems are similar, and that tumor-bearing mice are indeed a useful experimental model for studies on certain aspects of the cachectic process.

One may safely predict that the mechanisms of the cachectic process and its immunopathological implications will be unraveled only in small steps, with each step being almost irrelevant by itself and important only in the context. The major steps in the agenda are

- To identify abnormal biochemical patterns in different catabolic diseases and conditions,
- To identify abnormal biochemical parameters that are significantly correlated with mortality, disease progression, and weight loss,
- To prove cause-and-effect relationships by experimental intervention,

- To develop and optimize interventive strategies that may be suitable for clinical therapy, and
- To test new therapeutic strategies in clinical trials.

II. Elevated Venous Plasma Glutamate Levels as an Early and Possibly Universal Marker for Catabolic and Precatabolic Conditions

A. Evidence for an Impairment of Muscular Membrane Transport Activities

Virtually all diseases and conditions that are associated with skeletal muscle catabolism and that have been studied in this regard, including advanced malignancies (Dröge *et al.*, 1988a), HIV/SIV infection (Dröge *et al.*, 1988b; Eck *et al.*, 1989, 1991), sepsis (Roth *et al.*, 1985), and amyotrophic lateral sclerosis (Plaitakis and Caroscio, 1987), are associated with elevated plasma glutamate levels. At least in cancer, HIV, and SIV infection, the venous plasma glutamate levels were shown to increase before weight loss becomes detectable, i.e., in the precatabolic condition. In rhesus macaques, glutamate levels were shown to increase significantly within 2 weeks after inoculation of the simian immunodeficiency virus SIV_{251mac} (Eck *et al.*, 1991). The elevation of venous plasma glutamate levels may therefore be one of the earliest manifestations of the precachectic process.

Amino acid exchange studies on well-nourished (i.e., not yet overtly cachectic) cancer patients revealed that the increase of the postabsorptive venous glutamate level reflects the failure of the muscle tissue to take up glutamate from the circulation and is, in addition, associated with an impaired transport activity of the skeletal muscle tissue for other important metabolites such as glucose and ketone bodies (Hack *et al.*, 1996b).

B. Suggestive Evidence for an Abnormal Cysteine Catabolism and Glutathione Level in Skeletal Muscle Tissue

The major glutamate transport system in skeletal muscle cells is the Na^+-dependent X_{AG^-} system that is shared by aspartate (Horn, 1989; Low *et al.*, 1994; McGivan and Pastor-Anglada, 1994). As this membrane, transport is energetically driven by the electrochemical Na^+ gradient and as this, in turn, is linked via the Na^+/H^+ antiport to the intracellular pH, this glutamate transport activity is subject to inhibition by proton-generating processes. Another factor contributing to its pH dependency is the fact that this Na^+-dependent transport system also requires an antiport of OH^- or HCO_3^- (Bouvier *et al.*, 1992; Kanner, 1993). It is therefore reasonable to assume that the decreased glutamate transport activity is related to the increased glycolytic activity and lactate production that has been demonstrated in several different catabolic conditions, including cancer cachexia (Shaw et al., 1985; Shaw and Wolfe, 1987; Tayek, 1992) and sepsis (Roth et al., 1982; Roth, 1985). The glycolytic production of lactate and pyruvate causes an acidification of the skeletal muscle tissue. This acidification may be ameliorated by the fact that lactate and pyruvate are cotransported with protons across the plasma membrane and thereby remove the equivalent amount of protons generated. The acidification is aggravated, however, by the fact that the temporary increase of intracellular pyruvate causes an increased rate of cysteine/pyruvate transamination and consequently an increased cysteine catabolism into sulfate and protons. Increased intracellular sulfate levels indicative for an increased generation of protons have indeed been found in the skeletal muscle tissue of tumor-bearing mice (Hack et al., 1996a). This was associated with a decrease of the glutathione level in the skeletal muscle tissue, indicating that the cysteine catabolism was increased at the expense of glutathione biosynthesis in this tissue. There is, therefore, a possibility that the increase of the postabsorptive venous plasma glutamate level in cancer patients or HIV-infected persons may be an early indicator for an abnormal cysteine catabolism, a decrease of intracellular glutathione in the skeletal muscle tissue, and an impaired transport activity. Moreover, as the glycolytic pathway is suppressed by adequate levels of ATP and is generally not active in cells with an adequate mitochondrial oxidative metabolism, there is also a possibility that the elevated plasma glutamate level in the precachectic condition is indicative for a failure of the mitochondria to meet the energy requirements of the muscle tissue. In support of this hypothesis, a decreased ATP level has been found in the skeletal muscle tissue of patients with sepsis (Roth, 1985). An increased venous plasma glutamate level has also been found in healthy human subjects after a 4-week period with intensive anaerobic physical exercise programs which per definition exceeded the capacity of the oxidative metabolism and were therefore associated with high systemic lactate levels (Kinscherf et al., 1996).

C. Pathological Significance of Decreased Membrane Transport Activity and Elevated Venous Plasma Glutamate Levels

Whether the decrease of the skeletal muscle glutathione levels and/or the increase of the cysteine catabolism and its effect on the intracellular pH may play a causative role in the catabolic process remains to be determined. The impairment of the glutamate transport activity may be responsible, at least partly, for the decreased intracellular glutamate and/or glutamine levels that have been found in the skeletal muscle tissue of catabolic patients (Fürst et al., 1979; Askanazi et al., 1980; Roth et al., 1982). The uptake of glutamate from the circulation, the intracellular conversion of glutamate into glutamine, and the export of glutamine into the circulation are some of the most important biochemical functions of skeletal muscle tissue (Newsholme and Parry-Billings, 1990). In addition, the elevated venous glutamate levels may have a direct effect on the cysteine supply to cells of the immune system. Elevated extracellular concentrations of glutamate in the pathologically relevant range were found to inhibit the cellular uptake of cystine and to decrease intracellular cysteine and glutathione levels in macrophages and peripheral blood mononuclear cells (Eck and Dröge, 1989; Gmünder et al., 1991). It has therefore been proposed that an increased plasma glutamate level is a pathogenetic factor in its own right and is responsible for the decreased intracellular glutathione levels in the peripheral blood lymphocytes of HIV-infected patients (Eck et al., 1989) and for the decreased lymphocyte functions in advanced malignancies (Dröge et al., 1988a).

III. "Push" and "Pull" Mechanisms in Catabolic and Precatabolic Processes

A. Two Principal Mechanisms in the Catabolic Process

The hallmark of cachectic processes is an excessive urea production that proceeds at the expense of the skeletal muscle protein and which results in a negative nitrogen balance (Brennan, 1977; Shaw et al., 1985, 1988; Shaw and Wolfe, 1987). Because urea production occurs mainly in the liver, and catabolic processes are associated with biochemical changes in skeletal muscle tissue *and* liver, it has been an intriguing question whether the excessive urea production is primarily the consequence of a hepatic dysfunction that may drain indirectly the amino acid pool of the skeletal muscle tissue (i.e., a "pull" mechanism) or the consequence of a biochemical dysfunction in the skeletal muscle tissue (i.e., a "push" mechanism). Conditions with subnormal plasma amino acid levels should be indicative, by definition, for the "pull" mechanism, whereas conditions with higher than normal plasma amino acid levels should be indicative for the "push" mechanism. The available evidence suggests that both mechanisms are operating and may occur sequentially in the same disease process. As a rule, the "pull" condition is found in seemingly well-nourished patients with potentially cachectic diseases, i.e., in a stage that can be described as "precachectic." Although this condition may seem relatively benign, there is reason to believe that it triggers the progression to overt cachexia and, in the case of HIV infection, the breakdown of the immunological defense against the virus as discussed later.

B. Decreased Plasma Cystine Levels and Evidence for "Push" and "Pull" Mechanisms in Sepsis

Sepsis is a particularly well-studied model of cachexia. Patients with sepsis pass through a stage with a significant decrease of total plasma amino acid levels, including a decrease of plasma cystine, glutamine, and arginine levels (Roth *et al.*, 1985). In addition, significantly decreased glutamine levels have been found in skeletal muscle tissue by various authors (Fürst *et al.*, 1979; Askanazi *et al.*, 1980; Roth *et al.*, 1982). All of these findings are indicative for a "pull" mechanism. However, patients in the phase of severe skeletal muscle catabolism ("autocannibalism") were found to have increased plasma amino acid concentrations (Siegel *et al.*, 1979; Cerra *et al.*, 1980), indicative for a "push" mechanism. In either case, the amino acids glutamine and alanine were found to account for the majority of the amino acid nitrogen that was released from the skeletal muscle tissue and taken up by the splanchnic bed (Aulick and Wilmore, 1979; Duff *et al.*, 1979; Wilmore *et al.*, 1980; Souba and Wilmore, 1983).

C. Decreased Plasma Cystine and Cysteine Levels and Evidence for "Push" and "Pull" Mechanisms in HIV Infection

A significant decrease of plasma cystine and cysteine levels has also been found in HIV infection (Dröge *et al.*, 1988b; Eck *et al.*, 1989; Hortin *et al.*, 1994). A more detailed analysis of our data revealed that the most dramatic decrease of plasma cystine levels occurred in asymptomatic HIV-infected persons with CD4$^+$ counts <400 mm^{-3}, i.e., in the phase with the strongest decrease of the absolute CD4$^+$ T-cell number but without obvious weight loss (i.e., in the precachectic stage) (Table I). Using bioelectrical impedance analysis, Süttmann *et al.* (1991) found that the loss of body cell mass starts already in the asymptomatic stage, i.e., well before weight loss becomes detectable. The decrease of plasma cystine was associated with plasma glutamine and arginine levels significantly below the mean of healthy control subjects. The plasma cystine levels of symptomatic HIV-infected patients were, on the average, still lower than those of uninfected controls or infected persons with >400 mm^{-3} CD4$^+$ cells, but higher than the levels found in

TABLE I   Plasma Amino Acid Levels in HIV-Infected Persons

	n	Cystine (μM)	Glutamine (μM)	Arginine (μM)	Glutamate (μM)
Noninfected persons	145	56.8 ± 1.4	614 ± 11	89.7 ± 2.9	26.9 ± 1.2
HIV$^+$ without symptoms CD4$^+$ cells > 400 mm^{-3}	26	61.2 ± 3.4	485 ± 27	60.2 ± 3.0	81.0 ± 9.6
HIV$^+$ without symptoms CD4$^+$ cells < 400 mm^{-3}	14	40.3 ± 1.7	468 ± 26	70.8 ± 5.7	53.0 ± 5.7
HIV$^+$ with symptoms CD4$^+$ cells < 400 mm^{-3}	71	52.9 ± 1.9	468 ± 15	76.1 ± 8.0	70.0 ± 0.2

the intermediate asymptomatic stage with <400 mm^{-3} $CD4^+$ T cells. This increase in the later stages of the disease suggests the possibility that the catabolic process may shift, at least partly, from a "pull" to a "push" mechanism.

D. Decreased Plasma Cystine Levels and Evidence for "Push" and "Pull" Mechanisms in Malignant Diseases

Advanced cancer patients with massive cachexia have abnormally high plasma glutamate levels and often also relatively high plasma cystine, glutamine, and arginine levels indicative for a "push" mechanism of the catabolic process. This pattern was seen among others in a group of patients with liver cancer (Zhang and Pang, 1992) and in a group of advanced and strongly catabolic patients with gastrointestinal cancer, lung cancer, or cancer of the pancreas (V. Hack and W. Dröge, unpublished observation). However, at least one study on well-nourished patients with lung or breast cancer, i.e., on patients at a relatively early stage of the malignant disease, revealed that cystine, glutamine, and arginine levels were markedly decreased in comparison with healthy controls (Zhang and Pang, 1992). These findings suggest that "push" and "pull" mechanisms may also both be operating in malignant diseases and that the "pull" mechanism may be typical of the precatabolic stage. Significantly decreased plasma cystine levels have also been found in C57BL/6 mice bearing the syngeneic MCA 105 fibrosarcoma (Hack *et al.*, 1996a).

E. Abnormally Low Plasma Cystine and Glutamine Levels in Patients with Chronic Fatigue Syndrome (CFS)

The etiology and the pathogenetic mechanisms of CSF are poorly understood. By definition, these patients show a severe skeletal muscle dysfunction (fatigue), and they were also found to have significantly decreased plasma cystine and glutamine levels (Aoki *et al.*, 1993). In addition, they express an immunological dysfunction as discussed in Section IV,A.

F. Hypothetical Mechanism of the "Pull" Condition: Evidence for an Abnormal Cysteine and Glutathione Metabolism in Liver

Whereas the "push" mechanism of the catabolic process is still obscure, there are some important clues regarding the "pull" mechanism. Studies on patients with sepsis have already shown that the decrease of plasma cystine levels was associated with a decrease of hepatic cystine and taurine levels (Roth *et al.*, 1985). More recent studies on C57BL/6 mice bearing the syngeneic MCA 105 fibrosarcoma have shown that the decrease of the plasma cystine level was associated with a significant decrease of the intrahepatic sulfate level, suggesting that the proton-generating catabolism of cysteine into sulfate may be markedly decreased in the liver of the tumor-bearing host (Hack *et al.*, 1996a). As the intrahepatic pH is normally maintained by the ratio of hepatic urea production versus glutamine biosynthesis (Häussinger and Gerok, 1985), it is reasonable to assume that a decreased generation of protons will lead to an increased formation of carbamoylphosphate and urea biosynthesis at the expense of the glutamine pool (see scheme, Fig. 1). This hypothetical mechanism is supported by the experimental observation that the decrease of the hepatic sulfate levels in the tumor-bearing mice was associated with an increased hepatic urea level and with increased urea/glutamine and glutamate/glutamine ratios (Hack *et al.*, 1996a). To prove the cause-and-effect relationship between cysteine catabolism and hepatic urea levels, some of the tumor-bearing mice were treated with daily injections of cysteine. The results showed that the administration of cysteine reconstituted not only the hepatic sulfate level, but also decreased the hepatic urea level and normalized the urea/glutamine and glutamate/glutamine ratios.

Incidentally, the tumor-bearing mice showed a significant increase of the γ-glutamylcysteine synthetase activity, i.e., the first rate-limiting enzyme

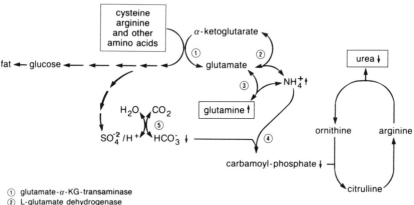

① glutamate-α-KG-transaminase
② L-glutamate dehydrogenase
③ glutamine synthetase / glutaminase
④ carbamoylphosphate synthase
⑤ carbonic anhydrase

FIGURE I   Role of cysteine in the feedback control of nitrogen disposal: A hypothetical model. The hypothesis states that the hepatic cysteine level determines the rate of its catabolic conversion into sulfate and protons. This, in turn, inhibits the conversion of amino groups into urea (↓) through carbamoylphosphate synthesis (↓) and the urea cycle. H^+-generating processes thus favor the conversion of NH_4^+ into glutamine (↑). Accordingly, the cysteine catabolism plays a role in conserving the glutamine pool. The hypothesis predicts that the availability of cysteine sets the threshold at which amino acids are being converted into glucose and fat. This threshold determines decisively the loss of nitrogen and contributes to the control of body cell mass and body fat.

of glutathione biosynthesis, suggesting that more cysteine is drained into the glutathione pool. The simplest interpretation for the enhanced production of urea in the "pull" condition is therefore that a decreased availability of cysteine in the liver leads to a decreased cysteine catabolism into sulfate and protons which, in turn, enhances urea biosynthesis at the expense of glutamine. The hepatic cysteine deficiency may be explained partly by an increased hepatic glutathione biosynthesis and partly by the increased cysteine catabolism in the skeletal muscle tissue (see Section II,B). The liver was previously shown to export glutathione disulfide mainly into the bile canaliculus (Akerboom and Sies, 1994a,b). Our own studies on the tumor-bearing mice revealed that the glutathione disulfide levels in the bile were significantly increased in comparison to the healthy controls. This suggests that cysteine is, at least partly, lost by an increased export of glutathione disulfide into the bile.

IV. Immunological Implications

A. Cysteine Deficiency and Immunological Dysfunction

In view of the important role of the cysteine supply and the intracellular glutathione level for lymphocyte functions, it has been hypothesized that the cysteine and glutathione deficiency in HIV-infected patients may contribute to the immunopathology of the disease (Dröge *et al.*, 1988b, 1992; Eck *et al.*, 1989). The almost complete clearance of the virus within a few weeks after primary infection (see Borrow *et al.*, 1994) and the relatively long asymptomatic period with relatively normal CD4[+] T-cell counts (Pantaleo *et al.*, 1993), plus the phenomenon of the long-term nonprogressors (Buchbinder *et al.*, 1994; Klein and van Baalen, 1995; Rinaldo *et al.*, 1995), are striking examples for the ability of the immune system to control this virus infection, at least in principle. The fact that the most profound decrease of plasma cystine levels coincides with the decrease of the CD4[+] T-cell numbers in asymptomatic HIV-infected persons (Table I), therefore, suggests the possibility that this decrease of the cystine levels may weaken the immune system so that the delicate balance between the virus and the immune system is tilted in favor of the virus. Other conditions with abnormally low cystine levels, including cancer and CFS, show clearly demonstrable immunological dysfunctions. It is a common finding that peripheral blood lymphocytes from patients with malignant diseases exhibit diminished proliferative responses to mitogens in comparison with healthy controls (Watkins, 1973; Wanebo *et al.*, 1975; Braun *et al.*, 1980; Collins *et al.*, 1980; Müller *et al.*, 1984; Yron *et al.*, 1986; Dröge *et al.*, 1988a). An even more profound suppression has been observed in various *in vitro* assays of the reactivity of tumor-infiltrating lymphocytes (Vose *et al.*, 1977; Tötterman *et al.*, 1980;

Miescher *et al.*, 1986). Persons with CFS were found to have, on the average, not only decreased plasma cystine and glutamine, but also a significantly decreased natural killer (NK) activity and relatively low numbers of CD16[+] cell numbers (Fc-γ receptor[+] cells) (reviewed in Aoki *et al.*, 1993). The decrease of NK activity was correlated with a decrease of antibody-dependent cell-mediated cytotoxicity (ADTC). In addition, a low CD4[+]/CD8[+] ratio has been observed in 20% or more of CFS patients (reviewed in Aoki *et al.*, 1993). These changes are indeed reminiscent of the cellular dysfunctions that are observed in HIV-infected patients before the CD4[+] T-cell numbers start to decline (Miedema *et al.*, 1988; Giorgi and Detels, 1989; Rosenberg and Fauci, 1989). Poli *et al.* (1985) have shown-that NK cells of HIV-infected persons are phenotypically and numerically normal but functionally defect. A qualitative defect of NK cells and LAK cells has also been reported (Fontana *et al.*, 1986; Ullum *et al.*, 1995; see also the review by Sirianni *et al.*, 1990). The loss of NK cell function may well be a decisive event in the switch from stable to progressive disease because NK cells were previously shown to play an important role in the defense against virus infections (Herberman and Ortaldo, 1981).

Finally, in a study on healthy human subjects, Kinscherf *et al.* (1994) found that persons with relatively low intracellular glutathione levels in the peripheral blood mononuclear cells (<20 nmol/mg protein) had significantly lower CD4[+] T-cell numbers than persons with approximately median levels of intracellular glutathione (20–30 nmol/mg protein). All of these phenomena, i.e., the immunological dysfunctions in cancer or CSF patients and the conspicuously low CD4[+] T-cell counts in persons with relatively low intracellular glutathione levels, may not seem very impressive in comparison with the severe destruction of the immune system in HIV infection (Pantaleo *et al.*, 1993). It would, therefore, be unreasonable to assume that the cysteine deficiency accounts for all aspects of the HIV-mediated immunopathology. However, even a relatively moderate impairment of immunological functions may be sufficient to have disastrous consequences in the course of the HIV disease by interfering with the initially successful battle between the host immune system and the virus. If so, this disastrous process may be prevented if the plasma cystine levels are corrected in time, as discussed in Section V.

B. Abnormal Glutathione Levels in HIV Infection

The cysteine-containing tripeptide glutathione is a limiting factor for the immune system to the extent that certain T-cell functions can be potentiated *in vivo* by the administration of glutathione (Dröge *et al.*, 1986). In a number of different experimental systems, the intracellular glutathione level of lymphocytes was shown to determine decisively the magnitude of immunological functions (reviewed in Dröge *et al.*, 1994). Because cysteine is a rate-limiting precursor for the glutathione biosynthesis in lymphoid cells

(Meister and Anderson, 1983), it was not unexpected to see that the decreased plasma cysteine and cystine levels of HIV-infected patients (Dröge *et al.*, 1988b; Eck *et al.*, 1989) were associated with a decrease of intracellular glutathione levels in the peripheral blood mononuclear cells, blood plasma, and epithelial lining fluid of the lung (Buhl *et al.*, 1989; Eck *et al.*, 1989; Roederer *et al.*, 1991). More recently, Aukrust *et al.* (1995) found abnormal glutathione/glutathione disulfide ratios but normal absolute glutathione levels in peripheral blood mononuclear cells of HIV-infected patients. The apparent discrepancy may be explained by the fact that the earlier findings were obtained mainly with patients at relatively early stages of the disease (Eck *et al.*, 1989), whereas Aukrust *et al.* (1995) may have studied more advanced patients, where the massive decrease of the plasma cystine level may have been largely reversed (see Section III,C).

C. Redox Regulation by Glutathione and Glutathione Disulfide

The changes of the intracellular glutathione and/or glutathione disulfide levels in HIV-infected patients (Eck *et al.*, 1989; Roederer *et al.*, 1991; Aukrust *et al.*, 1995) raise the question of how these changes may affect lymphocyte functions. The redox regulation of signal cascades and gene transcription is a relatively young area of research and is still not completely understood at the molecular level. It is clear, however, that glutathione disulfide has a profound effect on the DNA-binding activity of the nuclear factor-κB (NF-κB) in both intact cells and cell-free systems (Galter *et al.*, 1994; Mihm *et al.*, 1995). The transcription factor NF-κB is involved in the inducible transcription of several immunologically important genes, including those of the interleukin-2 receptor α-chain, tumor necrosis factor-α (TNF-α), major histocompatibility complex antigens, and c-fos (reviewed in Ullmann *et al.*, 1990; Baeuerle and Baltimore, 1991). The DNA-binding activity of NF-κB is inhibited in cell-free systems by physiologically relevant concentrations of GSSG even in the presence of a large excess of thiols. Moreover, three independent procedures that elevate GSSG levels in intact cells were found to inhibit DNA-binding activity and transactivating activity in intact cells (Galter *et al.*, 1994; Mihm *et al.*, 1995). Whereas NF-κB DNA-binding activity is inhibited by GSSG, it is enhanced by thiols, including dithiothreitol, cysteine, glutathione, and reduced thioredoxin (Matthews *et al.*, 1992; Okamoto *et al.*, 1992; Galter *et al.*, 1994). The latter is by far the most effective and possibly the only physiologically relevant thiol compound in this context. Thioredoxin was found to restore the DNA-binding activity of oxidized NF-κB *in vitro* and to augment gene expression from HIV-LTR in intact cells as shown with a corresponding reporter gene construct (Matthews *et al.*, 1992; Okamoto *et al.*, 1992). The induction of

NF-κB activation and nuclear translocation, in contrast, is strongly inhibited by thioredoxin (Schenk et al., 1994).

In contrast, the expression of a reporter gene under the control of the transcription factor activator protein-1 (AP-1), i.e., another immunologically important transcription factor, was shown to be strongly increased by procedures that elevate intracellular GSSG levels, indicating that the inhibitory effect of glutathione disulfide is selective (Galter et al., 1994). In contrast to NF-κB, AP-1 is almost exclusively localized in the nucleus, which contains considerably lower concentrations of GSSG. Because the DNA-binding activity of AP-1 in cell-free systems is not increased but rather decreased by GSSG, this finding suggests that GSSG modulates one of the components in the upstream signal cascade that induces AP-1 activity. Details of this regulatory effect, including the identification of the redox-sensitive component of the signal cascade and its biochemical modification by the oxidizing agent, remain to be investigated.

Because glutathione disulfide levels are determined largely by the availability of cysteine (Dröge et al., 1995), it makes sense that T lineage cells have generally a tightly controlled cysteine supply that is limited by an extremely weak membrane transport activity for cystine. Cystine is the quantitatively most important source of cysteine in blood plasma and in standard cell culture medium, and thus the limiting precursor for the biosynthesis of glutathione and GSSG. In T cells, the transport activity of cystine is more than 10-fold lower than that of cysteine, alanine, or arginine (Ishii et al., 1987, Gmünder et al., 1991, Lira et al., 1992). It is therefore conceivable that the delicate balance of the cysteine supply and of the intracellular glutathione and glutathione disulfide levels may be disturbed in HIV infection. HIV-infected persons were shown to have an increased expression of NF-κB controlled genes, including the genes of TNF-α, soluble IL-2 receptor (i.e., a truncated form of the IL-2 receptor α-chain), and β_2-microglobulin (reviewed in Dröge et al., 1992). The abnormal cytokine expression in HIV-infected persons is thus potentially a consequence of the abnormal redox regulation in these patients.

V. Therapeutic Intervention with a Cysteine Derivative: Effects of Long-Term Treatment with N-Acetylcysteine _____

Section IV,A discussed the hypothesis that the decrease of plasma cystine levels in HIV-infected patients may cause a (moderate) immunological dysfunction that tilts the balance between immune system and virus in favor of the virus and starts the catastrophic progression of the disease. Moreover, Section III,G described the hypothesis that the availability of cysteine may also regulate the loss of nitrogen from the amino acid reservoir in the (pre)catabolic condition. To test these hypotheses, we have looked for strategies to reconstitute the plasma cystine levels in these patients. Since 1988, we have collected longitudinal data on HIV-infected patients treated with the cysteine derivative N-acetylcysteine for periods up to 4 years. Data show that the plasma cystine and glutamine levels can indeed be substantially increased by this treatment in comparison with pretreatment values and with a large group of untreated HIV-infected patients (Figs. 2 and 3). CD4$^+$ T-cell numbers did, on the average, not change significantly after N-acetylcysteine treatment but remained essentially stable. This effect remains to be confirmed in a larger study. Excessive doses of N-acetylcysteine, however, may lead to higher than normal plasma cystine and cysteine levels and impose certain risks. The dose required to raise cystine and glutamine levels close to the mean of healthy individuals was found to vary in the course of the disease, suggesting a variable loss of cyst(e)ine. Data, therefore, suggest that HIV-infected patients should be treated with a cysteine derivative, but that the therapeutic dose may have to be adjusted according to individual needs. This requirement is analogous to the variable needs of diabetic patients with respect to insulin.

Section IV,A mentioned already the study of Kinscherf et al. (1994) on healthy human individuals showing that persons with intracellular glutathione levels of 20–30 nmol/mg protein had significantly higher numbers of

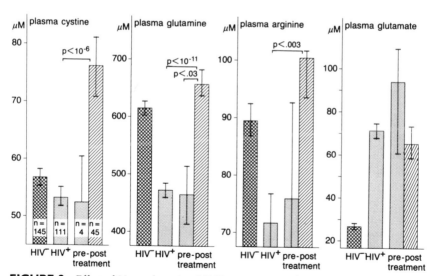

FIGURE 2   Effect of N-acetylcysteine (NAC) treatment on plasma amino acid patterns in HIV-infected individuals. Data show the mean plasma cystine, glutamine, arginine, and glutamate levels of 145 randomly selected healthy human subjects (HIV$^-$), 111 randomly selected HIV-infected persons without NAC treatment (HIV$^+$), and 4 HIV-infected persons before and after NAC treatment. Posttreatment data show the mean of a total of 45 longitudinal measurements of the 4 patients. Pretreatment data as well as the HIV$^-$ and HIV$^+$ data represent the mean of single measurements per person.

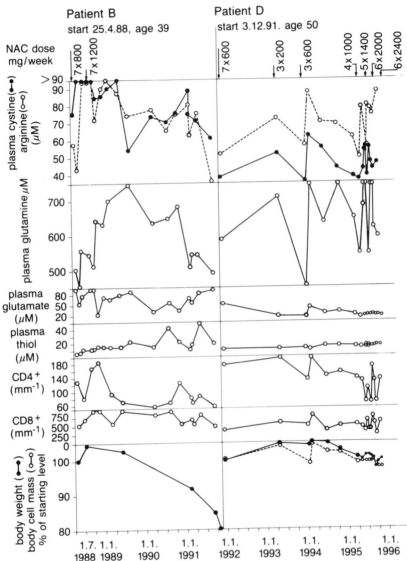

FIGURE 3   Longitudinal analysis of the effects of N-acetylcysteine (NAC) over >3-year periods. Longitudinal data include plasma amino acid levels, T-cell counts, body weight, and body cell mass from two HIV-infected persons observed over 4-year periods. The periods of NAC treatment and the doses used are indicated on the top of the figure. NAC treatment can cause a substantial increase of plasma cystine, glutamine, and arginine levels. However, the doses required to maintain plasma cystine, glutamine, and arginine levels close to the mean of healthy human subjects (i.e., > 50, > 600, and > 70 μM, respectively) may change considerably during the course of infection. The therapeutic dose of N-acetylcysteine may therefore have to be adjusted according to the individual needs of a given patient at a given time.

CD4$^+$ T cells than persons with lower glutathione levels. This study also showed that persons who changed during a 4-week observation period from the optimal to the suboptimal range (10–20 nmol/mg protein) experienced, on the average, a 30% decrease of CD4$^+$ T-cell numbers. This decrease was prevented by treatment with N-acetylcysteine (Kinscherf *et al.*, 1994).

VI. Concluding Remarks

The mechanisms of cachectic processes as they occur in HIV infection, cancer, and sepsis are complex and appear to involve at least two distinct mechanisms that can be described as "push" and "pull" mechanisms. The "push" mechanism appears to prevail in the phase of massive muscle cell catabolism, i. e., in the relatively late phase of the catabolic process. This process is believed to flood the system with amino acids and is therefore associated with normal or even elevated plasma amino acid levels. The "pull" mechanism, in contrast, is determined by a dysregulation at the site of the liver. Evidence suggests that an abnormally low cysteine level in the liver may cause an increased production of urea at the expense of the glutamine pool. The "pull" condition is therefore characterized by abnormally low plasma cystine and glutamine levels. The decrease of the plasma cystine level is explained partly by an increased rate of glutathione biosynthesis in the liver and partly by an increased cysteine catabolism in the skeletal muscle tissue. In the major catabolic diseases, including sepsis, HIV infection, and cancer, the "pull" mechanism appears to precede the "push" mechanism.

Because the decrease of plasma cystine levels in chronic fatigue syndrome and cancer patients is associated with a relatively mild immunological dysfunction, it seems unlikely that the cystine deficiency accounts for all aspects of the immunopathology of HIV infection. Nevertheless, even a relatively moderate impairment of lymphocyte functions may have disastrous consequences for the progression of the disease, if it tilts the initially stable balance between the immune system and the virus in favor of the virus.

Preliminary studies on N-acetylcysteine-treated HIV-infected patients indicate that the decrease of plasma cystine, glutamine, and arginine levels can be corrected by this agent. Anecdotal data also suggest that this strategy may slow or even prevent the progression of the disease. This possibility needs to be confirmed in a larger trial. The dose of N-acetylcysteine, however, may have to be adjusted according to the individual needs as judged by the plasma levels of cystine, glutamine, and acid-soluble thiol (i. e., cysteine). Placebo-controlled, double-blind trials with a constant and arbitrarily chosen dose of NAC may, therefore, not be adequate to evaluate the therapeutic potential of this drug.

Acknowledgment _____

The assistance of Mrs. I. Fryson in the preparation of this manuscript is gratefully acknowledged.

References _____

Akerboom, T. P. M., and Sies, H. (1994a). Transport of glutathione disulfide and glutathione S-conjugates in hepatocyte plasma membrane vesicles. *Methods Enzymol.* 233, 416–425.

Akerboom, T. P. M., and Sies, H. (1994b). Interorgan transport of glutathione and glutathione S-conjugates. *In* "Metabolic Aspects of Cell Toxicity" (P. Eyer, ed.), pp. 7–30, BI Wissenschaftsverlag, Mannheim.

Aoki, T., Miyakoshi, H., Usuda, Y., and Herberman, R. B. (1993). Low NK syndrome and its relationship to chronic fatigue syndrome. *Clin. Immunol. Immunopathol.* 69, 253–265.

Askanazi, J., Carpentier, Y. A., Michelsen, C. B., Elwyn, D. H., Furst, P., Kantrowitz, L. R., Gump, F. E., and Kinney, J. M. (1980). Muscle and plasma amino acids following injury. *Ann. Surg.* 192, 78–85.

Aukrust, P., Svardal, A. M., Müller, F., Lunden B., Berge, R. K., Ueland, P. M., and Froland, S. S. (1995). Increased levels of oxidized glutathione in CD4$^+$ lymphocytes associated with disturbed intracellular redox balance in human immunodeficiency virus type 1 infection. *Blood* 86, 258–267.

Aulick, L. H., and Wilmore, D. W. (1979). Increased peripheral amino acid release following burn injury. *Surgery* 85, 560–566.

Baeuerle, P. A., and Baltimore, D. (1991). *In* "Molecular Aspects of Cellular Regulation, Hormonal Control Regulation of Gene Transcription" P. Cohen and J. G. Foulkes, (eds.), pp. 423–446. Elsevier/North-Holland Biomedical Press, Amsterdam.

Borrow, P., Lewicki, H., Hahn, B. H., Shaw, G. M., and Oldstone, M. B. A. (1994). Virus-specific CD8$^+$ cytotoxic T lymphocyte activity associated with control of viremia in primary human immunodeficiency virus type 1 infection. *J. Virol.* 68, 6103–6109.

Bouvier, M., Szatkowski, M., Amato, A., and Attwell, D. (1992). The glial cell glutamate uptake carrier countertransports pH-changing anions. *Nature* 360, 471–474.

Braun, D. P., Cobleigh, M. A., and Harris, J. E. (1980). Multiple concurrent immunoregulatory effects in cancer patients with depressed PHA induced lymphocyte DNA synthesis. *Clin. Immunol. Immunopathol.* 17, 89–101.

Brennan, M. F. (1977). Uncomplicated starvation versus cancer cachexia. *Cancer Res.* 37, 2359–2364.

Buchbinder, S. P., Katz, M. H., Hessol, N. A., O'Malley, P. M., and Holmberg, S. D. (1994). Long-term HIV-1 infection without immunologic progression. *AIDS* 8, 1123–1129.

Buhl, R., Holroyd, K., Mastrangeli, A., Cantin, A. M., Jaffe, H. A., Wells, F. B., Saltini, C., and Crystal, R. G. (1989). Systemic glutathione deficiency in symptom-free HIV-soropositive individuals. *Lancet* Dec. 2, 1294–1298.

Cerra, F. B., Siegel, J. H., and Coleman, B. (1980). Septic autocannibalism: A failure of exogenous nutritional support. *Ann. Surg.* 192, 570–574.

Collins, P. B., Johnson, A. H., and Moriarty, M. (1980). T lymphocytes in human cancer. I. Mitogen-responsiveness of lymphocytes in cancer patients. *Ir. J. Med. Sci.* 149, 301–303.

Dröge, W., Eck, H.-P., Betzler, M., Schlag, P., Drings, P., and Ebert, W. (1988a). Plasma glutamate concentration and lymphocyte activity. *J. Cancer Res. Clin. Oncol.* 114, 124–128.

Dröge, W., Eck, H.-P., and Mihm, S. (1992). HIV-induced cysteine deficiency and T cell dysfunction: A rationale for treatment with N-acetylcysteine. *Immunol. Today* 13, 211–214.

Dröge, W., Eck, H.-P., Näher, H., Pekar, U., and Daniel, V. (1988b). Abnormal amino acid concentrations in the blood of patients with acquired immunodeficiency syndrome (AIDS) may contribute to the immunological defect. *Biol. Chem. Hoppe-Seyler* **369**, 143–148.

Dröge, W., Kinscherf, R., Mihm, S., Galter, D., Roth, S., Gmünder, H., Fischbach, T., and Bockstette, M. (1995). Thiols and the immune system. Effect of N-acetylcysteine on T cell system in human subjects. *Methods Enzymol.* **251**, 255–270.

Dröge, W., Pottmeyer-Gerber, C., Schmidt, H., and Nick, S. (1986). Glutathione augments the activation of cytotoxic T lymphocytes *in vivo*. *Immunobiology* **172**, 151–156.

Dröge, W., Schulze-Osthoff, K., Mihm, S., Galter, D., Schenk, H., Eck, H.-P., Roth, S., and Gmünder, H. (1994). Functions of glutathione and glutathione disulfide in immunology and immunopathology. *FASEB J.* **8**, 1131–1138.

Duff, J. H., Viidik, T., Marchuk, J. B., *et al.* (1979). Femoral arteriovenous amino acid differences in septic patients. *Surgery* **85**, 344–348.

Eck, H.-P., and Dröge, W. (1989). Influence of the extracellular glutamate concentration on the intracellular cyst(e)ine concentration in macrophages and on the capacity to release cysteine. *Biol. Chem. Hoppe-Seyler* **370**, 109–113.

Eck, H.-P., Gmünder, H., Hartmann, M., Petzoldt, D., Daniel, V., and Dröge, W. (1989). Low concentrations of acid soluble thiol (cysteine) in the blood plasma of HIV-1 infected patients. *Biol Chem Hoppe-Seyler* **370**, 101–108.

Eck, H.-P., Stahl-Hennig, C., Hunsmann, G., and Dröge, W. (1991). Metabolic disorder as an early consequence of simian immunodeficiency virus infection in rhesus macaques. *Lancet* **338**, 346–347.

Fontana, L., Sirianni, M. C., De Sanctis, G., Carbonari, M., Ensoli, B., and Aiuti, F. (1986). Deficiency of natural killer activity, but not of natural killer binding, in patients with lymphoadenopathy syndrome positive for antibodies to HTLV-III. *Immunobiology* **171**, 425–435.

Friedman, P. J. (1987). Is wasting itself lethal? A case-control prospective study. *Nutr. Res.* **7**, 707–717.

Fürst, P., Bergström, J., Chao, L., Larsson, J., Liljedahl, S.-O., Neuhäuser, M., Schildt, B., and Vinnars, E. (1979). Influence of amino acid supply on nitrogen and amino acid metabolism in severe trauma. *Acta Chir. Scand. Suppl.* **494**, 136–141.

Galter, D., Mihm, S., and Dröge W. (1994). Dinstinct effects of glutathione disulphide on the nuclear transcription factors κB and the activator protein-1. *Eur. J. Biochem.* **221**, 639–648.

Giorgi, J. V., and Detels, R. (1989). T cell subset alterations in HIV-infected homosexual men: NIAID multicenter AIDS cohort study. *Clin. Immunol. Immunopathol.* **52**, 10.

Gmünder, H., Eck, H.-P., and Dröge, W. (1991). Low membrane transport activity for cystine in resting and mitogenically stimulated human lymphocyte preparations and human T cell clones. *Eur. J. Biochem.* **201**, 113–117.

Grunfeld, C. (1991). Mechanisms of wasting in infection and cancer: An approach to cachexia in AIDS. In "Gastrointestinal and Nutritional Manisfestions of AIDS" (D. P. Kotler, ed.), pp. 207–229. Raven Press, New York.

Hack, V., Gross, A., Kinscherf, R., Bockstette, M., Fiers, W., Berke, G., and Dröge, W. (1996a). Abnormal glutathione and sulfate levels after interleukin-6 treatment and in tumor-induced cachexia. *FASEB J.* **10**, in press.

Hack, V., Stütz, O., Kinscherf, R., Schykowski, M., Kellerer, M., Holm, E., and Dröge, W. (1996b). Elevated venous glutamate levels in (pre)catabolic conditions result at least partly from a decreased glutamate transport activity. *J. Mol. Med.* **74**, 337–343.

Harnett, W. L. (1952) A survey of cancer in London. *Bri. Empire Cancer Campaign* **26**.

Häussinger, D., and Gerok, W. (1985). Hepatic urea synthesis and pH regulation: Role of CO_2, HCO_3^-, pH and the activity of carbonic anhydrase. *Eur. J. Biochem.* **152**, 381–386.

Herberman, R. B., and Ortaldo, J. R. (1981). Natural killer cells: Their role in defenses against disease. *Science* **214**, 24–30.

Horn, L. W. (1989). L-Glutamate transport in internally dialysed barnacle muscle fibres. *Am. J. Physiol.* 257, C442–C450.

Hortin, G. L., Landt, M., and Powderly, W. G. (1994). Changes in plasma amino acid concentrations in response to HIV-1 infection. *Clin. Chem.* 40, 785–789.

Ishii, T., Sugita, Y., and Bannai, S. (1987). Regulation of glutathione levels in mouse spleen lymphocytes by transport of cysteine. *J. Cell Physiol.* 133, 330–336.

Kanner, B. I. (1993). Glutamate transporters from brain: A novel neurotransmitter transporter family. *FEBS Lett.* 325, 95–99.

Kinscherf, R., Fischbach, T., Mihm, S., Roth, S., Hohenhaus-Sievert, E., Weiss, C., Edler, L., Bärtsch, P., and Dröge, W. (1994). Effect of glutathione depletion and oral N-acetyl-cysteine treatment on CD4+ and CD8+ cells. *FASEB J.* 8, 448–451.

Kinscherf, R., Hack, V., Fischbach, T., Friedmann, B., Weiss, C., Edler, L., Bärtsch, P., and Dröge, W. (1996). Low plasma glutamine in combination with high glutamate levels indicate risk for loss of body cell mass (BCM) in healthy individuals: The effect of N-acetyl-cysteine on BCM. *J. Mol. Med.,* in press.

Klein, MR., and van Baalen, C. A. (1995). Kinetics of Gag-specific cytotoxic T lymphocyte responses during the clinical course of HIV-1 infection: A longitudinal analysis of rapid progressors and long-term asymptomatics. *J. Exp. Med.* 181, 1356–1365.

Kotler, D. P., Wang, J., and Pierson, R. (1985). Body composition in patients with the acquired immunodeficiency syndrome. *Am. J. Clin. Nutr.* 42, 1255–1265.

Lawson, D. H., Richmond, A., Nixon, D. W., et al. (1982). Metabolic approaches to cancer cachexia. *Annu. Rev. Nutr.* 2, 277–301.

Lim, J.-S., Eck, H.-P., Gmünder, H., and Dröge, W. (1992). Expression of increased immunogenicity by thiol releasing tumor variants. *Cell. Immunol.* 140, 345–356.

Long, C. L., Crosby, F., Geiger, J. W., and Kinney, J. M. (1976). Parenteral nutrition in the septic patient: Nitrogen balance, limiting plasma amino acids, and calorie to nitrogen ratios. *Am. J. Clin. Nutr.* 29, 380–391.

Low, S. Y., Rennie, M. J., and Taylor, P. M. (1994). Sodium-dependent glutamate transport in cultured rat myotubes increases after glutamate deprivation. *FASEB J.* 8, 127–131.

Matthews, J. R., Wakasugi, N., Virelizier, J.-L., Yodoi, J., and Hay, R. T. (1992). Thioredoxin regulates the DNA binding activity of NF-κB by reduction of a disulphide bond involving cysteine 62. *Nucleic Acids Res.* 20, 3821–3830.

McGivan, J. D., and Pastor-Anglada, M. (1994). Regulatory and molecular aspects of mammalian amino acid transport. *Biochem. J.* 299, 321–334.

Meister, A., and Anderson, M. E. (1983). Glutathione. *Annu. Rev. Biochem.* 52, 711–760.

Miedema, F., Petit, A. J. C., Terpstra, F. G., Schattenkerk, J. K. M. E., de Wolf, F., Al, B. J. M., Roos, M., Lange, J. M. A., Danner, S. A., Gandsmit, J., and Schellekens, P. T. A. (1988). Immunological abnormalities in human immunodeficiency virus (HIV)-infected asymptomatic homosexual men. *J. Clin. Invest.* 82, 1908.

Miescher, S., Whiteside, T. L., Carrel, S., and von Fliedner, V. (1986). Functional properties of tumor-infiltrating and blood lymphocytes in patients with solid tumors: Effects of tumor cells and their supernatants on proliferative responses of lymphocytes. *J. Immunol.* 136, 1899–1907.

Mihm, S., Galter, D., and W. Dröge. (1995). Modulation of transcription factor NFκB activity by intracellular glutathione levels and by variations of the extracellular cysteine supply. *FASEB J.* 9, 246–252.

Müller, D. S., Manger, B., Zawatzky, R., Kirchner, H., and Kalden, J. R. (1984). Mitogen-induced pigg-interferon production in peripheral blood lymphocytes from patients with colorectal tumors. *Immunobiology* 166, 494–499.

Newsholme, E. A., and Parry-Billings, M. (1990). Properties of glutamine release from muscle and its importance for the immune system. *J. Parent. Ent. Nutr.* 14, 63S–67S.

Okamoto, T., Ogiwara, H., Hayashi, T., Mitsui, A., Kawabe, T., and Yodoi, J. (1992). Human thioredoxin/adult T cell leukemia-derived factor activates the enhancer binding protein of human immunodeficiency virus type 1 by thiol redox control mechanism. *Int. Immunol.* **4**, 811–819.

Pantaleo, G., Graziosi, C., and Fauci, A. S. (1993). The immunopathogenesis of human immunodeficiency virus infection. *N. Engl. J. Med.* **328**, 327–335.

Pisters, P. W., and Pearlstone, D. B. (1993). Protein and amino acid metabolism in cancer cachexia: Investigative techniques and therapeutic interventions. *Crit. Rev. Clin. Lab. Sci.* **30**, 223–272.

Plaitakis, A., and Caroscio, J. T. (1987). Abnormal glutamate metabolism in amyotrophic lateral sclerosis. *Ann. Neurol.* **22**, 575–579.

Poli, G., Introna, M., Zanaboni, F., Peri, G., Carbonari, M., Aiuti, F., Lazzarin, A., Moroni, M., and Mantovani, A. (1985). Natural killer cells in intravenous drug abusers with lymphadenopathy syndrome. *Clin. Exp. Immunol.* **62**, 128–135.

Rinaldo, C., Huang, X.-L., Fan, Z., Ding, M., Beltz, L., Panicali, D., Mazzara, G., Liebmann, J., Cattrill, M., and Gupta, P. (1995). High levels of anti-human immunodeficiency virus type 1 (HIV-1) memory cytotoxic T-lymphocyte activity and low viral load are associated with lack of disease in HIV-1-infected long-term nonprogressors. *J. Virol.* **69**, 5838–5844.

Roederer, M., Staal, F. J. T., Osada, H., Herzenberg, L. A., and Herzenberg, L. A. (1991) CD4 and CD8 T cells with high intracellular glutathione levels are selectively lost as the HIV infection progresses. *Int. Immunol.* **3**, 933–937.

Rosenberg, Z. F., and Fauci, A. S. (1989). The immunopathogenesis of HIV infection. *Adv. Immunol.* **47**, 377–431.

Roth, E. (1985). Untersuchungen zum Aminosäuren- und Proteinstoffwechsel bei kritisch Kranken. *Infusionstherapie* **12**, 270–280.

Roth, E., Funovics, J., Mühlbacher, F., Schemper, M., Mauritz, W., Sporn, P., and Fritsch, A. (1982). Metabolic disorders in severe abdominal sepsis: Glutamine deficiency in skeletal muscle. *Clin. Nutr.* **1**, 25–41.

Roth, E., Mühlbacher, F., Karner, J., Steininger, R., Schemper, M., and Funovics, J. (1985). Liver amino acids in sepsis. *Surgery* **97**, 436–442.

Schenk, H., Klein, M., Dröge, W., and Schulze-Osthoff, K. (1994). Dinstinct effects of thioredoxin and antioxidants on the activation of NFκB and AP-1. *Proc. Natl. Acad. Sci. U.S.A.* **91**, 1672–1676.

Shaw, J. H. F., and Wolfe, R. R. (1987). Glucose and urea kinetics in patients with early and advanced gastrointestinal cancer: The response to glucose infusion and TPN. *Surgery* **101**, 181–186.

Shaw, J. H., Humberstone, D. A., and Holdaway, C. (1988). Weight loss in patients with head and neck cancer: Malnutrition or tumor effect? *Aust. N. Z. J. Surg.* **58**, 505–509.

Shaw, J. H. F., Klien, and Wolfe, R. R. (1985). Assessment of alanine, urea and glucose interrelationships in normal subjects and in patients with sepsis using stable isotopic tracers. *Surgery* **97**, 557–567.

Siegel, J. H., Cerra, F. B., Coleman, B., *et al.* (1979). Physiological and metabolic correlations in human sepsis. *Surgery* **86**, 163–192.

Sirianni, M. C., Tagliaferri, F., and Aiuti, F. (1990). Pathogenesis of the natural killer cell deficiency in AIDS. *Immunol. Today* **11**, 81–82.

Souba, W. W., and Wilmore, D. W. (1983). Postoperative alteration of arteriovenous exchange of amino acids across the gastrointestinal tract. *Surgery* **94**, 342–350.

Süttmann, U., Hoogestraat, L., Ockenga, J., Coldewey, R., Schedel, I., Deicher, H., and Müller, M. J. (1991). Ernährungszustand und Immundefekt bei Patienten mit HIV-1-Infektion. *Infusionstherapie* **18**, 72–73.

Süttmann, U., Ockenga, J., Selberg, O., Hoogestraat, L., Deicher, H., and Muller, M. J. (1995). Incidence and prognostic value of malnutrition and wasting in human immunodeficiency virus-infected outpatients. *J. Acq. Immune Defic. Syndr. Hum. Retrovirol.* **8**, 239–246.

Tayek, J. A. (1992). A review of cancer cachexia and abnormal glucose metabolism in humans with cancer. *J. Am. Col. Nutr.* **11**, 445–456.

Tötterman, T. H., Parthenais, E., Häyry, P., Timonen, T., and Saksela, E. (1980). Cytological and functional analysis of inflammatory infiltrates in human malignant tumors. *Cell. Immunol.* **55**, 219–226.

Ullmann, K. S., Northrop, J. P., Verweij, C. L., and Crabtree, G. R. (1990). Transmission of signals from the T lymphocyte antigen receptor to the genes responsible for cell proliferation and immune function: the missing link. *Annu. Rev. Immunol.* **8**, 421–452.

Ullum, H., Gotzsche, P. C., Victor, J., Dickmeiss, E., Skinhoj, P., and Pedersen, B. K. (1995). Defective natural immunity: An early manifestation of human immunodeficiency virus infection. *J. Exp. Med.* **182**, 789–799.

Vose, B. M., Vanky, F., and Klein, E. (1977). Human tumor-lymphocyte interaction in vitro. V. Comparison of the reactivity of tumor-infiltrating blood and lymph-node lymphocytes with autologous tumor cells. *Int. J. Cancer* **20**, 895–902.

Wanebo, H. J., Jun, M. Y., Strong, E. W., and Oettgen, H. (1975). T cell deficiency in patients with squamous cell cancer of the head and neck. *Am. J. Surg.* **130**, 445–448.

Warren, S. (1932). The immediate causes of death in cancer. *Am. J. Med. Sci.* **184**, 610–615.

Watkins, S. M. (1973). The effects of surgery in lymphocyte transformation in patients with cancer. *Clin. Exp. Immunol.* **14**, 69–76.

Wilmore, D. W., Goodwin, C. W., Aulick, L. H., *et al.* (1980). Effect of injury and infection on visceral metabolism and circulation. *Ann. Surg.* **192**, 491–504.

Yron, I., Schickler, M., Fisch, B., Pinkas, H., Ovadia, J., and Witz, I. P. (1986). The immune system during the pre-cancer and the early cancer period: IL-2 production by PBL from postmenopausal women with and without endometrial carcinoma. *Int. J. Cancer* **38**, 331–338.

Zhang, P. C., and Pang, C. P. (1992). Plasma amino acid patterns in cancer. *Clin. Chem.* **38**, 1198–1199.

Charles S. Lieber

Mount Sinai School of Medicine (CUNY)
Alcohol Research and Treatment Center and G.I.-Liver-Nutrition Program
Bronx Veterans Affairs Medical Center
Bronx, New York 10468

Role of Oxidative Stress and Antioxidant Therapy in Alcoholic and Nonalcoholic Liver Diseases

Alcohol-induced oxidative stress is most severe in the liver because it is linked to the metabolism of ethanol, which occurs predominantly in that organ.

I. Metabolism of Ethanol

The hepatocyte contains three main pathways for ethanol metabolism, each located in a different subcellular compartment: (i) the alcohol dehydrogenase (ADH) pathway of the cytosol or the soluble fraction of the cell, (ii) the microsomal ethanol-oxidizing system (MEOS) located in the endoplasmic reticulum, and (iii) catalase located in the peroxisomes (Lieber, 1991) (Fig. 1). Each of these pathways produces a specific metabolic and toxic disturbance, and all three result in the production of acetaldehyde, a highly reactive metabolite.

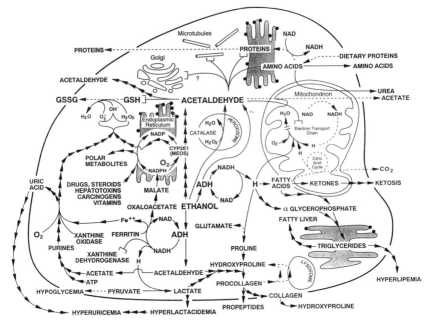

FIGURE 1   Oxidation of ethanol in the hepatocyte. Many disturbances in intermediary metabolism and toxic effects can be linked to (i) ADH-mediated generation of NADH; (ii) the induction of the activity of microsomal enzymes, especially the MEOS containing P4502E1 (CYP2E1); and (iii) acetaldehyde, the product of ethanol oxidation, GSH, reduced glutathione; GSSG, oxidized glutathione; ----, pathways that are depressed by ethanol; →→→→, stimulation or activation; ⊣, interference or binding. (From Lieber, 1994.)

A. Alcohol Dehydrogenase Pathway

The major portion of ethanol metabolism involves ADH. In this ADH-mediated oxidation, hydrogen is transferred from the substrate to the cofactor nicotinamide adenine dinucleotide (NAD), converting it to its reduced form (NADH), and acetaldehyde is produced (Fig. 1). Thus, the first step in the oxidation of ethanol generates an excess of reducing equivalents in the cytosol, primarily as NADH, with a marked shift in the redox potential of the cytosol, as measured by changes in the lactate/pyruvate ratio (Domschke *et al.*, 1974) using an approach first developed by Bücher and Klingenberg (1958). The reducing equivalents can also be transferred to NADP, and the increased NADPH can be utilized for synthetic pathways in the cytosol and microsomes.

Some of the hydrogen equivalents formed in this reaction are transferred from the cytosol into the mitochondria. The mitochondrial membrane is impermeable to NADH and the reducing equivalents are thought to enter the mitochondria via shuttle mechanisms such as the malate cycle (quantitatively, probably the most important), the fatty acid elongation cycle, and the α-glycerophosphate cycle. Normally, fatty acids are reduced via β-oxidation in the citric acid cycle of the mitochondria, which serves as a "hydrogen donor" for the mitochondrial electron transport chain. When ethanol is reduced, the generated hydrogen equivalents, which are shuttled into the mitochondria, supplant the citric acid cycle as source and the mitochondria are shifted to a more reduced redox state, as measured by changes in the ratio of β-hydroxybutyrate to acetoacetate (Domschke et al., 1974).

The altered redox state is responsible for a variety of metabolic abnormalities: the increased lactate/pyruvate ratio results in hyperlactacidemia because of both decreased utilization (Greenway and Lautt, 1990) and enhanced production of lactate by the liver. Hyperlactacidemia contributes to the acidosis and also reduces the capacity of the kidney to excrete uric acid, leading to secondary hyperuricemia (Lieber et al., 1962). Alcohol-induced ketosis (Lefevre et al., 1970) and enhanced purine breakdown (Faller and Fox, 1982) may also promote hyperuricemia. Another possible consequence of enhanced purine degradation is the increased production of activated oxygen species by xanthine oxidase (vide infra), as suggested by the protective effect of allopurinol against the alcohol-induced lipid peroxidation (Kato et al., 1990). Other metabolic consequences include enhanced lipogenesis, depressed lipid oxidation, and hypoglycemia (Lieber, 1992).

The increased NADH/NAD ratio also raises the concentration of α-glycerophosphate, which favors hepatic triglyceride accumulation by trapping fatty acids. In addition, excess NADH may promote fatty acid synthesis. Theoretically, enhanced lipogenesis can be considered a means for disposing of the excess hydrogen. The activity of the citric acid cycle is depressed, partly because of a slowing of the reactions of the cycle that require NAD; the mitochondria will use the hydrogen equivalents originating from ethanol instead of those derived from the oxidation of fatty acids that normally serve as the main energy source of the liver. Fatty acids of different sources can accumulate as triglycerides in the liver because of different metabolic disturbances: enhanced hepatic lipogenesis, decreased hepatic release of lipoproteins, increased mobilization of peripheral fat, enhanced hepatic uptake of circulating lipids, and, most important, decreased fatty acid oxidation, whether as a function of the reduced citric acid cycle activity secondary to the altered redox potential (vide supra) or as a consequence of permanent changes in mitochondrial structure and functions (Lieber, 1992), documented by breath analysis in alcoholics (Lauterburg et al., 1993).

A characteristic feature of liver injury in the alcoholic is the predominance of steatosis and other lesions in the perivenular zone, also called centrilobular or zone 3 of the hepatic acinus. The mechanism for this zonal selectivity of the toxic effects involves several distinct and not mutually exclusive mechanisms. The hypoxia hypothesis originated from the observation that liver slices from rats fed alcohol chronically consume more oxygen than those of controls. It was then postulated that the enhanced consumption of oxygen would increase the gradient of oxygen tension along the sinusoids to the extent of producing anoxic injury of perivenular hepatocytes (Israel *et al.*, 1975). Indeed both in human alcoholics (Kessler *et al.*, 1954) and in animals fed alcohol chronically (Jauhonen *et al.*, 1982; Sato *et al.*, 1983), decreases in either hepatic venous oxygen saturation (Kessler *et al.*, 1954) or PO_2 (Jauhonen *et al.*, 1982) and in tissue oxygen tension (Sato *et al.*, 1983) have been found during the withdrawal state. However, the changes in hepatic oxygenation found during the withdrawal state disappeared (Shaw *et al.*, 1977; Jauhonen *et al.*, 1982) or decreased (Sato *et al.*, 1983) when alcohol was present in the blood. Acute ethanol administration increased splanchnic oxygen consumption in naive baboons, but the consequences of this effect on oxygenation in the perivenular zone were offset by increased blood flow resulting in unchanged hepatic venous oxygen tension (Jauhonen *et al.*, 1982). In fact, ethanol induced an increase in portal hepatic blood flow (Stein *et al.*, 1973; Shaw *et al.*, 1977; Jauhonen *et al.*, 1982; Carmichael *et al.*, 1987; Lieber *et al.*, 1989). In cats (Greenway and Lautt, 1990) and in baboons (Lieber *et al.*, 1989) that were fed alcohol chronically, defective oxygen utilization rather than lack of blood oxygen supply characterized liver injury produced by high concentrations of ethanol. The low oxygen tension normally prevailing in perivenular zones exaggerates the redox shift produced by ethanol (Jauhonen *et al.*, 1982). Hypoxia, by increasing NADH, may in turn inhibit the activity of NAD^+-dependent xanthine dehydrogenase (XD), thereby favoring that of oxygen-dependent xanthine oxidase (XO) (Kato *et al.*, 1990). Purine metabolism via XO may lead to the production of oxygen radicals that can mediate toxic effects toward liver cells, including peroxidation. Physiological substrates for XO, hypoxanthine and xanthine, as well as AMP, significantly increased in the liver after ethanol, together with an enhanced urinary output of allantoin (a final product of xanthine metabolism). Allopurinol pretreatment resulted in 90% inhibition of XO activity, and also significantly decreased ethanol-induced lipid peroxidation (Kato *et al.*, 1990).

The main mechanism, however, whereby ADH-mediated ethanol metabolism affects oxidative stress is through the production of acetaldehyde (vide infra).

B. Microsomal Ethanol-Oxidizing System

I. Characterization of the MEOS and Its Role in Ethanol Metabolism

The first indication of an interaction of ethanol with the microsomal fraction of the hepatocyte was provided by the morphologic observation that alcohol feeding results in a proliferation of the smooth endoplasmic reticulum (SER) (Iseri *et al.*, 1964, 1966; Lane and Lieber, 1966). This increase in SER resembles that seen after the administration of a wide variety of hepatotoxins (Meldolesi, 1967), therapeutic agents (Conney, 1967), and some food additives (Lane and Lieber, 1967). Because most of the substances that induce a proliferation of the SER are metabolized, at least in part, by the cytochrome P450 enzyme system that is located on the SER, the possibility that alcohol may also be metabolized by similar enzymes was raised. Such a system was indeed demonstrated in liver microsomes *in vitro* and was found to be inducible by chronic alcohol feeding *in vivo* (Lieber and DeCarli, 1968) and was named the microsomal ethanol-oxidizing system (Lieber and DeCarli, 1968, 1970a). Its distinct nature was shown by (i) isolation of a P450-containing fraction from liver microsomes which, although devoid of any ADH or catalase activity, could still oxidize ethanol as well as higher aliphatic alcohols (e.g., butanol, which is not a substrate for catalase) (Teschke *et al.,* 1972, 1974) and (ii) reconstitution of ethanol-oxidizing activity using NADPH–cytochrome P450 reductase, phospholipid, and either partially purified or highly purified microsomal P450 from untreated (Ohnishi and Lieber, 1977) or phenobarbital-treated (Miwa *et al.,* 1978) rats. That chronic ethanol consumption results in the induction of a unique P450 was shown by Ohnishi and Lieber (1977) using a liver microsomal P450 fraction isolated from ethanol-treated rats. An ethanol-inducible form of P450, purified from rabbit liver microsomes (Koop *et al.,* 1982), catalyzed ethanol oxidation at rates much higher than other P450 isozymes, and also had an enhanced capacity to oxidize 1-butanol, 1-pentanol, and aniline (Morgan *et al.,* 1982), acetaminophen (Morgan *et al.,* 1983), CCl_4 (Morgan *et al.,* 1982), acetone (Ingelman-Sundberg and Johansson, 1984; Koop and Casazza, 1985), and N-nitrosodimethylamine (NDMA) (Yang *et al.,* 1985). The purified human protein (now called CYP2E1 or 2E1) was obtained in a catalytically active form, with a high turnover rate for ethanol and other specific substrates (Lasker *et al.,* 1987). MEOS has a relatively high K_m for ethanol (8–10 mM compared to 0.2–2 mM for hepatic ADH) but, contrasting with hepatic ADH, which is not inducible in primates as well as most other animal species, a fourfold induction of 2E1 was found in biopsies of recently drinking subjects using specific antibodies against this 2E1 and the Western blot technique (Tsutsumi *et al.,* 1989).

The molecular mechanism underlying 2E1 induction remains disputed. Investigations using rabbits (and some involving rats) appeared to have ruled out transcriptional activation of the 2E1 gene or stabilization of 2E1 mRNA as possible mechanisms since ethanol and similar agents (acetone, imidazole, 4-methylpyrazole, pyrazole, and pyridine) had little effect on 2E1 transcript content in liver (Song *et al.,* 1986, 1987; Khani *et al.,* 1987; Johansson *et al.,* 1988; Porter *et al.,* 1989). A posttranslational mechanism, namely protein stabilization, was thus proposed as (i) ethanol and imidazole were found to prevent the rapid decrease in 2E1 enzyme levels that occur in rat hepatocytes upon primary culture (Eliasson *et al.,* 1988) and (ii) acetone treatment was shown to prolong the *in vivo* half-life of 2E1 in rat liver by eliminating the fast-phase component associated with the normal degradation of the enzyme (Song *et al.*, 1989). Yet other studies support the role of enhanced *de novo* enzyme synthesis and/or increased mRNA in the 2E1 induction process. Kim and Novak (1990) observed increased rates of [^{14}C]leucine incorporation into 2E1 protein after treatment of rats with pyridine, a phenomenon attributed to the enhancement, by pyridine, of 2E1 mRNA translational efficiency (Kim *et al.*, 1990). Kubota *et al.* (1988) found that induction of 2E1 protein in hamsters by ethanol and pyrazole was associated with an increase in translatable 2E1 mRNA, whereas Diehl *et al.* (1991) described elevated levels of hepatic 2E1 mRNA in alcohol-treated rats. Enhanced levels of both hepatic 2E1 protein and mRNA were found in actively drinking patients (Takahashi *et al.*, 1993), indicating that mRNA stabilization and/or transcriptional activation is involved in ethanol-mediated 2E1 induction in humans. Furthermore, by measuring both the synthesis and the degradation of radiolabeled 2E1 protein at induced high steady-state levels in the rat, ethanol was found to stimulate rates of *de novo* 2E1 synthesis but had no effect on rates of enzyme degradation (Tsutsumi *et al.*, 1993). It is therefore reasonable to assume that the enhancement of *de novo* 2E1 synthesis noted in ethanol-treated rats results from increased steady-state levels of 2E1 message and/or the increased efficiency with which this mRNA is translated.

The dose of the inducer is also of importance because 2E1 induction appears to occur via two steps: (i) a posttranslational mechanism at low ethanol concentrations and (ii) an additional transcriptional one at high ethanol levels (Badger *et al.*, 1993; Ronis *et al.*, 1993).

The presence of 2E1 was also shown in extrahepatic tissues (Shimizu *et al.*, 1990) and in nonparenchymal cells of the liver, including Kupffer cells (Koivisto *et al.*, 1996). In rats, ethanol treatment caused a sevenfold increase in CYP2E1 content in Kupffer cells. It also increased the hydroxylation of *p*-nitrophenol, a relatively specific substrate for CYP2E1, demonstrating that the induced CYP2E1 was catalytically active. This reaction was significantly inhibited by anti-CYP2E1 IgG in both types of cells. Although CYP2E1 may not be the predominant pathway for ethanol metabolism in hepatocytes, it is possibly the major one in Kupffer cells. Thus, the induction of CYP2E1 by ethanol in these cells could cause significant changes in intracellular acetaldehyde concentrations, which may contribute to the development of alcoholic liver injury (Koivisto *et al.*, 1996).

2. Increased Xenobiotic and Retinoid Toxicity, Carcinogenicity, and Oxidative Stress in Alcoholics

Not infrequently, the metabolites produced in the microsomes are more toxic than the precursor compounds. Much of the medical significance of MEOS (and its ethanol-inducible 2E1) results not only from the oxidation of ethanol but also from the unusual and unique capacity of 2E1 to generate reactive oxygen intermediates, such as superoxide radicals (Fig. 1) (Dai *et al.*, 1993), and to activate many xenobiotic compounds to their toxic metabolites, often free radicals. This pertains, for instance, to carbon tetrachloride and other industrial solvents such as bromobenzene (Hetu *et al.*, 1983) and vinylidene chloride (Siegers *et al.*, 1983), as well as anesthetics such as enflurane (Tsutsumi *et al.*, 1990) and halothane (Takagi *et al.*, 1983). Ethanol also markedly increased the activity of microsomal low K_m benzene-metabolizing enzymes (Nakajima *et al.*, 1987) and aggravated the hemopoietic toxicity of benzene. Enhanced metabolism (and toxicity) also pertains to a variety of prescribed drugs, including isoniazid and phenylbutazone (Beskid *et al.*, 1980), and some over-the-counter medications such as acetaminophen (paracetamol, N-acetyl-p-aminophenol), all of which are substrates for or inducers of cytochrome P4502E1. Therapeutic amounts of acetaminophen (2.5 to 4 g per day) can cause hepatic injury in alcoholics. In animals given ethanol for long periods, hepatotoxic effects peak after withdrawal (Sato *et al.*, 1981) when ethanol is no longer competing for the microsomal pathway but when levels of the toxic metabolites are at their highest. Thus, alcoholics are most vulnerable to the toxic effects of acetaminophen shortly after the cessation of chronic drinking.

There is an association between alcohol misuse and an increased incidence of upper alimentary and respiratory tract cancers (Lieber *et al.*, 1986). Many factors have been incriminated, one of which is the effect of ethanol on enzyme systems involved in the cytochrome P450-dependent activation of carcinogens. This effect has been demonstrated with the use of microsomes derived from a variety of tissues, including the liver (the principal site of xenobiotic metabolism) (Garro *et al.*, 1981; Seitz *et al.*, 1981), the lungs (Garro *et al.*, 1981; Seitz *et al.*, 1981) and intestines (Seitz *et al.*, 1985, Shimizu *et al.*, 1990) (the chief portals of entry for tobacco smoke and dietary carcinogens, respectively), and the esophagus (Farinati *et al.*, 1985) (where ethanol consumption is a major risk factor in the development of cancer). Alcoholics are commonly heavy smokers, and a synergistic effect of alcohol consumption and smoking on cancer development has been described, as reviewed elsewhere (Lieber *et al.*, 1986). Indeed, long-term ethanol consumption was found to enhance the mutagenicity of tobacco-derived products (Lieber *et al.*, 1986).

Alcohol may also influence carcinogenesis in many other ways (Garro and Lieber, 1990), one of which involves vitamin A depletion. Indeed, alcoholic liver disease is associated with severe hepatic vitamin depletion, already at early stages (Leo and Lieber, 1983). Depressed hepatic levels of vitamin A were observed even when alcohol was given with diets containing large amounts of vitamin A (Sato and Lieber, 1981). New hepatic enzyme pathways of retinol metabolism, inducible by either ethanol or drug administration, have been discovered (Leo and Lieber, 1985; Leo *et al.*, 1987). Hepatic vitamin A depletion is associated with lysosomal lesions (Leo *et al.*,

1983) and decreased detoxification of NDMA (Leo et al., 1986). Although vitamin A deficiency might adversely affect the liver (Leo et al., 1983), an excess of vitamin A is also known to be hepatotoxic (Leo and Lieber, 1988). Long-term ethanol consumption enhances this effect, resulting in striking morphologic and functional alterations of the mitochondria (Leo et al., 1982), along with hepatic necrosis and fibrosis (Leo and Lieber, 1983). Hypervitaminosis A itself can induce fibrosis and even cirrhosis, as reviewed elsewhere (Leo et al., 1982). Thus, alcohol abuse narrows the therapeutic window for vitamin A, which hinders it therapeutic use.

There is also increased evidence that ethanol toxicity may be associated with an increased production of reactive oxygen intermediates. Numerous experimental data indicate that free radical mechanisms contribute to ethanol-induced liver injury. Increased generation of oxygen- and ethanol-derived free radicals occur at the microsomal level, especially through the intervention of the ethanol-inducible 2E1 (Nordmann et al., 1992). This induction is associated with proliferation of the endoplasmic reticulum (vide supra), which is accompanied by increased oxidation of NADPH with resulting H_2O_2 generation (Lieber and DeCarli, 1970b). There is also increased superoxide radical production. In addition, the 2E1 induction contributes to the well-known lipid peroxidation associated with alcoholic liver injury. DiLuzio (1964, 1968) was one of the first to report that ethanol produces increased lipid peroxidation in the liver and that the ethanol-induced fatty liver could be prevented by antioxidants. Lipid peroxidation correlated with the amount of CYP2E1 in liver microsomal preparations, and it could be inhibited by antibodies against 2E1 in control and ethanol-fed rats (Ekstrom and Ingelman-Sundberg, 1989; Castillo et al., 1992). Indeed, 2E1 is rather "leaky" and its operation results in a significant release of free radicals, including l-hydroxyethyl-free radical intermediates (Albano et al., 1987; Reinke et al., 1987), confirmed by detecting the hydroxyethyl radicals in vivo (Reinke et al., 1991). Ethanol can also be reduced by liver microsomes to acetaldehyde through a nonenzymatic pathway involving the presence of hydroxyl radicals originating from the iron-catalyzed degradation of H_2O_2 (Cederbaum, 1989). The production of ethanol-free radicals may be due to an oxidizing species bound to cytochrome P450 and abstracting a proton from the alcohol α-carbon (Albano et al., 1991). It is not known, however, whether hydroxyethyl-free radicals contribute to the damaging effects of ethanol. Hydroxyethyl radicals appear to be involved in the alkylation of hepatic proteins. An in vitro-produced hydroxyethyl radical forms stable adducts with albumin or fibrinogen (Clot et al., 1995), and patients with alcoholic cirrhosis have increased serum levels of both IgG and IgA reacting with proteins of liver microsomes incubated with ethanol and NADPH (Clot et al., 1995), which do not cross-react with the epitopes derived from acetaldehyde-modified proteins.

Most important, induction of the MEOS results in enhanced acetaldehyde production that, in turn, aggravates the oxidative stress directly as well as indirectly by impairing defense systems against it (vide infra).

C. Role of Catalase

Catalase is capable of oxidizing alcohol *in vitro* in the presence of an H_2O_2-generating system (Keilin and Hartree, 1945) (Fig. 1), and its interaction with H_2O_2 in the intact liver has been demonstrated (Sies and Chance, 1970). However, its role is limited by the small amount of H_2O_2 generated (Oshino *et al.*, 1973) and, under physiological conditions, catalase thus appears to play no major role in ethanol oxidation.

The catalase contribution might be enhanced if significant amounts of H_2O_2 become available through the β-oxidation of fatty acids in peroxisomes (Handler and Thurman, 1985). However, the peroxisomal enzymes do not oxidize short chain fatty acids such as octanoate; peroxisomal β-oxidation was observed only in the absence of ADH activity. In its presence the rate of ethanol metabolism is reduced by adding fatty acids (Williamson *et al.*, 1969) and, conversely, the β-oxidation of fatty acids is inhibited by NADH produced from ethanol metabolism via ADH (Williamson *et al.*, 1969). Similarly, the generation of reducing equivalents from ethanol by ADH in the cytosol inhibits H_2O_2 generation leading to significantly diminished rates of peroxidation of alcohols via catalase (Handler and Thurman, 1990). Various other results also indicated that peroxisomal fatty acid oxidation does not play a major role in alcohol metabolism (Inatomi *et al.*, 1989). Furthermore, when fatty acids were used by Handler and Thurman (1985) to stimulate ethanol oxidation, this effect was very sensitive to inhibition by aminotriazole, a catalase inhibitor. Therefore, if this mechanism were to play an important role *in vivo,* one would expect a significant inhibition of ethanol metabolism after aminothiazole administration *in vivo,* when physiologic amounts of fatty acids and other substrates for H_2O_2 generation are present. A number of studies, however, have shown that aminotriazole treatment has little, if any, effect on alcohol oxidation *in vivo,* as reviewed by Takagi *et al.* (1986) and Kato *et al.* (1987a,b).

Despite the considerable controversy that originally surrounded this issue, it is now agreed by the principal contenders involved that catalase cannot account for microsomal ethanol oxidation (Teschke *et al.*, 1977; Thurman and Brentzel, 1977). However, catalase could contribute to fatty acid oxidation. Indeed, long-term ethanol consumption is associated with increases in the content of a specific cytochrome (P4504A1) that promotes microsomal ω-hydroxylation of fatty acids, which may compensate, at least in part, for the deficit in fatty acid oxidation due to the ethanol-induced injury of the mitochondria (Lieber, 1992). Products of ω-oxidation also increase liver cytosolic fatty acid-binding protein (L-FABPc) content and peroxisomal β-oxidation (Kaikaus *et al.*, 1990), an alternate but modest pathway for fatty acid disposition (vide supra).

II. Acetaldehyde: Its Metabolism and Toxicity, Including Peroxidation

Acetaldehyde, the product of ethanol oxidation, is highly toxic and is rapidly metabolized to acetate, mainly by a mitochondrial low K_m aldehyde dehydrogenase (ALDH), the activity of which is significantly reduced by chronic ethanol consumption (Hasumura *et al.*, 1975). The decreased capacity of mitochondria of alcohol-fed subjects to oxidize acetaldehyde, associated with unaltered or even enhanced rates of ethanol oxidation (and therefore acetaldehyde generation because of MEOS induction: vide supra), results in an imbalance between production and disposition of acetaldehyde. The latter causes the elevated acetaldehyde levels observed after chronic ethanol consumption in humans (DiPadova *et al.*, 1987) and in baboons, which revealed a tremendous increase of acetaldehyde in hepatic venous blood (Lieber *et al.*, 1989), reflecting high tissue levels.

The toxicity of acetaldehyde is due, in part, to its capacity to form protein adducts, resulting in antibody production, enzyme inactivation, and decreased DNA repair (Espina *et al.*, 1988; Lieber *et al.*, 1989). It also is associated with a striking impairment of the capacity of the liver to utilize oxygen. Moreover, acetaldehyde promotes glutathione (GSH) depletion, free radical-mediated toxicity, and lipid peroxidation.

Indeed, acetaldehyde was shown to be capable of causing lipid peroxidation in isolated perfused livers (Müller and Sies, 1982, 1983). *In vitro*, metabolism of acetaldehyde via xanthine oxidase or aldehyde oxidase may generate free radicals, but the concentration of acetaldehyde required is much too high for this mechanism to be of significance *in vivo*. However, another mechanism used to promote lipid peroxidation is via GSH depletion. One of the three amino acids of this tripeptide is cysteine. Binding of acetaldehyde with cysteine and/or GSH may contribute to a depression of liver GSH (Shaw *et al.*, 1983). Rats fed ethanol chronically have significantly increased rates of GSH turnover (Morton and Mitchell, 1985). Acute ethanol administration inhibits GSH synthesis and produces an increased loss from the liver (Speisky *et al.*, 1985). GSH is selectively depleted in the mitochondria (Hirano *et al.*, 1992) and may contribute to the striking alcohol-induced alterations of that organelle. GSH offers one of the mechanisms for the scavenging of toxic-free radicals, as shown in Fig. 2, which also illustrates how the ensuing enhanced GSH utilization (and thus turnover) results in a significant increase in α-amino-n-butyric acid (Shaw and Lieber, 1980). Although GSH depletion per se may not be sufficient to cause lipid peroxidation, it is generally agreed upon that it may favor the peroxidation produced

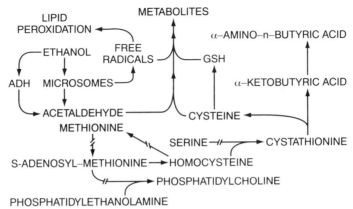

FIGURE 2 Link between accelerated acetaldehyde production and increased free radical generation by "induced" microsomes, resulting in enhanced lipid peroxidation. Metabolic blocks due to alcohol, folate deficiency, and/or alcoholic liver disease are illustrated with possible beneficial effects of GSH, its precursors (including S-adenosylmethionine), and phosphatidylcholine. (From Lieber, 1995b.)

by other factors. GSH has been shown to spare and potentiate vitamin E (Barclay, 1988); it is important in the protection of cells against electrophilic drug injury in general and against reactive oxygen species in particular, especially in primates, which are more vulnerable to GSH depletion than rodents (Shaw *et al.*, 1981). Iron overload may play a contributory role, as chronic alcohol consumption results in an increased iron uptake by hepatocytes (Zhang, 1993) and as iron exposure accentuates the changes of lipid peroxidation and in the glutathione status of the liver cell induced by acute ethanol intoxication (Valenzuela *et al.*, 1983). Lipid peroxidation is not only a reflection of tissue damage, it may also play a pathogenic role, e.g., by promoting collagen production (Geesin *et al.*, 1991; Tsukamoto, 1993).

III. Antioxidant Therapy

A. Carotenoids

Progress has been made concerning possible therapeutic effects of agents, including natural antioxidants, that oppose the alcohol-induced oxidative stress. Retinol is an antioxidant but it is a weak one and, as noted earlier, its use is complicated by its intrinsic hepatotoxicity, exacerbated by ethanol. Unlike retinol, its precursor β-carotene is considered to lack toxicity. Furthermore, in addition to acting as a retinol precursor, β-carotene is also an efficient quencher of singlet oxygen and can function as a radical-trapping antioxidant. β-Carotene also has been shown to have the potential of acting as a more efficient antioxidant than retinol. Carotenoids were found to inhibit free radical-induced lipid peroxidation (Krinsky and Deneke, 1982; Burton and Ingold, 1984; Krinsky, 1989). A mechanism whereby β-carotene may act as a lipid antioxidant has been provided by Burton (1989). β-Carotene was also shown to inhibit arachidonic acid oxidation (Halevy and Sklan, 1987). It may prevent lipid peroxidation by acting through specific enzyme inhibition. Indeed, as shown by Lomnitski *et al.* (1993), β-carotene inhibits the activity of lipoxygenase toward linoleate. Possible vitamin E–β-carotene interactions in the antioxidant defense system of the membrane have been assessed by Machlin and Bendich (1987) and a potential role in the prevention of cardiovascular disease has been reviewed (Gerster, 1991). Thus, β-carotene appears to play a significant role in the oxidant–antioxidant balance. Indeed, it has been shown in humans that administration of β-carotene reduces the level of circulating lipid peroxides (Mobarhan *et al.*, 1990). However, in a study in rats, Alam and Alam (1983) reported no change in either blood or tissue lipid peroxides following β-carotene ingestion of 180 mg/kg/day for a period of 11 weeks, and carotenoids did not protect against peroxidation in choline-deficient rats (Jenkins *et al.*, 1993). However, a study in guinea pigs noted a protective effect against *in vivo* lipid peroxidation when animals were pretreated with β-carotene (Kunert and Tappel, 1983). Furthermore, Palozza and Krinsky (1991) reported that β-carotene inhibited malondialdehyde production in a concentration-dependent manner and delayed the radical-initiated destruction of endogenous α- and γ-tocopherol in the rat, and Kim-Jun (1993) reported inhibitory effects of β-carotene on lipid peroxidation in mouse epidermis. As mentioned earlier, studies by Mobarhan *et al.* (1990) have shown that replenishing subjects with β-carotene can decrease the level of circulating lipid peroxides, but it was not reported whether this can be achieved in individuals who continue to drink and at a dose of β-carotene that has no toxicity in the presence of alcohol. One may also wonder whether the combination of alcohol and β-carotene, at the dose used for replenishment, may be useful in terms of preventing lipid peroxidation, without producing some signs of toxicity. At present, possible interactions of β-carotene with liver disease and/or alcohol are virtually uncharted but cannot be excluded as enhanced hepatic toxicity of β-carotene in the presence of ethanol has been observed with the possible existence of a defect in utilization and/or excretion associated with liver injury and/or alcohol abuse (Leo *et al.*, 1992, 1993). Furthermore, in men, heavy drinking was associated with a relative increase in serum β-carotene (Ahmed *et al.*, 1994), and relatively moderate drinking in women also has a similar effect (Forman *et al.*, 1995). It is noteworthy that epidemiologic studies have revealed that β-carotene supplementation may increase the incidence of pulmonary cancer and cardiovascular complications in smokers (Alpha-Tocopherol, β-Carotene and Cancer Prevention Study Group, 1994), an effect related to an interaction between β-carotene and alcohol (P. R. Taylor, personal communication, 1995). Thus, vitamin A or β-carotene supplementation must be used cautiously in alcoholics.

B. Methionine and S-Adenosylmethionine

As discussed earlier, one major antioxidant agent is the reduced form of GSH. However, therapeutic use of GSH itself is complicated by the fact that its replenishment through supplementation is hampered by its lack of penetration into the hepatocytes, except for its ethyl derivative, which is not very suitable for the treatment of alcoholic liver injury. Cysteine is one of the three amino acids of GSH, and the ultimate precursor of cysteine is methionine (Fig. 2); its deficiency in alcoholics has been incriminated and its supplementation has been considered for the treatment of alcoholic liver injury, but some difficulties have been encountered. Indeed, excess methionine was shown to have some adverse effects (Finkelstein and Martin, 1986), including a decrease in hepatic ATP. Horowitz *et al.* (1981) reported that the blood clearance of methionine after an oral load of this amino acid was slowed in patients with cirrhosis. Because about half of the methionine is metabolized by the liver, the previously mentioned observations suggest impaired hepatic metabolism of this amino acid in patients with alcoholic liver disease. To be utilized, methionine has to be activated to S-adenosylmethionine (SAMe) (Fig. 2). However, Duce *et al.* (1988) found a decrease in SAMe synthetase activity in cirrhotic livers. As a consequence, SAMe depletion ensues after chronic ethanol consumption (Lieber *et al.*, 1990). Potentially, such SAMe depletion may have a number of adverse effects. SAMe is the principal methylating agent in various transmethylation reactions important for nucleic acid and protein synthesis, as well as membrane fluidity and functions, including the transport of metabolites and the transmission of signals across membranes and maintenance of membranes. Thus, depletion of SAMe may promote the membrane injury documented in alcohol-induced liver damage (Yamada *et al.*, 1985). SAMe is not only the methyl donor in almost all transmethylation reactions, but it also plays a key role in the synthesis of polyamines and, as already alluded to, it provides a source of cysteine for GSH production, a major natural hepatoprotective agent, the lack of which promotes liver injury, in part through peroxidation (see earlier discussion).

Orally administered SAMe is a precursor for intracellular SAMe, both as unchanged SAMe and by the methionine it provides. Compared to methionine, the administration of SAMe has the advantage of bypassing the deficit in SAMe synthesis (from methionine) referred to earlier (Fig. 2). The usefulness of SAMe administration has been demonstrated in the baboon (Lieber *et al.*, 1990, 1994a) and in various clinical studies (Vendemiale *et al.*, 1989; Lieber, 1995a), some of which are still ongoing.

C. Vitamin E and Miscellaneous Antioxidants

α-Tocopherol, the major antioxidant in the membrane, is viewed as the "last line" of defense against membrane lipid peroxidation (McCay, 1985; Niki, 1987). Bjørneboe *et al.* (1987) reported a reduced hepatic α-tocopherol content after chronic ethanol feeding in rats receiving adequate amounts of vitamin E, as well as in the blood of alcoholics. Hepatic lipid peroxidation was significantly increased after chronic ethanol feeding in rats receiving a low vitamin E diet (Kawase *et al.*, 1989), indicating that dietary vitamin E is an important determinant of hepatic lipid peroxidation induced by chronic ethanol feeding. The lowest hepatic α-tocopherol was found in rats receiving a combination of low vitamin E and ethanol: both low dietary vitamin E and ethanol feeding significantly decreased hepatic α-tocopherol content, the latter, in part, because of increased conversion of α-tocopherol to α-tocopherylquinone (Kawase *et al.*, 1989). In patients with cirrhosis, diminished hepatic vitamin E levels have been observed (Leo *et al.*, 1993) (Fig. 3), as also shown by von Herbay *et al.* (1994). These deficient antioxidant defense systems, coupled with increased generation of acetaldehyde and oxygen radical by the ethanol-induced microsomes (Fig. 2), may contribute to liver damage via lipid peroxidation and also via enzyme inactivation (Dicker and Cederbaum, 1988). Effectiveness of vitamin E

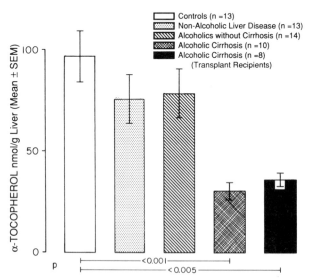

FIGURE 3   Effects of various liver diseases on total hepatic tocopherol levels. Only the two cirrhotic groups had significantly lower α-tocopherol levels (compared with α- and β-carotene and lycopene). (From Leo *et al.*, 1993.)

supplementation in the prevention of alcoholic liver injury is presently being evaluated.

Other antioxidant medications that have been proposed include (+)-cyanidanol-3, silymarin, selenium, and thioctic acid, but their beneficial effects still need confirmation (Seitz and Poschl, 1995).

D. Polyenyl- and Dilinoleoyl-Phosphatidylcholine

Oxidative stress is most likely to occur in patients with hyperlipidemia (Araujo *et al.*, 1995) and it is generally believed that the polyunsaturated lipids favor peroxidation. Indeed, because of their multiple double bond configuration, polyunsaturated fats are much more susceptible than saturated or monounsaturated ones to free radical peroxidation (Halliwell and Gutteridge, 1989). Surprisingly, however, some reports suggest that the opposite may occur. Effects of high monounsaturated and polyunsaturated fat diets on plasma lipoproteins and lipid peroxidation were studies in type 2 diabetes mellitus (Parfitt *et al.*, 1994). All indices of plasma lipid peroxidation in the diabetic group and lipid peroxides in the controls were significantly lower on these than on the baseline diet. It was postulated that because both high monounsaturated and polyunsaturated fat diets increase hepatic metabolism of low-density lipoproteins and shorten their circulating half-life, they may reduce lipid peroxidation, compared to high saturated fat diets; by this mechanism, polyunsaturated fat diets, may offset any increased susceptibility of polyunsaturated-enriched low-density lipoprotein to peroxidation. Similarly, whereas the malondialdehyde concentration in plasma increased with increasing blood lipids, it was inversely correlated to the proportion of linoleic acid in serum lipoprotein phospholipids, suggesting that oxidants and lipoprotein metabolism may be of greater importance for intravascular lipid peroxidation than the proportion of polyunsaturated fatty acids (PUFA) in the lipoprotein lipids. Furthermore, experimentally, studies in newborn rats demonstrated that lipid nutrition containing high concentrations of PUFA confers protection against pulmonary oxygen toxicity (Sosenko *et al.*, 1988, 1989). Specifically, newborn rat offspring of dams fed diets high in PUFA had elevated concentrations of PUFA in their lung lipids and significantly improved survival in hyperoxia compared with offspring of dams fed regular rat chow; conversely, newborn offspring of dams fed low PUFA, high saturated fatty acid diets were most susceptible to pulmonary oxygen toxicity. In addition, when Intralipid, derived from soybean oil and containing a high percentage of n-6 family PUFA, and linolenic acid, a n-3 family PUFA, were given for 3 weeks before and then throughout pregnancy and lactation, 1- and 5-day-old offspring of Intralipid diet-fed dams demonstrated significant increases in lung lipid n-6 family PUFA (plus elevated linolenic acid) compared to regular diet-fed offspring. Significantly improved hyperoxic survival rates were associated with these fatty acid changes. These findings supported the hypothesis that increasing lung PUFA content may provide increased O_2-free radical-scavenging capacity (Sosenko *et al.*, 1991). Similar results have been found with *in vitro* studies (Dennery *et al.*, 1990). However, there are also opposite results: cultured hamster fibroblast cells enriched with PUFA had increased susceptibility to the lethal effects of 95% oxygen (Spitz *et al.*, 1992).

In addition, evidence gathered in rodents and in nonhuman primates revealed striking antioxidant effects of a soybean extract rich in polyunsaturated lecithin, namely polyenylphosphatidylcholine (PPC), about half of which consists of dilinoleoylphosphatidylcholine (DLPC) (Lieber *et al.*, 1994a). Indeed, it was found that PPC prevents hepatic lipid peroxidation and attenuates associated injury induced by CCl_4 in rats (Aleynik *et al.*, 1995). Furthermore, PPC decreased oxidant stress and protected against alcohol-induced liver injury in the baboon (Lieber *et al.*, 1995). Using gas chromatography/mass spectroscopy (GC/MS), hepatic F_2-isoprostanes (F_2-IP), breakdown products of arachidonic acid peroxidation, were determined in liver needle biopsies. Whereas alcohol increased F_2-IP, PPC administration resulted in a 58% reduction. Moreover, as had been reported before (Lieber *et al.*, 1994a), the alcohol-induce septal fibrosis and cirrhosis were fully prevented by PPC under these conditions. In addition, a 54% decrease in circulating free F_2-IP was also observed in the plasma of the animals given PPC with ethanol. Alcohol feeding also significantly decreased GSH, an effect that was attenuated by PPC. As the phospholipid species of PPC are highly bioavailable (vide infra) and readily integrated in the liver membranes, they could act as scavengers of the excess O_2-free radicals and thereby prevent their toxic interaction with critical membrane polyunsaturated fatty acids. In a sense, they could act as some kind of radical "trap" or "sink." Furthermore, because peroxidation products are fibrogenic (Geesin *et al.*, 1991; Tsukamoto, 1993), their decrease after PPC could also explain, at least in part, the antifibrogenic property of the phospholipids.

PPC may also restore key enzyme activity: that of mitochondrial cytochrome oxidase *in vitro* and that of phosphatidylethanolamine *N*-methyltransferase (PEMT) *in vivo*. Indeed, the impairment of the mitochondria after chronic ethanol consumption includes a significant decrease in cytochrome oxidase activity, associated with a depletion in mitochondrial phosphatidylcholine; replenishment of the latter restored the cytochrome oxidase activity (Arai *et al.*, 1984). Concerning PEMT, one of its key functions is to utilize SAMe as a methyl donor in the methylation of phosphatidylethylanolamine to phosphatidylcholine (Fig. 2). However, it has been shown that chronic ethanol consumption is associated with a decrease of the activity in the enzymes, both in baboons (Lieber *et al.*, 1994b) and in men (Duce *et al.*, 1988). The reduction in PEMT activity may have some secondary effects on liver phospholipids. Indeed, a decrease in PEMT activity after alcohol may be responsible, at least in part, for the associated decrease in phospholipids (Lieber *et al.*, 1994a,b). Supplementation with polyenylphosphatidylcholine was shown to restore the PEMT activity (Lieber *et al.*, 1994b). Thus, PEMT depression after alcohol may exacerbate the hepatic phospholipid depletion and the associated membrane abnormalities, thereby promoting hepatic injury and triggering fibrosis, whereas PPC, by repleting hepatic phospholipids and normalizing PEMT activity, may contribute to the protection against alcoholic cirrhosis provided by PPC supplementation (Lieber *et al.*, 1992, 1994a) (Fig. 4).

PPC is especially suited to correct hepatic phospholipid depletion. Indeed, phospholipids rich in essential fatty acids have a high bioavailability. More than 50% of orally administered PPC is made biologically available for the organism either by intact absorption (lesser extent) or by reacylation of absorbed lysophosphatidylcholine (greater extent) (Fox, 1983). Pharmacokinetic studies in humans using $^3H/^{14}C$-labeled phosphatidylcholine showed the absorption to exceed 90% (Zierenberg and Grundy, 1982). Similar observations were made in animals (Lekim and Betzing, 1974; Parthasarathy *et al.*, 1974; Rodgers *et al.*, 1975). Although much of the PPC in the diet is degraded by pancreatic phopholipase A_2 (Arnesjo *et al.*, 1969), the products (1-acyl-lysophosphatidylcholine and fatty acids) are absorbed in the jejunum (Nilsson, 1968). Animal studies show that phosphatidylcholines recovered in intestinal lymph after feeding a fat-enriched diet with single fatty acids are highly enriched in both *sn*-1 and *sn*-2 positions with the same acyl groups that were fed (Patton *et al.*, 1984). Thus, it can be anticipated

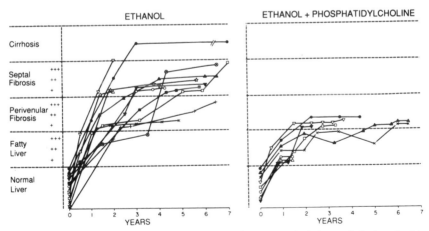

FIGURE 4   Sequential development of alcoholic liver injury in baboons fed ethanol with an adequate diet (left) and prevention of septal fibrosis and cirrhosis by supplementation with polyunsaturated phosphatidylcholine (PPC) (right). Liver morphology in animals pairfed control diets (with or without PPC) remained normal (not shown). (Data from Lieber *et al.*, 1994a.)

618   Charles S. Lieber

that during absorption of a diet enriched with 18 : 2 fatty acids, new phosphatidylcholines will be formed from dietary 18 : 2-lysophosphatidylcholine that will have an 18 : 2–18 : 2 composition. Various authors (Holz *et al.*, 1971; Lekim *et al.*, 1972; Lekim and Graf, 1976) reported PPC accumulation in the liver during the first 24 to 48 hr after administration. Indeed, all 18 : 2 phosphatidylcholines were present in the liver in significantly increased amounts in baboons fed PPC rich in DLPC (Lieber *et al.*, 1994a). It was also found in the baboon studies that, although the baseline value of DLPC was low, there was a significant increase of DLPC in the liver of the supplemented baboons (Lieber *et al.*, 1994a). Thus, PPC supplementation results in increased hepatic DLPC, which may be the active compound. Furthermore, in cultured stellate cells, PPC prevents acetaldehyde-mediated collagen accumulation (Li *et al.*, 1992), at least in part, by stimulating collagenase activity. The latter effect was mimicked by DLPC (Lieber *et al.*, 1994a).

In any event, whatever the mechanism of the hepatoprotective effect of PPC, it was manifest not only for the prevention, but also for the attenuation of preexisting liver fibrosis and cirrhosis (Ma *et al.*, 1996). Furthermore, PPC acts at prefibrotic steps: it attenuated hepatomegaly, fatty liver, and hyperlipemia in alcohol-fed rats (Navder *et al.*, 1996). The effectiveness of PPC for the prevention and/or treatment of liver cirrhosis is now being tested in a multicenter randomized trial in humans (VA Cooperative Study 391).

IV. Summary

The main pathway for the hepatic oxidation of ethanol to acetaldehyde proceeds via ADH and is associated with the reduction of NAD to NADH; the latter produces a striking redox change with various associated metabolic disorders. NADH also inhibits xanthine dehydrogenase activity, resulting in a shift of purine oxidation to xanthine oxidase, thereby promoting the generation of oxygen-free radical species. NADH also supports microsomal oxidations, including that of ethanol, in part via transhydrogenation to NADPH. In addition to the classic alcohol dehydrogenase pathway, ethanol can also be reduced by an accessory but inducible microsomal ethanol-oxidizing system. This induction is associated with proliferation of the endoplasmic reticulum, both in experimental animals and in humans, and is accompanied by increased oxidation of NADPH with resulting H_2O_2 generation. There is also a concomitant 4- to 10-fold induction of cytochrome P4502E1 (2E1) both in rats and in humans, with hepatic perivenular preponderance. This 2E1 induction contributes to the well-known lipid peroxidation associated with alcoholic liver injury, as demonstrated by increased rates of superoxide radical production and lipid peroxidation correlating with the amount of 2E1 in liver microsomal preparations and the inhibition of lipid peroxidation in liver microsomes by antibodies against 2E1 in control and ethanol-fed rats. Indeed, 2E1 is rather "leaky" and its operation results in a significant release of free radicals. In addition, induction of this microsomal system results in enhanced acetaldehyde production, which in turn impairs defense systems against oxidative stress. For instance, it decreases GSH by various mechanisms, including binding to cysteine or by provoking its leakage out of the mitochondria and of the cell. Hepatic GSH depletion after chronic alcohol consumption was shown both in experimental animals and in humans. Alcohol-induced increased GSH turnover was demonstrated indirectly by a rise in α-amino-n-butyric acid in rats and baboons and in volunteers given alcohol. The ultimate precursor of cysteine (one of the three amino acids of GSH) is methionine. Methionine, however, must be first activated to S-adenosylmethionine by an enzyme which is depressed by alcoholic liver disease. This block can be bypassed by SAMe administration which restores hepatic SAMe levels and attenuates parameters of ethanol-induced liver injury significantly such as the increase in circulating transaminases, mitochondrial lesions, and leakage of mitochondrial enzymes (e.g., glutamic dehydrogenase) into the bloodstream. SAMe also contributes to the methylation of phosphatidylethanolamine to phosphatidylcholine. The methyltransferase involved is strikingly depressed by alcohol consumption, but this can be corrected, and hepatic phosphatidylcholine levels restored, by the administration of a mixture of polyunsaturated phospholipids (polyenylphosphatidylcholine). In addition, PPC provided total protection against alcohol-induced septal fibrosis and cirrhosis in the baboon and it abolished an associated twofold rise in hepatic F_2-isoprostanes, a product of lipid peroxidation. A similar effect was observed in rats given CCl_4. Thus, PPC prevented CCl_4- and alcohol-induced lipid peroxidation in rats and baboons, respectively, while it attenuated the associated liver injury. Similar studies are ongoing in humans.

Acknowledgments

Original studies reviewed here were supported, in part, by NIH Grants AA05934 and AA07275 and by the Department of Veterans Affairs. Skillful typing of the manuscript by Ms. S. Dickerson is gratefully acknowledged.

References

Ahmed, S., Leo, M. A., and Lieber, C. S. (1994). Interactions between alcohol and β-carotene in patients with alcoholic liver disease. *Am. J. Clin. Nutr.* **60,** 430–436.
Alam, S. Q., and Alam, B. S. (1983). Lipid peroxide, α-tocopherol and retinoid levels in plasma and liver of rats fed diets containing β-carotene and 13-cis-retinoic acid. *J. Nutr.* **113,** 2608–2614.

Albano, E., Tomasi, A., Goria-Gatti, L., Persson, J. O., Terelius, Y., Ingelman-Sundberg, M., and Dianzani, M. U. (1991). Role of ethanol-inducible cytochrome P-450 (P450IIE1) in catalyzing the free radical activation of aliphatic alcohols. *Biochem. Pharmacol.* **411,** 1895–1902.

Albano, E., Tomasi, A., Goria-Gatti, L., Poli, G., Vannini, B., and Dianzani, M. U. (1987). Free radical metabolism of alcohols in rat liver microsomes. *Free Rad. Res. Commun.* **3,** 243–249.

Aleynik, S. I., Leo, M. A., Ma, X., Aleynik, M. K., and Lieber, C. S. (1995). Polyenylphosphatidylcholine (PPC) prevents hepatic lipid peroxidation and attenuates associated injury induced by CCl₄ in rats and by alcohol in baboons. *Gastroenterology* **108,** A1025.

Alpha-Tocopherol, β-carotene and Cancer Prevention Study Group. (1994). *N. Engl. J. Med.* **330,** 1029–1035.

Arai M., Gordon E. R., and Lieber, C. S. (1984). Decreased cytochrome oxidase activity in hepatic mitochondria after chronic ethanol consumption and the possible role of decreased cytochrome aa3 content and changes in phospholipids. *Biochim. Biophys. Acta.* **797,** 320–327.

Araujo, F. B., Barbosa, D. S., Hsin, C. Y., Maranhao, R. C., and Abdalla, D. S. P. (1995). Evaluation of oxidative stress in patients with hyperlipidemia. *Atherosclerosis* **117,** 61–71.

Arnesjö B., Nilsson, Å., Barrowman, J., and Borgström, B. (1969). Intestinal digestion and absorption of cholesterol and lecithin in the human: Intubation studies with a fat-soluble reference substance. *Scand. J. Gastroenterol.* **4,** 653–665.

Badger, T. M., Huang, J., Ronis, M., and Lumpkin, C. K. (1993). Induction of cytochrome P450 2E1 during chronic ethanol exposure occurs via transcription of the CYP 2E1 gene when blood alcohol concentrations are high. *Biochem. Res. Commun.* **190,** 780–785.

Barclay, L. R. (1988). The cooperative antioxidant role of glutathione with a lipid-soluble and a water-soluble antioxidant during peroxidation of liposomes initiated in the aqueous phase and in the lipid phase. *J. Biol. Chem.* **263,** 16138–16142.

Beskid, M., Bialck, J., Dzieniszewski, J., Sadowski, J., and Tlalka, J. (1980). Effect of combined phenylbutazone and ethanol administration on rat liver. *Exp. Pathol.* **18,** 487–491.

Bjørneboe, G. E. A., Bjørneboe, A., Hagen, B. F., Morland, J., and Drevon, C. A. (1987). Reduced hepatic α-tocopherol content after long-term administration of ethanol to rats. *Biochim. Biophys. Acta* **918,** 236–241.

Bjørneboe, G. E. A., Johnsen, J., Bjørneboe, A., Marklund, S. L., Skylv, N., Hoiseth, A., Bache-Wiig, J. E., Morland, J., and Drevon, C. A. (1988). Some aspects of antioxidant status in blood from alcoholics. *Alcoholism Clin. Exp. Res.* **12,** 806–810.

Borowsky, S. A., and Lieber, C. S. (1978). Interaction of methadone and ethanol metabolism. *J. Pharmacol. Exp. Ther.* **207,** 123–129.

Bücher, T. H., and Klingenberg, M. (1958). Wege des Wasserstoffs in der lebendigen Organisation. *Angew. Chem.* **70,** 552–570.

Burton, G. W. (1989). Antioxidant action of carotenoids. *J. Nutr.* **119,** 109–111.

Burton, G. W., and Ingold, K. U. (1984). β-carotene: An unusual type of lipid antioxidant. *Science* **224,** 569–573.

Carmichael, F. J., Saldivia, V., Israel, Y., McKaigney, J. P., and Orrego, H. (1987). Ethanol-induced increase in portal hepatic blood flow: Interference by anesthetic agents. *Hepatology* **97,** 89–94.

Castillo, T., Koop, D. R., Kamimura, S., Triadafilopoulos, G., and Tsukamoto. (1992). Role of cytochrome P-450 2E1 in ethanol-, carbon tetrachloride- and iron-dependent microsomal lipid peroxidation. *Hepatology* **16,** 992–996.

Cederbaum, A. I. (1989). Oxygen radical generation by microsomes: Role of iron and implications for alcohol metabolism and toxicity. *Free Rad. Biol. Med.* **7,** 559–567.

Clot, P., Bellomo, G., Tabone, M., Aricò, S., and Albano, E. (1995). Detection of antibodies against proteins modified by hydroxyethyl free radicals in patients with alcoholic cirrhosis. *Gastroenterology* **108,** 201–207.

Conney, A. H. (1967). Pharmacological implications of microsomal enzyme induction. *Pharmacol. Rev.* **19**, 317–366.

Dai, Y., Rashba-Step, J., and Cederbaum, A. I. (1993). Stable expression of human cytochrome P4502E1 in HepG2 cells: Characterization of catalytic activities and production of reactive oxygen intermediates. *Biochemistry* **32**, 6928–6937.

Dennery, P. A., Kramer, C. M., and Alpert, S. E. (1990). Effect of fatty acid profiles on the susceptibility of cultured rabbit tracheal epithelial cells to hyperoxic injury. *Am. J. Respir. Cell Mol. Biol.* **3**, 137–144.

Dicker, E., and Cederbaum, A. I. (1988). Increased oxygen radical-dependent inactivation of metabolic enzymes by liver microsomes after chronic ethanol consumption. *FASEB J.* **2**, 2901–2906.

Diehl, A. M., Bisgaard, H. C., Kren, B. T., and Steer, C. J. (1991). Ethanol interferes with regeneration-associated changes in biotransforming enzymes: A potential mechanism underlying ethanol's carcinogenicity? *Hepatology* **13**, 722–727.

DiLuzio, N. R. (1964). Prevention of the acute ethanol-induced fatty liver by the simultaneous administration of anti-oxidants. *Life Sci.* **3**, 113–119.

DiLuzio, N. R. (1968). The role of lipid peroxidation and antioxidants in ethanol-induced lipid alterations. *Exp. Mol. Pathol.* **8**, 394–402.

DiPadova, C., Worner, T. M., Julkunen, R. J. K., and Lieber, C. S. (1987). Effects of fasting and chronic alcohol consumption on the first pass metabolism of ethanol. *Gastroenterology* **92**, 1169–1173.

Domschke, S., Domschke, W., and Lieber, C. S. (1974). Hepatic redox state: Attenuation of the acute effects of ethanol induced by chronic ethanol consumption. *Life Sci.* **15**, 1327–1334.

Duce, A. M., Ortiz, P., Cabrero, C., and Mato, J. M. (1988). S-Adenosyl-L-methionine synthetase and phospholipid methyltransferase are inhibited in human cirrhosis. *Hepatology* **8**, 65–68.

Ekstrom, G., and Ingelman-Sundberg, M. (1989). Rat liver microsomal NADPH-supported oxidase activity and lipid peroxidation dependent on ethanol-inducible cytochrome P-450 (P-450IIE1). *Biochem. Pharmacol.* **38**, 1313–1319.

Eliasson, E., Johansson, I., and Ingelman-Sundberg, M. (1988). Ligand-dependent maintenance of ethanol-inducible cytochrome P-450 in primary rat hepatocyte cell cultures. *Biochem. Biophys. Res. Commun.* **150**, 436–443.

Espina, N., Lima, V., Lieber, C. S., and Garro, A. J. (1988). In vitro and in vivo inhibitory effect of ethanol and acetaldehyde on O^6-methylguanine transferase. *Carcinogenesis* **9**, 761–766.

Faller, J., and Fox, I. H. (1982). Evidence for increased urate production by activation of adenine nucleotide turnover. *N. Engl. J. Med.* **307**, 1598–1602.

Farinati, F., Zhou, Z., Bellah, J., Lieber, C. S., and Garro, A. J. (1985). Effect of chronic ethanol consumption on activation of nitrosopyrrolidine to a mutagen by rat upper alimentary tract, lung and hepatic tissue. *Drug Metab. Dispos.* **13**, 210–214.

Finkelstein, J. D., and Martin, J. J. (1986). Methionine metabolism in mammals: Adaptation to methionine excess. *J. Biol. Chem.* **261**, 1582–1587.

Forman, M. R., Beecher, G. R., Lanza, E., Reichman, M. E., Graubard, B. I., Campbell, Marr, T., Yong, L. C., Judd, J. T., and Taylor, P. R. (1995). Effect of alcohol consumption on plasma carotenoid concentrations in premenopausal women: A controlled dietary study. *Am. J. Clin. Nutr.* **62**, 131–135.

Fox, J. M. (1983). Polyene phosphatidylcholine: Pharmacokinetics after oral administration. *In* "Phospholipids and Atherosclerosis" (P. Avogaro, M. Macini, G., Ricci, and R. Paoletti, eds.). Raven Press, New York.

Garro, A. J., and Lieber, C. S. (1990). Alcohol and cancer. *Annu. Rev. Pharmacol. Toxicol.* **30**, 219–249.

Garro, A. J., Seitz, H. K., and Lieber, C. S. (1981). Enhancement of dimethylnitrosamine metabolism and activation to a mutagen following chronic ethanol consumption. *Cancer Res.* **41**, 120–124.

Geesin, J. C., Hendricks, L. J., Falkenstein, P. A., Gordon, J. S., and Berg, R. A. (1991). Regulation of collagen synthesis by ascorbic acid: Characterization of the role of ascorbate-stimulated lipid peroxidation. *Arch. Biochem. Biophys,* **290,** 127–132.

Gerster, H. (1991). Potential role of β-carotene in the prevention of cardiovascular disease. *Int. J. Vit. N Res.* **61,** 277–291.

Greenway, C. V., and Lautt, W. W. (1990). Acute and chronic ethanol on hepatic oxygen ethanol and lactate metabolism in cats. *Am. J. Physiol.* **258,** 411–418.

Halevy, O., and Sklan, D. (1987). Inhibition of arachidonic acid oxidation by beta-carotene, retinol and alpha-tocopherol. *Biochim. Biophys. Acta* **918,** 304–307.

Halliwell, B., and Gutteridge, J. M. C. (1989). "Free Radicals in Biology and Medicine," 2nd Ed., pp. 188–214. Clarendon Press, Oxford.

Handler, J. A., and Thurman, R. G. (1985). Fatty acid-dependent ethanol metabolism. *Biochem. Biophys. Res. Commun.* **133,** 44–51.

Handler, J. A., and Thurman, R. G. (1990). Redox interactions between catalase and alcohol dehydrogenase pathways of ethanol metabolism in the perfused rat liver. *J. Biol. Chem.* **265,** 1510–1515.

Hasumura, Y., Teschke, R., and Lieber, C. S. (1975). Hepatic microsomal ethanol oxidizing system (MEOS): Dissociation from reduced nicotinamide adenine dinucleotide phosphate-oxidase and possible role of form 1 of cytochrome P-450. *J. Pharmacol. Exp. Ther.* **194,** 469–474.

Hetu, C., Dumont, A., and Joly, J. G. (1983). Effect of chronic ethanol administration on bromobenzene liver toxicity in the rat. *Toxicol. Appl. Pharmacol.* **67,** 166–167.

Hetu, C., and Joly, J. G. (1985). Differences in the duration of the enhancement of liver mixed-function oxidase activities in ethanol-fed rats after withdrawal. *Biochem. Pharmacol.* **34,** 1211–1216.

Hirano, T., Kaplowitz, N., Tsukamoto, H., Kamimura, S., and Fernandez-Checa, J. C. (1992). Hepatic mitochondrial glutathione depletion and progression of experimental alcoholic liver disease in rats. *Hepatology* **6,** 1423–1427.

Holz, J., Wagner, H. (1971). Uber den einbau von intraduodenal appliziertem $^{14}C/^{32}P$-polyene-phosphatidylcholin in die leber von ratten und seine ausscheidung durch die galle. *Z. Naturforsch.* **26,** 1151–1158.

Horowitz, J. H., Rypins, E. B., Henderson, J. M., Heymsfield, S. B., Moffit, S. D., Bain, R. P., Chawla, R. K., Bleier, J. C., and Rudman, D. (1981). Evidence for impairment of transsulfuration pathway in cirrhosis. *Gastroenterology* **81,** 668–675.

Inatomi, N., Kato, S., Ito, D., and Lieber, C. S. (1989). Role of peroxisomal fatty acid beta-oxidation in ethanol metabolism. *Biochem. Biophys. Res. Commun.* **163,** 418–423.

Ingelman-Sundberg, M., and Johansson, I. (1984). Mechanisms of hydroxyl radical formation and ethanol oxidation by ethanol-inducible and other forms of rabbit liver microsomal cytochromes P-450. *J. Biol. Chem.* **259,** 6447–6458.

Iseri, O. A., Gottlieb, L. S., and Lieber, C. S. (1964). The ultrastructure of ethanol-induced fatty liver. *Fed. Proc.* **23,** 579.

Iseri, O. A., Lieber, C. S., and Gottlieb, L. S. (1966). The ultrastructure of fatty liver induced by prolonged ethanol ingestion. *Am. J. Pathol.* **48,** 535–555.

Israel, Y., Kalant, H., Orrego, H., Khanna, J. M., Videla, I., and Phillips, J. M. (1975). Experimental alcohol-induced hepatic necrosis: Suppression by propylthiouracil. *Proc. Natl. Acad. Sci. U.S.A.* **72,** 1137–1141.

Jauhonen, P., Baraona, E., Miyakawa, H., and Lieber, C. S. (1982). Mechanism for selective perivenular hepatotoxicity of ethanol. *Alcoholism Clin. Exp. Res.* **6,** 350–357.

Jenkins, M. Y., Sheikh, M. N., Mitchell, G. V., Grundel, E., Blakely, S. R., and Carter, C. J. (1993). Dietary carotenoids influenced biochemical but not morphological changes in adult male rats fed a choline-deficient diet. *Nutr. Cancer* **19,** 55–65.

Johansson, I. J., Ekstrom, G., Scholte, B., Puzycki, D., Jornvall, H., and Ingelman-Sundberg, M. (1988). Ethanol-fasting, and acetone-inducible cytochromes P-450 in rat liver: Regulation and characteristics of enzymes belonging to the IIB and IIE gene subfamilies. *Biochemistry* 27, 1925–1934.

Kaikaus, R. M., Chan, W. K., Lysenko, N., Ortiz, P., Montellano, D., and Bass, N. M. (1990). Induction of liver fatty acid binding protein (1-FABP) and peroxisomal fatty acid β-oxidation by peroxisome proliferators (PP) is dependent on cytochrome p-450 activity. *Hepatology* 12, A248.

Kato, S., Alderman, J., and Lieber, C. S. (1987a). Respective roles of microsomal ethanol oxidizing system (MEOS) and catalase in ethanol metabolism by deermice lacking alcohol dehydrogenase. *Arch. Biochem. Biophys.* 254, 586–591.

Kato, S., Alderman, J., and Lieber, C. S. (1987b). Ethanol metabolism in alcohol dehydrogenase deficient deermice is mediated by the microsomal ethanol oxidizing system, not by catalase. *Alcohol Alcoholism* (Suppl)1, 231–234.

Kato, S., Kawase, T., Alderman, J., Inatomi, N., and Lieber, C. S. (1990). Role of xanthine oxidase in ethanol-induced lipid peroxidation in rats. *Gastroenterology* 98, 203–210.

Kawase, T., Kato, S., and Lieber, C. S. (1989). Lipid peroxidation and antioxidant defense systems in rat liver after chronic ethanol feeding. *Hepatology* 10, 815–821.

Keilin, D., Hartree, E. F. (1945). Properties of catalase: Catalysis of coupled oxidation of alcohols, *Biochem J,* 39, 293–301.

Kessler, B. J., Lieber, J. B., Bronfin, G. J., and Sass, M. (1954). The hepatic blood flow and splanchnic oxygen consumption in alcohol fatty liver. *J. Clin. Invest.* 33, 1338–1345.

Khani, S. C., Zaphiropoulos, P. G., Fujita, V. S., Porter, T. D., Koop, D. R., and Coon, M. J. (1987). cDNA and derived amino acid sequence of ethanol-inducible rabbit liver cytochrome P-450 isozyme 3a (P450ALC). *Proc. Natl. Acad. Sci. U.S.A.* 84, 638–642.

Kim, S. G., and Novak, R. F. (1990). Induction of rat hepatic P450IIE1 (CYP 2E1) by pyridine: Evidence for a role of protein synthesis in the absence of transcriptional activation. *Biochem. Biophys. Res. Commun.* 166, 1072–1079.

Kim, S. G., Shehin, S. E., States, J. C., and Novak, R. F. (1990). Evidence for increased translational efficiency in the induction of P450IIE1 by solvents: Analysis of P450IIE1 mRNA polyribosomal distribution. *Biochem. Biophys. Res. Commun.* 172, 767–774.

Kim-Jun, H. (1993). Inhibitory effects of α- and β-carotene on croton oil-induced or enzymatic lipid peroxidation and hydroperoxide production in mouse skin epidermis. *Int. J. Biochem.* 25, 911–915.

Koivisto, T., Mishin, V. M., Mak, K. M., Cohen, P. A., and Lieber, C. S. (1996). Induction of cytochrome P-4502E1 by ethanol in rat Kupffer cells. *Alcoholism Clin. Exp. Res.* 20, 207–212.

Koop, D. R., and Casazza, J. P. (1985). Identification of ethanol-inducible P-450 isozyme 3a as the acetone and acetol monooxygenase of rabbit microsomes. *J. Biol. Chem.* 260, 13607–13612.

Koop, D. R., Morgan, E. T., Tarr, G. E., and Coon, M. I. (1982). P0!0!ication and characterization of a unique isozyme of cytochrome P-450 from liver microsomes of ethanol-treated rabbits. *J. Biol. Chem.* 257, 8472–8780.

Koop, D. R., and Tierney, D. J. (1990). Multiple mechanism in the regulation of ethanol-inducible cytochrome P450IIE1. *BioEssays* 12, 429–435.

Krinsky, N. I. (1989). Antioxidant functions of carotenoids. *Free Rad. Biol. Med.* 7, 617–635.

Krinsky, N. I., and Deneke, S. M. (1982). Interaction of oxygen and oxy-radicals with carotenoids. *J. Natl. Cancer Inst.* 69, 205–210.

Kubota, S., Lasker, J. M., and Lieber, C. S. (1988). Molecular regulation of ethanol inducible cytochrome P450-IIE1 in hamsters. *Biochem. Biophys. Res. Commun.* 150, 304–310.

Kunert, K. J., and Tappel, A. L. (1983). The effect of vitamin C on in vitro lipid peroxidation in guinea pigs as measured by pentane and ethanol production. *Lipids* 18, 271–274.

Lane, B. P., and Lieber, C. S. (1966). Ultrastructural alterations in human hepatocytes following ingestion of ethanol with adequate diets. *Am. J. Pathol.* 49, 593–603.

Lane, B. P., and Lieber, C. S. (1967). Effects of butylated hydroxytoluene on the ultrastructure of rat hepatocytes. *Lab. Invest.* **16**, 341–348.

Lasker, J. M., Raucy, J., Kubota, S., Bloswick, B. P., Black, M., and Lieber, C. S. (1987). Purification and characterization of human liver cytochrome P-450-ALC. *Biochem. Biophys. Res. Commun.* **148**, 232–238.

Lauterburg, B. H., Liang, D., Schwarzenbach, F. A., and Breen, K. J. (1993). Mitochondrial dysfunction in alcoholic patients as assessed by breath analysis. *Hepatology* **17**, 418–422.

Lefevre, A., Adler, H., and Lieber, C. S. (1970). Effect of ethanol on ketone metabolism. *J. Clin. Invest.* **49**, 1775–1782.

Lekim, D., and Betzing, H. (1974). The incorporation of essential phospholipids into the organs of intact and galactosamine intoxicated rats. *Drug Res.* **24**, 1217–1221.

Lekim, D., and Graf, E. (1976). Tierexperimentelle studien zur pharmakokinetik der 'essentiellen' phospholipide (EPL). *Arzneimittelforschung* **26**, 1772–1782.

Lekim, D., Betzing, H., Stoffel, W. (1972). Incorporation of complete phospholipid molecules in cellular membranes of rat liver after uptake from blood serum. *Z. Physiol. Chem.* **353S**, 929–946.

Leo, M. A., Arai, M., Sato, M., and Lieber, C. S. (1982). Hepatotoxicity of vitamin A and ethanol in the rat. *Gastroenterology* **82**, 194–205.

Leo, M. A., Kim, C. I., and Lieber, C. S. (1987). NAD^+-dependent retinol dehydrogenase in liver microsomes. *Arch. Biochem. Biophys.* **259**, 241–249.

Leo, M. A., Kim, C. I., Lowe, N., and Lieber, C. S. (1992). Interaction of ethanol with β-carotene: Delayed blood clearance and enhanced hepatotoxicity. *Hepatology* **15**, 883–891.

Leo, M. A., and Lieber, C. S. (1983). Hepatic fibrosis after long term administration of ethanol and moderate vitamin A supplementation in the rat. *Hepatology* **2**, 1–11.

Leo, M. A., and Lieber, C. S. (1985). New pathway for retinol metabolism in liver microsomes. *J. Biochem.* **260**, 5228–5231.

Leo, M. A., and Lieber, C. S. (1988). Hypervitaminosis A: A liver lover's lament. *Hepatology* **8**:412–417.

Leo, M. A., Lowe, N., and Lieber, C. S. (1986). Interaction of drugs and retinol. *Biochem. Pharmacol.* **35**, 3949–3953.

Leo, M. A., Rosman, A., and Lieber, C. S. (1993). Differential depletion of carotenoids and tocopherol in liver diseases. *Hepatology* **17**, 977–986.

Leo, M. A., Sato, M., and Lieber, C. S. (1983). Effect of hepatic vitamin A depletion on the liver in men and rats. *Gastroenterology* **84**, 562–572.

Li, J.-J., Kim, C.-I., Leo, M. A., Mak, K. M., Rojkind, M., and Lieber, C. S. (1992). Polyunsaturated lecithin prevents acetaldehyde-mediated hepatic collagen accumulation by stimulating collagenase activity in cultured lipocytes. *Hepatology* **15**, 373–381.

Lieber, C. S. (1988). The influence of alcohol on nutritional status. *Nutr. Rev.* **46**, 241–245.

Lieber, C. S. (1991). Hepatic, metabolic and toxic effects of ethanol: 1991 update. *Alcoholism Clin. Exp. Res.* **15**, 573–592.

Lieber, C. S. (1992). "Medical and Nutritional Complications of Alcoholism: Mechanisms and Management," p. 579. Plenum Press, New York.

Lieber C. S. (1995). Liver fibrosis: from pathogenesis to treatment. *In* "Advances in Hepatobiliary and Pancreatic Diseases" (D. F. de Pretis, ed.), pp 62–81. Kluwer Academic Press, Lancaster, UK.

Lieber, C. S. (1995a). Prevention and therapy with S-adenosyl-L-methionine and polyenylphosphatidyl-choline. *In* "Treatments in Hepatology" (V. Arroyo, J. Bosch, and J. Rodes, eds.), pp. 299–311. Masson, S. A., Barcelona.

Lieber, C. S., Baraona, E., Hernandez-Munoz, R., Kubota, S., Sato, N., Kawano, S., Matsumura, T., and Inatomi, N. (1989). Impaired oxygen utilization: A new mechanism for the hepatotoxicity of ethanol in sub-human primates. *J. Clin. Invest.* **83**, 1682–1690.

Lieber, C. S., Casini, A., DeCarli, L. M., Kim, C., Lowe, N., Sasaki, R., and Leo, M. A. (1990). S-Adenosyl-L-methionine attenuates alcohol-induced liver injury in the baboon. *Hepatology* **11**, 165–172.

Lieber, C. S., and DeCarli, L. M. (1968). Ethanol oxidation by hepatic microsomes: Adaptive increase after ethanol feeding. *Science* **162**, 917–918.

Lieber, C. S., and DeCarli, L. M. (1970a). Hepatic microsomal ethanol oxidizing system: In vitro characteristics and adaptive properties in vivo. *J. Biol. Chem.* **245**, 2505–2512.

Lieber, C. S., and DeCarli, L. M. (1970b). Reduced nicotinamide-adenine dinucleotide phosphate oxidase: Activity enhanced by ethanol consumption. *Science* **170**, 78–80.

Lieber, C. S., Garro, A., Leo, M. A., Mak, K. M., and Worner, T. M. (1986). Alcohol and cancer. *Hepatology* **6**, 1005–1019.

Lieber, C. S., Jones, D. P., Losowsky, M. S., and Davidson, C. S. (1962). Interrelation of uric acid and ethanol metabolism in man. *J. Clin. Invest.* **41**, 1863–1870.

Lieber, C. S., Lasker, J. M., DeCarli, L. M., Saeli, J., and Wojtowicz, T. (1988). Role of acetone, dietary fat, and total energy intake in the induction of the hepatic microsomal ethanol oxidizing system. *J. Pharmacol. Exp. Ther.* **247**, 791–795.

Lieber, C. S., Leo, M. A., Aleynik, S. I., Aleynik, M. A., and DeCarli, L. M. (1995). Polyenylphosphatidylcholine (PPC) decreases oxidant stress and protects against alcohol-induced liver injury in the baboon. *Hepatology* **22**, 225A.

Lieber, C. S., Li, J. I., Robins, S., DeCarli, L. M., Mak, K. M., and Leo, M. A. (1992). Dietary dilinoleoylphosphatidyl-choline (DLPC) is incorporated into liver phospholipids, protects against alcoholic cirrhosis, enhances collagenase activity and prevents acetaldehyde-induced collagen accumulation in cultured lipocytes. *Hepatology* **16**, 87A. [Abstract]

Lieber, C. S., Robins, S., Li, J., DeCarli, L. M., Mak, K. M., Fasulo, J. M., and Leo, M. A. (1994a). Phosphatidylcholine protects against fibrosis and cirrhosis in the baboon. *Gastroenterology* **106**, 152–159.

Lieber, C. S., Robins, S. J., and Leo, M. A. (1994b). Hepatic phosphatidylethanolamine methyltransferase activity is decreased by ethanol and increased by phosphatidylcholine. *Alcoholism Clin. Exp. Res.* **18**, 592–595.

Lomnitski, L., Bar-Natan, R., Sklan, D., and Grossman, S. (1993). The interaction between β-carotene and lipoxygenase in plant and animal systems. *Biochim. Biophys. Acta* **1167**, 331–338.

Machlin, L. J., and Bendich, A. (1987). Free radical tissue damage: Protective role of antioxidant nutrients. *FASEB J.* **1**, 441–445.

McCay, P. B. (1985). Vitamin E interaction with free radical and ascorbate. *Annu. Rev. Nutr.* **5**, 323–340.

Meldolesi, J. (1967). On the significance of the hypertrophy of the smooth endoplasmic reticulum in liver cells after administration of drugs. *Biochem. Pharmcol.* **16**, 125–131.

Misra, P. S., Lefevre, A., Ishii, H., Rubin, E., and Lieber, C. S. (1971). Increase of ethanol meprobamate and pentobarbital metabolism after chronic ethanol administration in man and in rats. *Am. J. Med.* **51**, 346–351.

Miwa, G. T., Levin, W., Thomas, P. E., and Lu, A. Y. H. (1978). The direct oxidation of ethanol by catalase- and alcohol dehydrogenase-free reconstituted system containing cytochrome P-450. *Arch. Biochem. Biophys.* **187**, 464–475.

Mobarhan, S., Bowen, P., Andersen, B., Evans, M., Stacewicz-Sapuntzakis, M., Sugerman, S., Simms, P., Lucchesi, D., and Friedman, H. (1990). Effects of β-carotene repletion of β-carotene absorption, lipid peroxidation, and neutrophil superoxide formation in young men. *Nutr. Cancer* **14**, 195–206.

Morgan, E. T., Koop, D. R., and Coon, M. J. (1982). Catalytic activity of cytochrome P-450 isozyme 3a isolated from liver microsomes of ethanol-treated rabbits. *J. Biol. Chem.* **257**, 13951–13957.

Morgan, E. T., Koop, D. R., and Coon, M. J. (1983). Comparison of six rabbit liver cytochrome P-450 isozymes in formation of a reactive metabolite of acetaminophen. *Biochem. Biophys. Res. Commun.* **112**, 8–13.

Morton, S., and Mitchell, M. C. (1985). Effects of chronic ethanol feeding on glutathione turnover in the rat. *Biochem. Pharmacol.* **34**, 1559–1563.

Müller, A., and Sies, H. (1982). Role of alcohol dehydrogenase activity and of acetaldehyde in ethanol-induced ethane and pentane production by isolated perfused rat liver. *Biochem. J.* **206**, 153–156.

Müller, A., and Sies, H. (1983). Inhibition of ethanol- and aldehyde-induced release of ethane from isolated perfused rat liver by pargyline and disulfiram. *Pharmacol. Biochem. Behav.* **18**, 429–432.

Nakajima, T., Okino, T., and Sato, A. (1987). Kinetic studies on benzene metabolism in rat liver: possible presence of three forms of benzene metabolizing enzymes in the liver. *Biochem. Pharmacol.* **36**, 2799–2804.

Navder, K. P., Baraona, E., and Lieber, C. S. (1996). Effects of polyenylphosphatidylcholine (PPC) on alcohol-induced fatty liver and hyperlipemia in rats. *Gastroenterology* **110**, 1275.

Niki, E. (1987). Interaction of ascorbate and α-tocopherol. *Ann. N.Y. Acad. Sci.* **493**, 186–199.

Nilsson, A. K. E. (1968). Intestinal absorption of lecithin and lysolecithin by lymph fistula rats. *Biochim. Biophys. Acta* **152**, 379–390.

Nordmann, R., Ribière, C., and Rouach, H. (1992). Implication of free radical mechanisms in ethanol-induced cellular injury. *Free Rad. Biol. Med.* **12**, 219–240.

Ohnishi, K., and Lieber, C. S. (1977). Reconstitution of the microsomal ethanol-oxidizing system: Qualitative and quantitative changes of cytochrome P-450 after chronic ethanol consumption. *J. Biol. Chem.* **252**, 7124–7131.

Oshino, N., Chance, B., Sies, H., and Bücher, T. (1973). The role of H_2O_2 generation in perfused rat liver and the reaction of catalase compound I and hydrogen donors. *Arch. Biochem. Biophys.* **154**, 117–131.

Palozza, P., and Krinsky, N. I. (1991). The inhibition of radical-initiated peroxidation of microsomal lipids by both α-tocopherol and β-carotene. *Free Rad. Biol. Med.* **11**, 407–414.

Parfitt, V. J., Desomeaux, K., Bolton, C. H., and Hartog, M. (1994). Effects of high monoun-saturated and polyunsaturated fat diets on plasma lipoproteins and lipid peroxidation in Type 2 diabetes mellitus. *Diab. Med.* **11**, 85–91.

Parthasarathy, S., Subbaiah, P. V., and Ganguly, J. (1974). The mechanism of intestinal absorption of phosphatidylcholine in rats. *Biochem. J.* **140**, 503–508.

Patton, G. M., Clark, S. B., Fasulo, J. M., and Robins, S. I. (1984). Utilization of individual lecithins in intestinal lipoprotein formation in the rat. *J. Clin. Invest.* **73**, 231–240.

Porter, T. D., Khani, S. C., and Coon, M. J. (1989). Induction and tissue-specific expression of rabbit cytochrome P450IIE1 and IIE2 genes. *Mol. Pharmacol.* **36**, 61–65.

Reinke, L., Lai, E. K., DuBose, C. M., McCay, P. B., and Janzen, E. G. (1987). Reactive free radical generation in vivo in heart and liver of ethanol-fed rats: Correlation with radical formation in vitro. *Proc. Natl. Acad. Sci. U.S.A.* **84**, 8223–8227.

Reinke, L. A., Kotake, Y., McCay, P. B., and Janzen, E. G. (1991). Spin trapping studies of hepatic free radicals formed following the acute administration of ethanol to rats: In vivo detection of l-hydroxyethyl radicals with PBN. *Free Rad. Biol. Med.* **11**, 31–39.

Rodgers, J. B., O'Brien, R. J., and Balint, J. A. (1975). The absorption and subsequent utilization of lecithin by the rat jejunum. *Am. J. Digest. Dis.* **20**, 208–211.

Ronis, M. J., Huang, J., Crouch, J., Mercado, C., Irby, D., Valentine, CR., Lumpkin, C. K., Ingelman-Sundberg, M., and Badger, T. M. (1993). Cytochrome P450 CYP 2E1 induction during chronic alcohol exposure occurs by a two-step mechanism associated with blood alcohol concentration in rats. *J. Pharmacol. Exp. Ther.* **264**, 944–950.

Sato, C., Matsuda, Y., and Lieber, C. S. (1981). Increased hepatotoxicity of acetaminophen after chronic ethanol consumption in the rat. *Gastroenterology* **80**, 140–148.

Sato, M., and Lieber, C. S. (1981). Hepatic vitamin A depletion after chronic ethanol consumption in baboons and rats. *J. Nutr.* **111**, 2015–2023.

Sato, N., Kamada, T., Kawano, S., Hayashin, N., Kishida, Y., Meren, H., Yoshihara, H., and Abe, H. (1983). Effect of acute and chronic ethanol consumption on hepatic tissue oxygen tension in rats. *Pharmacol. Biochem. Beh.* **18**, 443–447.

Seitz, H. K., and Poschl, G. (1995). Antioxidant drugs and colchicine in the treatment of alcoholic liver disease. In "Treatments in Hepatology" (V. Arroyo, J. Bosch, J. Rodes, eds.), pp. 271–276. Masson, S. A., Barcelona.

Seitz, H. K., Czygan, P., Waldherr, K., Veith, S., and Kommerell, B. (1985). Ethanol and intestinal carcinogenesis in the rat. *Alcohol* 2, 491–494.

Seitz, H. K., Garro, A. J., and Lieber, C. S. (1981). Enhanced pulmonary and intestinal activation of procarcinogens and mutagens after chronic ethanol consumption in the rat. *Eur. J. Clin. Invest.* 11, 33–38.

Shaw, S., Heller, E. A., Friedman, H. S., Baraona, E., and Lieber, C. S. (1977). Increased hepatic oxygenation following ethanol administration in baboon. *Proc. Soc. Exp. Biol. Med.* 156, 509–513.

Shaw, S., Jayatilleke, E., Ross, W. A., Gordon, E. R., and Lieber, C. S. (1981). Ethanol induced lipid peroxidation: Potentiation by long-term alcohol feeding and attenuation by methionine. *J. Lab. Clin. Med.* 98, 4171–425.

Shaw, S., and Lieber, C. S. (1980). Increased hepatic production of alpha-amino-n-butyric acid after chronic alcohol consumption in rats and baboons. *Gastroenterology* 78, 108–113.

Shaw, S., Rubin, K. P., and Lieber, C. S. (1983). Depressed hepatic glutathione and increased diene conjugates in alcoholic liver disease: Evidence of lipid peroxidation. *Digest Dis. Sci.* 28, 585–589.

Shimizu, M., Lasker, J. M., Tsutsumi, M., and Lieber, C. S. (1990). Immunohistochemical localization of ethanol-inducible P450IIE1 in the rat alimentary tract. *Gastroenterology* 99, 1044–1053.

Siegers, C. P., Heidbuchel, K., and Younes, M. (1983). Influence of alcohol, dithiocarb and (+)-catechin on the hepatotoxicity and metabolism of vinylidene chloride in rats. *J. Appl. Toxicol.* 3, 90–95.

Sies, H., and Chance, B. (1970). The steady state level of catalase compound I in isolated hemoglobin-free perfused rat liver. *FEBS Lett.* 11, 172–176.

Smith, T., DeMaster, E. G., Furne, J. K., Springfield, J., and Levitt, M. D. (1992). First-pass gastric mucosal metabolism of ethanol is negligible in the rat. *J. Clin. Invest.* 89, 1801–1806.

Song, B. J., Gelboin, H. V., Park, S. S., Yang, C. S., and Gonzalez, F. I. (1986). Complementary DNA and protein sequences of ethanol-inducible rat and human cytochrome P-450s: Transcriptional and post-transcriptional regulation of the rat enzyme. *J. Biol. Chem.* 261, 16689–16697.

Song, B. J., Matsunaga, T., Hardwick, J., Park, S. S., Veech, R. I., Yang, C. S., Gelboin, H. V., and Gonzalez, F. J. (1987). Stabilization of cytochrome P450j messenger ribonucleic acid in the diabetic rat. *Mol. Endocrinol.* 1, 542–547.

Song, B. J., Veech, R. I., Park, S. S., Gelboin, H. V., and Gonzalez, F. J. (1989). Induction of rat hepatic N-nitrosodimethylamine demethylase by acetone is due to protein stabilization. *J. Biol. Chem.* 264, 3568–3572.

Sosenko, I. R. S., Innis, S. M., and Frank, L. (1988). Polyunsaturated fatty acids and protection of newborn rats from oxygen toxicity. *J. Pediatr.* 112, 630–637.

Sosenko, I. R. S., Innis, S. M., and Frank, I. (1989). Menhaden fish oil, n-3 polyunsaturated fatty acids and protection of newborn rats from oxygen toxicity. *Pediatr. Res.* 25, 399–404.

Sosenko, I. R. S., Innis, S. M., and Frank, L. (1991). Intralipid increases lung polyunsaturated fatty acids and protects newborn rats from oxygen toxicity. *Pediatr. Res.* 30, 413–417.

Speisky, H., MacDonald, A., Giles, G., Orrego, H., and Israel, Y. (1985). Increased loss and decreased synthesis of hepatic glutathione after acute ethanol administration. *Biochem. J.* 225, 565–572.

Spitz, D. R., Kinter, M. T., Kehrer, J. P., and Roberts, R. J. (1992). The effect of monounsaturated and polyunsaturated fatty acids on oxygen toxicity in cultured cells. *Pediatr. Res.* 32, 366–372.

Stein, S. W., Lieber, C. S., Cherrick, G. R., Leevy, C. M., and Ablemann, W. H. (1963). The effect of ethanol upon systemic hepatic blood flow in man. *Am. J. Clin. Nutr.* **13**, 68–74.

Takagi, T., Ishii, H., Takahashi, H., Kato, S., Okuno, F., Ebihara, Y., Yamauchi, H., Nagata, Y., Tashiro M., and Tsuchiya, M. (1983). Potentiation of halothane hepatotoxicity by chronic ethanol administration in rat: An animal model of halothane hepatitis. *Pharmacol. Biochem. Behav.* **18**(Suppl. 1), 461–465.

Takagi, T., Alderman, J., Geller, J., and Lieber, C. S. (1986). Assessment of the role of non-ADH ethanol oxidation in vivo and in hepatocytes from deer mice. *Biochem. Pharmacol.* **35**, 3601–3606.

Takahashi, T., Lasker, J. M., Rosman, A. S., and Lieber, C. S. (1993). Induction of P450E1 in human liver by ethanol is due to a corresponding increase in encoding mRNA. *Hepatology* **17**, 236–245.

Teschke, R., Hasumura, Y., Joly, J. G., Ishii, H., and Lieber, C. S. (1972). Microsomal ethanol-oxidizing system (MEOS): Purification and properties of a rat liver system free of catalase and alcohol dehydrogenase. *Biochem. Biophys. Res. Commun.* **49**, 1187–1193.

Teschke, R., Hasumura, Y., and Lieber, C. S. (1974). Hepatic microsomal alcohol oxidizing system: Solubilization, isolation and characterization. *Arch. Biochem. Biophys.* **163**, 404–415.

Teschke, R., Matsuzaki, S., Ohnishi, K., DeCarli, L. M., and Lieber, C. S. (1977). Microsomal ethanol oxidizing system (MEOS): Current status of its characterization and its role. *Alcoholism Clin. Exp. Res.* **1**, 7–15.

Thurman, R. G., and Brentzel, H. J. (1977). The role of alcohol dehydrogenase in microsomal ethanol oxidation and the adaptive increase in ethanol metabolism due to chronic treatment with ethanol. *Alcoholism Clin. Exp. Res.* **1**, 33–38.

Tsukamoto, H. (1993). Oxidative stress, antioxidants, and alcoholic liver fibrogenesis. *Alcohol* **10**, 465–467.

Tsutsumi, M., Lasker, J. M., Shimizu, M., Rosman, A. S., and Lieber, C. S. (1989). The intralobular distribution of ethanol-inducible P450IIE1 in rat and human liver. *Hepatology* **10**, 437–446.

Tsutsumi, M., Lasker, J. M., Takahashi, T., and Lieber, C. S. (1993). *In vivo* induction of hepatic P4502E1 by ethanol: Role of increased enzyme synthesis. *Arch. Biochem. Biophys.* **304**, 209–218.

Tsutsumi, R., Leo, M. A., Kim, C., Tsutsumi, M., Lasker, J. M., Lowe, N., and Lieber, C. S. (1990). Interaction of ethanol with enflurane metabolism and toxicity: Role of P450IIE1. *Alcoholism Clin. Exp. Res.* **14**, 174–179.

Valenzuela, A., Fernandez, V., and Videla, L. A. (1983). Hepatic and biliary levels of flutathione and lipid peroxides following iron overload in the rat: Effect of simultaneous ethanol administration. *Toxicol. Appl. Pharmacol.* **70**, 87–95.

Vendemiale, G., Altomare, E., Trizzio, T., Le Grazzie, C., DiPadova, C., Salerno, T., Carrieri, V., and Albano, O. (1989). Effects of oral S-adenosyl-L-methionine on hepatic glutathione in patients with liver disease. *Scand. J. Gastroenterol.* **24**, 407–415.

von Herbay, A., de Groot, H., Hegi, U., Stremmel, W., Strohmeyer, G., and Sies, H. (1994). Low vitamin E content in plasma of patients with alcoholic liver disease, hemochromatosis and Wilson's disease. *J Hepatol.* **20**, 41–46.

Williamson, J. R., Scholz, R., Browning, E. T., Thurman, R. G., and Fukami, M. H. (1969). Metabolic effects of ethanol in perfused rat liver. *J. Biol. Chem.* **25**, 5044–5054.

Yamada, S., Mak, K. M., and Lieber, C. S. (1985). Chronic ethanol consumption alters rat liver plasma membranes and potentiates release of alkaline phosphatase. *Gastroenterology* **88**, 1799–1806.

Yang, C. S., Tu, Y. Y., Koop, D. R., and Coon, M. J. (1985). Metabolism of nitrosamines by purified rabbit liver cytochrome P-450 isozymes. *Cancer Res.* **45**, 1140–1145.

Zhang, H., Loney, L. A., and Potter, B. J. (1993). Effect of chronic alcohol feeding on hepatic iron status and ferritin uptake by rat hepatocytes. *Alcoholism Clin. Exp. Res.* **17**, 394–400.

Zierenberg, O., and Grundy, S. M. (1982). Intestinal absorption of polyenylphosphatidylcholine in man. *J. Lipid Res.* **23**, 1136–42.

Birgit Heller
Volker Burkart
Eberhard Lampeter
Hubert Kolb

Clinical Department
Diabetes Research Institute
Heinrich-Heine University Düsseldorf
D-40225 Düsseldorf, Germany

Antioxidant Therapy for the Prevention of Type I Diabetes

I. Introduction

Previous studies *in vitro* and in animal models indicate that reactive radicals contribute to the destruction of pancreatic islet cells in the pathogenesis of insulin-dependent diabetes mellitus (IDDM) (Rabinovitch *et al.*, 1992). During islet inflammation (insulitis), endothelial cells and infiltrating macrophages are thought to release cytotoxic amounts of nitric oxide (NO) (Kröncke *et al.*, 1991) and reactive oxygen intermediates (ROI) (Zweier *et al.*, 1988; Brenner *et al.*, 1993).

Islet cells were found to be highly susceptible to the toxic effects of chemically generated ROI (Malaisse *et al.*, 1982; Burkart *et al.*, 1992) and NO (Kröncke *et al.*, 1993). The reason for this unique sensitivity may be a deficiency of islet cells in radical defense (Grankvist *et al.*, 1981; Malaisse *et al.*, 1982). Therefore, it was suggested that these radicals are responsible for the destruction of insulin-producing β cells during insulitis.

Indeed, evidence shows that NO and ROI are involved in inflammatory and autoimmune-mediated tissue damage (Kolb and Kolb-Bachofen, 1992; Brenner et al., 1993). Activated macrophages spontaneously kill syngeneic islet cells but not cells isolated from thyroid or liver (Appels et al., 1989). Kröncke et al. (1993) showed that NO generated by activated macrophages plays an important role in the destruction of islet cells. In vitro studies showed that ROI may exert a similar extent of islet cell toxicity as NO (Burkart et al., 1992; Heller et al., 1994).

Based on these observations, various experimental approaches were performed to protect islet cells from ROI and NO by directly counteracting the toxic effects of the radicals. During the search for well-tolerated substances, the antioxidant dihydrolipoic acid was found to exert a considerable protective effect. The drug protected isolated rat islet cells from radical toxicity in a dual way. It decreased the toxicity of enzymatically generated ROI and suppressed NO production from macrophages (Burkart et al., 1993). In the nonobese diabetic (NOD) mouse, an animal model of human IDDM, lipoic acid also showed beneficial effects. The administration of the drug around the induction phase of diabetes resulted in a supression of the diabetes incidence from 60% in the control group to about 30% and in a suppression of inflammatory islet infiltration by immune cells (Faust et al., 1994). Taken together, these findings indicate that dihydrolipoic acid has limited protective effects with partial inhibition of islet inflammation and incomplete suppression of diabetes development. This led us to further elucidate the events involved in radical-induced islet cell death in order to develop an improved strategy for islet cell protection.

II. Mechanism of Reactive Oxygen Intermediates and Nitric Oxide Toxicity

Studies using defined chemical or enzymatic sources of NO or ROI and isolated islet cells have described the initial events of radical-induced islet cell death. DNA strand breaks and the activation of the nuclear enzyme poly(ADP-ribose)polymerase (PARP) are prominent early consequences of the radical-induced islet cell damage. PARP, a chromatin-associated enzyme, is involved in DNA repair and consumes large amounts of NAD^+ after excessive activation. The depletion of the intracellular NAD^+ pools leads to islet cell death which can be determined by the loss of cell membrane integrity several hours after exposure to radicals (Heller et al., 1994; Radons et al., 1994). In eukaryotic cells, two different classes of ADP-ribosylation reactions can be distinguished, namely mono- and poly(ADP-ribosyl)ation. Both pathways utilize NAD^+ as a source of ADP-ribose. To decide which type of (ADP-ribosyl)ation is involved in radical-induced islet cell death, pancreatic islet cells from normal Wistar rats or Balb/c mice were exposed to NO, released during the decomposition of sodium nitroprusside (NP) (Kallmann *et al.*, 1992) or to ROI generated during the oxidation of hypoxanthine (HX) by xanthine oxidase (XO) (Burkart *et al.*, 1992). Inhibitors of (ADP-ribosyl)ation were added 30 min prior to the initiation of radical treatment. After 18 hr of exposure, islet cell death was analyzed by the trypan blue exclusion assay and DNA strand breaks were determined using the *in situ* nick translation technique as described previously (Fehsel *et al.*, 1993). The activation of PARP was monitored by indirect immunofluorescence detecting the product of the enzymatic reaction, poly(ADP-ribose) (Heller *et al.*, 1994).

As shown in Fig. 1, radical-induced islet cell death could be largely prevented by the ADP-ribosylation inhibitors nicotinamide (NA) and 3-aminobenzamide (3-AB). However, NA could act as radical scavenger, as inhibitor of mono- and poly(ADP-ribosyl)ation, or as precursor in the biosynthesis of NAD^+. To identify the relevant inhibitory mechanism of NA on the radical-induced islet cell death, the effects of various more or less specific inhibitors of ADP-ribosylation were evaluated. The use of 3-AB and 4-amino-1,8-naphthalimide (NPH) (Banasik *et al.*, 1992), a potent novel inhibitor of poly(ADP-ribosyl)ation, showed the same inhibitory effect on radical-induced islet cell lysis, although these two compounds are not precursors of NAD^+. NPH, an inhibitor with high selectivity for PARP, protected islet cells to a similar extent as NA or 3-AB. In addition to poly(ADP-ribosyl)ation, an increasing number of enzymatic reactions is being described that also utilize NAD^+ as a substrate for the mono(ADP-ribosyl)ation of

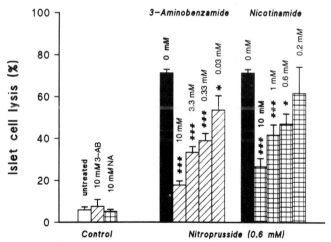

FIGURE 1   Protection of islet cells from NO-induced lysis by 3-aminobenzamide (3-AB) and nicotinamide (NA). Islet cells were exposed to 0.6 mM nitroprusside (NP) in the absence or presence of various concentrations of the drugs. After 18 hr of incubation, the lysis of the islet cells was determined by the trypan blue exclusion assay. Data show means ± SD from six determinations. ***$p < 0.001$ and *$p < 0.02$.

proteins (Moss and Vaughan, 1979). To determine which of these two types of reactions could play a functional role in islet cell lysis, inhibitors of poly(ADP-ribosyl)ation were used in parallel with mono(ADP-ribosyl)ation inhibitors (Loesberg *et al.*, 1990). The inhibitors of poly(ADP-ribosyl)ation could protect islet cells from radical toxicity, whereas the use of inhibitors of mono(ADP-ribosyl)ation was without effect (Radons *et al.*, 1994).

The question of whether NA is protective because of its antioxidant effect was answered by analysis of DNA strand breaks of radical-exposed islet cells in the absence versus presence of NA. In NA-treated cells there was only a slight decrease in the percentage of nuclei with DNA strand breaks after several hours, demonstrating that NA did not scavenge sufficient ROI to prevent DNA damage. The assumption of a causal relationship between PARP activation and islet cell death was finally tested directly by the examination of cells from mice with an inactivated PARP gene (Wang *et al.*, 1995).

The exposure of normal and PARP-deficient islet cells to increasing concentrations of NO and ROI revealed that PARP-deficient islet cells are more resistant to low doses of radicals than normal islet cells. Interestingly, at high concentrations of radicals, cell lysis also occurred in PARP-deficient cells. These findings indicate the existence of an alternative pathway of cell death that does not require PARP activation and NAD^+ depletion. Another issue resolved by the use of these cells was whether 3-AB exerts its protection against radical toxicity via PARP inhibition or by one of its other pharmacological properties. The prevention of radical-induced cytotoxicity by 3-AB occurred only in PARP containing islet cells and therefore identifies PARP as a target for the cytoprotective effect of 3-AB (Heller *et al.*, 1995). Taken together, the use of cells with a disrupted PARP gene enabled us to provide direct evidence for the causal relationship among PARP activation, NAD^+ depletion, and subsequent islet cell death. In the absence of PARP, a second, PARP-and NAD^+-independent pathway of cell death becomes evident. PARP-dependent and -independent forms of islet cell death were found to represent alternative responses of islet cells to the toxic immune mediators ROI and NO.

III. Intervention Studies in Animals

Based on the protective effect of NA on islet cells *in vitro*, this substance was used in different animal models of insulin-deficient diabetes. Streptozotocin (STZ) and alloxan (ALX) are widely used β-cell toxins (Rerup, 1970; Mordes and Rossini, 1981; Rayfield and Ishimura, 1987). Nonimmune-mediated diabetes can be induced by high doses of STZ, which damage islet cells directly via DNA damaging, PARP activation, and NAD^+ depletion, whereas the action of ALX may be independent of NAD$^+$ depletion. Insulin-deficient diabetes could also be induced by partial pancreatectomy (PPX). The effect of NA on diabetes development in these animal models is described in Table I. In all cases, NA showed a protective effect by preserving β-cell viability and promoting β-cell regeneration. In addition to the effect of NA in these nonimmune diabetes models, the protective effect of the substance was also demonstrated in immune-mediated diabetes. Rossini *et al.* (1977) observed the protective effect of NA in mice with diabetes induced by multiple, low-dose injections of STZ. Furthermore, the substance was tried in the spontaneously nonobese diabetic (NOD) mouse. The treatment of NOD mice from early life with high doses of NA was able to suppress the development of diabetes (Yamada *et al.*, 1982). Studies on the effect of NA in the BioBreeding (BB) rat, another model of spontaneous IDDM, remained inconclusive, with positive or negative outcome reported (Table I).

TABLE I   Effects of Nicotinamide (NA) on Nonimmune-Mediated Diabetes and Immune-Mediated Insulin-Dependent Diabetes Mellitus (IDDM) in Animal Models[a]

Model	NA dose (g/kg BW)	Effects of NA	Reference
Nonimmune-mediated IDDM			
HD SZ	0.3	Decreases plasma–glucose, normalizes plasma–insulin	Masiello *et al.* (1985)
	0.5	Protects against diabetes and β-cell destruction	Lazarus and Shapiro (1973)
ALX	0.9	Decreases diabetes incidence	Lazarow (1945); Lazarow *et al.* (1950)
PPX	0.5	Decreases glucose during IVGTT, induces islet hyperplasia	Yonemura *et al.* (1984)
Immune-Mediated IDDM			
LD STZ	0.5 ip	Decreases hyperglycemia	Rossini *et al.* (1977)
NOD	0.5 ip	Protects against IDDM, decreases insulitis and sialitis	Yamada *et al.*, (1982)
BB	0.5 o	Slight, nonsignificant decrease of IDDM	Kolb *et al.* (1989)
BB	0.5 o	Decreases IDDM incidence from 72 to 35%	Sarri *et al.* (1989)
BB	0.5 ip	Slight, nonsignificant decrease of IDDM	Hermitte *et al.* (1989)

Abbreviations: BB, BioBreeding rat; BW, body weight; HD STZ, high-dose streptozotocin; ip, intraperitoneally; IVGTT, intravenous glucose tolerance test; LD STZ, low-dose streptozotocin; NOD, nonobese diabetic mouse; o, orally; PPX, partial pancreatectomy.
[a] Modified from Mandrup-Poulson *et al.* (1993).

IV. Intervention Studies in Humans _____

A. Trials at Onset of Diabetes

Based on the encouraging results of the animal studies, a number of clinical trials were initiated in which IDDM patients with a disease duration ranging from 1 week to a maximum of 5 years received daily doses of 20 mg up to 3 g NA/kg body weight. The aim of these studies was the protection of the pancreatic β cells from the ongoing autoimmune attack to preserve the residual insulin secretory capacity. At present, the most reliable results on the therapeutic effects of NA over a period of 12 months may derive from a meta-analysis of 10 randomized clinical trials (Mendola et al., 1989; Pozzilli et al., 1989, 1994, 1995; Chase et al., 1990; Ilkova et al., 1991; Guastamacchia et al., 1992, Lewis et al., 1992; Taboga et al., 1994; Staman et al., 1995) of which 5 are placebo controlled (Mendola et al., 1989; Chase et al., 1990; Guastamacchia et al., 1992; Lewis et al., 1992; Pozzilli et al., 1995). In general, the combined analysis of data from the terminated trials and from the intermediate evaluation of the ongoing studies with a total number of 211 NA-treated patients and 166 patients in the control group indicates a protective effect of NA treatment on residual β-cell function (P. Pozzilli et al., in press). After 1 year of follow-up, the patients of the control groups showed a decline of the basal and stimulated secretion of C-peptide to about 50% of the value at the entry into the study. In the same period, NA-treated patients showed only a slight decrease in the stimulated C-peptide secretion, but, interestingly, these patients were able to keep their baseline C-peptide release, and after the first year of follow-up it was significantly ($p < 0.005$) higher than in the control group. However, there was no significant difference between the two groups in their insulin requirement for metabolic control, as judged from the HbA1c concentrations. In summary, these data indicate a certain β-cell protective capacity of NA as judged from the preservation of residual β-cell function after 1 year of IDDM manifestation or later.

B. Trials for Prevention

NA protects β cells from inflammatory cell death but may not interfere with the inflammation itself. Consequently, clinically more significant effects may be anticipated when the drug is introduced during the long prediabetic stage, with still sufficient β-cell mass present. A first large open trial was started in Auckland, New Zealand, by studying school children who were found positive for islet cell antibodies (ICA) (Elliot and Pilcher, 1991). Similar to study designs where vaccination strategies were tested, all schools of the district were randomly allocated to a control or an intervention group. ICA testing was offered to 32,000 children ages 5–7 years and 20,195 accepted. Out of these children, 152 were identified as ICA positive and to be at increased risk of diabetes. These children were treated with 1 g/day of a sustained release preparation of NA (Endur-Amide, Innovite, Oregon). The incidence of IDDM was significantly ($p < 0.05$) decreased by approximately 50% in the tested and treated group as compared to 48,335 control pupils of the same age or the 13,463 children offered to treat but not tested (mean observation time: 4.2, 5.2, and 4.2 years; incidence: 7.1, 21.0, and 17.7/105, respectively).

Subsequently, two double-blind trials to prevent diabetes by NA have been initiated. The first trial, "Deutsche Nikotinamid Interventions Studie" (DENIS), recruited 3- to 12-year-old siblings of diabetic children with positive ICA levels higher than 20 JDF (Juvenile Diabetes Foundation) units (Haastert and Giani, 1993; Lampeter, 1993). This subgroup of first-degree relatives was chosen because of the particularly high risk of IDDM in young siblings. To date, 50 children have entered the trial, after random allocation to placebo or NA treatment. The dose of 1.2 g NA/m^2 body surface/day is split into two doses (Endur-Amide, sustained release preparation). Interim analysis did not reveal any side effects of treatment. The study aims at the inclusion of 74 probands but will be sequentially statistically analyzed after every 5 observed diabetes manifestations. First results are expected by the end of 1997. A different proband group is being recruited by the "European Nicotinamide Diabetes Intervention Trial" (ENDIT). In this study, first-degree relatives of 5–40 years of age with ICA ≥ 20 JDF units and manifestation of diabetes before 20 years in the index family member are being included. Dosage and study design are similar to the DENIS trial. Recruitment of probands is exepted to be completed by the end of 1996 and the first interim analysis is due in late 1997.

V. Concluding Remarks

The results of the studies described in this chapter suggest that different species of radicals are involved in islet inflammation and β-cell destruction. Hence any "antioxidant therapy" must not only target "classical" oxygen radicals, such as superoxide anions, hydroxyl radicals, and singlet oxygen, but also nitric oxide and related compounds. Instead of radical scavenging, the first large-scale clinical trials of diabetes prevention aim at supporting the defenses of β cells.

References

Appels, B., Burkart, V., Kantwerk-Funke, G., Funda, J., Kolb-Bachofen, V., and Kolb, H. (1989). Spontaneous cytotoxicity of macrophages against pancreatic islet cells. *J. Immunol.* **142**, 3803–3808.

Banasik, M., Komura, H., Shimoyama, M., and Ueda, K. (1992). Specific inhibitors of poly (ADP-ribose) synthetase and mono(ADP-ribosyl)transferase. *J. Biol. Chem.* **25,** 1569–1575.

Brenner, H.-H., Burkart, V., Rothe, H., and Kolb, H. (1993). Oxygen radical production is increased in macrophages from diabetes prone BB rats. *Autoimmunity* **15,** 93–98.

Burkart, V., Koike, T., Brenner, H.-H., Imai, Y., and Kolb, H. (1993). Dihydrolipoic acid protects pancreatic islet cells from inflammatory attack. *Agents Actions* **38,** 60–65.

Burkart, V., Koike, T., Brenner, H.-H., and Kolb, H. (1992). Oxygen radicals generated by the enzyme xanthine oxidase lyse rat pancreatic islet cells in vitro. *Diabetologia* **35,** 1028–1034.

Chase, H. P., Butler-Simon, N., Garg, S., McDuffie, M., Hoops, S. L., and O'Brien, D. (1990). A trial of nicotinamide in newly diagnosed patients with type 1 (insulin-dependent) diabetes. *Diabetologia* **33,** 114–146.

Elliot, R. B., and Pilcher, C. C. (1991). Prevention of diabetes in normal school children *Diab. Res. Clin. Pract.* **14** (Suppl. 1), 85.

Faust, A., Burkart, V., Ulrich, H., Weischer, C. H., and Kolb, H. (1994). Effect of lipoic acid on cyclophosphamide-induced diabetes and insulitis in non-obese diabetic mice. *Int. J. Immunopharmacol.* **16,** 61–66.

Fehsel, K., Jalowy, A., Sun, Q., Burkart, V., Hartmann, B., and Kolb, H. (1993). Islet cell DNA is a target of inflammatory attack by nitric oxide. *Diabetes* **42,** 496–500.

Grankvist, K., Marklund, S. L., and Talledal, I. B. (1981). CuZn-superoxide dismutase, Mn-superoxid dismutase, catalase and glutathione peroxidase in pancreatic islets and other tissues of the mouse. *Biochem. J.* **199,** 393–398.

Guastamacchia, E., Ciampolillo, A., Lollino, G., Caragiulo, L., De Robertis, O., Lattanzi, V., and Giorgino, R. (1992). Effetto della terapia con nicotinamide sull'induzione della durata della remissione clinica in diabetici di typo 1 all'esordio sottoposti a terapia insulinicaottimizzata mediante microinfusore. *II Diabete (Suppl. 1)* Marzo, 210.

Haastert, B., and Giani, G. (1993). On the sequential design of the Deutsche Nikotinamid Interventionsstudie-DENIS. *Diabete-Metab.* **19** (1 Pt 2), 100–104.

Heller, B., Bürkle, A., Radons, J., Fengler, E., Jalowy, A., Müller, M., Burkart, V., and Kolb, H. (1994). Analysis of oxygen radical toxicity in pancreatic islets at the single cell level. *Biochem. Biophys. Hoppe-Seyler* **375,** 597–602.

Heller, B., Wang, Z.-Q., Wagner, E., Radons, J., Bürkle, A., Fehsel, K., Burkart, V., and Kolb, H. (1995). Inactivation of the poly(ADP-ribose)polymerase gene affects oxygen radical and nitric oxide toxicity in islet cells. *J. Biol. Chem.* **270,** 11176–11180.

Hermitte, L., Vialettes, B., Atlef, N., Payan, M. J., Doll, N., Scheinmann, A., and Vague, Ph. (1989). High dose nicotinamide fails to prevent diabetes in BB rats. *Autoimmunity* **5,** 79–86.

Ilkova, H., Gorpe, U., Kadioglu, P., Ozyazar, M., and Bagriacik, N. (1991). Nicotinamide in type 1 diabetes mellitus of recent onset: A double blind, placebo controlled trial. *Diabetologia* **34** (Suppl. 2), A179.

Kallmann, B., Burkart, V., Kröncke, K.-D., Kolb-Bachofen, V., and Kolb, H. (1992). Toxicity of chemically generated nitric oxide towards pancreatic islet cells can be prevented by nicotinamide. *Life Sci.* **51,** 671–678.

Kolb, H., and Kolb-Bachofen, V. (1992). Nitric oxide: A pathogenetic factor in autoimmunity. *Immunol. Today* **13,** 157–160.

Kolb, H., Schmidt, M., and Kiesel, U. (1989). Immunomodulary drugs in type 1 diabetes. *In* "Immunotherapy of Type 1 Diabetes and Selected Autoimmune Diseases" (G. S. Eisenbarth, ed.), pp. 111–122. CRC Press, Boca Raton, FL.

Kröncke, K.-D., Brenner, H.-H., Rodriguez, M.-L., Etzkorn, K., Noack, E. A., Kolb, H., and Kolb-Bachofen, V. (1993). Pancreatic islet cells are highly susceptible toward the cytotoxic effects of chemically generated nitric oxide. *Biochem. Biophys. Acta* **1182,** 221–229.

Kröncke, K.-D., Kolb-Bachofen, V., Betschick, B., Burkart, V., and Kolb, H. (1991). Activated macrophages kill pancreatic syngeneic islet cells via arginine-dependent nitric oxide generation. *Biochem. Biophys. Res. Commun.* 175, 752–758.

Lampeter, E. F. (1993). Intervention with nicotinamide in pre-type 1 diabetes: The Deutsche Nikotinamid Interventionsstudie-DENIS. *Diabete-Metab.* 19(1 Pt 2); 105–109.

Lazarow, A. (1945). Protection against alloxan diabetes. *Anat. Rec.* 97, 353.

Lazarow, A., Liambies, J., and Tausch, A. J. (1950). Protection against diabetes with nicotinamide. *J. Lab. Clin. Med.* 38, 249–258.

Lazarus, S., and Shapiro, S. H. (1973). Influence of nicotinamide and pyridine nucleotides on streptozotocin and alloxan induced pancreatic B cell cytotoxicity. *Diabetes* 22, 499–506.

Lewis, C. M., Canafax, D. M., Sprafka, J. M., and Barbosa, J. J. (1992). Double-blind randomized trial of nicotinamide on early onset diabetes. *Diabetes Care* 15, 121–123.

Loesberg, C., van Rooij, H., and Smets, L. A. (1990). Meta-iodobenzylguanidin (MIBG), a novel high-affinity substrate for cholera toxin that interferes with cellular mono(ADP-ribosylation). *Biochim. Biophys. Acta* 1037, 92–99.

Malaisse, W. J., Malaisse-Lagae, F., Sener, A., and Pipeleers, D. G. (1982). Determinants of the selective toxicity of alloxan to the pancreatic B-cell. *Proc. Natl. Acad. Sci. U.S.A.* 79, 927–930.

Mandrup-Poulson, T., Reimers, J. I., Andersen, H. U., Pociot, F., Karlsen, A. E., Bjerre, U., and Nerup, J. (1993). Nicotinamide treatment in the prevention of insulin-dependent diabetes mellitus. *Diabetes/Metab. Rev.* 9, 295–309.

Masiello, P., Cubeddu, T. L., Frosina, G., and Bergamini, E. (1985). Protective effect of 3-aminobenzamide, an inhibitor of poly(ADP-ribose) synthetase, against streptozotocin-induced diabetes. *Diabetologia* 28, 683–686.

Mendola, G., Casamitjana, R., and Gomis, R. (1989). Effect of nicotinamide therapy upon β-cell function in newly diagnosed type 1 (insulin-dependent) diabetic patients. *Diabetologia* 32, 160–162.

Mordes, J. P., and Rossini, A. A. (1981). Animal models of diabetes. *Am. J. Med.* 70, 353–360.

Moss, J., and Vaughan, M. (1979). Activation of adenylate cyclase by choleragen. *Annu. Rev. Biochem.* 48, 581–600.

Pozzilli, P., Browne, P. D., and Kolb, H. (1996). Meta-analysis of nicotinamide treatment in patients with recent onset insulin dependent diabetes. *Diabetes Care,* in press.

Pozzilli, P., on behalf of the IMDIAB Study group. (1994). Randomized trial comparing nicotinamide and nicotinamide plus cyclosporin in recent onset insulin-dependent diabetes. *Diabetes Med.* 17, 897–900.

Pozzilli, P., Visalli, N., Ghirlanda, G., Manna, R., and Andreani, D. (1989). Nicotinamide increases C-peptide secretion in patients with recent onset type 1 diabetes. *Diabetes Med.* 6, 316–321.

Pozzilli, P., Vasalli, N., Signore, A., et al. (1995). Double blind trial of nicotinamide in recent onset insulin dependent diabetes mellitus. *Diabetologia* 38, 848–852.

Rabinovitch, A., Suarez, W. L., Thomas, P. D., Strynadka, K., and Simpson, I. (1992). Cytotoxic effects of cytokines on rat islets: Evidence for involvement of free radicals in lipid peroxidation. *Diabetologia* 35, 409–413.

Radons, J., Heller, B., Bürkle, A., Hartmann, B., Rodriguez, M. L, Kröncke, K.-D., Burkart, V., and Kolb, H. (1994). Nitric oxide toxicity in islet cells involves poly(ADP-ribose) polymerase activation and concomitant NAD⁺ depletion. *Biochem. Biophys. Res. Commun.* 199, 1270–1277.

Rayfield, E. J., and Ishimura, K. (1987). Environmental factors and insulin-dependent diabetes mellitus. *Diabetes/Metab. Rev.* 3, 925–957.

Rerup, C. C. (1970). Drugs producing diabetes through damage of the insulin secreting cells. *Pharmacol. Rev.* 22, 485–520.

Rossini, A. A., Like, A. A., Chick, W. L., Appel, M. C., and Cahill, G. F., Jr. (1977). Studies of streptozotocin-induced insulitis and diabetes. *Proc. Natl. Acad. Sci. U.S.A.* 74, 2485–2489.

Sarri, Y., Mendola, J., Ferrer, J., and Gomis, R. (1989). Preventive effects of nicotinamide administration on spontaneous diabetes of BB rats. *Med. Sci. Res.* **17**, 987–988.

Staman, I., Dinccag, N., Karsidag, K., Ozer, E., Altuntas, Y., and Yilmaz, M. T. (1995). The effect of nicotinamide in recent onset type 1 diabetes regarding the level of beta cell reserve. *Klinik Gelisim* **8**, 3882–3886.

Taboga, C., Tonutti, L., and Noacco, C. (1994). Residual beta cell activity and insulin requirements in insulin-dependent diabetic patients treated from the beginning with high doses of nicotinamide: A two year follow-up. *Recent Progr. Med.* **85**, 513–516.

Wang, Z.-Q., Auer, B., Stingl, L., Berghammer, H., Haidacher, D., Schweiger, M., and Wagner, E. (1995). Mice lacking ADPRT and poly(ADPribosyl)ation develop normally but are susceptible to skin disease. *Genes Dev.* **9**, 509–520.

Yamada, K., Nonaka, K., Hanafusa, T., Miyazaki, A., Toyoshima, H., and Tarui, S. (1982). Preventive and therapeutic aspects of large dose nicotinamide injections on diabetes associated with insulitis: An observation in non-obese diabetic (NOD) mice. *Diabetes* **31**, 749–753.

Yonemura, Y., Takishima, T., Miwa, K., Miyazadi, I., Yamamoto, H., and Okamoto, H. (1984). Amelioration of diabetes mellitus in partially depancreatized rats by poly(ADP-ribose) synthetase inhibitors. *Diabetes* **33**, 401–404.

Zweier, J. I., Kuppusamy, P., and Lutty, G. A. (1988). Measurement of endothelial cell free radical generation: Evidence for a central mechanism of free radical injury in postischaemic tissues. *Proc. Natl. Acad. Sci. U.S.A.* **85**, 4046–4050.

Karin Scharffetter-Kochanek

Experimental and Clinical Photodermatology
Department of Dermatology
University of Cologne
50931 Köln, Germany

Photoaging of the Connective Tissue of Skin: Its Prevention and Therapy

I. Introduction

The skin is always in contact with oxygen and is increasingly exposed to ultraviolet (UV) irradiation. Therefore, the risk of photo-oxidative damage of the skin induced by reactive oxygen species (ROS) has increased substantially (Darr and Fridovich, 1994). The term reactive oxygen species not only collectively includes oxygen-centered radicals such as the superoxide anion ($O_2 \cdot^-$) and the hydroxyl radical ($OH \cdot$), but also some nonradical species, such as hydrogen peroxide (H_2O_2) and singlet oxygen ($^1O_2$), among others, all being produced in the skin upon UV irradiation. A complex antioxidant defense system has evolved in the skin and protects against ROS (Fridovich, 1989; Shindo et al., 1993). However, UV-generated ROS substantially compromise the enzymatic and nonenzymatic antioxidant defense of the skin (Witt et al., 1993; Biesalski et al., 1996), thus tilting the balance toward a prooxidant state (Sies, 1986, 1991). The resulting oxidative stress causes damage to cellular components and changes the pattern of gene expression, finally leading to skin pathology such as nonmelanoma and melanoma skin cancers, phototoxicity, and photoaging (Oikarinen *et al.*, 1985; Oikarinen and Kallioinen, 1989; Epstein, 1989; Gallagher *et al.*, 1989; Urbach, 1989; Henriksen *et al.*, 1990; Kligman, 1992; Scharffetter-Kochanek *et al.*, 1992, 1995b).

This chapter focuses on the molecular mechanisms of the UV-induced connective tissue damage of the skin, with particular emphasis on the involvement of ROS in cutaneous photodamage, its prevention, and therapy.

II. The Nature of Photoaging

Several intrinsic and extrinsic factors contribute to the complex phenomenon of aging. Chronological aging affects the skin in a manner similar to other organs (Uitto, 1986). Superimposed on this innate process, photoaging is related to severe UV-induced damage of the dermal connective tissue. Increasing evidence shows that these two processes, chronological aging and photoaging, have different biological, biochemical, and molecular mechanisms (Oikarinen, 1990), including differences in the histological picture, the content of various extracellular matrix proteins, the formation of cross-links within the collagen molecule (Bailey *et al.*, 1974; Yamauchi *et al.*, 1991), and the capacity of fibroblasts to organize extracellular matrix molecules (Marks *et al.*, 1990). Photoaging in humans becomes apparent as a final stage after several decades of chronic sun exposure. This protracted evolution of chronic sun-induced damage of the connective tissue and the impossibility to assess the involved wavelengths and the cumulative dose have substantially hampered the studies on chronic UV effects on the connective tissue and its metabolism in humans. During the last decade, the hairless skin mouse (skh-1), a fully immunocompetent mouse, proved to be a relevant model in photoaging research with similar connective tissue alterations as found in actinically damaged human skin. Most important, the process of photoaging occurs over a shorter period of time (Kligman *et al.*, 1982, 1983, 1985). Clinically, dermal photoaging is characterized by wrinkle formation, loss of recoil capacity, increased fragility of the skin with blister formation, and impaired wound healing. At the histological level, a loss of mature collagen, a distinct basophilic appearance of collagen ("basophilic degeneration"), and greatly increased deposition of glycosaminoglycans, as well as an increase in fragmented elastic fibers (Mitchell, 1967; Kligman, 1969), are constant features in cutaneous photodamage (Chen *et al.*, 1986). Biochemically, quantitative and qualitative alterations of dermal extracellular matrix proteins such as elastin (Smith *et al.*, 1962; Braverman and Fonferko, 1982; Uitto, 1986), glycosaminoglycans (Sams and Smith, 1961; Smith *et al.*, 1962), and interstitial collagen (Trautinger *et al.*, 1989; Lever and Schaumburg-Lever, 1990; Schwartz *et al.*, 1991) are involved. Using a transgenic mouse line, the activity of the elastin promotor-driven reporter CAT gene substantially increased upon UV irradiation (Bernstein *et al.*, 1995). In contrast, collagen type I, which belongs to a family of closely related but genetically distinct proteins (van der Rest and Garrone, 1991) providing the dermis with tensile strength and stability, has been found to be diminished in photoaged skin (Oikarinen *et al.*, 1985; Trautinger *et al.*, 1989). Because the enzymatic capacity for extracellular matrix synthesis and its degradation reside in the dermal fibroblast, much effort has been devoted to studying these fibroblast-controlled processes. In addition to UV-affected posttranslational modification of the newly synthesized collagen molecule (Johnston *et al.*, 1984; Oikarinen *et al.*, 1985), it has been shown that various matrix metalloproteinases (MMPs), responsible for the breakdown of dermal interstitial collagen and other connective tissue components (Table I), are dose dependently induced *in vitro* and *in vivo* by UVA and UVB irradiation (Scharffetter *et al.*, 1991; Petersen *et al.*, 1992; Scharffetter-Kochanek *et al.*, 1992; Herrmann *et al.*, 1993, Koivukangas *et al.*, 1994; Brenneisen *et al.*, 1996; Herrmann *et al.*, 1996). The family of MMPs is growing and comprises at least 10 members (Table I). Whereas MMP-1 cleaves collagen type I, MMP-2 is able to degrade basement membrane compounds, including collagen type IV and type VII. MMP-3 reveals the broadest substrate specifity for proteins, such as collagen type IV, proteoglycans, fibronectin, and laminin (Liotta *et al.*, 1991; Senior *et al.*, 1991; Woessner, 1991; Matrisian, 1992). As to their proteolytic activity, UV-induced MMPs may contribute to the dissolution of the basement membrane and dermal structural proteins, leading to blister formation and photoaging. In fact, blister formation following PUVA treatment (UVA + 8-methoxypsoralene) (Heidbreder, 1980; Friedmann *et al.*, 1987) or after tanning on sun beds (Epstein *et al.*, 1973; Farr *et al.*, 1988) has been observed.

TABLE I   Matrix Metalloproteinases (MMP) and Their Substrates

Designation	Size (kDa)	Substrate
MMP-1 (interstitial collagenase)	52, 57	Collagen type I, II, III
MMP-8 (neutrophil collagenase)	75	Collagen type I, II, III
MMP-2 (72-kDA type IV collagenase)	72	Collagen type IV, V, VIII, gelatin, fibronectin
MMP-9 (92-kDa type IV collagenase)	92	Collagen type IV, V, gelatin
MMP-3 (stromelysin-1)	57, 60	Collagen type III, IV, V, gelatin, proteoglycans, laminin, fibronectin
MMP-10 (stromelysin-2)	53	Collagen type III, IV, V, fibronectin, gelatin
MMP-7 (matrilysin)	28	Fibronectin, gelatin
MMP-11 (stromelysin-3)	51	Unknown
MMP-13 (collagenase-3)	48	Collagen type
MMP-12 (metalloelastase)	?	Unknown

The activity of all MMPs is inhibited by a special class of tissue inhibitors of metalloproteinases (TIMPs). Similar to MMPs, TIMPs are synthesized by ordinarily resident fibroblasts.

Prior to discussing the role of distinct reactive oxygen species in the up regulation of matrix metalloproteinases with subsequent connective tissue degradation, some biophysical aspects of UV irradiation and current concepts concerning the UV generation of ROS and their prominent biochemical properties will be presented.

III. Role of Ultraviolet-Generated Oxygen Species in Cutaneous Photoaging

A. Penetration of Different Spectra in the Skin and Generation of ROS

The depth of penetration of UV irradiation into the skin determines the site of reactive oxygen generation and is clearly dependent on the wavelength of UV and on the dose as reviewed by Bryce (1993). In contrast to UVB irradiation which is mainly absorbed by the epidermis, UVA irradiation penetrates into the dermis, making the fibroblast an accessible target even in the deep dermis. Certainly, the transmission also depends on the thickness of the stratum corneum, its state of hydration, and the pigmentation of the epidermis. A standardized skin specimen (Bruls et al., 1984) will transmit 19% of the intensity at 365 nm, 9.5% at 313 nm, 2.4% at 297 nm, and 0.27% at 290 nm. While it is well accepted that low levels of reactive oxygen species are continuously produced in vivo and are involved in signal transduction pathways, cell activation, differentiation, and growth control, there is accumulating indirect evidence for the damaging effect of higher concentrations of ROS generated in vivo following UVA and UVB irradiation of the skin (Black, 1987; Taira et al., 1992; Dalle-Carbonare and Pathak, 1994; Jurkiewicz and Buettner, 1994; Jurkiewicz et al., 1995). In addition to direct absorption of UVB photons by DNA and subsequent structural changes, the generation of reactive oxygen species following irradiation with UVA and UVB has been reported requiring the absorption of photons by endogenous photosensitizer molecules. Photosensitized oxidations occur as type I or type II reactions (Foote, 1991). Mechanistically, the absorption of UV photons by a sensitizer results in its electronically excited state. Photosensitizers include riboflavin, porphyrins, quinones, and bilirubin in mammalian cells (Rosenstein et al., 1983). The excited sensitizer subsequently reacts with another substrate (type I reaction) or with oxygen (type II reaction). The resulting products of type I reactions are radicals or radical ions, whereas type II reactions produce reactive oxygen species, including $O_2 \cdot^-$, OH·, and $^1O_2$. $O_2 \cdot^-$ may also be formed from other sources, e.g., when chelated Fe^{3+} is reduced and undergoes autoxidation (Halliwell and Gutteridge, 1990). $O_2 \cdot^-$ produced by phagocytes plays an important role in the killing of bacteria (Babior and Woodman, 1990) and, if overproduced or not correctly dismutated, may play a role in the pathogenesis and severity of central nervous system injuries and of neurologic degenerative disorders (Hall, 1994). Superoxide dismutases convert $O_2 \cdot^-$ to H_2O_2. H_2O_2 is able to cross all membranes easily. However, neither of these species reacts directly with DNA (Halliwell and Aruoma, 1991). Therefore, H_2O_2 and $O_2 \cdot^-$ are thought to participate in the generation of more dangerous species such as the hydroxyl radical (OH·). This can happen *in vitro* and *in vivo* by two related although different mechanisms. $O_2 \cdot^-$ can reduce Fe(III) or Cu(II), and the subsequent Fe(II) or Cu(I) can reduce H_2O_2, finally resulting in the generation of the hydroxyl radical (OH·) (Darr and Fridovich, 1994). *In vivo*, $O_2 \cdot^-$ enhances a release of Fe(II) from [4Fe-4S] clusters of dehydratases, and the released Fe(II) subsequently reduces H_2O_2 to HO^- and HO· (Liocher and Fridovich, 1994). In addition, $O_2 \cdot^-$ is able to release Fe(II) from ferritin. This has also been reported following UV irradiation and most likely is due to the effect of UV-generated $O_2 \cdot^-$ (Boyer and McCleary, 1987; Biemond *et al.*, 1988. Using electron spin resonance (ESR), OH· has been detected in skin upon UV irradiation (Jurkiewicz and Buettner, 1994).

Singlet oxygen is a particularly damaging molecule with a short half-life. UV irradiation generates singlet oxygen by energy transfer from a photosensitizer molecule to ground-state oxygen. In addition to endogenous photosensitizers, the skin has access to an increasing number of exogenous photosensitizers in cosmetics, medications, drugs, plants, and industrial emissions. Singlet oxygen and other reactive oxygen species have been implicated in the pathogenesis of the severe connective tissue damage in photodermatologic disorders, including drug-induced phototoxicity, porphyrias, and photoaging (Epstein *et al.*, 1973; Oikarinen *et al.*, 1985; Oikarinen and Kallioinen, 1989; Kligman, 1992).

B. Mechanism of Connective Tissue Damage by ROS

Because ROS may prove particularly relevant to future developments of UV protective agents for the skin, there have been efforts to better define the involvement of distinct oxygen species in the upregulation of MMPs which are responsible for the connective tissue degradation in photoaging, tumor invasion, and metastasis. For this purpose, fibroblast monolayer cultures were subjected to various ROS-generating systems or to UV irradiation at different wavelengths and spectra. Using nontoxic concentrations of chemicals which increase or specifically inhibit the activity of ROS-detoxifying enzymes and iron chelators blocking the Fenton reaction, distinct ROS have been increased peri- or intracellularly.

Exposure of cultured human fibroblasts to singlet oxygen, generated in a dark reaction by thermodecomposition of the endoperoxide of the sodium salt of 3,3'-(1,4-naphtylidene) dipropionate (NDPO$_2$), induced collagenase (MMP-1) mRNA steady-state levels in a dose-dependent manner (Scharffetter-Kochanek et al., 1993). The increase in collagenase expression after singlet oxygen exposure generated with 3 mM NDPO$_2$ was equivalent to that observed with UVA at a dose of 200–300 kJ/m^2 and was developed in a similar time course. In contrast, mRNA levels of TIMP-1, the specific inhibitor of metalloproteinases, remained unchanged. Indirect evidence for the role of singlet oxygen in the UVA induction of collagenase comes from studies using enhancers or quenchers of singlet oxygen. Accordingly, incubation in deuterium oxide, an enhancer of the life span of singlet oxygen, led to an additional increase in the steady-state levels of collagenase mRNA after exposure to NDPO$_2$ or UVA irradiation. In contrast, sodium azide, a potent quencher of singlet oxygen, almost totally abrogated the induction of collagenase after the exposure of fibroblasts to NDPO$_2$ or UVA irradiation. Similar results were obtained on the protein level (Wlaschek et al., 1995). The importance of singlet oxygen in photoaging becomes particularly clear in patients suffering from porphyria cutanea tarda. These patients show substantially accelerated photoaging and more frequent and severe blister formation compared to UV-exposed individuals. Biochemically, an increase in photosensitizing porphyrins, mainly of uroporphyrin in the skin, has been reported (Schaefer et al., 1991).

Photoexcitation of uroporphyrin I enhances the formation of singlet oxygen. In order to simulate this metabolic disorder, fibroblast monolayer cultures have been subjected to a combined treatment with uroporphyrin and subsequent irradiation (320–460 nm). This combined treatment resulted in a much stronger induction of MMPs compared to irradiation alone (Herrmann et al., 1996). This is in line with the clinical experience of a substantial improvement of photoaging after a therapeutical decrease of abnormally elevated porphyrins by means of chloroquin, further underlining the central role of singlet oxygen in dermal photoaging and its particular relevance to the future development of UV protective agents for the skin. Because even high doses of singlet oxygen quenchers could not completely inhibit the UVA-induced collagenase, it was concluded that other ROS may be involved. In order to define distinct ROS other than singlet oxygen that are possibly involved in the upregulation of MMPs, paraquat, a redox-cycling agent known for its capacity to increase the intracellular concentration of the superoxide anion, has been used. A time-dependent increase in steady-state mRNA levels for interstitial collagenase (MMP-1) with a maximal increase of eight-fold was observed at 72 hr following exposure of fibroblasts to paraquat at nontoxic concentrations (Brenneisen et al., 1996). Interestingly, an induction of the corresponding protein was only detected at 24 hr with no protein at later time points suggesting an uncoupling of transcription and translation events. The isolated and combined inhibition of glutathione peroxidase, catalase, and the Fenton reaction by buthionine sulfoximine (BSO), aminotriazole (ATZ), and iron chelators such as desferrioxamine (DFO) or N-2-hydroxybenzyl-N'-benzylethylenediamine diacetic acid (HBED), respectively, enhances MMP-1 mRNA levels up to sixfold after paraquat treatment or UVA irradiation compared to paraquat-treated or UVA-irradiated controls. Although inhibition of the Cu,Zn superoxide dismutase by diethyldithiocarbamate (DDC) diminished the steady-state MMP-1 mRNA level after paraquat treatment, MMP-1 mRNA levels were increased after UVA irradiation (Brenneisen et al., 1996). These results indicate that following paraquat treatment, H_2O_2 may be responsible for the induction of MMP-1. Following UVA irradiation in addition to $^1O_2$, both H_2O_2 and $O_2\cdot^-$ appear to play a role in the induction of MMP-1 synthesis (Wlaschek et al., 1995; G. Herrmann et al., unpublished results). In contrast, following UVB irradiation, the iron chelators, particularly DFO or HBED, are able to inhibit the upregulation of MMP-1 mRNA levels by 60% compared to the UVB-irradiated control, thus pointing to the importance of the Fenton reaction and possibly OH· (P. Brenneisen and K. Briviba, unpublished results). The involvement of transitional metals in accelerating photoaging is further supported by in vivo studies (Bisset et al., 1991). Chronic exposure of hairless mice (skh-1) with suberythemal doses of UVB results in an increased level of nonheme iron in the skin. A similar increase in nonheme iron was observed in sun-exposed human skin compared to nonexposed body sites. Topical application of certain iron chelators (for details see Section IV,A) to hairless mice skin substantially delayed the onset of UVB-induced histological alterations such as hyperplasia and fragmentation of elastin fibers and basophilic degeneration of collagen. These in vitro and in vivo results and the delay in photoaging by antioxidant strategies provide strong evidence for an important role of UV-generated ROS in the upregulation and activation of matrix-degrading metalloproteinases. In addition, reactive oxygen species generated either by UV irradiation or by inflammatory cells present in photodamaged skin are able to directly attack and destroy extracellular matrix proteins (Monboisse and Borel, 1992; Baker, 1994).

Taken together, recent work has identified distinct ROS involved in connective tissue damage. However, further studies are required to outline preventive strategies that may enhance the rational development of protective agents for photoaging.

C. Signal Transduction Pathway in the ROS-Induced Upregulation of MMPs

Due to their short half-lives of 10^{-5} to 10^{-8} sec, ROS represent ideal signaling molecules and, in this capacity, are involved in several physiological processes. A UV-induced extracellular factor(s) from human fibroblasts has been reported to communicate the UV response to nonirradiated cells (Schorpp *et al.*, 1984). Furthermore, much information has accumulated on a group of signaling peptides comprising growth factors, cytokines, and interleukins that had been implicated in tissue remodeling and degradation in a variety of physiological and pathological processes. The exact nature of the factors and their relation to ROS have now been unraveled, at least in part. Using antisense strategies, function-blocking antibodies, and bioassays, evidence for a UVA-induced cytokine network consisting of interleukin-1α, interleukin-1β, and interleukin-6, which induced interstitial collagenase (MMP-1) via interrelated autocrine loops, has been provided. While an early peak of IL-1 bioactivity at 1-hr postirradiaton is responsible for the induction of IL-6 and together with IL-6 leads to an increase in MMP-1, the latter posttranscriptionally controls synthesis and release of IL-1 and thus perpetuates the UV response (Wlaschek *et al.*, 1993, 1994). Similarly, Krämer and co-workers (1993) showed that the UVC irradiation of HeLa cells induced the synthesis and release of IL-1α and basic fibroblast growth factor, which together stimulate the synthesis of collagenase (MMP-1). Obviously, the UV induction of collagenase (MMP-1) is mediated by different cytokines, depending on the UV spectra and possibly on the cell type.

Evidence suggests that the UVA generation of singlet oxygen and H_2O_2 precedes the synthesis and release of IL-1α, IL-1β, and IL-6. Accordingly, UV-generated ROS are rather initial intermediates in the complex signaling cascade, inducing the release and sythesis of IL-1α and IL-6, finally leading to the upregulation of matrix metalloproteinases and possibly connective tissue degradation. Also, Meier and co-workers (1989) showed that IL-1 stimulates the release of ROS by human fibroblasts. Hence, ROS may represent ideal targets for pharmacological intervention.

Another line of investigation concerns the redox regulation of transcription factors, such as AP-1, which transactivate the interstitial collagenase (MMP-1) gene, resulting in its enhanced expression. Most of this work, however, has been done with epithelial cell lines, and it is unclear whether the redox regulation of AP-1 also plays a role in human dermal fibroblasts and how the cytokine network would fit into this complex signaling cascade.

In HeLa cells, the redox regulation of AP-1 DNA-binding activity is mediated by a conserved cysteine residue, which is localized in the DNA-binding domain of the AP-1 protein (Abate and Curran, 1990). Substitution of Cys-154 in Fos and Cys-272 in Jun with a serine residue leads to enhanced DNA-binding activity and a loss of redox control (Abate *et al.*, 1990). In a pro-oxidant state, the critical cysteine residue is converted to a state that does not effectively bind to DNA. The cysteine residue is in close contact with DNA, and treatment with reducing agents restores its DNA-binding activity. Evidence shows that AP-1 binding is regulated by a tightly controlled redox cascade involving the redox factor-1 (Ref-1) and thioredoxin (Abate *et al.,* 1990; Xanthoudakis and Curran, 1992). Confocal immunocytology has identified Ref-1 as being localized in the nucleus. Furthermore, immunodepletion analysis with function-blocking antibodies against Ref-1 suggests that Ref-1 is a major AP-1 redox activity in HeLa cell nuclear extracts. Interestingly, in addition to its redox-controlling properties, Ref-1 possesses an apurinic/apyrimidinic (AP) endonuclease DNA repair activity, suggesting a link among the regulation of transcription factors, oxidative signaling, and repair processes of oxidative DNA damage in human cells (Xanthoudakis *et al.,* 1992). It remains to be established whether Ref-1 plays a similarly essential role in fibroblasts and whether it is induced on UV irradiation.

IV. Prevention and Therapy of Connective Tissue Damage

A. Prevention of UV-Induced Connective Tissue Damage

As shown earlier, ROS play a major role in the UV-dependent upregulation of matrix metalloproteinases. Hence, a decrease of the ROS load following UV irradiation by efficient sunscreens and/or antioxidants represents promising strategies to prevent or minimize cutaneous photoaging. Support for this view comes from experiments with the hairless skh-1 mouse model, showing that the concomitant application of sunscreens prior and during chronic UVB or UVA irradiation effectively prevented connective tissue damage (Kligman *et al.,* 1982, 1983, 1985; Bissett *et al.,* 1987; Plastow *et al.,* 1987). Furthermore, application of sunscreens on an already photodamaged skin stopped further damage and even promoted the naturally occurring repair capacity of the skin. The repair zone directly beneath the epidermis revealed new collagen formation (Kligman *et al.,* 1982, 1983).

A variety of ROS-scavenging or quenching compounds effectively delayed the onset of wrinkling and histological changes in cutaneous photoaging in the hairless mouse model. This has been shown for topically applied compounds, such as α-tocopherol, ascorbic acid, and propyl gallate (Bissett *et al.,* 1990a,b). Interestingly, a topical application of α-tocopherol effectively prevents the UVA reduction of interstitial collagen type I and type III in hairless mice (Trautinger *et al.,* 1989). Using electron paramagnetic resonance spectroscopy, it has been shown that α-tocopherol sorbate is more efficient in decreasing the UV-induced radical flux in the skin compared to α-tocopherol acetate (Jurkiewicz *et al.,* 1995).

Conjugated dienes, known for their reactivity with singlet oxygen, were effective in preventing the UVB-induced wrinkling (Bissett *et al.,* 1990a). However, dietary supplements with β-carotene, a potent singlet oxygen quencher, did not prevent cutaneous photoaging (Kligman and Mathews-

Roth, 1990). In contrast, β-carotene effectively prevented phototoxicity in patients suffering from erythropoetic protoporphyria (EPP) (Mathews-Roth et al., 1970). Furthermore, all-*trans*-β carotene serum levels of patients suffering from EPP increased substantially during continuous treatment with β-carotene (von Laar et al., 1995). Hydroxyl radicals are generated from hydrogen peroxide with ferrous iron via the Fenton reaction. Iron appears to play a central role in skin photodamage via the enhanced formation of hydroxyl radicals. The topical application of certain iron chelators, such as 1,10-phenanthroline and 2,2'-dipyridylamine, substantially delayed the onset of wrinkling. In contrast, nonchelating analogs did not provide any photoprotection (Bissett et al., 1991).

Even though available studies suggest a protective/therapeutic potential for antioxidants in photoaging, no study has investigated the additive photoprotective effect of combinations of antioxidant substances on UV-induced connective tissue damage. Also, no comprehensive biochemical data are available for a protective effect of antioxidants on connective tissue damage in humans.

A rational design of antioxidants for topical and systemical application finally will depend on our understanding of the molecular mechanisms and the identification of distinct reactive oxygen species involved in cutaneous photoaging.

B. Therapy of UV-Induced Photodamage

As early as 1982, Kligman and co-workers observed that within 3 months after the cessation of chronic UV irradiation, a subepidermal zone of new collagen formation spontaneously occurred in hairless mice. A similar repair zone has also been observed in human photodamaged skin. The concept of these investigators initiated multiple studies regarding the possibility to enhance the repair of photodamaged connective tissue by pharmacologic agents.

In addition to retinoids, which accelerate the spontaneous repair capacity of the connective tissue, α hydroxy acids (AHAs) and related compounds may have beneficial effects on photodamaged skin, although their molecular mode of action is poorly understood. Comprehensive reviews are available (Griffith and Voorhees, 1995; Scott and Yu, 1995).

Retinoids are derived from all-*trans*-retinol (tretinoin), known as vitamin A, (Ross, 1992), and play major roles in the regulation of cell proliferation, differentiation, and production of matrix during morphogenesis, wound healing, and vision (Tabin, 1991). Topical application of tretinoin on human skin results, at least in part, in the formation of 13-*cis*-retinoic acid (isotretinoin), among other stereoisomers (Frickel, 1984; Duell et al., 1992; Levin et al., 1992). Both tretinoin and 13-*cis*-retinoic acid are effective in the reversal of cutaneous photodamage. Morphologically, an increase in

the width of the subepidermal repair zone (Chen *et al.*, 1992) and an effacement of permanent wrinkles (Bryce *et al.*, 1988) in photoaged skin have
been reported. Biochemically, retinoic acids induce collagen synthesis, as
has been shown by an increase in collagen type I and type III steady-state
mRNA levels (Kim *et al.*, 1992), by an increase in the incorporation of
radioactive proline into collagen hydroxyproline in organ cultures (Chen *et
al.*, 1992), and by an increase in the aminopropeptide of collagen type III
as measured by radioimmunoassays and immunohistochemistry (Schwartz
et al., 1991). Interestingly, stimulation of collagen synthesis only occurs in
chronically photodamaged skin and not in skin of age-matched control mice
(Chen *et al.*, 1992; Kim *et al.*, 1992). Furthermore, retinoic acids are able
to inhibit matrix-degrading enzymes (Edwards *et al.*, 1987).

Apart from translocation of the elastotic material into the lower dermis,
no changes have been found in the desmosin content quantitatively reflecting
elastin, following retinoic acid treatment (Bryce *et al.*, 1988). Although
retinoic acids have antioxidant activity (Hiramatsu and Packer, 1990) and
may provide protection from UV penetration by increasing epidermal thickness, the molecular mechanism underlying the repair capacity of retinoic
acids is thought to be due to an enhanced epidermal release of transforming
growth factor-β (TGF-β) (Fisher *et al.*, 1992; Kim *et al.*, 1992). The family
of TGF-β currently comprises more than five closely related members, each
encoded by a distinct gene. TGF-β is particularly important for the remodeling and organization of connective tissue. TGF-β stimulates the production
and deposition of extracellular matrix proteins and exhibits chemotactic
effects on human dermal fibroblasts (Sporn and Roberts, 1986). Indirect
evidence suggests that retinoic acids stimulate the processing of preformed
TGF-β protein which is released from the epidermis and modulates the
connective tissue metabolism of dermal fibroblasts (Fisher *et al.*, 1992; Kim
et al., 1992).

Topical tretinoin application also reverses, at least partly, the cutaneous
photodamage in human skin as reported by several studies, among them
double-blind, vehicle-controlled studies (Cordero, 1983; Kligman *et al.*,
1986, Weiss *et al.*, 1988; Ellis *et al.*, 1990).

Although a substantial improvement of coarse wrinkles and histologi-
cally an increased thickness of the epidermis are well-documented effects of
retinoic acids on human photodamaged skin, less quantitative and biochemical studies are available. A preliminary report on six patients claimed that
treatment of photoaged skin with topical tretinoin increased the number of
anchoring fibrils (collagen type VII); however, the number of anchoring
fibrils in individuals without any photodamage has not been established,
nor has the study been repeated with a statistically more reliable number
of patients (Woodley *et al.*, 1990). Using an antibody that recognizes a 19-
aminoacid portion of the aminopeptide cleavage site of human procollagen
type I, Griffiths and co-workers (1993) showed immunohistologically that

collagen type I formation was substantially diminished in photodamaged skin compared to sun-protected skin and that treatment of photodamaged skin with tretinoin produced an 80% increase in collagen type I formation compared to the vehicle control. This study was carefully performed, but the value of quantification of immunostaining is still controversially discussed. Although retinoids are widely used in the United States for the treatment of cutaneous photodamage, this indication has not been approved by the FDA.

The second group of chemical substances that have become major determinants in skin care are α hydroxy acids. AHAs are composed of glycolic acid, lactic acid, and citric acid, also called fruit acids. The molecular mechanism of their beneficial effect on photodamaged skin is not understood. A clinical histometric study (Ditre *et al.*, 1996) of topical effects of several AHAs on photoaged skin has documented an increase in skin thickness, most likely due to an increase in dermal glycosaminoglycans. However, there are no studies on the effect of AHAs on the connective tissue metabolism of the skin.

Further understanding of the molecular mechanism of photoaging and the development of repair pharmaceuticals, combined with strategies including application of effective sunscreens and antioxidant agents, hold definite promise for the protection of the connective tissue damage in sun-exposed skin.

References

Abate, C., Patel, L., Rauscher, F. J., III, and Curran, T. (1990). Encounter with Fos and Jun on the road to AP-1. *Science* **249,** 1157–1161.
Abate, C., and Curran, T. (1990). Redox regulation of Fos and Jun DNA-binding activity in vitro. *Semin. Cancer. Biol.* **1,** 19–26.
Bailey, A. J., Robins, S. P., and Balian, G. (1974). Biological significance of the intermolecular crosslinks of collagen. *Nature* **251,** 105–109.
Baker, M. S. (1994). Free radicals and connective tissue damage. *In* "Free Radicals Damage and Its Control" (C. A. Rice-Evans and R. H. Burdon, eds.). Elsevier, Amsterdam.
Barbior, B. M., and Woodman, R. C. (1990). Chronic granulomatous disease. *Semin. Hematol.* **27,** 247–259.
Bernstein, E. F., Brown, D. B., Urbach, F., Forbes, D., Del Monaco, M., Wu, M., Katchman, S. D., and Uitto, J. (1995). Ultraviolet radiation activates the human elastin promoter in transgenic mice: A novel in vivo and in vitro model of cutaneous photoaging. *J. Invest Dermatol.* **105,** 269–273.
Biemond, P., Swakk, A. J. G., Eijk, H. G., and Koster, J. F. (1988). Superoxide dependent iron release from ferritin in inflammatory diseases. *Free Rad. Biol. Med.* **4,** 185–198.
Biesalski, H. K., Hemmes, C., Hopfenmüller, W., Schmid, C. and Gollnick, H. P. M. (1996). Effects of controlled exposure of sunlight on plasma and skin levels of β-carotene. *Free Rad. Res.,* in press.
Bissett, D. L., Chatterjee, R., and Hannon, D. P. (1990a). Photoprotective effect of superoxide scavenging antioxidants against ultraviolet radiation-induced chronic skin damage in the hairless mouse. *Photodermatol. Photoimmunol. Photomed.* **7,** 56–62.

Bissett, D. L., Chatterjee, R., and Hannon, D. P. (1991). Chronic ultraviolet radiation-induced increase in skin iron and the photoprotective effect of topically applied iron chelators. *Photochem. Photobiol.* **54**, 215–223.

Bissett, D. L., Hannon, D. P., and Orr, T. V. (1987). An animal model of solar-aged skin: Histological, physical and visible changes in UV-irradiated hairless mouse skin. *Photochem. Photobiol.* **46**, 367–378.

Bissett, D. L., Majeti, S., Fu, J.-J. L., McBride, J. F., and Wyder, W. E. (1990b). Photoprotective effect of topically applied hexadienes against ultraviolet radiation-induced chronic skin damage in the hairless mouse. *Photodermatol. Photoimmunol. Photomed.* **7**, 63–67.

Black, H. (1987). Potential involvement of free radical reactions in ultraviolet light-mediated cutaneous damage. *Photochem. Photobiol.* **46**, 213–221.

Boyer, R. F., and Cleary, C. J. (1987). Superoxide ion as a primary reductant in ascorbate-mediated ferritin iron release. *Free Rad. Biol. Med.* **3**, 389–395.

Bravermann, I. M., and Fonferko, E. (1982). Studies in cutaneous aging. I. The elastic fiber network. *J. Invest. Dermatol.* **78**, 434–443.

Brenneisen, P., Briviba, K., Wenk, J., Wlaschek, M., and Scharffetter-Kochanek, K. (1996). Hydrogen peroxide (H_2O_2) mediates the transcriptional activity of collagenase/MMP-1, tissue inhibitor of MMP-1 (TIMP-1) and interleukin-1 in human dermal fibroblasts in vitro after applying the herbicide paraquat. *Free Rad. Biol. Med.* in press.

Brenneisen, P., Oh, J., Wlaschek, M., Wenk, J., Briviba, K., Hommel, C., Herrmann, G., Sies, H., and Scharffetter-Kochanek, K. (1996). UVB-wavelength dependence for the regulation of two major matrix-metalloproteinases and their inhibitor TIMP-1 in dermal human fibroblasts. *Photochem. Photobiol.* in press.

Bruls, W. A. G., Slaper, H., van der Leun, J. C., and Berrens, L. (1984). Transmission of human epidermis and stratum corneum as a function of thickness in the ultraviolett and visible wavelengths. *Photochem. Photobiol.* **40**, 485–494.

Bryce, G. F., Bogdan, N. J., and Brown, C. C. (1988). Retinoic acids promote the repair of dermal damage and the effacement of wrinkles in the UVB-irradiated hairless mouse. *J. Invest. Dermatol.* **91**, 175–180.

Bryce, G. F. (1993). The effects of UV radiation on skin connective tissue. In "Oxidative Stress in Desmatology" (J. Fuchs and L. Packer, eds.), pp. 105–125. Marcel Dekker, Inc., New York.

Chen, S., Kiss, I., and Tramposch, K. M. (1992). Effects of all-trans retinoic acid on UVB-irradiated and non-irradiated hairless mouse skin. *J. Invest. Dermatol.* **2**, 248–254.

Chen, V. L., Fleischmajer, R., Schwartz, E., Palia, M., and Timpl, R. (1986). Immunochemistry of elastotic material in sun-damaged skin. *J. Invest. Dermatol.* **87**, 334–337.

Cordero, A., Jr. (1983). La vitamina A acida en la piel senil. *Actual. Terap. Dermatol.* **6**, 49–54.

Dalle-Carbonare, M., and Pathak, M. A. (1992). Skin photosensitizing agents and the role of oxygen species in photoaging. *Photochem. Photobiol.* B **14**, 105–124.

Darr, D., and Fridovich, I. (1994). Free radicals in cutaneous biology. *J. Invest. Dermatol.* **102**, 671–675.

Ditre, C. M., Griffin, T. D., Murphy, G. F., *et al.* (1996). The effects of alpha hydroxy acids (AHAs) on photoaged skin: A pilot clinical histological and ultrastructural study. Submitted for publication.

Duell, E. A., Aström, A., Griffiths, C. E. M., Chambon, P., and Voorhees, J. J. (1992). Human skin levels of retinoic acid and cytochrome P-450-derived 4-hydroxyretinoic acid after topical application of retinoic acid in vivo compared to concentrations required to stimulate retinoic acid receptor-mediated transcription in vitro. *J. Clin. Invest.* **90**, 1269–1274.

Edwards, D. R., Murphy, G., Reynolds, J. J., *et al.* (1987). Transforming growth factor β modulates the expression of collagenase and metalloproteinase inhibitor. *EMBO J.* **6**, 1899–1904.

Ellis, C. E., Weiss, J. S., Hamilton, T. A., Headington, J. T., Zelickson, A. S., and Voorhees, J. J. (1990). Substained improvement with prolonged topical tretinoin (retinoic acid) for photoaged skin. *J. Am. Acad. Dermatol.* **23**, 629–637.

Epstein, J. H. (1989). Photomedicine. *In* "The Science of Photobiology" (K. C. Smith, ed.), pp. 155–192. Plenum Press, New York.

Epstein, J. H., Tuffanelli, D. L., and Epstein W. L. (1973). Cutaneous changes in the Porphyrias. *Arch. Dermatol.* 107, 689–698.

Farr, P. M., Marks, J. M., Diffey, B. L., and Ince P. (1988). Skin fragility and blistering due to use of sunbeds. *BMJ* 296, 1708–1709.   •

Foote, C. S. (1991). Definition of Type I and Type II photosensitized oxidation. *Photochem. Photobiol.* 54, 659.

Frickel, F. (1984). Chemistry and physical properties of retinoids. *In* "The Retinoids" (M. B. Sporn and A. B. Roberts, eds.), Vol. 1, pp. 8–45. Academic Press, San Diego.

Fridovich, I. (1989). Superoxide dismutase: An adaption to a paramagnetic gas. *J. Biol. Chem.* 264, 7761–7764.

Friedmann, P. S., Coburn, P., Dahl M. G. C., Diffey, B. L., Ross, J., Ford, G. P., Parker, S. C., and Bird, P. (1987). PUVA-induced blisters, complement deposition and damage to dermoepidermal junction. *Arch. Dermatol.* 123, 1471–1477.

Fisher, G. J., Tavakkol, A., Griffiths, C. E. M., Elder, J. T., Zhang, Q.-Y., Finkel, L., Danielpour, D., Glick, A. B., Higley, H., Ellingsworth, L., and Voorhees, J. J. (1992). Differential modulation of transforming growth factor-β1 expression and mucin deposition by retionic acid and sodium lauryl sulfate in human skin. *J. Invest. Dermatol.* 98, 102–108.

Gallagher, R. P., Elwood, J. M., and Yang, C. P. (1989). Is chronic sunlight exposure important in accounting for increase in melanoma incidence? *Int. J. Cancer* 44, 813–815.

Griffiths, C. E. M, Russman, A. N., Majumudar, G., Singer, R. S., Hamilton, T. A., and Voorhees, J. J. (1993). Restoration of collagen formation in photodamaged human skin by tretinoin (retinoic acid). *N. Engl. J. Med.* 329, 530–535.

Griffiths, C. E. M., and Voorhees, J. J. (1995). Effects of retionoids on photodamage skin. *In* "Photodamage" (B. A. Gilchrest, ed.). Blackwell Science.

Hall, E. D. (1994). Free radicals in central nervous system injury. *In* "Free Radical Damage and Its Control" C. A. Rice-Evans and R. H. Burdon, eds.). Elsevier, Amsterdam.

Halliwell, B., and Aruoma, O. I. (1991). DNA damage by oxygen-derived species: Its mechanism and measurement in mammalian systems. *FEBS Lett.* 281, 9–19.

Halliwell, B., and Gutteridge, J. M. C. (1990). Role of free radicals and catalytic metal ions in human disease: An overview. *Methods Enzymol.* 186, 1–85.

Heidbreder, G. (1980). Lokalisierte Blasen bei Fotochemotherapie-Eine akrobulloese Photodermatose. *Z. Hautkr.* 55, 84–99.

Henriksen, T., Dahlback, A., Larsen, S. H. H., and Moan, J. (1990). Ultraviolet-radiation and skin cancer-effect of an ozone layer depletion. *Photochem. Photobiol.* 51, 579–582.

Herrmann, G., Wlaschek, M., Bolsen, K., Prenzel, K., Goerz, G., and Scharffetter-Kochanek, K. (1996). Pathogenic implication of matrix-metalloproteinases (MMPs) and their counteracting inhibitor TIMP-1 in the cutaneous photodamage of human porphyria cutanea tarda (PCT). *J. Invest. Dermatol.*, in press.

Herrmann, G., Wlaschek, M., Lange, T. S., Prenzel, K., Goerz, G., and Scharffetter-Kochanek, K. (1993). UVA irradiation stimulates the synthesis of various matrix-metalloproteinases (MMPs) in cultured human fibroblasts. *Exp. Dermatol.* 2, 92–97.

Hiramatsu, M., and Packer, L. (1990). Antioxidant activity of retinoids. *Methods Enzymol.* 190, 273–280.

Johnston, K. J., Oikarinen, A. I., Lowe, N. J., Clark, J. G., and Uitto, J. (1984). Ultraviolet irradiation-induced connective tissue changes in the skin of hairless mice. *J. Invest. Dermatol.* 82, 587–590.

Jurkiewicz, B. A., Bissett, D. L., and Buettner, G. R. (1995). Effect of topically applied tocopherol on ultraviolet radiation-mediated free radical damage in skin. *J. Invest. Dermatol.* 104, 484–488.

Jurkiewicz, B. A., and Buettner, G. R. (1994). Ultraviolet-light-induced free radical formation in skin: An electron paramagnetic resonance study. *Photochem. Photobiol.* 59, 1–4.

Kim, H.-J., Bogdan, N. J., D'Agostaro, L. J., Gold, L. I., and Bryce, G. F. (1992). Effect of topical retinoic acids on the levels of collagen mRNA during the repair of UVB-induced dermal damage in the hairless mouse and the possible role of TGF-β as a mediator. *J. Invest. Dermatol.* **98**, 359–363.

Kligman, A. M. (1969). Early destructive effect of sunlight on human skin. *J. Am. Med. Assoc.* **210**, 2377–2380.

Kligman, L. H. (1986). Effect of all-trans-retinoic acid on the dermis of hairless mice. *J. Am. Acad. Dermatol.* **15**, 779–784.

Kligman, L. H. (1992). UVA induced biochemical changes in hairless mouse skin collagen: A contrast to UVB effects. *In* "Biological Responses to Ultraviolet A Radiation" (F. Urbach, ed.), pp. 209–216. Valdemar, Overland Park.

Kligman, L. H., Akin, F. J., and Kligman, A. M. (1982). Prevention of ultraviolet damage to the dermis of hairless mice by sunscreens. *J. Invest. Dermatol.* **78**, 181–189.

Kligman, L. H., Akin, F. J., and Kligmann, A. M. (1983). Sunscreens promote the repair of ultraviolet radiation-induced dermal damage. *J. Invest. Dermatol.* **81**, 98–102.

Kligman, L. H., Akin, F. J., and Kligman, A. M. (1985). The contributions of UVA and UVB to connective tissue damage in hairless mice. *J. Invest. Dermatol.* **84**, 272–276.

Kligman, A. M., Grove, G. L., Hirose, R., and Leyden, J. J. (1986). Topical tretionin for photoaged skin. *J. Am. Acad. Dermatol.* **15**, 836–859.

Kligman, L. H., and Mathews-Roth, M. M. (1990). Dietary β-carotene and 13-cis-retinoic acid are not effective in preventing some features of UVB-induced dermal damage in hairless mice. *Photochem. Photobiol.* **51**, 733–735.

Koivukangas, V., Kalliloinen, M., Autio-Harmainen, H., and Oikarinen, A. (1994). UV irradiation induces the expression of gelatinases in human skin in vivo. *Acta Derm. Venereol.* **74**, 279–282.

Krämer, M., Sachsenmaier, C., Herrlich, P., and Rahmsdorf, H. J. (1993). UV irradiation-induced interleukin-1 and basic fibroblast growth factor synthesis and release mediate part of the UV response. *J. Biol. Chem.* **268**, 6734–6741.

Lever, W. F., and Schaumburg-Lever, G. (1990). "Histopathology of the Skin," 7th Ed., pp. 298–300. Lipppincott, Philadephia.

Levin, A. A., Sturzenbecker, I. J., Kazmer, S., Bosakowski, T., Huselton, C., Allenby, G., Speck, J., Kratzeisen, Gl., Rosenberger, M., Lovey, A., and Grippo, J. F. (1992). 9- Cis retinoic acid stereoisomer binds and activates the nuclear receptor RXRα. *Nature* **355**, 359–361.

Liocher, S. L., and Fridovich, I. (1994). The role of $O_2{}^-$ in the production of HO$\cdot$: In vitro and in vivo. *Free Rad. Biol. Med.* **16**, 29–33.

Liotta, L. A., Steeg, P. S., and Stetler-Stevenson. W. G. (1991). Cancer metastasis and angiogenesis: An imbalance of positive and negative regulation. *Cell* **64**, 327–336.

Marks, M. W., Morykwas, M. J., and Wheathly, M. J. (1990). Fibroblast-mediated contraction in actinically exposed and actinically protected aging skin. *Plast. Reconstr. Surg.* **86**, 255–259.

Mathews-Roth, M. M., Pathak, M. A, Fitzpatrick, T. B., Harber, L. C., and Kass, E. H. (1970). Beta-Carotene as a photoprotective agent in erythropoetic protoporphyria. *N. Engl. J. Med.* **282**, 1231–1234.

Matrisian, L. M. (1992). The matrix degrading metalloproteinases. *BioEssays* **14**, 455–463.

Meier, B., Radeke, H. H., Selle, S., Younes, M., Sies, H., Resch, K., and Habermehl, G. G. (1989). Human fibroblasts release reactive oxygen species in response to interleukin-1 or tumour necrosis factor-α. *Biochem. J.* **263**, 539–545.

Mitchell, R. E. (1967). Chronic solar elastosis: A light and electron microscopic study of the dermis. *J. Invest. Dermatol.* **48**, 203–220.

Monboisse, J. C., and Borel, J. P. (1992). Oxidative damage to collagen. *In* "Free Radicals and Aging" (I. Emerit and B. Chance, eds.), pp. 323–327. Birkhäuser Verlag, Basel.

Oikarinen, A. (1990). The aging of skin: Chronoaging versus photoaging. *Photodermatol. Photoimmunol. Photomed.* **7**, 3–4.

Oikarinen, A., and Kallioinen, M. (1989). A biochemical and immunohistochemical study of collagen in sun-exposed and protected skin. *Photodermatology* 6, 24–31.

Oikarinen, A., Karvonen, J., Uitto, J., and Hannuksela, M. (1985). Connective tissue alterations in skin exposed to natural and therapeutic UV-radiation. *Photodermatology* 2, 15–26.

Petersen, M. J., Hansen, C., and Craig, S. (1992). Ultraviolet A irradiation stimulates collagenase production in cultured human fibroblasts. *J. Invest. Dermatol.* 99, 440–440.

Plastow, S. R., Lovell, C. R., and Young, A. R. (1987). UVB-induced collagen changes in the skin of the hairless albino mouse. *J. Invest. Dermatol.* 88, 145–148.

van der Rest, M., and Garrone, R. (1991). Collagen family of proteins. *FASEB J.* 5, 2814–2823.

Rosenstein, B. S., Ducore, J. M., and Cummings, S. W. (1983). The mechanism of bilirubin photosensitzed DNA strand breakage in human cells exposed to phototherapy light. *Mutat. Res.* 112, 397–406.

Ross, A. C. (1992). Cellular metabolism and activation of retinoids: Roles of cellular retinoid-binding proteins. *FASEB J.* 7, 317–327.

Sams, W. M., and Smith, J. G. (1961). The histochemistry of chronically sun-damage skin. *J. Invest. Dermatol.* 37, 447–452.

Schaefer, T., Scharffetter, K., Bolsen, K., Jugert, F., Lehmann, P., Merk, H. F., and Goerz, G. (1991). Effect of UVASUN on porphyrin metabolism and P-450 isoenzymes in hexachlorbenzene-induced porphyric rats. *In* "Metabolic Disorders and Nutrition Correlated with Skin" (B. J., Vermeer, K. D., Wuepper, W. A. Van Vloeten, H. Baart de la Faille, and J. G. van der Schroeff, eds.), pp. 106–115.

Scharffetter, K., Wlaschek, M, Hogg, A., Bolsen, K., Schothorst, A., Goerz, G. Krieg, T., and Plewig, G. (1991). UVA irradiation induces collagenase in human dermal fibroblasts in vitro and in vivo. *Arch. Dermatol. Res.* 283, 506–511.

Scharffetter-Kochanek, K., Goldermann, R., Lehmann, P., Hölzle, E. and Goerz, G. (1995a). PUVA therapy in disabling pansclerotic morphoea of children. *Br. J. Dermatol.* 132, 827–839.

Scharffetter-Kochanek, K., Wlaschek, M., Bolsen, K., Herrmann, G., Lehmann, P., Goerz, G., Mauch, C., and Plewig, G. (1992). Mechanisms of cutaneous photoaging. *In* "The Environmental Threat of the Skin," (G. Plewig and R. Marks, eds.), pp. 72–82. *Martin Dunitz Publishers,* London.

Scharffetter-Kochanek, K., Wlaschek, M., Brenneisen, P., Wenk, J., Herrmann, G., Hommel, C., and Krieg, T. (1995b). Pathopysiologie der degenerativen Bindegewebserkrankungen am Beispiel der Photoalterung der Haut. *In* "Dermatologie" (B. Tebbe, S. Goerdt, and C. E. Orfanos, eds.). Thieme.

Scharffetter-Kochanek, K., Wlaschek, M., Briviba, K., and Sies, H. (1993). Singlet oxygen induces collagenase expression in human skin fibroblasts. *FEBS Lett.* 331, 304–306.

Schorpp, M., Mallick, U., Rahmsdorf, H. J., and Herrlich, P. (1984). UV-induced extracellular factor from human fibroblasts communicates the UV response to nonirradiated cells. *Cell* 37, 861–868.

Schwartz, E., Criuckshank, F. A., Mezick, J. A., and Kligman, L. H. (1991). Topical all-trans retinoic acid stimulates collagen synthesis in vivo. *J. Invest. Dermatol.* 96, 975–978.

Schwartz, E., Cruickshank, F. A., Perlish, J. S., and Fleischmajer, R. (1989). Alterations in dermal collagen in ultraviolet irradiated hairless mice. *J. Invest. Dermatol.* 93, 142–146.

Scott, E. J., and Yu, R. J. (1995). Actions of alpha hydroxy acis on skin compartments. *J. Geriatr. Dermatol.* 3 (Suppl.A), 19A–25A.

Senior, R. M., Griffin, G. L., Fliszar, C. J., Shapiro, S. D., Goldberg, G. I., and Welgus, H. G. (1991). Human 92- and 72-kilodalton type IV collagenases are elastases. *J. Biol. Chem.* 266, 7870–7875.

Shindo, Y., Witt, E., and Packer, L. (1993). Antioxidant defense mechanisms in murine epidermis and dermis and their responses to ultraviolet light. *J. Invest. Dermatol.* 100, 260–265.

Sies, H. (1986). Biochemistry of oxidative stress. *Angewandte Chem.* 25, 1058–1071.

Sies, H. (1991). "Oxidative stress: Oxidants and antioxidants." Academic Press, New York.

Smith, J. G., Davidson, E. A., Sams, W. M., and Clark, R. D. (1962). Alterations in human dermal connective tissue with age and chronic sun damage. *J. Invest. Dermatol.* **39**, 347–350.

Sporn, M. B., and Roberts, A. B. (1986). Peptide growth factors and inflammation, tissue repair and cancer. *J. Clin. Invest.* **78**, 329–332.

Sporn, M. B., and Roberts, A. B. (1989). Transforming growth factor beta. *JAMA* **262**, 938–941.

Tabin, C. J. (1991). Retinoids, homeoboxes and growth factors toward molocular models for limb development. *Cell* **66**, 199–217.

Taira, J., Mimura, K., Yoneya, T., Hagi, A., Murakami, A., and Makino, K. (1992). Hydroxyl radical formation by UV-irradiated epidermal cells. *J. Biochem.* **111**, 693–695.

Thomsen, K. (1989). Solarium pseudoporphyria. *Photodermatology* **6**, 61–62.

Trautinger, F., Trenz, A., Raff, M., and Kokoschka, E. M. (1989). Influence of UV radiation on dermal collagen content in hairless mice. *Arch. Dermatol. Res.* **281**, 144.

Uitto, J (1986). Connective tissue biochemistry of the aging dermis: Age-related alterations in collagen and elastin. *Dermatol. Clin.* **4**, 433–446.

Urbach, F. (1989). Potential effects of altered solar ultraviolett radiation on human skin cancer. *Photochem. Photobiol.* **50**, 507–514.

van der Rest, M., and Garrone, R. (1991). Collagen family of proteins. *FASEB J.* **5**, 2814–2823.

von Laar, J., Stahl, W., Bolsen, K., Goerz, G., and Sies, H. (1995). β-Carotene serum levels in patients with erythropoetic protoporphyria on treatment with the synthetic all-trans isomer or a natural isomeric mixture of β-carotene. *Photochem. Photobiol.* **99**, 1–999.

Woessner, J. F., Jr. (1991). Matrix metalloproteinases and their inhibitor in connective tissue remodelling. *FASEB J.* **5**, 2145–2154.

Weiss, J. S., Ellis, C. N., Headington, J. T., Tincoff, Hamilton, T. A., and Voorhess, J. J. (1988). Topical tretinoin improves photoaged skin. *JAMA* **259**, 527–532.

Witt, E. H., Motchnik, P., and Packer, L. (1993). Evidence for UV light as an oxidative stressor in skin. "Oxidative Stress in Dermatology" (J. Fuchs and L. Packer, eds.), pp. 29–47. Marcel Dekker, Inc., New York.

Wlaschek, M., Bolsen, K., Hermann, G., Schwarz, A., Wilmroth, F., Heinrich, P. C., Goerz, G., and Scharffetter-Kochanek, K. (1993). UVA-induced autocrine stimulation of fibroblast-derived-collagenase by IL-6: A possible mechanism in dermal photodamage? *J. Invest. Dermatol.* **101**, 164–168.

Wlaschek, M., Briviba, K., Stricklin, G. P., Sies, H., and Scharffetter-Kochanek, K. (1995). Singlet oxygen may mediate the ultraviolet A induced synthesis of interstitial collagenase. *J. Invest. Dermatol.* **104**, 194–198.

Wlaschek, M., Heinen, G., Poswig, A., Schwarz, A., Krieg, T., and Scharffetter-Kochanek, K. (1994). UVA-induced stimulation of fibroblast-derived collagenase/MMP-1 by interelated loops of interleukin-1 and interleukin-6. *Photochem. Photobiol.* **59**, 550–556.

Woodley, D. T., Zelickson, A. S., Briggaman, R. A., Hamilton, T. A., Weiss, J. S., Ellis, C. N., and Voorhees, J. J. (1990). Treatment of photoaged skin with topical tretinoin increases epidermal-dermal anchoring fibrils. *JAMA* **263**, 3057–3059.

Xanthoudakis, S., and Curran, T. (1992). Identification and characterization of REF-1, a nuclear protein that facilitates AP-1 DNA binding activity. *EMBO J.* **11**, 653–665.

Xanthoudakis, S., Miao, G., Wang, F., Pan, Y.-C. E., and Curran, T. (1992). Redox activation of FOS-Jun DNA binding activity is mediated by a DNA repair enzyme. *EMBO J.* **11**, 3323–3335.

Yamauchi, M., Prisayanh, P., Haque, Z., and Woodley, D. T. (1991). Collagen cross-linking in sun-exposed and unexposed sites of aged human skin. *J. Invest. Dermatol.* **97**, 938–941.

Susan T. Mayne

Department of Epidemiology and Public Health
Yale University School of Medicine and Yale Cancer Center
New Haven, Connecticut 06520

Antioxidant Nutrients and Cancer Incidence and Mortality: An Epidemiologic Perspective

I. Introduction

The global burden of cancer is overwhelming, with more than 7.6 million newly diagnosed cases of cancer in 1985 (Muir, 1995). Lung cancer is the most common tumor worldwide (Muir, 1995). Patterns of cancer vary regionally, with a preponderance of cancers of the lung, breast, large bowel, uterine corpus, and prostate in developed countries, contrasted by a preponderance of cervical, mouth/pharynx, esophageal, and liver cancers in the developing world (Muir, 1995). These conspicuous differences in site distribution for the developed versus developing countries are expected to diminish gradually, following changes such as vaccination against hepatitis B virus and increased use of tobacco products in developing countries (Muir, 1995). Thus, a "globocancer" pattern is likely to emerge, with tobacco-related cancers being of particular concern.

Smoking exposes lipids, proteins, and nucleic acids in the lungs and upper aerodigestive tract to oxidants. Lung tissue, in particular, is exposed

657

to a relatively high concentration of oxygen by virtue of its role in oxygen-ation of the body. Thus, oxidative stress may be important in the etiology of tobacco-related cancers and lung cancer. However, all cells are exposed to oxidative stress, and thus oxidation, and free radicals, may be important in carcinogenesis at multiple tumor sites.

A. Role of Oxidation in Cancer

The hypothesis that free radicals may be involved in carcinogenesis is based primarily on observations that many carcinogens are free radicals, are converted to free radicals *in vivo,* stimulate the production of free radicals, or are products of biological free radical reactions. Also, many antioxidants have been shown to inhibit carcinogenesis in a variety of animal models. Evidence suggests that free radicals may play a role in the initiation of carcinogenesis. For example, free radical metabolites of many known carcin-ogens may initiate carcinogenesis by direct DNA damage (Demopoulos *et al.,* 1980). Also, by-products of lipid peroxidation, including malondialde-hyde and lipid peroxides, are mutagenic and may be carcinogenic (Kotin and Falk, 1963; Mukai and Goldstein, 1976). Lipid peroxidation compromises cellular membranes, which may affect the activity of membrane-bound en-zymes, many of which are involved in the activation and detoxification of carcinogens.

Radicals also may be important in tumor promotion and/or progression. A common property of tumor promoters may be their ability to produce oxygen radicals. Examples include unsaturated fats, hydrogen peroxide, phorbol esters, phenobarbital, and radiation (Ames, 1983). The mechanisms by which radicals and active oxygen may be involved in promotion are unknown. However, one likely explanation is effects resulting from mem-brane lipid peroxidation. Membranes play a role in cellular communication and provide a supportive structure for several critical regulatory enzymes. Evidence also implicates oxidation in the process of apoptosis (programmed cell death), with bcl-2 apparently functioning in an antioxidant capacity (Hockenbery *et al.,* 1993).

Humans have evolved with antioxidant defense systems to protect against free radicals (Machlin and Bendich, 1987). Nutrients such as vitamin E play a critical role in our antioxidant defense systems. Moreover, a number of anti-oxidant enzyme systems require nutrients for activity, e.g., selenium-dependent glutathione peroxidase. Despite these and other antioxidants, some free radicals still escape to cause damage. Because our antioxidant defense systems are not completely efficient, it has been proposed that increas-ing dietary antioxidants may be important in diminishing the cumulative ef-fects of oxidative damage over the long human life span (Halliwell, 1994).

This chapter considers the nutrients β-carotene, vitamins C and E, and selenium. While part of the antioxidant defense system, these nutrients may not always act as antioxidants and may have actions unrelated to antioxidant activity (Olson, 1996). However, for the sake of simplicity, the term "antioxidant nutrient" is used to refer to these nutrients.

B. Overview of Epidemiologic Studies of Antioxidant Nutrients and Cancer

Epidemiologic studies of antioxidant nutrients and cancer can be divided broadly into two categories: observational studies and intervention trials. Observational studies examine the association between antioxidant nutrient intake and cancer incidence, or between blood or tissue levels of antioxidant nutrients and cancer incidence. While these studies have contributed enormously to the literature regarding antioxidants and cancer, the interpretation of these studies is difficult for many reasons. First, fruits and vegetables, primary dietary sources of β-carotene and vitamin C, are complex mixtures containing other vitamins, minerals, fibers, and numerous phytochemicals. Thus, people who consume more β-carotene and vitamin C also consume more of these other substances, many of which may have cancer preventive properties either related or unrelated to antioxidant activity (Dragsted *et al.*, 1993). Also, individuals who consume more antioxidant-rich diets, or who choose to self-supplement with antioxidant nutrients, may also consume less dietary fat, or may be more health-conscious in other ways as compared to individuals who consume relatively few antioxidant nutrients. Observational epidemiologic studies of antioxidant nutrients and human health must therefore be interpreted cautiously, as it is entirely possible that effects observed may result from dietary or other factors correlated with antioxidant nutrient intake rather than from antioxidant nutrients themselves.

Due to the limitations inherent in observational studies of antioxidant nutrients and cancer, supporting data from intervention studies are important for causal inference. Intervention trials of antioxidant nutrients and health consist primarily of trials using supplemental β-carotene, vitamin C, vitamin E, and selenium, either alone or in combinations. Randomized, placebo-controlled intervention trials, unlike the observational studies, generally are not subject to bias and/or are confounding. However, intervention trials are extraordinarily expensive, difficult to conduct, and are of necessity relatively short in duration (typically 5–10 years). Consequently, the interpretation of intervention trials of antioxidant nutrients and cancer is also complex, as detailed further later.

II. Antioxidant Nutrients and Cancer Prevention ─────────────

A. Observational Studies

β-Carotene, vitamin C, and vitamin E have been found to be inversely associated with risks of various cancers in numerous observational studies, reviewed elsewhere (Block, 1991; Byers and Perry, 1992; Das, 1994; Mayne, 1996). Much of the interest in antioxidant nutrients and cancer prevention has concerned the tobacco-related cancers, such as lung cancer. A recent review of dietary factors and lung cancer shows that fruit and vegetable intake, the primary dietary sources of β-carotene, vitamin C, and other antioxidants such as glutathione, is inversely associated with lung cancer risk in men and women from various countries; in smokers, ex-smokers, and never smokers; and for all histologic types (Ziegler et al., 1996). Whether or not protective effects are due to carotenoids, vitamin C, or other micronutrients in fruits and vegetables is unclear. In general, studies suggest that there is roughly a 50–100% increase in lung cancer risk for persons below the (approximate) 25th percentile of carotene intake as compared with those above the (approximate) 75th percentile (Byers, 1994). Low levels of vitamin C intake have been associated with increased lung cancer risk in some but not all studies (Block, 1991; Byers, 1994; Ziegler et al., 1996). Intake of vitamin E, unlike vitamin C and β-carotene, is difficult to measure by dietary questionnaire because vitamin E is present in a wide variety of foods and the concentration of vitamin E in a given foodstuff is variable. Thus, usual vitamin E intake can best be measured either by assessing levels in blood or by assessing prior/current use of vitamin E supplements. Epidemiologic studies using these approaches are somewhat inconsistent, even within a given tumor site, as reviewed elsewhere (Das, 1994). For example, of six prospective studies that measured α-tocopherol in prediagnostic blood samples and accrued at least 25 lung cancers, only three found inverse associations (Ziegler et al., 1996).

Selenium intake, like that of vitamin E, is difficult to measure by questionnaire as the selenium content of an individual food is determined primarily by soil selenium levels, which vary regionally. Consequently, selenium status is best measured in human tissues (blood, nail clippings). Lung cancer studies using this approach have had mixed results, with some suggesting protective effects of higher selenium levels, some suggesting no effect, and some suggesting enhancement of carcinogenesis (Ziegler et al., 1996). Notably, the two largest studies of selenium and lung cancer were conducted in populations with a low to average consumption of selenium, with both reporting an advantage with higher selenium status (Ziegler et al., 1996).

Another antioxidant found in foods that has received limited epidemiologic attention is glutathione. Concentrations of glutathione vary in foods (Jones et al., 1992), and studies in humans indicate that orally administered glutathione (a tripeptide) is absorbed intact by buccal and intestinal mucosal cells. The first epidemiologic study of glutathione was a case control study of dietary glutathione and oral/pharyngeal cancer risk (Flagg et al., 1994). The study found that glutathione intake from fruit, and from vegetables commonly consumed raw, was inversely associated with risk. However, the effect could not be distinguished from the more general protective effect of fruits and raw vegetable consumption in the study population. This study underscores the fact that fruits and vegetables contain many substances with antioxidant activity and that it will be difficult, if not impossible, to make inferences about a given compound based on data gleaned from observational studies of dietary patterns.

There are considerably more observational studies of lung cancer and diet than of other tumor sites; however, protective associations with fruit and vegetable intake have been found with many other tumor sites. For example, increased consumption of fruits and vegetables was found to be significantly protective in 132 of 170 dietary studies reviewed (Block *et al.*, 1992). In addition to lung cancer, apparent protective effects of fruits and vegetables were most consistently noted for oral cavity, pharynx, and larynx (13 of 13 studies), esophagus (15 of 16 studies), stomach (17 of 19 studies), and cervix (7 of 8 studies). Tobacco is an important risk factor for all of these tumor sites, with the exception of stomach cancer. In contrast, the protective effects of fruits and vegetables are somewhat less consistent for colorectal cancers (20 of 27 studies; 3 indicated harmful effects), breast cancers (8 of 14 studies), and prostate cancers (4 of 14 studies; 2 indicated harmful effects) (Block *et al.*, 1992).

B. Intervention Trials

Observational studies thus suggest that dietary intake of the antioxidant nutrients β-carotene, vitamin C, vitamin E, and selenium may play a role in the prevention of certain cancers, particularly tobacco-related cancers, in humans. Moreover, observational studies have failed to find any consistent adverse effects associated with increased intake and/or levels of antioxidant nutrients. These observations, along with supporting data from *in vitro* and animal carcinogenesis studies, served as the impetus for a number of intervention trials designed to determine formally whether or not antioxidant supplementation was of benefit in terms of reducing malignancy or premalignancy. Results are now available from several of these trials.

1. Colon/Rectum

Some of the first trials of antioxidant nutrients examined the effect of supplementation on the recurrence of colorectal polyps, which are thought to be part of the carcinogenic continuum for colon and rectal cancers. DeCosse and colleagues (1975) reported on the effect of ascorbic acid (3 g/ day for 4–13 months) on rectal polyps in five patients with familial polyposis. Rectal polyps regressed completely in two, partially in two, and increased in one patient following vitamin C supplementation. In 1982, Bussey and co-workers completed a randomized, double-blinded trial of ascorbic acid in polyposis coli ($n = 36$ total patients). The investigators evaluated polyps every 3 months over an 18-month duration. Supplementation with pharmacological doses of vitamin C reduced polyps; however, the effect was statistically significant at only one time point (9 months, $p < 0.03$). Larger randomized trials have failed to demonstrate efficacy for either vitamin C or E in polyp recurrence, in both sporadic cases (McKeown-Eyssen *et al.*, 1988) or in patients with familial adenomatous polyposis (DeCosse *et al.*, 1989).

The effect of supplemental β-carotene on polyp recurrence has also been studied. Fifteen milligrams of β-carotene/day did not reduce the recurrence of colorectal adenomas (Kikendall *et al.*, 1991). In the largest polyp prevention trial that has been completed, β-carotene (25 mg/day), the combination of 1 g of vitamin C plus 400 mg vitamin E, or the three nutrients together did not reduce the incidence of new colorectal adenomas (Greenberg *et al.*, 1994). The relative risk for β-carotene was 1.01 (95% CI = 0.85–1.20); for vitamins C and E it was 1.08 (95% CI = 0.91–1.29).

While antioxidant nutrient supplementation has not been found to have efficacy in the prevention of colorectal adenomas, a more promising approach for chemoprevention in this setting may be dietary intervention. The National Cancer Institute is currently coordinating a trial to determine whether a low-fat, high-fiber, vegetable- and fruit-enriched diet will decrease the recurrence rate of large bowel adenomatous polyps (Freedman and Schatzkin, 1992). The trial is a multicenter, randomized, controlled trial involving approximately 2000 men and women with a 4-year follow-up period. This intervention would be expected to enhance antioxidant status not only with regard to antioxidant nutrients, but also with regard to many nonnutrient substances, such as some of the phytochemicals, which appear to have antioxidant activity (Dragsted *et al.*, 1993). Promising results from this trial, if obtained, would argue strongly for dietary approaches to cancer prevention, as opposed to antioxidant nutrient supplementation, at least for this tumor site.

2. Oral Cavity

In contrast to the disappointing results of the trials of antioxidant nutrients in colonic premalignancy are results of trials of antioxidant nutrients in oral premalignancy. Oral leukoplakia is a precancerous lesion, and oral micronuclei are an indicator of genotoxic damage to the oral epithelium. Studies in the 1980s demonstrated that supplemental β-carotene reduces the frequency of oral micronuclei significantly (Stich *et al.*, 1984, 1985, 1988). Seven trials have investigated the effects of supplemental β-carotene, alone or in combination with other agents, on regression of oral leukoplakia. Five nonrandomized studies reported response rates ranging from 44 to 71% (Garewal *et al.*, 1990; Malaker *et al.*, 1991; Toma *et al.*, 1992; Kaugars *et al.*, 1994; Garewal *et al.*, 1995). One randomized study of β-carotene alone and in combination with retinol (Stich *et al.*, 1988) and another study of the combination of β-carotene, retinol, and vitamin E (Zaridze *et al.*, 1993) reported treatment advantages with supplementation. Vitamin E as a single agent has also been reported to regress oral leukoplakia (Benner *et al.*, 1993).

The promising results obtained in trials with antioxidant nutrients and oral leukoplakia suggest that trials of antioxidant nutrients in head and neck cancer prevention should be done. Three such trials are ongoing (Mayne *et al.*, 1992; Bairati *et al.*, 1994; Toma *et al.*, 1994). All are designed to determine whether supplemental β-carotene, alone (Mayne *et al.*, 1992; Toma *et al.*, 1994) or in combination with vitamin E (Bairati *et al.*, 1994), reduces the incidence of second malignant cancers of the oral cavity, pharynx, larynx, esophagus, and lung in patients who have been "cured" of an early stage head or neck cancer. The results of these studies will be followed with interest.

3. Lung

The Tyler (Texas) Chemoprevention Trial randomized 755 asbestos workers to receive β-carotene (50 mg/day) and retinol (25,000 IU every other day) versus placebo to see if the nutrient combination could reduce the prevalence of atypical cells in sputum. After a mean intervention period of 58 months, there was no difference in the two groups in the prevalence of sputum atypia (McLarty *et al.*, 1995). In another randomized, placebo-controlled trial, 14 weeks of supplemental β-carotene (20 mg/day) reduced micronuclei counts significantly in sputum from heavy smokers (Van Poppel *et al.*, 1992). β-Carotene was not found, however, to reduce oxidative damage as measured by urinary excretion of 8-oxo-7,8-dihydro-2'-deoxyguanosine (Van Poppel *et al.*, 1995), suggesting that antioxidant activity was not responsible for the reduction in micronuclei. The relevance of micronuclei to lung carcinogenesis is unknown.

Two trials using lung cancer as a primary end point have now been completed. The first involved 29,133 males age 50–69 years old from Finland (Alpha-Tocopherol, β-Carotene Cancer Prevention Study Group, 1994) who were heavy cigarette smokers at entry (average one pack/day for 36 years). The study design was a two-by-two factorial with participants randomized to receive supplemental β-carotene (20 mg/day), α-tocopherol (50 mg/day), the combination, or placebo for 5 to 8 years. No reduction in lung cancer incidence was observed among men receiving α-tocopherol (RR = 0.98, 95% CI = 0.86–1.12). Unexpectedly, participants receiving β-carotene (alone or in combination with α-tocopherol) had a statistically significant 18% increase in lung cancer incidence (RR = 1.18; 95% CI = 1.03–1.36) and an 8% increase in total mortality (RR = 1.08, 95% CI = 1.01–1.16) relative to participants receiving placebo. Supplemental β-carotene did not appear to affect the incidence of other major cancers occurring in this population.

The finding of an increased incidence of lung cancer in β-carotene-supplemented smokers has now been replicated in another major trial. Investigators of CARET (Carotene and Retinol Efficacy Trial) held a press conference on January 18, 1996, to announce that the intervention component of CARET was being terminated nearly 2 years early. CARET is a multicenter lung cancer prevention trial of supplemental β-carotene (30 mg/day) plus retinol (25,000 IU/day) versus placebo in asbestos workers and smokers (Goodman et al., 1992a). According to Dr. Gilbert Omenn, the lead investigator of CARET, the intervention was stopped early because interim analyses of data indicated that should the trial have continued for its planned duration, it is highly unlikely that the intervention would have been found to be beneficial, given the results as of late 1995. Furthermore, the interim results indicated that the supplemented group was developing more lung cancer, not less, consistent with the results of the Finnish trial. Overall, lung cancer incidence was increased by 28% in the supplemented subjects (RR = 1.28; 95% CI = 1.04–1.57) and total mortality was also increased (RR = 1.17, 95% CI = 1.03–1.33). The increase in lung cancer following β-carotene supplementation was observed for current but not former smokers.

Major findings of one additional trial, the Physicians' Health Study of supplemental β-carotene versus placebo in 22,071 male U.S. physicians, were also released at the same press conference. Dr. Charles Hennekens, lead investigator of the Physicians' Health Study, announced that there was no significant effect—positive or negative—of 12 years of supplementation of β-carotene (50 mg every other day) on cancer or cardiovascular disease. The lack of an effect of long-term supplementation of β-carotene on lung cancer incidence in this cohort is noteworthy. Although the cohort contained only 11% current smokers at entry (approximately 2500 smokers), intervention and follow-up lasted for 12 years as compared to only 4 years in CARET (Hennekens et al., 1996; Omenn et al., 1996).

In contrast, more encouraging results for lung cancer prevention come from an esophageal and gastric cancer prevention trial in China (see later). The relative risk of death from lung cancer was 0.55 (95% CI = 0.26–1.14) among those receiving the combination of β-carotene, α-tocopherol, and selenium (Blot et al., 1994). However, this result is not statistically significant, based on only 31 total lung cancer deaths.

A clear mechanism to explain the apparent enhancement of lung carcinogenesis by supplemental β-carotene in smokers has yet to emerge. Some mechanisms that have been proposed to date include inhibition of the intestinal absorption of other nutrients by daily large doses of β-carotene and a possible pro-oxidant effect of β-carotene in the damaged lungs of long-term heavy smokers (Olson, 1996). While mechanistic studies continue, it is prudent to recommend that heavy smokers, particularly those from well-nourished populations, should avoid high-dose supplements of β-carotene for lung cancer chemoprevention.

4. Esophagus/Stomach

Certain regions of China have strikingly high incidence rates of esophageal and gastric cancers; moreover, blood levels of various micronutrients are consistently low in these populations. These observations have led to two esophageal and/or gastric cancer prevention trials in China involving antioxidant nutrients. The first trial was conducted in residents from the general population (Blot *et al.*, 1993). Nearly 30,000 men and women ages 40–69 took part in the study, which tested the efficacy of four different nutrient combinations at inhibiting the development of esophageal and gastric cancers. The nutrient combinations included retinol plus zinc, riboflavin plus niacin, ascorbic acid plus molybdenum, and the combination of β-carotene, selenium, and α-tocopherol. After a 5-year intervention period, those who were given the combination of β-carotene, vitamin E, and selenium had a 13% reduction in total cancer deaths (RR = 0.87; 95% CI = 0.75–1.00), a 4% reduction in esophageal cancer deaths (RR = 0.96; 95% CI = 0.78–1.18), and a 21% reduction in gastric cancer deaths (RR = 0.79; 95% CI = 0.64–0.99). None of the other nutrient combinations reduced gastric or esophageal cancer deaths significantly in this trial. The finding that vitamin supplements reduced cancer deaths in this population provides compelling data supporting the concept of cancer prevention with antioxidant nutrients; however, the applicability of these results for populations with adequate nutritional status and for other tumor sites may be limited. Also, it is unclear which nutrient(s) (β-carotene, vitamin E, or selenium) was responsible for the observed protection.

The other Linxian trial was done to determine whether a multivitamin/multimineral preparation plus β-carotene (15 mg) reduced esophageal and gastric cardia cancers in 3318 residents with esophageal dysplasia (Li *et al.*, 1993). Cumulative esophageal/gastric cardia death rates after the 6-year intervention period were 8% lower (RR = 0.92, 95% CI = 0.67–1.28), esophageal cancer mortality was 16% lower (RR = 0.84, 95% CI = 0.54–1.29), and total cancer mortality was 4% lower (RR = 0.96, 95% CI = 0.71–1.29) in the supplemented group. Stomach cancer mortality, however, was 18% higher (RR = 1.18, 95% CI = 0.76–1.85) in the supplemented group. None of the results were statistically significant.

In addition to the Chinese trials, a trial from Uzbekistan used a factorial design to study the combination of β-carotene, retinol, and α-tocopherol, with and without riboflavin, in subjects with chronic esophagitis (Zaridze *et al.*, 1993). The risk of progression or no change versus regression was nonsignificantly decreased by 34% in those receiving retinol, β-carotene, and α-tocopherol (OR = 0.66; 95% CI = 0.37–1.16) versus those who did not receive these agents.

5. Bladder

Lamm *et al.* (1994) randomized 65 patients with biopsy-confirmed transitional cell carcinoma of the bladder to a multivitamin (recommended dietary allowance or RDA levels) alone or supplemented with 40,000 IU retinol, 100 mg pyridoxine, 2000 mg ascorbic acid, 400 units of α-tocopherol, and 90 mg zinc. The 5-year estimate of tumor recurrence was

91% in the RDA arm versus 41% in the megadose arm ($p = 0.0014$). These results are promising in that the intervention was essentially nontoxic, with only one patient (3%) requiring dose reduction for mild stomach upset.

6. Uterine Cervix

The only randomized trial involving an antioxidant nutrient in cervical neoplasia that has been published was a trial from The Netherlands that randomized women with a histological diagnosis of cervical dysplasia to either 10 mg β-carotene/day or placebo. After 3 months of intervention, there was no detectable effect of supplemental β-carotene on the regression and progression of cervical dysplasia (DeVet et al., 1991). In contrast are promising findings from a nonrandomized phase II intervention trial of cervical dysplasia (Meyskens and Manetta, 1995) that reported a 70% response rate with 6 months of supplementation with 30 mg β-carotene/day. Results of other large placebo-controlled trials of β-carotene in cervical dysplasia should be forthcoming in the next few years.

7. Skin

Greenberg et al. (1990) demonstrated that supplementation with 50 mg β-carotene/day for 5 years did not reduce the occurrence of new skin cancers (relative rate = 1.05; 95% CI = 0.91–1.22) in 1805 persons with a previous nonmelanoma skin cancer.

III. Antioxidant Nutrients and Cancer Survival

Although a tremendous amount of research has been done in the area of antioxidant nutrients and cancer prevention, a dearth of information exists regarding the effects of antioxidant nutrients and other dietary factors on survival in cancer patients. The lack of research in this area is in many ways surprising, in that animal studies have demonstrated that certain antioxidants, such as β-carotene, can regress existing tumors (Schwartz and Shklar, 1988). Also, individuals diagnosed with cancer often seek nutritional "therapies" despite the fact that sound scientific research either supporting or refuting efficacy is not available.

Studies of antioxidant nutrients and cancer survival, like those of cancer incidence, can take the form of observational studies or trials. Randomized, placebo-controlled trials of antioxidant supplements in cancer patients have not been done. As for observational studies of dietary patterns, several studies have found significant associations between dietary patterns at or before diagnosis of breast cancer and survival. Ingram (1994) reported that β-carotene consumption, as estimated at the time of cancer diagnosis, was the dietary variable most significantly associated with improved survival in Australian women with breast cancer. Only 1 death occurred in the tertile of women with the highest consumption of β-carotene compared to 8 and 12 deaths in the intermediate and lowest consumption groups, respectively. This study did not reassess dietary patterns during the follow-up period, therefore it is unknown whether dietary change following diagnosis of breast cancer reduced the risk of death.

Using a slightly different experimental design, Jain *et al.* (1994) examined dietary data obtained from a cohort of disease-free women who were participating in the National Breast Screening Study in Canada. During the follow-up period, 678 women who had completed a diet history were diagnosed with breast cancer from 1982 to 1992, and 83 of these women died. There was a lower risk of dying of breast cancer in the highest quartiles of β-carotene intake (hazard ratio = 0.48; 95% CI = 0.23–0.99) and vitamin C intake (hazard ratio = 0.43; 95% CI = 0.21–0.86); both vitamins showed a significant dose–response relationship. Saturated fat intake increased the risk of dying of breast cancer (hazard ratio = 1.50; 95% CI = 1.08–2.08). These effects persisted even after adjusting for known prognostic factors. The authors concluded that "more attention should be paid to premorbid dietary habits in relation to breast cancer prognosis. Further studies, however, need to be done with full ascertainment of dietary changes prior to and subsequent to diagnosis."

Rohan *et al.* (1993) reported that upper levels of intake of β-carotene and vitamin C at diagnosis were associated with reduced risk of mortality in Australian women with breast cancer, although the association was not significant. Holm *et al.* (1993) assessed the association between dietary intake at the time of diagnosis and treatment failure (contralateral breast cancer, locoregional recurrence, or distant metastasis) in Swedish women with breast cancer. No association was found between dietary habits and treatment failure for women with estrogen receptor-poor cancers. In women having tumors rich in estrogen receptors, however, some dietary associations were evident. Women who did not fail ($n = 119$) consumed significantly less polyunsaturated fat ($p = 0.04$), 22% more carotenoids ($p > 0.05$), and 11% more vitamin C ($p > 0.05$) than women who failed ($n = 30$). Vitamin E intake, however, was nonsignificantly lower in women who remained disease free, perhaps related to a reduced consumption of vegetable oils. In a study of Danish women with breast cancer, Ewertz *et al.* (1991) reported that the "risk of dying was slightly decreased for frequent consumption of vegetables." This literature as a whole, combined with a growing literature implicating dietary fat in poorer survival from breast cancer (Cohen *et al.*, 1993), suggests that an optimal balance between plant and animal products could be an important determinant of breast cancer survival.

Studies of dietary patterns and survival from cancers other than breast are more limited. Goodman *et al.* (1992b) reported that increased vegetable consumption was associated with longer survival in female lung cancer patients, but not in males. The adjusted median survival times for women from the highest to the lowest quartiles of vegetable intake were 33, 21, 15, and 18 months, respectively (p for trend $= 0.03$). Higher fruit consumption was also associated with longer survival among women (p for trend $= 0.02$). Antioxidant and other nutrients may also improve tolerance to chemotherapy and radiation (Jaakkola *et al.*, 1992).

IV. Antioxidants and Cancer: Emerging Issues

As described earlier, observational studies suggest that antioxidant nutrients hold promise for the prevention of cancer incidence and possibly improvements in survival. Results from intervention trials of antioxidant nutrients and cancer prevention, however, are in striking contrast to the observational studies, with the exception of the Chinese trial (Blot *et al.*, 1993, 1994). An understanding of these apparently discordant observations is critically needed. The trial results should not be overinterpreted, as the results may apply only to populations with similar baseline nutritional status and risk factors, to a single tumor site, and to the specific intervention used (combinations versus single agents). Negative trials also should not necessarily override positive results from observational studies, as the antioxidant nutrients may have been given too late in the carcinogenic process, may have been given for an inadequate duration, may have been given to a population with near-optimal antioxidant status, or may not have been provided in a form that maximizes the effectiveness of a given antioxidant nutrient.

Given this, how can we proceed to better understand the potential of antioxidant nutrients for the prevention of human cancers? First, results from other ongoing and/or completed trials of vitamin E, β-carotene, and selenium are needed. One such trial is the Heart Protection Study, a British trial of vitamins E and C and β-carotene in at least 20,000 subjects, approximately 8000 of whom have been randomized as of early 1996.

Second, researchers should seek explanations for the apparently discordant findings of observational studies versus intervention trials. Pressing research issues include studies of interactions of carotenoids with themselves and with other phytochemicals and mechanistic studies of the actions of β-carotene in lung carcinogenesis and cardiovascular disease. Other priority areas include effects of foods versus supplements in animal models; the pharmacology of antioxidant nutrients, particularly β-carotene; and development and validation of measures of antioxidant intake/antioxidant capacity.

One experimental approach for assessing the cancer preventive effects of fruits and vegetables versus their component antioxidant nutrients is to perform animal carcinogenesis studies, where animals are fed fruit and/or vegetable supplemented diets, or fruit/vegetable extracts, as opposed to single antioxidant supplements. This approach was used by Root *et al.* (1994), who reported promising effects of an orange juice extract in a rodent aflatoxin hepatocarcinogenesis model. While food extracts may require more effort to prepare as compared to commercially available antioxidant supplements, the effort may be worthwhile in that this approach allows for the testing of antioxidants in naturally occurring combinations, and in their natural form. Most commercial supplements of β-carotene, for example, are predominantly in the all-*trans* configuration, although carotenoids in nature are in *cis* as well as *trans* configurations. Thus, comparative data of foods/food extracts versus supplements may lead to a greater understanding of antioxidants and cancer prevention.

Another issue relates to the possibility that the efficacy of a nutrient may be a function of baseline nutritional status and, ultimately, the plasma and tissue levels achieved by a given intervention. With regard to β-carotene, populations seem to differ in their plasma response to a given dose, perhaps related to differences in fat intake or differences in bioavailability of supplements from different manufacturers (reviewed in Mayne, 1996). These wide discrepancies in plasma response to a given oral dose of β-carotene emphasize our lack of understanding of the pharmacokinetics of β-carotene. However, this information may be quite relevant in that some investigators have argued that there are threshold levels for antioxidant nutrients and that no protective benefit may be achieved with supplementation if the plasma concentration already exceeds the risk threshold level (Gey *et al.*, 1993).

The third issue concerns interactions of supplements (e.g., pharmacological doses of nutrients) with other nutrients (antioxidant and other). A few reports in the literature have suggested that pharmacological doses of β-carotene may adversely affect vitamin E levels in blood (Xu *et al.*, 1992; Mobarhan *et al.*, 1994) or tissues (Mobarhan *et al.*, 1994). Larger studies (Goodman *et al.*, 1994; Nierenberg *et al.*, 1994) do not support an adverse effect of long-term supplemental β-carotene on vitamin E levels in blood, although data are lacking regarding effects on tissue stores of vitamin E.

Limited evidence suggests an adverse effect of supplemental β-carotene on plasma levels of other carotenoids, particularly lutein (Micozzi *et al.*, 1992; Kostic *et al.*, 1995). In contrast to these relatively small, short-term studies, Wahlqvist *et al.* (1994) reported that supplementation of 20 mg β-carotene/day for 24 months not only elevated plasma α-carotene levels (211% in men and 166% in women), but also lycopene levels (176% in men; lycopene levels were elevated nonsignificantly in women; total $n =$ 224). Lutein/zeaxanthin levels were unaffected by β-carotene supplementation in this study. Omenn *et al.* (1993) reported that supplemental β-carotene plus retinol increased the blood levels of α-carotene in participants in the Asbestos Workers Pilot Study for CARET ($n = 721$).

A final area in need of research attention, antioxidant indices, centers around the fact that different antioxidants operate in different regions of the cell (lipid versus aqueous environments), coupled with the fact that antioxidant interactions clearly occur. Vitamin C functions to maintain vitamin E in its reduced and functional form as an antioxidant, and glutathione functions to reduce oxidized vitamin C back to its functional reduced form (Jones *et al.,* 1995). Thus, it may be illogical to examine effects of isolated antioxidants. Some epidemiologic evidence also supports the concept of antioxidant interactions being important in disease prevention. For example, Salonen *et al.* (1985) conducted a prospective study of serum micronutrients and cancer. The relative risk for cancer (all sites) was 5.8 for individuals in the lowest tertile of serum Se (relative to the highest tertile); the relative risk was elevated to 11.4 for individuals in the lowest tertile of serum Se *and* serum vitamin E (relative to individuals in the highest tertile of serum Se and vitamin E). Many investigators have generated indices of antioxidant intake or of antioxidant levels in blood. While the concept of an antioxidant index is appealing, there is no uniform approach for creating an antioxidant index reflective of dietary intake or plasma levels of multiple antioxidant nutrients, nor is there a standard approach for measuring "total antioxidant activity" in biological specimens such as plasma. Assays of antioxidant activity are dependent on the source of oxidative stress, as well as the target oxidizable substrate (Halliwell and Gutteridge, 1995). This suggests a need for development and validation of meaningful indices of antioxidant intake and status.

V. Conclusions and Policy Implications _____

Supplementation with any single antioxidant nutrient, or limited combinations of antioxidant nutrients, for cancer prevention at this time is not generally recommended in that randomized trials of antioxidant nutrient supplements have not demonstrated chemopreventive efficacy, with the exception of a trial in a Chinese population with notable micronutrient deficiencies (Blot *et al.,* 1993). Moreover, our limited understanding of the pharmacology of and nutrient interactions with antioxidant supplements, such as β-carotene, suggests caution in the use of any antioxidant supplements, other than in research settings. Recommendations to increase consumption of fruits and vegetables, which contain several antioxidant nutrients, including vitamin C, carotenoids, and glutathione, are not controversial. Fruits and vegetables, however, are poor sources of selenium, which is concentrated in seafood, meats, and cereals, and α-tocopherol, which is concentrated in vegetable oils. Thus, should these nutrients be found to reduce cancer incidence in ongoing trials, then supplementation or fortification may be the only practical avenues for increasing the intake of these nutrients.

Acknowledgment ───────────────────────────────────

Support for the author's research on antioxidants and cancer prevention from the U.S. National Cancer Institute, Grants CA42101 and CA64567, is gratefully acknowledged.

References ──────────────────────────────────────

Alpha-Tocopherol, β-Carotene Cancer Prevention Study Group. (1994). The effect of vitamin E and beta carotene on the incidence of lung cancer and other cancers in male smokers. *N. Engl. J. Med.* 330, 1029–1035.

Ames, B. N. (1983). Dietary carcinogens and anti-carcinogens: Oxygen radicals and degenerative diseases. *Science* 221, 1256–1264.

Bairati, I., Roy, J., Gelinas, M., Brochet, F., Nabid, A., Tetu, B., Masse, B., and Meyer, F. (1994). Beta-carotene and alpha-tocopherol chemoprevention of second primary malignancies in head and neck cancer patients. *In* "Second International Conference on Antioxidant Vitamins and Beta-carotene in Disease Prevention," Berlin, October, Abstract P-54.

Benner, S. E., Winn, R. J., Lippman, S. M., Poland, J., Hansen, K. S., Luna, M. A., and Hong, W. K. (1993). Regression of oral leukoplakia with α-tocopherol: A community clinical oncology program chemoprevention study. *J. Natl. Cancer Inst.* 85, 44–47.

Block, G. (1991). Vitamin C and cancer prevention: The epidemiologic evidence. *Am. J. Clin. Nutr.* 53, 270S–282S.

Block, G., Patterson, B., and Subar, A. (1992). Fruit, vegetables, and cancer prevention: A review of the epidemiological evidence. *Nutr. Cancer* 18, 1–29.

Blot, W. J., Li, J.-Y., Taylor, P. R., Guo, W., Dawsey, S., Wang, G.-Q., Yang, C. S., Zheng, S.-F., Gail, M., Li, G.-Y., Yu, Y., Liu, B.-Q., Tangrea, J., Sun, Y.-H., Liu, F., Fraumeni, J. F., Jr., Zhang, Y.-H., and Li, B. (1993). Nutrition intervention trials in Linxian, China: Supplementation with specific vitamin/mineral combinations, cancer incidence, and disease-specific mortality in the general population. *J. Natl. Cancer Inst.* 85, 1483–1492.

Blot, W. J., Li, J.-Y., Taylor, P. R., and Li, B. (1994). Lung cancer and vitamin supplementation: Letter to the editor. *N. Engl. J. Med.* 331, 614.

Bussey, H. J. R., DeCosse, J. J., Deschner, E. E., Eyers, A. A., Lesser, M. L., Morson, B. C., Ritchie, S. M., Thomson, J. P. S., and Wadsworth, J. (1982). A randomized trial of ascorbic acid in polyposis coli. *Cancer* 50, 1434–1439.

Byers, T. (1994). Diet as a factor in the etiology and prevention of lung cancer. *In* "Epidemiology of Lung Cancer" (J. M. Samet, ed.), pp. 335–352. Dekker, New York.

Byers, T., and Perry, G. (1992). Dietary carotenes, vitamin C, and vitamin E as protective antioxidants in human cancers. *Annu. Rev. Nutr.* 12, 139–159.

Cohen, L. A., Rose, D. P., and Wynder, E. L. (1993). A rationale for dietary intervention in postmenopausal breast cancer patients: An update. *Nutr. Cancer* 19, 1–10.

Das, S. (1994). Vitamin E in the genesis and prevention of cancer: A review. *Acta Oncol.* 33, 615–619.

DeCosse, J. J., Adams, M. B., Kuzma, J. F., LoGerfo, P., and Condon, R. E. (1975). Effect of ascorbic acid on rectal polyps of patients with familial polyposis. *Surgery* 78, 608–612.

DeCosse, J. J., Miller, H. H., and Lesser, M. L. (1989). Effect of wheat fiber and vitamin C and E on rectal polyps in patients with familial adenomatous polyposis. *J. Natl. Cancer Inst.* 81, 1290–1297.

Demopoulos, H. B., Pietronigro, D. D., Flamm, E. S., and Seligman, M. L. (1980). The possible roles of free radical reactions in carcinogenesis. *J. Environ. Pathol. Toxicol.* 3, 273–303.

DeVet, H. C. W., Knipschild, P. G., Willebrand, D., Schouten, H. J. A., and Sturmans, F. (1991). The effect of beta-carotene on the regression and progression of cervical dysplasia: A clinical experiment. *J. Clin. Epidemiol.* **44**, 273–283.

Dragsted, L. O., Strube, M., and Larsen, J. C. (1993). Cancer protective factors in fruits and vegetables: Biochemical and biological background. *Pharmacol. Toxicol.* **72** (Suppl.), s116–s135.

Ewertz, M., Gillanders, S., Meyer, L., and Zedeler, K. (1991). Survival of breast cancer patients in relation to factors which affect the risk of developing breast cancer. *Int. J. Cancer* **49**, 526–530.

Flagg, E. W., Coates, R. J., Jones, D. P., Byers, T. E., Greenberg, R. S., Gridley, G., McLaughlin, J. K., Blot, W. J., Haber, M., Preston-Martin, S., Schoenberg, J. B., Austin, D. F., and Fraumeni, J. F., Jr. (1994). Dietary glutathione intake and the risk of oral and pharyngeal cancer. *Am. J. Epidemiol.* **139**, 453–465.

Freedman, L. S., and Schatzkin, A. (1992). Sample size for studying intermediate endpoints within intervention trials or observational studies. *Am. J. Epidemiol.* **136**, 1148–1159.

Garewal, H., Meyskens, F., Katz, R. V., Friedman, S., Morse, D. E, Alberts, D., and Girodias, K. (1995). Beta-carotene produces sustained remissions in oral leukoplakia: Results of a 1 year randomized, controlled trial. *Proc. Am. Soc. Clin. Oncol.* **14**, 496. [Abstract]

Garewal, H. S., Meyskens, F. L., Killen, D., Reeves, D., Kiersch, T. A., Elletson, H., Strosberg, A., King, D., and Steinbronn, K. (1990). Response of oral leukoplakia to beta-carotene. *J. Clin. Oncol.* **8**, 1715–1720.

Gey, K. F. (1993). Prospects for the prevention of free radical disease, regarding cancer and cardiovascular disease. *Br. Med. Bull.* **49**, 679–699.

Goodman, G. E., Metch, B. J., and Omenn, G. S. (1994). The effect of long-term beta-carotene and vitamin A administration on serum concentrations of alpha-tocopherol. *Cancer Epidemiol. Biomark. Prev.* **3**, 429–432.

Goodman, G. E., Omenn, G. S., and CARET Coinvestigators and Staff. (1992a). Carotene and retinol efficacy trial: Lung cancer chemoprevention trial in heavy cigarette smokers and asbestos-exposed workers. *In* "The Biology and Prevention of Aerodigestive Tract Cancers" (G. R. Newell, and W. K. Hong, eds.), pp. 137–140. Plenum Press, New York.

Goodman, M. T., Kolonel, L. N., Wilkens, L. R., Yoshizawa, C. N., Le Marchand, L., and Hankin, J. H. (1992b). Dietary factors in lung cancer prognosis. *Eur. J. Cancer* **28**, 495–501.

Greenberg, E. R., Baron, J. A., Stukel, T. A., Stevens, M. M., Mandel, J. S., Spencer, S. K., Elias, P. M., Lowe, N., Nierenberg, D. W., Bayrd, G., Vance, J. C., Freeman, D. H., Jr., Clendenning, W. E., Kwan, T., and the Skin Cancer Prevention Study Group. (1990). A clinical trial of beta carotene to prevent basal cell and squamous cell cancers of the skin. *N. Engl. J. Med.* **323**, 789–795.

Greenberg, E. R., Baron, J. A., Tosteson, T. D., Freeman, D. H., Jr., Beck, G. J., Bond, J. H., Colacchio, T. A., Coller, J. A., Frankl, H. D., Haile, R. W., Mandel, J. S., Nierenberg, D. W., Rothstein, R., Snover, D. C., Stevens, M. M., Summers, R. W., van Stolk, R. U., and the Polyp Prevention Study Group. (1994). A clinical trial of antioxidant vitamins to prevent colorectal adenoma. *N. Engl. J. Med.* **331**, 141–147.

Halliwell, B. (1994). Free radicals, antioxidants, and human disease: Curiosity, cause, or consequence? *Lancet* **344**, 721–724.

Halliwell, B., and Gutteridge, J. M. C. (1995). The definition and measurement of antioxidants in biological systems: Letter to the editor. *Free Rad. Biol. Med.* **18**, 125–126.

Hennekens, C. H., Buring, J. E., Manson, J. E., Stampfer, M., Rosner, B., Cook, N. R., Belanger, C., LaMotte, F., Gaziano, J. M., Ridker, P. M., Willett, W., and Peto, R. (1996). Lack of effect of long-term supplementation with beta carotene on the incidence of malignant neoplasms and cardiovascular disease. *N. Engl. J. Med.* **334**, 1145–1149.

Hockenbery, D. M., Oltvai, Z. N., Yin, X.-M., Milliman, C. L., and Korsmeyer, S. J. (1993). Bcl-2 functions in an antioxidant pathway to prevent apoptosis. *Cell* **75**, 241–251.

Holm, L.-E., Nordevang, E., Hjalmar, M.-L., Lidbrink, E., Callmer, E., and Nilsson, B. (1993). Treatment failure and dietary habits in women with breast cancer. *J. Natl. Cancer Inst.* **85**, 32–36.

Ingram, D. (1994). Diet and subsequent survival in women with breast cancer. *Br. J. Cancer* **69**, 592–595.

Jaakkola, K., Lahteenmaki, P., Laakso, J., Harju, E., Tykka, A., and Mahlberg, K. (1992). Treatment with antioxidant and other nutrients in combination with chemotherapy and irradiation in patients with small-cell lung cancer. *Anticancer Res.* **12**, 599–606.

Jain, M., Miller, A. B., and To, T. (1994). Premorbid diet and the prognosis of women with breast cancer. *J. Natl. Cancer Inst.* **86**, 1390–1397.

Jones, D. P., Coates, R. J., Flagg, E. W., Eley, J. W., Block, G., Greenberg, R. S., Gunter, E. W., and Jackson, B. (1992). Glutathione in foods listed in the National Cancer Institute's health habits and history food frequency questionnaire. *Nutr. Cancer* **17**, 57–75.

Jones, D. P., Kagan, V. E., Aust, S. D., Reed, D. J., and Omaye, S. T. (1995). Impact of nutrients on cellular lipid peroxidation and antioxidant defense system. *Fund. Appl. Toxicol.* **26**, 1–7.

Kaugars, G. E., Silverman, S., Jr., Lovas, J. G. L., Brandt, R. B., Riley, W. T., Dao, Q., Singh, V. N., and Gallo, J. (1994). A clinical trial of antioxidant supplements in the treatment of oral leukoplakia. *Oral Surg. Oral Med. Oral Pathol.* **78**, 462–468.

Kikendall, J. W., Mobarhan, S., Nelson, R., Burgess, M., and Bowen, P. E. (1991). Oral beta carotene does not reduce the recurrence of colorectal adenomas. *Am. J. Gastroenterol.* **36**, 1356. [Abstract]

Kostic, D., White, W. S., and Olson, J. A. (1995). Intestinal absorption, serum clearance and interactions between lutein and β-carotene when administered to human adults in separate or combined oral doses. *Am. J. Clin. Nutr.* **62**, 604–610.

Kotin, P., and Falk, H. L. (1963). Organic peroxides, hydrogen peroxide, epoxides, and neoplasia. *Radiat. Res. Suppl.* **3**, 193–211.

Lamm, D. L., Riggs, D. R., Shriver, J. S., vanGilder, P. F., Rach, J. F., and DeHaven, J. I. (1994). Megadose vitamins in bladder cancer: A double-blind clinical trial. *J. Urol.* **151**, 21–26.

Li, J.-Y., Taylor, P. R., Li, B., Dawsey, S., Wang, G.-Q., Ershow, A. G., Guo, W., Liu, S.-F., Yang, C. S., Shen, Q., Wang, W., Mark, S. D., Zou, X.-N., Greenwald, P., Wu, Y.-P., and Blot, W. J. (1993). Nutrition intervention trials in Linxian, China: Multiple vitamin/ mineral supplementation, cancer incidence, and disease-specific mortality among adults with esophageal dysplasia. *J. Natl. Cancer Inst.* **85**, 1492–1498.

Machlin, L. J., and Bendich, A. (1987). Free radical tissue damage: Protective role of antioxidant nutrients. *FASEB J.* **1**, 441–445.

Malaker, K., Anderson, B. A., Beecroft, W. A., and Hodson, D. I. (1991). Management of oral mucosal dysplasia with beta-carotene and retinoic acid: A pilot cross-over study. *Cancer Detect. Prev.* **15**, 335–340.

Mayne, S. T. (1996). Beta-carotene, carotenoids, and disease prevention in humans. *FASEB J.* **10**, 690–701.

Mayne, S. T., Zheng, T., Janerich, D. T., Goodwin, W. J., Jr., Fallon, B. G., Cooper, D. L., Friedman, C. D. (1992). A population-based trial of β-carotene chemoprevention of head and neck cancer. *In* "The Biology and Prevention of Aerodigestive Tract Cancers" (G. R. Newell and W. K. Hong, eds.), pp. 119–127. Plenum Press, New York.

McKeown-Eyssen, G., Holloway, C., Jazmaji, V., Bright-See, E., Dion, P., and Bruce W. R. (1988). A randomized trial of vitamins C and E in the prevention of recurrence of colorectal polyps. *Cancer Res.* **48**, 4701–4705.

McLarty, J. W., Holiday, D. B., Girard, W. M., Yanagihara, R. H., Kummet, T. D., and Greenberg, S. D. (1995). Beta-carotene, vitamin A and lung cancer chemoprevention: Results of an intermediate endpoint study. *Am. J. Clin. Nutr.* **62**(Suppl.), 1431S–1438S.

Meyskens, F. L., Jr., and Manetta, A. (1995). Prevention of cervical intraepithelial neoplasia and cervical cancer. *Am. J. Clin. Nutr.* **62**(Suppl.), 1417S–1419S.

Micozzi, M. S., Brown, E. D., Edwards, B. K., Bieri, J. G., Taylor, P. R., Khatchik, F., Beecher, G. R., and Smith, J. C., Jr. (1992). Plasma carotenoid response to chronic intake of selected foods and beta-carotene supplements in men. *Am. J. Clin. Nutr.* **55**, 1120–1125.

Mobarhan, S., Shiau, A., Grande, A., Kolli, S., Stacewicz-Sapuntzakis, M., Oldham, T., Liao, Y., Bowen, P., Dyavanapalli, M., Kazi, N., McNeal, K., and FroAmmel, T. (1994). β-Carotene supplementation results in an increased serum and colonic mucosal concentration of β-carotene and a decrease in alpha-tocopherol concentration in patients with colonic neoplasia. *Cancer Epidemiol. Biomark. Prev.* **3**, 501–505.

Muir, C. S. (1995). International Patterns of Cancer. *In* "Cancer Prevention and Control" (P. Greenwald, B. S. Kramer, and D. L. Weed, eds.), pp. 37–68. Dekker, New York.

Mukai, F. H., and Goldstein, B. D. (1976). Mutagenicity of malondialdehyde, a decomposition product of peroxidized polyunsaturated fatty acids. *Science* **191**, 868–869.

Nierenberg, D. W., Stukel, T. A., Mott, L. A., and Greenberg, E. R. (1994). Steady-state serum concentration of alpha tocopheral not altered by supplementation with oral beta carotene. *J. Natl. Cancer Inst.* **86**, 117–120.

Olson, J. A. (1996). Benefits and liabilities of vitamin A and carotenoids. *J. Nutr.* **126** (Suppl.), 1208S–1212S.

Omenn, G. S., Goodman, G. E., Thornquist, M. D., Balmes, J., Cullen, M. R., Glass, A., Keogh, J. P., Meyskens, F. L., Jr., Valanis, B., Williams, J. H., Jr., Barnhart, S., and Hammar, S. (1996). Effects of combination of beta carotene and vitamin A on lung cancer and cardiovascular disease. *N. Engl. J. Med.* **334**, 1150–1155.

Omenn, G. S., Goodman, G. E., Thornquist, M. D., Rosenstock, L., Barnhart, S., Gylys-Colwell, I., Metch, B., and Lund, B. (1993). The Carotene and Retinol Efficacy Trial (CARET) to prevent lung cancer in high risk populations: Pilot study with asbestos-exposed workers. *Cancer Epidemiol. Biomark. Prev.* **2**, 389–396.

Rohan, T. E., Hiller, J. E., and McMichael, A. J. (1993). Dietary factors and survival from breast cancer. *Nutr. Cancer* **20**, 167–177.

Root, M., Chung, J., and Parker, R. S. (1994). Inhibition of hepatocarcinogenesis in rats by a carotenoid-rich fraction of orange juice. *In* "Second International Conference on Antioxidant Vitamins and Beta-Carotene in Disease Prevention," Berlin, October. Abst. P-67.

Salonen, J. T., Salonen, R., Lappetlainen, R., Maenpaa, P. H., Alfthan, G., and Puska, P. (1985). Risk of cancer in relation to serum concentrations of selenium and vitamins A and E: Matched case-control analysis of prospective data. *Br. Med. J. Clin. Res. Ed.* **290**, 417–420.

Schwartz, J., and Shklar, G. (1988). Regression of experimental oral carcinomas by local injection of beta-carotene and canthaxanthin. *Nutr. Cancer* **11**, 35–40.

Stich, H. F., Hornby, A. P., and Dunn, B. P. (1985) A pilot beta-carotene intervention trial with Inuits using smokeless tobacco. *Int. J. Cancer* **36**, 321–327.

Stich, H. F., Rosin, M. P., Hornby, P., Mathew, B., Sankaranarayanan, R., and Nair, M. K. (1988). Remission of oral leukoplakias and micronuclei in tobacco/betel quid chewers treated with beta-carotene and with beta-carotene plus vitamin A. *Int. J. Cancer* **42**, 195–199.

Stich, H. F., Rosin, M. P., and Vallejera, M. O. (1984). Reduction with vitamin A and beta-carotene administration of proportion of micronucleated buccal mucosal cells in Asian betel nut and tobacco chewers. *Lancet* **1**, 1204–1206.

Toma, S., Benso, S., Albanese, E., Palumbo, R., Cantoni, E., Nicolo, G., and Margiante, P. (1992). Treatment of oral leukoplakia with beta-carotene. *Oncology* **49**, 77–81.

Toma, S., Bonelli, G., Cortesina, G., Gandolfo, G., Mira, E., Sartorie, A., Radaelli De Zinis, L., Vincenti, M., and Palumbo, R. (1994). Beta-carotene in head and neck tumors chemoprevention. *In* "Second International Conference on Antioxidant Vitamins and Beta-Carotene in Disease Prevention," Berlin, October. Abst P-70.

Van Poppel, G., Kok, F. J., and Hermus, R. J. (1992). Beta-carotene supplementation in smokers reduces the frequency of micronuclei in sputum. *Br. J. Cancer* **66**, 1164–1168.

Van Poppel, G., Poulsen, H., Loft, S., and Verhagen, H. (1995). No influence of beta carotene on oxidative DNA damage in male smokers. *J. Natl. Cancer Inst.* **87**, 310–311.

Wahlqvist, M. L., Wattanapenpaiboon, N., Macrae, F. A., Lambert, J. R., MacLennan, R., Hse-Hage, B. H.-H., and Australian Polyp Prevention Project Investigators. (1994). Changes in serum carotenoids in subjects with coloretal adenomas after 24 months of β-carotene supplementation. *Am. J. Clin. Nutr.* **60**, 936–943.

Xu, M. J., Plezia, P. M., Alberts, D. S., Emerson, S. S., Peng, Y. M., Sayers, S. M., Liu, Y., Ritenbaugh, C., and Gensler, H. L. (1992). Reduction in plasma or skin alpha-tocopherol concentration with long-term oral administration of beta-carotene in humans and mice. *J. Natl. Cancer Inst.* **84**, 1559–1565.

Zaridze, D., Evstifeeva, T., and Boyle, P. (1993). Chemoprevention of oral leukoplakia and chronic esophagitis in an area of high incidence of oral and esophageal cancer. *Ann. Epidemiol.* **3**, 225–234.

Ziegler, R. G., Mayne, S. T., and Swanson, C. A. (1996). Nutrition and lung cancer. *Cancer Causes Control* **7**, 157–177.

Index

Contents of Previous Volumes

Volume 29B

Volume 32

Signal Sorting by G-Protein-Linked Receptors
Graeme Milligan

Regulation of Phospholipase A_2 Enzymes: Selective Inhibitors and
Their Pharmacological Potential
Keith B. Glaser

Platelet Activating Factor Antagonists
James B. Summers and Daniel H. Albert

Pharmacological Management of Acute and Chronic
Bronchial Asthma
Michael K. Gould and Thomas A. Raffin

Anti-Human Immunodeficiency Virus Immunoconjugates
Seth H. Pincus and Vladimir V. Tolstikov

Recent Advances in the Treatment of Human Immunodeficiency
Virus Infections with Interferons and Other Biological
Response Modifiers
Orjan Strannegård

Advances in Cancer Gene Therapy
Wei-Wei Zhang, Toshiyoshi Fujiwara, Elizabeth A. Grimm, and Jack A. Roth

Melanoma and Melanocytes; Pigmentation, Tumor Progression,
and the Immune Response to Cancer
Setaluri Vijayasaradhi and Alan N. Houghton

High-Density Lipoprotein Cholesterol, Plasma Triglyceride, and
Coronary Heart Disease: Pathophysiology and Management
Wolfgang Patsch and Antonio M. Gotto, Jr.

Neurotransmitter-like Actions of l-DOPA
Yoshimi Misu, Hiroshi Ueda, and Yoshio Goshima

New Approaches to the Drug Treatment of Schizophrenia
Gavin P. Reynolds and Carole Czudek

Membrane Trafficking in Nerve Terminals
Flavia Valtorta and Fabio Benfenati

Inhaled Nitric Oxide Therapy of Pulmonary Hypertension and Respiratory Failure in Premature and Term Neonates
Steven H. Abman and John P. Kinsella

Clinical Applications of Inhaled Nitric Oxide in Children with Pulmonary Hypertension
David L. Wessel and Ian Adatia

Volume 35

Interactions between Drugs and Nutrients
C. Tschanz, W. Wayne Stargel, and J. A. Thomas

Induction of Cyclo-Oxygenase and Nitric Oxide Synthase in Inflammation
Ian Appleton, Annette Tomlinson, and Derek A. Willoughby

Current and Future Therapeutic Approaches to Hyperlipidemia
John A. Farmer and Antonio M. Gotto, Jr.

In Vivo Pharmacological Effects of Ciclosporin and Some Analogues
Jean F. Borel, Götz Baumann, Ian Chapman, Peter Donatsch, Alfred Fahr, Edgar A. Mueller, and Jean-Marie Vigouret

Mono-ADP-ribosylation: A Reversible Posttranslational Modification of Proteins
Ian J. Okazaki and Joel Moss

Activation of Programmed (Apoptotic) Cell Death for the Treatment of Prostate Cancer
Samuel R. Denmeade and John T. Isaacs

Reversal of Atherosclerosis with Therapy: Update of Coronary Angiographic Trials
Howard N. Hodis

Unnatural Nucleotide Sequences in Biopharmaceutics
Lawrence A. Loeb

Pharmacology of the Neurotransmitter Release Enhancer Linopirdine (DuP 996), and Insights into Its Mechanism of Action
Simon P. Aiken, Robert Zaczek, and Barry S. Brown